Biomaterials, Bioengineering and Sustainability

Volume 7

There is an urgent need to address the current paradigm shifts in bioengineering for Human Health aiming the creation of breakthrough tissue engineered products, manufacturing technologies and effective regenerative treatments for tackling different diseases/disorders in a personalized manner. Yet, the excessive costs and wastes related to the development and production of biomaterials and advanced therapy medicinal products, and the increasing use of plastics in cell culture methods and animal derived reagents has recognized the importance of decreasing the direct carbon footprint and thus, implement sustainable principles and solutions in the innovation ecosystem. The main goal of the volumes in Biomaterials, Bioengineering and Sustainability series is to catapult and consolidate new concepts and solutions towards the development of the next-generation of sustainable and eco-friendly biomaterials and tissue engineering and regenerative medicine approaches. Each volume will focus on the latest developments dealing with the identification of new sources of sustainable or recycled biomaterials, providing ideas for green technologies and methods that can be applied for biomaterials advanced processing and scaffolding strategies, and applications in biofabrication, tissue engineering, regenerative medicine, and drug delivery systems. It also aims to include the exploitation of renewable and sustainable source of human cells applied for cell therapies or in combination with sustainable biomaterials. The develop complex in vitro 3D/4D models and dynamic cell culture systems will be other subjects to be further explored from a sustainable perspective. This series aims to attract the contributions of leading experts in bioengineering, cell biology, materials engineering, and environmental sciences.

Sudarshan Singh
Popat Mohite • Deepanjan Datta
Editors

Biomaterial-based Additive Manufacturing in Tissue Engineering and Regeneration

Springer

Editors
Sudarshan Singh
Office of Research Administration
Chiang Mai University
Mueang Chiang Mai, Thailand

Popat Mohite
AETs St. John Institute of Pharmacy and Research
Palghar, Maharashtra, India

Deepanjan Datta
Manipal College of Pharmaceutical Sciences
Manipal Academy of Higher Education
Manipal, Karnataka, India

ISSN 2731-751X ISSN 2731-7528 (electronic)
Biomaterials, Bioengineering and Sustainability
ISBN 978-3-031-96069-7 ISBN 978-3-031-96070-3 (eBook)
https://doi.org/10.1007/978-3-031-96070-3

This Springer imprint is published by the registered company Springer Nature Switzerland AG
The registered company address is: Gewerbestrasse 11, 6330 Cham, Switzerland

Contents

Introduction to Biomaterials in Design of Regenerative Medicine

1

Sagar R. Pardeshi, Amol D. Gholap, Nitin K. Sen, Prashant J. Shinde, and Prakash D. Khandagale

Abstract

Recent advancements in biomaterials have driven tissue engineering and regenerative medicine forward, overcoming limitations of traditional methods in tissue repair and restoration. These biomaterials, classified according to their polymer composition, possess tailored properties ideal for a wide range of regenerative applications. By ensuring biocompatibility and biodegradability, these materials avoid issues of immune rejection and toxicity. Central to tissue engineering, three-dimensional scaffolds provide an optimal environment for cell growth and proliferation. Innovative approaches such as hydrogels and cryogels offer porous structures that support cell migration and tissue regeneration. Whether sourced from natural or synthetic origins, these matrices find extensive use across various tissue engineering domains. Biomaterials demonstrate notable effectiveness in cardiovascular, orthopedic, and dental implant procedures, highlighting their adaptability in medical contexts. Additionally, integrating stem cells with biomaterials presents a promising avenue in regenerative medicine, offering viable alternatives to autologous and allogenic sources. This chapter underscores the versatility of biomaterials in meeting diverse clinical needs. It concludes by emphasizing the importance of ongoing research and innovation in advancing

S. R. Pardeshi (✉) · A. D. Gholap · P. J. Shinde
Department of Pharmaceutics, AETs St. John Institute of Pharmacy and Research, Palghar, Maharashtra, India
e-mail: sagarp@sjipr.edu.in

N. K. Sen
Department of Pharmacology, AETs St. John Institute of Pharmacy and Research, Palghar, Maharashtra, India

P. D. Khandagale
Department of Pharmacognosy, AETs St. John Institute of Pharmacy and Research, Palghar, Maharashtra, India

S. Singh et al. (eds.), *Biomaterial-based Additive Manufacturing in Tissue Engineering and Regeneration*, Biomaterials, Bioengineering and Sustainability 7, https://doi.org/10.1007/978-3-031-96070-3_1

biomaterial-based solutions for regenerative medicine, providing insights into future directions in the field.

Keywords

Biomaterials · Tissue engineering · Regenerative medicine · Bioprinting · Personalized medicine

1.1 Introduction

The exploration for innovative solutions for betterment of healthcare and aging has emerged as a global challenge. Many strategies and drug delivery carrier systems are developed for effective human therapeutics. The repair and regeneration of damaged tissues or organs involving restoration of the functions are essential for the wellbeing of humans [113]. Dr. YC Fung first proposed the concept of tissue engineering for a help handicapped program meeting in 1987. At the initial phase of tissue engineering, it was considered for production for biological substitute which can be restored or maintained for improvisation of the tissue function. Later, tissue engineering was applied in biological, engineering, and chemical sectors for repair, regeneration along with the restoration of the living tissue for regeneration of the tissues belonging to the skin, bone along with cartilage [25]. A single lineage of tissue has been regenerated by using tissue engineering while a complex human system consists of the multitude of lineages which requires simultaneous type of regeneration. Thus, regenerative engineering is involved in the integration of material science with the help of developmental cell biology. This field has progressed from individual tissue repair to that of the regeneration of complex tissue or organ systems. The generation of the induced pluripotent stem cells is done based on the proper understanding of the adult stem cells along with the adult cells. An advance into the biomaterials can provide manipulation of the biological processes while with the help of the nanotechnology tool the broader applications with targeted action of the tissue engineered materials can be possible in it [20, 116].

The advancement of tissue engineering and regenerative medicine consists of combining cells, growth factors and scaffold for tissue regeneration or the replacement of damaged tissue or diseased ones. Tissue engineering involves a group of therapy-like cell-based therapy, gene therapy and the immunomodulation for better tissue or organ regenerations as presented in Fig. 1.1. The tissue engineering and regeneration is one of the multidisciplinary sciences which involves basic sciences, biomechanics, cell biology and medical sciences for better use of tools required for organ repair, tissue regeneration or reconstruction [11, 20]. At present on a global platform, there is an increase in the demand for organ replacement and tissue engineering can address the issues raised the same way. Tissue regeneration involves a three-dimensional (3D) cell or the biomaterial complex which provides a similar function to that of the living tissue or organ and can be used for the regeneration of the tissue or the organ. The complex formation is required for cell growth support,

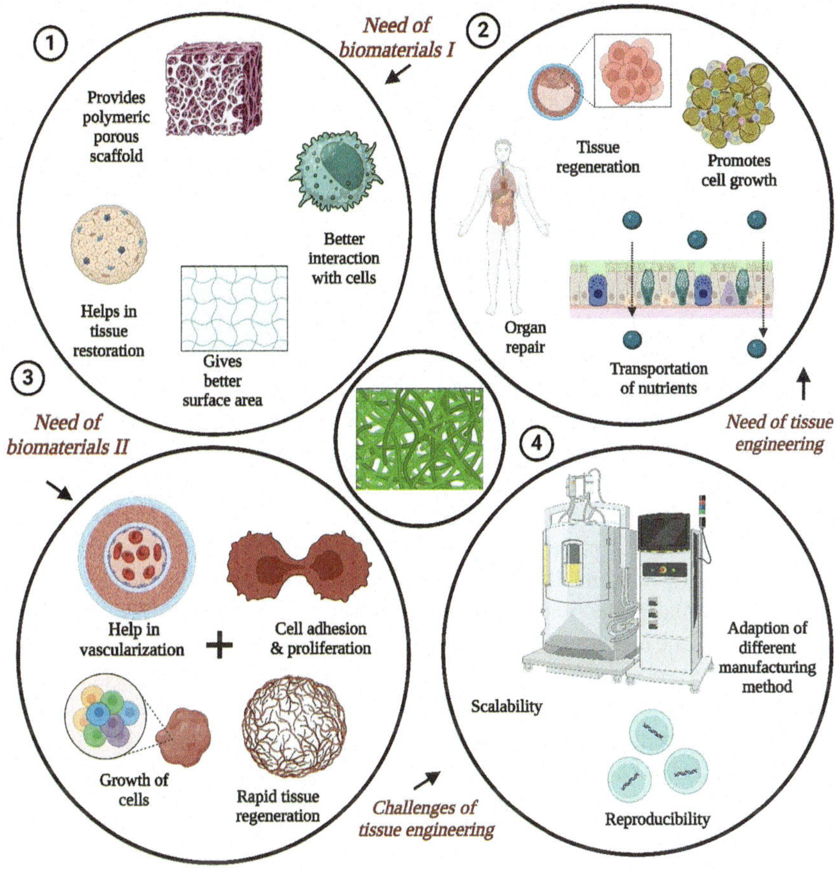

Fig. 1.1 Summary of biomaterial used in tissue engineering involves role of biomaterial, need of tissue engineering and challenges of tissue engineering. (Created by using Biorender.com)

gas exchange, waste transportation and nutrients. The implementation of the tissue engineering and regeneration involves three different methods including implantation of the cell seeded biomaterials into the body for repair as well as tissue regeneration or organ regeneration [20, 118].

The second method is dealing with the transplantation of stem cells or cell systems while the third system consists of biomaterial systems which can be implanted into the body resulting in the process of tissue integration. Despite the several advantages provided of the tissue engineering and regeneration still some questions are required for to be addressed including selection of cell sources, development into tissue specific type of materials, specialized bioreactors along with the process of the construction of the complex organs. The progress of the tissue engineering field is related to the amount of similarity achieved by the tissue engineering method for organ/tissue development along with the healing as compared to the natural process for the same. Thus, for wide applications and better patient compliance stability, security and effectiveness of the tissue engineered materials required to be checked for use of the same in clinical therapeutics [12, 20, 66].

Polymeric porous scaffolds are required for favorable environment regarding the tissue restoration ensuring better interaction among with biomaterials with the cells. The auto graft method is involved in the same based on the availability of the donor tissue, pain, bleeding, and risk related to the infectious diseases in the patients. Different properties are offered by the biomaterials used for tissue regeneration. It provides adequate mechanical and physical properties to 3D nexus for better micro-architecture and micro-textures in surface topography with good porosity and inter-connected pores in it. It provides better surface area to the volume ratio for effective vascularization, growth of cells and mass transfer involved into it [20, 49].

The lesser degree of inflammation with good amount of safety can be ensured through biocompatible biomaterials into tissue engineering and regenerations. The bioactivity of the selected biomaterials will be ensured about the cell interaction for cell adhesion and proliferation. It also shares effective contact among the cells and their prolonged period of migration. With the help of this the scaffold can be able to include the biological molecules on their surface for effective cell adhesion and assist in the growth stimulating substances like growth factors for quick regeneration. The optimal protection and nutrition along with mimicking of native tissue required for tissue regeneration. The bio-absorption property of biomaterials ensures the new tissue replacement into the available space while the versatility of the biomaterials provides adaptation for different manufacturing techniques. The wide applications of the tissue regeneration to the large tissues required scalability, reproducibility and accessibility properties which can be acquired through selection of the appropriate biomaterials into the tissue engineering and regeneration [4, 20].

1.2 Essential Requirements in Designing Biomaterials for Tissue Engineering and Regenerative Medicine Applications

1.2.1 Surface Properties

The complex web of proteins and chemicals known as the extracellular matrix (ECM) supports cells within tissues structurally and sends signals to them, which is essential for tissue regeneration and repair. The biomaterial utilized in therapeutic applications needs to be biodegradable in order to help restore the natural ECM and facilitate the successful integration of seeded cells into local tissue [38]. These materials allow for cell infiltration and tissue remodelling as they break down. In tissue engineering, surface characteristics affect how cells engage biomaterials. Roughness, wettability, functional groups, and topography are critical surface attributes [115]. Regulated roughness encourages neurite development, and roughness has an impact on cell adhesion, proliferation, and differentiation [17]. By boosting tissue integration, cell reactions, and protein adsorption, wetting can be optimized to increase biocompatibility. Neurite extension is guided, and cell behaviour is modulated by functional groups on the exterior [53]. Furthermore, surface morphology, which includes characteristics at the micro- and nanoscale, can facilitate neurite alignment and cell orientation, hence promoting the development of functioning neural networks. Both natural and synthetic polymers can be used to create scaffolds, and each has special properties that should be thoroughly studied before use [96]. In a study on tissue engineering for spinal cord injury, a group of researchers [17], examined the effects of different biomaterials—chitosan, poly (ε-caprolactone), and poly (L-lactic acid)—on neural cells. They found that biomaterial surface properties significantly influenced cell behaviour, including attachment, morphology, and differentiation. Specifically, poly (ε-caprolactone) and poly (L-lactic acid) supported attachment and maturation of neural cells, while promoting neuronal differentiation and network development [65]. These findings underscore the critical role of selecting biomaterials with suitable surface properties for enhancing cell interactions and optimizing therapeutic outcomes in SCI treatment.

1.2.2 Mechanical Requirement

Research is now being done on improving novel biomaterials chemically and physically so they can interact with living cells. Whether synthetic or hybrid, these materials need to be engineered to work well with biological systems in both in vitro and in vivo settings [13]. The particular use of the scaffold will determine the type of biomaterial to use and how to design it. Tensile strength, compressibility, elasticity, plasticity, and hardness are important mechanical attributes. For instance, hydroxyapatite and tricalcium phosphate ceramics are excellent for bone repair [22]. The bio-ceramic scaffold's mechanical characteristics should be similar to the physiological site where it is intended to be inserted, and its rate of degradation should correspond with the necessary functional surface for successful tissue regeneration. In addition, the varied rates of bone regeneration in different age groups need to be considered, since older persons typically regenerate at a slower rate than younger adults [107]. This factor is essential when creating scaffolds to guarantee that they promote the best possible tissue regeneration in a range of age groups. It is imperative to guarantee the mechanical stability of implanted scaffold constructions and their capacity to tolerate loads and stresses throughout the process of reconstruction. In biological structures, scaffold stability is affected by several variables, including temperature, flexibility, stress, strength, and material absorption that is linked to chemical deterioration [55]. Therefore, evaluating several crucial attributes is necessary when choosing a suitable biomaterial:

Elastic behaviour: measures the reaction to compressive or tensile stress. A linear relationship between stress (load) and strain (displacement) characterizes this reversible behaviour [23].

Plastic behaviour: once compressive or tensile stress applied uniaxial exceeds yield strength, permanent deformation occurs [37].

- *Tensile strength:* the highest stress that a substance can withstand before breaking.
- *Ductility:* the maximum plastic strain a material can withstand prior to failure.
- *Toughness:* the amount of energy needed to break a material with a given volume.
- *Flexural behaviour:* the relationship between flexural stress and strain in reaction to compressive or tensile tension perpendicular to the bar is known as flexural behaviour [41].

In addition, a biomaterial's mechanical characteristics are directly affected by swelling, porosity, pore size, shape, orientation, and interaction, all of which have a consequence on the mechanical behavior of materials [84]. The effectiveness of substrates in tissue engineering depends on striking a balance between mechanical qualities and a porous structure that permits cell penetration and vascularization [67]. Furthermore, cellular regrowth, polarization, and intracellular signaling are strongly impacted by the mechanical rigidity, surface roughness, and physical stimulation that the scaffold's three-dimensional microstructure provides [115].

1.2.3 Biological Requirements

Efficient 3D biomimetic scaffolds for tissue engineering depend on the production of appropriate scaffolds that promote cell growth and differentiation while emulating the physiological functions of extracellular matrix proteins [108]. Making sure the scaffold is biocompatible is crucial to avoiding ectopic calcifications and unfavorable immunological reactions in vivo. To improve cell adhesion, movement, growth, and differentiation, biomaterial surfaces should have exceptional chemical characteristics [47]. According to [61], scaffolds utilized in tissue engineering ought to be non-toxic in order to prevent sensitive or inflammatory reactions. A major factor in determining the implant's effectiveness and the material's level of biocompatibility is how the body reacts to it. Two main aspects influencing biocompatibility are tissue reactions to the material and the substance's internal deterioration[92]. In order to facilitate the development of new tissue, biodegradability must be controlled. Tissues respond to the implant surface when a biomaterial implant is inserted into the body. Two of the body's reactions to implants are: (1) the development of a fibrous capsule surrounding the implant surface; and (2) thrombosis, or blood coagulation, after platelets adhering to the implant surface [10]. The biomaterial utilized in the implants determines the kind of response.

Biomaterials, based on the body's responses, can be classified into three main groups:

- *Bioinert:* materials that do not interact with the biological environment and remain unchanged after implantation.
- *Bioactive:* materials that interact with the biological environment, promoting bonding with surrounding tissues.
- *Bioresorbable:* materials designed to degrade and be absorbed by the body over time, facilitating tissue regeneration without leaving a permanent foreign object [58].

Additionally, the mechanical properties of the scaffold should be tailored to match the specific tissue requirements. This includes ensuring appropriate porosity to allow for nutrient and waste exchange, as well as mechanical stability to withstand physiological loads [50]. Proper vascularization within the scaffold is also crucial for supplying nutrients and removing metabolic waste, thus supporting cell viability and function. These biological and mechanical considerations are vital for the successful application of scaffolds in tissue engineering and regenerative medicine [51].

Bioinert substances hardly engage connective tissue capsules or adjacent tissues. Over the implant surface, attachments rarely form. Zirconia, polymethyl methacrylate (PMMA), alumina, titanium, and stainless steel are a few examples of bioinert materials [24]. Conversely, bioactive substances attach to the surrounding soft tissue and bone to generate adhesions. Around the implant, new tissue grows and firmly fuses with the implant surface. Bio glass, glass ceramics, synthetic substances, and hydroxyapatite (HAp) are a few kinds of bioactive materials [105]. Following implantation, bioresorbable substances are intended to disintegrate and be substituted by new tissue. Heness and Ben-Nissan cite polyglycolic acid, polylactic acid, protein composites, and tricalcium phosphate as examples of these materials. Because bioactive substances integrate well with surrounding soft tissue and bone, they are frequently chosen over bioinert substances for implant fabrication. Bioresorbable polymers have garnered a lot of attention lately. Because they encourage the repair of both tissue function and anatomical structure, such scaffolds are thought to be more appropriate for patient therapy [100]. The risk of thrombosis and fibrosis is decreased by the entry of new tissue into the resorbed location. Better long-term results are another benefit that bioactive and bioresorbable materials provide, in addition to their integrability [64]. By facilitating the gradual shift from implant to native tissue, bioresorbable scaffolds reduce the risks associated with long-term implants. Tissue engineering and regenerative medicine can benefit from the use of such substances because their use is in line with the body's natural healing processes [109]. This approach enhances patient outcomes by reducing the need for secondary surgeries to remove the implant and promoting seamless tissue regeneration.

1.3 Categories of Biomaterials for Regenerative Biomedicine

1.3.1 Natural Polymeric-Based Biomaterials

Biomaterials made from natural and synthetic polymers are widely used in the regeneration of tissues as matrices that support cells [97]. Because these substances are chemically and biologically like ECM, they are especially appealing. Biocompatibility, an acceptable microstructure, mechanical strength, a reasonable degradation rate, and the capacity to support cellular activities that reinstate metabolic activity are all desirable characteristics for a material used in tissue engineering [98]. Naturally, derived biomaterials create a biological environment conducive to interaction with the body's metabolic systems. They offer advantages such as reduced inflammation and immunological reactions, low cytotoxicity, and excellent biocompatibility [9]. A diverse range of polymers can be extracted from living organisms and classified into three main groups based on their chemical structures: polysaccharides, proteins, and nucleic acids. Polysaccharides include hyaluronic acid (HA), chondroitin sulfate, chitin, chitosan, alginates, and cellulose. Proteins, which play crucial roles in biological processes, include elastin, silk, collagen, and fibrinogen. Both polysaccharides and proteins are extensively used as scaffolds in tissue engineering and in drug delivery systems [21].

Biotechnology through microorganism fermentation or enzymatic processes offers innovative methods for producing natural polymers. Despite these advancements, animals, plants, and algae remain primary sources for polymer extraction [91]. Manipulating these natural materials into porous scaffolds typically requires more complex techniques compared to synthetic polymers. Methods such as using suitable cross-linkers are essential to address the mechanical and physical limitations of natural polymers, although cross-linkers can introduce cell toxicity, posing a challenge for their utilization [101]. Natural biomaterials must also meet clinical requirements for processing into various forms, such as nano or microparticles, to control release applications or to be structured into two- or three-dimensional forms [9]. While these materials boast excellent biocompatibility, they often suffer from poor mechanical and thermal properties. In contrast, synthetic polymers can be optimized for these desired properties [5]. Therefore, combining synthetic and natural polymers, along with selecting appropriate cross-linkers, can help overcome these challenges, enhancing the overall performance of the biomaterials.

1.3.2 Synthetic Polymer-Based Biomaterials

Synthetic biodegradable polymers are a prominent group of materials widely used in tissue engineering. These polymers support cell attachment, proliferation, and differentiation, facilitating the reconstruction of tissue defects [39]. Unlike natural biopolymers, synthetic polymers have simpler structures, allowing for precise control over their properties. The degradation rates of synthetic polymers can be tailored by adjusting factors such as the mixing ratio, molecular weight of components, and other parameters, ensuring that they match the tissue regeneration rate [95]. Examples of synthetic polymers used in tissue engineering include polyanhydrides, polyesters, polyphosphazenes, poly (glycerol sebacate), and polyurethanes [39]. These materials can be engineered to possess desirable characteristics for specific applications. For instance, polyanhydrides are highly hydrophobic and degrade in a controlled manner from the surface inward, making them suitable for drug delivery applications due to their favourable degradation pattern [83]. Polyesters, formed from a dicarboxylic acid and a diol, are notable for their ability to degrade through hydrolysis of ester bonds. The rate of degradation can be fine-tuned by modifying the polymer concentration, structure, and molecular weight [117]. Key examples of synthetic biopolymers in this category include polylactide, polyglycolide (PGA), their copolymer poly (lactide-co-glycolide) (PLGA), and poly (ε-caprolactone) (PCL). These materials are particularly suitable for applications in bone and cartilage regeneration due to their favourable properties [68]. Polyphosphazenes are another group of synthetic biopolymers characterized by an inorganic backbone of alternating phosphorus and nitrogen atoms, with each phosphorus atom bonded to two organic substituents. The properties of polyphosphazenes can be tailored by altering these organic groups, making them highly versatile [18]. These polymers are well-suited for use in tissue engineering scaffolds and drug delivery systems due to their customizable nature and high molecular weight. Together, these synthetic polymers offer a range of properties that can be engineered to meet specific needs in tissue engineering, from supporting cell growth to providing controlled degradation rates that match tissue regeneration processes [82].

1.3.3 Ceramic-Based Biomaterials

Ceramic-based biomaterials are pivotal for hard tissue regeneration due to their similarity to the mineral components of natural bones and teeth, as well as their ability to mimic mechanical properties [110]. These materials are known for their osteo-conductivity and bone-bonding capabilities, making them ideal for use in bone and dental regenerative medicine. However, their intrinsic fragility and the complexity of their manufacturing processes present significant limitations for clinical applications [56]. Common ceramics with good biocompatibility include alumina, zirconia, titania, hydroxyapatite (HA), and calcium silicate. Despite their potential, there are challenges in the clinical use of ceramic biomaterials, particularly related to the inadequate long-term bonding of ceramic bioimplants to adjacent bones [2]. Additionally, their inherent fragility makes them unsuitable for repairing large bone defects. Innovations in ceramic materials design aim to enhance osteoinductive and osteoconductive properties, which could revolutionize orthopedic implants. Developing ceramics that can form strong bonds with bones while overcoming their fragility is crucial for advancing their clinical application in orthopedic and dental regenerative medicine [43].

1.3.4 Metal Based Biomaterials

Metals have long been used in surgical procedures due to their excellent mechanical properties, making them ideal for load-bearing applications like knee and hip reconstructions. Stainless steel (SS) is cost-effective, corrosion-resistant, and strong, but suffers from poor wear resistance [114]. To address this, cobalt-chromium (Co-Cr) alloys are used, offering superior wear and corrosion resistance. However, both materials have drawbacks, such as nickel-related allergenic potential and ion leaching in Co-Cr alloys [111]. Recent advancements have led to the development of nickel-free alloys and improved formulations to enhance their performance in medical applications, particularly in joint prostheses and artificial disc replacements.

Moreover, Titanium (Ti)-based implants are commonly used in orthopaedic applications due to their favourable properties, such as a moderately high elastic modulus of around 110 GPa and lower density compared to stainless steel (SS) and cobalt-chromium (Co-Cr) alloys [6]. Ti-based implants offer superior corrosion resistance due to the formation of a protective inert oxide layer on their surface, which minimizes ion release. Additionally, these implants exhibit excellent osseointegration, enhancing their long-term performance over traditional SS implants [1]. Alloys like Ti-6Al-4V have been developed to improve titanium's properties, though the release of aluminium and vanadium ions can be cytotoxic. Consequently, research continues to focus on creating titanium alloys that maintain these advantages without the associated health risks [88].

1.4 Applications in Regenerative Medicines

1.4.1 Cardiovascular Devices and Implants

Biomaterials play a crucial role in modern medical devices and systems, including synthetic skin, drug delivery systems, tissue cultures, hybrid organs, synthetic blood vessels, artificial hearts, and more specialized applications like cardiac pacemakers and orthopedic implants [35, 87, 94]. Among these, cardiovascular biomaterials (CB) are particularly significant, where their use hinges on blood compatibility and integration with the surrounding environment upon implantation [19, 34, 81]. When selecting CB, two key considerations are paramount:

- *Physical and mechanical features:* This includes factors like strength, deformation characteristics, fatigue resistance, friction and wear resistance, flow dynamics, and pressure handling capabilities, all crucial for engineering cardiovascular implants as presented in Fig. 1.2 [36, 78, 79].
- *Biocompatibility:* This refers to the material's compatibility with living tissue, ensuring minimal adverse reactions, and promoting successful integration. Biocompatibility is rigorously evaluated through a range of in vivo and in vitro experiments [13]

Cardiac implants, such as replacement valves, illustrate the diversity in biomaterial applications. Mechanical valves, categorized into types like caged ball, single tilting disk, and bileaflet, offer durability but necessitate lifelong anticoagulation therapy [106]. Biological valves, sourced from human or animal tissue (homograft or heterograft bioprosthetic valves, respectively), provide better hemodynamics but have limited durability [93, 99, 112]. Additionally, cardiovascular implants include stents and stent grafts, which are used to restore blood flow through arteries [44]. Stents can be balloon-expandable or self-expanding, each tailored to different procedural needs for restoring vessel patency and ensuring optimal long-term outcomes [8].

Metals have long been integral to biomedical applications, particularly in cardiovascular medicine where stainless steel, cobalt-chromium (CoCr) alloys, and titanium (Ti) alloys are extensively used due to their strength, stiffness, corrosion

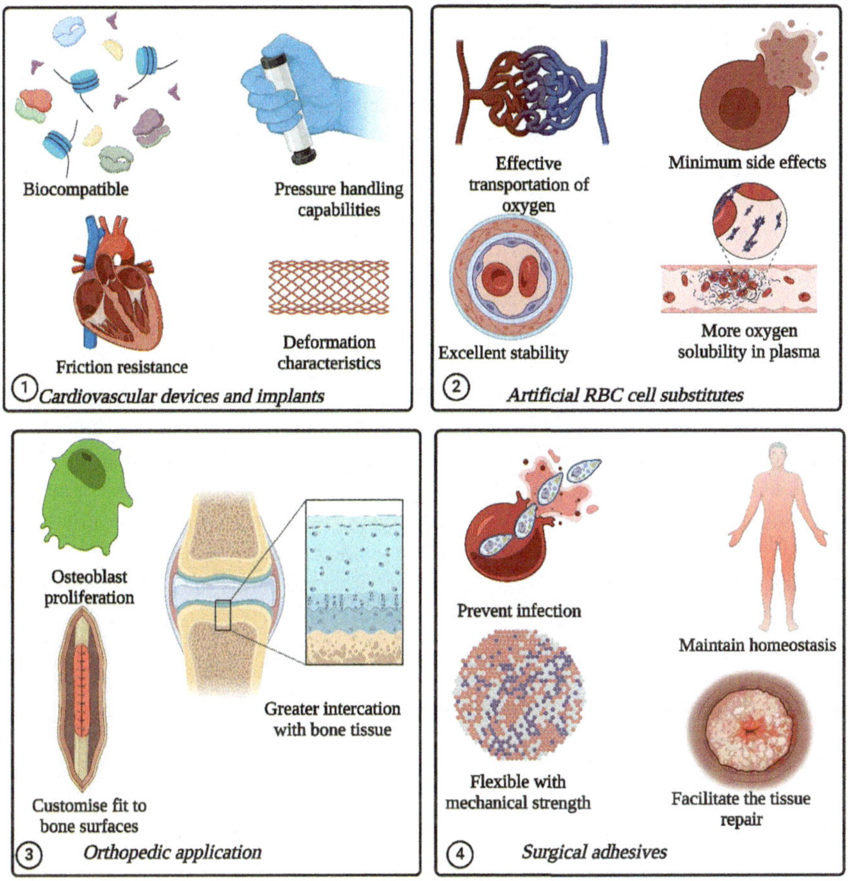

Fig. 1.2 Summary of biomaterial application through tissue engineering containing (1) Cardiovascular devices and implants (2) Artificial RBC cell substitutes (3) Orthopedic application (4) Surgical adhesives for prevention of infection, maintenance of homeostasis, better flexibility of mechanical strength and tissue repair. (Created by using Biorender.com)

resistance, and biocompatibility [16]. Cobalt alloys are dense and non-ferromagnetic, making them suitable for coronary stents, while titanium alloys, including nitinol with shape memory properties, are favored for self-expanding stents. The evolution from bare metal stents (BMS) to drug-eluting stents (DES) aimed to improve outcomes by delivering antiproliferative drugs, although polymer use in DES can lead to complications [85]. Bioabsorbable stents, such as magnesium alloys, offer a promising alternative, designed to degrade over time, promote healing, and minimize long-term medication requirements while ensuring biocompatibility and adequate structural support during arterial remodeling [7].

1.4.2 Artificial RBCs Cell Substitutes

Artificial red blood cell substitutes are researched for their potential in blood replacement therapies, particularly in perioperative hemo-dilution and haemorrhagic blood loss resuscitation. Unlike natural blood, these substitutes are designed solely for oxygen and carbon dioxide transport throughout the body [45]. The ideal artificial blood product must be safe, compatible with various blood types, capable of effectively transporting and releasing oxygen, and shelf stable [30–32]. It aims to supplement or replace homologous and autologous transfusions, in conjunction with erythropoietin therapy [40]. Currently, two main types of artificial blood substitutes are in development: perfluorocarbons (PFCs) and haemoglobin-based products. PFCs are chemically inert fluorine-substituted hydrocarbons that enhance oxygen solubility in plasma without reacting with gases [46]. They maintain dissolved oxygen stability regardless of temperature, pH, or other physiological factors, ensuring a linear oxygen dissociation curve z. This characteristic facilitates efficient oxygen transfer from red blood cells to tissues, particularly advantageous for organ perfusion [33, 59, 76]. While homologous blood transfusions remain cost-effective and clinically effective with minimal side effects, ongoing research aims to develop artificial blood products that offer superior safety and efficacy without adverse effects, enhancing medical care delivery [14].

1.4.3 Orthopaedic Applications/Dental Implants/ Cartilage Implants

In orthopaedic and dental implant applications, alongside metallic implants, naturally derived matrices such as hyaluronic acid, chitosan, collagen, and fibrinogen have been utilized in bone tissue engineering [26, 28, 29, 74]. However, these matrices can be challenging to sterilize and may provoke immune responses in recipients [27]. Synthetic polymers like poly (alpha-hydroxy acids), polypropylene fumarate, and polyethylene glycol offer advantages in fabrication flexibility but can degrade into potentially toxic components. Bioactive inorganic materials like bioactive glass, hydroxyapatite, porous coralline, and tricalcium phosphate show promise due to their ability to enhance bone tissue integration [3].

Current research aims to enhance scaffold bio-functionality by incorporating osteo-inductive cues to promote osteoblast proliferation. Cartilage tissue engineering employs scaffolds made from natural polymers such as agarose, alginate, chitosan, and synthetic polymers like poly (lactic acid), polyglycolic acid, and their copolymers [60, 71, 77]. Dental implants are categorized into endosteal (embedded within bone tissue) and subperiosteal (placed on bone surfaces) systems. Endosteal implants, including root forms and blades, are fixed into bone, while subperiosteal implants are customized to fit bone surfaces. Common materials for dental implants include titanium and its alloys, as well as surface coatings like hydroxyapatite, enhancing biocompatibility and integration with bone tissue [57].

1.4.4 Surgical Sutures/Surgical Adhesives

Surgical sutures and adhesives are essential tools in modern medicine, used extensively in various surgical procedures to close wounds, approximate tissues, and promote healing. They are crucial for achieving haemostasis, preventing infection, and facilitating tissue repair with minimal scarring. These materials have evolved significantly over time, from natural origins to advanced synthetic compositions, offering a range of properties tailored to specific medical needs [104]. Surgical adhesives are essential in various medical procedures for wound closure and tissue healing, offering a less invasive alternative to traditional methods like sutures, staples, and metallic grafts [69, 70]. These adhesives are used in applications such as closing lacerations, vessel, and intestine anastomoses, and attaching skin flaps to prevent fluid accumulation [103]. The ideal surgical adhesive must balance mechanical strength and flexibility to support tissue healing without being too strong or brittle. Current materials used in surgical adhesives include cyanoacrylates, albumin and glutaraldehyde, polyethylene glycol (PEG), polyurethane, and fibrin [102]. Cyanoacrylates, first used in 1959, cure rapidly in the presence of moisture and are effective for emergency wound closure, providing tensile strength like absorbable sutures. However, they are unsuitable for high-tension areas, mucosal surfaces, and areas with dense hair growth. Products like Indermil®, Histoacryl®, and LiquiBand® have been approved for skin re-approximation and act as microbial barriers, showing comparable performance to sutures and fibrin glues with higher patient satisfaction [89].

Despite the effectiveness of traditional methods, they often result in secondary tissue damage, infection, fluid leakage, and poor cosmetic outcomes, particularly in intricate surgeries like vascular anastomosis, nerve repair, or ocular surgeries [78]. Metallic grafts used in orthopedic surgeries also face issues such as modulus mismatch and aseptic loosening, leading to poor bone healing [72, 73]. Tissue adhesives offer a promising alternative, spreading over the contact area to eliminate stress localization and facilitate load transfer between fractured surfaces [62]. Studies indicate that cyanoacrylate adhesives perform well compared to sutures, with no significant differences in wound complications or cosmetic outcomes. Indermil® is noted for its faster application and ease of use, while Histoacryl Blue® has demonstrated superior performance in wound closure compared to fibrin glue. These adhesives cause minimal tissue damage and resorb over time, making them effective and reliable options for tissue reconstruction and wound healing [90].

1.4.5 Hydrogels for Neurodegenerative Diseases

Regenerative medicine for CNS injuries aims to replace or regenerate cells, tissues, or organs. It employs two strategies: delivering stem cells for integration with host tissue and administering drugs to stimulate endogenous cell regeneration [54]. However, the CNS's blood-brain barrier (BBB) and blood-cerebrospinal fluid barrier (BCSFB) limit the effectiveness of traditional therapies, requiring high doses or frequent administration, which can increase toxicity. Invasive methods like BBB disruption and direct CNS drug delivery are used to overcome this but can cause additional damage [80]. Consequently, novel delivery methods, such as injectable hydrogels, are being researched for their potential to treat CNS disorders more effectively and safely [74]. Because of their high porosity and watery nature, hydrogels colloid gels with water as the dispersion medium—closely mimic biological tissue and facilitate waste and nutrition exchange [42]. Hydrogel frameworks are made from synthetic and natural polymers and are utilized for tissue regeneration [63, 75]. To facilitate nerve cell regeneration, hydrogels made of substances such as collagen, hyaluronic acid (HA), chitosan, self-assembled peptides, and decellularized ECMs are employed [86]. These hydrogels are frequently coupled with medications, bioactive substances, or cells to improve therapeutic effects. Research on brain injuries has placed a great deal of emphasis on traumatic brain injury (TBI), which is a major cause of mortality and disability for people of all ages [15]. After the original trauma, there is frequently secondary injury that breaks down the blood-brain barrier (BBB) and causes significant inflammatory reactions. This leads to ongoing neuronal damage. Recent studies indicate that TBI-related brain injury is primarily caused by reactive oxygen species (ROS) generated during secondary damage. Conventional treatments, such as ROS scavengers and growth factors, face limitations due to their inability to effectively cross the BBB [48]. Innovative approaches are being explored to address these challenges. For instance, Qian et al. developed hydrogels that respond to the post-TBI microenvironment and eliminate ROS, incorporating curcumin to facilitate neuron regeneration. Additionally, stem cell replacement therapy (CRT) shows promise for TBI treatment when combined with injectable hydrogels that create a supportive microenvironment for transplanted stem cells [52]. Lu et al., created a hydrogel scaffold with a high-water content that is perfect for encapsulating human umbilical cord mesenchymal stem cells (hUC-MSCs). The scaffold is composed of hyaluronic acid (HA) and sodium alginate (SA). According to in vivo experiments, HA/SA hydrogels help TBI patients regain their nerve functions by supporting the survival of hUC-MSCs and encouraging the regeneration of endogenous nerve cells. In a similar vein, Sultan created hydrogels based on silk fibroin to efficiently encapsulate human mesenchymal stem cells (hMSCs) and treat brain damage [52].

1.5 Conclusion and Future Perspectives

In conclusion, biomaterials have revolutionized regenerative medicine by providing versatile platforms for tissue engineering and therapeutic interventions. They encompass natural polymers mimicking the extracellular matrix and synthetic materials with precise mechanical properties, crucial for supporting cell growth, guiding tissue regeneration, and enhancing medical device functionality. Future advancements will likely focus on advanced functionalization with bioactive molecules, growth factors, and therapeutic agents, alongside the development of smart biomaterials responsive to physiological cues. Bioprinting and 3D biomanufacturing techniques will further enable the creation of complex tissues, while biohybrid implants integrating biomaterials with living cells promise dynamic interaction with the host environment. Personalized medicine approaches, regulatory considerations, and ethical implications will also shape the field's trajectory, ensuring biomaterial-based therapies continue to innovate and translate into improved clinical outcomes globally.

References

1. Abd-Elaziem W, Darwish MA, Hamada A, Daoush WM (2024) Titanium-based alloys and composites for orthopedic implants applications: a comprehensive review. Mater Des 241:112850. https://doi.org/10.1016/j.matdes.2024.112850
2. Al-Amin M, Abdul Rani AM, Abdu Aliyu AA, Bryant MG, Danish M, Ahmad A (2020) Bio-ceramic coatings adhesion and roughness of biomaterials through PM-EDM: a comprehensive review. Mater Manuf Process 35(11):1157–1180. https://doi.org/10.1080/10426914.2020.1772483
3. Alizadeh-Osgouei M, Li Y, Wen C (2019) A comprehensive review of biodegradable synthetic polymer-ceramic composites and their manufacture for biomedical applications. Bioact Mater 4:22–36. https://doi.org/10.1016/j.bioactmat.2018.11.003
4. Al-Taani B, Khanfar MAI, Alsuod OA (2018) Enhancement of the release of curcumin by the freeze drying technique using inulin and neusilin as carriers. Int J Appl Pharm 10(3):42–48. https://doi.org/10.22159/ijap.2018v10i3.24429
5. Arif U, Haider S, Haider A, Khan N, Alghyamah AA, Jamila N, Khan MI, Almasry WA, Kang I-K (2019) Biocompatible polymers and their potential biomedical applications: a review. Curr Pharm Des 25(34):3608–3619. https://doi.org/10.2174/1381612825999191011105148
6. Bandyopadhyay A, Mitra I, Goodman SB, Kumar M, Bose S (2023) Improving biocompatibility for next generation of metallic implants. Prog Mater Sci 133:101053. https://doi.org/10.1016/j.pmatsci.2022.101053
7. Boccafoschi F, Fusaro L, Cannas M (2018) Immobilization of peptides on cardiovascular stent. In: Functionalised cardiovascular stents. Elsevier, pp 305–318. https://doi.org/10.1016/B978-0-08-100496-8.00016-0
8. Borhani S, Hassanajili S, Ahmadi Tafti SH, Rabbani S (2018) Cardiovascular stents: overview, evolution, and next generation. Prog Biomater 7(3):175–205. https://doi.org/10.1007/s40204-018-0097-y
9. Brovold M, Almeida JI, Pla-Palacín I, Sainz-Arnal P, Sánchez-Romero N, Rivas JJ, Almeida H, Dachary PR, Serrano-Aulló T, Soker S, Baptista PM (2018) Naturally-derived biomaterials for tissue engineering applications, pp 421–449. https://doi.org/10.1007/978-981-13-0947-2_23
10. Carnicer-Lombarte A, Chen S-T, Malliaras GG, Barone DG (2021) Foreign body reaction to implanted biomaterials and its impact in nerve neuroprosthetics. Front Bioeng Biotechnol 9. https://doi.org/10.3389/fbioe.2021.622524
11. Castañeda-Rodríguez S, González-Torres M, Ribas-Aparicio RM, Del Prado-Audelo ML, Leyva-Gómez G, Gürer ES, Sharifi-Rad J (2023) Recent advances in modified poly (lactic acid) as tissue engineering materials. J Biol Eng 17(1):21. https://doi.org/10.1186/s13036-023-00338-8
12. Chae S, Cho D-W (2023) Biomaterial-based 3D bioprinting strategy for orthopedic tissue engineering. Acta Biomater 156:4–20. https://doi.org/10.1016/j.actbio.2022.08.004
13. Chandra P, Yoo JJ, Lee SJ (2015) Biomaterials in regenerative medicine. In: Translational regenerative medicine. Elsevier, pp 151–167. https://doi.org/10.1016/B978-0-12-410396-2.00013-X
14. Charbe NB, Castillo F, Tambuwala MM, Prasher P, Chellappan DK, Carreño A, Satija S, Singh SK, Gulati M, Dua K, González-Aramundiz JV, Zacconi FC (2022) A new era in

oxygen therapeutics? From perfluorocarbon systems to haemoglobin-based oxygen carriers. Blood Rev 54:100927. https://doi.org/10.1016/j.blre.2022.100927
15. Chen Y, Lin J, Yan W (2022) A Prosperous application of hydrogels with extracellular vesicles release for traumatic brain injury. Front Neurol 13. https://doi.org/10.3389/fneur.2022.908468
16. Cui Y-W, Wang L, Zhang L-C (2024) Towards load-bearing biomedical titanium-based alloys: from essential requirements to future developments. Prog Mater Sci 144:101277. https://doi.org/10.1016/j.pmatsci.2024.101277
17. da Silva VA, Bobotis BC, Correia FF, Lima-Vasconcellos TH, Chiarantin GMD, De La Vega L, Lombello CB, Willerth SM, Malmonge SM, Paschon V, Kihara AH (2023) The impact of biomaterial surface properties on engineering neural tissue for spinal cord regeneration. Int J Mol Sci 24(17):13642. https://doi.org/10.3390/ijms241713642
18. Deng M, Kumbar SG, Wan Y, Toti US, Allcock HR, Laurencin CT (2010) Polyphosphazene polymers for tissue engineering: an analysis of material synthesis, characterization and applications. Soft Matter 6(14):3119. https://doi.org/10.1039/b926402g
19. Desai N, Pande S, Gholap AD, Rana D, Salave S, Vora LK (2024) Regulatory processes involved in clinical trials and intellectual property rights around vaccine development. In: Advanced vaccination technologies for infectious and chronic diseases. Elsevier, pp 279–309. https://doi.org/10.1016/B978-0-443-18564-9.00008-4
20. Ding D, Kundukad B, Somasundar A, Vijayan S, Khan SA, Doyle PS (2018) Design of mucoadhesive PLGA microparticles for ocular drug delivery. ACS Appl Bio Mater 1(3):561–571. https://doi.org/10.1021/acsabm.8b00041
21. Dong C, Lv Y (2016) Application of collagen scaffold in tissue engineering: recent advances and new perspectives. Polymers 8(2):42. https://doi.org/10.3390/polym8020042
22. Dorozhkin SV (2011) Biocomposites and hybrid biomaterials based on calcium orthophosphates. Biomatter 1(1):3–56. https://doi.org/10.4161/biom.1.1.16782
23. Duck FA (1990) Mechanical properties of tissue. In: Physical properties of tissues. Elsevier, pp 137–165. https://doi.org/10.1016/B978-0-12-222800-1.50009-7
24. Eftekhar Ashtiani R, Alam M, Tavakolizadeh S, Abbasi K (2021) The role of biomaterials and biocompatible materials in implant-supported dental prosthesis. Evid Based Complement Alternat Med 2021:1–9. https://doi.org/10.1155/2021/3349433
25. Esdaille CJ, Washington KS, Laurencin CT (2021) Regenerative engineering: a review of recent advances and future directions. Regen Med 16(5):495–512. https://doi.org/10.2217/rme-2021-0016
26. Faiyazuddin M, Sophia A, Ashique S, Gholap AD, Gowri S, Mohanto S, Karthikeyan C, Nag S, Hussain A, Akhtar MS, Bakht MA, Ahmed MG, Rustagi S, Rodriguez-Morales AJ, Salas-Matta LA, Mohanty A, Bonilla-Aldana DK, Sah R (2023) Virulence traits and novel drug delivery strategies for mucormycosis post-COVID-19: a comprehensive review. Front Immunol 14. https://doi.org/10.3389/fimmu.2023.1264502
27. Fan L, Ren Y, Emmert S, Vučković I, Stojanovic S, Najman S, Schnettler R, Barbeck M, Schenke-Layland K, Xiong X (2023) The use of collagen-based materials in bone tissue engineering. Int J Mol Sci 24(4):3744. https://doi.org/10.3390/ijms24043744
28. Gholap AD, Bhowmik DD, Deshmukh AY, Hatvate NT (2023a) Quintessential impact of dendrimer bioconjugates in targeted drug delivery, pp 257–302. https://doi.org/10.1016/bs.coac.2023.02.009
29. Gholap AD, Gupta JS, Kamandar PA, Banchhod GV, Hatvate NT (2023b) Antibody-drug conjugates for cancer therapy: an up-to-date review on the chemistry and pharmacology, pp 105–190. https://doi.org/10.1016/bs.coac.2023.02.007
30. Gholap AD, Hatvate NT, Khuspe PR, Mandhare TA, Kashid P, Gaikwad VD (2023c) Proteomics in oncology: retrospect and prospects. In: Novel technologies in biosystems, biomedical & drug delivery. Springer Nature Singapore, pp 243–269. https://doi.org/10.1007/978-981-99-5281-6_10
31. Gholap AD, Hatvate NT, Sawant SD, Khemka PN, Vora LK (2023d) Unlocking the potential of nanobiohybrids to combat environmental pollution. In: Nanobiohybrids for advanced

wastewater treatment and energy recovery. IWA Publishing, pp 169–190. https://doi.org/10.2166/9781789063592_0169
32. Gholap A, Pardeshi S, Giram P (2023e) Preparation, antibacterial and antiviral activity measurements and detection methods. In: Antibacterial and antiviral functional materials, vol 1. ACS Publications, pp 33–64. https://doi.org/10.1021/bk-2023-1458.ch002
33. Gholap AD, Said RP, Pawar RD, Ambore GS, Hatvate NT (2023f) Importance of carbohydrate-drug conjugates in vaccine development: a detailed review, pp 191–256. https://doi.org/10.1016/bs.coac.2023.03.002
34. Gholap AD, Sawant SD, Sayyad SF, Hatvate NT, Chavan M, Rojekar S, Faiyazuddin M (2023g) Food biopackaging for human benefits: status and perspectives. In: Biobased packaging materials. Springer Nature Singapore, pp 1–35. https://doi.org/10.1007/978-981-99-6050-7_1
35. Gholap AD, Pardeshi SR, Hatvate NT, Dhorkule N, Sayyad SF, Faiyazuddin M, Khalid M (2024a) Environmental implications and nanotechnological advances in octocrylene-enriched sunscreen formulations: a comprehensive review. Chemosphere 358:142235. https://doi.org/10.1016/j.chemosphere.2024.142235
36. Gholap AD, Uddin MJ, Faiyazuddin M, Omri A, Gowri S, Khalid M (2024b) Advances in artificial intelligence for drug delivery and development: a comprehensive review. Comput Biol Med 178:108702. https://doi.org/10.1016/j.compbiomed.2024.108702
37. Gleiter H (1996) Microstructure. In: Physical metallurgy. Elsevier, pp 843–942. https://doi.org/10.1016/B978-044489875-3/50014-4
38. Gu X (2021) Biodegradable materials and the tissue engineering of nerves. Engineering 7(12):1700–1703. https://doi.org/10.1016/j.eng.2021.10.011
39. Guo B, Ma PX (2014) Synthetic biodegradable functional polymers for tissue engineering: a brief review. SCIENCE CHINA Chem 57(4):490–500. https://doi.org/10.1007/s11426-014-5086-y
40. Haldar R, Gupta D, Chitranshi S, Singh MK, Sachan S (2019) Artificial blood: a futuristic dimension of modern day transfusion sciences. Cardiovasc Hematol Agents Med Chem 17(1):11–16. https://doi.org/10.2174/1871525717666190617120045
41. Häneke T, Sahara M (2022) Progress in bioengineering strategies for heart regenerative medicine. Int J Mol Sci 23(7):3482. https://doi.org/10.3390/ijms23073482
42. Hasanzadeh E, Seifalian A, Mellati A, Saremi J, Asadpour S, Enderami SE, Nekounam H, Mahmoodi N (2023) Injectable hydrogels in central nervous system: unique and novel platforms for promoting extracellular matrix remodeling and tissue engineering. Mater Today Bio 20:100614. https://doi.org/10.1016/j.mtbio.2023.100614
43. He P, Sahoo S, Ng KS, Chen K, Toh SL, Goh JCH (2013) Enhanced osteoinductivity and osteoconductivity through hydroxyapatite coating of silk-based tissue-engineered ligament scaffold. J Biomed Mater Res A 101A(2):555–566. https://doi.org/10.1002/jbm.a.34333
44. Hoffmann G, Lutter G, Cremer J (2008) Durability of bioprosthetic cardiac valves. Dtsch Arztebl Int. https://doi.org/10.3238/arztebl.2008.0143
45. Khan F, Singh K, Friedman MT (2020) Artificial blood: the history and current perspectives of blood substitutes. Discover 8(1):e104. https://doi.org/10.15190/d.2020.1
46. Kim J-H, Jung E-A, Kim J-E (2024) Perfluorocarbon-based artificial oxygen carriers for red blood cell substitutes: considerations and direction of technology. J Pharm Investig 54(3):267–282. https://doi.org/10.1007/s40005-024-00665-y
47. Krishani M, Shin WY, Suhaimi H, Sambudi NS (2023) Development of scaffolds from bio-based natural materials for tissue regeneration applications: a review. Gels 9(2):100. https://doi.org/10.3390/gels9020100
48. Li Q, Shao X, Dai X, Guo Q, Yuan B, Liu Y, Jiang W (2022) Recent trends in the development of hydrogel therapeutics for the treatment of central nervous system disorders. NPG Asia Mater 14(1):14. https://doi.org/10.1038/s41427-022-00362-y
49. Liu S, Yu J-M, Gan Y-C, Qiu X-Z, Gao Z-C, Wang H, Chen S-X, Xiong Y, Liu G-H, Lin S-E, McCarthy A, John JV, Wei D-X, Hou H-H (2023) Biomimetic natural biomaterials for tissue

engineering and regenerative medicine: new biosynthesis methods, recent advances, and emerging applications. Mil Med Res 10(1):16. https://doi.org/10.1186/s40779-023-00448-w
50. Loh QL, Choong C (2013) Three-dimensional scaffolds for tissue engineering applications: role of porosity and pore size. Tissue Eng Part B Rev 19(6):485–502. https://doi.org/10.1089/ten.teb.2012.0437
51. Lovett M, Lee K, Edwards A, Kaplan DL (2009) Vascularization strategies for tissue engineering. Tissue Eng Part B Rev 15(3):353–370. https://doi.org/10.1089/ten.teb.2009.0085
52. Lu P, Ruan D, Huang M, Tian M, Zhu K, Gan Z, Xiao Z (2024) Harnessing the potential of hydrogels for advanced therapeutic applications: current achievements and future directions. Signal Transduct Target Ther 9(1):166. https://doi.org/10.1038/s41392-024-01852-x
53. Luo J, Walker M, Xiao Y, Donnelly H, Dalby MJ, Salmeron-Sanchez M (2022) The influence of nanotopography on cell behaviour through interactions with the extracellular matrix—a review. Bioact Mater 15:145–159. https://doi.org/10.1016/j.bioactmat.2021.11.024
54. Mao AS, Mooney DJ (2015) Regenerative medicine: current therapies and future directions. Proc Natl Acad Sci 112(47):14452–14459. https://doi.org/10.1073/pnas.1508520112
55. Miri Z, Haugen HJ, Loca D, Rossi F, Perale G, Moghanian A, Ma Q (2024) Review on the strategies to improve the mechanical strength of highly porous bone bioceramic scaffolds. J Eur Ceram Soc 44(1):23–42. https://doi.org/10.1016/j.jeurceramsoc.2023.09.003
56. Mishchenko O, Yanovska A, Kosinov O, Maksymov D, Moskalenko R, Ramanavicius A, Pogorielov M (2023) Synthetic calcium–phosphate materials for bone grafting. Polymers 15(18):3822. https://doi.org/10.3390/polym15183822
57. Mitra J, Tripathi G, Sharma A, Basu B (2013) Scaffolds for bone tissue engineering: role of surface patterning on osteoblast response. RSC Adv 3(28):11073. https://doi.org/10.1039/c3ra23315d
58. Moghadasi K, Mohd Isa MS, Ariffin MA, Mohd Jamil MZ, Raja S, Wu B, Yamani M, Bin Muhamad MR, Yusof F, Jamaludin MF, Ab Karim MSB, Abdul Razak BB, Yusoff NB (2022) A review on biomedical implant materials and the effect of friction stir based techniques on their mechanical and tribological properties. J Mater Res Technol 17:1054–1121. https://doi.org/10.1016/j.jmrt.2022.01.050
59. Mohanto S, Biswas A, Gholap AD, Wahab S, Bhunia A, Nag S, Ahmed MG (2024) Potential biomedical applications of terbium-based nanoparticles (TbNPs): a review on recent advancement. ACS Biomater Sci Eng 10(5):2703–2724. https://doi.org/10.1021/acsbiomaterials.3c01969
60. More MP, Pardeshi SR, Pardeshi CV, Sonawane GA, Shinde MN, Deshmukh PK, Naik JB, Kulkarni AD (2021) Recent advances in phytochemical-based nano-formulation for drug-resistant cancer. Med Drug Discov 10:100082. https://doi.org/10.1016/j.medidd.2021.100082
61. Moztarzadeh S, Mottaghy K, Sefat F, Samadikuchaksaraei A, Mozafari M (2019) Nanoengineered biomaterials for lung regeneration. In: Nanoengineered biomaterials for regenerative medicine. Elsevier, pp 305–323. https://doi.org/10.1016/B978-0-12-813355-2.00013-2
62. Naar Z, Adányi N, Bata-Vidács I, Tömösközi-Farkas R, Tömösközi-Farkas R (2017) Nanoencapsulation technologies for the food and nutraceutical industries S.M. Jafari Nanoencapsulation technologies for the food and nutraceutical industries 125 London Wall, London EC2Y 5AS, United Kingdom: Academic Press, Elsevier Science Publishing Co. Acta Aliment 46(3):390–394. https://doi.org/10.1556/066.2017.46.3.16
63. Naik JB, Pardeshi SR, Patil RP, Patil PB, Mujumdar A (2020) Mucoadhesive micro-/nano carriers in ophthalmic drug delivery: an overview. BioNanoScience 10(3):564–582. https://doi.org/10.1007/s12668-020-00752-y
64. Nappi F, Nenna A, Larobina D, Martuscelli G, Singh SSA, Chello M, Ambrosio L (2021) The use of bioactive polymers for intervention and tissue engineering: the new frontier for cardiovascular therapy. Polymers 13(3):446. https://doi.org/10.3390/polym13030446
65. Nikolova MP, Chavali MS (2019) Recent advances in biomaterials for 3D scaffolds: a review. Bioact Mater 4:271–292. https://doi.org/10.1016/j.bioactmat.2019.10.005

66. Ntege EH, Sunami H, Shimizu Y (2020) Advances in regenerative therapy: a review of the literature and future directions. Regen Ther 14:136–153. https://doi.org/10.1016/j.reth.2020.01.004
67. O'Brien FJ (2011) Biomaterials & scaffolds for tissue engineering. Mater Today 14(3):88–95. https://doi.org/10.1016/S1369-7021(11)70058-X
68. Okamoto M, John B (2013) Synthetic biopolymer nanocomposites for tissue engineering scaffolds. Prog Polym Sci 38(10–11):1487–1503. https://doi.org/10.1016/j.progpolymsci.2013.06.001
69. Pardeshi S, More M, Patil P, Pardeshi C, Deshmukh P, Mujumdar A, Naik J (2021a) A meticulous overview on drying-based (spray-, freeze-, and spray-freeze) particle engineering approaches for pharmaceutical technologies. Dry Technol 39(11):1447–1491. https://doi.org/10.1080/07373937.2021.1893330
70. Pardeshi S, Patil P, Rajput R, Mujumdar A, Naik J (2021b) Preparation and characterization of sustained release pirfenidone loaded microparticles for pulmonary drug delivery: spray drying approach. Dry Technol 39(3):337–347. https://doi.org/10.1080/07373937.2020.1833213
71. Pardeshi S, Damiri F, Zehravi M, Joshi R, Kapare H, Prajapati MK, Munot N, Berrada M, Giram PS, Rojekar S, Ali F, Rahman MH, Barai HR (2022a) Functional thermoresponsive hydrogel molecule to material design for biomedical applications. Polymers 14(15):3126. https://doi.org/10.3390/polym14153126
72. Pardeshi SR, Kole EB, Kapare HS, Chandankar SM, Shinde PJ, Boisa GS, Salgaonkar SS, Giram PS, More MP, Kolimi P, Nyavanandi D, Dyawanapelly S, Junnuthula V (2022b) Progress on thin film freezing technology for dry powder inhalation formulations. Pharmaceutics 14(12):2632. https://doi.org/10.3390/pharmaceutics14122632
73. Pardeshi SR, More MP, Patil PB, Mujumdar A, Naik JB (2022c) Statistical optimization of voriconazole nanoparticles loaded carboxymethyl chitosan-poloxamer based in situ gel for ocular delivery: in vitro, ex vivo, and toxicity assessment. Drug Deliv Transl Res 12(12):3063–3082. https://doi.org/10.1007/s13346-022-01171-0
74. Pardeshi SR, Deshmukh NS, Telange DR, Nangare SN, Sonar YY, Lakade SH, Harde MT, Pardeshi CV, Gholap A, Deshmukh PK, More MP (2023a) Process development and quality attributes for the freeze-drying process in pharmaceuticals, biopharmaceuticals and nanomedicine delivery: a state-of-the-art review. Futur J Pharm Sci 9(1):99. https://doi.org/10.1186/s43094-023-00551-8
75. Pardeshi CV, Kothawade RV, Markad AR, Pardeshi SR, Kulkarni AD, Chaudhari PJ, Longhi MR, Dhas N, Naik JB, Surana SJ, García MC (2023b) Sulfobutylether-β-cyclodextrin: a functional biopolymer for drug delivery applications. Carbohydr Polym 301:120347. https://doi.org/10.1016/j.carbpol.2022.120347
76. Pardeshi SR, More MP, Pardeshi CV, Chaudhari PJ, Gholap AD, Patil A, Patil PB, Naik JB (2023c) Novel crosslinked nanoparticles of chitosan oligosaccharide and dextran sulfate for ocular administration of dorzolamide against glaucoma. J Drug Deliv Sci Technol 86:104719. https://doi.org/10.1016/j.jddst.2023.104719
77. Pardeshi SR, Nikam A, Chandak P, Mandale V, Naik JB, Giram PS (2023d) Recent advances in PLGA based nanocarriers for drug delivery system: a state of the art review. Int J Polym Mater Polym Biomater 72(1):49–78. https://doi.org/10.1080/00914037.2021.1985495
78. Pardeshi S, Pillai D, Giram P, Pardeshi C, Gholap A, Naik J (2024) Chapter 15: nanocarriers for ocular drug delivery. In: Nanocarriers for ocular drug delivery. De Gruyter, pp 469–500. https://doi.org/10.1515/9783111320847-015
79. Parit S, Manchare A, Gholap AD, Mundhe P, Hatvate N, Rojekar S, Patravale V (2024) Antibody-drug conjugates: a promising breakthrough in cancer therapy. Int J Pharm 659:124211. https://doi.org/10.1016/j.ijpharm.2024.124211
80. Partridge B, Eardley A, Morales BE, Campelo SN, Lorenzo MF, Mehta JN, Kani Y, Mora JKG, Campbell E-OY, Arena CB, Platt S, Mintz A, Shinn RL, Rylander CG, Debinski W, Davalos RV, Rossmeisl JH (2022) Advancements in drug delivery methods for the treatment of brain disease. Front Vet Sci 9. https://doi.org/10.3389/fvets.2022.1039745

81. Patil A, Pardeshi S, Kapase M, Patil P, More M, Dhole S, Kole E, Deshmukh P, Gholap A, Mujumdar A, Naik J (2024) Continuous preparation of sustained release vildagliptin nanoparticles using tubular microreactor approach. Dry Technol 42(4):661–673. https://doi.org/10.1080/07373937.2023.2298778
82. Place ES, George JH, Williams CK, Stevens MM (2009) Synthetic polymer scaffolds for tissue engineering. Chem Soc Rev 38(4):1139. https://doi.org/10.1039/b811392k
83. Poetz KL, Shipp DA (2016) Polyanhydrides: synthesis, properties, and applications. Aust J Chem 69(11):1223. https://doi.org/10.1071/CH16144
84. Prasadh S, Wong RCW (2018) Unraveling the mechanical strength of biomaterials used as a bone scaffold in oral and maxillofacial defects. Oral Sci Int 15(2):48–55. https://doi.org/10.1016/S1348-8643(18)30005-3
85. Raikar AS, Priya S, Bhilegaonkar SP, Somnache SN, Kalaskar DM (2023) Surface engineering of bioactive coatings for improved stent hemocompatibility: a comprehensive review. Materials 16(21):6940. https://doi.org/10.3390/ma16216940
86. Revete A, Aparicio A, Cisterna BA, Revete J, Luis L, Ibarra E, Segura González EA, Molino J, Reginensi D (2022) Advancements in the use of hydrogels for regenerative medicine: properties and biomedical applications. Int J Biomater 2022:1–16. https://doi.org/10.1155/2022/3606765
87. Rojekar S, Gholap AD, Togre N, Bhoj P, Haeck C, Hatvate N, Singh N, Vitore J, Dhoble S, Kashid S, Patravale V (2024) Current status of mannose receptor-targeted drug delivery for improved anti-HIV therapy. J Control Release 372:494–521. https://doi.org/10.1016/j.jconrel.2024.06.002
88. Ronoh K, Mwema F, Dabees S, Sobola D (2022) Advances in sustainable grinding of different types of the titanium biomaterials for medical applications: a review. Biomed Eng Adv 4:100047. https://doi.org/10.1016/j.bea.2022.100047
89. Sagar P, Prasad K, Lalitha RM, Ranganath K (2015) Cyanoacrylate for intraoral wound closure: a possibility? Int J Biomater 2015:1–6. https://doi.org/10.1155/2015/165428
90. Sanders L, Nagatomi J (2014) Clinical applications of surgical adhesives and sealants. Crit Rev Biomed Eng 42(3–4):271–292. https://doi.org/10.1615/CritRevBiomedEng.2014011676
91. Satchanska G, Davidova S, Petrov PD (2024) Natural and synthetic polymers for biomedical and environmental applications. Polymers 16(8):1159. https://doi.org/10.3390/polym16081159
92. Sefat F, Raja TI, Zafar MS, Khurshid Z, Najeeb S, Zohaib S, Ahmadi ED, Rahmati M, Mozafari M (2019) Nanoengineered biomaterials for cartilage repair. In: Nanoengineered biomaterials for regenerative medicine. Elsevier, pp 39–71. https://doi.org/10.1016/B978-0-12-813355-2.00003-X
93. Shabib Akhtar M, Chandrasekaran K, Saminathan S, Rajalingam SR, Mohsin N, Awad Alkarem Ahmed KA, Alhazmi Y, Walbi IA, Abdel-Wahab BA, Gholap AD, Faiyazuddin M, Sundaram G (2024) Nanoengineered chitosan functionalized titanium dioxide biohybrids for bacterial infections and cancer therapy. Sci Rep 14(1):3705. https://doi.org/10.1038/s41598-024-52847-1
94. Shaikh MS, Faiyazuddin M, Khan MS, Pathan SK, Syed IJ, Gholap AD, Akhtar MS, Sah R, Mehta R, Sah S, Bonilla-Aldana DK, Luna C, Rodriguez-Morales AJ (2024) Chikungunya virus vaccine: a decade of progress solving epidemiological dilemma, emerging concepts, and immunological interventions. Front Microbiol 15. https://doi.org/10.3389/fmicb.2024.1413250
95. Shiva S, Asuwin Prabu RG, Bajaj G, John AE, Chandran S, Kumar VV, Ramakrishna S (2023) A review on the recent applications of synthetic biopolymers in 3D printing for biomedical applications. J Mater Sci Mater Med 34(12):62. https://doi.org/10.1007/s10856-023-06765-9
96. Simitzi C, Ranella A, Stratakis E (2017) Controlling the morphology and outgrowth of nerve and neuroglial cells: the effect of surface topography. Acta Biomater 51:21–52. https://doi.org/10.1016/j.actbio.2017.01.023
97. Singh S, Chunglok W (2022a) Biopolymers towards green and sustainable development. Bentham Science Publishers. https://doi.org/10.2174/97898150793021220101

98. Singh S, Chunglok W (2022b) Potential sources of biodegradable polymers. In: Biopolymers towards green and sustainable development. Bentham Science Publishers. https://doi.org/10.2174/97898150793021220101
99. Singh T, Kumar VR, Gholap AD, Sanap G, Anjaiah B, Roja K (2016) In vitro antipyrial activity of psidium guajava leaf extract. Eur J Pharm Med Res 3(1):312–315
100. Song R, Murphy M, Li C, Ting K, Soo C, Zheng Z (2018) Current development of biodegradable polymeric materials for biomedical applications. Drug Des Devel Ther 12:3117–3145. https://doi.org/10.2147/DDDT.S165440
101. Suamte L, Tirkey A, Babu PJ (2023) Design of 3D smart scaffolds using natural, synthetic and hybrid derived polymers for skin regenerative applications. Smart Mater Med 4:243–256. https://doi.org/10.1016/j.smaim.2022.09.005
102. Syukri DM, Nwabor OF, Singh S, Ontong JC, Wunnoo S, Paosen S et al (2020) Antibacterial-coated silk surgical sutures by ex situ deposition of silver nanoparticles synthesized with *Eucalyptus camaldulensis* eradicates infections. J Microbiol Methods 174:105955
103. Syukri DM, Nwabor OF, Singh S, Voravuthikunchai SP (2021) Antibacterial functionalization of nylon monofilament surgical sutures through in situ deposition of biogenic silver nanoparticles. Surf Coat Technol 413:127090
104. Syukri DM, Singh S, Nwabor OF, Ontong JC, Dejyong K, Sunghan J et al (2024) Prevention of post-operative bacterial colonization on mice buccal mucosa using biogenic silver nanoparticles-coated nylon sutures. Regen Eng Transl Med 10(5):1–15
105. Szwed-Georgiou A, Płociński P, Kupikowska-Stobba B, Urbaniak MM, Rusek-Wala P, Szustakiewicz K, Piszko P, Krupa A, Biernat M, Gazińska M, Kasprzak M, Nawrotek K, Mira NP, Rudnicka K (2023) Bioactive materials for bone regeneration: biomolecules and delivery systems. ACS Biomater Sci Eng 9(9):5222–5254. https://doi.org/10.1021/acsbiomaterials.3c00609
106. Taghizadeh B, Ghavami L, Derakhshankhah H, Zangene E, Razmi M, Jaymand M, Zarrintaj P, Zarghami N, Jaafari MR, Moallem Shahri M, Moghaddasian A, Tayebi L, Izadi Z (2020) Biomaterials in valvular heart diseases. Front Bioeng Biotechnol 8. https://doi.org/10.3389/fbioe.2020.529244
107. Thangavel M, Elsen Selvam R (2022) Review of physical, mechanical, and biological characteristics of 3D-printed bioceramic scaffolds for bone tissue engineering applications. ACS Biomater Sci Eng 8(12):5060–5093. https://doi.org/10.1021/acsbiomaterials.2c00793
108. Turnbull G, Clarke J, Picard F, Riches P, Jia L, Han F, Li B, Shu W (2018) 3D bioactive composite scaffolds for bone tissue engineering. Bioact Mater 3(3):278–314. https://doi.org/10.1016/j.bioactmat.2017.10.001
109. Vach Agocsova S, Culenova M, Birova I, Omanikova L, Moncmanova B, Danisovic L, Ziaran S, Bakos D, Alexy P (2023) Resorbable biomaterials used for 3D scaffolds in tissue engineering: a review. Materials 16(12):4267. https://doi.org/10.3390/ma16124267
110. Vaiani L, Boccaccio A, Uva AE, Palumbo G, Piccininni A, Guglielmi P, Cantore S, Santacroce L, Charitos IA, Ballini A (2023) Ceramic materials for biomedical applications: an overview on properties and fabrication processes. J Funct Biomater 14(3):146. https://doi.org/10.3390/jfb14030146
111. Viennot S, Dalard F, Lissac M, Grosgogeat B (2005) Corrosion resistance of cobalt-chromium and palladium-silver alloys used in fixed prosthetic restorations. Eur J Oral Sci 113(1):90–95. https://doi.org/10.1111/j.1600-0722.2005.00190.x
112. Vora LK, Gholap AD, Hatvate NT, Naren P, Khan S, Chavda VP, Balar PC, Gandhi J, Khatri DK (2024) Essential oils for clinical aromatherapy: a comprehensive review. J Ethnopharmacol 330:118180. https://doi.org/10.1016/j.jep.2024.118180
113. Vyas J, Shah I, Singh S, Prajapati BG (2023) Biomaterials-based additive manufacturing for customized bioengineering in management of otolaryngology: a comprehensive review. Front Bioeng Biotechnol 11:1234340
114. Vyas J, Raytthatha N, Singh S, Prajapati BG, Mohite P, Munde S (2024) Sustainable sources of raw materials as substituting biomaterials for additive manufacturing of dental implants: a review. Perio Implant Res 8(1):3

115. Wang F, Cai X, Shen Y, Meng L (2023a) Cell–scaffold interactions in tissue engineering for oral and craniofacial reconstruction. Bioact Mater 23:16–44. https://doi.org/10.1016/j.bioactmat.2022.10.029
116. Wang Y, Wang Z, Dong Y (2023b) Collagen-based biomaterials for tissue engineering. ACS Biomater Sci Eng 9(3):1132–1150. https://doi.org/10.1021/acsbiomaterials.2c00730
117. Wang Y, van Putten R-J, Tietema A, Parsons JR, Gruter G-JM (2024) Polyester biodegradability: importance and potential for optimisation. Green Chem 26(7):3698–3716. https://doi.org/10.1039/D3GC04489K
118. Wei Q, Zhou J, An Y, Li M, Zhang J, Yang S (2023) Modification, 3D printing process and application of sodium alginate based hydrogels in soft tissue engineering: a review. Int J Biol Macromol 232:123450. https://doi.org/10.1016/j.ijbiomac.2023.123450

Biomaterials for Regenerative Medicine: Historical Perspectives and Current Trends

2

Ayush Madan, Abhik Paul, Krishnendu Adhikary, Anas Islam, Sumel Ashique, Radheshyam Pal, Masoumeh Bagheri, and Naheed Mojgani

Abstract

Biomaterials have undergone substantial development in regenerative medicine, beginning as inert supports and progressing to bioactive materials that drive tissue regeneration. The purpose of these biomaterials is to imitate the natural

A. Madan
Department of Biotechnology, School of Research and Technology, People's University, Bhopal, Madhya Pradesh, India

Institute of Tropical Aquaculture and Fisheries, Universiti Malaysia Terengganu, Terengganu, Malaysia

A. Paul
Department of Pharmaceutical Technology, NSHM Knowledge Campus, Kolkata Group of Institutions, Kolkata, West Bengal, India

Department of Pharmaceutical Technology, Jadavpur University, Kolkata, West Bengal, India

K. Adhikary
Department of Interdisciplinary Science, Centurion University of Technology and Management, R.Sitapur, Odisha, India

A. Islam
Faculty of Pharmacy, Integral University, Lucknow, Uttar Pradesh, India

S. Ashique (✉)
Department of Pharmaceutical Technology, Bharat Technology, Uluberia, West Bengal, India

R. Pal
Department of Pharmaceutics, Bengal College of Pharmaceutical Sciences & Research, Durgapur, West Bengal, India

M. Bagheri · N. Mojgani
Biotechnology Department, Razi Vaccine and Serum Research Institute—Agricultural Research, Education and Extension Organization, Markazi Province, Iran

S. Singh et al. (eds.), *Biomaterial-based Additive Manufacturing in Tissue Engineering and Regeneration*, Biomaterials, Bioengineering and Sustainability 7,
https://doi.org/10.1007/978-3-031-96070-3_2

extracellular matrix to encourage cell adhesion, migration, proliferation, and differentiation. There has been a change toward an integrated strategy involving material researchers, biology researchers, and physicians, with a primary emphasis on tissue engineering and regenerative medicine, according to historically significant perspectives. They play an essential part in bone tissue engineering by facilitating bone regeneration through various scaffolds, including hydrogels, nanofiber scaffolds, and composite scaffolds printed using 3D printing technology. To facilitate cell proliferation and differentiation to replicate the extracellular matrix, scaffolds are synthesized from natural or synthetic polymers to create biomaterials for use in tissue engineering. Developing highly biodegradable frameworks that include antimicrobial qualities is now a topic that is being pursued to accelerate tissue regeneration and prevent infections. The multidisciplinary advancements in biomaterials continue to propel innovation in regenerative medicine. These advancements highlight the significance of cell-biomaterial interactions and the harmonious degradation of biomaterials for the successful regeneration of tissue. To summarize, the development of biomaterials in regenerative medicine represents a multidisciplinary strategy intending to develop implants that speed up the process of tissue regeneration. This chapter will provide the perspectives on biomaterials in regenerative medicine.

Keywords

Biomaterials · Regenerative medicine · Drug-delivery systems · Scaffolds · Natural polymers · Synthetic polymers

2.1 Introduction

Regenerative means replacing or rejuvenating damaged human cells, tissues, organs, or muscles. Other than this, regenerative medicines help bracing the human body's natural healing process to cure damaged cells and tissues. It is a medicinal stream where scientists apply life science and engineering together for clinical therapies. Several cells, primarily stem cells and primary cells, have been effectively used to regenerate organs and tissues [1, 2]. However, in practical cases, it has been shown that only a tiny percentage of cells can be embedded at the target site, and a significant percentage of cells die within the initial hour of delivery. To minimize cell death and promote cell engraftment, a combination of cells and biomaterials has been developed [2, 3]. Biomaterial components have a pivotal role in regenerative medical applications such as artificial organs, cell-based therapies, living tissues, etc., and heavily rely on the capacity of biomaterial. Biomaterials are not new; instead, their existence has been around us for many years, and recent advancements in materials science are focused on developing biomaterials, particularly for tissue engineering that are highly bioactive, biodegradable, and capable of interacting with tissues on multiple levels [4, 5]. Biomaterials can be considered "tools" that help achieve a successful therapeutic outcome independently or in conjunction with other things. These therapeutic outcomes can be made positively by precisely manipulating the physiochemical characteristics of such materials. Biodegradable materials, particularly polymers like poly-lactic acid or poly-glycolic acid, do not create toxic elements when they dissolve or break down. On the other hand, non-biodegradable materials can produce harmful toxins when they break down or can create long-term issues by being persistent in the body for an infinite amount of time. Picking biodegradable components for surgical implants minimizes the likelihood of chronic inflammation and reduces the procedure of monitoring the patient regularly [6]. While designing biomaterials for regenerative medicine, some factors, such as biomaterial's capacity to sustain cell survival, optimal cellular function upon transplantation, and enhancement of autologous tissue growth in situ, etc. must be considered [7]. Biobased and biodegradable polymers provide an extensive range of engineering prospects, including regulated rates of degradation, modified mechanical attributes, and surface qualities that complement human physiology. To support the generation of new tissue, a biomaterial framework must provide structure, mechanical support, and cellular-level design. Various manufacturing techniques have been explored to produce biomaterials that mimic the physiochemical properties of target tissues. To enhance different aspects of biomaterial scaffolds, a range of composite biomaterials has been developed by combining natural and synthetic bioactive materials [2]. The technical and commercial development of biomaterials involves numerous mathematical applications, along with physical, chemical, and biological principles, aimed at enhancing the performance of biologically based industrial processes. Further research into biomaterial medicines in the medical field, through in vitro and in vivo studies, is necessary to determine their therapeutic potential, evaluate safety and efficacy, test the final product, assess clinical performance and human safety, and obtain regulatory approval based on satisfactory product performance [5]. This chapter seeks to provide a comprehensive overview of the history and current trends of biomaterials in regenerative medicine.

2.2 Biomaterials and Regenerative Medicine

Using bio-based and bio-active materials for medical purposes is not a new example; instead, in history, it has been found that ancient people used to apply such materials as medical tools. Findings from archaeological research revealed that materials utilized for replacing lost human bones and teeth include wood, corals, shells, and many metals. Other examples can be considered as wooden dental implants used by Neanderthal men, using gold and silver for dental repair, using iron threads to repair bone fractures, etc. Another instance of a lifelong foreign material injection into the dermis is tattooing, which has been around for nearly 5000 years. Around 4000 years ago, the Egyptian people used linen threads to heal wounds, and they even set the most significant example of bio-based substances in medicines through "mummification". Surprisingly, the heads of large ants were used in South Africa and India to hold two edges of wounds. In ancient Greece, golden wires had been used by surgeons and physicians as ligatures. Inca physicians used Golden plates to repair cranial fractures and restore bone tissue. Also, ancient Mayan civilizations developed prosthetic teeth out of seashells, through which osseointegration was achieved in that era [8]. Most primitive alternate substances were bioinert or physiologically inert; they interact with surrounding tissues less effectively and even be more harmful to people [9]. As metals were used for replacements and surgeries, all those metals were tested on various animal tissues where some metals showed a discoloration effect, some showed inadequate mechanical features and metals like steel and iron showed quick corrosion. Studies of substances that are essential for the creation of implants and prostheses were conducted after World War II in the urge for the treatment of innumerable casualties. As the heart is considered a pump, in 1812, Le Gallois put forward the idea of replacing the heart with an artificial pump, and he also stated that organs can also be preserved by circulating blood through them. Later, between 1910 and 1920, modern metallic, polymeric materials, and ceramic were created, though these were used for military purposes. Thus, substances like Teflon, silicones, nylons, polyurethanes, methacrylates, etc., started using for medical purposes by surgeons [2]. Therefore, scientists and researchers have been trying to replace or repair human tissue or organs with foreign substances for centuries, and they met possibilities and failure simultaneously.

Around the sixteenth century, some plastic surgeons, including Gaspare Tagliacozzi, had successfully replaced missing noses by using autogenous skin flaps. These real-life or practical surgeries or operations had been done without any constraints related to problems and limitations of material science. At that time, researchers or surgeons were unaware of factors like sterilization, inflammation, immunological reaction, etc. [8]. However, all these research and successes have pointed out that the human body can change to tolerate alien substances. This is how the path for biomaterials and their application in regenerative medicines started, and it is continuing. It was only approximately 150 years ago that medical professionals began to systematically investigate how the body behaves when the insertion of foreign molecules occurs.

2.3 Characteristics of Biomaterials

The development of regenerative medicines is dependent on cell activity as the innate repairing capability of harmed or injured tissues is minimal for cells to grow in number, migrate, and differentiate. The development of regenerative medicines is impossible without using biomaterials, which offer practical ways to improve cell biology both *in vivo* and *in vitro*. The primary function or aim of biomaterials is to replace or substitute damaged tissue or cells, and this can be accomplished by integrating the chemical, physical, and biological properties of biomaterials. Some required characteristics of biomaterials are briefly mentioned below.

2.3.1 Biocompatibility

In terms of biology, biocompatibility is the phenomenon of accepting non-living foreign substances by living cells or tissues. It is essential that an implanted biomaterial does not adversely affect the surrounding tissues or host organs, as it is intended for use in close contact with living tissues. Along with this, regular cellular activity should not be altered or affected by the implanted material. It is possible for a biocompatible material to be partially "inert"; in actuality, it all depends on how the host reacts. The biocompatibility feature is dynamic in nature; for instance, a biomaterial that is biocompatible for bone implants may not work for skin or tissue replacement.

2.3.2 Non-toxicity

The main reason implants and other regenerative medicines are replaced from non-biodegradable to biodegradable is 'toxicity'. Non-biodegradable materials cause long-lasting complications by breaking down into byproducts, which either directly affect the human body or create toxins by reacting with targeted cells. In the context of a living creature, a well-constructed biological substance should fulfil its function without endangering other cells, organs, or the body as an entire unit. Therefore, the toxic effects of biomaterials should be examined before they are implanted.

2.3.3 Host Response

A broad spectrum of biomaterials has not yet reached clinical applications as they affect patient health, which is brought on by the host immunological response. The host tissue or cells are sensitive to the materials and can cause some life-long complications. Most elements are never inert, and the way a foreign substance interacts with the host tissue has a significant impact on the therapeutic outcome of a biomaterial. The interaction between a host site and biomaterials often relies on the duration and location of implantation [10]. The debasement of the implanted or induced materials activates the immune system.

2.3.4 Mechanical Property

Along with biocompatibility, no-toxicity, and host response, another prominent feature of a biomaterial is its mechanical feature. Focusing on the mechanical features of a biomaterial is crucial as it is liable to preserve the three-dimensional porosity structure to promote cell adhesion, migration, infiltration, and proliferation [11]. Prior to implementation, mechanical qualities are included in the design of biomaterials. When implanted into the living part, a material undergoes several phenomena such as shear, stress, strain, corrosion, etc.

2.4 Different Sources of Biomaterials

Regenerative medicine seeks to repair and restore the structure and function of damaged tissues and organs through biomaterials. These biomaterials can be sourced from various natural and synthetic origins, each with unique properties that influence their suitability for different regenerative applications. Understanding the different sources of biomaterials is crucial for developing effective regenerative strategies.

2.4.1 Natural Sources of Biomaterials

Natural biomaterials, sourced from living organisms, inherently offer biocompatibility and bioactivity. Natural polymers, obtained from biological origins such as plants, animals, and microorganisms, are particularly suitable for tissue engineering due to their excellent biocompatibility and bioactivity. Examples of these natural polymers include collagen, gelatin, chitosan, hyaluronic acid, and alginate. Collagen, the most abundant protein in the human body, provides structural support and enhances cell adhesion, making it a favored material for scaffolds in regenerative medicine. Gelatin, derived from collagen, retains many of collagen's properties and is utilized in wound dressings and drug delivery systems. Chitosan, a polysaccharide from chitin, possesses antimicrobial properties and is applied in tissue engineering and wound healing. Hyaluronic acid, which is found in connective tissues, is known for its viscoelastic properties and is used in osteoarthritis treatment and tissue engineering. Alginate, extracted from seaweed, forms hydrogels that are used for cell encapsulation and drug delivery [12].

2.4.1.1 Extracellular Matrix Components

The extracellular matrix (ECM) offers both structural support and biochemical signals that facilitate cell attachment, proliferation, and differentiation. Natural ECM components like collagen, fibrin, and elastin closely resemble the microenvironment of native tissues. Collagen, the body's most abundant protein, is particularly favored in tissue engineering because of its biocompatibility and its capacity to enhance cell adhesion and proliferation.

2.4.1.2 Polysaccharides

Polysaccharides such as chitosan, hyaluronic acid, and alginate are widely employed in regenerative medicine. Chitosan, derived from chitin, exhibits antimicrobial properties, and supports wound healing. Hyaluronic acid is essential for cell signaling and tissue hydration. Alginate forms hydrogels capable of encapsulating cells and bioactive substances, making it suitable for applications in drug delivery and tissue engineering.

2.4.1.3 Proteins

Proteins like silk fibroin and gelatin are used in regenerative medicine. Silk fibroin, derived from silkworm cocoons, is biocompatible and mechanically robust, making it suitable for scaffolds. Gelatin, derived from collagen, is biodegradable and promotes cell adhesion and growth.

2.4.1.4 Extracellular Vehicles

Extracellular vehicles (EVs) are tiny vehicles released by cells that facilitate intercellular communication. They contain bioactive molecules such as proteins, lipids, and nucleic acids, making them promise for applications in regenerative medicine. For instance, EVs sourced from mesenchymal stem cells have demonstrated capabilities in enhancing tissue regeneration and alleviating inflammation.

2.4.2 Synthetic Sources of Biomaterials

Synthetic biomaterials are created through chemical processes and can be customized to possess specific characteristics (Fig. 2.1). Synthetic polymers, manufactured materials with adjustable properties, can mimic the extracellular matrix (ECM) found in tissues. These polymers offer benefits such as mechanical strength, control over degradation, and ease of fabrication. Examples of synthetic polymers employed in regenerative medicine include poly (lactic-co-glycolic acid) (PLGA), polyethylene glycol (PEG), and polycaprolactone (PCL). PLGA, a biodegradable polymer, is widely utilized in drug delivery and tissue engineering due to its predictable degradation and compatibility with biological systems. PEG, a hydrophilic polymer, can be modified to improve cell attachment and drug release capabilities [13]. PCL, a biodegradable polyester, is used in scaffolds for tissue engineering and drug delivery applications.

2.4.2.1 Polymers

Poly (lactic-co-glycolic acid) (PLGA), polyethylene glycol (PEG), and polycaprolactone (PCL) are extensively utilized in regenerative medicine as synthetic polymers. PLGA is biodegradable and has tunable degradation rates, making it suitable for drug delivery. PEG is hydrophilic and can be modified to improve its biocompatibility. PCL is biocompatible and has mechanical properties like human tissues.

2.4.2.2 Ceramics

Ceramics are inorganic materials known for their biocompatibility and osteoconductive properties, making them appropriate for bone regeneration. Common ceramics utilized in regenerative medicine include hydroxyapatite (HA), calcium phosphate, and bio glass. Synthetic ceramics like hydroxyapatite (HA) and tricalcium phosphate (TCP) are favored for bone regeneration due to their similarity to the mineral composition of bone. HA, specifically, promotes bone formation as it mimics the main mineral found in bone and is often used in bone grafts and implant coatings [14]. TCP is biocompatible and gradually resorbs, making it suitable for

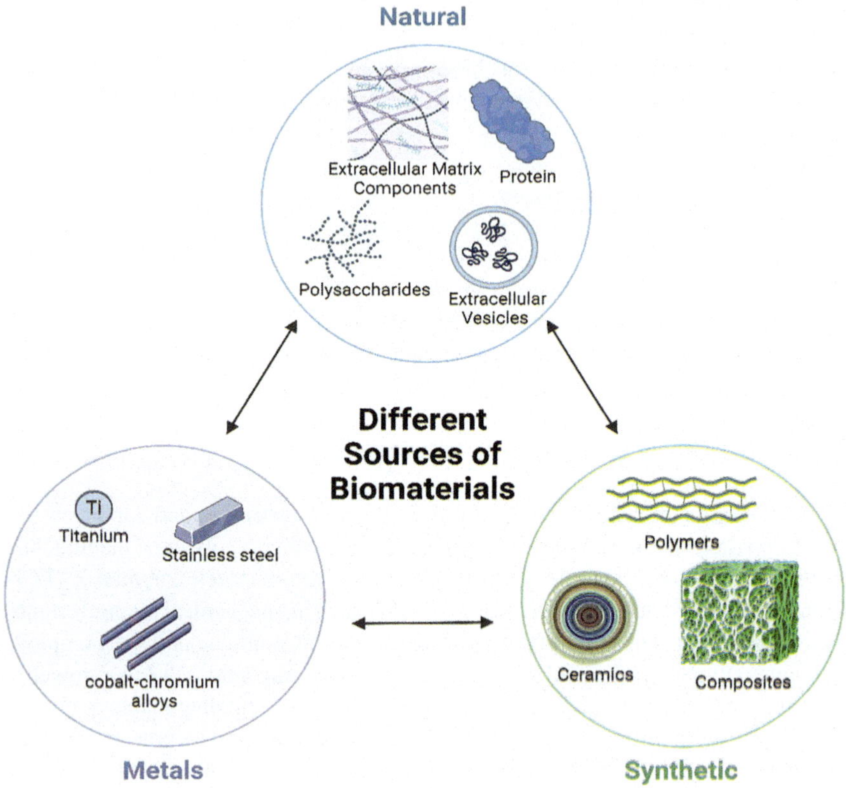

Fig. 2.1 Different sources of biomaterials. (Created by Biorender.com)

bone grafting. Calcium phosphate ceramics such as tricalcium phosphate (TCP) and biphasic calcium phosphate (BCP) are also utilized in bone regeneration due to their resemblance to natural bone minerals [15]. Bio glass, a bioactive glass, bonds with bone tissue and stimulates new bone growth, making it valuable for bone grafts and scaffolds in tissue engineering [16].

2.4.2.3 Composites

Composite biomaterials combine natural and synthetic components to achieve synergistic properties. Composites are materials that blend two or more types of biomaterials to attain desired properties. For instance, composite scaffolds may integrate natural polymers with ceramics or metals to enhance both mechanical strength and bioactivity. These composite biomaterials find wide application in regenerative medicine, including areas such as bone regeneration, cartilage repair, and wound healing [17]. For example, scaffolds combining collagen and PLGA combine collagen's biocompatibility with PLGA's robust mechanical properties, making them well-suited for tissue engineering purposes.

2.4.2.4 Metals

They are another source of biomaterials for regenerative medicine, particularly in orthopedic and dental applications. Metals offer high mechanical strength and durability, making them suitable for load-bearing implants. Common metals used in regenerative medicine include titanium, stainless steel, and cobalt-chromium alloys. Titanium is biocompatible and corrosion-resistant, making it an ideal material for dental implants and joint replacements [18]. Stainless steel is widely used in orthopedic implants due to its strength and corrosion resistance [19]. Cobalt-chromium alloys are also used in orthopedic implants and have excellent wear resistance. Overall, biomaterials derived from natural and synthetic sources, as well as metals, ceramics, and composites, play a critical role in regenerative medicine. These biomaterials offer a wide range of properties and advantages for tissue engineering and regenerative therapies. By understanding the unique properties of each biomaterial source, researchers can continue to develop innovative strategies for regenerative medicine.

2.5 Types of Biomaterials in Regenerative Medicine

Biomaterials play a crucial role in regenerative medicine by providing essential support for tissue regeneration and repair. They are broadly classified into three main categories: polymers (both synthetic and natural), metals, and ceramics (including carbon-based materials, glass-ceramics, and glasses). Each category possesses distinct properties and applications that contribute to the progression of regenerative medicine.

2.5.1 Polymers: Synthetic and Natural

Polymers, whether synthetic or natural, are highly versatile materials used extensively in regenerative medicine and other biomedical applications. These materials are composed of large, repeating units known as monomers, forming long chains with diverse properties and functionalities.

2.5.1.1 Synthetic Polymers

Synthetic polymers are engineered materials designed to meet specific requirements for various medical applications. They offer controllable properties such as degradation rates, mechanical strength, and chemical composition, making them highly adaptable for different biomedical uses.

2.5.1.1.1 Poly (Methyl Methacrylate) (PMMA)

PMMA is a hydrophobic polymer extensively utilized in orthopedic applications as bone cement because of its outstanding mechanical properties and biocompatibility. Its high transparency also makes it suitable for use in intraocular lenses (IOLs) and rigid contact lenses.

2.5.1.1.2 Poly (Glycolic Acid) (PGA) and Poly (Glycolic-Co-lactic Acid) (PLGA)

PGA is frequently employed in absorbable surgical sutures, whereas PLGA finds application in resorbable surgical sutures, drug delivery systems, and orthopedic fixation devices. PLGA's degradation products lactic acid and glycolic acid—are non-toxic, which improves its appropriateness for biomedical uses [20].

2.5.1.1.3 Poly (Hydroxyethyl Methacrylate) (PHEMA): PHEMA Is Used to Produce Soft Contact Lenses

To maintain dimensional stability, it is cross-linked with ethylene glycol dimethacrylate (EGDMA). Adding trace amounts of Poly methacrylic acid enhances the wettability of the lenses, improving comfort and usability.

2.5.1.1.4 Polypropylene (PP)

PP is an isotactic crystalline polymer valued for its strong rigidity and tensile strength, which render it suitable for applications in surgical sutures and hernia repair. Its biocompatibility and mechanical characteristics make it a preferred option for these medical uses.

2.5.1.1.5 Poly (Vinyl Alcohol) (PVA)

PVA is used in biomedical applications such as tubing and blood storage bags, including blood transfusion tubes and dialysis tubing. Its biocompatibility and chemical stability are crucial for these applications.

2.5.1.1.6 Polyethylene (PE) and Polyethylene Terephthalate (PET)

PE is used in various medical devices due to its durability and biocompatibility. PET is used in vascular grafts and other implantable devices due to its strength and resistance to biodegradation [21].

2.5.1.1.7 Silicone Rubber (SR)

SR is widely used in medical devices due to its flexibility, biocompatibility, and stability. It is used in catheters, tubing, and implants [22].

2.5.1.1.8 Polytetrafluoroethylene (PTFE)

PTFE, known for its low friction and high chemical resistance, is used in vascular grafts and other implantable devices. Its inertness and stability make it suitable for long-term applications in the body.

2.5.1.1.9 Poly (Vinyl Chloride) (PVC)

PVC is utilized in diverse medical applications such as blood bags and tubing because of its flexibility, durability, and ease of manufacturing. The addition of plasticizers further enhances its flexibility, making it well-suited for medical device applications.

2.5.1.1.10 Poly (Ethylene Glycol) (PEG)

PEG is used in drug delivery systems and as a surface modifier to improve the biocompatibility of other materials. Its hydrophilicity and non-toxicity make it suitable for various biomedical applications.

2.5.1.2 Natural Polymers

Natural polymers originate from biological sources and offer excellent biocompatibility and bioactivity, making them ideal for regenerative medicine applications.

2.5.1.2.1 Alginate

Alginate is a natural polysaccharide made up of repeating units of L-guluronic acid and D-mannuronic acid. When it encounters calcium ions, alginate forms a three-dimensional gel that is employed to encapsulate growth factors or specific types of cells. Alginate gels have been shown to support chondrocyte phenotype maintenance, making them suitable for cartilage regeneration. Alginate gels also provide a supportive environment for chondrocyte storage, with cells surviving up to 8 months in these gels [12].

2.5.1.2.2 Collagen

Collagen, a fundamental element in human skin and other connective tissues, is well-suited for applications in skin regeneration. Combining bovine collagen with human autologous keratinocytes and fibroblasts has demonstrated encouraging outcomes in treating severe burn injuries. Collagen is also employed in bilayer artificial skin constructs, like Integra, which integrates collagen with glycosaminoglycans to facilitate skin regeneration.

2.5.1.2.3 Hyaluronic Acid (Hyaluronan)

Hyaluronic acid is employed in tissue regeneration; however, its high solubility in water and rapid clearance from host tissues restrict its application. Chemical modifications, such as esterification with benzyl alcohol to form HYAFF 11, enhance its stability and make it suitable for cartilage repair and other regenerative applications [23].

2.5.1.2.4 Fibrin

Fibrin is a natural component involved in wound healing. Fibrin gels are used as scaffolds for tissue regeneration, but their fast biodegradation due to fibrinolysis is a limitation. Modified fibrin gels, such as Tissucol, provide longer-lasting scaffolds for extracellular matrix development by chondrocytes.

2.5.1.2.5 Other Natural Polymers

Cellulose, agarose, chitosan, and gelatin are utilized as scaffold materials in regenerative medicine, providing diverse properties such as biocompatibility and the ability to promote cell growth and tissue regeneration.

2.5.2 Metals

Metals are extensively utilized in biomedical applications for their robust mechanical strength, durability, and malleability. However, they can pose challenges such as limited biocompatibility and susceptibility to corrosion in physiological conditions.

2.5.2.1 Stainless Steels (316, 316L)

Stainless steels are employed in orthopedic implants, surgical instruments, and dental devices because of their strength and ability to resist corrosion. The low-carbon variant (316L) offers improved corrosion resistance, making it suitable for long-term implants.

2.5.2.2 Titanium and Titanium Alloys

Titanium and its alloys, such as Ti-6Al-4V, are favored for implants because they exhibit exceptional biocompatibility, corrosion resistance, and favorable mechanical properties. Titanium implants find widespread use in dental, orthopedic, and cardiovascular applications [24].

2.5.2.3 Cobalt-Chromium Alloys

These alloys are utilized in joint replacements and dental prosthetics because of their superior wear resistance and strength. They provide durable solutions for load-bearing implants [25].

2.5.2.4 Tantalum and Silver

Tantalum is chosen for orthopedic and dental implants because it possesses outstanding biocompatibility and corrosion resistance. Silver, known for its antimicrobial properties, is used in wound dressings and coatings for medical devices to prevent infections [26].

2.5.2.5 Vitallium

Vitallium, an alloy composed of cobalt, chromium, and molybdenum, is employed in orthopedic and dental implants owing to its exceptional strength and biocompatibility. It provides durable and long-lasting solutions for various medical applications [22].

2.5.3 Ceramics

Ceramics, encompassing carbons, glass-ceramics, and glasses, are inorganic and non-metallic materials known for their excellent biocompatibility and resistance to corrosion. These attributes render them highly suitable for a wide range of biomedical applications.

2.5.3.1 Hydroxyapatite (HAp)

Hydroxyapatite (HAp) is a calcium phosphate ceramic that is essential for bone tissue regeneration. Its chemical similarity to natural bone mineral makes it an ideal material for bone grafts and coatings for metal implants to enhance osseointegration.

2.5.3.2 Aluminum Oxide

Aluminum oxide is used in dental implants, joint replacements, and bone fillers due to its high hardness, wear resistance, and biocompatibility. It provides a durable and long-lasting solution for load-bearing applications [27].

2.5.3.3 Calcium Phosphates

Calcium phosphate ceramics, such as tricalcium phosphate and biphasic calcium phosphate, are used in bone grafts and scaffolds for bone regeneration. They provide bioactive surfaces that support bone cell attachment and proliferation [14].

2.5.3.4 Bioglass

Bioglass is a type of bioactive glass that promotes bone bonding and regeneration. It is used in bone grafts, coatings for metal implants, and dental applications due to its ability to form a bond with bone tissue [28].

Carbon-Based Materials Carbon materials, including carbon fibers and carbon nanotubes, are used in biomedical applications for their strength, biocompatibility, and electrical conductivity. They are used in implants, prosthetics, and as scaffolds for tissue engineering.

2.5.3.5 Glass-Ceramics

Glass-ceramics are materials that combine the properties of glasses and ceramics. They are used in dental applications, bone grafts, and coatings for implants due to their biocompatibility and mechanical strength [29].

2.5.3.6 Composite Materials

Ceramic composites, including metal-ceramic composites, leverage the strength of metals alongside the biocompatibility of ceramics. These materials are utilized in orthopedic implants and dental prosthetics, and other biomedical applications where both strength and biocompatibility are required.

2.6 Applications of Biomaterials in Regenerative Medicine

Biomaterials play a crucial role in regenerative medicine, serving as scaffolds that facilitate tissue repair and regeneration. Their applications span various fields, including tissue engineering, gene therapy, and controlled drug release. Table 2.1 briefly presents the application of various biomaterials as regenerative medicine.

Table 2.1 Applications of biomaterials in regenerative medicine

Class of biomaterial	Examples	Applications	Properties and benefits	Challenges	References
Synthetic polymers	PMMA (Poly (methyl methacrylate))	Bone cement, intraocular lenses (IOLs), hard contact lenses	Excellent light transmittance, biocompatibility	Hydrophobic nature requires surface modifications for improved cell interactions	[30]
	PLGA (Poly (glycolic-co-lactic acid))	Resorbable surgical sutures, drug delivery systems, orthopedic fixation devices	Biodegradable, non-toxic degradation products (lactic and glycolic acid)	Rapid degradation rate can be a limitation in long-term applications	[31]
	PHEMA (Poly (hydroxyethyl methacrylate))	Soft contact lenses	Good optical clarity, biocompatible	Requires cross-linking for dimensional stability	[30]
	PP (Polypropylene)	Surgical sutures, hernia repair	High rigidity, good tensile strength	Limited biocompatibility without surface treatment	[32]
Natural polymers	Alginate	Cell encapsulation, cartilage regeneration	Biocompatible, forms gels with calcium ions	Rapid degradation, limited mechanical strength	[33]
	Collagen	Skin regeneration, wound healing	Excellent biocompatibility supports cell adhesion and growth	Risk of immunogenicity, variability in natural sources	[34]
	Hyaluronic acid	Scaffold for tissue regeneration, drug delivery	Biocompatible, supports cell migration	High water solubility, rapid clearance from the body	[35]
	Fibrin	Wound healing, tissue engineering scaffolds	Biocompatible, supports cell growth	Rapid degradation, requires modification to prolong stability	[36]

Metals	Titanium and its alloys (Ti-6Al-4V)	Orthopedic implants, dental implants	High strength, corrosion resistance, biocompatibility	Risk of metal ion release, potential for allergic reactions	[37]
	Stainless steel (316L)	Bone plates, screws, cardiovascular stents	High mechanical strength, corrosion resistance	Susceptibility to corrosion in long-term implantation	
	Cobalt-chromium alloys	Joint replacements, dental implants	High wear resistance, strength	High modulus mismatch with bone, potential for metal ion release	[38]
Ceramics	Hydroxyapatite (HAp)	Bone grafts, dental implants	Excellent osteoconductivity, biocompatibility	Brittle nature, low impact resistance	[39]
	Bioactive glass	Bone regeneration, dental repair	Stimulates bone growth, biocompatible	Processing and fabrication difficulties	[31]
	Alumina (aluminum oxide)	Hip prostheses, dental implants	High wear resistance, biocompatible	Brittle, limited tensile strength	[39]
	Zirconia	Dental crowns, orthopedic implants	High strength, fracture toughness, biocompatible	Processing complexities, high cost	[40]

2.6.1 Polymers in Regenerative Medicine

2.6.1.1 Synthetic Polymers

Polymeric biomaterials find extensive applications in tissue regeneration due to their adjustable mechanical properties and biocompatibility. For instance, PMMA is widely employed as bone cement and in manufacturing intraocular lenses (IOLs) because of its exceptional light transmittance and mechanical characteristics [17]. PLGA, another synthetic polymer, is utilized in resorbable surgical sutures and drug delivery systems. Its breakdown products, lactic acid, and glycolic acid, are non-toxic, making it suitable for temporary medical uses [31]. However, the rapid degradation rate of PLGA can restrict its applicability in long-term medical applications.

2.6.1.2 Natural Polymers

Natural polymers like alginate, collagen, and hyaluronic acid are highly promising in regenerative medicine. Alginate, comprising L-guluronic acid and D-mannuronic acid, forms hydrogels when exposed to calcium ions, making it suitable for applications such as cell encapsulation and cartilage regeneration [12]. Collagen, a crucial component of the extracellular matrix, is extensively used for skin regeneration and wound healing due to its excellent biocompatibility [34]. However, challenges arise from the natural variability and potential immunogenicity associated with collagen.

2.6.2 Metals in Regenerative Medicine

Metals like titanium and its alloys, stainless steel, and cobalt-chromium alloys are widely utilized in orthopedic and dental implants because of their superior mechanical strength and resistance to corrosion. Titanium and its alloys are renowned for their exceptional biocompatibility and ability to integrate seamlessly with bone tissue through osseointegration, making them ideal for joint replacements and dental implants [41]. However, the risk of metal ion release and allergic reactions is a concern.

2.6.3 Ceramics in Regenerative Medicine

Ceramics such as hydroxyapatite (HAp) and bioactive glass are employed in bone regeneration and dental applications because of their outstanding biocompatibility and ability to promote bone growth (osteo-conductivity). Hydroxyapatite, which closely mimics the mineral structure of bone, is commonly utilized in bone grafts and dental implants [39]. Bioactive glass stimulates bone growth and repair but presents challenges in processing and fabrication due to its brittle nature [31].

2.7 Biomaterials in Public Health Emergencies: Innovations and Applications

Biomaterials play a critical role in addressing public health emergencies by offering innovative solutions for the prevention, diagnosis, treatment, and management of health crises. These materials, which include polymers, metals, and ceramics, provide the necessary functionality and versatility to support various medical and healthcare needs during emergencies such as pandemics, natural disasters, and bioterrorism events. This essay explores the diverse applications of biomaterials in public health emergencies, highlighting their significance in enhancing healthcare delivery and improving patient outcomes.

2.7.1 Polymers in Public Health Emergencies

Polymers, both synthetic and natural, are extensively used in public health emergencies due to their versatility, biocompatibility, and ability to be engineered for specific applications. They play crucial roles in personal protective equipment (PPE), medical devices, and drug delivery systems.

2.7.1.1 Personal Protective Equipment (PPE)

The COVID-19 pandemic underscored the importance of PPE, such as masks, gloves, and gowns, in controlling the spread of infectious diseases. Polymers like polypropylene (PP) are widely used in the manufacture of N95 respirators and surgical masks due to their excellent filtration efficiency and breathability. Polyethylene (PE) and polyvinyl chloride (PVC) are commonly used in gloves and gowns, providing barrier protection against pathogens and hazardous materials [42].

2.7.1.2 Medical Devices

During public health emergencies, there is an increased need for medical devices such as syringes, catheters, and components for ventilators. Polymers like polycarbonate (PC) and polyether ether ketone (PEEK) are utilized in these devices due to their robustness, durability, and biocompatibility. Also, biodegradable polymers such as polylactic acid (PLA) and poly (lactic-co-glycolic acid) (PLGA) are used in making absorbable surgical sutures and drug delivery systems. This reduces the necessity for additional surgeries to remove devices and promotes faster patient recovery [22].

2.7.1.3 Drug Delivery Systems

Effective drug delivery is essential during public health emergencies to ensure timely and targeted treatment. Polymers like PLGA are used to develop controlled-release drug delivery systems, providing sustained therapeutic effects, and reducing dosing frequency. Hydrogel-based systems, composed of polymers such as poly (ethylene glycol) (PEG) and polyvinyl alcohol (PVA), are used for localized drug delivery, particularly in wound care and tissue regeneration [43].

2.7.2 Metals in Public Health Emergencies

Metals and metal alloys are indispensable in the development of durable and reliable medical devices and implants. Their applications in public health emergencies span from surgical instruments to diagnostic equipment.

2.7.2.1 Surgical Instruments

Stainless steel and titanium are commonly used in surgical instruments due to their high strength, corrosion resistance, and sterilizability. These materials are crucial in emergency surgeries and trauma care, ensuring precision and reliability during critical procedures [24].

2.7.2.2 Diagnostic Equipment

Metals such as gold and silver find application in diagnostic devices like lateral flow assays and biosensors because of their exceptional electrical conductivity and chemical stability. These diagnostic tools are vital for rapid and accurate detection of pathogens during outbreaks, enabling prompt public health responses [44].

2.7.2.3 Implants and Prosthetics

Titanium and its alloys are extensively used in orthopedic implants and dental prosthetics for their biocompatibility and ability to integrate with bone tissue. In emergencies involving traumatic injuries, these implants provide durable solutions for bone repair and reconstruction [45].

2.7.3 Ceramics in Public Health Emergencies

Ceramics and glass-ceramics offer unique properties such as high biocompatibility, resistance to corrosion, and ability to withstand high temperatures. These materials are used in various medical applications, from bone grafts to medical imaging.

2.7.3.1 Bone Grafts and Scaffolds

Hydroxyapatite (HAp), a ceramic material, is widely used in bone grafts and scaffolds due to its similarity to natural bone minerals. In emergencies involving bone fractures or defects, HAp-based materials support bone regeneration and healing. Additionally, bioactive glasses, which can bond to bone tissue, are used to enhance the osteo-conductivity of implants [46].

2.7.3.2 Medical Imaging

Ceramics like barium titanate and lead zirconate titanate (PZT) are used in medical imaging devices such as ultrasound transducers and X-ray detectors. These materials provide high sensitivity and resolution, essential for accurate diagnosis and monitoring of medical conditions during public health emergencies.

2.7.3.3 Dental Applications

Ceramics such as zirconia and alumina are used in dental restorations, including crowns and bridges. Their high strength and aesthetic properties make them ideal for dental emergencies, ensuring functional and cosmetic restoration [47].

2.7.4 Integrated Approaches

The convergence of biomaterials with advanced technologies such as nanotechnology, tissue engineering, and regenerative medicine holds promise for transforming public health emergency responses. For example, nanomaterials can augment the properties of biomaterials, offering enhanced mechanical strength, antimicrobial activity, and capabilities for drug delivery.

2.7.4.1 Nanotechnology

Nanomaterials, such as silver nanoparticles and carbon nanotubes, can be integrated into biomaterials to impart antimicrobial properties, making them effective in preventing infections. These nanocomposites are used in wound dressings, coatings for medical devices, and PPE to reduce the risk of pathogen transmission [48].

2.7.4.2 Tissue Engineering and Regenerative Medicine

The use of biomaterials in tissue engineering and regenerative medicine offers promising solutions for repairing and regenerating damaged tissues. Scaffold materials, including hydrogels and biopolymers, are engineered to provide the necessary support and bioactive cues for cell growth and tissue formation. In emergencies involving severe injuries or tissue loss, these regenerative approaches can significantly improve patient outcomes.

2.7.4.3 3D Printing and Customization

3D printing technology allows for the rapid production of customized medical devices and implants using biomaterials. This capability is particularly valuable in public health emergencies, where there is a need for tailored solutions to meet specific patient needs. For example, 3D-printed PPE and medical devices can be produced quickly to address shortages and enhance the effectiveness of emergency responses [49, 50].

2.8 Advantages and Limitations of Biomaterials in Regenerative Medicine

Biomaterials are pivotal in regenerative medicine, providing numerous benefits for tissue engineering and repair. Nonetheless, they also present certain drawbacks that must be carefully assessed in their utilization. Recognizing these advantages and limitations is crucial for advancing effective regenerative therapies. This article delves into a comprehensive exploration of the strengths and weaknesses of

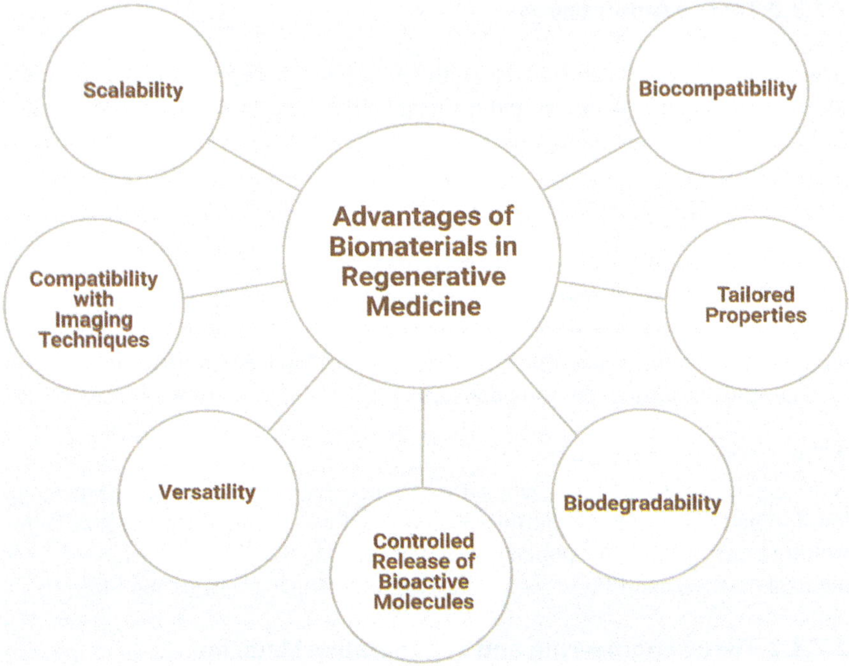

Fig. 2.2 Advantages of biomaterials in regenerative medicine. (Created by using Biorender.com)

biomaterials in regenerative medicine, leveraging current research and literature in the field. Advantages of biomaterials in regenerative medicine are illustrated in Fig. 2.2.

2.8.1 Biocompatibility

A key benefit of biomaterials is their capability to interact harmoniously with biological systems, avoiding negative immune responses. Biocompatible biomaterials support essential processes such as cell adhesion, proliferation, and differentiation, thereby promoting effective tissue regeneration [50].

2.8.2 Tailored Properties

Biomaterials can be tailored to possess precise mechanical, chemical, and biological characteristics, enabling their suitability for diverse regenerative applications. For instance, scaffolds can be engineered to emulate the mechanical attributes of natural tissues, offering ideal support for cellular growth and tissue maturation.

2.8.3 Biodegradability

Many biomaterials are designed to degrade over time, allowing for the gradual integration of regenerated tissue with the surrounding environment. This property reduces the risk of long-term complications and the need for surgical removal of implants [51].

2.8.4 Controlled Release of Bioactive Molecules

Biomaterials serve as carriers for bioactive molecules such as growth factors, cytokines, and drugs. By controlling the release of these molecules, biomaterials can augment tissue regeneration processes through mechanisms like facilitating cell migration, proliferation, and differentiation.

2.8.5 Versatility

Biomaterials are highly versatile, offering a broad spectrum of materials and fabrication methods. This versatility enables the creation of tailored solutions for various applications in regenerative medicine, such as bone regeneration, cartilage repair, and wound healing.

2.8.6 Compatibility with Imaging Techniques

Biomaterials can be designed to be compatible with different imaging methods, including MRI, CT scans, and ultrasound. This compatibility allows for non-invasive monitoring of tissue regeneration processes, enabling real-time assessment of treatment outcomes.

2.8.7 Scalability

Biomaterials can be produced in large quantities using scalable manufacturing processes. This scalability is essential for the commercialization of regenerative therapies, ensuring their availability and affordability for a wide range of patients [51].

Whereas the limitations of biomaterials in regenerative medicine includes.

2.8.8 Immune Response

Although biomaterials are biocompatible, they may still induce an immune response in certain individuals. This response can lead to inflammation, fibrosis, and rejection of the biomaterial, compromising its effectiveness in tissue regeneration [52].

2.8.9 Degradation Rate Variability

The degradation rate of biomaterials can vary depending on factors such as the implantation site, local environment, and material composition. This variability can affect the integration of the regenerated tissue and may require additional interventions to control the degradation process [51].

2.8.10 Lack of Standardization

The field of regenerative medicine lacks standardized protocols for biomaterial development and testing. This lack of standardization can result in variability in the quality and performance of biomaterials, which may impede their broad clinical adoption.

2.8.11 Mechanical Properties

Creating biomaterials that accurately replicate the mechanical properties of native tissues is difficult despite engineering efforts. Mismatched mechanical properties can lead to implant failure or tissue damage, especially in load-bearing applications.

2.8.12 Cost

The development and production of biomaterials can be costly, particularly for complex materials and fabrication techniques. This cost can limit access to regenerative therapies, especially in resource-limited settings [51].

2.8.13 Long-Term Stability

Some biomaterials may degrade too quickly, leading to a loss of structural integrity and functionality over time. Ensuring the long-term stability of biomaterials is essential for their success in regenerative medicine.

2.8.14 Regulatory Challenges

The regulatory approval process for biomaterials in regenerative medicine can be lengthy and complex. This process can delay the clinical translation of promising biomaterials, limiting their impact on patient care [53]. Overall, biomaterials provide significant benefits for tissue engineering and regenerative medicine, including biocompatibility, customizable properties, and controlled release of bioactive substances. However, they also present challenges such as potential immune reactions, variability in degradation rates, and the lack of standardized manufacturing processes. Overcoming these limitations through ongoing research and innovation is crucial for advancing regenerative medicine and enhancing patient outcomes.

2.9 Challenges of Biomaterials for Regenerative Medicine

Nevertheless, despite the substantial advancements in regenerative engineering, the use of these techniques in clinical settings has been restricted. Moreover, there are substantial hurdles that must be surmounted for future expansions, including ethical as well as biological considerations [54]. For instance, tissue growth in laboratory conditions remains constrained due to the lengthy maturation period required in bioreactors, inadequate structural stability for surgical manipulation, and vulnerability to oxygen deprivation until vascularization occurs. Upon implantation, all engineered cellular structures undergo a certain level of oxygen deficiency, known as hypoxia. The present challenge is in the difficulty of constructing a continuous network of small blood vessels that can extend to a connectable vascular segment. Consequently, our ability to generate tissues with increased thickness is significantly impeded. As a result, the rapid formation of blood vessels remains a significant obstacle that impedes the advancement of regenerative engineering. The public and the medical community have great aspirations for regenerative medicine or tissue engineering, even though there are just a few confirmed clinical research involving humans in this discipline. Considering the vast amount of basic research that has been carried out in this field, it is advantageous to take into consideration the elements that are responsible for the delayed progress of tissue engineering in clinical applications. The search for perfect bioinks is an important area of research in tissue engineering. The primary focus of this research is on the development of materials that can satisfy all the criteria. Among the criteria are the enhancement of cell adhesion, the guarantee of efficient printability, and the facilitation of cell displacement [8]. This effort underscores the current reliance on specific components like gelatin and highlights a significant deficiency in the variety of hydrogels tailored for specific applications, such as cartilage repair, as evidenced by several research studies.

Despite notable advancements in the field, a significant hurdle remains in enhancing hydrogels to accurately mimic the mechanical properties of various tissues. Attributes such as strength, electrical conductivity, and durability over extended periods are particularly challenging to achieve. Furthermore, the inherent viscoelastic qualities of hydrogels give rise to issues connected with aging and creep. These complications include the possibility of deformation that changes over time as a response of continuous stress or strain [2]. There is a potential risk that this action could compromise the hydrogel's ability to provide sustained support for cell attachment, migration, and proliferation, all critical for effective tissue repair and regeneration. In order to address these concerns, it is vital to devise a strategy that has been meticulously thought out and that successfully blends mechanical qualities with biological activities in order to fulfill the needs of various cell kinds and the functions that they play. Also, it is essential to ensure that the materials used in clinical applications have a long lifespan. Another significant challenge is accurately regulating the behavior of stem cells, which includes their capacity to multiply, specialize in specific cell types, and effectively merge with the tissues of the body. In the context of healing and regeneration processes, these elements play a crucial role. They involve stimulating cell mobility and proliferation, mimicking signals from the natural extracellular matrix, and seamlessly integrating with the body's tissues. Properly managing immune responses and enhancing the bioactivity of hydrogel-based implants and grafts are essential for the success of regenerative therapies [55]. The goal of replicating the complex composition and capabilities of biological tissues, like cartilage, is one that is both important and challenging. The objective is not only to replicate the physical qualities of tissues, but also to maintain the viability of cells and make certain that the structures that are formed are able to properly integrate with the tissues that are in the surrounding areas of the physical body.

Continuous attempts are being made to improve hydrogels to obtain increased capabilities in controlled and targeted medication delivery (also known as "controlled drug delivery"). The purpose of this initiative is to develop pharmaceutical delivery systems that can deliver medications in two stages or correctly dispensing growth factors and other medicinal ingredients. In the field of regenerative medicine, the objective is to maximize the efficacy of treatment while simultaneously minimizing adverse effects, with the end aim of improving the results for patients. Hydrogel technology is undergoing constant research and development, which indicates a determination to transcend current restrictions. Therefore, advancements in tissue engineering and regenerative medicine are being facilitated [56].

2.10 Current Advancement of Biomaterials for Regenerative Medicine

Regenerative medicine may be studied in three primary ways: (1) cell therapy; (2) repair and development of cells via biological or synthetic materials; (3) scaffolds for cell seeding. The human body is composed of several cell kinds, each with distinct purposes. For cell treatment, fresh, healthy cells are introduced into sick areas. It may employ undifferentiated or developed stem cells, depending on the circumstances. On the one hand, patient tissues collect differentiated endogenous primary cells that are prepared for implantation without modification—just growing. Even for organs like the liver that have a high capacity for replication in vivo, these cells are difficult to get in vitro due to the lack of the microenvironment necessary for proliferation. Even though these cells do not seem to be associated with rejection or severe inflammatory responses, their use will decrease in the future. Alternatively, if stem cells (SCs) are not encouraged to transform into a specific cell type, they could proliferate and self-renew. There are several methods to get SCs. Stem cells that are autologous, allogeneic, or xenogeneic originate from patients, people, or animals, in that order. Isolated from all adult body tissues, ASCs work to repair minor tissue damage by restoring tissue function. Mesenchymal stem cells (MSCs) derived from bone marrow are extensively researched and play a significant role. Using various culture techniques, they can develop into several cell types to treat disorders of the bone, cartilage, nervous system, muscles, heart, blood, and gastrointestinal tract. Aspirated from an embryo's inner cell mass at the blastocyst stage (5 days after conception), Embryonic stem cells (ESCs) can proliferate extensively while maintaining their pluripotency until they are directed to differentiate into specific cell types from all three embryonic germ layers. To create human ESCs, extra embryos from in vitro fertilization may be used [57]. Hyaluronic acid is the most used natural biomaterial. It is an anti-aging component present in numerous skincare products and injectable facial fillers. Researchers successfully demonstrated laparoscopic bladder augmentation in minipigs using an acellular tissue matrix derived from porcine intestine and small intestinal submucosa. This was carried out to illustrate how useful the process is. The ideal biomaterial should be biocompatible and degrade at a pace that aligns with the rate of regeneration, but it should also not produce any deleterious byproducts, obstruct the process of regeneration, or induce inflammation or obstruction. Another important property of the material is its porosity, which allows nutrients and trash to be exchanged. It is quite difficult to correctly achieve this property, and 3D bio-printers are the ideal solution to this problem. The research team illustrated that by enclosing functional nanoparticles within a biocompatible hydrogel, they could bio print a three-dimensional liver-like device that could detoxify blood. By employing dynamic optical projection stereolithography, intricate three-dimensional microstructures can be printed within seconds. Among these structures are the blood arteries. Printing vasculature is essential to ensure the delivery of nutrients and oxygen in tissue-engineered organs such as the liver and kidneys. Without vascular networks, these organs would be non-functional in clinical applications. Bio fabrication technology utilizes a photo-induced solidification process for this purpose. Using soft, biocompatible hydrogels that are alive, this method produces a single, continuous layer of solid structure with each iteration. This is achieved by illuminating a region of a solution containing cells and photosensitive biopolymers. Several hours may pass if another 3D bio fabrication technique, such two-photon photopolymerization, is used to create a three-dimensional object. Finally making an appearance is Organovo, a medical start-up that aims to provide bio printed organs—like livers—for transplantation and surgical procedures [58].

Biomaterials have continuously expanded during the last 50 years, bringing new ideas and lucrative subfields. The field's present state and prospective future advances are included in the review. Recent developments in biomaterials that control biological processes and promote healing were presented. The newest biomaterials include biologically inspired materials that imitate natural processes, signaling group immobilization, controlled synthetic materials for medication and cell carriers, surface modification for protein adsorption, and 3D designs for unique diagnostic patterns [59]. 2D approaches are used in traditional stem cell culture. To nourish and differentiate cells, in this culture technique, cells are grown on a feeder layer and supplemented with tissue culture media containing growth factors or cytokines. The following are some of the disadvantages of two-dimensional cultures: (1) flattening and elongating of the cells; (2) inadequate differentiation and formation of cell junctions; (3) abnormally high rates of proliferation; and (4) significant differences in phenotypic and gene expression. These problems have been tried to be addressed by recent 2D cultural advancements. Biomaterial modifications for tissue culture may change the fate of cells [60]. However, 3D cultures methods more accurately mimic *in vivo* environments. Numerous studies have shown that 3D culture techniques enhance cell morphology, proliferation, differentiation, and responsiveness to stimuli. These cultures can be scaffold-based or anchorage-independent, utilizing specialized 3D platforms. Recent advances in techniques like electrospinning and 3D bioprinting can further enhance scaffold-based approaches for producing intricate structures. Models of cell culture should replicate all aspects of the natural environment of cells, such as nourishment, oxygen, waste product elimination, and microenvironment. 3D culture has a few drawbacks despite its advantages over 2D culture, including an uneven supply of nutrients, growth hormones, and oxygen, which might render cells far from the matrix surface dormant. Rather than focusing on mature quiescent ECM–cell interactions, many tissue-engineered structures aim to create regenerative culture models. Future development must handle the greater costs, differences in experiment replication, and data interpretation associated with this sort of culture. Scaffold-based anchorage-dependent culture uses elaborate multilayer scaffolds to make simple extracellular-like matrices. Scaffold selection is influenced by cellular composition of the target tissue and physical factors that provide structural stability [61]. Complex structure construction has been altered by three-dimensional bioprinting. Transplantable devices' non-functional vasculature impedes successful in vivo applications and clinical translation. Unique 3D anchorage-independent techniques include magnetic forces, hand-drop, low attachment vessel, and magnetic levitation. The low attachment plate technique makes use of ultra-low attachment culture containers. With anchorage-independent techniques, cells cluster, sphericalize, and produce extracellular matrix. The most common technique for producing cartilage is spheroid culture. Magnetic levitation is used to suspend cultured cells that include magnetic nanoparticles. This approach is reproducible and decreases spheroid core necrosis. These conditions maintain the properties and quiescence of stem cells for potential therapeutic applications [62]. In terms of anchorage-independent culture, organoids are interesting. Tissue culture is used in making organoids to self-organize and renew 3D cultures. There have been reports of organoid cultures in the renal, eye, brain, stomach, and lung. Recent study demonstrated the ability to replicate tissue architecture in adipospheres formed from leukocytes, endothelial cells, and stem cells obtained from adipose tissue [63]. This technique produced differentiated unilocular adipocytes and endothelial structures resembling lumen-containing vessels. Cell self-aggregation without an attachment wall is used in the hand-drop method. Spheroid size is controlled by the volume of the cell solution. Static cultures are less effective than this cell culture. For cardiovascular tissue engineering, researchers have looked at the conversion of adipocytes to cardiomyocytes. The hand-drop method increased the differentiation yield of retinal progenitors from adipocytes by inducing these cells to respond to glutamate neurotransmitters. This approach has also been applied in other preclinical studies for bone healing, cardiac tissue regeneration, and cartilage repair [64, 65].

2.11 Clinical Significance of Biomaterials for Regenerative Medicine

Several clinical applications, particularly in regenerative medicine, have advanced patient care and established new standards, although widespread clinical use of bioengineered organs on a large scale is still years away. Currently, only thin, and/or avascular tissues such as skin, cartilage, heart tissue, and liver tissue have benefited from these advancements. It is essential to assess a new polymeric biomaterial's cytotoxicity on a model cell line before considering it for use in regenerative medicine, whether it be an implant device or biomaterial. Any harmful effect that a drug causes at the cellular level, such as cellular permeability alterations or death, is referred to as cytotoxicity [66]. Any biomaterial deemed "toxic" will either directly or indirectly harm cells. Testing for in vitro cytotoxicity is required for biomaterials intended for use in medical implants. The cytotoxicity profile is examined, and then further testing is done to determine how biocompatible the implant is. Test material might be added to cells directly or removed and then added during assay procedures [67]. Three techniques may be used to assess a material's biocompatibility: elution, agar diffusion, or direct touch. In regenerative medicine and tissue engineering, the direct contact method is frequently utilized. This method involves using a near-confluent monolayer of L-929 (mammalian fibroblast) cells in the experiment. Culture media is replaced with fresh media on the culture plates. Biomaterial specimens or implants are placed carefully on the plates and then incubated for 24 h at 37 °C in a humid environment. After incubation, cells are typically stained using fluorescent dyes such as fluorescein diacetate (FDA) or cytochemical stains like hematoxylin and eosin. Toxicology is shown by the absence of marked cells underneath and surrounding the substance. The MTT test, also known as 3-(4,5-dimethyl thiazol-2-yl)-2,5-diphenyltetrazolium bromide, is another direct contact assay. In this assay, cells are cultured on the biomaterial for several days. The MTT reagent is introduced after the medium has been removed from the wells for the experiment. In this experiment, after 4 h of incubation, living cells change the tetrazolium salt MTT into purple formazan. The optical density at 570 nm may be used to quantify the colorful complex produced by solubilizing formazan crystals in DMSO. To create skin tissue for wound healing, fibroblasts, keratinocytes, and scaffold were used in the first clinical use of bioengineered tissue in 1980. Ever then, studies on artificial skin have been conducted to mimic the process of natural healing and hasten the healing of wounds and restoration of function. If there are limited autologous options, these substitutes may be used in lieu of conventional skin transplants. Several FDA-approved brands and kinds of bioengineered skin substitutes have gained popularity in recent years as a wound healing option for patients. Further, in previously incurable injuries, skin has been successfully regenerated by some bioengineered treatments. Organogenesis and material origin—autologous, allogenic, or xenogeneic—are the two main classifications for the many bioengineered skin substitutes. A class's therapeutic significance varies according on the depth of the wound and whether temporary or permanent therapy is required [68]. Acellular skin substitutes like Integra R. replace the dermis and epidermis with collagen and a silicone membrane. These substitutes act as a barrier to protect against infection and fluid loss while promoting the body's natural wound healing process through the inclusion of growth factors, cytokines, and dermal matrix components. They are particularly effective for treating superficial burns and wounds [67, 69, 70].

A scaffold layer or two that is seeded with autologous or allogeneic cells makes up cellular skin replacements, which are more complex [71]. They expedite healing and may eventually fully rebuild injured tissue, much like their acellular counterparts. There are no restrictions on the availability of donors for commonly used allogeneic cellular skin replacements, such as Derma graft. Autologous cellular skin replacements or split thickness skin transplants are often used to replace these transient wound coverings. Epicil R, Epidex R, and Tiscover R (A-Skin) are the commercially available autologous options. To provide appropriate cells, grafts need the involvement of the epidermis and superficial dermis, which complicates their creation. The stability and durability of native skin are absent from cultured epidermal autografts, notwithstanding the possibility of providing long-term wound coverage. Thus, they are mostly used in situations when it is necessary to replace just the keratinocyte layer or when donor skin is hard to come by. Unlike skin, cartilage is devoid of robust lymphatics, blood arteries, and nerves. Usually, in vivo healing is impeded. It is difficult to repair faults when there is not enough supply of chondrocytes. Common injuries were formerly treated using invasive, difficult, and inaccurate techniques, which gave rise to novel bioengineered substitutes. Joint and tracheal replacements using bioengineered cartilage are the two clinical applications being researched. The only successful cartilage replacements are small joint bioengineered replacements. For clinical usage, Lee et al. produced anatomically developed, vascularized tibial condyles. Larger joints such as the hip and knee are difficult to scale up due to vascular and load bearing constraints. Scaffolds for the regeneration of cartilage tissue were created using a hybrid technique that combined electrospinning and inkjet printing, resulting in improved mechanical and biological properties. Even though bioengineered cartilage repair has been available for around 30 years, the mechanical properties, zonal organization, and ECM composition of natural tissue cannot be replicated by existing techniques. Bioengineered tracheal transplants were initially hailed as a pioneering clinical application of tissue engineering. However, ethical concerns and unfavorable outcomes have since cast doubts on their widespread adoption. The same research team reported the first three successful tissue-engineered airway transplants, each using a different method. Cardiovascular disease has a widespread impact on several parts of the body and is a significant cause of illness and death worldwide. Pharmaceuticals and surgery have enhanced therapeutic options, yet they remain inadequate. This is relevant to both acquired and congenital disorders. For example, it would be optimal to have a substitute part that could expand alongside the child in instances of juvenile cardiac illnesses, such as anomalies in the septum. Nevertheless, it is more advantageous to develop a dense patch that can contract and transmit electrical currents in an adult who has had a heart attack. It's evident that over the last two decades, scaffolds and stem cells have played pivotal roles in cardiovascular repair, addressing a range of needs with advancements in alternative components. While the heart possesses a limited number of resident stem cells with inherent regenerative capabilities, other sources like hematopoietic, mesenchymal, and induced pluripotent stem cells have proven more beneficial [72]. Matsumara's work, for instance, focused on bone marrow-derived stem cells and endothelial cells for seeding scaffolds, leading to the development of autografts for vascular tissue engineering. This technique holds significant promise for treating blocked arteries, dysfunctional blood vessels, and replacing hemodialysis access grafts.

The process of creating replacement components such as blood vessels and heart valves in a controlled, in vitro environment involves isolating and cultivating cells alongside constructing supportive structures known as scaffolds. This process is particularly demanding when using autologous stem cells, requiring substantial time and effort. Moreover, the ability to generate whole organs or thick tissues is constrained by the lack of interconnected blood vessel systems. As a result, injectable treatments have been devised, including direct stem cell injections, or using carrier scaffolds. Injectable treatments offer the advantage of less invasive delivery methods like trans coronary infusions or trans-endocardial injections [73].

Initial studies in rat models of myocardial infarction demonstrated significant improvements in viability using a fibrin carrier for injectable myoblasts, due to fibrin's inherent framework containing relevant growth factors. Subsequent innovations include calcium-crosslinked alginate delivered via trans-coronary infusion using percutaneous catheters to reverse left ventricular remodeling in pig models of myocardial infarction [74].

Most documented scaffolds are hydrogel-based due to their structural similarity to the natural extracellular matrix and their ability to create a conducive environment for cell growth [67]. Currently, the bladder stands as the only fully bioengineered human organ, owing to its thin, hollow structure. Yoo et al. achieved the first successful canine model in 1998 by seeding autologous cells onto a decellularized bladder submucosa. This resulted in the formation of normal bladder tissue histologically and functionally indistinguishable from native tissue, contrasting with acellular matrices that evolved into aberrant and dysfunctional tissue over time. This success quickly translated to human subjects within a year, utilizing biodegradable scaffolds seeded with autologous cells, which continue to show effective outcomes even after 15 years. This study is now approaching the final stages of human clinical trials with the aim of securing approval from the Food and Drug Administration, employing the same procedures and manufacturing methods as in the initial investigations [75].

2.12 Challenges and Future Directions

While biomaterials have significantly advanced regenerative medicine, challenges remain. Synthetic polymers may provoke immune responses or have degradation products that are not fully biocompatible. Metals can suffer from corrosion and mechanical mismatch with natural tissues. Ceramics, despite their biocompatibility, often have low impact resistance and processing difficulties. Future research is focused on developing new materials and improving existing ones to address these challenges. Advances in nanotechnology, biomimetic materials, and bioactive coatings are promising areas for enhancing the performance and functionality of biomaterials. Combining different types of biomaterials to create hybrid systems that leverage the strengths of each component is another promising strategy.

2.13 Conclusion

Tissue engineering advancements are merging with progress in biomaterials science, stem cell biology, and growth factors to shape the rapidly advancing field of regenerative engineering. It has already generated clinically applicable technologies that may enhance patient outcomes and is poised to further transform the surgical and medical domains. Surgeons need to familiarize themselves with some of its fundamental elements, since it is anticipated to influence patient treatment, to some extent, soon. The cutting-edge field of tissue engineering is creating new avenues for understanding the promise of regenerative medicine. In the near future, tissue engineering and nanotechnology might completely transform medical care. Tissue engineering relies on the cultivation of cells on scaffolds created from natural or artificial membranes composed of polymers. The benefits of synthetic and natural biopolymers have shown to be significantly different. The cytokines and growth factors found in natural membranes support cell survival, but they may also trigger immune defense reactions. The design of biomaterial scaffolds that modify the immune response may consider breakthroughs in cell and developmental biology that identify important immunological mediators of repair and regeneration. This new field of study is known as biomaterials-directed regenerative immunology.

References

1. Levi B, Longaker MT (2011) Concise review: adipose-derived stromal cells for skeletal regenerative medicine. Stem Cells 29:576–582
2. Vyas J, Shah I, Singh S, Prajapati BG (2023) Biomaterials-based additive manufacturing for customized bioengineering in management of otolaryngology: a comprehensive review. Front Bioeng Biotechnol 11:1234340
3. Braghirolli DI, Steffens D, Pranke P (2014) Electrospinning for regenerative medicine: a review of the main topics. Drug Discov Today 19:743–753
4. Dee KC, Puleo DA, Bizios R (2002) An introduction to tissue-biomaterial interactions. Wiley
5. Chandra P, Yoo JJ, Lee SJ (2015) Biomaterials in regenerative medicine: challenges in technology transfer from science to process development. In: Translational regenerative medicine. Academic, Cambridge, MA
6. Trucillo P (2024) Biomaterials for drug delivery and human applications. Materials 17:456
7. Sekuła M, Zuba-Surma EK (2018) Biomaterials and stem cells: promising tools in tissue engineering and biomedical applications. In: Biomaterials in regenerative medicine. IntechOpen
8. Todros S, Todesco M, Bagno A (2021) Biomaterials and their biomedical applications: from replacement to regeneration. Processes 9:1949
9. Qu H, Fu H, Han Z, Sun Y (2019) Biomaterials for bone tissue engineering scaffolds: a review. RSC Adv 9:26252–26262

10. Mariani E, Lisignoli G, Borzì RM, Pulsatelli L (2019) Biomaterials: foreign bodies or tuners for the immune response? Int J Mol Sci 20
11. Ullah S, Chen X (2020) Fabrication, applications and challenges of natural biomaterials in tissue engineering. Appl Mater Today 20:100656
12. Singh S, Chunglok W (2022) Biopolymers towards green and sustainable development. Bentham Science Publishers. https://doi.org/10.2174/97898150793021220101
13. An H, Zhu L, Shen J et al (2020) Self-healing PEG-poly(aspartic acid) hydrogel with rapid shape recovery and drug release. Colloids Surf B: Biointerfaces 185:110601. https://doi.org/10.1016/j.colsurfb.2019.110601
14. Dorozhkin SV (2010) Calcium orthophosphates as bioceramics: state of the art. J Funct Biomater 1:22–107
15. Bohner M (2010) Design of ceramic-based cements and putties for bone graft substitution. Eur Cell Mater 20. https://doi.org/10.22203/eCM.v020a01
16. Rahaman MN, Day DE, Sonny Bal B et al (2011) Bioactive glass in tissue engineering. Acta Biomater 7:2355–2373
17. Gaharwar AK, Peppas NA, Khademhosseini A (2014) Nanocomposite hydrogels for biomedical applications. Biotechnol Bioeng 111. https://doi.org/10.1002/bit.25160
18. Vyas J, Raytthatha N, Singh S, Prajapati BG, Mohite P, Munde S (2024) Sustainable sources of raw materials as substituting biomaterials for additive manufacturing of dental implants: a review. Perio Implant Res 8(1):3
19. Amini AR, Laurencin CT, Nukavarapu SP (2012) Bone tissue engineering: recent advances and challenges. Crit Rev Biomed Eng 40. https://doi.org/10.1615/CritRevBiomedEng.v40.i5.10
20. Makadia HK, Siegel SJ (2011) Poly lactic-co-glycolic acid (PLGA) as biodegradable controlled drug delivery carrier. Polymers (Basel) 3. https://doi.org/10.3390/polym3031377
21. Hutmacher DW (2000) Scaffolds in tissue engineering bone and cartilage. Biomaterials 21. https://doi.org/10.1016/S0142-9612(00)00121-6
22. Ratner BD, Hoffman AS, Schoen FJ, Lemons JE (2004) Biomaterials science: an introduction to materials in medicine, 2nd edn. Academic
23. Hemshekhar M, Thushara RM, Chandranayaka S et al (2016) Emerging roles of hyaluronic acid bioscaffolds in tissue engineering and regenerative medicine. Int J Biol Macromol 86:917–928. https://doi.org/10.1016/j.ijbiomac.2016.02.032
24. Geetha M, Singh AK, Asokamani R, Gogia AK (2009) Ti based biomaterials, the ultimate choice for orthopaedic implants—a review. Prog Mater Sci 54
25. Long M, Rack HJ (1998) Titanium alloys in total joint replacement—a materials science perspective. Biomaterials 19:1621–1639
26. Lansdown ABG (2006) Silver in health care: antimicrobial effects and safety in use. Curr Probl Dermatol 33:17
27. Brüggemann D (2013) Nanoporous aluminium oxide membranes as cell interfaces. J Nanomater 2013. https://doi.org/10.1155/2013/460870
28. Hench LL (2006) The story of bioglass®. J Mater Sci Mater Med 17(11):967–978
29. Chevalier J, Gremillard L, Virkar AV, Clarke DR (2009) The tetragonal-monoclinic transformation in zirconia: lessons learned and future trends. J Am Ceram Soc 92. https://doi.org/10.1111/j.1551-2916.2009.03278.x
30. Hoffman AS (2012) Hydrogels for biomedical applications. Adv Drug Deliv Rev 64:18–23. https://doi.org/10.1016/j.addr.2012.09.010
31. Slaughter BV, Khurshid SS, Fisher OZ et al (2009) Hydrogels in regenerative medicine. Adv Mater 21:3307–3329
32. Orhan ZD, Ciğerim L (2024) Evaluation of the effect of polybutester and polypropylene sutures on complications after impacted lower third molar surgery. Appl Sci 14:1448. https://doi.org/10.3390/app14041448
33. Sharma R, Malviya R, Singh S, Prajapati B (2023) A critical review on classified excipient sodium-alginate-based hydrogels: modification, characterization, and application in soft tissue engineering. Gels 9(5):430

34. Karageorgiou V, Kaplan D (2005) Porosity of 3D biomaterial scaffolds and osteogenesis. Biomaterials 26:5474–5491
35. Prè ED, Conti G, Sbarbati A (2016) Hyaluronic acid (HA) scaffolds and multipotent stromal cells (MSCs) in regenerative medicine. Stem Cell Rev Rep 12:664–681. https://doi.org/10.1007/s12015-016-9684-2
36. Jayakumar R, Prabaharan M, Sudheesh Kumar PT et al (2011) Biomaterials based on chitin and chitosan in wound dressing applications. Biotechnol Adv 29:322–337
37. Branquinho MV, Ferreira SO, Alvites RD et al (2021) In vitro and in vivo characterization of PLLA-316L stainless steel electromechanical devices for bone tissue engineering—a preliminary study. Int J Mol Sci 22:7655. https://doi.org/10.3390/ijms22147655
38. Fratila A, Jimenez-Marcos C, Mirza-Rosca JC, Saceleanu A (2023) Mechanical properties and biocompatibility of various cobalt chromium dental alloys. Mater Chem Phys 304:127867. https://doi.org/10.1016/j.matchemphys.2023.127867
39. Bhardwaj N, Kundu SC (2010) Electrospinning: a fascinating fiber fabrication technique. Biotechnol Adv 28:325–347
40. Cionca N, Hashim D, Mombelli A (2017) Zirconia dental implants: where are we now, and where are we heading? Periodontol 73:241–258. https://doi.org/10.1111/prd.12180
41. Chen FM, Liu X (2016) Advancing biomaterials of human origin for tissue engineering. Prog Polym Sci 53:86–168
42. Chellamani KP, Veerasubramanian D, Vignesh Balaji RS (2013) Surgical face masks: manufacturing methods and classification. J Acad Ind Res 2:320–324
43. Peppas NA (2004) Intelligent therapeutics: biomimetic systems and nanotechnology in drug delivery. Adv Drug Deliv Rev 56:1529–1531
44. Liu W, Worms I, Slaveykova VI (2020) Interaction of silver nanoparticles with antioxidant enzymes. Environ Sci Nano 7. https://doi.org/10.1039/c9en01284b
45. Elias CN, Lima JHC, Valiev R, Meyers MA (2008) Biomedical applications of titanium and its alloys. JOM 60:46–49
46. Hench LL (1991) Bioceramics: from concept to clinic. J Am Ceram Soc 74. https://doi.org/10.1111/j.1151-2916.1991.tb07132.x
47. Piconi C, Maccauro G (1999) Zirconia as a ceramic biomaterial. Biomaterials 20:1–25
48. Rai M, Yadav A, Gade A (2009) Silver nanoparticles as a new generation of antimicrobials. Biotechnol Adv 27:76–3
49. Lee Ventola C (2014) Medical applications for 3D printing: current and projected uses. P and T 39:704
50. Williams DF (2008) On the mechanisms of biocompatibility. Biomaterials 29. https://doi.org/10.1016/j.biomaterials.2008.04.023
51. O'Brien FJ (2011) Biomaterials & scaffolds for tissue engineering. Mater Today 14:88–95. https://doi.org/10.1016/S1369-7021(11)70058-X
52. Anderson JM, Rodriguez A, Chang DT (2008) Foreign body reaction to biomaterials. Semin Immunol 20:86–100
53. Song X, Tang Z, Liu W et al (2022) Biomaterials and regulatory science. J Mater Sci Technol 128:221–227. https://doi.org/10.1016/j.jmst.2022.04.018
54. Williams DF (2019) Challenges with the development of biomaterials for sustainable tissue engineering. Front Bioeng Biotechnol 7. https://doi.org/10.3389/fbioe.2019.00127
55. Ashammakhi N, Ghavaminejad A, Tutar R et al (2022) Highlights on advancing frontiers in tissue engineering. Tissue Eng Part B Rev 28:633–664
56. Goldenberg D, McLaughlin C, Koduru SV, Ravnic DJ (2021) Regenerative engineering: current applications and future perspectives. Front Surg 8:731031
57. Imberti B, Casiraghi F, Cugini D et al (2011) Embryonic stem cells, derived either after in vitro fertilization or nuclear transfer, prolong survival of semiallogeneic heart transplants. J Immunol 186:4164–4174. https://doi.org/10.4049/jimmunol.1000654
58. Grounds MD (2018) Obstacles and challenges for tissue engineering and regenerative medicine: Australian nuances. Clin Exp Pharmacol Physiol 45. https://doi.org/10.1111/1440-1681.12899

59. Wubneh A, Tsekoura EK, Ayranci C, Uludağ H (2018) Current state of fabrication technologies and materials for bone tissue engineering. Acta Biomater 80:1–30
60. Swanson WB, Omi M, Woodbury SM et al (2022) Scaffold pore curvature influences MSC fate through differential cellular organization and YAP/TAZ activity. Int J Mol Sci 23. https://doi.org/10.3390/ijms23094499
61. Edgar L, McNamara K, Wong T et al (2016) Heterogeneity of scaffold biomaterials in tissue engineering. Materials 9, 332
62. Srivastava GK, Martinez-Rodriguez S, Md Fadilah NI et al (2024) Progress in wound-healing products based on natural compounds, stem cells, and microRNA-based biopolymers in the European, USA, and Asian markets: opportunities, barriers, and regulatory issues. Polymers (Basel) 16. https://doi.org/10.3390/polym16091280
63. Ebrahimi B, Mokhtari T, Ghaffari N et al (2024) Acellular spinal cord scaffold containing quercetin-encapsulated nanoparticles plays an anti-inflammatory role in functional recovery from spinal cord injury in rats. Inflammopharmacology 32:2505–2524. https://doi.org/10.1007/s10787-024-01478-z
64. Ma L, Liang X, Yu S, Zhou J (2022) Expression, characterization, and application potentiality evaluation of recombinant human-like collagen in Pichia pastoris. Bioresour Bioprocess 9. https://doi.org/10.1186/s40643-022-00606-3
65. Stricher M, Vigneron P, Delbecq F et al (2024) The microalga *Volvox carteri* as a cell supportive building block for tissue engineering. Mater Today Bio 25. https://doi.org/10.1016/j.mtbio.2024.101013
66. Das P, Mukherjee T, Adhikary K, Mohanty S (2024) Progressive approaches in adipose radio imaging: cancer utilization and necessity for advancements. Infect Disord Drug Targets 24. https://doi.org/10.2174/0118715265314722240508092646
67. Adhikary K, Banerjee P, Barman S et al (2024) Nutritional aspects, chemistry profile, extraction techniques of lemongrass essential oil and it's physiological benefits. J Am Nutr Assoc 43:183–200
68. Mukherjee T, Das T, Basak S et al (2024) Mucormycosis during COVID-19 era: a retrospective assessment. Inf Med 3:100112. https://doi.org/10.1016/j.imj.2024.100112
69. Mukherjee T, Mohanty S, Kaur J et al (2024) Exploring small-molecule inhibitors targeting MAPK pathway components: focus on ERK, MEK1, and MEK2 kinases in cancer treatment. Chem Biol Lett 11:659. https://doi.org/10.62110/sciencein.cbl.2024.v11.659
70. Bhattacharya K, Dey R, Sen D et al (2024) Polycystic ovary syndrome and its management: in view of oxidative stress. Biomol Concepts 15:20220038.
71. Posinasetty B, Soni M, Pande SD et al (2024) Enhancing single-cell trajectory inference and microbial data intelligence. In: Microbial data intelligence and computational techniques for sustainable computing. Microorganisms for sustainability. Springer Singapore
72. Banerjee P, Adhikary K, Sarkar R et al (2022) Prion diseases: a rare group of neurodegenerative disorders. In: Viral, parasitic, bacterial, and fungal infections: antimicrobial, host defense, and therapeutic strategies. Academic
73. Adhikary K (2021) An updated review on nanomaterials for biomedical advancements: concepts and applications. Biosci Biotechnol Res Commun 14. https://doi.org/10.21786/bbrc/14.4.9
74. Banerjee P, Adhikary K, Chatterjee A et al (2021) Digestion and gut microbiome. In: Nutrition and functional foods in boosting digestion, metabolism and immune health. Academic
75. Sharma P, Kumar P, Sharma R et al (2019) Tissue engineering; current status & futuristic scope. J Med Life 2019. https://doi.org/10.25122/jml-2019-0032

Overview of Additive Manufacturing for Tissue Engineering and Regeneration

3

Md. Jasim Uddin, Safiya Mohammed Saeed, Samiha Binte Sadiq, Rehnuma Binta Hafiz, Jahida Yeasmin, Farjana Yesmin, and Amol D. Gholap

Abstract

Additive manufacturing (AM) has emerged as a game-changing technology in the field of tissue engineering and regeneration, providing remarkable precision to produce complex structures. Tissue engineering and regeneration have undergone a revolutionary change owing to AM, which allows for the accurate creation of complex tissue constructs with biomimetic architectures. This overview delves into the principles, biomaterials, design factors, and uses of additive manufacturing in tissue engineering. The benefits and drawbacks of several AM methods, including stereolithography, fused deposition modeling, and bioprinting, are covered. For bioprinting applications, the formulation of bioinks and the selection of biomaterials natural, synthetic, and hybrid materials are stressed as critical steps. The incorporation of cells and bioactive substances into AM structures is described, demonstrating the possibility for mimicking native tissue microenvironments and directing biological behavior. Furthermore, key concerns for scaffold design include resolution, porosity, and mechanical qualities, with an emphasis on maximizing these factors to resemble native tissue architecture and enable tissue regeneration. Tissue engineering-specific design concerns

M. J. Uddin (✉)
Department of Pharmaceutical Technology, Faculty of Pharmacy, Universiti Malaya, Kuala Lumpur, Malaysia

School of Pharmacy, Brac University, Dhaka, Bangladesh
e-mail: j.uddin@um.edu.my

S. M. Saeed · S. B. Sadiq · R. B. Hafiz · J. Yeasmin · F. Yesmin
Drug Delivery and Therapeutics Lab, Dhaka, Bangladesh

A. D. Gholap
Department of Pharmaceutics, AETs St. John Institute of Pharmacy and Research, Palghar, Maharashtra, India

S. Singh et al. (eds.), *Biomaterial-based Additive Manufacturing in Tissue Engineering and Regeneration*, Biomaterials, Bioengineering and Sustainability 7,
https://doi.org/10.1007/978-3-031-96070-3_3

and computer-aided design (CAD) methods are addressed, including patient-specific design methodologies and scaffold architecture optimization. The chapter explores the many uses of AM in tissue engineering, from the creation of skin replacements and vascular tissue structures to the regeneration of bone and cartilage, allowing for patient-specific treatment strategies. Along with prospects and developing technology in the sector, difficulties including biocompatibility, regulatory barriers, and scalability for clinical translation are also addressed. It illustrates the revolutionary potential of AM in enhancing tissue engineering and regeneration via case studies, eventually opening the door to the creation of novel regenerative medicine solutions. It aims to provide researchers, clinicians, and biomedical engineers with a thorough understanding of capabilities and potential applications in tissue engineering and regeneration, paving the way for novel therapeutic interventions and better patient outcomes.

Keywords

Additive manufacturing · 3D printing · Computer-aided design · Selective laser sintering · Scaffold · Vascular graft

Abbreviations

AM	Additive manufacturing
CAD	Computer-aided design
AM3DP	Additive Manufacturing/3D Printing
ASME	American Society of Mechanical Engineers
SLA	Stereolithography
UV	Ultraviolet
STL	Standard Tessellation Language
SLS	Selective laser sintering
DMLS	Direct Metal Laser Sintering
EBM	Electron beam melting
DIW	Direct ink writing
ECM	Extracellular matrix
PLA	Poly (lactic acid)
PGA	Poly (glycolic acid)
PLGA	Poly (lactic-co-glycolide)
PCL	Poly(caprolactone)
HA	Hydroxyapatite
PEG	Polyethylene glycol
CAM	Computer-aided manufacturing
CT	Computed tomography
MRI	Magnetic resonance imaging
CFD	Computational fluid dynamics
FEA	Finite element analysis

OCT	Optical coherence tomography
AI	Artificial intelligence
SGBR	Scaffold-guided bone regeneration
TEVG	Tissue-engineered vascular graft
dECM	Decellularized extracellular matrix
OOCs	Organ-on-a-chip
BioMEMS	Biological Micro-Electro-Mechanical Systems
3D	Three-dimensional
3DP	3D printing technologies
4DP	Four-dimensional printing
5DP	Five-dimensional printing
6DP	Six-dimensional printing

3.1 Introduction

With the use of cutting-edge technology, three-dimensional products are now possible to construct from digital models thanks to 3D printing. Additive manufacturing (AM) builds objects layer by layer utilizing materials like plastic, metal, or biological matter, in contrast to traditional manufacturing processes that involve extracting material from a larger block or sculpting a shape [1]. Almost 230,000,000 major surgical operations are performed globally each day [2]. Treatment commonly involves reconstructing, repairing, or replacing one or more impaired tissues or organs. Tissue, engineering in conjunction with AM has shown promise in the treatment of damaged organs and tissues by creating customized replacements that preserve, enhance, or repair tissue function [3]. The way we design and produce things could be dramatically changed by this technology, which offers benefits including increased flexibility, speed, and customization possibilities combined with reduced costs and waste. As a result, it is becoming more important in manufacturing and could transform several other industries in the future [4]. Taking in consideration of this chapter, looking into biomedical applications specifically, there are several benefits of 3D printing: (a) lowered waste levels; (b) economical; (c) improved capabilities; (d) adaptability in design; (e) personalization; (f) quick fabrication and prototyping; (g) reduction of recurrent surgery; (h) support for bone ingrowth; (i) better implant attachment; (j) enhanced cell proliferation; (k) restoration of normal joint kinematics; (l) strong mechanical qualities, and healing of (m) osteochondral defects [5, 6].

Because scaffold acts as a template for tissue regeneration, it is necessary for the fabrication of functional tissues or organs. AM is very useful for reconstructive surgery for facial injury since it may be used to create scaffolds with anatomical geometries individual to each patient. It can help in building such a template layer by layer. Tissue engineering scaffolds will be utilized increasingly frequently in clinical settings as research and practical translation advance. Nowadays, compounds that promote bone ingrowth are often used to treat wounds during reconstructive surgery [7]. Extrusion of materials AM is one of the most popular technologies for creating scaffolds for tissue engineering because it enables the extrusion of a wide range of materials, such as hydrogels, polymers, and ceramic pastes. The procedure is often called fusion deposition modeling, or FDM, for thermoplastics. The filaments emerge from a nozzle in material extrusion AM, and they are placed in relation to one another in a pattern that the user chooses. A totally interconnected network of pores can be used to construct complex geometries and porous objects that are not achievable with traditional production techniques like injection molding or machining [8]. Because porosity promotes the formation of blood vessels, nutrition diffusion, oxygen transport, and waste removal all critical processes for the regeneration of fully functional tissues it is particularly significant in scaffold construction.

Gariboldi and Best [9] presented four geometric levels for scaffolds used in tissue engineering. Surface topography, pore size and shape, porous networks, and macroscopic pore arrangement, which could include structural architecture with a functional gradient—are some of these layers. The next three levels strongly depend on the scaffolding geometry, which is determined by the orientation and arrangement of filaments, whereas the first level deals with surface characteristics of 10 μm or less and is usually governed by selecting the material and processing method. "Position and orientation of filaments" in this context refers to the precise geometric placements of filaments within a scaffold in relation to one another, including the spacing between filaments and the angle of placement of filaments on different levels. It is possible to alter these parts of the scaffold geometry without affecting the overall outward scaffold geometry design [7].

Numerous scaffold geometry characteristics, including mechanical properties, pore volume fraction, pore size, pore shape, and functional gradients of these qualities, can be controlled by the placement and orientation of filaments. These are all necessary components for tissue engineering scaffolds. Importantly, several studies have shown that the scaffold's shape affects a range of biological processes, including tissue growth, cell division, and cell seeding [10–12].

3.2 Evolution of AM in Healthcare

Charles Hull created additive manufacturing, also known as rapid prototyping, freeform manufacturing, or three-dimensional (3D) printing, in the 1980s. This state-of-the-art digital manufacturing technique creates physical models of complex shapes and structures by stacking on materials [13]. As a result, it makes it possible to produce structures more quickly and cheaply than with the traditional manufacturing approach. When used to product innovation, additive manufacturing can reduce lead times by 90% and manufacturing costs by up to 70%.

Compared to other disciplines where AM is utilized, the proportion of AM committed to medicine stays around 22%, according to a series of studies done between 2020 and 2022. The significance of creating fresh trends and enhancing medical applications is discussed in this study. The Additive Manufacturing/3D Printing (AM3DP) Medical Annual Report [14], which was recently released by the

American Society of Mechanical Engineers (ASME), highlights the important role that AM will play in the medical industry in 2020 and 2021. Where the greatest advances have been made are in the fields of prototypes and anatomical models. However, because of their enormous potential and development, more specialized fields like prosthesis design and fabrication, orthosis, bone tissue reconstruction, organ fabrication, dental applications, and medical device manufacturing have been taken into consideration in this research [15].

AM technology has been studied in a range of medical contexts since it was originally introduced in the 1980s, including as scaffolds for tissue engineering of patient-customized devices, research labs, and medicine delivery systems. According to the aforementioned trends, medical professionals should have a basic understanding of 3D printing technologies (3DP) in order to choose the method that would work best for them when applying AM to treat a certain medical condition. Potential solutions may include a revolutionary staff technique to address medical demands, an AM application, or a 3D-printed product. The creation of artificial organs, implants, prosthetics, orthoses, new surgical instruments, and surgical planning are examples of creative solutions that are found. Moreover, treatments open new possibilities for controlling or regulating the rates at which the body of the patient interacts with outside substances. Examples include the use of more suitable materials and methods to improve bone growth and the regulation of medication dosages [16]. In order to address issues whose constraints cannot be overcome; new AM fields are emerging. The most recent advancements in this subject, including bioprinting and other state-of-the-art printing techniques like four-dimensional printing (4DP), five-dimensional printing (5DP), and six-dimensional printing (6DP), are therefore summarized in this article. These can open up new avenues for study and research interest, which might result in cutting-edge applications that provide solutions and drive the demand for improvements in additive manufacturing techniques and a never-ending cycle of growth [14].

3.3 Importance of AM in Tissue Engineering

Tissue engineering combined with AM has emerged as a promising technique for healing injured tissues and organs by producing patient-specific substitutes that maintain, improve, or restore tissue function. Advances in the wide field of tissue engineering have resulted in the creation of a unique collection of tissue replacement components and application methods. To create new or improved tissues in the lab, one can combine biologically active substances, synthetic extracellular matrices (sometimes referred to as "scaffolds"), stem cells, biomaterials, growth and differentiation agents, and biomimetic environments. The requirement for ever-more complicated functioning, biomechanical stability, and vascularization in laboratory-grown tissues meant for transplantation is one of the main concerns facing tissue engineering today [17]. The capacity of additive manufacturing to produce a wide range of products from various materials is gaining prominence. Of all the industrial and academic uses of additive manufacturing, printing methods are perhaps the most common. In addition to electronic functional materials, it is believed that the next scientific and technological revolution will entail the two- and three-dimensional printing of a wide range of materials, including metals, polymers, and ceramics. Notably, these technologies are being used in many medical applications. Tissue-engineering benefits from printers' ability to provide a precise, computer-controlled, repeatable process for creating human tissues and organs. The three-dimensional printing of scaffolds, nanomaterials (including cells), and cellular materials lacking scaffolds in tissue engineering could be very helpful in the production of functional tissues and organs [18].

AM have also been utilized to give aligned cell development for better tissue morphogenesis and to achieve spatial modulation of skeletal muscle growth hormones, enabling directed tissue growth. AM is ideally suited to reproducing the difficult change that occurs across several orders of magnitude of mechanical characteristics within tiny length scales across diverse orthopedic contacts. Even though AM already offers significant advantages in the production of customized metallic and ceramic implants for orthopedic stabilization, the integration of cellular and material investigations in bioprinting promises to yield sophisticated biosynthetic grafts for regeneration as well as restoration of tissue deficits as opposed to just stabilization [18]. Though bioinks, a mixture of different cell types, including biological molecules. In combination with polymers or ceramics—have undergone research to enhance the hypothetical potential of AM devices in regenerative medicine, the primary application of 3D printing has been the production of scaffolds devoid of cells. A significant shift from conventional 3D printing techniques is required to enable AM of biological commodities since cells react adversely to strong chemicals, extreme heat, and strong UV light. This means that there is a limited range of substances and equipment accessible for graft creation via bioprinting [19].

3.4 Fundamentals of AM Techniques

3.4.1 3D Printing

3D bioprinting creates physiologically mimicked complex tissue structures by entailing layer by layer accumulation of biomaterials incorporated with cells within a predefined structural framework which is utilized in various ways in regenerative medication and tissue engineering. It is a well-constructed integration of tissue engineering, live cells and controlled motor systems. The physiologically mimicked complex tissue structures are constructed into scaffold form by regulated mechanical, permeability and porosity as well as biomedical devices [20, 21]. Moreover, 3D printing is foreseen to be an immaculate discovery in further future due to its sustainable features including increased affordability, minimal post-processing and material wastages, capacity to reutilize, reuse plastics, lower emissions and nonetheless the ability to create detailed and well-balanced geometrical construction. 3D printed materials are lightweight components with superior strength [22, 23]. AM is an essential procedure for the basic construction of a particular object in 3D printing. Firstly, the essential and fundamental blueprint of the object is made by computer software connected with a 3D printer for the transaction of the blueprint to the 3D printer. Secondly, the 3D printer interprets the blueprint in one single two-dimensional layer each time and applies one layer on top of the other to produce the desired product [23].

3.4.1.1 Stereolithography

Stereolithography (SLA) is defined as a three-dimensional printing method in which the stereolithographic machine helps liquid photopolymers to be embedded into 3D objects. There are essential components of stereolithography such as laser source, computer, perforated table, ultraviolet (UV) curable photopolymer liquid for process control. Firstly, the procedure is initiated by generating a Standard Tessellation Language (STL) file containing necessary geometrical information of a particular object which is converted through a CAD file [24]. Secondly, 3D printers begin to operate, and the liquid tank is immersed beneath the perforated table after they have read the STL file type. Interaction between the liquid polymer and the table occurs through the perforations as it descends. As a result, the photo liquid polymer is instantly hardened as soon as the liquid enters contact with the table due to striking the upper surface with a UV laser. Again, the table descends to produce a layer-by-layer geometry with the base layer serving as the starting point for the fusion of each subsequent layer. The component which is generated via three-dimensional printing is immersed in a secondary resin after the final layer is finished in order to separate it derived from the liquid polymer. After the implementation of this procedure, resin binds all of the layers together with rigidity and the 3D printed model is then ready for baking in a UV curing oven with a predefined temperature. As all of the layers are solidified or stiffened, the rigidity extends as well as the appropriate exterior area finish is achieved, which is **shown in** Fig. 3.1, which is a diagrammatic representation of SLA process as initially designed by Hull, which shaped the foundation of modern SLA to date [25]. According to Hull, utilization of photosensitive materials being cured by ultraviolet (UV) light, the 3D objects are created layer by layer.

3.4.1.2 Fused Deposition Modeling

This process involves pushing out thermoplastic filament layer by layer after it has dried to the point of melting in order to create three-dimensional objects. The usable support base of the 3D printers for FDM is arranged to move the filament vertically with a certain degree of freedom. A companion extruder joined with the filament heats the bottom plate till the freezing temperature as well as uses a nozzle to extrude the filament gradually by layers to create the desired object. Consequently, the extruder can move in all axis directions. During the deposition procedure, the adjoining layers combine, and it is completed by the extruder. As a result, the final product is dipped in resin depending on the required surface exterior [26, 27]. This is illustrated in Fig. 3.2.

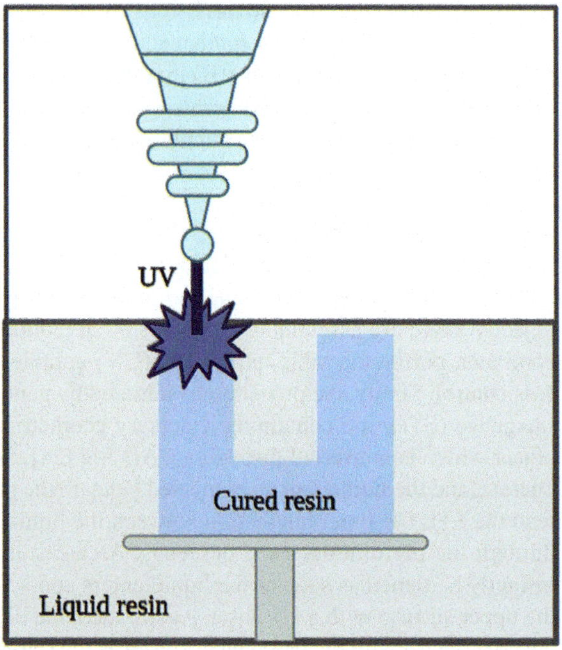

Fig. 3.1 Basic SLA process. (Created by using Biorender.com)

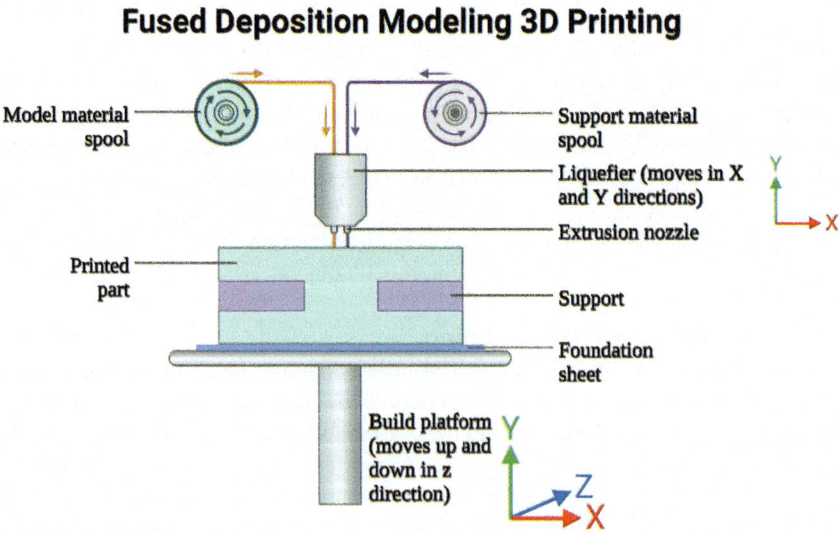

Fig. 3.2 It illustrates about the fused deposition modeling 3D printing process used for development of 3D printed material. (Created by using Biorender.com)

3.4.1.3 Selective Laser Sintering

Solid 3D scaffold structures are formed by powder sintering and utilizing lasers as power sources via selective laser sintering (SLS). This procedure is applicable for printing on several diverse materials such as polymeric materials, metals as well as earthenware bricks. Additionally, scaffolds which are generated by composites and polymeric biomaterials via SLS are also applicable in the myriad implementation in tissue engineering [28]. For generating complex as well as detailed geometrical structures, successive layers of powdered material are fused together over one another with SLS. CO_2 or nitrogen lasers aid in the solidification of layers on the surface exterior and amalgamation to produce objects by using the chemical compound powder made of metal, glass, ceramics, thermoplastic etc. Particularly, metal powder assisting in the production of objects is referred to as Direct Metal Laser Sintering (DMLS). The SLS printer is comprised of two compartments. Powder is transferred from the first compartment into the second compartment in which the procedure of manufacturing takes place. Moreover, a decreased temperature than the melting point of the comparable material is used for heating. Layers surfacing the powder are formed by the leveler or roller at the upper portion [29]. Standard SLS procedure is consists of three core stages: (a) preheating phase: The layer-bed is heated up till it achieves the predetermined temperature (called the layer-bed temperature) during this phase and this temperature is maintained throughout the component construction operation. To save laser activity and prevent other component deformities in the cooling process, the layer-bed temperature is maintained just below the polymer's softening temperature. (b) construction stage: This is the primary stage of the fabrication process involving multiple processes. The support structure is first lowered to collect the powder particles that are being propagated by the spreading blade or the roller. The layer of particles is then melted by the laser beam following the programmed route. The component is finally gradually cooled to the bed temperature in order for it to solidify. (c) phase of cooling: After turning off the heat source, the powder bed gradually cools to the extraction temperature of the component [30].

3.4.1.4 Inkjet Bioprinting

Bioprinting is one of the significant methods to generate bioengineered structures (i.e. AM) of biological and biologically related materials using computer-aided transfer and build-up processes. Inkjet bioprinting is a non-contact printing method that dispenses droplets of dilute liquids using thermal, piezoelectric, or microvalve processes. Traditional inkjet process used by desktop inkjet printers uses in inject bioprinting technology, which uses single droplets to pattern a substrate. Inkjet bioprinting uses the thermal or acoustic delivery of bioink. Inkjet bioprinters can be designed to print materials other than bioinks. It constructs three dimensional structures by dispensing a set amount of bioink through a needle with air pressure or with the help of mechanical pistons [31]. It plays an important role in tissue engineering by enabling the development of sophisticated cellular scaffolds with high resolution, form fidelity, and cell survival. Achieving these characteristics, it faces some challenge, like it requiring bioinks that are biocompatible, printable, and biodegradable. A pressure pulse produced by an electrical heater or piezoelectric actuator is used in inkjet bioprinting to force the bioink droplet onto the surfaces [32].

3.4.1.5 Electron Beam Melting

This method involves creating three-dimensional objects by melting metal powder using a powerful electron beam. In magnetic field, an electron beam generates a stream of electrons that melt layers of powdered metal to build an object, and it must meet the precise parameters according to CAD model. Electron beam melting (EBM) works on a similar principle to SLM technology. EBM is a new and advanced method of additive manufacturing that can handle very reactive materials with high melting points [33]. In a vacuum room, production is conducted to prevent oxidation, otherwise highly reactive materials could be damaged. EBM creates high-strength parts that take advantage of the intrinsic qualities of the metals utilized in the process, avoiding impurities that may form when casting metals or using other methods of fabrication. Here, electron beam radiation is used to melt the powder at a voltage range of 30–60 kV. In high-vacuum conditions, EBM is performed which reduces oxidation, atmospheric interference, and associated contamination. EBM offers the unique ability to create metallic components utilizing an alloyed powder mixture to achieve novel mechanical and functional qualities. EBM is suited for strategic applications such as outer spacecraft and medicinal implants. It is used to print parts for aerospace, automotive, defense, petrochemical, biomedical implants, and medical applications [34].

3.4.1.6 Direct Ink Writing

Direct ink writing (DIW) is a printing method that uses continuous filament ink stream extrusion and computer-controlled deposition to quickly create 3D structures layer by layer without needing costly tools or masks [35]. DIW is a 3D printing technology in which viscous ink is extruded from a syringe under pneumatic pressure. This extrusion-based technology allows for the printing of various polymers, such as epoxy and photocurable acrylic resin, as well as reinforcements such as silicon carbide whiskers, nylon fibers, and carbon fibers. It is a user-friendly technology that has been successfully used in the biomedical industry for tissue engineering applications. The DIW technique prints the 3D object by moving the syringe horizontally and vertically. In this procedure, the initial step is to mix the resin and curing agent before feeding them into the syringe as ink. After extrusion, the ink is cured using UV induction or thermal heating [36]. The techniques used by the DIW are divided into two groups: droplet-based approaches and filamentary-based approaches. Different designs of ink, including polymer melts, highly shear-thinning colloidal suspensions, waxes, and concentrated polyelectrolyte complexes, have been employed in DIW processes. The inks could solidify either through temperature, liquid evaporation, gelation, or solvent-induced phase change [37].

3.5 Bioprinting

It is a significant technology in tissue regeneration of living cells to produce reconstructive tissue where bioinks and biomaterials are mixed with cells. The 3D bioprinting uses AM techniques. One major benefit of bioprinting is the consistent spread of cells, which can be a challenge when seeding cells onto large tissue scaffolds. Current bioprinting approaches rely on inkjet, extrusion, or laser AM technology [38].

Traditional 3D printing techniques have advanced with the development of sacrificial resin molds for the construction of 3D scaffolds from biological materials. Successful tissue regeneration has been linked to cell seeding, scaffolding, and construction technologies. 3D printing and electrospinning methods can help make scaffolds quickly and accurately for cellularization using vascular progenitors or mixes of mature human vascular cells, creating bioengineered blood vessels [39]. The progress in 3D bio-printing is driven by the increasing demand for new ways to make tissues and even organs, since current conventional methods can't fabricate materials with the necessary properties like structural, mechanical, and biological properties such as biocompatibility and complexity [40].

3.6 Bioprinting Technologies: Advantages and Limitations

Bioprinting generates critical interconnected tissue structures by utilizing hydrogel bioinks. According to a study, a vascular constructed system is bio-printed by GelMA which enhances endothelial monolayer formation, the quantity of healthy cells and metabolic transportation. Interconnected tissue structures are generated by thermally reversible gelation printed biomaterials embedded with cells and blood vessels. Recently, hydrogel has been utilized in the bioprinting of aortic valves according to a study which has reported an increased cell survival rate. Yet, mechanical properties of bioink materials fall short of clinical requirements due to their lack of elasticity and flexibility. Moreover, the production of cartilage tissue engineering scaffolds with ceramics and nanomaterials has been facilitated by the widespread utilization of bioprinting techniques. Consequently, bioink produced by incorporating delicate human chondrocytes with a combination of nano-fibrillated cellulose and alginate has demonstrated excellent rheological characteristics such as shear thinning and increased quantity of healthy cells in an experimental 3D culture study. Also, the bioink was effectively able to generate different physiological shapes associated with cartilages due to the high plasticity of the ink [41]. Since many long-term illnesses and burn injuries result in permanent skin damage, bioprinting technology is especially important for creating skin substitutes that can be transplanted to replace damaged skin. LAB method incorporated and printed bio-skin printing has been able to successfully create skin substitutes which is proved by observing healing of skin wounds of mice. Also, the bio-skin has effectively adhered, proliferated and differentiated the tissue surrounding the skin wound. Bioprinted skin tissue incorporated with keratinocytes and fibroblasts has shown essential morphological and biological characteristics [42]. Bioprinting techniques have been recently utilized to generate similar structures of liver hepatic cord such as canaliculi to extend the duration of liver cellular activity and functionality. This method utilized in various clinical trials has been observed to be helpful in further medical and biological research, as well as to help identify drug toxicities [43]. A significant disadvantage of inkjet bioprinters is the narrow selection of bioink materials. To be ejected out of the nozzle's small orifice, the bioink material requires for being liquid and consisting of the right viscosity which should be enough to maintain the specified formation but not so much that it clogs the nozzle. The usage of customized 3D printing technology is thought to result in several regulatory obstacles with regard to the oversight of printed products. Nonetheless, to ensure the sustainable advancement of 3D printing technology, it is imperative to ensure the management of laws and regulations for ensuring the sustainable advancement of bioprinting technology. The primary technological obstacles are the scarcity of proper bioinks which can be utilized to achieve biological function as well as consist of good mechanical strength and biocompatibility [44].

3.7 Biomaterials for AM in Tissue Engineering

Tissue engineering relies significantly on biomaterials. They are intended to offer an architectural foundation similar to the original extracellular matrix, encouraging cell proliferation and, eventually, tissue regeneration. Bone and cartilage are two separate tissues with different compositions and mechanical properties [45]. The study of biomaterials that can promote regenerative processes by successfully conveying cell populations and therapeutic agents, as well as providing structural scaffolding that confers suitable mechanical qualities to tissues, is an important aspect of tissue engineering. Furthermore, the biomaterial should disintegrate at a rate proportional to the formation of new tissue at the implantation site. Among a multiplicity of applications, the musculoskeletal tissues of bone and cartilage have sparked significant interest from researchers. There are two types of biomaterials, such as natural biomaterials and synthetic biomaterials [46]. These are further explored in this section.

3.7.1 Natural Biomaterials

For decades, naturally generated biomaterials have been used in a variety of regenerative medical applications [47]. They are typically the first choice when selecting a functional and useful biomaterial [48]. From renewable resources, such as plants, animals, and microorganisms, natural biomaterials originate and have a wide range of distinct but complicated ingredients, microstructures, and physiological features [49]. In their natural environment, such materials provide biological support for cell adhesion and growth, as well as a variety of functions. Thus, natural biomaterials can be used for specific applications in tissue engineering and regenerative medicine [50]. Natural materials are classified into different types based on their origin. Some examples include proteins like collagen, gelatin, silk, and fibrin; some examples include carbohydrates like cellulose, chitin/chitosan, alginate, and agarose; some other examples include glycosaminoglycans like hyaluronic acid, chondroitin sulfate, dermatan sulfate, heparan sulfate, and keratan sulfate [51].

Recently, more complex biomaterials have developed, as in the case of cell/organs-derived matrices [52]. They all have some unique features, including biodegradability, biocompatibility, and remodeling that enhanced the scientific interest in developing tissue engineering technologies around these biomaterials [53]. Some significant examples of natural biomaterials are explained below.

3.7.1.1 Collagen

Collagen is the primary structural protein found in the majority of animal tissues, and It is necessary to maintain the biological and structural integrity of the extracellular matrix (ECM) and provide physical support for cells and tissues. Fibroblasts produce collagen, which controls the cellular shape, adhesion, migration movement, and differentiation [54]. Collagen is mainly seen in fibrous tissues such as skin, tendons, and ligaments as well as in other tissues that require both strength and flexibility, such as blood vessels, bones, cartilage, corneas, the gut, intervertebral discs, and dentin. Collagen makes up about 25–35% of the protein content of the human body [55]. Currently, 29 collagen isoforms have been described. Because of its importance in biology, collagen has been a focal point of tissue engineering research. Collagen is easy to extract and purify on a large scale. Furthermore, collagen is biodegradable, biocompatible, and has no cytotoxic properties. Collagen structures have been utilized for different purposes, like heart valve replacements, blood vessel implants, medicine delivery systems, eye treatments, and nerve repair. Also, small capsules with collagen layers have been developed in three-dimensional structures for building soft tissues. In liver tissue engineering, using collagen-coated silicone structures is a key method for making healthy 3D liver cell cultures [56].

3.7.1.2 Gelatin

Through the hydrolysis of collagen, gelatin is formed. It is a biodegradable and fully absorbable biopolymer. It has been widely employed and has demonstrated numerous advantages over its parent protein because of its biological nature, excellent solubility in aqueous systems, and widespread commercial availability at a reasonable price [57]. Gelatin-derived biomaterials have multiple applications, including cardiovascular, bone, skeletal muscle, and hepatic tissue engineering, as well as wound healing and injectable fillers. As a biomaterial, gelatin has a wide variety of applications. Gelatin formulation alterations have been demonstrated in stem cell research to influence stem cell fate in injectable cell-based therapies [58].

3.7.1.3 Alginate

It is derived from brown seaweed, and it is a naturally occurring anionic polymer. It has good biocompatibility, low toxicity, and low cost. Alginate can be easily processed into three-dimensional scaffolding materials. Alginate hydrogels are one of the most popular compositions. Through numerous cross-linking techniques, alginate hydrogels can be produced, and their structural similarities to the ECM of living tissues provide a wide range of wound healing management applications [57].

3.7.2 Synthetic Biomaterials

The material surrounding cells, known as the extracellular matrix, consists of a combination of proteins and polysaccharides that form a structured network to support cells. Naturally generated polymers have the potential to improve cell adhesion and function, and they are popular for scaffolds. However, the complex structural composition of natural polymers, their immunogenicity, and the risk of spreading pathogens prompted the development of synthetic polymers as scaffolding materials. Synthetic polymers typically have a regulated structure, improved manufacturing flexibility, and no immunological problems. The most frequently used environmentally friendly synthetic polymers for 3D scaffolds in tissue engineering are saturated poly (lactic acid) (PLA) and poly (glycolic acid) (PGA), as well as poly (lactic-co-glycolide) (PLGA) copolymers [59]. Some significant synthetic biomaterials, such as: polycaprolactone Poly (lactic-co-glycolic acid) is explained below:

3.7.2.1 Poly(Caprolactone)

In tissue engineering and biomedical uses, Poly(caprolactone) (PCL) is utilized. In normal conditions, microorganisms, hydrolytic, enzymatic, and intracellular processes can all breakdown PCL [60]. It is a semi-crystalline material with a low glass transition temperature of about −60 °C. Therefore, it remains elastic and has a high material permeability, which allows materials to pass through easily under physiological conditions. However, PCL breaks down more slowly compared to PLA, PGA, and PLGA, which makes it less suitable for common tissue regeneration, but it is more suitable for long-term implants and drug-delivery systems [61].

3.7.2.2 Poly Lactic-Co-Glycolic Acid

Poly lactic-co-glycolic acid is a type of biodegradable material made from combining lactic acid and glycolic acid to form a polymer. It is widely utilized in medicines and medical engineering materials due to its biocompatibility, non-toxicity, and high plasticity [62]. The study of the latest technology in the field shows that there are new methods for creating scaffolds and materials that can help make biomimetic PLGA substrates. These biomimetic PLGA substrates are able to modulate cell interaction for improved substitution, restoration, or enhancement of bone tissue function [63].

3.7.3 Hybrid and Composite Biomaterials

Hybrid and composite biomaterials offer an advantage by blending materials to create enhanced properties [64]. For example, combining synthetic polymers can result in materials, with biocompatibility strong mechanical properties and controlled degradation rates [65]. Natural polymers like collagen, gelatin and alginate create an environment for cell attachment and growth while artificial polymers, like polycaprolactone and poly (lactic co glycolic acid) contribute to the structural strength and adjustability [66]. Because ceramic polymer composites can mimic the amount of minerals of bones and promote bone regeneration, they are generating interest in the field of cells engineering [67]. Mixing hydroxyapatite (HA), a bone mineral, with polymer materials improves their capacity to promote bone formation and growth. These composite materials can also be customized with growth factors or bioactive substances to encourage reactions. In addition, nanomaterials like carbon nanotubes, graphene and nanoparticles have been integrated into composites to provide characteristics [68–70]. These nanomaterials can enhance strength, electrical conductivity and biological functions making them appropriate for use in applications like neural tissue engineering and bio-sensing [71].

The way scaffolds are designed architecturally is just as important, as the materials used in tissue engineering. Additive manufacturing methods like 3D printing allow for creating scaffold designs customized to each patient including specific porosity, pore sizes and interconnections. These characteristics are crucial, for supporting cell movement transporting nutrients and promoting tissue growth [72].

One exciting avenue, in manufacturing for tissue engineering involves exploring production methods that merge various manufacturing processes or blend advanced manufacturing with traditional techniques. For instance, the fusion of 3D printing and electrospinning can create scaffolds displaying intricate macro and micro architectures closely resembling the level structure found in natural tissues [73]. Moreover, incorporating elements, like growth factors, cells or decellularized extracellular matrix into the compositions has displayed encouraging outcomes in tissue engineering endeavors. These bioactive elements can be enclosed within the biomaterial during the printing phase. Added after fabrication facilitating tissue regeneration and functional recovery [74]. Even though there has been advancement, in creating composite biomaterials for AM in tissue engineering there are still some hurdles to overcome. These challenges involve enhancing biocompatibility, refining control over material degradation rates and refining the manufacturing processes for accuracy and efficiency. Moreover, incorporating materials like those, with self-healing or shape memory capabilities could greatly improve the effectiveness of tissue engineered structures [75].

An exciting avenue, in this domain involves the advancement of self-assembling biomaterials of creating structures that imitate the intricacy of natural tissues [76]. These materials use interactions and supramolecular chemistry to construct designs from the ground up [77]. To illustrate hydrogels based on peptides can autonomously organize into networks to the extracellular matrix offering a nurturing setting for cell development and specialization. Moreover, incorporating stimuli-responsive elements into biomaterial formulations has become a hot topic. Such material reacts to external agents like temperature, pH, or light exposure, providing an ability to regulate certain elements' characteristics and performance [78]. For instance, Thermo-responsive polymers, which are utilized in an in-situ tissue engineering system, may change from a solid to a gel state at human body temperature.

Furthermore, the combined use of current bio-fabrication techniques like bio-printing and bio-assembly with biomaterials has produced a novel method for creating complex tissue structure. These methods enable the exact placement of cells, biomolecules, and biomaterials into pre-planned patterns, helping in the restoration of the complex heterogeneity and structure of the initial tissues [79]. To conclude, the advancement of hybrid and composite biomaterials for AM in tissue engineering allows to create tissue constructs uniquely tailored to desired needs. The use of different materials or their combinations allows them to mimic the attributes of the native cell's environment. Their use should facilitate enhanced cell survival, growth and differentiation [80].

3.8 Bioinks: Properties and Formulation

In the diverse focus of tissue engineering, elements of biology, engineering, and materials science are combined to generate functioning substitutes for organs and cells that have been harmed by illness or injury. AM or the utilization of 3D printing in Tissue engineering is becoming essential. Because it facilitates the creation of multidimensional and patient-specific architectural models and material with control over its composition. This is particularly possible with the use of bioinks. Bioinks are substances that help support cell life, functioning, and division, while further providing the mechanical and ideal structural orientation for the destined use [81].

3.8.1 Properties of Bioinks

The bioinks must satisfy these additional requirements: biocompatibility, printability, mechanical properties, and biodegradability.

3.8.1.1 Biocompatibility

The materials need to be non-toxic and stimulate cell survival, adhesion, and differentiation. The bioinks, when implanted, should not evoke dangerous immune and inflammatory reactions.

3.8.1.2 Printability

The inks should have suitable rheological properties to achieve proper viscosity, shear-thinning, and quick gelation or cross-linking to create the constructs accurately and keep their shape and efficacy during and after the printing process [82].

3.8.1.3 Mechanical Properties

The structural features of the inks should be appropriate to the native tissue, including the stiffness, elasticity, and strength properties.

Biodegradability The bioinks usually need to be degradable to be replaced with new tissue as the implant degrades.

3.8.1.4 Cell Adhesion

In addition, a bioink must allow adhesion, migration, and spreading of cells, thus enabling the creation of the engineered tissue and their fusion with the surrounding tissue [83].

3.8.2 Formulation of Bioinks

Polymers, both organic and artificial, can be utilized as bioinks to their compositions and hybrids. These biomaterials are selected according to the requirements for the printing of the target tissue.

3.8.2.1 Natural Polymers

Proteins Commonly used proteins in bioinks include collagen, gelatin, silk fibroin, and fibrin [84]. Proteins are favorable to be employed as bioinks since they are biocompatible, biodegradable, and cell-binding proteins [85].

Polysaccharides They are frequently used as bioinks because of their excellent physiological compatibility with ECM and the capacity to steering cell-matrix interactions. The bioinks of polysaccharide include the following: (alginate, chitosan, hyaluronic acid, agarose).

3.8.2.2 Synthetic Polymers

Polymer or repeat unit that makes up the backbone of the polymer can be defined by a chemical structure. A few instances of synthetic polymers in hydrogels consist of, poly glycolic acid, and their copolymers. These polymers have high and increasing applications in hydrogels due to biodegradability and tunable mechanical properties [86].

Many alternative examples are polymers like poly caprolactone PCL which is a polyester that decomposes naturally excellent mechanical performance that is compatible with various substances and biomolecules. Polyethylene glycol (PEG) is another broad hydrophilic polymer utilized to modify the hydrogel's characteristics or to enhance cell viability and degrade the polymer at specific speeds.

3.8.3 Composites and Hybrids

Composites can be developed by combining various both organic and artificial polymers to gain from their distinct features: biocompatibility, high mechanics, and the possibility of regulating degradation:

Hybrid bioinks can include other elements, such as nanoparticles, growth factors, or ECM proteins, to improve specific characteristics or functions [87].

Formulation Strategies

Photo-crosslinking: to produce hydrogels, the use of a crosslinking agent is needed, turning the bioinks from a printable to a less viscous form and back after printing. The most common strategies to date are UV or visible light, with the corresponding photochemistry used to initiate the crosslinking reaction. This method allows for the manufacturing of hydrogels with a variety of mechanical properties and susceptibility to degradation [88, 89].

Ionic crosslinking: this approach utilizes divalent cations, for example calcium or barium ions bind with the anionic polysaccharides like alginates to quickly form a stable hydrogel [90].

Thermal gelation: Solvo-gel transitions based on polymers that are responsive to temperature may be used in the printing of bioinks. For example, gelatin or Pluronic-based materials that can gel at AMD temperatures may be used to print, with the crucial occurring *in situ*.

3.8.4 Enzymatic Crosslinking

Protein-based bioinks can be crosslinked in a relatively mild and bio-friendly manner by enzymes. Transglutaminase or thrombin crosslinked gelatin or fibrin form stable hydrogels, but with mild conditions.

Solvent-Based Systems Polymers can be dissolved in biocompatible solvents, such as dimethyl sulfoxide or N-methyl-2-pyrrolidone, and subsequently solidified through their evaporation or precipitation solvent-based systems [91].

The choices of bioinks as well as strategies of formulation and optimization is crucial. For achieving desired properties and performance in cells engineering [92]. The research in this area is also focused on developing advanced bioinks with advanced functionality, printability, and tissue-mimicking properties, as well as discovering new biomaterials and formulating techniques [93].

3.9 Design Considerations and Computer-Aided Design in Tissue Engineering

3.9.1 Computer-Aided Design Software for Tissue Engineering Applications

Computer assisted design software can assist with the procedure of the bone tissue engineering and generation of 3D bone scaffolds with simple geometrical structures. It has many applications in the healthcare industry, including therapeutics, personalized medications, pharmacogenetics, orthodontics and medical devices. 3D printing is a conventional modeling method that utilizes computer-aided design. CAD modeling has made extensive usage as it can quickly create many models [94]. A secondary development method of CAD software is utilized to design scaffolds with internal and external structures. Effective CAD technique has been studied to enhance the ability of designing and integration of a convoluted network of channels under dispute of a set of variable parameters in a short period of time via combination of CAD software and object associated programming. The output data includes a CAD file of a manageable size that can be quickly transformed into an STL format for the manufacturing of the product [95]. 3D printing utilizes software for segmentation and computer-aided design that layers 2D medical pictures into 3D models one after the other as printable digital files. Furthermore, 3D bioprinting provides precision and customizability since Cells are stacked exactly by utilizing CAD and produced by 3D printers to imitate the extracellular matrix of the patient. Recently, the medical sector made extensive usage of CAD software increasing the productivity by generating 3D structural framework which adheres to precise geometry, both internal and external and ensures minimal structural porosity to prevent problems. CAD software creates digital 3D structural frameworks that can be manually altered to verify details, eliminating imperfections and fixing automated computational errors [96]. Computer-aided design techniques make a variety of biomaterials usable. Any 3D organ or missing portion of any organ can be precisely generated with the help of CAD software. Certain characteristics of living organs could be included in the CAD software generated 3D model. CAD software can be used to create scaffold geometry for irregularly shaped tissue defects detected via MRI and CT scans. The CAD model of the scaffold-based product can be generated by utilizing free software [97]. CAD is widely usable technology in the engineering, architectural and manufacturing sectors to generate accurate 2D and 3D models of physical components by increasing structural efficiency and enabling simplistic modifications. Also, the enhancement in its ability occurs due to integration of 3D printing and computer-aided manufacturing (CAM). Computer-aided design generates scaffolds by utilizing 3D scans of particular biofabrication products facilitating the production of controlled-release scaffolds for tissue engineering [98].

3.9.2 Design Optimization for Tissue Constructs

Optimization strategies significantly improve the performance and function of tissue constructs fabricated using AM. Using advanced optimization techniques, simulation tools, and computational modeling, it is possible for researchers to forecast how constructed tissues will behave in different settings and optimize distinctive performance trait configurations. The tools are used to optimize characteristics such as mechanical properties, nutrient transport, cell layout and tissue integration. Ultimately, this leads to better tissue construct design [99].

In the field of design optimization, a popular method involves utilizing topological optimization algorithms. These algorithms are used to determine the way to distribute materials efficiently within a designated design area while meeting requirements. This approach enables the development of structures that're strong and durable resembling the complex structures seen in natural tissues. Topological optimization algorithms have been effectively utilized to enhance the create of scaffolds, over bone tissue engineering. This has led to the creation of structures with characteristics and better nutrient flow [100].

Design optimization also involves incorporating patient data utilizing techniques for medical imaging like computed tomography (CT) or magnetic resonance imaging (MRI) to integrate anatomical information unique to each patient, into the design process. This method permits the development of customized tissue structures designed to fulfill the needs of each patient, potentially enhancing clinical results, and minimizing the chances of rejection or adverse effects [101].

Researchers have not worked on enhancing the structural design but have also dedicated efforts to improve the bioactivity and functionality of tissue constructs. One strategy includes integrating substances, growth factors or biomolecular gradients into the design of the construct. These bioactive elements can be strategically placed within the construct to encourage cellular reactions, like cell growth, specialization, or tissue repair [102].

Optimizing the flow of nutrients and oxygen within tissue constructs has been greatly enhanced using computational fluid dynamics (CFD) simulations, which are considered valuable tools, in this field. By replicating the movement of fluids and the transfer of substances scientists can pinpoint the factors needed to guarantee sufficient nutrient supply and oxygen delivery to cells throughout the construct, thereby enhancing cell viability and tissue formation. Moreover, researchers have utilized mathematical modeling and simulation methods to enhance the attributes of tissue construct. These models can forecast how stress is distributed, how deformation occurs and the overall mechanical strength of constructs, under loading scenarios. This enables the creation of constructs, possessing mechanical qualities tailored for specific tissue applications [103].

The enhancement of tissue construct design goes beyond mechanical aspects; it also encompasses optimizing bio-manufacturing processes. Scientists have investigated employing techniques to fine tune process parameters like printing speed, temperature, and material composition. This optimization aims To attain the intended characteristics of constructs and enhance the efficiency and consistency of the manufacturing process. Using computational models and experimental techniques it is possible to identify the optimal conditions for tissue development [104].

Designing tissue constructs poses a challenge due to the changes in the tissue development process. As cells grow, transform and create matrices the characteristics of the construct can evolve. To tackle this issue scientists have utilized dynamic modeling methods to simulate how the properties of the construct change over time and adjust the design accordingly. For instance, sophisticated computational models have been created to forecast how mechanical properties and nutrient transport within tissue engineered evolve as the tissue matures [105].

In design optimization a different method involves combining modeling with finite element analysis (FEA) to forecast how tissue constructs behave under normal loading conditions. By mimicking how the construct reacts, scientists can enhance the design to guarantee functionality mechanical properties and minimize the risk of failure or deformation during implantation and tissue regeneration [106].

Progress in imaging methods like computed tomography and microscopy has played a role in improving the design of tissue constructs. These methods permit researchers to see and analyze the structure and layout of constructs, which helps validate models and enhance design parameters. Moreover, real time monitoring of tissue development within constructs is possible through in situ imaging techniques like optical coherence tomography (OCT) and confocal microscopy enabling adjustments to the design or manufacturing process as needed [107]. Moreover, combining manufacturing with cutting edge technologies, like bioprinting and microfluidics has introduced opportunities for enhancing design in tissue engineering. Bioprinting facilitates placement of cells, biomaterials, and bioactive substances, within tissue frameworks allowing for the development of varied structures. Additionally integrating systems into tissue frameworks enables regulation of nutrient delivery, oxygen supply and biochemical signals to enhance the microenvironment conducive to tissue growth [108].

A growing field in design improvement involves combining machine learning and Artificial intelligence (AI) methods [109]. By utilizing datasets that include data and computational simulations these techniques can uncover correlations and trends between design factors and construction performance. Subsequently optimization AI-based algorithms can be used to navigate design realms and pinpoint solutions that might not be easily identifiable using conventional optimization approaches.

Various strategies and methods for enhancing the design of tissue constructs produced through manufacturing are explored in this section. The mentioned sources offer information on topological optimization algorithms, patient-specific design, integration of bioactive elements, computational fluid dynamics simulations, optimization of mechanical properties fine tuning process parameters optimization and the application of machine learning and artificial intelligence techniques [110]. Furthermore, an overview provided in the introduction and a review on manufacturing within tissue engineering give an understanding of the field. Furthermore, this section delves into objective optimization strategies, dynamic modeling approaches, biomechanical modeling coupled with finite element analysis, imaging methods for validation and monitoring purposes optimizing biological and biochemical factors, for better outcomes as well as combining AM with bioprinting and microfluidics to optimize design. These references shed light on the methods and approaches employed to refine tissue construct designs [111].

3.9.3 Scaffold Architecture Considerations

The word scaffold is used to describe 3D materials that help cells rebuild tissues and organs by providing a favorable environment. Scaffolds are polymeric biomaterials, giving structural support for cell adhesion and subsequent tissue growth. When scaffolds are used for tissue engineering, researchers frequently face a wide range of options [112]. The goal is to create scaffolds that can offer regeneration signals to cells and mimic natural behavior. Scientists are working to create methods for manufacturing scaffolds which are very close to the natural environment, have multi-functional qualities, and are cost-effective and clinically useful [113].

The scaffolds' structure is the most significant aspect. Interconnected pores and high porosity enable cell adhesion, facilitating cell growth, proliferation, and differentiation, three-dimensional tissue regeneration and waste diffusion as well as scaffold breakdown products. The pore size must be large enough to allow cell motility as well as being tiny enough for cell attachment to the scaffold. The scaffold should be break-down within the same time of tissue regeneration. An ideal scaffold for bone tissue would be osteo-conductive, biodegradable, and have appropriate mechanical qualities. It must be able to distribute cells, construct scaffolds in unusual forms, and, of course, be commercially successful [12]. Tissue engineering strives to create artificial constructs such as 3D scaffolds containing suitable cells that can be implanted directly in vivo to stimulate and direct the development of new tissues. The human body is a delicate system, thus, materials for 3D scaffolds must be biocompatible, easily sterilizable, and have good mechanical properties [114]. Features and functions of scaffolds are given below:

Architecture: it should allow void volume for vascularization, new tissue development, and remodeling in order to facilitate host tissue integration following implantation. The biomaterials should be treated to provide a porous structure that allows for effective nutrition and metabolite transfer. To maintain the scaffold's mechanical stability, biomaterials should be treated to provide a porous structure that allows effective transfer of nutrition and metabolites. Furthermore, the biomaterials should degrade at the same rate as the new matrix produced by the developing tissue [115].

Cyto- and tissue compatibility: scaffolds should facilitate the attachment, growth, and differentiation of either extraneously applied or endogenous cells during in vitro culture and in vivo implantation. The scaffolds' biomaterials must be compatible with the biological components of the created tissues as well as natural cells in the host tissue [116].

Bioactivity: It can actively interact with the biological components of newly created tissues to assist and regulate their functions. Biomaterials may contain biological cues, for example, cell-adhesive ligands, to improve adhesion, or physical cues, such as topography, to alter cell shape and alignment [117]. The scaffold can also function as a delivery system or reservoir for external growth-stimulating signals such as growth factors, which can accelerate regeneration. Here, the biomaterials must be compatible with the biomolecules and amenable to an encapsulating approach for controlled release of the biomolecules while retaining bioactivity [118].

Mechanical properties: For scaffolding, intrinsic mechanical properties of biomaterials are used, and their post-processing qualities should be compatible with the host tissue. Recent research in mechanobiology has highlighted the significance of a scaffold's mechanical qualities for seeded cells. When traction pressures are applied to a substrate, several mature cell types, including epithelial cells, fibroblasts, muscle cells, and neurons, detect the substrate's stiffness and exhibit distinct morphology and adhesive properties [119]. This mechano-sensitivity has also been shown in the differentiation of MSC, where the stiffness of the agarose gel determines how likely they are to differentiate. The hMSC can differentiate into nerve cells, muscle cells, or bone cells depending on stiffness that approximates those of the brain, muscle, and bone tissues, respectively [120].

Many scaffolds are created from hydrogels. A hydrogel is a 3D flexible network made of natural or synthetic polymers that are insoluble in water. Examples include polyethylene glycol and alginate-based hydrogels. Hydrogels can resemble live tissues by responding to variations in ambient factors such as pH, temperature, and electric field. They are suitable for minimally invasive delivery and in situ gelling at body temperature [121]. It can be used for bone regeneration, and one of its main advantages is the ability to encapsulate cells and perform chemical biofunctionalization. For tissue engineering hydrogels are used as matrices because of their porosity. Cells can develop and expand there, medicines can be released, and nutrients and waste may pass through them. Another advantage is the ease of modification with adhesion ligands. Self-cross-linking hydrogels, such as water-soluble chitosan and oxidised hyaluronic acid, are employed as cell transporters since they do not require any chemical cross-linking agents [122].

3.9.4 Patient-Specific Design Approaches

3.9.4.1 Scaffold-Guided Bone Regeneration

Scaffold-guided bone regeneration (SGBR) carries growth factors or bone graft and physically guides tissue regeneration by serving as a template for the host cells and the transplanted bone graft eventually leading to tissue ingrowth and physiological remodeling. It provides a favorable environment in response to the highly dynamic tissue processes during bone regeneration that occurred due to disruption of local soft tissue integrity and normal vascular function and surgical distortion of bone marrow. Appropriate setting for bone regeneration occurs due to prolonged degradation of scaffold involving a series of procedures such as controlled bleeding, inflammation, generation of contemporary blood vessels and osseous tissue as well as bone remodeling [123].

3.9.4.2 Selective Laser Sintering in New Bone Formation, Fibroblast Cell Attachment, and Growth

Selective laser sintering generates composite scaffolds consisting of polycaprolactone and hydroxyapatite. It stimulates the generation of new bone indicating great potential for orthopedic and reconstructive usages. In a study of vascular tissue engineering, Selective laser sintering scaffolds constructed from polyurethane have demonstrated excellent bone structures, effective diffusion and the capacity to sustain fibroblast cell attachment and growth [124].

3.9.4.3 3D CT for Targeting Organ and Tissues

The three-dimensional form of a target organ or tissue can now be automatically replicated through by utilizing 3D CT and additional imaging methods. This is accomplished by using the patient's three-dimensional volumetric data from imaging during the scanning exam to utilize reverse engineering to create a 3D digital model that precisely matches. The size and form of the flaw when the patient undergoes a pre surgery magnetic resonance imaging or CT scan [124].

3.9.4.4 Tissue-Engineered Vascular Graft

The main advantage of the tissue-engineered vascular graft (TEVG) it's the ability for biodegradability which creates a fully self-generated vascular intermediary expanding as well as changing alongside with its host and terminating the need for repeated surgeries to enlarge the graft because of somatic overgrowth [125]. Clinical and TEVG engineering collaboration has been researched to generate patient-specific models for guiding clinical decision and forecast results. Significant advancements in both fields logically lead to the incorporation of patient-specific modeling and tailored regenerative implants to create an optimal patient-specific TEVG [126].

3.9.4.5 Cell-Based Tissue Engineering Techniques

The possibility of customized cell-based tissue engineering associated methods to repair damaged IVD as well as effectively restore full disc function is gaining attention for the treatment of lower back pain and intervertebral disc degradation. It incorporates bioink consisting of elastin and silk fibroin hydrogel to bio-print patient-specific replacements which imitates the structural framework of intervertebral disc with the precise measurement of size and shape, porosity, mechanical characteristics and biodegradability. These personalized implants accelerate post-surgical recovery and support the restoration of spine functionality which can be useful in clinical aspects [127].

3.9.4.6 Decellularized Extracellular Matrix Scaffolds

Scaffolds made of decellularized extracellular matrix have been utilized in the treatment of irregularly large soft tissue injuries have demonstrated the capacity to regulate the immune response and promote healing. Though it provides probable solution for the regeneration of functional tissue, acellular decellularized extracellular matrix (dECM) scaffolds are critical to fabricate into integrated complex structural frameworks. Decellularized extracellular matrix scaffolds have been demonstrated to regulate the immune response to initiate wound healing and improve tissue regeneration in in-vivo clinical trials. In addition of providing a structural framework for the penetration of macrophages and progenitor cells, the dECM scaffolds accumulates growth factors, extracellular vesicles initiating the development of skeletal muscle, generation of blood vessels, and transduction of nerves in paralyzed body part [128].

3.9.4.7 Tissue Engineering Technique in Rhinoplasty, Auricular Reconstruction, and the Treatment of Microtia

Virtual surgery simulation software is used to measure the reconstruction of a patient's nose during a pre-surgical rhinoplasty. For personalized nasal implants composed of biodegradable materials, facial geometries can be digitally measured by 3D printing technology such as CAD or CAM software to fit the patient's unique form. The post-surgical facial model which is generated via computer graphics as well as provided by the virtual simulation software permits assumption of the nose shape and its balance upon particular facial structure with least amount of error and minimal complications. Moreover, this technique is also effective for human auricular reconstruction and the treatment of congenital ear malformation such as microtia [129].

3.10 Applications of AM in Tissue Engineering

3.10.1 Bone Tissue Engineering

The development of bone tissue engineering has been made possible by AM, commonly referred to as 3D printing. Modern technology provides benefits over traditional manufacturing techniques by enabling the fabrication of individualized implants with unique forms. We are going to look at the use of manufacturing in tissue engineering in this talk, with a particular emphasis on how it is applied in bone tissue engineering [130]. The applications of the AM are presented in Fig. 3.3.

Scaffold Creation By supporting attachment of cells, division, and tissue development, scaffolds are useful in tissue engineering. AM techniques make the production possible of scaffolds with customized designs, pore sizes and interconnected structures. These features are essential for promoting cell movement facilitating exchange, waste removal and stimulating blood vessel formation within the engineered structure [131].

Custom Implants for Patients One significant advantage of manufacturing is its ability to create patient implants. This technology in bone tissue engineering allows to produce implants based on each patient's data obtained from imaging methods like computed tomography or magnetic resonance imaging. These personalized implants can closely match the patient's bone structure resulting in fitment, reduced surgical time and potentially improved healing and integration [132].

Incorporation of Bioactive Materials Various bioactive substances, like growth factors, medications or biomolecules can be integrated into the scaffold structure using manufacturing methods during production. Incorporating these bioactive

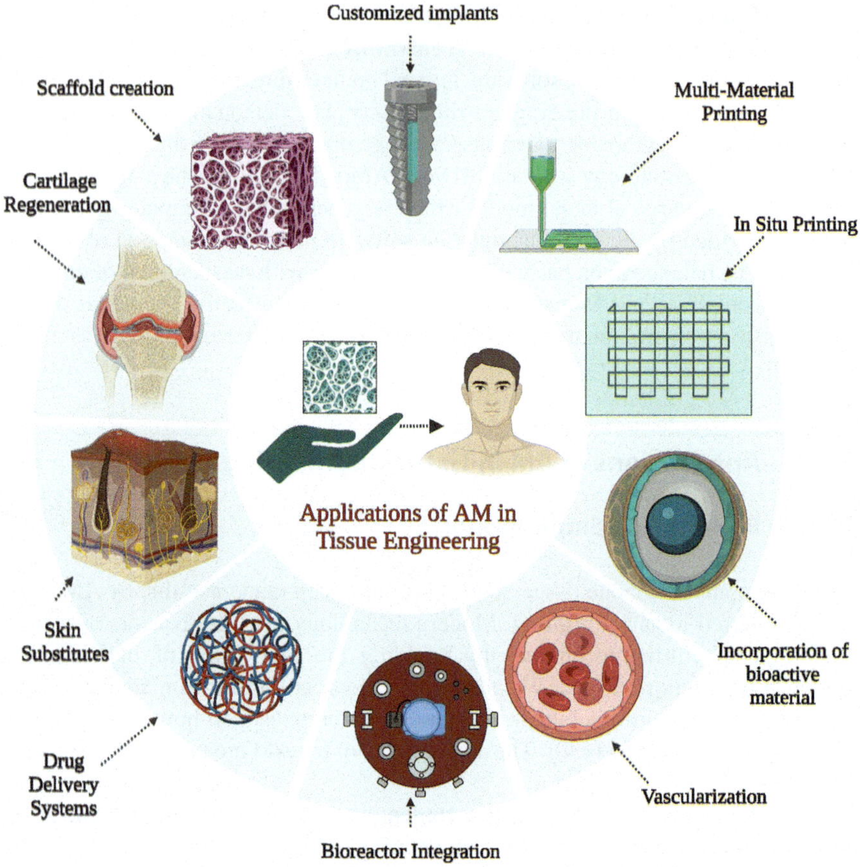

Fig. 3.3 Applications of AM in tissue engineering. (Created by using Biorender.com)

components aids in promoting the differentiation of cells, growth and tissue renewal ultimately improving the effectiveness of the engineered structure [133].

Multi-material Printing Certain methods in manufacturing enable the printing of materials, at the same time making it possible to produce composite scaffolds with diverse mechanical and biological characteristics. This feature can be utilized to replicate the structure and makeup of natural bone tissue.

In Situ Printing AM also allows for the option of, on site printing, where the framework or implant is created directly at the location of the injury or defect. This method reduces the requirement for surgery. Could speed up recovery by offering instant structural support and administering bioactive substances straight to the affected area [134].

Vascularization One of the hurdles in bone tissue engineering involves ensuring blood vessel formation within the created structure. AM methods enable the production of supports featuring networks of blood vessels. These vascular pathways help transport oxygen and nutrients and eliminate waste materials promoting the growth and health of cells, in the engineered tissue [135].

Bioreactor Integration AM can work alongside bioreactor systems, which offer a regulated setting, for the growth and maturation of tissues. Designed 3D printed scaffolds can be created with characteristics to seamlessly fit into bioreactors facilitating the use of mechanical forces, like fluid flow or compression to improve the growth and performance of artificial bone tissue.

Drug Delivery Systems AM has the possibility to be used in producing drug delivery systems, purpose is bone tissue engineering. 3D printed scaffolds could be custom designed with controlled release mechanisms to administer substances, like growth factors or anti-inflammatory medications aiming to enhance bone regeneration and recovery.

There are still hurdles to overcome, such as guaranteeing that the materials used remain compatible with the body over the term and fine tuning their characteristics. Continued research and partnerships among engineers, material scientists and medical professionals are crucial for propelling this field [136].

3.10.2 Skin Substitutes

AM plays a role in skin tissue engineering by creating scaffolds that offer support for tissue to grow and develop resulting in the development of tissue. Unlike methods, AM allows to produce scaffolds with precise designs and customizable features, like porosity and mechanical properties. Various methods have been studied for creating scaffolds, like deposition modeling stereolithography and bioprinting. FDM builds up layers of a polymer while SLA uses a process that turns resins into solid structures. Bioprinting involves integrating living cells and biomaterials into the printing procedure allowing for the development of tissue constructs [115].

Researchers have also utilized manufacturing for producing skin grafts and wound dressings. They have delved into bioprinting to craft layered skin structures resembling natural skin composition. These structures can integrate cell varieties, including keratinocytes and fibroblasts along with biomaterials such as collagen as well as hyaluronic acid to help the healing of wounds and regeneration of tissues. The capacity to produce structures unique to each sufferer is a significant advantage of AM in skin tissue engineering. Researchers can create customized scaffolds or skin grafts that exactly match the patient's wound geometry by utilizing computer-aided design software and imaging data. This individualized strategy helps quicken the recovery process and enhance treatment outcomes [137].

Researchers have also utilized manufacturing for producing skin grafts and wound dressings. They have delved into bioprinting to craft layered skin structures resembling natural skin composition. These structures can integrate cell varieties, including keratinocytes and fibroblasts along with biomaterials such as collagen as well as hyaluronic acid to help the healing of wounds and regeneration of tissues. The capacity to produce structures unique to each sufferer is a significant advantage of AM in skin tissue engineering. Researchers can create customized scaffolds or skin grafts that exactly match the patient's wound geometry by utilizing computer-aided design software and imaging data [138]. This individualized strategy helps quicken the recovery process and enhance treatment outcomes.

There are several challenges utilizing AM in skin tissue engineering. The primary difficulty is the choice of biodegradable and biocompatible materials, which are suited to printing and capable of supporting cell growth and tissue formation [139]. Lately there have been progressions in utilizing manufacturing, for the area of skin tissue engineering. For example, research conducted by Cubo and colleagues in 2017 showcased the development of skin replacement by integrating bioprinting and electrospinning methods [140]. This structure included a layer crafted from bioprinter fibroblasts and an artificial epidermis produced through electrospinning mirroring the characteristics and roles of skin.

A significant advancement involves utilizing 3D bioprinting to create skin grafts that include networks. Scientists at the Wake Forest Institute for Regenerative Medicine have effectively printed a skin graft with blood vessels by combining skin cells and a hydrogel based bioink. This method is designed to tackle the issue of transporting nutrients and oxygen in tissue structures, which's crucial for ensuring their long-term survival and integration.

Therefore, AM has become a tool in tissue engineering showing applications in creating skin substitutes. By allowing the production of personalized scaffolds, skin grafts and wound coverings this technology could transform the treatment of burns, persistent wounds, and other skin conditions. However further research is needed to tackle issues like choosing materials, maintaining cell health and incorporating features such as vascularization. Moving forward collaborative efforts among engineers, biologists and medical professionals will play a part in advancing the domain and applying these cutting-edge technologies [141].

3.10.3 Cartilage Regeneration

In the fields of tissue engineering and regenerative medicine, 3D printing has quickly progressed as an additive manufacturing technique. By simultaneously building 3D artificial implants or complex biological tissues integrated with living cells, extracellular matrices, and other bioactive materials, this technology overcomes the limitations of cartilage tissue engineering. The ability to distribute 3D bioprinted cells, hydrogels, and active ingredients spatially and hierarchically in accordance with the necessary 3D functions is the main advantage of 3D bioprinting in cartilage tissue engineering. The structure of interconnected pores and large surface area created by 3D bioprinting, in contrast to conventional solvent hydrogels, facilitates gas and nutrient exchanges as well as intercellular communication and seed cell attachment, growth, and regeneration, thereby significantly promoting cartilage tissue regeneration [38]. When chondrocyte factors are expressed, hyaluronic acid acts as a three-dimensional scaffold to promote chondrocyte growth. The 3D fibrin-alginate hydrogel, along with chondrocyte and hyaluronic acid supplements, can hasten glycosaminoglycan accumulation and cell proliferation when cartilage matrix is embedded within it. By encasing chondrocytes in a mixture of PCL and alginate, gelatin-hydroxyapatite is enzymatically cross-linked, enhancing the scaffold's ability to promote stem cell cartilage differentiation and successfully repair injured tissue. Implementation of collagen-alginate bioink for 3D bioprinted cartilage tissue engineering has demonstrated exceptional mechanical and expansion properties as well as the preservation of chondrocyte phenotype. Moreover, 3D printed PEG-alginate hydrogel composites can generate a higher quantity of healthy cells than even natural cartilage which consists of agarose, alginate, GelMA and bioink. Tissue engineering replicates the natural cartilage environment as well as creates viable grafts for cartilage regeneration by the combination of cells and biomaterial scaffolds to create a tissue framework [142]. The characteristics of such tissue frameworks have been engineered both in vitro and in vivo, demonstrating that dynamic mechanical compression promotes matrix synthesis, initiates chondrocyte differentiation, and boosts cell viability and proliferation. Consequently, these advantageous effects are most likely the result of a combination of interstitial flow-initiated and compression-initiated mass transport enhancement, mechanical factors and their transduction of gene expression. Constructions subjected to mechanical loading in biomaterials exhibited compressive characteristics similarly those of natural cartilage due to the applied force determining the characteristics of chondrocyte deposition in the matrix. Structural bone scaffolds are utilized in order to encourage cellular longevity and adherence, expedite bone remodeling, provide osteoconductive institutional instruction as well as function as recipients of antimicrobial agents and growth stimuli as well as treatments via genetic engineering [143].

3.10.4 Vascular Tissue Engineering

Scaffold based fabrication allows vascular cells to be distributed rationally in particular area by creating growth microenvironments for cell propagation and adhesion, localization and colonization. Moreover, 3D printed scaffolds can be constructed to match the vascular anatomy of the host for enhancement of the host vasculature and promotion of anastomosis amid the blood vessels that were designed and their comparable host vascular structures with incorporated physiological signals regarding target vascular anastomosis. In order to create fresh vascular tissues in vivo, vasculature scaffolds can be generated through 3D printing in vivo [144].

Firstly, 3D reconstruction is utilized via various ways such as microcomputed tomography, magnetic resonance imaging or parametric modeling to create vascular scaffold models. Secondly, a combination of 3D printing technologies are used to prepare scaffolds based on bioinks. Thirdly, the implanted vessel being grafted needs to withstand the host reaction which frequently causes thrombus formation, inflammation and graft failure. In order to facilitate regenerative integration with the host tissue, remodeling events must ultimately occur. These processes include EC migration and the establishment of cell-cell contact between the graft and host ECs, SMC migration across the interface blurring the host-graft interface and circumferential graft ECM rearrangement restoring mechanical homeostasis [145]. Increased tensile mechanical properties, decreased thrombogenicity and regenerative remodeling for successful integration are the primary requirements for designing a durable 3D printed artery. Arteries need to be both tensile and compliant enough to withstand rupture after abruptly splitting off the aorta and allowing highly pulsatile pressure flow. Three widely used techniques are involved in creating tubular structures using biomaterials which are sheet rolling, tubular molding, and direct scaffolding. Cell-derived graft made from an annular glass mold has demonstrated the effective potential in a model of replacement of the lamb artery in clinical trials [146].

3.10.5 Organ-on-Chip Models

Organ-on-a-chip (OOCs) engineering has grabbed considerable focus from the various pharmaceutical industries, legislative bodies as well as military services. The principal aim of organ-on-a-chip technology is to duplicate some aspects of human physiology which are parenchymal cell organization at the tissue level such as heart, liver, skeletal muscle, tumors, multicellular vascular or epithelial interfaces and the pharmacokinetics of elimination associated organs. Beside combining different cell types, organ-on-a-chip technology also includes engineering substances such as cellular spatial confinement or the addition of sensors and microfluidic channels [147]. The main technique used in organ-on-a-chip technologies is the fusion of proportionately pre-differentiated cells to imitate the structural and functional ability of native tissues. Additionally, different 3D constructs can be integrated by organ-on-a-chip technology into a dynamic circulation system for imitating cellular characteristics within various tissues and organs. Organ-on-a-chip acts as a good potential model due to its enhanced suitability, efficiency and implications for determining pathophysiology as well as suitable platform for disease modeling and drug discovery [148]. Two important methods in which microfluidic chips permeate the tissue framework with uncontrolled fluid behavior and connections, whereas BioMEMS (Biological Micro-Electro-Mechanical Systems) allows for precise control over cellular microenvironments under biological conditions. They are potent new tools that bridge the gap between human disease and animal models of translation. Because they can replicate the structural, physiological, microenvironmental and biological functions of human organs on microfluidic chips, organ-on-a-chip particularly holds their potential to replace animal trials in the future [149]. Many single-organ chips have been created to study adverse drug reactions and look into disease mechanisms for generating more sustainable OOCs. Apart from their emphasis on emulating the functions of individual organs, single-OOCs have also drawn attention for their ability to integrate multiple organ units and produce more dependable data. A high degree of control attained by using OOCs promotes the creation of a unique cellular environment in culture. Various studies have employed cell printing and patterning for combining 3D bioprinting and OOCs [150]. Encapsulated fibroblasts via decellularized extracellular matrix and polycaprolactone framework incorporated with endothelial cells have generated imitation of in-vivo conditions and progressively initiate endothelial regeneration. The vascularized airway-on-a-chip models consist of the capability of accurately replicating pathophysiological reactions to the stimulation in vitro via 3D cell printing. Furthermore, OOC engineering has the potential to lead 3D printing technology by generating advanced technology incorporating cell line and bioactive materials for performing similar tissue functionality to that of humans [151].

3.11 Challenges and Future Directions

3.11.1 Biocompatibility and Tissue Integration

The first step in the bioprinting process is to determine the anatomical structure of the target tissue using an appropriate imaging technology. To enable the printing process to create layers on top of each other, the image is then converted using advanced software into a CAD drawing of cross-sectional layers. Then, using a specific printing method—such as inkjet 3D bioprinting, micro-extrusion 3D bioprinting, laser-assisted 3D biological printing, or stereolithography—the bioprinting apparatus creates the tissue. It does this by using a variety of printing materials, including scaffold, bioink, and other additive components. Tissue viability, security, and deposition accuracy can vary depending on the method used [152]. Subsequently, the generated tissue undergoes post-processing within a bioreactor, which replicates the in vivo conditions necessary to maintain tissue viability during the maturation stage. Ozbolat and Dababneh gave an excellent review on the process of 3D bioprinting. The main goal of 3D bioprinting is to implement a method for 3D operational complicated organ fabrication as a source for tissue grafts, whole organ transplants, and animal-substitute models for medication evaluation. Bioprinting is a relatively new technology, but it is developing quickly due to the volume of research being done in areas like printing engineering, tissue engineering, and cell sciences. The field of bioprinting has developed from traditional 2D printing on paper to 3D printing, which includes the use of extrusion and stereolithography for non-biological materials and inkjet and laser printing for biological materials [153]. It follows that the engineering aspect's greater level of sophistication than the bioink materials technology is not surprising. Though each printing process is designed to produce materials free of biological matter, there are still a number of issues with material compatibility when using bioink in place of other construction materials.

For instance, low viscosity bioink is needed for inkjet printing. This complicates the deposition of extracellular matrix and several efficient, extremely viscous hydrogels. The use of stereolithography-based bioprinting for tissue printing is further limited by the high UV radiation intensity required for the cross-linking process and the protracted post-processing needs [154]. On the other hand, the most difficult issue that is still impeding the advancement of bioprinting technology is bioink development. The optimal characteristics of bioink should fulfil the physical and biological material needs to facilitate cellular behaviors that are similar to those observed in vivo, including maturation, migration, differentiation, and proliferation. The viscosity, structural strength, printing capability, deterioration, and functionality are the physical attributes. There are several bioink compositions available to satisfy the demands of particular printing technologies. Viscosity of the bioink is an important parameter that must always be optimized in order to modify tissue structure stability, cell encapsulation effectiveness, and bioink flow. Hydrogels' hydrophilicity, low cytotoxicity, and biocompatibility have made them very promising for the creation of bioinks. They can also build networks of polymers that give them an ECM-like structure [155].

3.11.2 Regulatory and Ethical Considerations

Currently, there are three distinct regulatory domains in Europe: biologics, medical devices, and pharmaceuticals. Since tissue engineering products frequently consist of both cells and a supporting structure, they are frequently hybrid in character. Certain products can be approved as medical devices, while others must be approved as medicinal products. In Europe, new regulatory frameworks addressing these concerns have been seen [156]. The rationale for the difficulties in coming to a regulatory agreement on this matter comes from a UK survey. The authors blame these issues on how closely tied to and overlapping these issues are with other technologies, such as xenotransplantation. Regulatory agencies are unable to manage it independently for this reason The ethical debates surrounding this and related research topics (such as the ethics of organ transplantation and the stem cell controversy) further complicate regulation. The survey previously cited illustrates how a particular technique can be perceived as either "pure" or "polluted" by the same social actor using autologous cartilage transplantation as an example [157].

Tissue engineering is largely affected by two fundamental trends in the European Union. These include Directive 2004/23/EC, which was adopted by the European Parliament in 2004 and establishes quality and security standards for the acquisition and distribution of human tissues, as well as a proposed regulation on tissue-engineered products. The latter had its premiere in Brussels in 2004 and was developed with assistance from the European Commission's DG Enterprise [158].

3.11.3 Scaling-up for Clinical Translation

A significant development in the field of medical technology, scaling up AM for clinical translation heralds a new era of individualized healthcare solutions. This project represents the fusion of state-of-the-art AM methods with the complex requirements of healthcare settings, and it has the power to drastically alter patient treatment. Now, doctors can design implants specifically for each patient, prostheses, in addition to medical equipment by means of unmatched precision and customization thanks to AM. This technology's scalability promotes cost-effectiveness and streamlines production processes, making customized medical interventions more accessible to a wider range of people [159]. Furthermore, the adaptability of AM permits the creation of complicated forms as well as complex geometries that smoothly fit among the various anatomical nuances observed in clinical practice. The potential applications range from personalized medicine delivery systems to orthopedic implants that are designed to fit the unique bone structures of each patient to a wide range of potentially revolutionary uses. Moreover, quick prototyping and iteration are made possible by AM's iterative nature, which speeds up the development and improvement of medical equipment and treatments [160]. But as we move forward with clinical translation, issues like standardization, biocompatibility of materials, and regulatory compliance become more and more pressing, calls for interdisciplinary cooperation and stringent quality control procedures. Despite these obstacles, AM holds unquestionably great promise for clinical translation, with the prospective to revolutionize the approach by which healthcare is dispensed by elevating patient-centered, precise, and individualized interventions to the forefront of medical practice [161].

3.11.4 Emerging Technologies and Innovations

The most recent developments in AM have shown great efficacy as viable remedies for the present shortcomings in medical therapies. Numerous investigations have been carried out, offering prospects for subsequent uses predicated on the results of continuing trials. The introduction of new technology has led to a rise in interest in improving medical processes, according to recent trends. Notable attempts are being made to create materials that spark fresh research and lead to the identification of innovative uses. Because 5DP and 6DP technologies are very new, they haven't made much of an impact [14]. On the other hand, they are being used to create techniques and procedures that satisfy medical needs as needed. It is imperative to emphasize the critical role that raw materials play in innovation, particularly in the domains of implants and tissue engineering. The most difficult part of these research topics is creating an imprint material that is comparable to the body's needs in terms of biocompatibility, mechanical resistance, and stimuli responsiveness [162].

Thus, ME fields had a spike in interest in 2020 and 2021 whereas VP saw the development of new materials, which led to a spike in their utilization in 2022—far exceeding the major one in the years that followed [14]. The expansion of the bibliographic base verified the growth, since the majority of the content dealt with the creation and preparation of biocompatible materials for use in organisms.

3.12 Conclusion

AM has become a game-changer because it provides unmatched accuracy and customization for the production of intricate tissue constructions in tissue engineering and regeneration. This overview highlights the noteworthy advancements in the area by providing a thorough analysis of various AM methods, biomaterials, and design concerns. The wide range of uses for AM, from the restoration of bone as well as cartilage, to development of skin substitutes and vascular tissues, demonstrate its adaptability and promise to tackle critical challenges facing regenerative medicine. The path to broad clinical acceptance is hampered by issues with scalability, regulatory compliance, and biocompatibility, notwithstanding the encouraging developments. But further research and technology advancements are setting the stage for a time when AM-enabled tissue engineering will be able to offer customized, efficient solutions for demands unique to individual patients. In order to overcome current obstacles and comprehend the complete ability of AM in tissue engineering and regeneration, it will be imperative to integrate developing technologies with multidisciplinary collaboration. With further research and development, AM has the potential to be a key player in the developing area of regenerative medicine, which might eventually enhance the health of patients and bring about a revolution in the healthcare sector.

References

1. Vyas J, Shah I, Singh S, Prajapati BG (2023) Biomaterials-based additive manufacturing for customized bioengineering in management of otolaryngology: a comprehensive review. Front Bioeng Biotechnol 11:1234340
2. Weiser TG, Regenbogen SE, Thompson KD, Haynes AB, Lipsitz SR, Berry WR, Gawande AA (2008) An estimation of the global volume of surgery: a modelling strategy based on available data. Lancet 372:139–144. https://doi.org/10.1016/S0140-6736(08)60878-8
3. Vyas J, Raytthatha N, Singh S, Prajapati B (2024) Next-generation computational automation-based additive manufacturing of pharmaceuticals: an approach to fabricate precise medicine. In: Handbook of 3D printing in pharmaceutics. CRC Press, pp 153–175
4. Vyas J, Singh S, Shah I, Prajapati BG (2023) Potential applications and additive manufacturing technology-based considerations of mesoporous silica: a review. AAPS PharmSciTech 25(1):6
5. Mamo HB, Adamiak M, Kunwar A (2023) 3D printed biomedical devices and their applications: a review on state-of-the-art technologies, existing challenges, and future perspectives. J Mech Behav Biomed Mater 143:105930. https://doi.org/10.1016/j.jmbbm.2023.105930
6. Rehman M, Yanen W, Mushtaq RT, Ishfaq K, Zahoor S, Ahmed A, Kumar MS, Gueyee T, Rahman MM, Sultana J (2023) Additive manufacturing for biomedical applications: a review on classification, energy consumption, and its appreciable role since COVID-19 pandemic. Prog Addit Manuf 8:1007–1041. https://doi.org/10.1007/s40964-022-00373-9

7. Gleadall A, Visscher D, Yang J, Thomas D, Segal J (2018) Review of additive manufactured tissue engineering scaffolds: relationship between geometry and performance. Burn Trauma 6. https://doi.org/10.1186/s41038-018-0121-4
8. Loh QL, Choong C (2013) Three-dimensional scaffolds for tissue engineering applications: role of porosity and pore size. Tissue Eng Part B Rev 19:485–502. https://doi.org/10.1089/ten.teb.2012.0437
9. Gariboldi MI, Best SM (2015) Effect of ceramic scaffold architectural parameters on biological response. Front Bioeng Biotechnol 3. https://doi.org/10.3389/fbioe.2015.00151
10. Sobral JM, Caridade SG, Sousa RA, Mano JF, Reis RL (2011) Three-dimensional plotted scaffolds with controlled pore size gradients: effect of scaffold geometry on mechanical performance and cell seeding efficiency. Acta Biomater 7:1009–1018. https://doi.org/10.1016/j.actbio.2010.11.003
11. Roohani-Esfahani S-I, Newman P, Zreiqat H (2016) Design and fabrication of 3D printed scaffolds with a mechanical strength comparable to cortical bone to repair large bone defects. Sci Rep 6:19468. https://doi.org/10.1038/srep19468
12. Abbasi N, Hamlet S, Love RM, Nguyen N-T (2020) Porous scaffolds for bone regeneration. J Sci Adv Mater Devices 5:1–9. https://doi.org/10.1016/j.jsamd.2020.01.007
13. Kumar R, Kumar M, Chohan JS (2021) The role of additive manufacturing for biomedical applications: a critical review. J Manuf Process 64:828–850. https://doi.org/10.1016/j.jmapro.2021.02.022
14. Amaya-Rivas JL, Perero BS, Helguero CG, Hurel JL, Peralta JM, Flores FA, Alvarado JD (2024) Future trends of additive manufacturing in medical applications: an overview. Heliyon 10:e26641. https://doi.org/10.1016/j.heliyon.2024.e26641
15. Kholgh Eshkalak S, Rezvani Ghomi E, Dai Y, Choudhury D, Ramakrishna S (2020) The role of three-dimensional printing in healthcare and medicine. Mater Des 194:108940. https://doi.org/10.1016/j.matdes.2020.108940
16. Sufaru I-G, Macovei G, Stoleriu S, Martu M-A, Luchian I, Kappenberg-Nitescu D-C, Solomon SM (2022) 3D printed and bioprinted membranes and scaffolds for the periodontal tissue regeneration: a narrative review. Membranes (Basel) 12:902. https://doi.org/10.3390/membranes12090902
17. Thomas D, Singh D (2019) Novel techniques of engineering 3D vasculature tissue for surgical procedures. Am J Surg 218:235–236. https://doi.org/10.1016/j.amjsurg.2018.06.004
18. Miar S, Shafiee A, Guda T, Narayan R (2018) Additive manufacturing for tissue engineering. In: 3D printing and biofabrication. Springer, Cham, pp 1–52
19. Ahn S, Lee H, Bonassar LJ, Kim G (2012) Cells (MC3T3-E1)-laden alginate scaffolds fabricated by a modified solid-freeform fabrication process supplemented with an aerosol spraying. Biomacromolecules 13:2997–3003. https://doi.org/10.1021/bm3011352
20. MacDonald E, Wicker R (2016) Multiprocess 3D printing for increasing component functionality. Science (80-) 353. https://doi.org/10.1126/science.aaf2093
21. Gopinathan J, Noh I (2018) Recent trends in bioinks for 3D printing. Biomater Res 22. https://doi.org/10.1186/s40824-018-0122-1
22. Lee J-Y, An J, Chua CK (2017) Fundamentals and applications of 3D printing for novel materials. Appl Mater Today 7:120–133. https://doi.org/10.1016/j.apmt.2017.02.004
23. Jandyal A, Chaturvedi I, Wazir I, Raina A, Ul Haq MI (2022) 3D printing—a review of processes, materials and applications in industry 4.0. Sustain Oper Comput 3:33–42. https://doi.org/10.1016/j.susoc.2021.09.004
24. Wei L, Yang C, Wu W (2023) Recent advances of stereolithographic 3D printing enabled by photon upconversion technology. Curr Opin Green Sustain Chem 43:100851. https://doi.org/10.1016/j.cogsc.2023.100851
25. Huang J, Qin Q, Wang J (2020) A review of stereolithography: processes and systems. PRO 8:1138. https://doi.org/10.3390/pr8091138
26. Verma N, Awasthi P, Gupta A, Banerjee SS (2023) Fused deposition modeling of polyolefins: challenges and opportunities. Macromol Mater Eng 308. https://doi.org/10.1002/mame.202200421

27. Morales MA, Maranon A, Hernandez C, Michaud V, Porras A (2023) Colombian sustainability perspective on fused deposition modeling technology: opportunity to develop recycled and biobased 3D printing filaments. Polymers (Basel) 15:528. https://doi.org/10.3390/polym15030528
28. Song Y, Ghafari Y, Asefnejad A, Toghraie D (2024) An overview of selective laser sintering 3D printing technology for biomedical and sports device applications: processes, materials, and applications. Opt Laser Technol 171:110459. https://doi.org/10.1016/j.optlastec.2023.110459
29. Tang H, Zhang S, He L, Yang Z, Liu T (2024) 3D printing of high-stiffness and high-strength glass fiber reinforced PEEK composites by selective laser sintering. Compos Part A Appl Sci Manuf 187:108470. https://doi.org/10.1016/j.compositesa.2024.108470
30. Karanwad T, Jorvekar SB, Mandal S, Borkar RM, Banerjee S (2024) Additive manufacturing of SmartEx QD 100 designed oral three-dimensional printlets containing isoniazid for immediate gastric release by selective laser sintering. Mol Pharm 21:5272–5284. https://doi.org/10.1021/acs.molpharmaceut.4c00693
31. Wu Y, Zhang Y, Yan M, Hu G, Li Z, He W, Wang X, Abulimit A, Li R (2024) Research progress on the application of inkjet printing technology combined with hydrogels. Appl Mater Today 36:102036. https://doi.org/10.1016/j.apmt.2023.102036
32. Bernasconi R, Invernizzi GP, Gallo Stampino E, Gotti R, Gatti D, Magagnin L (2023) Printing MEMS: application of inkjet techniques to the manufacturing of inertial accelerometers. Micromachines 14:2082. https://doi.org/10.3390/mi14112082
33. Uçak N, Çiçek A, Aslantas K (2022) Machinability of 3D printed metallic materials fabricated by selective laser melting and electron beam melting: a review. J Manuf Process 80:414–457. https://doi.org/10.1016/j.jmapro.2022.06.023
34. Palmquist A, Jolic M, Hryha E, Shah FA (2023) Complex geometry and integrated macro-porosity: clinical applications of electron beam melting to fabricate bespoke bone-anchored implants. Acta Biomater 156:125–145. https://doi.org/10.1016/j.actbio.2022.06.002
35. Coffigniez M, Gremillard L, Balvay S, Lachambre J, Adrien J, Boulnat X (2021) Direct-ink writing of strong and biocompatible titanium scaffolds with bimodal interconnected porosity. Addit Manuf 39:101859. https://doi.org/10.1016/j.addma.2021.101859
36. Baniasadi H, Ajdary R, Trifol J, Rojas OJ, Seppälä J (2021) Direct ink writing of aloe vera/cellulose nanofibrils bio-hydrogels. Carbohydr Polym 266:118114. https://doi.org/10.1016/j.carbpol.2021.118114
37. Shahzad A, Lazoglu I (2021) Direct ink writing (DIW) of structural and functional ceramics: recent achievements and future challenges. Compos Part B Eng 225:109249. https://doi.org/10.1016/j.compositesb.2021.109249
38. Vanaei S, Parizi MS, Vanaei S, Salemizadehparizi F, Vanaei HR (2021) An overview on materials and techniques in 3D bioprinting toward biomedical application. Eng Regen 2:1–18. https://doi.org/10.1016/j.engreg.2020.12.001
39. Gu Z, Fu J, Lin H, He Y (2020) Development of 3D bioprinting: from printing methods to biomedical applications. Asian J Pharm Sci 15:529–557. https://doi.org/10.1016/j.ajps.2019.11.003
40. Rossi A, Pescara T, Gambelli AM, Gaggia F, Asthana A, Perrier Q, Basta G, Moretti M, Senin N, Rossi F et al (2024) Biomaterials for extrusion-based bioprinting and biomedical applications. Front Bioeng Biotechnol 12. https://doi.org/10.3389/fbioe.2024.1393641
41. Koçak E, Yıldız A, Acartürk F (2021) Three dimensional bioprinting technology: applications in pharmaceutical and biomedical area. Colloids Surf B: Biointerfaces 197:111396. https://doi.org/10.1016/j.colsurfb.2020.111396
42. Tan B, Gan S, Wang X, Liu W, Li X (2021) Applications of 3D bioprinting in tissue engineering: advantages, deficiencies, improvements, and future perspectives. J Mater Chem B 9:5385–5413. https://doi.org/10.1039/D1TB00172H
43. Zandrini T, Florczak S, Levato R, Ovsianikov A (2023) Breaking the resolution limits of 3D bioprinting: future opportunities and present challenges. Trends Biotechnol 41:604–614. https://doi.org/10.1016/j.tibtech.2022.10.009

44. Yakimova MS, Aboushanab SAS, Ivantsova MN, Kamel M (2020) Overview on 3D bioprinting technology: potentials and current challenges. AIP Conf Proc 2313:080034
45. Germaini M-M, Belhabib S, Guessasma S, Deterre R, Corre P, Weiss P (2022) Additive manufacturing of biomaterials for bone tissue engineering—a critical review of the state of the art and new concepts. Prog Mater Sci 130:100963. https://doi.org/10.1016/j.pmatsci.2022.100963
46. Singh S, Chunglok W (2022) Biopolymers towards green and sustainable development. Bentham Science Publishers. https://doi.org/10.2174/97898150793021220101
47. Bothiraja C, Gholap AD, Shaikh KS, Pawar AP (2014) Investigation of ethyl cellulose microsponge gel for topical delivery of eberconazole nitrate for fungal therapy. Ther Deliv 5:781–794. https://doi.org/10.4155/tde.14.43
48. Gholap AD, Rojekar S, Kapare HS, Vishwakarma N, Raikwar S, Garkal A, Mehta TA, Jadhav H, Prajapati MK, Annapure U (2024) Chitosan scaffolds: expanding horizons in biomedical applications. Carbohydr Polym 323:121394. https://doi.org/10.1016/j.carbpol.2023.121394
49. Brovold M, Almeida JI, Pla-Palacín I, Sainz-Arnal P, Sánchez-Romero N, Rivas JJ, Almeida H, Dachary PR, Serrano-Aulló T, Soker S et al (2018) Naturally-derived biomaterials for tissue engineering applications. Adv Exp Med Biol 1077:421–449
50. Gholap AD, Kapare HS, Pagar S, Kamandar P, Bhowmik D, Vishwakarma N, Raikwar S, Garkal A, Mehta TA, Rojekar S et al (2024) Exploring modified chitosan-based gene delivery technologies for therapeutic advancements. Int J Biol Macromol 260:129581. https://doi.org/10.1016/j.ijbiomac.2024.129581
51. Huang G, Li F, Zhao X, Ma Y, Li Y, Lin M, Jin G, Lu TJ, Genin GM, Xu F (2017) Functional and biomimetic materials for engineering of the three-dimensional cell microenvironment. Chem Rev 117:12764–12850. https://doi.org/10.1021/acs.chemrev.7b00094
52. Gholap AD, Gupta J, Kamandar P, Bhowmik DD, Rojekar S, Faiyazuddin M, Hatvate NT, Mohanto S, Ahmed MG, Subramaniyan V et al (2024) Harnessing nanovaccines for effective immunization—a special concern on COVID-19: facts, fidelity, and future prospective. ACS Biomater Sci Eng 10:271–297. https://doi.org/10.1021/acsbiomaterials.3c01247
53. Ullah S, Chen X (2020) Fabrication, applications and challenges of natural biomaterials in tissue engineering. Appl Mater Today 20:100656. https://doi.org/10.1016/j.apmt.2020.100656
54. Li Y, Liu Y, Li R, Bai H, Zhu Z, Zhu L, Zhu C, Che Z, Liu H, Wang J et al (2021) Collagen-based biomaterials for bone tissue engineering. Mater Des 210:110049. https://doi.org/10.1016/j.matdes.2021.110049
55. Pachence JM (1996) Collagen-based devices for soft tissue repair. J Biomed Mater Res 33:35–40. https://doi.org/10.1002/(SICI)1097-4636(199621)33:1<35::AID-JBM6>3.0.CO;2-N
56. Amirrah IN, Lokanathan Y, Zulkiflee I, Wee MFMR, Motta A, Fauzi MB (2022) A comprehensive review on collagen type I development of biomaterials for tissue engineering: from biosynthesis to bioscaffold. Biomedicine 10:2307. https://doi.org/10.3390/biomedicines10092307
57. Sharma R, Malviya R, Singh S, Prajapati B (2023) A critical review on classified excipient sodium-alginate-based hydrogels: modification, characterization, and application in soft tissue engineering. Gels 9(5):430
58. Malrautu P, Laha A, Ramakrishna S (2021) Gelatin nanofibers in drug delivery systems and tissue engineering. Eng Sci. https://doi.org/10.30919/es8d527
59. Gentile P, Chiono V, Carmagnola I, Hatton P (2014) An overview of poly(lactic-co-glycolic) acid (PLGA)-based biomaterials for bone tissue engineering. Int J Mol Sci 15:3640–3659. https://doi.org/10.3390/ijms15033640
60. Pawar R, Pathan A, Nagaraj S, Kapare H, Giram P, Wavhale R (2023) Polycaprolactone and its derivatives for drug delivery. Polym Adv Technol 34:3296–3316. https://doi.org/10.1002/pat.6140
61. Shah SR, Modi CD, Singh S, Mori DD, Soniwala MM, Prajapati BG (2024) Recent advances in additive manufacturing of polycaprolactone-based scaffolds for tissue engineering

applications: a comprehensive review. In: Regenerative engineering and translational medicine. Springer, pp 1–20
62. Gholap AD, Sawant SD, Sayyad SF, Hatvate NT, Chavan M, Rojekar S, Faiyazuddin M (2023) Food biopackaging for human benefits: status and perspectives. In: Biobased packaging materials. Springer Nature Singapore, Singapore, pp 1–35
63. Lu Y, Cheng D, Niu B, Wang X, Wu X, Wang A (2023) Properties of poly (lactic-co-glycolic acid) and progress of poly (lactic-co-glycolic acid)-based biodegradable materials in biomedical research. Pharmaceuticals 16:454. https://doi.org/10.3390/ph16030454
64. Rojekar S, Gholap AD, Togre N, Bhoj P, Haeck C, Hatvate N, Singh N, Vitore J, Dhoble S, Kashid S et al (2024) Current status of mannose receptor-targeted drug delivery for improved anti-HIV therapy. J Control Release 372:494–521. https://doi.org/10.1016/j.jconrel.2024.06.002
65. Pardeshi S, Gholap A, More M, Togre N, Rebello N, Giram P (2023) Dendrimers based antibacterial and antiviral materials. In: Antibacterial and antiviral functional materials, vol 1. ACS Publications, pp 139–169
66. Las-Casas B, Dias IKR, Yupanqui-Mendoza SL, Pereira B, Costa GR, Rojas OJ, Arantes V (2023) The emergence of hybrid cellulose nanomaterials as promising biomaterials. Int J Biol Macromol 250:126007. https://doi.org/10.1016/j.ijbiomac.2023.126007
67. Gholap AD, Hatvate NT, Sawant SD, Khemka PN, Vora LK (2023) Unlocking the potential of nanobiohybrids to combat environmental pollution. In: Nanobiohybrids for advanced wastewater treatment and energy recovery. IWA Publishing, pp 169–190
68. Gholap AD, Said RP, Pawar RD, Ambore GS, Hatvate NT (2023) Importance of carbohydrate-drug conjugates in vaccine development: a detailed review. In: Comprehensive analytical chemistry. Elsevier, pp 191–256
69. Gholap AD, Gupta JS, Kamandar PA, Banchhod GV, Hatvate NT (2023) Antibody-drug conjugates for cancer therapy: an up-to-date review on the chemistry and pharmacology. In: Comprehensive analytical chemistry. Elsevier, pp 105–190
70. Gholap AD, Bhowmik DD, Deshmukh AY, Hatvate NT (2023) Quintessential impact of dendrimer bioconjugates in targeted drug delivery. In: Comprehensive analytical chemistry, vol 103. Elsevier, pp 257–302
71. Casarin M, Todesco M, Fontanella CG, Morlacco A, Dal Moro F, Bagno A (2013) Hybrid materials for tissue repair and replacement: another frontier in biomaterial exploitation focusing on cardiovascular and urological fields. Processes 2023:11. https://doi.org/10.3390/pr11072013
72. Kennedy SM, Vasanthanathan A, Jeen Robert RB, Amudhan K (2024) Advancements and prospects of polymer-based hybrid composites for bone plate applications. Polym Technol Mater 63:68–87. https://doi.org/10.1080/25740881.2023.2274564
73. Xing J, Zhang M, Liu X, Wang C, Xu N, Xing D (2023) Multi-material electrospinning: from methods to biomedical applications. Mater Today Bio 21:100710. https://doi.org/10.1016/j.mtbio.2023.100710
74. Dzobo K, Motaung KSCM, Adesida A (2019) Recent trends in decellularized extracellular matrix bioinks for 3D printing: an updated review. Int J Mol Sci 20:4628. https://doi.org/10.3390/ijms20184628
75. Liu H, Xing F, Yu P, Lu R, Ma S, Shakya S, Zhou X, Peng K, Zhang D, Liu M (2024) Biomimetic fabrication bioprinting strategies based on decellularized extracellular matrix for musculoskeletal tissue regeneration: current status and future perspectives. Mater Des 243:113072. https://doi.org/10.1016/j.matdes.2024.113072
76. Gholap AD, Khuspe PR, Pardeshi SR, Uddin MJ, Das U, Hatvate NT, Rojekar S, Giram P, Khalid M, Choonara YE et al (2024) Achieving optimal health with host-directed therapies (HDTs) in infectious diseases—a new horizon. Adv Ther. https://doi.org/10.1002/adtp.202400169
77. Gholap A, Pardeshi S, Giram P (2023) Preparation, antibacterial and antiviral activity measurements and detection methods. In: Antibacterial and antiviral functional materials. ACS Publications, pp 33–64

78. Nah SH, Lee Y, Yu K, Chi Y, Lee H, Chen B, Patel M, Wang K, Yang S (2024) Moisture absorbing and water self-releasing from hybrid hydrogel desiccants. Adv Funct Mater 34. https://doi.org/10.1002/adfm.202313881
79. Castillo-Henríquez L, Castro-Alpízar J, Lopretti-Correa M, Vega-Baudrit J (2021) Exploration of bioengineered scaffolds composed of thermo-responsive polymers for drug delivery in wound healing. Int J Mol Sci 22:1408. https://doi.org/10.3390/ijms22031408
80. Sarwan T, Kumar P, Choonara YE, Pillay V (2020) Hybrid thermo-responsive polymer systems and their biomedical applications. Front Mater 7. https://doi.org/10.3389/fmats.2020.00073
81. Socci MC, Rodríguez G, Oliva E, Fushimi S, Takabatake K, Nagatsuka H, Felice CJ, Rodríguez AP (2023) Polymeric materials, advances and applications in tissue engineering: a review. Bioengineering 10:218. https://doi.org/10.3390/bioengineering10020218
82. Benwood C, Chrenek J, Kirsch RL, Masri NZ, Richards H, Teetzen K, Willerth SM (2021) Natural biomaterials and their use as bioinks for printing tissues. Bioengineering 8:27. https://doi.org/10.3390/bioengineering8020027
83. Gungor-Ozkerim PS, Inci I, Zhang YS, Khademhosseini A, Dokmeci MR (2018) Bioinks for 3D bioprinting: an overview. Biomater Sci 6:915–946. https://doi.org/10.1039/C7BM00765E
84. Gholap AD, Hatvate NT, Khuspe PR, Mandhare TA, Kashid P, Gaikwad VD (2023) Proteomics in oncology: retrospect and prospects. In: Novel technologies in biosystems, biomedical & drug delivery. Springer Nature Singapore, Singapore, pp 243–269
85. Chen XB, Fazel Anvari-Yazdi A, Duan X, Zimmerling A, Gharraei R, Sharma NK, Sweilem S, Ning L (2023) Biomaterials/bioinks and extrusion bioprinting. Bioact Mater 28:511–536. https://doi.org/10.1016/j.bioactmat.2023.06.006
86. Ashammakhi N, Ahadian S, Xu C, Montazerian H, Ko H, Nasiri R, Barros N, Khademhosseini A (2019) Bioinks and bioprinting technologies to make heterogeneous and biomimetic tissue constructs. Mater Today Bio 1:100008. https://doi.org/10.1016/j.mtbio.2019.100008
87. Ilyas R, Zuhri M, Norrahim M, Misenan M, Jenol M, Samsudin S, Nurazzi N, Asyraf M, Supian A, Bangar S et al (2022) Natural fiber-reinforced polycaprolactone green and hybrid biocomposites for various advanced applications. Polymers (Basel) 14:182. https://doi.org/10.3390/polym14010182
88. Gholap AD, Sayyad SF, Hatvate NT, Dhumal VV, Pardeshi SR, Chavda VP, Vora LK (2023) Drug delivery strategies for avobenzone: a case study of photostabilization. Pharmaceutics 15:1008. https://doi.org/10.3390/pharmaceutics15031008
89. Gholap AD, Pardeshi SR, Hatvate NT, Dhorkule N, Sayyad SF, Faiyazuddin M, Khalid M (2024) Environmental implications and nanotechnological advances in octocrylene-enriched sunscreen formulations: a comprehensive review. Chemosphere 358:142235. https://doi.org/10.1016/j.chemosphere.2024.142235
90. Zhang Q, Yan K, Zheng X, Liu Q, Han Y, Liu Z (2024) Research progress of photo-crosslink hydrogels in ophthalmology: a comprehensive review focus on the applications. Mater Today Bio 26:101082. https://doi.org/10.1016/j.mtbio.2024.101082
91. Zennifer A, Manivannan S, Sethuraman S, Kumbar SG, Sundaramurthi D (2022) 3D bioprinting and photocrosslinking: emerging strategies & future perspectives. Biomater Adv 134:112576. https://doi.org/10.1016/j.msec.2021.112576
92. Parit S, Manchare A, Gholap AD, Mundhe P, Hatvate N, Rojekar S, Patravale V (2024) Antibody-drug conjugates: a promising breakthrough in cancer therapy. Int J Pharm 659:124211. https://doi.org/10.1016/j.ijpharm.2024.124211
93. GhavamiNejad A, Ashammakhi N, Wu XY, Khademhosseini A (2020) Crosslinking strategies for 3D bioprinting of polymeric hydrogels. Small 16. https://doi.org/10.1002/smll.202002931
94. Su X, Wang T, Guo S (2021) Applications of 3D printed bone tissue engineering scaffolds in the stem cell field. Regen Ther 16:63–72. https://doi.org/10.1016/j.reth.2021.01.007
95. Top N, Şahin İ, Gökçe H, Gökçe H (2021) Computer-aided design and additive manufacturing of bone scaffolds for tissue engineering: state of the art. J Mater Res 36:3725–3745. https://doi.org/10.1557/s43578-021-00156-y
96. Chinnasami H, Dey MK, Devireddy R (2023) Three-dimensional scaffolds for bone tissue engineering. Bioengineering 10:759. https://doi.org/10.3390/bioengineering10070759

97. Fay CD (2020) Computer-aided design and manufacturing (CAD/CAM) for bioprinting. Methods Mol Biol 2140:27–41
98. Zaszczyńska A, Moczulska-Heljak M, Gradys A, Sajkiewicz P (2021) Advances in 3D printing for tissue engineering. Materials (Basel) 14:3149. https://doi.org/10.3390/ma14123149
99. Bardini R, Di Carlo S (2024) Computational methods for biofabrication in tissue engineering and regenerative medicine—a literature review. Comput Struct Biotechnol J 23:601–616. https://doi.org/10.1016/j.csbj.2023.12.035
100. Imran R, Al Rashid A, Koç M (2022) Review on computational modeling for the property, process, product and performance (PPPP) characteristics of additively manufactured porous magnesium implants. Bioprinting 28:e00236. https://doi.org/10.1016/j.bprint.2022.e00236
101. Huang H-M, Shih Y-Y (2014) Pushing CT and MR imaging to the molecular level for studying the "omics": current challenges and advancements. Biomed Res Int 2014:1–17. https://doi.org/10.1155/2014/365812
102. Nikolova MP, Chavali MS (2019) Recent advances in biomaterials for 3D scaffolds: a review. Bioact Mater 4:271–292. https://doi.org/10.1016/j.bioactmat.2019.10.005
103. Hoyle HW, Stenger CML, Przyborski SA (2022) Design considerations of benchtop fluid flow bioreactors for bio-engineered tissue equivalents in vitro. Biomater Biosyst 8:100063. https://doi.org/10.1016/j.bbiosy.2022.100063
104. Xie Z, Gao M, Lobo AO, Webster TJ (2020) 3D bioprinting in tissue engineering for medical applications: the classic and the hybrid. Polymers (Basel) 12:1717. https://doi.org/10.3390/polym12081717
105. Mirshafiei M, Rashedi H, Yazdian F, Rahdar A, Baino F (2024) Advancements in tissue and organ 3D bioprinting: current techniques, applications, and future perspectives. Mater Des 240:112853. https://doi.org/10.1016/j.matdes.2024.112853
106. Song Y, Shao E, Bíró I, Baker JS, Gu Y (2022) Finite element modelling for footwear design and evaluation: a systematic scoping review. Heliyon 8:e10940. https://doi.org/10.1016/j.heliyon.2022.e10940
107. Keklikoglou K, Arvanitidis C, Chatzigeorgiou G, Chatzinikolaou E, Karagiannidis E, Koletsa T, Magoulas A, Makris K, Mavrothalassitis G, Papanagnou E-D et al (2021) Micro-CT for biological and biomedical studies: a comparison of imaging techniques. J Imaging 7:172. https://doi.org/10.3390/jimaging7090172
108. Shelmerdine SC, Simcock IC, Hutchinson JC, Aughwane R, Melbourne A, Nikitichev DI, Ong J, Borghi A, Cole G, Kingham E et al (2018) 3D printing from microfocus computed tomography (micro-CT) in human specimens: education and future implications. Br J Radiol 20180306. https://doi.org/10.1259/bjr.20180306
109. Gholap AD, Uddin MJ, Faiyazuddin M, Omri A, Gowri S, Khalid M (2024) Advances in artificial intelligence for drug delivery and development: a comprehensive review. Comput Biol Med 178:108702. https://doi.org/10.1016/j.compbiomed.2024.108702
110. Wu C, Fang J, Entezari A, Sun G, Swain MV, Xu Y, Steven GP, Li Q (2021) A time-dependent mechanobiology-based topology optimization to enhance bone growth in tissue scaffolds. J Biomech 117:110233. https://doi.org/10.1016/j.jbiomech.2021.110233
111. Wu J, Zhang Y, Lyu Y, Cheng L (2023) On the various numerical techniques for the optimization of bone scaffold. Materials (Basel) 16:974. https://doi.org/10.3390/ma16030974
112. Zielińska A, Karczewski J, Eder P, Kolanowski T, Szalata M, Wielgus K, Szalata M, Kim D, Shin SR, Słomski R et al (2023) Scaffolds for drug delivery and tissue engineering: the role of genetics. J Control Release 359:207–223. https://doi.org/10.1016/j.jconrel.2023.05.042
113. Krishani M, Shin WY, Suhaimi H, Sambudi NS (2023) Development of scaffolds from bio-based natural materials for tissue regeneration applications: a review. Gels 9:100. https://doi.org/10.3390/gels9020100
114. Chocholata P, Kulda V, Babuska V (2019) Fabrication of scaffolds for bone-tissue regeneration. Materials (Basel) 12:568. https://doi.org/10.3390/ma12040568
115. Suamte L, Tirkey A, Barman J, Jayasekhar Babu P (2023) Various manufacturing methods and ideal properties of scaffolds for tissue engineering applications. Smart Mater Manuf 1:100011. https://doi.org/10.1016/j.smmf.2022.100011

116. Adel IM, ElMeligy MF, Elkasabgy NA (2022) Conventional and recent trends of scaffolds fabrication: a superior mode for tissue engineering. Pharmaceutics 14:306. https://doi.org/10.3390/pharmaceutics14020306
117. Kim HS, Kumbar SG, Nukavarapu SP (2021) Biomaterial-directed cell behavior for tissue engineering. Curr Opin Biomed Eng 17:100260. https://doi.org/10.1016/j.cobme.2020.100260
118. Wang F, Cai X, Shen Y, Meng L (2023) Cell–scaffold interactions in tissue engineering for oral and craniofacial reconstruction. Bioact Mater 23:16–44. https://doi.org/10.1016/j.bioactmat.2022.10.029
119. Hussain Z, Mehmood S, Liu X, Liu Y, Wang G, Pei R (2024) Decoding bone-inspired and cell-instructive cues of scaffolds for bone tissue engineering. Eng Regen 5:21–44. https://doi.org/10.1016/j.engreg.2023.10.003
120. Yi B, Xu Q, Liu W (2022) An overview of substrate stiffness guided cellular response and its applications in tissue regeneration. Bioact Mater 15:82–102. https://doi.org/10.1016/j.bioactmat.2021.12.005
121. El-Sherbiny IM, Yacoub MH (2013) Hydrogel scaffolds for tissue engineering: progress and challenges. Glob Cardiol Sci Pract 2013:38. https://doi.org/10.5339/gcsp.2013.38
122. El-Husseiny HM, Mady EA, Hamabe L, Abugomaa A, Shimada K, Yoshida T, Tanaka T, Yokoi A, Elbadawy M, Tanaka R (2022) Smart/stimuli-responsive hydrogels: cutting-edge platforms for tissue engineering and other biomedical applications. Mater Today Bio 13:100186. https://doi.org/10.1016/j.mtbio.2021.100186
123. Liu J, Kerns DG (2014) Mechanisms of guided bone regeneration: a review. Open Dent J 8:56–65. https://doi.org/10.2174/1874210601408010056
124. Turnbull G, Clarke J, Picard F, Riches P, Jia L, Han F, Li B, Shu W (2018) 3D bioactive composite scaffolds for bone tissue engineering. Bioact Mater 3:278–314. https://doi.org/10.1016/j.bioactmat.2017.10.001
125. Breuer T, Jimenez M, Humphrey JD, Shinoka T, Breuer CK (2023) Tissue engineering of vascular grafts: a case report from bench to bedside and back. Arterioscler Thromb Vasc Biol 43:399–409. https://doi.org/10.1161/ATVBAHA.122.318236
126. Durán-Rey D, Crisóstomo V, Sánchez-Margallo JA, Sánchez-Margallo FM (2021) Systematic review of tissue-engineered vascular grafts. Front Bioeng Biotechnol 9. https://doi.org/10.3389/fbioe.2021.771400
127. Shah DD, Raghani NR, Chorawala MR, Singh S, Prajapati BG (2023) Harnessing three-dimensional (3D) cell culture models for pulmonary infections: state of the art and future directions. Naunyn Schmiedeberg's Arch Pharmacol 396(11):2861–2880
128. Zhang X, Chen X, Hong H, Hu R, Liu J, Liu C (2022) Decellularized extracellular matrix scaffolds: recent trends and emerging strategies in tissue engineering. Bioact Mater 10:15–31. https://doi.org/10.1016/j.bioactmat.2021.09.014
129. Yi H-G, Choi Y-J, Jung JW, Jang J, Song T-H, Chae S, Ahn M, Choi TH, Rhie J-W, Cho D-W (2019) Three-dimensional printing of a patient-specific engineered nasal cartilage for augmentative rhinoplasty. J Tissue Eng 10:204173141882479. https://doi.org/10.1177/2041731418824797
130. Haleem A, Javaid M, Khan RH, Suman R (2020) 3D printing applications in bone tissue engineering. J Clin Orthop Trauma 11:S118–S124. https://doi.org/10.1016/j.jcot.2019.12.002
131. Zhang Q, Zhou J, Zhi P, Liu L, Liu C, Fang A, Zhang Q (2023) 3D printing method for bone tissue engineering scaffold. Med Nov Technol Devices 17:100205. https://doi.org/10.1016/j.medntd.2022.100205
132. Mobarak MH, Islam MA, Hossain N, Al Mahmud MZ, Rayhan MT, Nishi NJ, Chowdhury MA (2023) Recent advances of additive manufacturing in implant fabrication—a review. Appl Surf Sci Adv 18:100462. https://doi.org/10.1016/j.apsadv.2023.100462
133. Echeverria Molina MI, Malollari KG, Komvopoulos K (2021) Design challenges in polymeric scaffolds for tissue engineering. Front Bioeng Biotechnol 9. https://doi.org/10.3389/fbioe.2021.617141
134. Do A, Khorsand B, Geary SM, Salem AK (2015) 3D printing of scaffolds for tissue regeneration applications. Adv Healthc Mater 4:1742–1762. https://doi.org/10.1002/adhm.201500168

135. Nazir A, Gokcekaya O, Md Masum Billah K, Ertugrul O, Jiang J, Sun J, Hussain S (2023) Multi-material additive manufacturing: a systematic review of design, properties, applications, challenges, and 3D printing of materials and cellular metamaterials. Mater Des 226:111661. https://doi.org/10.1016/j.matdes.2023.111661
136. Claudia Spataru M, Simona Baltatu M, Victor Sandu A, Vizureanu P (2024) General trends on biomaterials applications: advantages and limitations. In: Novel biomaterials for tissue engineering. IntechOpen
137. Shah D, Dave B, Chorawala MR, Prajapati BG, Singh S, Elossaily GM et al (2024) An insight on microfluidic organ-on-a-chip models for PM2. 5-induced pulmonary complications. ACS Omega 9(12):13534–13555
138. Giannitelli SM, Accoto D, Trombetta M, Rainer A (2014) Current trends in the design of scaffolds for computer-aided tissue engineering. Acta Biomater 10:580–594. https://doi.org/10.1016/j.actbio.2013.10.024
139. Hölzl K, Lin S, Tytgat L, Van Vlierberghe S, Gu L, Ovsianikov A (2016) Bioink properties before, during and after 3D bioprinting. Biofabrication 8:032002. https://doi.org/10.1088/1758-5090/8/3/032002
140. Cubo N, Garcia M, del Cañizo JF, Velasco D, Jorcano JL (2016) 3D bioprinting of functional human skin: production and in vivo analysis. Biofabrication 9:015006. https://doi.org/10.1088/1758-5090/9/1/015006
141. Olejnik A, Semba JA, Kulpa A, Dańczak-Pazdrowska A, Rybka JD, Gornowicz-Porowska J (2022) 3D bioprinting in skin related research: recent achievements and application perspectives. ACS Synth Biol 11:26–38. https://doi.org/10.1021/acssynbio.1c00547
142. Little C, Kulyk W, Chen X (2014) The effect of chondroitin sulphate and hyaluronic acid on chondrocytes cultured within a fibrin-alginate hydrogel. J Funct Biomater 5:197–210. https://doi.org/10.3390/jfb5030197
143. Anderson DE, Johnstone B (2017) Dynamic mechanical compression of chondrocytes for tissue engineering: a critical review. Front Bioeng Biotechnol 5. https://doi.org/10.3389/fbioe.2017.00076
144. Wang P, Sun Y, Shi X, Shen H, Ning H, Liu H (2021) 3D printing of tissue engineering scaffolds: a focus on vascular regeneration. Bio-Design Manuf 4:344–378. https://doi.org/10.1007/s42242-020-00109-0
145. Liu N, Zhang X, Guo Q, Wu T, Wang Y (2022) 3D bioprinted scaffolds for tissue repair and regeneration. Front Mater 9. https://doi.org/10.3389/fmats.2022.925321
146. Ozdemir S, Yalcin-Enis I, Yalcinkaya B, Yalcinkaya F (2022) An investigation of the constructional design components affecting the mechanical response and cellular activity of electrospun vascular grafts. Membranes (Basel) 12:929. https://doi.org/10.3390/membranes12100929
147. Ma C, Peng Y, Li H, Chen W (2021) Organ-on-a-chip: a new paradigm for drug development. Trends Pharmacol Sci 42:119–133. https://doi.org/10.1016/j.tips.2020.11.009
148. Singh D, Mathur A, Arora S, Roy S, Mahindroo N (2022) Journey of organ on a chip technology and its role in future healthcare scenario. Appl Surf Sci Adv 9:100246. https://doi.org/10.1016/j.apsadv.2022.100246
149. Pattanayak P, Singh SK, Gulati M, Vishwas S, Kapoor B, Chellappan DK, Anand K, Gupta G, Jha NK, Gupta PK et al (2021) Microfluidic chips: recent advances, critical strategies in design, applications and future perspectives. Microfluid Nanofluid 25:99. https://doi.org/10.1007/s10404-021-02502-2
150. Deng S, Li C, Cao J, Cui Z, Du J, Fu Z, Yang H, Chen P (2023) Organ-on-a-chip meets artificial intelligence in drug evaluation. Theranostics 13:4526–4558. https://doi.org/10.7150/thno.87266
151. Junka R, Quevada K, Yu X (2020) Acellular polycaprolactone scaffolds laden with fibroblast/endothelial cell-derived extracellular matrix for bone regeneration. J Biomed Mater Res Part A 108:351–364. https://doi.org/10.1002/jbm.a.36821

152. Bishop ES, Mostafa S, Pakvasa M, Luu HH, Lee MJ, Wolf JM, Ameer GA, He T-C, Reid RR (2017) 3-D bioprinting technologies in tissue engineering and regenerative medicine: current and future trends. Genes Dis 4:185–195. https://doi.org/10.1016/j.gendis.2017.10.002
153. Cui H, Nowicki M, Fisher JP, Zhang LG (2017) 3D bioprinting for organ regeneration. Adv Healthc Mater 6. https://doi.org/10.1002/adhm.201601118
154. Ghosh S, Yi H-G (2022) A review on bioinks and their application in plant bioprinting. Int J Bioprinting 8:612. https://doi.org/10.18063/ijb.v8i4.612
155. Fang Y, Guo Y, Liu T, Xu R, Mao S, Mo X, Zhang T, Ouyang L, Xiong Z, Sun W (2022) Advances in 3D bioprinting. Chin J Mech Eng Addit Manuf Front *1*:100011. https://doi.org/10.1016/j.cjmeam.2022.100011
156. Thom RL, Cronin AJ (2023) Legal and regulatory challenges for emerging regenerative medicine solutions for diabetes. Transplantation 108(5):1072–1079. https://doi.org/10.1097/TP.0000000000004797
157. Brown N, Faulkner A, Kent J, Michael M (2006) Regulating hybrids: 'making a mess' and 'cleaning up' in tissue engineering and transpecies transplantation. Soc Theory Health 4:1–24. https://doi.org/10.1057/palgrave.sth.8700062
158. Pirnay J-P, Vanderkelen A, De Vos D, Draye J-P, Rose T, Ceulemans C, Ectors N, Huys I, Jennes S, Verbeken G (2013) Business oriented EU human cell and tissue product legislation will adversely impact member states' health care systems. Cell Tissue Bank 14:525–560. https://doi.org/10.1007/s10561-013-9397-6
159. Singh AB (2024) Transforming healthcare: a review of additive manufacturing applications in the healthcare sector. In: Proceedings of the IECMA 2024, September 10. MDPI, Basel Switzerland, p 2
160. Alzoubi L, Aljabali AAA, Tambuwala MM (2023) Empowering precision medicine: the impact of 3D printing on personalized therapeutic. AAPS PharmSciTech 24:228. https://doi.org/10.1208/s12249-023-02682-w
161. Đorđević S, Gonzalez MM, Conejos-Sánchez I, Carreira B, Pozzi S, Acúrcio RC, Satchi-Fainaro R, Florindo HF, Vicent MJ (2022) Current hurdles to the translation of nanomedicines from bench to the clinic. Drug Deliv Transl Res 12:500–525. https://doi.org/10.1007/s13346-021-01024-2
162. Tamayo JA, Riascos M, Vargas CA, Baena LM (2021) Additive manufacturing of Ti6Al4V alloy via electron beam melting for the development of implants for the biomedical industry. Heliyon 7:e06892. https://doi.org/10.1016/j.heliyon.2021.e06892

Advances in Additive Manufacturing in Tissue Engineering and Regeneration

4

Shekhar V. Kokate, Yogesh V. Ushir, Akanksha M. Kamble, and Sudarshan Singh

Abstract

Additive manufacturing (AM), or 3D printing, has emerged as a transformative force in tissue engineering and regenerative medicine. This chapter explores the convergence of AM and tissue engineering, emphasizing its revolutionary impact on regenerative medicine. AM enables precise fabrication of scaffolds with tailored mechanical properties, geometries, and bio-functionalities that mimic native tissue structures. Technologies like stereolithography, extrusion-based printing, and inkjet printing offer unique benefits for scaffold creation using materials such as natural polymers, synthetic polymers, and hybrid biomaterials. AM has diverse applications in tissue engineering, including bone, cartilage, and skin regeneration. In bone tissue engineering, AM facilitates customized implants and osteoconductive scaffolds for individual patient needs. Cartilage regeneration benefits from bio-printed constructs and hydrogels that address cartilage defects, while skin tissue engineering utilizes AM to develop cell-seeded scaffolds, 3D-printed dressings, and patches for wound healing, highlighting its adaptability and potential for personalized medicine. Integrating AM with advanced biomaterials, such as bioinks and decellularized extracellular matrices, enhances scaffold bioactivity and biocompatibility, promoting cell adhesion, proliferation, and differentiation. Additionally, incorporating growth factors, genes, and cells into AM scaffolds allows precise control over tissue regeneration, while AM technologies support organ-on-a-chip platforms and

S. V. Kokate (✉) · Y. V. Ushir · A. M. Kamble
SMBT College of Pharmacy, Nashik, Maharashtra, India

S. Singh
Office of Research and Administration, Chiang Mai University, Mueang Chiang Mai, Thailand

Faculty of Pharmacy, Chiang Mai University, Chiang Mai, Thailand

S. Singh et al. (eds.), *Biomaterial-based Additive Manufacturing in Tissue Engineering and Regeneration*, Biomaterials, Bioengineering and Sustainability 7,
https://doi.org/10.1007/978-3-031-96070-3_4

patient-specific implants for personalized medical applications and drug screening. Challenges like biocompatibility, mechanical properties, vascularization, and regulatory barriers remain, but advanced techniques such as multi-material printing, vascularization strategies, 4D printing, and bioactive agents offer solutions. Future directions include new biomaterials, technological advancements, clinical translation, commercialization, and ethical considerations to further drive innovation and clinical adoption in tissue engineering.

Keywords

Additive manufacturing · 3D printing · Tissue engineering · Tissue regeneration · Bioprinting · Biomaterials

Abbreviations

AM	Additive manufacturing
TE	Tissue Engineering
RM	Rigenerative Medicine
UV	Ultra Visible
VISCAM RP	Visual Interactive Simulation and Computer-Aided Manufacturing for Rapid Prototyping
MAGICS	Materialise Automated Geometries for Integrated Cross-Sections.
DoF	degrees of freedom
SLA	Stereolithography
PB	powder bed
MSCs	mesenchymal stem cells
BG	bioactive glass
PVP	Poly vinyl pyrrolidone
PCL	poly ε-caprolactone
PGA	polyglycolic acid
PLA	polylactic acid
PU	polyurethane
PLA	poly lactic acid
PLGA	polylactic-co-glycolic acid
PHA	polyhydroxyalkanoates
PHB	poly-3-hydroxybutyrate
PHO	poly-3-hydroxyoctanoate
PEG	polyethylene glycol
P(3)HB	poly 3-hydroxybutyrate
TCP	Tricalcium phosphate
HA	Hydroxyapatite
FEA	Finite Element Analysis.
CT	Computed Tomography
IVD	In Vitro Diagnostic

FDM	Fused Deposition Modelling
HyA	Hyaluronic Acid
ECM	Extracellular Matrix
GelMA	Gelatin methacrylate
HAMA	hyaluronic acid methacrylate
VEGF	Vascular Endothelial Growth Factor
DLP	Digital Light Processing
MCDM	Multi-Criteria Decision Making
BTE	bone tissue engineering
MMAM	Multi-material additive manufacturing
PEEK	Polyether Ether Ketone
ABS	Acrylonitrile Butadiene Styrene
PET	Polyethylene Terephthalate
MIT	Massachusetts Institute of Technology
BMPs	Bone morphogenetic proteins
IGF-1	insulin-like growth factor 1
bFGF	Basic fibroblast growth factor
TGF-β	transforming growth factor
GFs	beta growth factors
PSI	Patient-Specific Implant
DICOM	Digital Imaging and Communications in Medicine
NC	Nano cellulose
CNC	Cellulose nano-crystals
NFC	Nano-fibrillate cellulose
ALG	Alginate
TCP	tri-calcium phosphate
IP	intellectual property
RPT	Rapid Prototyping Technology
3D	3 dimension
β-TCP	Beta-tricalcium phosphate

4.1 Introduction

A tissue engineering [TE] and regenerative medicine [RM] is biomedical fields that apply fundamental concepts to create practical solutions. The modern term "tissue engineering" originated in 1987 for medicine. The consensus definition: "Tissue Engineering is the application of the principles and methods of engineering and life sciences toward the fundamental understanding of structure-function relationships in normal and pathologic mammalian tissue and the development of biological substitutes to restore, maintain, or improve function." Creating artificial tissues, organs, and actual beings has been a myth throughout history.

In recent decades, this vision has become more viable and is now being used in clinical care. Scientific advancements have made to creation to new tissues and organs from small living specimens. However, the future influence on clinical care remains uncertain. To anticipate the future for TE and RM and for previews the vital visions and activities. TE and RM, that build complex organisms from smaller components, are profoundly ingrained in people's imaginations. Throughout history, there has been a shift in our understanding of how to make and regenerate complex organs and organisms, reflecting humanity's cultural history. Throughout history, males have been believed to be capable of achieving independence without sexual intimacy. Greek mythology, such as the creation of Prometheus exemplifies the concept of producing live creatures from non-living materials. The Biblical account of Eve being formed from ribs belonging to Adam is well-known illustration of this principle. In modern times, this is considered hybrid cloning. Abundance is an example in literature. The arts reflect humanity's drive to establish self-sufficient individuals or groups and Individuals' most insignificant components. Individuals' social, cultural, and scientific backgrounds shape their conceptions of life-creation measures. As science and medicine advanced, many stories, reports, artworks, and films explored the possibility of creating life using "scientific" methods. In addition to mythological, religious, and fictional accounts, individuals have pioneered real methods for generating, healing, or regenerating bodily parts. Tissue engineering has been linked to advancements in clinical medicine (reconstructive surgery, prosthetics, transplantation, micro-surgery) and biology (biochemistry, genetics, cell and molecular biology) [1].

Additive Manufacturing became a breakthrough technology in healthcare, offering novel solutions across a spectrum of medical applications. Bio-printing's key contribution is in personalized medical devices like implants and prosthetics, custom-fit to individual anatomy, boosting comfort and function. Additionally, AM helps create patient-specific models of anatomy based on medical imaging data, empowering surgeons with invaluable preoperative planning tools and enabling them to rehearse complex procedures with greater accuracy. Beyond device manufacturing, AM allows for personalized drug delivery systems, improving therapeutic outcomes and reducing side effects. In bio-printing, scientists are crafting living tissues and organs, opening doors for regenerative medicine and transplants. Dental applications, surgical instruments, and educational tools further underscore the versatility and impact of AM in healthcare, revolutionizing treatment paradigms and fostering innovation [2].

This book chapter digs into the variety of uses and transformative power in 3D printing in biomedicine and healthcare, highlighting its adaptability and influence across various sectors. Also, it will embark on a comprehensive journey through the evolving field of additive manufacturing and its transformative impact on tissue engineering. Beginning with the fundamentals, we will explore various AM techniques like stereolithography, fused deposition modeling, selective laser sintering, and inkjet bio-printing, highlighting their unique capabilities and applications. In orthopaedics and prosthetics, it reshapes the landscape by enabling the production of tailor-made implants and prosthetic devices, thereby ensuring optimal fit and enhanced patient outcomes. Dentistry undergoes a transformation with the introduction of 3D-printed dental implants, crowns, and orthodontic appliances, offering patients personalized solutions to address their oral health needs. In tissue engineering, it assumes a pivotal role in creating organoids, implants, and scaffolds for bone regeneration, significant advances in regeneration medicine. Bio-printing improves surgical precision with precise tools and virtual planning and advances pharmaceuticals through tailored drug delivery. 3D printing is revolutionizing medical practices, optimizing procedures, and enhancing patient care across various applications. We'll cover materials like polymers and ceramics for tissue engineering, and explore innovations like multi-material and 4D printing with case studies of custom implants and bio-printed tissues. Additionally, advent for 3D-printed ingestible tablets represents a ground-breaking approach to drug delivery, enhancing medication administration and patient compliance. Overall, this book emphasizes the transformational potential for 3-D printing, underscoring its role for driving innovation, ultimately improving patient care across diverse biomedical applications [3].

Tissue engineering has undergone a revolutionary leap forward, propelled by cutting-edge techniques like 3D bioprinting, this breakthrough technology allows for the accurate manufacture of skin-like structures with remarkable accuracy. Gone are the days of rudimentary approaches; we now wield a formidable arsenal of bioprinting technologies, each more formidable than the last. From the intricate precision of inkjet bioprinting to the versatile process of extrusion-based methods, we stand on the brink of a new era where the very building blocks of life can be orchestrated with unparalleled finesse. But the marvels don't end there. Polymer scaffolds, once mere supports, have evolved into architectural wonders that not only provide mechanical reinforcement but also serve as nurturing sanctuaries for cellular development. Picture-perfect replicas of natural tissues are now within our grasp, thanks to the relentless march of progress in laser technology and stereolithography. These advancements have bestowed upon us the power to breathe life into bio-printed tissues with such precision and resilience that they rival their organic counterparts in both form and function. Indeed, the landscape of tissue engineering has been reshaped beyond recognition, with each stride forward revealing new vistas of possibility. The dream of regenerating damaged tissues and organs is no longer confined to the realm of science fiction but has emerged as a tangible reality, promising hope and healing to countless lives. Addressing the challenges facing AM in tissue engineering, from biocompatibility and mechanical properties to regulatory hurdles and clinical translation. In this golden age of biotechnology, where the boundaries of what we can achieve continue to expand, the future holds infinite promise for the advancement of human health and well-being [4].

4.2 Fundamentals of Additive Manufacturing

AM methods create models that fuse, sintering/polymerizing material in predefined layers without the need for tools. This allows for the production of intricate geometries, including detailed interiors, impossible with traditional machining or molding (8). AM operates by slicing CAD data with specialized software, with layer thickness ranging from 10 to 200 μm depending on machine parameters. The visibility of layers during printing impacts the final output quality. The "staircase effect" refers to how layer thickness affects surface orientation in additive manufacturing. Thinner layers create smoother surfaces but increase processing time and resolution demands. In AM, layers are stacked on the Z-axis. After treating each layer, new substance is coated. In resin based systems, the traversal end smooth the resins; in powdered systems, placed powder distributed by roller/wiper; and in certain systems, the material is dropped through nozzle that delivers what is needed. Because the recoating interval is greater than the layer time required for processing. Numerous components can be built simultaneously during the time it takes to recoat a single layer in additive manufacturing. Software like Smart Space, VISCAM RP and MAGICS, helps position and orient parts optimally for assembly. Fragile parts sometimes require structural support during the construction process, which AM systems supply using specialized components. Thin, pointed teeth serve as typical support structures, allowing for quick removal after printing [5]. This approach can result in the creation of components that are lighter and exhibit superior performance compared to assemblies. It also reduces assembly time, repair duration, shop floor complexity, inventory of replacement parts, and the need for tooling, leading to cost savings across the entire process [6].

4.3 Materials Used in AM for Tissue Engineering

4.3.1 Synthetic Polymers

Synthetic polymers are made under laboratory circumstances using hydrocarbon building components [7]. Several synthetic polymers possess the necessary characteristics for biodegradable scaffolds, including polyurethanes, polyacetals, and polyanhydrides. Among them, synthetic aliphatic poly-esters like poly (ε-caprolactone) (PCL), polyglycolic acid (PGA), polylactic acid (PLA), and their copolymers are widely utilized in biodegradable tissue engineering. These polymers undergo hydrolytic degradation through de-esterification, yielding easily removable monomers like lactic acid and glycolic acid, making them excellent options for tissue scaffold materials. The most frequently utilized synthetic polymers' like polyurethane (PU), poly (lactic acid), polylactic-co-glycolic acid (PLGA), poly (glycolic acid) (PGA), and polycaprolactone (PCL). These polymers boast high biocompatibility and are produced using renewable methods. Biodegradable and bioresorbable qualities of these polymers have been extensively studied and applied in clinical products. However, drawbacks exist, such as the low cytocompatibility and biological invincibility of pure PLA, which can lead to decreased cellular contact and tissue regeneration. However, it is worth noting that they are frequently copolymerized or modified for specific use [8].

Synthetic polymers, like natural polymers, deteriorate and lose mechanical qualities, which might be problematic if tissue growth is inadequate [9]. Additionally, degradation by-products can be hazardous. Other alterations include the incorporation of synthetic polymers using materials such as aliphatic polyester group polyhydroxyalkanoates (PHA), that comprised of poly-3-hydroxybutyrate (PHB) and poly-3-hydroxyoctanoate (PHO). Such polymers, primarily produced by microbes under controlled imbalanced conditions, but they can also be synthesized. Polyanhydrides are the sole FDA-approved surface-eroding biomaterials, yet their challenging manufacturing process and low mechanical strength restrict their broad usage. On the flip side, polyethylene glycol (PEG) is a commonly used polymer known for its soft, gel-like properties, particularly in medication delivery and wound treatment due to its cross-linked hydrogel structure [10].

4.3.2 Natural Polymers

Natural polymers that are acquired from livelihood or biological sources, like plants, animals, microbes, and biological systems [7, 10]. These polymers are alginate, collagen, fibrin, gelatin, Hyaluronic Acid (HyA), and chitosan is well-suited for cell attachment owing to their bio-compatibility, bio-chemical similarity for Extracellular Matrix (ECM), with the presence of Structural patterns for binding. Collagen and HyA is integral elements for ECM, acting as Receptors facilitating cell adhesion, and are currently under extensive research scrutiny. Natural polymers degrade through enzymatic pathways, allowing cells to reconstruct the scaffold, while manufactured polymers degrade by hydrolysis. Natural polymers have less mechanical qualities than synthetic polymers and ceramics, making them unsuitable for osteochondral techniques that require instant load bearing capability. Additionally, concerns regarding immunogenic response, refining process, restricted supply, and variability between batches could constrain scalability of application in clinical settings for scaffolds established on natural polymers [9].

Carbohydrates encompass biopolymers such as arabinoxylan, starch and chitosan alongside proteins like keratin and gelatin as well as PHA and [P(3HB)]. Chitosan is a well-known member of the group known as polysaccharide carbohydrate family. Biosustainability, biodegradability and compatible are desired in many companies [11]. Although collagen offers superior biocompatibility by closely resembling the in vivo environment, its weak mechanical properties and potential for immunogenic responses pose limitations. Nonetheless, NP possess the benefit of integrating sites for cell recognition and adhesion, like the arginine-glycine-aspartate pattern, which was initially identified in natural polymers like fibrin and collagen [10].

4.3.3 Ceramics and Bio-ceramic

Ceramics based primarily on calcium phosphate, which include hydroxyapatite (HA) as well as Tricalcium phosphate (TCP) coatings, are applied onto metal implants so that enhance Bone fusion along with Bone growth facilitation. This is facilitated by the formation of a layer resembling bone apatite on the surface of the coating, which helps prevent implant loosening. While reducing requirement for further revision surgery [9]. Hydroxyapatite (HA) is a bio-ceramic extensively utilized as a supporting framework material in BTE due to excellent bio-compatibility and similarity of genuine material of bone. It is frequently coupled with biopolymers/bio-ceramics to improve binding reactions as well as mechanical characteristics during the additive manufacturing procedures. Beta-tricalcium phosphate (β-TCP) is well-suited for craniofacial Irregularities due to exceptional biodegradation, durability, ability to chemically bond with tissues of bones with varying environmental stressors. The primary challenge with β-TCP lies in maintaining its sintering temperature at 1100 °C. Beyond this temperature, β-TCP converts to α-TCP, which is capable of dissolution and undergoes chemical transformation less stable than β-TCP [12].

Bio-ceramics are often employed in the restoration and replacement of wounded tissues due to several benefits, including exact chemical composition, which plays an important contribution to the integration of soft and hard tissues. Zirconia ceramics outperform alumina ceramics in terms of fracture toughness and flexural strength. Using ceramics, instead of typical metals and polymer derivatives results in decreased rate of wear at articulated surfaces [13]. To enhance the biocompatibility and biomimetic properties of scaffolds, these polymers are commonly mixed in solid form with other polymers to customize breakdown with physical characteristics. Moreover, these modifications, are incorporating HyA, ceramics particularly hydroxyapatite have shown to enhance the bioactivity of several polymers, enabling more efficient utilization in tissue engineering [10]. In bone and even osteochondral tissue engineering ceramics made of bioactive agents are utilized, because of similar composition bone mineral and enhanced biomineralization features [9].

4.4 Techniques of Additive Manufacturing

4.4.1 Stereolithography

The technique of stereolithography (SLA), which employs UV light, is one of several methods for creating 3D structures. This approach is the earliest and still extensively used. Charles Hull patented this technique in 1986 as shown in Fig. 4.1 [6, 14–17]. These devices use a radiation source. It is a multilayer technique that employs selected photoinitiated cured low-molecular-weight additives, prepolymer and photoinitiators. The liquid photopolymer can be selectively solidified using either focused UV beam light or mask based irradiation. Either concentrated UV beam light/mask based irradiation. Stereolithography-based bioprinting has a high build velocity and precision to make it ideal for organ 3D bioprinting applications [14, 17, 18]. Hull's proposal for modern stereolithography sparked four generations of technological improvement and overview is shown in Fig. 4.2. Researchers have used numerous technologies to create physical systems that improve stereolithography performance.

Stereolithography boosts several advantages. It excels in achieving high accuracy and intricate detail in final products, while also delivering a smooth surface

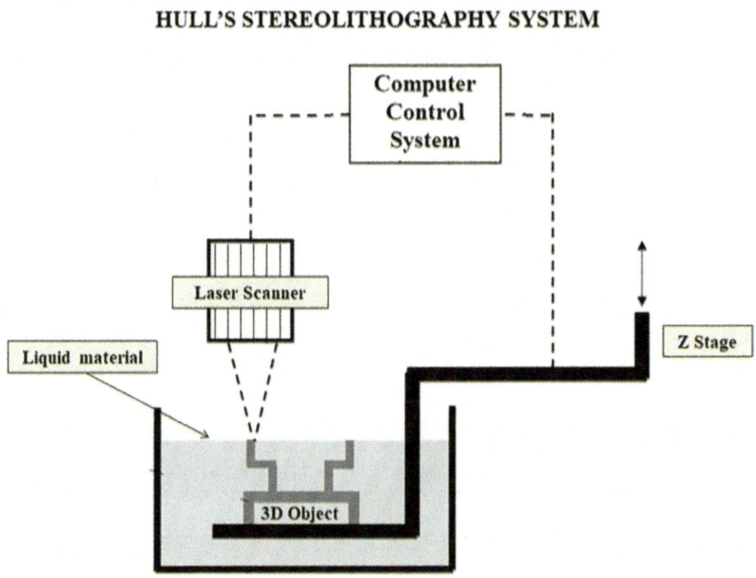

Fig. 4.1 Hull's stereolithography system. (Reproduce with permission from [14] under CCBY 3.0)

finish that minimizes the necessity for extensive post-processing. Its versatility extends to crafting intricate and complex geometries with ease. Moreover, it supports a broad spectrum of materials suitable for various applications, offering flexibility in material selection. Additionally, rapid prototyping capabilities allow for swift design iteration, facilitating efficient development and refinement of prototypes. Among the drawbacks they are expensive of device, harmful effects of lights and photoinitiators. Stereolithography comes with its share of drawbacks. It faces limitations when it comes to producing large-sized objects, often necessitating post-processing to eliminate excess resin and attain the desired aesthetics. Material options are comparatively restricted compared to alternative 3D printing techniques. Additionally, managing liquid resins can be messy and demands safety precautions. Moreover, the cost of stereolithography machines tends to be relatively high, adding to the investment required for utilizing this technology [15].

Stereolithography finds diverse applications across various industries. It is also utilized to generate molds for the fabrication of implants for cranial surgeries utilizing data from CT imaging [19]. It is extensively utilized in prototyping and product development, as well as for creating medical and dental models. The process is also employed in crafting custom jewellery and fashion accessories, architectural models, art pieces, and sculptures. Moreover, it serves in rapid tooling and manufacturing aids, educational models, and the production of engineering and aerospace components, along with consumer electronics parts [17].

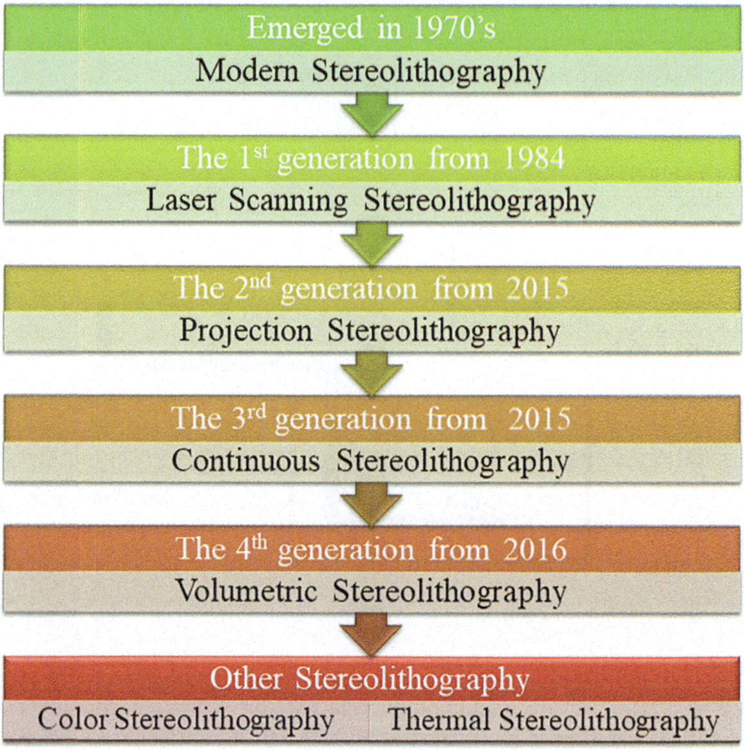

Fig. 4.2 Overview of stereolithography system. (Reproduce with permission from [14] under CCBY 3.0)

4.4.2 Fused Deposition Modelling

Fused deposition modelling (FDM), i.e. recognized by material extrusion additive manufacturing, and utilizes polymers in the form of filament as its raw material. Typically, filament undergoes heating, so that it reaches a "State of liquefaction" before it is pressed out by the machine's spout. The spout head is capable of movement in 3 DoF that accurately position the pressed out polymer onto building surface based on the instructions encoded in the G-code. In FDM the filament is continuously fed through the machine's extruder and nozzle via 2 rollers rotating opposing ways. This is then layered onto the building surface successively till its required form and size of the product is obtained [20]. It is a widely adopted additive manufacturing method for producing prototypes and functional components using fundamental engineering polymers [20, 21].

The procedure involves heating plastic filaments as feedstock and extruding them by a spout tip to form layer's onto a platform, enabling the direct construction of items layered one by one from a computer model [22]. It is a highly dynamic yet cost-effective approach for a variety of manufacturing phases: from prototyping through mass production. FDM is a cutting-edge manufacturing method for thermoelectric materials [23]. RPT has the unique capacity to precisely fit prosthesis to a patient based on his particular proportions. Implants for many bodily parts, including hips, knees, and spines, have been proposed. RPT can create physical models of atomic structures, which may aid in diagnosis, surgical planning, and manufacturing personalized implants [24].

4.4.3 Selective Laser Sintering

Selective laser sintering is (SLS) lasers-based technique for AM that can quickly print both metallic and non-metallic in three dimensions physical objects layer-by-layer straight from CAD models [25]. Industrial 3D printing technology utilizes a powder bed [PB] for construction of the 3D object. Thus, instead employing a spray solution, SLS employs a laser beam to fuse the powdered particles' together. Throughout process of printing, the laser is guided to trace a precise pattern onto the surface of the powder bed as shown Fig. 4.3. Upon completion of the first layer, a roller evenly distributes a fresh layer of powder atop the preceding one. The object is systematically built one by one layer, and retrieved by PB. An industrial three-dimensional printing technology like PB employs a powder bed to fabricate the 3-D

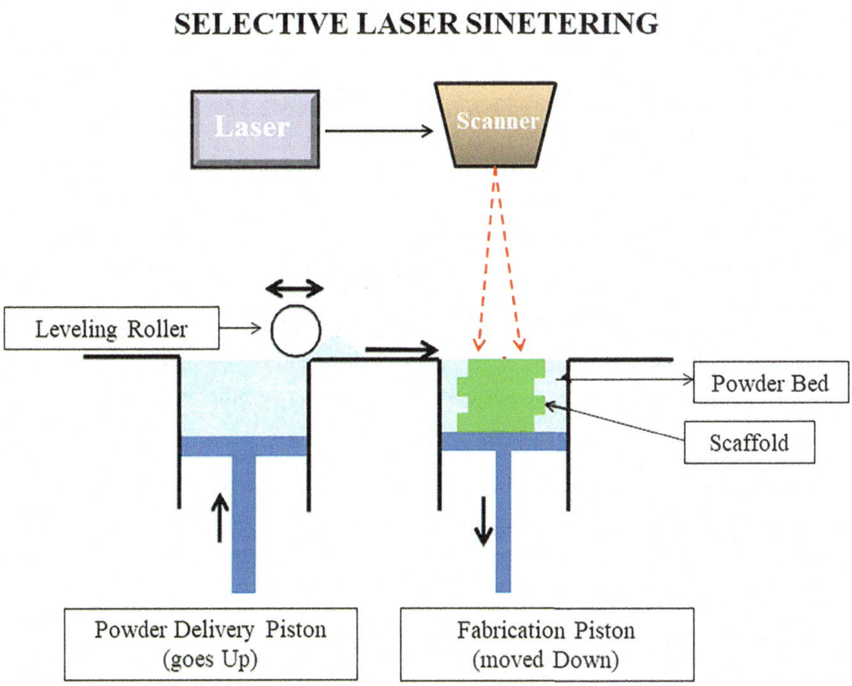

Fig. 4.3 Schematic illustration of selective laser sintering. (Reproduce with permission from [27] under CCYB 3.0)

object; SLS utilizes laser to fuse the powder particles instead of relying on a spray solution [26].

However, the primary materials in the form that are used in this method are of polymers, metals, ceramics, alloys, which necessitate high-temperature sintering along with high-energy lasers. Due to such demanding printing conditions, they have impeded the adopting such manufacturing method within p'ceuticals industry. It is widely acknowledged as high-energy input of laser can deteriorate in quality, value, or conditions of p'ceuticals industry if utilized as base materials. Consequently, SLS printing has mainly been employed in the healthcare to fabricate scaffolds for TE or medication administration devices, with the drug added post-printing. Currently, there have been no reported studies on Formulations containing drugs and prepared using SLS [25, 28].

4.4.4 Inkjet Bio-printing

Inkjet-based bio-printers, also referred as droplet-based bioprinters [29]. Hewlett-Packard Company invented inkjet printing as a two-dimensional printing process in the 1970s. It wasn't until 2003 that the first patent for a modified inkjet printer for bioprinting live cells was granted. This technology was originally used for 3D printing and it was one of the early AM technologies [30]. It offers precise control over pattern size and droplet deposition through computer-controlled technology and also employs individual droplets to imprint a substrate, akin to desktop inkjet printers IP enables the controlled positioning of biological and synthetic substances like proteins, enzymes, and cells, which can be costly [31, 32]. It employs thermal, piezoelectric/microvalve techniques to dispense droplets of diluted liquids without direct contact. Thermal and piezoelectric materials are commonly used to fabricate structures. Bioprinting with live systems often uses prepolymer, which may or may not contain cells. Thermal inkjet printers expels droplets through spouts by generating elevated pressure pocket of trapped air through heating, reaching temperatures of up to 300 °C [30, 33].

Thermal inkjet printing method utilizes a heating component to spray drops. The element quickly heats the surrounding ink, causing bubbles to form in the pressure cavity. Gasification pressure causes certain droplets to conquer surface tension and exit the extrusion nozzle. When the voltage is turned off, the bioink immediately cools and returns into the pressure chambers. Droplets can be squeezed continually through applying and releasing voltage. Piezoelectric inkjet printing method utilizes piezoelectric substances to spray droplets. Applying electricity to both ends of a piezoelectric device causes it to flex and squeeze out droplets near the nozzle. When the voltage is withdrawn, the piezoelectric element recovers [29].

It finds extensive use in bio-fabrication and biomedical applications. Scientists have utilized it to print human mesenchymal stem cells (MSCs) within a PEGDMA scaffold containing BG and HA polymer. The bio-printed scaffolds distributed cells evenly and mitigated the accumulation of cells caused by gravity, a challenge faced by other scaffold fabrication methods. Thermal inkjet bioprinting stands out as an efficient and cost-effective printing technique, although it still faces various hurdles. It proves particularly suitable for bioprinting applications [30].

4.5 Advanced Techniques and Innovations

4.5.1 Multi-material Printing

Multi-material additive manufacturing (MMAM) may enhance the usefulness and complexity of 3-D printed constituents can be diversified for incorporating multi-materials or various blends of the same material. MMAM provides potential to designing sophisticated, functional, personalized, and high-value goods with enhanced features. Combinations of materials can be produced using an AM technique to increase characteristics. By integrating materials at different scales, components can have customized optical, electrical, mechanical, thermal, and multifunctional capabilities. MMAM can be crafted from a diverse array of materials. The materials being investigated fall into either the similar category (for instance polymer-polymer)/2 distinct types (for instance, polymer-metal). The materials under study belong to either the similar group (for instance, poly-poly) or 2 different groups (e.g., poly-metal). According to the reviewed literary works, materials such as polymers and metals have been examined. Polymers like PLA, PEEK, ABS, as well as PET have been merely studied for their power, stability, also utility [34].

Equipment limitations in multi-material 3D printing primarily involve the decrease in the available printing space and the adjustment of extrusion heads calibration. Accommodating extra extrusion heads reduces build space, particularly in direct drive systems, which are ideal for flexible filaments. This limitation remains unavoidable without altering the printer frame. Some kits use the Bowden extrusion variant to save space and weight, but it's often incompatible with flexible materials. Maintaining precise alignment of multiple extrusion heads is another challenge, requiring time-consuming adjustments. Switching between extrusion heads creates boundary interfaces, disrupting printing and impacting mechanical performance. Part design must address this by overlaying materials at boundaries. Controlling filament residence time in the hot end is difficult in multi-material printing, leading to polymer drooling. Minimizing this requires temperature control, filament selection, and retracting filament when inactive. Chemical composition differences between materials can cause thermal expansion discrepancies, resulting in part distortion. Careful material selection, preferably with a common base material, is essential to mitigate these issues [35].

4.5.2 Vascularization Strategies

Vascular Endothelial Growth Factor [VEGF] initiates vascularization. Vascularization has a vital function in both ossification and subsequent bone development and remodelling processes. Given the significance in bone healing as well as bone formation is of vascularization, and is a evident It's essential to consider this aspect in bone-tissue-engineering [36]. Angiogenic growth factors help creates new blood vessels from an existing vascular network. Vasculogenesis, on the other hand, occurs when endothelial progenitor cells produce new blood vessels from the ground up. Angiogenic growth factors are often mixed along factors of osteogenic that improve bone-tissue development used in BTE. Advantages include technological simplicity, the capability to merge angiogenic and osteogenic growth factors, along with a broad selection of commercially accessible angiogenic growth factors, presents advantages. However, challenges include the necessity to fine-tune the drug release mechanism, the limited half-life of certain angiogenic growth agents, and the time lag in achieving sufficient blood supply post-implantation [36].

Cell-based approaches include co-culturing endothelial cells along with the desired cell type and prevascularization in both vivo and in vitro settings. While these approaches can create functional vasculature, concerns remain about cell culture time and anastomosis with host vessels [37]. Bio-fabrication of vessel networks approaches have Endothelial cells are actually bioprinted at high density using bioinks like fibrin or collagen, forming lines, branched structures similar to genuine blood arteries. Microchannels are bioprinted using sacrificial fugitive bioinks. A surgical strategy for vascularization requires time for the substituted tissue to be vascularized via the host is a significant concern with all of the techniques mentioned above. To generate stable structures that integrates reliably within the host tissue, this process is essential for supplying nutrition to these cells through neovascularization. Angiogenesis is an event in which neovascularization occurs. Vasculogenesis occurs when host vessels multiply and penetrate the implant, indicating the development of vessels within it. It is technically difficult [36].

4.5.3 4D Printing in Tissue Engineering

4D printing is creative manufacturing technique which integrates 3-D printing as an additional dimension: time. This enables the desired changes in shape, size, and functionality. The concept was initially introduced by Tibett at Massachusetts Institute of Technology (MIT). It is a dynamic technology that uses biomaterials containing or not containing cells to create complicated tissue engineering constructions. 4D bioprinting differs from 3D bioprinting in terms of the material and stimulus employed. The application of stimuli provides degrees of control the process, with materials are capable to respond to 1 or more stimuli are referred to as stimuli-responsive material [38]. 4D bio-fabrication is able to carry out 3 approaches: scaffold manufacturing; scaffold production; and scaffold production concurrently by cell seeding-containing material, and then followed by material transformation. Hence, it's crucial to consider the interaction of live cells with the material, stimulation, and/or material alteration. Materials utilized in bio-fabrication should be non-toxic, bio-compatible, and conducive to cell attachment along with proliferation. Cells are then seeded first to materials trans-formation, it's imperative that stimuli and transformations don't affect vitality or cell-type properties. In the last principle, bioprinting manufacturing should be compatible with cell vitality. Consequently, cell-traction forces depend on cell phenotype, adhesion, density, necessitating adjustment that controls conformations within structures those are printed [39].

The implications of 4-D printing in BTE are: (1) Injectable thermo-sensitive hydrogels for 4-D bone tissue re-generation: These hydrogels can inject to uneven defective regions and turn into gel-like substance at the temperature of the body. (2) Shape-transformation-based 4-dimensional printing for bone tissue: Shape memory scaffolds may change size to fill holes, allowing for tailored repairing in bone-defect. (3) 4-D printing for bone-tissue in a bio-mimetic microenvironment: By using changed topologies, a 4-D printed bio-mimetic scaffold may facilitate the development of newly made bone tissue functionally and increase stem cell osteogenesis, thus boosting the synthesis of new bone tissue [40].

4.5.4 Integration of Bioactive Agents and Growth Factors

Commonly growth-factors utilized in tissue-engineering consist of BMPs, IGF-1, bFGF, VEGF, TGF-β, are critical in bone-tissue engineering. Additionally, certain cytokines, which are proteins released by cells to regulate the actions of other cells, can also function as growth factors (GFs). They play crucial role in bio-active coatings, because these are responsible for osteo-conduction and osseo-integration. Bioactive coatings govern the release of medicinal medicines. Encapsulating GFs prevents burst discharge [41]. Administering different growth factors concurrently amplifies the inherent bone-healing mechanism. This approach can retain enough bioactive components over the interval required To stimulate osteogenesis, using minimal doses of growth factors is preferred to avoid side effects associated with excessive doses [42]. GFs are employed in many therapies to regenerate tissues, including musculoskeletal, neuronal, hepatic, vascular, and hybrid systems. In tissue engineering, the goal is to repair tissues or restore their function alongside regenerative medicines, it's necessary to recreate the microenvironments present in the natural extracellular matrix serve as the benchmark for tissue regeneration. Exogenous bioactive signals can aid in mimicking these conditions. Transporting growth factors to biological compartments is facilitated through carriers and drug delivery systems. is a potential option for improving bioavailability, reducing administration frequency, and expanding effective routes of administration [43]. Appropriate bioactive signals, which include inorganic minerals, growth factors, and medicines, can be coated onto the surface or disseminated throughout a biomaterial to give osteoconductive/osteoinductive indicators that can increase bone formation and its integration with the surrounding tissue of the host [44].

4.6 Applications: Additive Manufacturing in Tissue Engineering

4.6.1 Bone Tissue Engineering

4.6.1.1 Customized Implants

A tailored implant method is devised to reduce bone loss while maximizing fit and "usability". The procedure combines CT, image processing, stereolithography, FEA, and investment casting to design and fabricate a close-to-ideal match among the sufferer bones and component of the implant. Solid models allow for implant customization using various design tools and patient data. The data needed for implant design varies, but typically includes patient age, weight, activity, and other factors. Specific formulations and designs get selected for achieving a near-perfect fit with minimal removal of bone, taking into account the patient's bone density and natural bone design. Traditional and unorthodox design formulas are applied to Customized implants significantly surpass "standard" implants due to their tailored design and robust construction reduces need for natural bone removal [45]. Customizing implants based on CT data is a practical and interesting approach. Creating physical models from scanned data has the potential to revolutionize medical procedures [45]. Standard and customized knee implants are presented in Fig. 4.4.

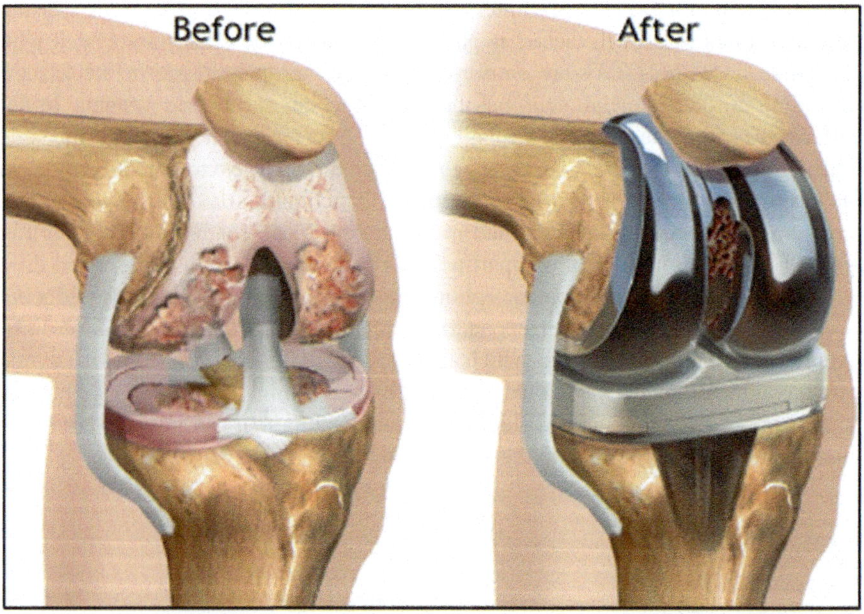

Fig. 4.4 Standard and customized knee implants. (Reproduce with permission from [46] CCYB)

4.6.1.2 Osteoconductive Scaffolds

Scaffolds usually constructed by polymeric bio-materials, which offers structural support for cell adhesion as well as subsequent tissue development. Majorly 4 scaffolding techniques for tissue engineering has emerged. These scaffolding approaches are useful guides for tissue engineering for complicated tissues like IVD and can be utilized in combination [47]. Natural polymeric scaffolds mimic the ECM consists of proteins, polysaccharides, and polynucleotides, are considered most similar to original tissue as they show osteoinductive properties. However, despite their good compressive and bending strength, CPCs, including HA, have lower fracture toughness and tensile strength compared to natural bone [48]. Furthermore, tissue-specific concerns such as the level of injured, distinctive structural as well as functional connection, different tissues compositions, and connections in IVD requires specialized observation [47]. 3D scaffolds have the capability to guide the differentiation of a particular progenitor cell population into a specific lineage, either through the wholesale integration of soluble substances inside the scaffolds or through exogenous introduction of chemicals, hormones, other growth factors within the culture medium. Metal scaffolds, such as titanium, are biocompatible and ideal for hard-tissue applications, including as the development for rat dental pulp progenitor cells towards odontoblast cells [49]. The scaffold promotes the development of cells that have been sown within its porous structure or have migrated from surrounding tissue [50].

4.6.2 Cartilage Tissue Engineering

4.6.2.1 Bio-printed Constructs

Cartilage, a sleek tissue characterized by low chondrocyte density and no blood vessels or nerves, encases ends within the joint bones. The 3-D bio-printing of scaffolds containing live cells for potential engineering of cartilage tissue presents a fascinating avenue for exploration. The development of bioinks plays a crucial role in enabling precise and efficient cell bioprinting. Bioinks, which are cell preparations compatible with automated biomanufacturing techniques, typically comprise hydrogel formulations. They too integrate cell-free biomaterial inks, such as exosomes, which show potential for enhancing tissue repair. Exosome-loaded 3D materials hold promise as therapeutic products devoid of cells for tissue repair and can be engineered for targeted drug delivery in cartilage repair. While traditional bioprinting involves layer-by-layer deposition of bioinks, emerging technologies inject bioinks into a suspension while printing, offering opportunities to tackle biological challenges beyond medicine. However, challenges persist in designing Developing appropriate bioinks and intricate tissue fabrication, particularly in preserving cell viability and ensuring structural integrity during printing. The primary challenge ahead lies in developing biomimetic cartilage structures capable of inducing dependent on specific regions cartilage specialization with extracellular matrix accumulation. Future progressions in 3-D printing technology, including Cartilage tissue containing MSCs with both dual-factor gradient and release structure, hold promise for enhancing regeneration and repair of cartilage within the body [51].

4.6.2.2 Hydrogels for Cartilage Regeneration

Hydrogels are the most popular bio-inks due to its hydrophilic chains with high absorbability of water, customizable topology's, designable sizes, and biological property serve as versatile tools for drugs and cells transportation in TE along with drug delivery. Bio-printing hydrogels can be categorized into two varieties: hydrogels based on natural polymers and those based on synthetic polymers. Natural polymers such as gelatin, HyA, cellulose, silk, alginate, starch, chitosan and fibroin are crucial components for Generating ECM: conditions for cells, are commonly utilized in bio-printing hydrogels for tissue regeneration. Synthetic polymers must generally be combined with natural polymers in order to adjust degradation also enhance biological properties through appropriate physical-chemical processes for bioprinting applications [52]. In this field, PEG hydrogels and GelMA: gelatin methacrylate combined with HAMA: hyaluronic acid methacrylate has employed, similar to other hydrogel-based ideas [53].

4.6.3 Skin Tissue Engineering

4.6.3.1 Cell-Seeded Scaffolds

Scaffold cell-seeding represents a crucial step in standardizing and uniformly maturing tissue structures. This is especially important into the field of additively built frameworks, because excessive porosity and the size of the pores frequently restrict preservation of the dosing a solution is resulting in suboptimal seeding efficacy and asymmetrical cell scattering. For overcome such challenge, this chapter outlines a straightforward yet efficient cell seeding method involving preincubation of the scaffold in 100% serum for 1 h, resulting in consistent seeding outcomes. We demonstrate an example of concept using extremely permeable molten electro written polycaprolactone scaffolds as a cellular transporter. To validate seeding and provide quality control of the cellularized construct, assays such as the pigment alizarin red, Sirius red, and the immunostaining are recommended due to the significance of cell count, dissemination, and differentiation inside the structure of the scaffold [54].

Proper positioning of a cell in a 3D scaffold is crucial for angiogenesis. Proper cell positioning in a 3D scaffold is crucial for angiogenesis, after selecting the appropriate cell source. After the precise selection and implantation of designated cells in their designated sites, guiding their differentiation becomes another significant hurdle. This process involves utilizing a range of factors, including biological components like growth factors, as well as mechanical and electrical elements such as scaffold stiffness, durability, and even gravitational force application. Hence, crafting an ideal scaffold/hydrogel integrated with the correct cellular source is critical. To support implanted cells for strong adhesion, development, and differentiation, precise physical and mechanical stimuli must be modelled. Regulating cell movement and expansion, especially for the endothelial cells, pericytes, and even macrophages, is pivotal for fostering blood vessel development within engineered tissues. Mechanical variables play a crucial role in determining cell destiny and commitment. Therefore, they must be thoroughly studied and linked to angiogenesis [53].

4.6.3.2 3D-Printed Dressings and Patches

In the exploration of wound care, 3D-printed multicomponent biocomposite materials have been utilized to create hydrogel wound dressings. These dressings are designed to carry specific compounds such as small molecules, proteins and metal nanoparticles, allowing for CR at the wound location. Specifically, hydrogel wound dressings embedded with silver nano-particles with anti-bacterial properties and the growth factor that promotes the development of blood vessels i.e. VEGF were utilized to examine how wounds in a mouse model responded physiologically. By comparing them with control dressings, researchers observed improvements in the formation of granulation tissue and different levels of density of vascular, which were influenced by the growth factor's release pattern [55].

Electric, pH, Temperature, light intensity, magnetic fields, biological molecules are all examples of stimuli that smart hydrogels can respond to. Smart hydrogels are composed of intelligent cross-linked networks capable of reorganizing themselves from a collapsed to an expanded state. Additionally, they feature intelligent surfaces that alter their hydrophilicity in response to specific stimuli. In certain instances, Smart macromolecules, whether in linear or solubilized form, undergo a transition from a single-phase to a two-phase state, leading to the creation of reversible sol-gel hydrogels. Both Digital Light Processing (DLP) and SLA offer advantages in production for such wound dressings. Wound dressings fabricated through SLA/DLP processes via 3D printing can incorporate Nanoparticles with antibacterial properties and antibiotic agent. This stage of printing is meticulously regulated to fabricate the required component. The DLP technique achieves this by printing successive layers, incrementally raising the platform to match the desired part thickness [56]. FITC-peptides were utilized as markers in peptide release assays. FITC-peptide was inserted onto lyophilized thin filament patches via chemical coupling. Fluorescence microscopy was used to examine peptide release at various time periods after incubating in the experiments; 3-D-printed patches' containing peptides of FITC and are immersed in hundred ml of PBS at 37 °C. Through chemical coupling, these patches demonstrated significantly enhanced release of drug efficiency comparing with Absorption through physical methods. As a result, miniature patches of filaments are employed in subsequent trials [57].

4.7 Case Studies

4.7.1 Patient-Specific Bone Implants

All models were made using the Arburg freeformer 200-3X and the Additive Powder Fusion.

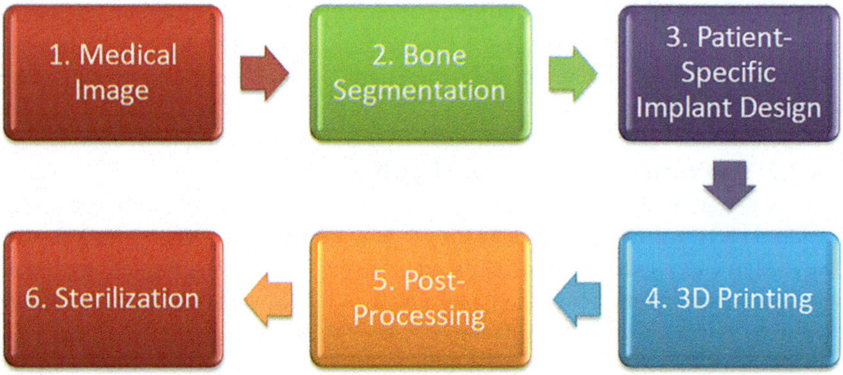

Fig. 4.5 General flowchart of patient-specific design

The patient-specific creation and manufacturing process for mandibular-alveolar bone re-generation starts with imaging data of the patient by CT. Next, important anatomical components are segmented from the data. Using the sorted data, a Patient-Specific Implant (PSI) developed. This PSI is furtherly manufactured in 3D using medical-grade raw material pellets. Subsequently, removal of water-soluble support-structures was made by using an ultrasonic bath. Finally, packaging and sterilization of implant is done [58] (Fig. 4.5).

Each implant design was based on patient CT scans, processed digitally after being saved in DICOM format. Materialize Mimics software extracted bone structures from these Scans are conducted for 3D printing. The quality of the substrate prior to the optimization of surface modification techniques are through cytotoxicity, non-toxicity study, as well as microstructural analysis. Surface modification explored two methods: functional polymeric granules to enhance implant properties, and a biocompatible coating with polymeric microspheres and a hydrogel carrier for prolonged growth factor release. Results showed β-TCP-containing granules improved bone cell growth, but further porosity optimization may be needed. The biocompatible coating, still in development, showed promise in biocompatibility and growth factor release. A prototype mandibular implant was produced, but further testing is needed to extend growth factor release, including from microspheres within the PVP coating [59].

4.7.2 Bio-printed Cartilage Constructs

Cellulose The polymer with the highest abundance found on Earth, is a polysaccharide comprised of units of D-glucopyranose linked by β-1,4 glycosidic bonds. Its physical properties are dictated by the existence of 3-hydroxyl groups (–OH) at positions [C-2], [C-3], [C-6]. Extracted from sources such as cotton, wood, algae, hemp, potato tubers, cellulose-based materials, including nanocellulose (NC), maybe categorized in cellulose nano-crystals (CNC) and nano-fibrillated cellulose

(NFC). NC, especially NFC, exhibits promising properties for applications in 3D bioprinting and tissue engineering with Non-Newtonian shear thinning behavior and high mechanical strength. Despite its resistance to degradation in the body, functionalization with cellulase enzymes can aid in breaking down β-1,4 linkages. NC blends have been utilized successfully in bioprinting cartilage tissue, offering improved mechanical properties, stability, and structural integrity.

A bioink composed of NFC and ALG, successfully 3D bioprinting anatomically-shaped cartilage models like an ear of human and a meniscus of sheep. ALG/NFC with the ratio of $30/70_{w/w}$ demonstrated the greatest resistance to compression. However, high ALG concentrations compromised mechanical properties and cell viability due to poor diffusion. By printing an open-porosity auricular construct with NFC/ALG, inducing redifferentiation of human chondrocytes. Henriksson et al. developed CELLINK-H, a bioink combining NC and HA, showing advantages in adipose tissue engineering over 2D culture systems. CELLINK® bioink demonstrated excellent structural integrity and promoted cartilage synthesis in vivo. NFC/ALG bioink utilized to bioprint human cartilage in mice, exhibiting favorable proliferation and ECM deposition after a 60-day period [60].

4.7.3 3D-Printed Skin Substitutes

In 2009, the first study on skin bioprinting used actual human dermal fibroblasts and a collagen hydrogel [61]. In 2009, Lee and colleagues explained a customized or unrestricted in form for fabrication method to create the multi-layered skin-substitute using type-I collagen sourced from tails of rats, along with human fibroblasts and human keratinocytes. Although only evaluated in vitro, the study found that cell proliferation was effective on both planar and nonplanar surfaces, supporting the utilization of 3D printing for on-demand manufacturing of skin transplants has been explored. Binder and colleagues were pioneers in testing 3-D ink-jet-Printed fibroblasts from human and keratin cells on wounds of full thickness in mice with athymic disease. Their investigation established the viability of healing wounds, exhibiting improvements outcomes compared to untreated allogeneic implant controls. The development of reliable bioinks is paramount for bioprinted tissue engineering, alongside advancements in fabrication technologies [62]. While numerous available products in the market and in clin-trials, achieving optimal skin growth requires replication of all skin traits and functions. Challenges such as scaling up production, creating a distinct junction between outermost and innermost skin appendages, and attaining mechanical properties similar to natural skin remain to be addressed [63].

Skin bioprinting focuses on advanced materials are crucial for accurate printing and accelerated wound healing. Alongside the biofunctionality and mechanical characteristics of bioink, such as viscoelasticity, the printing technology significantly influences the macrostructure of the resulting construct through engineering principles. 3D bioprinting for skin regeneration, covering technology, bioink composition, and cell-laden constructions are studied. Researchers have added keratinocytes, fibroblasts, and adding stromal cells from stem cells to bio ink enhances regenerative medicine efforts. These results in cell-laden device, once printed, can be applied to a wound either immediately or after maturation. Researchers have devised hybrid coculture-system on 3-D layer structures to replicate the double layered epidermal/dermal structure of skin that is normal [61].

4.8 Challenges and Limitations

4.8.1 Biocompatibility and Material Selection

Biocompatibility is a complex process influenced by a variety of parameters such as biofunctionality, bioinertia, bioactivity, and biostability [64]. Biocompatible materials are the primary design criteria for any scaffold. Initially, biocompatibility focused on long-term implanted materials that were minimally reactive to biological systems [9]. Compatibility with cells, immunological response, prevention of thrombosis, and site-specific properties tailored to the recipient's implantation site are all important aspects of a scaffold's biocompatibility. Biocompatible scaffolds facilitate bone cell adhesion, infiltration, and healthy proliferation while mitigating toxicity and immunological reactions in the in vivo system, implanted materials interact with the host response by forming layers on their surface, with macrophages located internally, connective tissue with fibroblasts externally [64]. Identifying acceptable binders for different product kinds is a key task. Poor mechanical qualities, size distribution, limited range of materials, texture, and color, etc. [65].

In the Indian market, numerous producers of varying quality use identical polymer raw material or filament. Using the MCDM method to optimize a polymer raw material based on experimental results from different manufacturers for the same filament is a novel approach. Relying solely on online ratings to select the optimal polymer raw material can be challenging. Polymers are available in various forms, including powders, granules, filaments, or resins, adding complexity in the process of making decisions. Material extrusion contains the use of thermoplastic polymers. Fusibility occurs when heated. It is innovative to utilize MCDM approaches to select the best polymers for the user from polymers with comparable applications [66]. Bioactive ceramics exhibit favorable interacting with the tissue of tissue to support integration and re-generation. Due to the chemical and mechanical qualities, they are only effective for bone regeneration [67].

4.8.2 Mechanical Properties of AM Construct

Biomaterials may undergo mechanical degradation throughout their production, storage, and utilization phases. Environmental factors throughout the printing procedure can induce resulting from tension, shear forces and compression. Addressing the challenge involves matching the properties which are mechanical for synthetic materials of natural tissues within the human body. Optimal functioning of support systems, whether in engineering structures or bone within the human body, hinges on possessing suitable mechanical characteristics. Tensile, strain, compressive stresses, bending moment, torque, shear force, with other mechanical properties significantly influence the suitability of materials for various applications [68].

4.8.3 Vascularization and Tissue Integration

Vascularization has an important role in both ossification and subsequent bone growth and remodelling processes. Given the crucial role of vascularization throughout bone healing and formation, it is evident that must be looked in (BTE: bone tissue engineering). In practice of clinical, bone abnormalities maybe addressed through autologous/Bone grafts from a donor of the same species, Using bone from the patient's own body substitution being the preferred method. Bone defects of a size that exceeds the body's capacity for natural regeneration, which are relatively common, can arise from various causes such as trauma, malignancy, osteomyelitis, and non-uniform fractures can be potential causes of critical-size bone defects, which are quite prevalent [36].

Tissue engineering along with regenerative medicine have led to advancements in tissue healing, organ replacement, and the development of vitro tissue models. However, this has made a demand for vascularized-tissue-constructs. Long-term as well as size viability of these structures may influence through their vascular network functionality and complexity. In vitro, tissue-engineered-constructs face size limitations due to passive diffusion. Nevertheless, bio-reactors may facilitate Transporting nutrients and gases through dynamic culture a condition, addressing these constraints, allowing for longer-term survivability of tissue constructs. AM methods offered automation, reproducibility, scale-up, and standardization, but they can also be integrated with traditional biomanufacturing processes to create sophisticated, vascularized tissue constructions for clinical translation [9].

4.8.4 Regulatory and Clinical Translation Challenges

3D bioprinting poses distinct risks compared to standard 3D printing, necessitating unique policy concerns. The distinctions encompass philosophical and bioethical challenges, as well as practical risk, biosafety, and security concerns. Regulatory problems for 3D printed devices stem from automated fabrication processes. Fabrication processes can cause unique surface, edge, and interlayer flaws that must be addressed. Regulating digital processing, including computer-aided design and software systems, is essential for modern production. Some argue that existing laws are sufficient for 3D printed objects, minimizing the regulatory complexity connected with the manufacturing process. The process's new capabilities, such as personalization and decentralization, have the potential to destabilize regulations. Legal questions arise around liability for defective 3D-printed products, including who may be held responsible. Without precedent, it's unclear how legislation will address these problems [69]. Bioprinting technology can create vascular channels faster, but recreating a complex and multifunctional vascular network remains challenging and time-consuming. Additional options for improving vascularization include growth agents, The practice of co-culturing endothelial cells alongside various cell types, coupled with the utilization of biomaterials [70].

Despite decades of study in tissue engineering, just a few clinically licensed therapies are based on its principles. Tissue engineering for illness treatment has regulatory challenges, as well as biological and engineering design limits, limiting its widespread usage. Understanding regulatory constraints is crucial for scientists designing new therapeutics for clinical use, even if it is not often the primary priority when developing new technologies in the lab [71]. Successfully transitioning a 3D printed tissue scaffold from laboratory settings to clinical applications involves a range of factors, encompassing clinical and commercial considerations which impact the advancement regarding tissue engineered products that hinges on understanding and addressing various factors [9].

4.9 Discussion of Recent Advancements in 3DP

The on-going advancement of biomaterials is a continuous process aimed at creating safer, more dependable, cost-effective, and physiologically compatible substitutes for damaged or diseased human tissues. Individuals employed materials from their local habitats to replace tissues lost due to damage or sickness in the first generation of biomaterial development. The initial wave of biomaterial development aimed to create materials that were inert when introduced in the body. Resorbable polymers are another type of second-generation biomaterials. Currently, in the third era of biomaterial research, biomaterials are being developed that promote or hinder specific cell functions [72]. The commonly used bio-materials are utilized to simulate the matrix of cells in tissues engineering, biological medicine, organ-on-a-chip technologies, and systems for drug delivery. Bioceramics and metallic biomaterials are two frequently employed types in bioprinting. Biomaterials of Calcium phosphate like hydroxyl-apatite (HA), tri-calcium phosphate (TCP), and bio-glass 45S5, as well as bioactive glasses, exhibit great biocompatibility with closely resemble the makeup of bone, teeth. Consequently, they are often utilized to fabricate porous frameworks enabling tissue and bone restoration. Metallurgical Biomaterial, such as Titanium-based ceramics is typically manufactured using powder-bed fusion methods. Three promising types of biomaterials for next-generation additive manufacturing include bio-polymers can be used to print high-resolution tissue engineering scaffolds, hydro gels for cell the printing process, and multimaterial architectures [73].

Advancements in AM technologies have diversified materials, improved printing techniques for complex geometries, increased speed and efficiency, and enhanced scalability. Integrated design and simulation tools optimize parts for AM, while hybrid and multi-material printing enable varied functionalities. In healthcare, bioprinting offers potential for tissue engineering and personalized medical treatments. These advancements are driving innovation, efficiency, and new applications across industries, positioning AM as a cornerstone of future manufacturing. The field of healthcare, bioprinting technologies have advanced significantly, enabling the formation of living organs and tissues, through fabrication. These developments show potential for therapeutic regeneration and medication development, testing, and personalized medical treatments. Rapid tooling, lightweight construction, material economy, quicker lead time, and capacity to create complicated structures are significant advantages over traditional manufacturing procedures.

The ISO/ASTM 52900 standard classifies AM in 7 methods: powder-bed fusion, binder-jetting, material extrusion and jetting, energy-directed deposition, vat-polymerization, sheet-lamination. Among the widely utilized 3D printing technologies are FDM, LOM, SLA, SLM, and SLS [74, 75].

Nano-medicines (NNMs) aim to address inflammatory conditions by targeting their underlying biological mechanisms. However, the process of translating NNMs into clinical use is both expensive and time-consuming. Clinical development of NNMs involves addressing biological challenges, scaling up manufacturing, ensuring Biological compatibility and safety of Health, navigating IP: intellectual-property considerations, adhering to government rules, and assessing affordability compared to existing medications. One significant hurdle contributing to delays is the complexity of arranging collaboration and licensing agreements, which can be addressed to expedite progress. Moreover, the formulation's intricacy plays a crucial role in the clinical application of NNMs. Comprehensive toxicology studies is essential so verify the safety of NNMs for human use before they can be successfully translated into clinical practice [76]. Addressing the various attributes of bio-inks, including their origin, variability between batches, and tissue-specific properties, presents significant challenges that must be overcome to enable successful commercialization and clinical application [77].

Understanding how new nanomaterial's are absorbed, distributed, metabolized, and excreted in the body is crucial for predicting their toxicological effects. Special toxic study assessments within models of animals are necessary to evaluate both short- and long-term toxicity. New intellectual property (IP) policies and protocols required for simplification of the journey from innovation to commercialisation. For accelerate clinical trials of promising novel NNMs, consider establishing or coordinating Labs and institutions with experience in characterization of NNM algorithms platforms, conducting preclinical research for legislative filing, scaling up laboratory production of nanomaterials, and designing clinical trials [76].

The cluster focuses on the ethical concerns of animal testing. Animals may be used as cell sources or for tissue engineering research purposes. Laboratory animals are used for illness modelling, medicine and product development, and tissue engineering research. To validate the utility of translational animal models, experts emphasize the importance of considering anatomical, physiological, and functional factors. Experiments on animals may cause unexpected pain or incapacity. To limit injury to animals, researchers must utilize humane treatment methods. The question of identifying and owning autologous tissues and cells is a significant concern.

Questions raised include whether autologous means autonomous, if donors have rights to donated cells with altered biological features, and if we truly own our bodies and cells [78]. Moral and religious considerations may arise regarding. The application of decellularized tissues from humans: particularly cadaveric cells, as scaffold elements. As a result, synthetic scaffolding, animal-derived structures, and decellularized cells from human donors have become preferred possibilities. The problem with ownership and commodity creates difficulties regarding whoever can claim rights of ownership over donor and modified tissues, including the tissue donation, the person acquiring the transplant, and those involved with the donor production. Additionally, there is on-going discussion regarding the potential benefits of claiming intellectual property rights over TE, including patenting the technology and its products, and the conditions under which such rights are advantageous [79].

4.10 Conclusion

Advancements in Additive Manufacturing have revolutionized tissue engineering by enabling the creation of complex structures with precise material management. Techniques such as Stereolithography, Fused Deposition Modelling, and Inkjet Bioprinting offer flexibility in crafting personalized implants and scaffolds for regenerating bone, cartilage, and skin. Despite significant progress, challenges persist in ensuring biocompatibility, achieving mechanical strength, and promoting vascularization. Innovations like multi-material printing, vascularization strategies, and 4D printing hold promise for overcoming these obstacles.

The development of advanced bioinks, particularly those blending Nano Cellulose and Alginate, has profound implications for tissue engineering. These bioinks enhance the mechanical properties and cell viability in bioprinted tissues, allowing for high-precision, anatomically accurate constructs. This level of control supports the construction of complex tissues and improves cell differentiation and extracellular matrix deposition, making artificial tissues more similar to natural ones.

While current 4D printing technologies enable basic shape transformations, future advancements must incorporate dual or multiple shape-changing mechanisms to replicate more complex natural processes. Making bioprinters accessible in clinical settings and refining biomaterials for physiological conditions are crucial next steps. Optimized post-processing techniques, including crosslinking and sterilization, are necessary to ensure functional and mechanical performance. Close collaboration with clinicians and understanding regulatory requirements are essential for translating these innovations into practical therapeutic solutions. The ongoing research and technological progress hold the potential to significantly enhance regenerative therapies and improve patient outcomes.

References

1. Meyer U (2009) The history of tissue engineering and regenerative medicine in perspective. In: Fundamentals of tissue engineering and regenerative medicine. Springer
2. Vyas J, Raytthatha N, Singh S, Prajapati B (2024) Next-generation computational automation-based additive manufacturing of pharmaceuticals: an approach to fabricate precise medicine. In: Handbook of 3D printing in pharmaceutics. CRC Press, pp 153–175
3. Vyas J, Singh S, Shah I, Prajapati BG (2023) Potential applications and additive manufacturing technology-based considerations of mesoporous silica: a review. AAPS PharmSciTech 25(1):6
4. Shah SR, Modi CD, Singh S, Mori DD, Soniwala MM, Prajapati BG (2024) Recent advances in additive manufacturing of polycaprolactone-based scaffolds for tissue engineering applications: a comprehensive review. Regen Eng Transl Med 11(1):1–20
5. Vyas J, Shah I, Singh S, Prajapati BG (2023) Biomaterials-based additive manufacturing for customized bioengineering in management of otolaryngology: a comprehensive review. Front Bioeng Biotechnol 11:1234340
6. Rosen DW (2014) What are principles for design for additive manufacturing? Proc Int Conf Prog Addit Manuf 0(May 2024):85–90. https://doi.org/10.3850/978-981-09-0446-3_027
7. Singh S, Chunglok W (2022) Biopolymers towards green and sustainable development. Bentham Science Publishers. https://doi.org/10.2174/97898150793021220101
8. Liu F, Wang X (2020) Synthetic polymers for organ 3D printing. Polymers 12(8):1–34. https://doi.org/10.3390/polym12081765
9. Vyas C (2019) Development of a multi-material and multi-scale 3D bioprinted scaffold for osteochondral tissue engineering. The University of Manchester
10. Mohite P, Shah SR, Singh S, Rajput T, Munde S, Ade N et al (2023) Chitosan and chito-oligosaccharide: a versatile biopolymer with endless grafting possibilities for multifarious applications. Front Bioeng Biotechnol 11:1190879
11. Singh S, Nwabor OF, Syukri DM, Voravuthikunchai SP (2021) Chitosan-poly (vinyl alcohol) intelligent films fortified with anthocyanins isolated from *Clitoria ternatea* and *Carissa carandas* for monitoring beverage freshness. Int J Biol Macromol 182:1015–1025
12. Damiri F, Fatimi A, Musuc AM, Santos ACP, Paszkiewicz S, Idumah CI et al (2024) Nano-hydroxyapatite (nHAp) scaffolds for bone regeneration: preparation, characterization and biological applications. J Drug Deliv Sci Technol 95:105601
13. Huang J, SM (2022) Tissue engineering using ceramics and polymers, 3rd edn. Woodhead Publishing Series in Biomaterials
14. Huang J (2021) A review of stereolithography: processes and systems. Processes 8(September 2020):01–16. https://doi.org/10.3390/pr8091138
15. Soori M (2019) Applications of additive manufacturing to tissue engineering (April, 2019). [Online]. Available: https://www.researchgate.net/publication/332570568
16. Sharma S (2019) A review of recent developments on stereolithography. Int J Eng Res Technol 8(08):180–185
17. Lakkala P, Ram S, Bandari S, Repka M (2023) Additive manufacturing technologies with emphasis on stereolithography 3D printing in pharmaceutical and medical applications: a review. Int J Pharm X 5(December 2022):100159. https://doi.org/10.1016/j.ijpx.2023.100159
18. Joel AT, Najmon C, Raeisi S (2019) 2—review of additive manufacturing technologies and applications in the aerospace industry. In Additive manufacturing for the aerospace industry. Elsevier, pp 7–31. [Online]. Available: https://doi.org/10.1016/B978-0-12-814062-8.00002-9
19. Melchels FPW, Feijen J, Grijpma DW (2010) A review on stereolithography and its applications in biomedical engineering. Biomaterials 31(24):6121–6130. https://doi.org/10.1016/j.biomaterials.2010.04.050
20. Mwema FM, Akinlabi ET (2020) Basics of fused deposition modelling (FDM). In: Fused deposition modeling, SpringerBriefs in applied sciences and technology, pp 1–15. https://doi.org/10.1007/978-3-030-48259-6_1

21. Mogan J et al (2023) Fused deposition modelling of polymer composite: a progress. Polymers 15(1). https://doi.org/10.3390/polym15010028
22. Masood SH (2014) 10.04—advances in fused deposition modeling. Compr Mater Process:69–91. [Online]. Available: https://doi.org/10.1016/B978-0-08-096532-1.01002-5
23. Matthew Montez PLM, Willis K, Rendler H, Marshall C, Rubio E, Rajak DK, Rahman MH (2022) 5—fused deposition modeling (FDM): processes, material properties, and applications. In: Tribology of additively manufactured materials. fundamentals, modeling, and applications, Elsevier series on tribology and surface engineering, pp 137–163. [Online]. Available: https://doi.org/10.1016/B978-0-12-821328-5.00005-6
24. Singh R, Garg HK (2021) Fused deposition modeling—a state of art review and future applications. Elsevier Ltd. https://doi.org/10.1016/B978-0-12-803581-8.04037-6
25. Sathiya P, Varatharajulu M, Bhuvanesgh Kumar M (2021) Selective laser sintering. In: Advances in additive manufacturing processes, 2021st edn. Bentham Science Publishers, pp 28–47
26. Goel A, Bourell D (2009) Electrochemical infiltration of laser sintered preforms with metals. In: 20th annual international solid freeform fabrication symposium, SFF 2009, January 2009, pp 332–341
27. Rider P, Kačarević ŽP, Alkildani S, Retnasingh S, Schnettler R, Barbeck M (2018) Additive manufacturing for guided bone regeneration: a perspective for alveolar ridge augmentation. Int J Mol Sci 19(11):3308. https://doi.org/10.3390/ijms19113308
28. Fabrizio Fina AWB, Goyanes A, Gaisford S (2017) Selective laser sintering (SLS) 3D printing of medicines. Int J Pharm 529(1–2):285–293
29. Salve RK (2020) An overview of inkjet bioprinting. Int J Futur Gener Commun Netw 13(2):651–656
30. Huang Y, Zhang X, Gao G, Yonezawa T, Cui X (2017) 3D bioprinting and the current applications in tissue engineering. Biotechnol J 17:1–16. https://doi.org/10.1002/biot.201600734
31. Rider P, Peri Ž, Alkildani S, Retnasingh S, Barbeck M (2018) Bioprinting of tissue engineering scaffolds. J Tissue Eng 9:1–16. https://doi.org/10.1177/2041731418802090
32. Saunders RE, Derby B (2014) Inkjet printing biomaterials for tissue engineering: bioprinting. Int Mater Rev 59(8):430–448. https://doi.org/10.1179/1743280414Y.0000000040
33. Wu CA, Zhu Y, Woo YJ (2023) Advances in 3D bioprinting: techniques, applications, and future directions for cardiac tissue engineering. Bioengineering 10(7). https://doi.org/10.3390/bioengineering10070842
34. Nazir A, Gokcekaya O, Masum K, Ertugrul O, Jiang J (2023) Multi-material additive manufacturing: a systematic review of design, properties, applications, challenges, and 3D printing of materials and cellular metamaterials materials & design multi-material additive manufacturing: a systematic review of desi. Mater Des 226(February):111661. https://doi.org/10.1016/j.matdes.2023.111661
35. Lopes LR, Silva AF, Carneiro OS (2018) Multi-material 3D printing: the relevance of materials affinity on the boundary interface performance. Addit Manuf 23(March):45–52. https://doi.org/10.1016/j.addma.2018.06.027
36. Simunovic F, Finkenzeller G (2021) Vascularization strategies in bone tissue engineering. Cells 10(7):1–17. https://doi.org/10.3390/cells10071749
37. Lovett M et al (2014) Vascularization strategies for tissue engineering. Tissue Eng Part B Rev 15(August):353–370. https://doi.org/10.1089/ten.TEB.2009.0085
38. Zeenat L, Zolfagharian A, Sriya Y, Sasikumar S, Bodaghi M (2023) 4D printing for vascular tissue engineering: progress and challenges. Adv Mater Technol 8(23):1–21. https://doi.org/10.1002/admt.202300200
39. Saska S, Pilatti L, Blay A, Shibli JA (2021) Bioresorbable polymers: advanced materials and 4D printing for tissue engineering. Polymers 13(4):1–24. https://doi.org/10.3390/polym13040563
40. Wan Z, Zhang P, Liu Y, Lv L, Zhou Y (2020) Four-dimensional bioprinting: current developments and applications in bone tissue engineering. Acta Biomater 101(December):26–42. https://doi.org/10.1016/j.actbio.2019.10.038

41. Bjelić D, Finšgar M (2021) The role of growth factors in bioactive coatings. Pharmaceutics 13(7). https://doi.org/10.3390/pharmaceutics13071083
42. Oliveira ÉR et al (2021) Advances in growth factor delivery for bone tissue engineering. Int J Mol Sci 22(2):1–33. https://doi.org/10.3390/ijms22020903
43. Chen FM, Zhang M, Wu ZF (2010) Toward delivery of multiple growth factors in tissue engineering. Biomaterials 31(24):6279–6308. https://doi.org/10.1016/j.biomaterials.2010.04.053
44. Aldana AA, Bauer J, Baker MB (2022) Synthetic biomaterials. Tissue Eng (Third Edition) 42(2):173–212. https://doi.org/10.1016/B978-0-12-824459-3.00006-8
45. Hosni YA, Harrysson O (2015) Design and manufacturing of customized implants, May 2002
46. Knee joint replacement. MedlinePlus. National Library of Medicine (US), Bethesda. [Online]. Available: https://medlineplus.gov/ency/article/002974.htm
47. Chan BP, Leong KW (2008) Scaffolding in tissue engineering: general approaches and tissue-specific considerations. Eur Spine J 17(Suppl):4. https://doi.org/10.1007/s00586-008-0745-3
48. Ghassemi T, Shahroodi A, Ebrahimzadeh MH, Mousavian A, Movaffagh J, Moradi A (2018) Current concepts in scaffolding for bone tissue engineering. Arch Bone Jt Surg 6(2):90–99. https://doi.org/10.22038/abjs.2018.26340.1713
49. Hosseinkhani M, Mehrabani D, Karimfar MH, Bakhtiyari S, Manafi A, Shirazi R (2014) Tissue engineered scaffolds in regenerative medicine. World J Plast Surg 3(1):3–7. [Online]. Available: http://www.ncbi.nlm.nih.gov/pubmed/25489516%0A; http://www.pubmedcentral.nih.gov/articlerender.fcgi?artid=PMC4236978
50. Garg T (2011) Scaffold: tissue engineering and regenerative medicine. Int Res J Pharm 2(12):648–659. https://doi.org/10.1002/9781119719809.ch57
51. Huang J et al (2021) 3D bioprinting of hydrogels for cartilage tissue engineering. Gels 7(3):1–15. https://doi.org/10.3390/gels7030144
52. Han X, Chang S, Zhang M, Bian X, Li C, Li D (2021) Advances of hydrogel-based bioprinting for cartilage tissue engineering. Front Bioeng Biotechnol 9(September):1–14. https://doi.org/10.3389/fbioe.2021.746564
53. Shokrani H, Shokrani A, Sajadi SM, Seidi F (2022) Cell-seeded biomaterial scaffolds: the urgent need for unanswered accelerated angiogenesis. Int J Nanomedicine 17(March 2022):1035–1068
54. Blaudez F, Vaquette C, Ivanovski S (2023) Cell seeding on 3D scaffolds for tissue engineering and disease modeling applications. In: Oral biology. Methods in molecular biology, vol 2588. Humana Press, New York, pp 473–483. https://doi.org/10.1007/978-1-0716-2780-8_28
55. Moien Alizadehgiashi EK, Nemr CR, Chekini M, Ramos DP, Mittal N, Ahmed SU, Khuu N, Kelley SO (2021) Multifunctional 3D-printed wound dressings. ACS Nano 17(July). https://doi.org/10.1021/acsnano.1c04499
56. Tsegay BH, Elsherif M (2022) Smart 3D printed hydrogel skin wound bandages: a review. Polymers (Basel) 14(5). https://doi.org/10.3390/polym14051012
57. Gaopeng Guan WL, Lv Q, Liu S, Jiang Z, Zhou C (2022) 3D-bioprinted peptide coupling patches for wound healing. Mater Today Bio 13. https://doi.org/10.1016/j.mtbio.2021.100188
58. Maintz M et al (2024) Patient-specific implants made of 3D printed bioresorbable polymers at the point-of-care: material, technology, and scope of surgical application. 3D Print Med 10(1):1–10. https://doi.org/10.1186/s41205-024-00207-0
59. Major R et al (2020) Patient specific implants for jawbone reconstruction after tumor resection. Colloids Surf B: Biointerfaces 193(March). https://doi.org/10.1016/j.colsurfb.2020.111056
60. Szychlinska MA, Bucchieri F, Fucarino A, Ronca A, D'Amora (2022) Three-dimensional bioprinting for cartilage tissue engineering: insights into naturally-derived bioinks from land and marine sources. J Funct Biomater 13(3). https://doi.org/10.3390/jfb13030118
61. Fayyazbakhsh F, Leu MC (2020) A brief review on 3D bioprinted skin substitutes. Procedia Manuf 48:790–796. https://doi.org/10.1016/j.promfg.2020.05.115
62. Maniţă PG, García-Orue I, Santos-Vizcaíno E, Hernández RM, Igartua M (2020) 3D bioprinting of functional skin substitutes for chronic wound treatment: from current achievements to future goals. SSRN Electron J 14. https://doi.org/10.2139/ssrn.3751557

63. Chelladurai Karthikeyan Balavigneswaran VM, Selvaraj S, Vasudha TK, Iniyan S (2023) Tissue engineered skin substitutes: a comprehensive review of basic design, fabrication using 3D printing, recent advances and challenges. Biomater Adv 153. https://doi.org/10.1016/j.bioadv.2023.213570
64. Anitha Radhakrishnan RPS, Sreekumaran S, Anjali S, Resmi R (2022) Chapter 20—emerging strategies in bone tissue engineering. In: Tissue engineering: current status and challenges, pp 469–492. https://doi.org/10.1016/B978-0-12-824064-9.00013-7
65. Shahrubudin N, Koshy P, Alipal J, Kadir MHA, Lee TC (2020) Challenges of 3D printing technology for manufacturing biomedical products: a case study of Malaysian manufacturing firms. Heliyon 6(4):1–14. https://doi.org/10.1016/j.heliyon.2020.e03734
66. Raja S, John Rajan A (2023) Challenges and opportunities in additive manufacturing polymer technology: a review based on optimization perspective. Adv Polym Technol 2023. https://doi.org/10.1155/2023/8639185
67. Miramini S, Fegan KL, Green NC, Espino DM, Zhang L, Thomas-Seale LEJ (2020) The status and challenges of replicating the mechanical properties of connective tissues using additive manufacturing. J Mech Behav Biomed Mater 103:1–54. https://doi.org/10.1016/j.jmbbm.2019.103544
68. Olamide Omiyale B, Abiodun Rasheed A, Omoboyode Akinnusi R, Olumide Olugbade T (2023) Influence of mechanical properties of biomaterials on the reconstruction of biomedical parts via additive manufacturing techniques: an overview. Biotechnol Biosensors Biomater Tissue Eng 2022(June):1–11. https://doi.org/10.5772/intechopen.104465
69. Mladenovska T et al (2023) The regulatory challenge of 3D bioprinting the regulatory challenge of 3D bioprinting. Regen Med 18(8):659–674. https://doi.org/10.2217/rme-2022-0194
70. Ke D, Murphy SV (2018) Current challenges of bioprinted tissues towards clinical translation. Tissue Eng Part B Rev:1–32. https://doi.org/10.1089/ten.TEB.2018.0132
71. Webber MJ, Khan OF, Sydlik SA, Tang BC, Avenue M (2016) HHS public access. Ann Biomed Eng 43(3):641–656. https://doi.org/10.1007/s10439-014-1104-7.A
72. Narayan RJ (2016) The next generation of biomaterial development. Adv Process Biomater 368(1917):1831–1837. https://doi.org/10.1098/rsta.2010.0001
73. Chang J et al (2018) Advanced material strategies for next-generation additive manufacturing. Materials 11(1). https://doi.org/10.3390/ma11010166
74. Pérez M, Carou D, Rubio EM, Teti R (2020) Current advances in additive manufacturing. Procedia CIRP 88(March):439–444. https://doi.org/10.1016/j.procir.2020.05.076
75. Verma AS, Bisht A, Kant S, Bahl S (2022) Recent advances in additive manufacturing for current challenges, materials and their applications. Indian J Eng Mater Sci 29(3):331–344. https://doi.org/10.56042/ijems.v29i3.65761
76. Hua S, de Matos MBC, Metselaar JM, Storm G (2018) Current trends and challenges in the clinical translation of nanoparticulate nanomedicines: pathways for translational development and commercialization. Front Pharmacol 9(JUL):1–14. https://doi.org/10.3389/fphar.2018.00790
77. Vijayavenkataraman S (2023) 3D bioprinting: challenges in commercialization and clinical translation. J 3D Printing Med 7(3). https://doi.org/10.2217/3dp-2022-0026
78. Azharuddin NSM, Aa'zamuddin Ahmad Radzi M, Azhim A, Sha'ban M, Hashi AA (2022) Ethical issues in tissue engineering: a systematic review. Malays J Med Health Sci 18(1):275–289
79. De Kanter AFJ, Jongsma KR, Verhaar MC, Bredenoord AL (2023) The ethical implications of tissue engineering for regenerative purposes: a systematic review. Tissue Eng Part B Rev 29(2):167–187. https://doi.org/10.1089/ten.teb.2022.0033

5 Advanced Smart Biomaterials and Constructs for Hard Tissue Engineering and Regeneration Through Additive Manufacturing

Amol D. Gholap, Pankaj R. Khuspe, Swapnil A. Phade, Dipali V. Mane, Navnath T. Hatvate, Abhijeet V. Puri, and Satish Rojekar

Abstract

Hard tissue engineering has been transformed by the combination of smart biomaterials and additive manufacturing methods, opening up new possibilities for regenerative medicine. With an emphasis on additive manufacturing, this chapter offers a thorough review of sophisticated smart biomaterials and constructions used in hard tissue engineering. It covers the basics of smart biomaterials and how they are used to regenerate hard tissues like cartilage and bone. Smart biomaterials are materials that can react to external impetus including mechanical forces, temperature, and pH. To replicate the natural tissue milieu, intricate structures requiring exact control over shape and porosity are made possible using additive manufacturing. Scaffolds that facilitate cell attachment, proliferation, and differentiation are made using a variety of processes, such as stereolithography, selective laser sintering, and extrusion-based printing. To improve their regeneration qualities, these scaffolds are frequently packed with intelligent biomaterials like hydrogels, nanoparticles, and growth factors. Recent developments in the subject are highlighted in this chapter, including the creation of bioactive scaffolds that can promote tissue growth and healing. Additionally,

A. D. Gholap · A. V. Puri
AETs St. John Institute of Pharmacy and Research, Palghar, Maharashtra, India

P. R. Khuspe · S. A. Phade · D. V. Mane
Shriram Shikshan Sanstha's College of Pharmacy, Solapur, Maharashtra, India

N. T. Hatvate
Institute of Chemical Technology, Marathwada Campus, Jalna, Maharashtra, India

S. Rojekar (✉)
Center for Translational Medicine and Pharmacology, Department of Pharmacological Sciences, Icahn School of Medicine at Mount Sinai, New York, NY, USA

S. Singh et al. (eds.), *Biomaterial-based Additive Manufacturing in Tissue Engineering and Regeneration*, Biomaterials, Bioengineering and Sustainability 7,
https://doi.org/10.1007/978-3-031-96070-3_5

it talks about how additive manufacturing procedures can be used to manufacture patient-specific implants that will improve treatment outcomes by utilizing imaging techniques like computed tomography (CT) and magnetic resonance imaging (MRI). The use of smart biomaterials for drug delivery is also explored in this chapter, allowing for the regulated release of therapeutic substances to support tissue regeneration. Even with these developments, there are still issues to be resolved, such as the need for better smart biomaterials' mechanical and biocompatibility qualities as well as the scalability and affordability of additive printing methods. In order to advance the area and integrate these technologies into clinical applications, it will be imperative to tackle these obstacles.

Keywords

Advanced smart biomaterials · Hard tissue engineering · Regeneration · Additive manufacturing · 3D printing · Stimuli-responsive materials · Hydrogels · Nanocomposites

Abbreviations

FDA	Food and Drug Administration
EMA	European Commission and the European Medicines Agency
AM	Additive manufacturing
HA	Hydroxyapatite
NIAID	National Institute of Allergy and Infectious Diseases
FGFs	Fibroblast growth factors
BMPs	Bone morphogenetic proteins.
TNF	Tumor necrosis factor
MSCs	mesenchymal stem cells
PCL	Poly(caprolactone)
SLA	stereolithography
SLS	Selective laser sintering
FDM	Fused deposition modeling
EBM	Electron beam melting
CT	Computed tomography
MRI	Magnetic resonance imaging
ECM	Extracellular matrix
iPSCs	induced pluripotent stem cells
SEM	Scanning electron microscopy
micro-CT	micro-computed tomography
SMPs	Shape memory polymers
SMAs	Shape memory alloys
PLGA	Poly(lactic-co-glycolic acid)
PLA	Polylactic acid.

5.1 Introduction

The intersection of biology, medicine, and materials science is where the quickly expanding subject of "hard tissue engineering" lies. Its main objective is to create novel methods for the regeneration or repair of diseased or damaged hard tissues, like cartilage and bone [65]. The development of new biomaterials and cutting-edge tissue engineering methods has been greatly aided by the interdisciplinary approach, which has also improved our comprehension of the intricate processes guiding tissue formation and regeneration [133, 134]. Creating biomaterials that can replicate the target tissue's extracellular matrix (ECM) for regeneration is a major challenge in hard tissue engineering. In addition to giving cells structural support, the ECM is essential for controlling a variety of cell behaviors, including as adhesion, migration, proliferation, and differentiation. For tissue engineering to be effective, biomaterials must be biocompatible, biodegradable, and have mechanical characteristics similar to those of the original tissue [55].

Different bioactive compounds, including growth factors, peptides, and proteins, can be added to the scaffold to increase the bioactivity of biomaterials [46, 50, 85]. These substances can promote cellular reactions, including differentiation, adhesion, and proliferation, which improves tissue regeneration [45, 48, 93]. Furthermore, by boosting their biocompatibility and bioactivity, surface changes like the addition of nanoparticles or coatings can further enhance the performance of biomaterials [63]. The application of stem cells, which can differentiate into diverse cell types like chondrocytes and osteoblasts, which create cartilage and bone, respectively, is essential to the study of hard tissue engineering. Stem cells can be extracted from several sources, including bone marrow, fat tissue, and induced pluripotent stem cells (iPSCs), in order to facilitate tissue regeneration. These stem cells can then be seeded into biomaterial scaffolds. A number of variables, including the intended tissue type, scalability, and ethical issues, influence the choice of stem cell source [122].

Hard tissue engineering has advanced thanks in large part to developments in imaging and characterization methods as well as biomaterials and stem cells. Researchers can see the content and structure of manufactured tissues at great resolution thanks to techniques like scanning electron microscopy (SEM) and micro-computed tomography (micro-CT), which offer important insights into the process of tissue regeneration [5]. In general, hard tissue engineering field exhibits significant potential for creating innovative treatments aimed at restoring and revitalizing impaired or ailing hard tissues. Researchers are getting closer to creating safe and efficient treatments for diseases including osteoporosis, osteoarthritis, and bone fractures by fusing developments in biomaterials, stem cell biology, and imaging technologies [15]. The components of tissue engineering are presented in Fig. 5.1.

Hard tissue engineering has advanced significantly thanks to smart biomaterials, which provide cutting-edge approaches to healing and restoring harmed or diseased tissues. These materials have special qualities that enable them to react to outside stimuli like light, pH, temperature, or certain proteins. This allows for exact control over the process of tissue regeneration [9].

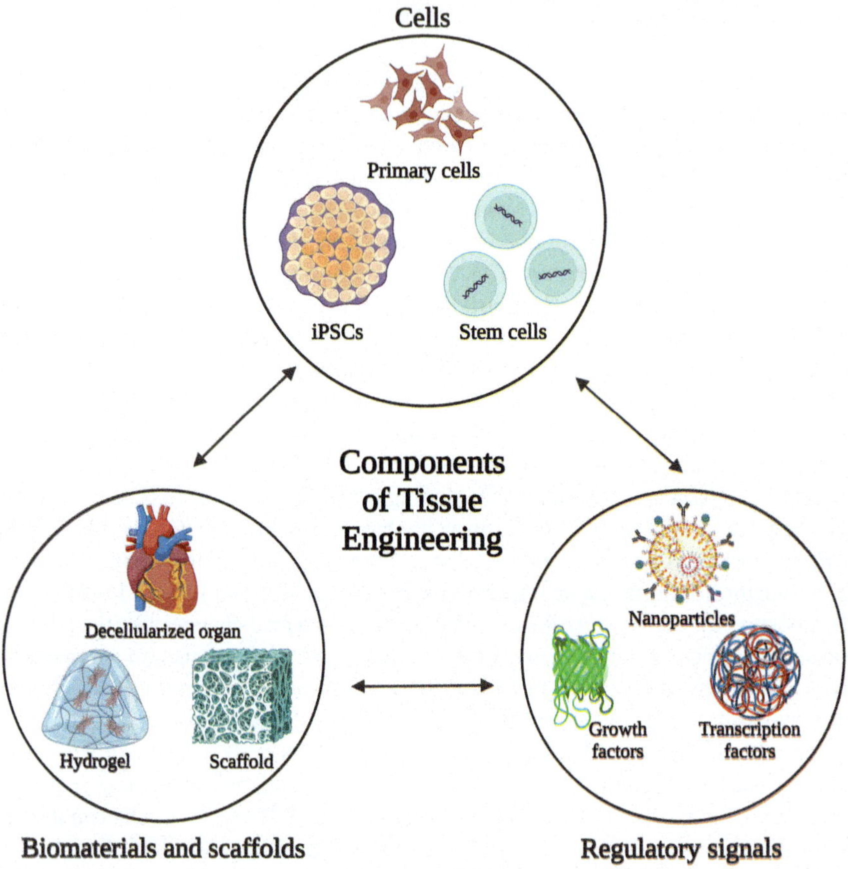

Fig. 5.1 Summary of components of tissue engineering consist of cells, biomaterials, scaffold, and regulatory signals including growth factors, transcription factors and nanoparticles. (Created by using Biorender.com)

The creation of biomaterials that can replicate the intricate structure and functionalities of natural tissues is one of the main obstacles in tissue engineering. By providing dynamic and interactive platforms that can actively interact with the surrounding biological environment, smart biomaterials address this problem. These materials provide mechanical strength, promote cell adhesion and proliferation, induce tissue regeneration, and can be engineered to control the release of bioactive substances [72]. The capacity of smart biomaterials to react to stimuli in the surrounding environment is one of their main advantages [47–49]. For instance, pH-sensitive biomaterials can undergo modifications in their physical and chemical characteristics in response to shifts in pH levels, which often indicate tissue inflammation or infection. The controlled release of medications or growth factors in reaction to physiological changes in the body is made possible by thermo-responsive biomaterials, which may alter their structural characteristics in response to temperature variations [63, 70].

The capacity of smart biomaterials to offer spatiotemporal control over the process of tissue regeneration is another crucial feature. Growth factor-releasing microsphere-embedded scaffolds, for instance, can release growth factors continuously to support tissue growth and cell differentiation at particular scaffold regions [142]. The creation of complex tissues with discrete structural and functional sections depends on this spatial regulation [42, 46, 47]. Moreover, intelligent biomaterials can be engineered to communicate with tissues or cell types. Biomaterials can be functionalized, for instance, by adding peptides or ligands unique to target cell types, which bind to receptors on the target cell's surface and encourage adhesion and growth. By reducing off-target effects, this tailored method maximizes the biomaterial's ability to promote tissue regeneration [138]. The statistics of tissue engineering use are presented in Fig. 5.2.

Smart biomaterials are essential for tissue engineering research and development in addition to their function in fostering tissue regeneration [37, 41, 92]. These materials are useful resources for investigating disease progression, tissue formation, and cell behavior in a laboratory environment [40, 43, 44]. Smart biomaterials provide a platform that mimics the natural tissue environment, enabling researchers to gain deeper insights into biological processes and aiding in the creation of novel treatments for various hard tissue diseases [88]. All things considered, smart biomaterials offer unparalleled control and accuracy in tissue regeneration, thereby heralding in a paradigm shift in the field of hard tissue engineering [11, 98, 115]. These

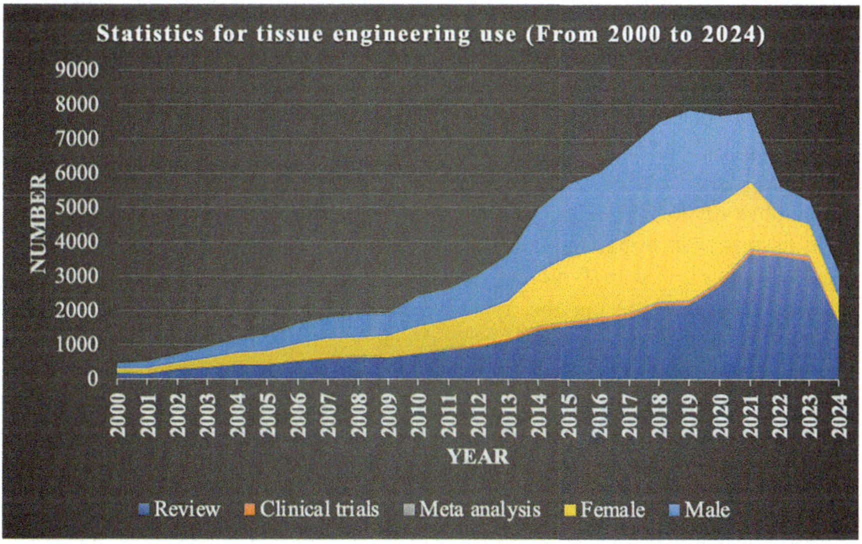

Fig. 5.2 The statistics of tissue engineering applications from the year 2000 to 2024 including number of review articles published, number of clinical trials conducted, number of meta-analyses published and gender wise data. (Data obtained from pubmed.com)

materials have enormous potential for studying tissue formation and disease mechanisms, as well as for the development of cutting-edge treatments for bone and cartilage regeneration. Smart biomaterials are predicted to play a major role in transforming the treatment of hard tissue illnesses as research in this area progresses [107].

5.1.1 Regenerative Medicine Approaches

Regenerative medicine techniques cover a wide range of tactics intended to replace, regenerate, or restore organs and tissues in order to treat illness or damage. This chapter examines the most recent developments in regenerative medicine, including tissue engineering, gene editing, and stem cell therapies, as well as how they are being used in other medical specialties. Because of their extraordinary capacity to develop into a wide variety of cell types, stem cells have become a highly promising tool in regenerative medicine [82]. Adult stem cells, iPSCs, and embryonic stem cells (ESCs) are among the stem cell types used in therapeutics. Early-stage embryos provide the source of stem cells, which can develop into any kind of body cell. Since iPSCs are reprogrammed, adult cells produced from the patient's own cells, there is less chance of rejection than with ESCs. The body contains adult stem cells in a variety of tissues, and they can develop into several cell types. Treatments for numerous ailments, such as diabetes, heart disease, and spinal cord injury, have showed promise using stem cell therapy [146].

Regenerative medicine has undergone a revolution thanks to gene editing tools like CRISPR-Cas9. These technologies allow for precise alterations to the genome. Researchers can possibly fix genetic flaws or introduce therapeutic genes by editing specific genes linked to disease or injury using CRISPR-Cas9. By repairing the underlying genetic abnormalities, gene editing has demonstrated promise in the treatment of genetic illnesses like sickle cell anemia and cystic fibrosis. Gene editing can also be used to create cells for regenerative medicine treatments, like boosting stem cells' ability to regenerate or altering immune cells for cancer immunotherapy [1, 73].

In tissue engineering, living organs or tissues are grown in the lab for use in transplantation or repair. Typically, this method entails seeding cells onto biomaterial scaffolds and offering the appropriate cues such as growth hormones and mechanical stimulation to encourage the development and maturation of tissue [72]. For use in organ repair, wound healing, and disease modeling, tissue engineering has been utilized to construct a range of tissues, including skin, cartilage, and blood vessels. Tissue engineering has come a long way because to developments in biomaterials, stem cell biology, and biofabrication methods, which open up new avenues for regenerative medicine [91].

Approaches from regenerative medicine have been used with encouraging outcomes of therapeutic contexts [108, 112]. Certain diseases like heart disease, stroke, and spinal cord injuries have been treated with stem cell therapy; some patients have seen notable improvements in their quality of life and ability to function [90].

Several clinical trials are underway to evaluate the safety and efficacy of gene editing, which has demonstrated potential in the treatment of genetic disorders [33, 66, 67]. Skin grafts for burn patients, cartilage implants for joint repair, and vascular grafts for cardiovascular disease patients have all been made using tissue engineering techniques. Even with the advancements in regenerative medicine, several obstacles still need to be overcome. These include the immune system's tendency to reject newly implanted cells or tissues, moral concerns about using embryonic stem cells, and the capacity of these treatments to be scaled up for general application [148]. The creation of customized treatments based on each patient's unique genetic composition, the creation of more complex tissues and organs using biomaterials and bioengineering methods, and the fusion of regenerative medicine techniques with other therapeutic modalities like immunotherapy and precision medicine are some of the future directions in regenerative medicine. All things considered, regenerative medicine has enormous potential to transform the way a variety of illnesses and wounds are treated, giving both patients and medical professionals fresh hope [82].

5.1.2 Significance of Additive Manufacturing in Creating Complex Tissue Constructs

The subject of tissue engineering could undergo a significant transformation due to the breakthrough technology known as additive manufacturing, or 3D printing. Tissue engineering uses additive manufacturing because of its capacity to construct intricate three-dimensional structures precisely and sharply [95]. The capacity of conventional manufacturing methods, such molding and casting, to produce complicated geometries which are sometimes needed to replicate the intricate architecture of native tissues which is limited. By constructing structures layer by layer, additive manufacturing gets over this restriction and makes it possible to fabricate scaffolds with exact control over both their internal and external aspects [87].

The capacity of additive manufacturing to tailor scaffolds based on anatomical information unique to a patient is another important advantage [28, 30, 100, 131]. Personalized medicine has the potential to transform regenerative medicine by creating implants that are anatomically specific to each patient. It may lower the possibility of implant rejection and enhance overall therapeutic results, resulting in more efficient and customized care [62]. In addition, doctors can rehearse difficult procedures before operating on real patients by using additive manufacturing to produce patient-specific models for surgical planning and simulation. Not only can additive printing produce intricate shapes, but it also gives scaffold builders a wide range of material options. Using a variety of biomaterials in additive manufacturing methods, such as hydrogels, polymers, and ceramics, researchers can customize the mechanical, biological, and degrading characteristics of the scaffolds to satisfy particular tissue engineering needs. Moreover, bioactive substances like medications and growth hormones can be included into the scaffold via additive manufacturing, allowing for better tissue regeneration through spatiotemporal control over their release [4].

The development of vascularized tissue constructs is one of the most exciting uses of additive manufacturing in tissue engineering [31, 34, 101, 127]. Vascularization is essential to the long-term viability and proper operation of synthetic tissues because it gives the scaffold's cells a way to receive nutrients and oxygen. Scaffolds with complex circulatory networks that resemble the innate vasculature of tissues can be created by additive manufacturing [21]. The repair and regeneration of sizable tissues and organs is just one of the many possible uses for these vascularized tissue constructions. Bioprinting, which is the process of depositing living cells and biomaterials to form functioning tissues, is another field in which additive manufacturing is making great progress. Bioprinting provides fine control over the arrangement of cells inside the scaffold, enabling the construction of intricate tissue structures. By allowing the creation of organs tailored to each patient and completely compatible with the recipient's immune system, this technique holds the potential to completely transform the practice of organ transplantation [132].

Additive manufacturing in tissue engineering has many benefits, but there are still a number of issues that need to be resolved. These include the creation of novel biomaterials with improved bioactivity and printability, the fine-tuning of printing parameters to increase scaffold qualities, and the technology's scalability for use in clinical settings [114]. However, additive manufacturing is set to play a major part in the production of complex tissue constructs for a wide range of biomedical applications, ultimately enhancing patient outcomes and quality of life. This is due to continuous research and technology advancements [10].

5.2 Smart Biomaterials for Hard Tissue Engineering

5.2.1 Definition and Characteristics of Smart Biomaterials

A class of materials known as "smart biomaterials" has special characteristics that allow them to react to particular environmental stimuli. The ultimate objective of these materials is to increase the efficacy of biomedical devices and treatments by means of controlled interactions with biological systems. Thanks to developments in materials science, nanotechnology, and biotechnology, the field of smart biomaterials has grown significantly in recent years [88]. The capacity of smart biomaterials to react to external stimuli, such as variations in pH, temperature, light, or the presence of biomolecules, is what makes them smart. This responsiveness makes it possible to activate physiological reactions, modify mechanical qualities, or release medications or bioactive chemicals under control. For instance, in reaction to pH variations, pH-sensitive biomaterials can change their structure, enabling the targeted administration of medications to particular tissues or cells [13].

Another key aspect of smart biomaterials is their biocompatibility, which refers to their ability to interact with biological systems without causing adverse reactions

[29, 102–105]. For biomedical devices and treatments to be safe and effective, biocompatible materials are necessary. The ECM, which is the natural environment in which cells live, has characteristics that smart biomaterials are frequently made to resemble. Smart biomaterials can enhance cell adhesion, proliferation, and differentiation, hence boosting tissue regeneration and repair, thanks to their capacity to mimic the extracellular matrix [3].

Smart biomaterials are biocompatible and responsive, but they also have customizable qualities that enable researchers to modify them to suit certain application needs [32, 35, 36, 77–79]. For example, smart biomaterials' mechanical characteristics can be changed to more closely resemble those of natural tissues, guaranteeing appropriate support and functionality. In a similar vein, premature degradation or tissue rejection can be avoided by regulating the rate at which smart biomaterials degrade to correspond with the rate at which tissue regenerates. Numerous biomedical applications, such as medication delivery, tissue engineering, and medical diagnostics, make use of smart biomaterials [10]. In drug delivery, smart biomaterials are employed to encapsulate drugs and release them in a controlled manner, enhancing the effectiveness of the medication while reducing side effects. Smart biomaterials are used in tissue engineering to build scaffolds that promote tissue regeneration and cell development, allowing diseased or damaged tissues to be repaired. Smart biomaterials are utilized in medical diagnostics to identify particular infections or biomarkers, enabling the early diagnosis and treatment of illnesses [64].

One strategy that shows promise for increasing the efficacy of biomedical devices and therapies is the use of smart biomaterials. They are useful tools for a variety of biological applications due to their biocompatibility, adjustable characteristics, and capacity to respond to specific stimuli. Smart biomaterials are anticipated to become more significant in the creation of novel and enhanced biomedical technologies as this field of study progresses [71].

5.2.2 Types of Stimuli-responsive Materials Used in Hard Tissue Engineering

A class of materials known as shape memory polymers (SMPs) can "remember" and revert to its original shape in response to external stimuli like heat, light, or pH changes. SMPs have special benefits in hard tissue engineering because of their capacity for significant, reversible deformations [6]. Delivery methods that limit physical harm are made possible by the ability of these materials to create scaffolds that, when compressed or bent, return to their original structure. Because of this characteristic, SMPs are very helpful in developing implants that can adjust to the intricate geometries of bone defects, offering a more personalized fit for better tissue regeneration. To improve their functionality in tissue engineering applications, SMPs can also be functionalized with bioactive compounds or nanoparticles [117].

Hydrogels, which consist of three-dimensional networks of hydrophilic polymers, have the capacity to absorb and hold considerable volumes of water. Tissue regeneration and cell proliferation are greatly aided by these materials because they closely mimic the extracellular matrix (ECM) present in real tissues. Moreover, hydrogels employed in hard tissue engineering can be made to react differently to different stimuli, like light, temperature, and pH. For instance, in reaction to pH variations, which are frequently suggestive of infection or inflammation, pH-responsive hydrogels might release medications or growth hormones. Temperature-responsive hydrogels are perfect for applications involving controlled drug administration or cell encapsulation because they can change their mechanical characteristics and structure in an irreversible manner in response to temperature fluctuations [81].

Materials called nanocomposites are made of nanoparticles mixed with a polymer matrix to produce materials with improved mechanical, biological, or electrical properties. Nanocomposites are employed in hard tissue engineering to strengthen scaffolds and implants mechanically, hence enhancing tissue regeneration support [13]. Furthermore, adding nanoparticles to nanocomposites can give them stimulus-responsive qualities, which can be used for mechanical reinforcement or controlled drug release in response to stimuli. For instance, when subjected to external magnetic fields, nanocomposites containing magnetic nanoparticles can generate heat or experience mechanical deformation; this ability might be used to improve drug release or promote cell development [111].

Other stimuli-responsive materials used in hard tissue engineering include shape memory alloys (SMAs), which are derived from metal alloys like nickel-titanium and display shape memory properties comparable to those of SMPs [117]. These materials are employed in addition to hydrogels, nanocomposites, and SMPs. Because SMAs can withstand significant deformations and are biocompatible, they are utilized in orthopedic implants. In hard tissue engineering, electrically conductive materials, including carbon nanotubes or conductive polymers, can also be employed to improve the electrical characteristics of tissues or promote cell development. In tissue engineering applications by making it easier to create scaffolds and implants that can actively interact with their biological surroundings, these materials present a unique chance to improve their efficacy [121].

5.2.3 Applications of Smart Biomaterials in Bone and Cartilage Regeneration

To support tissue repair and function restoration, bone and cartilage regeneration are intricate processes that frequently call for the usage of cutting-edge biomaterials. Because smart biomaterials may react to particular environmental stimuli, they provide special benefits for these kinds of applications. With the goal of regenerating bone and cartilage, they are being used to create scaffolds that closely resemble the composition and capabilities of natural tissues [139]. These scaffolds offer a structure for the attachment, growth, and differentiation of cells, which eventually results in the production of new cartilage or bone. Tightly controlled porosity, mechanical characteristics, and rates of degradation can be achieved in scaffolds made of smart biomaterials, opening new possibilities for customized approaches to tissue regeneration. For instance, scaffolds composed of shape memory polymers can adapt their shape in response to particular stimuli, improving tissue integration and offering a more precise fit for bone abnormalities [143].

Drug delivery techniques for bone and cartilage repair also employ smart biomaterials. In order to encourage tissue regeneration and lessen inflammation, these materials can be designed to release bioactive substances in a regulated manner, such as growth factors or anti-inflammatory medications. Hydrogels, for instance, can be used to encapsulate growth hormones and release them in response to stimuli, like temperature or pH changes. The prolonged supply of bioactive molecules to the site of tissue regeneration is made possible by this regulated release, which enhances the therapeutic effect [150]. Smart biomaterials are used in tissue engineering methods for bone and cartilage regeneration in addition to scaffolds and drug delivery systems. To create functioning, implantable tissues, tissue engineering combines cells, growth agents, and biomaterials. In this process, smart biomaterials are essential because they offer an environment that is conducive to the proliferation of cells and the formation of tissues. Nanocomposite scaffolds, for instance, can release bioactive chemicals to promote cell growth and differentiation and offer mechanical support for tissue regeneration [64].

Overall, smart biomaterials provide distinct advantages for bone and cartilage regeneration, offering tailored solutions for tissue repair and restoration. By combining advanced materials science with tissue engineering principles, researchers are developing innovative strategies to improve the outcomes of bone and cartilage regeneration therapies. Smart biomaterials are anticipated to become more significant in the creation of novel therapies for diseases of the bone and cartilage as this field of study progresses [25].

5.2.4 Biofabrication Strategies

Tissue engineering has undergone a revolution because to biofabrication techniques, which have made it possible to create complex organs and tissues for regenerative medicine. By precisely depositing cells, biomaterials, and bioactive chemicals, 3D bioprinting a revolutionary technology—makes it possible to construct intricate three-dimensional structures. With the use of this technology, tissues that closely resemble natural tissues can be created with complex topologies and customized characteristics [96]. Because bio-printed structures may be engineered to enhance cell adhesion, proliferation, and differentiation, they are perfect for tissue regeneration. A range of biomaterials, including hydrogels, bioinks, and decellularized matrices, are being investigated by researchers in an effort to improve the mechanical qualities and biocompatibility of bioprinted tissues [24]. Through the creation of microscale models of human organs on a chip, researchers are able to replicate the mechanical and physiological characteristics of the human body by utilizing organ-on-a-chip technology. These microfluidic devices offer a platform for drug screening, disease modeling, and personalized treatment since they can mimic the actions of organs like the liver, heart, and lungs. Because organ-on-a-chip technology produces more precise and predictive models than conventional cell culture techniques, it holds the potential to completely transform drug discovery and development [116].

With the advent of 4D printing, 3D printing now contains a temporal component that allows structures to adapt their form or function to outside stimuli. This technology makes use of smart materials that can reversibly adapt in response to light, pH, and temperature, such as hydrogels and shape memory polymers. In the field of tissue engineering, 4D printing has promise for producing dynamic and adaptable structures that will enable the creation of self-assembling scaffolds, programmed drug delivery systems, and biomimetic tissues that react to bodily stimuli [137].

Tissue engineering uses biofabrication techniques to create vascularized tissues, skin replacements, and organoids, among other things. The creation of vascular networks within constructs can be achieved by biofabrication techniques, which are essential for the life and functionality of created tissues. Bioprinted skin substitutes have demonstrated encouraging outcomes in skin regeneration and wound healing. Organ-on-a-chip technology can be used to produce miniature organs called organoids, which can be used to evaluate drug reactions in a more physiologically appropriate model and explore disease causes [141]. The integration of vascular networks into large-scale tissues, the scalability of biofabrication processes, and the regulatory approval of bio-fabricated products are only a few of the problems that persist in biofabrication despite its remarkable advances. The development of multi-material bioprinting methods, the design of completely functional organs for transplantation, and the use of AI and machine learning to tissue engineering are some of the future directions in biofabrication. All things considered, tissue engineering and regenerative medicine could greatly benefit from biofabrication techniques, which could open up new treatment options for a variety of illnesses and wounds [145].

5.3 Innovative Additive Fabrication Approaches for Hard Tissue Regeneration

Stereolithography (SLA) is a well-liked method because of its high precision and resolution in hard tissue engineering. In SLA, a liquid photopolymer resin is selectively cured using a laser to create a 3D structure layer by layer. With this method, scaffolds with exact geometries that resemble the intricate structure of bone or cartilage can be made. SLA is very useful for producing scaffolds with intricate details, which makes it perfect for uses requiring a high degree of precision [136]. Because of its excellent resolution, stereolithography is a great method for producing scaffolds with complex geometries and fine details. By layer by layer selectively curing a liquid photopolymer resin using a laser, SLA can attain this high resolution. However, because each layer needs to be cured separately, SLA can be slower than other AM methods. The material compatibility of SLA is also restricted because it needs photopolymer resins that have laser curing capabilities [144].

Another popular additive manufacturing method in hard tissue engineering is selective laser sintering (SLS). SLS creates three-dimensional structures by selectively fusing powdered materials, such ceramics or polymers, layer by layer with a high-power laser. SLS's versatility in material selection which includes biocompatible polymers and ceramics which allows for the creation of scaffolds with precisely engineered mechanical characteristics. Scaffolds with regulated porosity and mechanical strength for the regeneration of bone and cartilage are frequently made using SLS. Material compatibility is improved by selective laser sintering, which can employ a variety of powdered materials, such as metals, ceramics, and polymers. This makes it possible to create scaffolds for hard tissue engineering that have specific mechanical characteristics. In addition, SLS provides comparatively good resolution when compared to other powder-based AM methods. SLS's speed, however, may be constrained by the lengthy sintering process, which is particularly true for intricate designs [113].

Widely used additive manufacturing technology fused deposition modeling (FDM) is renowned for its affordability, ease of usage, and adaptability. Thermoplastic filaments are heated and extruded via a nozzle to create a three-dimensional structure in FDM. Even while FDM might not provide as much resolution as SLA or SLS, it's still a useful technique for creating scaffolds for the regeneration of bone and cartilage and for quick prototyping [133, 134]. Scaffolds with tunable mechanical characteristics and geometries can be made with FDM. The benefits of fused deposition modeling include its affordability, ease of usage, and adaptability. To do this, FDM uses heated nozzles to extrude thermoplastic filaments, which solidify to build three-dimensional structures. Because the extrusion process is continuous, FDM can be faster than other AM methods. Nevertheless, FDM has a resolution limit because the nozzle size dictates the smallest feature size that may be produced. Furthermore, because FDM requires thermoplastic filaments that can be melted and extruded, its material compatibility is limited [135].

An electron beam is used in the additive manufacturing process known as electron beam melting (EBM) to melt and fuse metal particles into three-dimensional structures. EBM is frequently used to precisely produce metal implants for purposes involving hard tissue engineering. EBM is a useful technology for designing customized implants for bone and cartilage regeneration because it can be used to generate specific mechanical properties with intricate geometries in metal implants. Applications in hard tissue engineering, electron beam melting provides benefits in terms of material compatibility and resolution. EBM does this by melting and fusing metal powders with an electron beam, enabling the customized mechanical qualities with intricate geometries in metal implants production [84]. When considering alternative metal AM methods, like laser powder bed fusion, EBM may obtain a comparatively high resolution. EBM's pace, however, may be constrained by the lengthy melting and fusing process which particularly for big implants [16].

Another additive manufacturing method utilized in hard tissue engineering is inkjet 3D printing, which builds up a 3D structure using material droplets deposited from a print head. The ability to add cells and growth factors to scaffolds made using inkjet 3D printing is a special benefit that enables the development of scaffolds that actively encourage tissue regeneration. This method is very helpful for building scaffolds for cartilage regeneration, since the inclusion of cells is necessary for the development of functional tissue in these cases. Compatibility with materials and the capacity to add cells and growth factors to scaffolds for hard tissue creation are two benefits of inkjet 3D printing [80]. This is accomplished by inkjet 3D printing, which builds out a 3D structure using material droplets deposited from a print head. This makes it possible to build scaffolds that actively encourage tissue regeneration. However, because the lowest feature size achievable depends on the droplet size, inkjet 3D printing may have resolution limitations. Furthermore, compared to other AM methods, the deposition process of inkjet 3D printing might be comparatively slow, which can have an impact on its pace [147].

All things considered, these additive manufacturing methods provide special benefits for applications in hard tissue engineering, enabling scientists and medical professionals to design scaffolds and implants with specific qualities for tissue regeneration. It is anticipated that as the area develops, additive manufacturing will become more significant in the creation of novel therapies for problems involving the bones and cartilage, ultimately leading to better patient outcomes and quality of life [147]. Regarding resolution, speed, and material compatibility, every additive manufacturing method has its own benefits and drawbacks for applications involving hard tissue engineering. When choosing an AM technique for a particular application, researchers and clinicians need to carefully analyze these parameters, considering the requirements of the tissue regeneration process as well as the desired scaffold qualities. It's anticipated that new methods with enhanced resolution, speed, and material compatibility for hard tissue engineering will surface as additive manufacturing develops [126].

5.3.1 Recent Advancements in Additive Manufacturing for Hard Tissue Engineering

Hard tissue engineering has advanced significantly as a result of recent developments in additive manufacturing (AM), which present new possibilities for the creation of intricate structures with improved characteristics. The creation of new materials with a focus on tissue regeneration is known as AM in the field of hard tissue engineering. When compared to conventional biomaterials, these materials have better mechanical qualities, degradation profiles, and biocompatibility [97]. For instance, scientists have produced biodegradable polymers that can be utilized to build scaffolds with specific rates of disintegration, enabling the release of bioactive compounds under control. Furthermore, adding nanoparticles to these materials can improve their mechanical characteristics and encourage the adherence and proliferation of cells [69].

A noteworthy progression in additive manufacturing for hard tissue engineering is the enhancement of printing methodologies to get superior resolution and precision. For instance, scientists have created novel printing techniques like bioprinting and multi-material printing, which make it easier to create scaffolds with intricate geometries and hierarchical architectures. By depositing several materials simultaneously during a printing process, multi-material printing makes it possible to create scaffolds with various mechanical and biological characteristics. In contrast, bioprinting offers new opportunities for tissue regeneration by using living cells and biomaterials to build functional tissues [83].

New applications and methods have been developed in AM for hard tissue engineering recently, in addition to innovative materials and printing methods. For instance, AM has been utilized by researchers to fabricate implants for bone and cartilage regeneration that are unique to each patient, enhancing surgical results and lowering implant rejection rates [83]. Scaffolds with bioactive coatings that can encourage tissue regeneration and lower the risk of infection have also been made using AM. These developments could completely transform the area of hard tissue engineering and open up new avenues for the management of a range of bone and cartilage conditions [99]. New materials, enhanced printing methods, and inventive applications have been the main topics of recent AM research effort for hard tissue engineering. The creation of bioactive compounds that can actively encourage tissue regeneration is one field of study. For instance, scientists have created compounds that, when released under regulated conditions, can stimulate the production of new tissues and the multiplication of cells. These materials present novel avenues for treating diseases where effective outcomes depend on stimulating tissue regeneration, such as osteoarthritis and bone fractures [120].

The development of novel printing processes that can raise the resolution and precision of printed scaffolds is a further field of study. For instance, researchers have looked into the application of high-resolution printing methods like two-photon polymerization, which precisely polymerizes a photosensitive resin using a concentrated laser beam [94]. With the use of this approach, scaffolds with micron-scale features can be created, facilitating the creation of structures that resemble real tissues architecture. The application of 3D bioprinting, which entails the deposition of living cells and substances to generate functioning tissues, has been investigated by other researchers. By opening up new possibilities for the development of intricate tissues and organs, this technique has the potential to completely transform the area of tissue engineering [76]. Recent research on additive manufacturing for hard tissue engineering has not only concentrated on improving materials and printing methods but also on creating novel applications and strategies. Creating implants tailored to each patient to promote bone and cartilage repair is one field of research. Researchers can make 3D models of a patient's anatomy using sophisticated imaging methods like CT and MRI scans. These models can then be used to build implants that closely fit each patient's individual anatomy. Compared to conventional implants, these implants provide superior fit and functionality, which benefits patients [87].

The creation of scaffolds with bioactive coatings that can encourage tissue regeneration and lower the risk of infection is another field of study. To lower the risk of infections connected to implants, scientists have created coatings that produce antimicrobial peptides, which can kill germs on contact. The application of coatings that release growth factors or other bioactive chemicals to promote the development of new tissues and cells has been investigated by various researchers. These coatings present fresh opportunities to enhance the functionality of scaffolds and implants, resulting in better patient outcomes [128]. Hard tissue engineering has advanced dramatically as a result of recent developments in additive manufacturing, opening the door for the creation of intricate structures with improved characteristics. By creating new materials, refining printing methods, and investigating innovative uses, scientists are laying the groundwork for the creation of novel treatments for a range of bone and cartilage problems. It is anticipated that additive manufacturing will become more and more important in the treatment of different diseases as this field of study develops, eventually leading to better patient outcomes and quality of life [2].

5.4 Applications of Smart Biomaterials and Additive Manufacturing in Hard Tissue Engineering

5.4.1 Fabrication of Bioactive Scaffolds with Controlled Release of Growth Factors and Drugs

Hard tissue engineering has undergone a revolution because to smart biomaterials and additive manufacturing, which have produced ground-breaking approaches to tissue regeneration and repair [65]. The ability of smart biomaterials to react to certain stimuli in the surrounding environment offers special benefits for the development of scaffolds and implants that closely resemble the dynamic properties of original tissues. Conversely, additive manufacturing makes it possible to fabricate intricate structures with extreme control and precision, enabling the development of implants that are specially made to meet the demands of each patient [87].

These scaffolds can be designed to react to physiological stimuli, including variations in pH, temperature, or the presence of biomolecules. This enables the release of medications or growth hormones to encourage tissue regeneration in a regulated manner. Furthermore, scaffolds with customized mechanical properties that better assist tissue growth and repair can be made using smart biomaterials [26].

To develop scaffolds and implants for hard tissue engineering, additive manufacturing processes like stereolithography, selective laser sintering, and fused deposition modeling provide special benefits. For example, SLA is perfect for producing complex structures because it can produce scaffolds with a high degree of precision and resolution. On the other hand, SLS enables the creation of scaffolds with customized mechanical qualities by utilizing a broad variety of materials, like biocompatible ceramics as well as polymers. FDM is a cost-effective and versatile method that may be used for quick prototyping and scaffold fabrication with configurable geometries, even if it does not give the same level of resolution as SLA or SLS [61, 140].

New therapeutic options for a range of bone and cartilage ailments are made possible by the combination of smart biomaterials and additive manufacturing. Through the integration of the distinct characteristics of intelligent biomaterials with the accuracy of additive manufacturing, scientists and medical professionals may produce personalized implants and scaffolds that facilitate tissue repair and enhance patient results. It is anticipated that additive manufacturing and smart biomaterials will become more significant as this field of study develops and new treatments for hard tissue illnesses are created [27].

5.4.2 Use of Imaging Techniques for Patient-Specific Implants

The process of fabricating bioactive scaffolds that allow for the regulated release of medications and growth factors is complex and requires careful consideration of a number of criteria, such as material selection, scaffold design, inclusion of bioactive compounds, and release kinetics characterization [125]. The way bioactive scaffolds are designed affects how well they work to promote tissue regeneration. The ECM of the target tissue must be mimicked in scaffold design to provide an environment that is conducive to cell adhesion, proliferation, and differentiation. Cell behavior and the diffusion of nutrients are influenced by the scaffold's architecture, which includes pore size, porosity, and interconnectivity. For tissue to develop and change over time, the scaffold also needs to be biodegradable [8].

For bioactive scaffolds to function well and be biocompatible, the materials used are essential. Biodegradable polymers with variable breakdown rates, such poly(caprolactone) (PCL), poly(lactic-co-glycolic acid) PLGA, and polylactic acid (PLA), are widely utilized because of their biocompatibility. To process these polymers and produce scaffolds with the required structure and qualities, 3D printing or electrospinning might be used. The mechanical characteristics or bioactivity of the scaffold can also be improved by using other materials, such as hydrogels and ceramics [19].

Bioactive scaffolds can be modified to improve tissue regeneration by adding growth factors and medications. BMPs and PDGFs are two examples of growth factors that are essential for promoting cell division and proliferation. These molecules can be distributed throughout the scaffold by encapsulating them in hydrogels or biodegradable microspheres. Antibiotics and other anti-inflammatory drugs can also be used to prevent infection and lessen inflammation. To guarantee their efficacy and reduce negative effects, the release of these bioactive compounds needs to be meticulously regulated [18].

It is necessary to evaluate the release kinetics of bioactive compounds from scaffolds to comprehend their behavior in vivo. The release kinetics can be affected by several variables, including the kind of material, the way it is included, and the size of the molecules. The quantification of bioactive compounds released over time can be achieved by several techniques such as spectrophotometry, chromatography, and imaging. Optimizing the design of bioactive scaffolds and forecasting their therapeutic efficacy require a thorough understanding of release kinetics [56]. Although the research has made great strides, there are still several obstacles to overcome in the creation of bioactive scaffolds with regulated release properties. A sustained release of bioactive chemicals over a long time is a challenge because a fast release may have unintended consequences and be less effective. To increase the control and duration of release, researchers are looking into new materials and fabrication processes. Making sure the encapsulated compounds remain stable and bioactive during processing and release presents another difficulty. More stable and efficient formulations have been created as a result of recent developments in drug delivery technologies and biomaterials [60].

The process of creating bioactive scaffolds that allow for the regulated release of medications and growth factors is intricate and needs properly thought-out scaffold design, material selection, inclusion of bioactive compounds, and release kinetics characterization. Novel scaffolds with enhanced therapeutic efficacy have been developed because of recent developments in the sector, opening new avenues for tissue regeneration and regenerative medicine. It is anticipated that bioactive scaffolds will become more crucial in the treatment of several tissue injuries and illnesses as this field of study develops [63].

5.4.3 Incorporation of Cells and Biologics into Scaffolds for Enhanced Tissue Regeneration

One important tactic in tissue engineering to improve tissue regeneration is the incorporation of cells and biologics into scaffolds. Tissue engineering can make use of differentiated cells, progenitor cells, and stem cells, among other cell types. Because of their immunomodulatory qualities and capacity to develop into many cell types, stem cells in particular are appealing. Mesenchymal stem cells (MSCs) and induced pluripotent stem cells (iPSCs) are two prominent types of these stem cells. Tissue regeneration can also be assisted by progenitor cells, which are partially differentiated cells with a restricted ability for proliferation. For tissue regeneration, differentiated cells such as osteoblasts or chondrocytes which can be employed when a particular cell type is needed [14].

Scaffolds can include cells by a variety of techniques, including as surface modification, encapsulation, and seeding. Directly seeding cells onto the scaffold surface enables them to adhere and multiply. This process is known as seeding. The process of encapsulation entails placing the cells inside a hydrogel or microsphere and then inserting them into the scaffold. The process of surface modification entails altering the scaffold's surface to encourage cell adhesion and growth. Depending on the particular application and cell type, each approach offers pros and cons [75].

By encouraging cell division, proliferation, and ECM synthesis, the integration of cells into scaffolds can greatly improve tissue regeneration. Additionally, cells can stimulate angiogenesis and regulate the immune system, which improves tissue vascularization and integration. The kind of cell used and the way it is incorporated can affect how well the scaffold works to encourage tissue regeneration. For instance, because stem cells may develop into multiple cell types and release growth factors that aid in tissue repair, it has been demonstrated that they improve tissue regeneration [110].

To improve tissue regeneration, biologics like growth factors and cytokines can also be added to scaffolds. Growth factors, such as bone morphogenetic proteins (BMPs) and fibroblast growth factors (FGFs), are important mediators of cell division and proliferation. Tumor necrosis factor (TNF) and interleukins are examples of cytokines that can influence the immune system and encourage tissue remodeling. Biologics can improve scaffolds' therapeutic efficacy and encourage more effective tissue regeneration [129].

Notwithstanding the noteworthy advancements in the integration of cells and biologics into scaffolds, several obstacles still need to be addressed. Making sure the cells integrated into the scaffold survive and continue to function is one challenge. Factors like the availability of oxygen and nutrients, mechanical pressures, and interactions with the scaffold material can all have an impact on the survival of cells. In an effort to increase cell survival and function in scaffolds, researchers are investigating novel approaches like as co-culture systems and bioreactors [20].

A regulated release of biologics from the scaffold is another difficulty. Biologics should be released gradually to avoid unwanted side effects, as prolonged release is necessary to encourage tissue regeneration. Innovative drug delivery technologies and biomaterials are being developed by researchers to enable the controlled release of biologics from scaffolds. All things considered, adding cells and biologics to scaffolds has enormous potential to improve tissue regeneration and advance tissue engineering. It is anticipated that additional research in this field will result in the creation of novel treatments for a variety of tissue ailments [106].

5.5 Clinical Trials and Patents Details

The clinical trials and patents details for tissue engineering and regenerative medicines are tabulated in Tables 5.1 and 5.2.

5.6 Key Challenges and Future Directions

Even though the use of cells and biologics in scaffolds for tissue engineering has advanced significantly, there are still many issues that need to be resolved. New approaches are being looked into in order to overcome these obstacles and improve tissue regeneration even more [58]. Achieving a uniform distribution of cells throughout the scaffold is one of the primary challenges. Cells frequently tend to gather or cluster in specific places of the scaffold, which might result in an uneven distribution of cells and possibly reduce the scaffold's efficacy. In order to obtain a more uniform distribution of cells within the scaffold, researchers are investigating novel approaches such as 3D bioprinting and cell patterning techniques. These methods provide you exact control over where the cells are placed, which produces results that are more reliable and consistent [119].

Assuring the survival and functionality of the integrated cells within the scaffold presents another difficulty. Cell viability and function can be impacted by variables like pH levels, mechanical pressures, and the availability of oxygen and nutrients. New scaffold designs and growing conditions are being developed by researchers to provide a more conducive environment for the cells. Perfusion bioreactors, for

Table 5.1 The clinical trials updates for tissue engineering and regenerative medicines

NCT number	Study status	Conditions	Sponsor	Phases	Start date	Completion date
NCT01834326	TERMINATED	Disorder of oral mucous membrane	Stephen E. Feinberg	PHASE2	2014-04	15-01-2019
NCT04747574	UNKNOWN	SARS-CoV-2	Tel-Aviv Sourasky Medical Center	PHASE1	25-09-2020	25-03-2021
NCT03174522	TERMINATED	Peripheral arterial disease	Ixaka Ltd	PHASE3	25-04-2017	13-02-2023
NCT04319848	UNKNOWN	Mild to moderate corneal endothelial decompensation	Singapore Eye Research Institute	PHASE1	01-12-2017	2021–02
NCT05700526	ENROLLING_ BY_INVITATION	Musculoskeletal deformity	Istituto Ortopedico Rizzoli	NA	01-06-2022	30-06-2026
NCT04134676	COMPLETED	Chronic ulcer	Sukma Skin Treatment	PHASE1	01-06-2019	10-06-2020
NCT04881240	RECRUITING	Acute lymphoblastic leukemia, in relapse	St. Jude Children's Research Hospital	PHASE1	14-02-2024	2027-06
NCT04066881	ENROLLING_ BY_INVITATION	HIV infections	MRC/UVRI and LSHTM Uganda Research Unit	PHASE2	15-12-2020	31-12-2024
NCT03563495	COMPLETED	Cleft lip and palate	Cairo University	NA	2015–02	2017-09
NCT03111238	TERMINATED	Peripheral arterial disease	Ixaka Ltd	PHASE3	05-04-2017	31-03-2021
NCT02770209	UNKNOWN	Cartilage diseases	Chinese PLA General Hospital	NA	2012-02	2016-06
NCT02315703	COMPLETED	Healthy	Janssen Vaccines & Prevention B.V.	PHASE1|PHASE2	22-12-2014	28-03-2022
NCT04185064	COMPLETED	Shoulder injuries	Encore Research Inc.	NA	01-09-2020	17-05-2023
NCT05081921	COMPLETED	Osteoarthritis of knee	Jagiellonian University	PHASE1|PHASE2	05-01-2022	20-02-2024
NCT03844061	RECRUITING	Systemic sclerosis	Hospital for Special Surgery, New York	PHASE2	29-07-2019	30-06-2026
NCT02417311	UNKNOWN	Cardiomyopathy	UniversitÃ¤tsklinikum Hamburg-Eppendorf		2014-06	2019-06
NCT02346747	ACTIVE_NOT_ RECRUITING	Ovarian cancer|ovarian neoplasms	Gradalis, Inc.	PHASE2	2015-02	2025-12
NCT02942277	COMPLETED	Malaria	National Institute of Allergy and Infectious Diseases (NIAID)	PHASE1	21-10-2016	15-07-2020

Data obtained from https://clinicaltrials.gov/search?intr=tissue%20engineered%20group

Table 5.2 The patent details for hard tissue engineering

Patent ID	Patent title	Inventor/author	Publication date	Grant date
AU-2009201588-B2	Growth factor modified protein matrices for tissue engineering	Jeffrey A. Hubbell, Anna Jen, Matthias Lutolf, Jason Schense	24-02-2011	24-02-2011
US-9023972-B2	Polyesters, methods of making polyesters and uses therefor	Hunghao Chu, Yadong Wang, Zhengwei You	05-05-2015	05-05-2015
AU-2018266193-B2	Medical and dental integrated multiphasic biomaterials for single or multi-tissue reconstruction/regeneration	Erfan DASHTIMOGHADAM, Farahnaz FAHIMIPOUR, Lobat TAYEBI	28-03-2024	28-03-2024
US-10729534-B2	Methods of treating a musculoskeletal tissue interface injury using decellularized composite tissue bioscaffolds	James Chang, Colin Woon, Hung Pham	04-08-2020	04-08-2020
US-11497829-B2	Method for manufacturing support for regenerating core-shell structured hard tissue and support for regenerating core-shell structured hard tissue manufactured thereby	Hui Suk YUN, Naren RAJA, Jong Man Lee	15-11-2022	15-11-2022
US-9040070-B2	Material for induction of hard tissue regeneration	Shinichi Okamoto, Yasuhiko Tabata	26-05-2015	26-05-2015
US-6730252-B1	Methods for fabricating a filament for use in tissue engineering	Swee Hin Teoh, Dietmar Werner Hutmacher, Kim Cheng Tan, Kock Fye Tam, Iwan Zein	04-05-2004	04-05-2004
US-6376742-B1	In vivo tissue engineering with biodegradable polymers	Richard J. Zdrahala, Ivanka J. Zdrahala	23-04-2002	23-04-2002
CA-2410637-C	Shape memory polymers seeded with dissociated cells for tissue engineering	Andreas Lendlein, Ralf Knischka, Karl Kratz	10-04-2007	10-04-2007
EP-1494730-B1	Composition for hard tissue augmentation	Enrico Zamparo, Aaldert Rens Molenberg, Daniel Fehr	18-01-2012	18-01-2012
US-11364323-B2	Combination grafts for tissue repair or regeneration applications	Nicholas Elian, Sean M. O'Connell, William K. Boss, Jr.	21-06-2022	21-06-2022
US-8992534-B2	Method and system for cutting knee joint using robot	Chun Tek Lee, Sung Hwan Yoon, Oh Myoung Kwon, Joon Sik Park, Hang Jae Lee, Mi Ro Kang, Masei Marty Trabish, Jai Karan	31-03-2015	31-03-2015
US-11980374-B2	Methods and apparatus for performing discectomy	Damien Kahmer	14-05-2024	14-05-2024
US-9131597-B2	Electrosurgical system and method for treating hard body tissue	Richard J. Taft, Danielle Seybold	08-09-2015	08-09-2015

(continued)

Table 5.2 (continued)

Patent ID	Patent title	Inventor/author	Publication date	Grant date
US-7963967-B1	Bone preparation tool	Michael Woods	21-06-2011	21-06-2011
US-9510853-B2	Tissue resection and treatment with shedding pulses	Nikolai Aljuri, Surag Mantri, Luis Baez, George Surjan, Michael W. Sasnett, Jonathan Foote	06-12-2016	06-12-2016
US-11464536-B2	Automated image-guided tissue resection and treatment	Nikolai Aljuri, Surag Mantri, George Surjan, Michael W. Sasnett, Jonathan Foote	11-10-2022	11-10-2022
AU-2007245488-B2	Implant for tissue engineering	Kajal K. Mallick, James O. Meredith	19-04-2012	19-04-2012
US-9232937-B2	Rearchitecting the spine	Neville Alleyne	12-01-2016	12-01-2016
US-9795686-B2	Biomaterials comprising hyaluronic acid binding peptides and bifunctional biopolymer molecules for hyaluronic acid retention and tissue engineering applications	Janice Lee, Jennifer H. Elisseeff, Shimon A. Unterman	24-10-2017	24-10-2017
EP-1032319-B1	Anterior-posterior femoral resection guide with set of detachable collets	Michael J. Pappas	06-12-2006	06-12-2006
US-10524822-B2	Image-guided eye surgery apparatus	Nikolai Aljuri, Surag Mantri	07-01-2020	07-01-2020
US-9968463-B2	Devices and methods for tissue engineering	James Jenq Liu	15-05-2018	15-05-2018
US-9050176-B2	At least partially resorbable reticulated elastomeric matrix elements and methods of making same	Arindam Datta, Lawrence P. LAVELLE, Craig Friedman, Balakrishna Haridas	09-06-2015	09-06-2015
US-11642849-B2	In vivo live 3D printing of regenerative bone healing scaffolds for rapid fracture healing	Venu G. Varanasi, Azhar Ilyas, Philip Roger Kramer, Taha Azimaie, Pranesh B. Aswath, Tugba Cebe	09-05-2023	09-05-2023
EP-3119930-B1	Micro-fiber webs of poly-4-hydroxybutyrate and copolymers thereof produced by centrifugal spinning	David Martin, Said Rizk	13-06-2018	13-06-2018
US-11497583-B2	Methods of designing and manufacturing customized dental prosthesis for periodontal or osseointegration and related systems	Ruedger Rubbert, Ernst-Ulrich Gerhard Berndt, Jakob Sebastian Marquard, Hauke Schmidt-Martens, Lea Ellermeier Nesbit	15-11-2022	15-11-2022
US-9925301-B2	Methods of producing and using silk microfibers	David L. Kaplan, Biman B. Mandal	27-03-2018	27-03-2018

US-9672302-B2	Producing a three-dimensional model of an implant	Howard David Dean, Krishnamoorthy Subramanyan, Alexandros T. Moullas, Robert A. Ratcheson	06-06-2017	06-06-2017
US-8105086-B2	Medical and dental biomaterial and method of use for the same	Saeed Asgary	31-01-2012	31-01-2012
US-11844879-B2	Articles of poly(butylene succinate) and copolymers thereof	Simon F. Williams, Said Rizk, David P. Martin, Skander Limem, Kai Guo, Amit Ganatra, German Oswaldo Hohl Lopez	19-12-2023	19-12-2023
US-10004587-B2	Macroporous 3-D scaffolds for tissue engineering	Fan Yang, Li-Hsin Han, Xinming Tong	26-06-2018	26-06-2018
US-8337487-B2	Reticulated elastomeric matrices, their manufacture and use in implantable devices	Arindam Datta, Craig Friedman, Daniel Klempner, Aisa Sendijarevic	25-12-2012	25-12-2012
EP-2694124-B1	Biocompatible and biodegradable gradient layer system for regenerative medicine and for tissue support	Thorsten Steinberg, Pascal Tomakidi, Simon Schulz, Marco Angarano, Rolf MÃ¼lhaupt, Martin Fabritius	04-01-2017	04-01-2017
US-10201640-B2	Ultrafine electrospun fibers of poly-4-hydroxybutyrate and copolymers thereof	David P. Martin, Kai Guo, Said Rizk, Simon F. Williams	12-02-2019	12-02-2019
US-10844338-B1	Automated tissue engineering system	Timothy J. N. Smith, Sydney M. Pugh, Martin R. Pecaric, Rupert Hagg, Roberto Tommasini, Yves Larcher, Lowell D. Misener	24-11-2020	24-11-2020
CA-2969429-C	Methods of orienting multifilament yarn and monofilaments of poly-4-hydroxybutyrate and copolymers thereof	Amit Ganatra, Fabio Felix, Bhavin Shah, Matthew Bernasconi, Said Rizk, David P. Martin, Simon F. Williams	27-10-2020	27-10-2020
US-8772456-B2	Fibrous protein fusion and use thereof in the formation of advanced organic/inorganic composite materials	David L. Kaplan, Jia Huang, Cheryl Wong Po Foo, Rajesh Naik, Anne George	08-07-2014	08-07-2014
AU-2022200696-B2	Decellularised cell wall structures from plants and fungus and use thereof as scaffold materials	Charles Michel CUERRIER, Ryan Joseph HICKEY, Daniel J. MODULEVSKY, Andrew Edward PELLING	28-09-2023	28-09-2023
US-8784893-B2	Polymer formulations for delivery of bioactive agents	George Daniloff, David Gravett, Robert C. Spiro	22-07-2014	22-07-2014

Data obtained from https://patents.google.com/?q=(Hard+tissue+engineering)&oq=Hard+tissue+engineering

instance, can give the cells an uninterrupted supply of nutrients and oxygen, enhancing their survival and functionality within the scaffold [119]. Another difficulty in tissue engineering is regulating the discharge of biologics from the scaffold. Growth factors and cytokines are examples of biologics that are essential for tissue regeneration, yet their quick release might have unfavorable effects. Researchers are creating innovative biomaterials and drug delivery technologies to achieve regulated release of biologics from the scaffold. These systems ensure that biologics are effective in stimulating tissue regeneration by enabling their sustained release throughout time [58]. Future developments in the field of integrating biologics and cells into scaffolds for tissue engineering could lead to the creation of more specialized and targeted treatments. Therapeutics that are customized to meet the needs of specific patients are being developed because to developments in tissue engineering and stem cell research. Researchers can build more effective and less harmful tailored medicines by employing the patient's own cells and creating scaffolds that imitate the original tissue environment [5, 14, 17].

The incorporation of sophisticated imaging methods and computational modeling into scaffold design represents another avenue for future development. Researchers are able to create scaffolds that closely resemble the original tissue architecture thanks to the accurate assessment of tissue composition and structure made possible by advanced imaging techniques like CT and MRI. Furthermore, by utilizing computer modeling to predict scaffold behavior in vivo, scientists can improve the results of tissue regeneration by fine-tuning their designs [130].

The incorporation of cells and biologics in scaffolds for tissue engineering has tremendous potential for developing novel treatments for a range of tissue ailments, despite ongoing hurdles. Researchers are opening up new avenues and tackling difficult problems in order to develop more individualized and successful therapies that will benefit patients with illnesses and injuries to their tissues [68].

5.6.1 Improving Cellular Compatibility and Mechanical Properties of Smart Biomaterials

To successfully be used in tissue engineering and regenerative medicine, smart biomaterials must have improved mechanical qualities and biocompatibility. This chapter examines several approaches and methods for improving these qualities, including as surface modification, the addition of nanoparticles, and the creation of composite materials. Surface modification is a popular strategy to increase the biocompatibility of smart biomaterials [12]. Researchers can improve cell adhesion, proliferation, and differentiation by altering the material's surface. Plasma treatment is a frequently used technique for surface modification that improves the material's interaction with tissues and cells by adding functional groups to its surface. The surface characteristics of intelligent biomaterials can also be altered by other techniques including chemical grafting and biomolecule immobilization [53].

Adding nanoparticles to smart biomaterials can enhance their biocompatibility and mechanical characteristics. The strength and stiffness of a material can be improved by adding nanoparticles, such as silica or hydroxyapatite (HA), which makes the material more appropriate for load-bearing applications. Additionally, to increase the material's bioactivity and encourage tissue regeneration, nanoparticles can be functionalized with bioactive chemicals like growth hormones or antibiotics. The addition of nanoparticles can, however, present certain difficulties as well, such as toxicity and aggregation problems that need to be properly handled [7].

Smart biomaterials' mechanical characteristics and biocompatibility can be enhanced by the use of composite materials, which are made up of two or more materials having distinct features. For instance, to improve the mechanical strength and durability of composite materials, researchers have created a biodegradable polymer matrix reinforced with fibers or nanoparticles. Furthermore, the ability of composite materials to release bioactive substances in a controlled manner might greatly enhance their therapeutic efficiency [52]. Smart biomaterials can be made more bioactive and tissue regeneration can be encouraged by adding biological molecules, such as cytokines or growth factors. These molecules can be enclosed within the material matrix or fixed onto the material's surface, enabling regulated release over time. By distributing these chemicals directly to the location of injury or defect, smart biomaterials can promote cell proliferation and differentiation and aid in tissue regeneration [118]. Creating smart biomaterials with a structure and set of characteristics like biological tissues is another way to enhance their mechanical and biocompatibility characteristics. Smart biomaterials can enhance tissue regeneration by promoting cell adhesion, proliferation, and differentiation by imitating the structure and content of original tissues. For instance, scientists have created scaffolds with improved mechanical qualities and better interaction with neighboring tissues by modeling the structure of bone or cartilage [59].

In the end, improving smart biomaterials' mechanical characteristics and biocompatibility is crucial to their successful application in tissue engineering and regenerative medicine. Through utilization of techniques including surface modification, nanoparticle integration, composite material development, and bioinspired design, scientists may optimize the functionality of intelligent biomaterials and provide novel treatments for an extensive array of tissue ailments [54].

5.6.2 Scalability and Cost-Effectiveness of Additive Manufacturing Techniques

When it comes to the adoption and application of additive manufacturing techniques, scalability and cost-effectiveness are crucial, especially in sectors like aerospace, automotive, and healthcare. Maintaining the uniformity and quality of printed items as the size of the production batch grows is one of the fundamental issues in scaling up AM processes. At scale, elements like machine dependability, material qualities, and build time become increasingly important. To address these issues, scientists and business experts are investigating a range of approaches, such as process monitoring and control systems implementation, new material development, and printing parameter optimization [109].

Maintaining part quality and consistency at scale requires optimizing printing parameters including layer thickness, printing speed, and laser power. To forecast the impacts of various printing settings on part qualities and optimize them for large-scale manufacturing, researchers are employing computational modeling and simulation. Manufacturers can increase process efficiency and cut waste by knowing how these parameters impact the finished part. Enhancing scalability can also involve creating new materials especially made for additive manufacturing. For instance, scientists are working on creating metal alloys and high-strength polymers that work well with additive manufacturing techniques and can endure the strains of mass production. By enhancing part performance and lowering the requirement for post-processing, these materials can increase AM's economic viability in large-scale manufacturing [57].

Ensuring the quality and consistency of printed parts at scale can be facilitated by the implementation of process monitoring and control systems. These systems monitor important process parameters in real-time and adjust guarantee that parts fulfill quality standards using sensors and data analytics. Manufacturers can lower the possibility of errors and increase overall process efficiency by putting these measures in place. Although additive manufacturing has the potential to be a cost-effective manufacturing technique, several factors, including as material costs, machine costs, and post-processing requirements, may affect this. Scholars and business experts are investigating a range of approaches to lower these expenses and improve AM's competitiveness in comparison to conventional manufacturing techniques [124].

The cost-effectiveness of AM can be significantly influenced by material costs. For additive manufacturing techniques, researchers are creating new materials that are more economical and effective. Recycled materials and bio-based polymers, for instance, can lower material prices without sacrificing part quality. Further lowering material waste and expenses is possible by optimizing material consumption through design for AM. The cost-effectiveness of AM can also be impacted by machine expenses, which include the price of AM equipment and maintenance. New AM technologies that are more cost-effective and efficient for mass production are being developed by researchers. Large-format AM machines, for instance, can print larger pieces, which minimizes the quantity of separate parts that must be manufactured and put together. Furthermore, labor expenses related to AM processes might be decreased by advancements in automation and robotics [123].

AM costs may increase due to post-processing requirements such surface finishing and heat treatment. Researchers are creating novel additive manufacturing techniques including binder jetting and laser sintering that require less post-processing. Additionally, by guaranteeing that parts fulfill quality standards throughout printing, advancements in in-situ monitoring and control can lessen the requirement for post-processing. All things considered, scalability with economy is important element to be considered when adopting and using AM techniques for mass production. Researchers and industry professionals can make AM more competitive with traditional manufacturing methods and unlock its full potential for mass production by lowering material, machine, and post-processing costs, developing new materials, implementing process monitoring and control systems, and optimizing printing parameters [74].

5.7 Regulatory Challenges and Clinical Translation

To move these technologies from the lab to the clinic, regulatory issues, clinical translation of smart biomaterials, and additive manufacturing in hard tissue engineering are important factors to consider for the same. Hard tissue engineering additive manufacturing and smart biomaterials face a complicated regulatory environment that differs from nation to nation. The Food and Drug Administration (FDA) in the US controls the creation and distribution of medical devices, including those made using biomaterials and additive manufacturing techniques. The European Commission and the European Medicines Agency (EMA) oversee regulating tissue-engineered goods and medical devices in Europe. Similar obligations are assigned to regulatory organizations in other nations [149].

Demonstrating the safety and effectiveness of smart biomaterials and additive manufacturing technology is a major obstacle to regulatory approval. To evaluate biocompatibility, mechanical characteristics, and degradation rates, extensive preclinical investigations are needed. These studies should include both in vitro and in vivo testing. Furthermore, it is frequently necessary to conduct clinical trials to assess the efficacy and safety of these technologies in humans. Making sure that additive manufacturing technologies and smart biomaterials adhere to legal requirements for quality assurance and production procedures is another difficulty. This involves making certain that the production processes are repeatable and consistent, as well as that the materials employed in these technologies are of the highest caliber. Regulatory approval depends on quality control procedures like impurity testing and sterility assurance [23].

Researchers and experts in the industry are using a variety of approaches to help accelerate the clinical translation of additive manufacturing technologies and smart biomaterials in the field of hard tissue engineering to overcome these obstacles. To make sure the technologies fulfill regulatory standards, one tactic is to work with regulatory bodies early in the development process. This can assist in the early identification of potential regulatory obstacles and the development of solutions. Undertaking extensive preclinical research is an additional tactic to produce solid data regarding the safety and effectiveness of smart biomaterials and additive manufacturing technology. This entails doing biomechanical testing to evaluate the materials' mechanical qualities as well as testing the materials in appropriate animal models. Additionally, to assess how well these technologies integrate with native tissues, researchers are using sophisticated imaging methods like CT and MRI scans [86].

To prove that smart biomaterials and additive manufacturing techniques are safe and effective for use in people, clinical trials are necessary. Usually, these trials are carried out in stages, beginning with small-scale research to gauge safety and progressing to larger studies to gauge effectiveness. Regulatory agencies must approve clinical trials, and they must follow stringent ethical and scientific guidelines. Finding enough patients to take part in clinical trials is one of the challenges involved. This can be especially difficult in cases of uncommon illnesses or ailments. To get over this problem, researchers are looking into creative recruitment techniques include working with patient advocacy organizations and creating patient registries. Making sure the clinical studies are appropriately planned and carried out to produce dependable and repeatable outcomes presents another difficulty. This involves minimizing bias and guaranteeing the validity of the data by utilizing blinding, randomization, and standardized processes. Additionally, in order to evaluate the data and reach relevant findings, researchers are employing sophisticated statistical techniques [89].

Smart biomaterials and additive manufacturing technologies need to go through regulatory scrutiny before being approved, following the conclusion of clinical trials. To verify the safety and effectiveness of the technologies, a lot of data and documentation must be submitted, which can make the procedure drawn out and complicated. Before approving something, regulatory bodies could need more research or data. Smart biomaterials and additive manufacturing technologies can be commercialized and made accessible to patients if they receive regulatory approval. However, due to national variations in healthcare systems and payment policies, market access might be difficult. Scholars and practitioners in the field are striving to enlighten legislators and healthcare providers about the advantages of these technologies and promote their incorporation into payment guidelines [51].

Researchers, business, and regulatory authorities must work together to address regulatory obstacles, clinical translation of smart biomaterials, and additive manufacturing in hard tissue engineering. Through tackling these obstacles and implementing techniques to expedite clinical translation, scientists can quicken the advancement and commercialization of these innovations while also enhancing patient outcomes [22].

5.8 Conclusion

Hard tissue engineering's use of smart biomaterials and additive manufacturing is developing quickly, and considerable strides are being made in expanding clinical translation and resolving regulatory obstacles. The techniques of additive manufacturing are becoming more practical for large-scale production due to their increasing scalability and cost-effectiveness. Smart biomaterials' mechanical qualities and biocompatibility are being improved by techniques like surface modification, nanoparticle inclusion, and composite material synthesis. This is opening the door for their wider application in tissue engineering. Even with these developments, there are still several obstacles to overcome before additive manufacturing may become more economically viable. These include the demand for additional printing parameter optimization, the creation of new materials, and a reduction in the amount of post-processing that is needed. Key factors in the clinical translation of additive manufacturing and smart biomaterials are market accessibility and regulatory approval. In hard tissue engineering, the status of smart biomaterials and additive manufacturing today has enormous potential for individualized therapies and regenerative medicine. By offering specialized solutions that encourage tissue regeneration and enhance patient outcomes, these technologies have the potential to completely transform the way that a variety of tissue problems are treated.

Researchers can improve tissue regeneration and provide more potent treatments for a variety of tissue injuries and illnesses by adding cells and biologics into scaffolds. It is anticipated that future studies and developments in additive manufacturing and smart biomaterials will propel tissue engineering forward. Technological developments in stem cell research, biomaterials, and 3D printing will propel the creation of novel therapeutics customized to the specific requirements of each patient. In addition to enhancing patients' quality of life, these therapies will lower healthcare expenses by offering more effective and focused care. Regenerative medicine and customized therapies are expected to be major developments in the future thanks in large part to smart biomaterials and additive manufacturing. Through tackling regulatory obstacles, enhancing scalability and cost-efficiency, and propelling clinical translation, these technologies will persist in propelling innovation in tissue engineering and fundamentally altering the trajectory of tissue diseases treatment.

References

1. Akram F, Haq I ul, Sahreen S et al (2022) RETRACTED: CRISPR/Cas9: a revolutionary genome editing tool for human cancers treatment. Technol Cancer Res Treat 21. https://doi.org/10.1177/15330338221132078
2. Akshaya S, Rowlo PK, Dukle A, Nathanael AJ (2022) Antibacterial coatings for titanium implants: recent trends and future perspectives. Antibiotics 11:1719. https://doi.org/10.3390/antibiotics11121719
3. Al-Zyoud W, Haddadin D, Hasan SA et al (2023) Biocompatibility testing for implants: a novel tool for selection and characterization. Materials (Basel) 16:6881. https://doi.org/10.3390/ma16216881
4. Ariz A, Tasneem I, Bharti D et al (2021) Is additive manufacturing of patient-specific implant is beneficial for orthopedics. Apollo Med. https://doi.org/10.4103/am.am_20_20
5. Ashammakhi N, GhavamiNejad A, Tutar R et al (2022) Highlights on advancing frontiers in tissue engineering. Tissue Eng Part B Rev 28:633–664. https://doi.org/10.1089/ten.teb.2021.0012
6. Bhattacharya S, Prajapati BG, Singh S (2023) A critical review on the dissemination of PH and stimuli-responsive polymeric nanoparticular systems to improve drug delivery in cancer therapy. Crit Rev Oncol Hematol 185:103961
7. Bin RA, Haque M, Islam SMM, Uddin Labib KMR (2024) Nanotechnology-enhanced fiber-reinforced polymer composites: recent advancements on processing techniques and applications. Heliyon 10:e24692. https://doi.org/10.1016/j.heliyon.2024.e24692
8. Bitar KN, Zakhem E (2014) Design strategies of biodegradable scaffolds for tissue regeneration. Biomed Eng Comput Biol 6:BECB.S10961. https://doi.org/10.4137/BECB.S10961
9. Bonferoni MC, Caramella C, Catenacci L et al (2021) Biomaterials for soft tissue repair and regeneration: a focus on Italian research in the field. Pharmaceutics 13:1341. https://doi.org/10.3390/pharmaceutics13091341
10. Bose S, Ke D, Sahasrabudhe H, Bandyopadhyay A (2018) Additive manufacturing of biomaterials. Prog Mater Sci 93:45–111. https://doi.org/10.1016/j.pmatsci.2017.08.003
11. Bothiraja C, Gholap AD, Shaikh KS, Pawar AP (2014) Investigation of ethyl cellulose microsponge gel for topical delivery of eberconazole nitrate for fungal therapy. Ther Deliv 5:781–794. https://doi.org/10.4155/tde.14.43
12. Cai S, Wu C, Yang W et al (2020) Recent advance in surface modification for regulating cell adhesion and behaviors. Nanotechnol Rev 9:971–989. https://doi.org/10.1515/ntrev-2020-0076

13. Cecen B, Hassan S, Li X, Zhang YS (2024) Smart biomaterials in biomedical applications: current advances and possible future directions. Macromol Biosci 24. https://doi.org/10.1002/mabi.202200550
14. Chan BP, Leong KW (2008) Scaffolding in tissue engineering: general approaches and tissue-specific considerations. Eur Spine J 17:467–479. https://doi.org/10.1007/s00586-008-0745-3
15. Chen F-M, Liu X (2016) Advancing biomaterials of human origin for tissue engineering. Prog Polym Sci 53:86–168. https://doi.org/10.1016/j.progpolymsci.2015.02.004
16. Davoodi E, Montazerian H, Mirhakimi AS et al (2022) Additively manufactured metallic biomaterials. Bioact Mater 15:214–249. https://doi.org/10.1016/j.bioactmat.2021.12.027
17. De Chiara F, Ferret-Miñana A, Fernández-Costa JM, Ramón-Azcón J (2024) The tissue engineering revolution: from bench research to clinical reality. Biomedicine 12:453. https://doi.org/10.3390/biomedicines12020453
18. De Witte T-M, Fratila-Apachitei LE, Zadpoor AA, Peppas NA (2018) Bone tissue engineering via growth factor delivery: from scaffolds to complex matrices. Regen Biomater 5:197–211. https://doi.org/10.1093/rb/rby013
19. Dwivedi R, Kumar S, Pandey R et al (2020) Polycaprolactone as biomaterial for bone scaffolds: review of literature. J Oral Biol Craniofacial Res 10:381–388. https://doi.org/10.1016/j.jobcr.2019.10.003
20. Dzobo K, Thomford NE, Senthebane DA et al (2018) Advances in regenerative medicine and tissue engineering: innovation and transformation of medicine. Stem Cells Int 2018:1–24. https://doi.org/10.1155/2018/2495848
21. Elomaa L, Yang YP (2017) Additive manufacturing of vascular grafts and vascularized tissue constructs. Tissue Eng Part B Rev 23:436–450. https://doi.org/10.1089/ten.teb.2016.0348
22. Evans CH (2011) Barriers to the clinical translation of orthopedic tissue engineering. Tissue Eng Part B Rev 17:437–441. https://doi.org/10.1089/ten.teb.2011.0228
23. Fang Y, Guo Y, Liu T et al (2022) Advances in 3D bioprinting. Chinese J Mech Eng Addit Manuf Front 1:100011. https://doi.org/10.1016/j.cjmeam.2022.100011
24. Fang W, Yang M, Wang L et al (2023) Hydrogels for 3D bioprinting in tissue engineering and regenerative medicine: current progress and challenges. Int J Bioprinting 9:759. https://doi.org/10.18063/ijb.759
25. Farag MM (2023) Recent trends on biomaterials for tissue regeneration applications: review. J Mater Sci 58:527–558. https://doi.org/10.1007/s10853-022-08102-x
26. García A, Cabañas MV, Peña J, Sánchez-Salcedo S (2021) Design of 3D scaffolds for hard tissue engineering: from apatites to silicon mesoporous materials. Pharmaceutics 13:1981. https://doi.org/10.3390/pharmaceutics13111981
27. Germaini M-M, Belhabib S, Guessasma S et al (2022) Additive manufacturing of biomaterials for bone tissue engineering—a critical review of the state of the art and new concepts. Prog Mater Sci 130:100963. https://doi.org/10.1016/j.pmatsci.2022.100963
28. Gholap AD (2012a) Artificial viral envelopes: emerging vehicle for gene delivery. Pharmatutor-Art-1439 1:1–3
29. Gholap AD (2012b) Translational nanomedicine: status assessment and opportunities. Int J Res Pharmaceut Biomed Sci 4:1454–1463
30. Gholap AD, Dighe SB (2012a) Role of sebum in human health & acne. Inven Rapid Cosmeceuticals [ISSN 0976-3864] 1:1–3
31. Gholap AD, Dighe SB (2012b) Dandruff: pathogenesis and treatment. Inven Rapid Cosmeceuticals [ISSN 0976-3864] 1:1–4
32. Gholap AD, Musmade DS (2011) Advances in the field of nanooncology: an overview. Inven Impact NDDS [ISSN 0976-3791] 4:1–3
33. Gholap AD, Rajurkar VG (2012) Phonophoresis: ultrasound enhanced transdermal drug delivery. Inven Rapid NDDS [ISSN 0976-3791] 1:1–3
34. Gholap AD, Tupe D (2012) Potential applications of botanical extract. Inven Rapid Ethnopharmacol [ISSN 0976-3805] 1:1–4
35. Gholap AD, Borude SS, Mahajan AM (2011a) Smart crystals technology: a review. Pharmacology 1:238–243

36. Gholap AD, Borude SS, Mahajan AM, Gholap MAD (2011b) Aquasomes: a potential drug delivery carrier. Pharmacology 1:230–237
37. Gholap A, Pardeshi S, Giram P (2023a) Preparation, antibacterial and antiviral activity measurements and detection methods. In: Antibacterial and antiviral functional materials. ACS Publications, pp 33–64
38. Gholap AD, Bhowmik DD, Deshmukh AY, Hatvate NT (2023b) Quintessential impact of dendrimer bioconjugates in targeted drug delivery. In: Bioconjugated materials part 2 applications in drug delivery, vaccine formulations and important conjugates for cancer therapy. Elsevier, pp 257–302
39. Gholap AD, Gupta JS, Kamandar PA et al (2023c) Antibody-drug conjugates for cancer therapy: an up-to-date review on the chemistry and pharmacology. In: Bioconjugated materials part 2 applications in drug delivery, vaccine formulations and important conjugates for cancer therapy. Elsevier, pp 105–190
40. Gholap AD, Hatvate NT, Khuspe PR et al (2023d) Proteomics in oncology: retrospect and prospects. In: Novel technologies in biosystems, biomedical & drug delivery. Springer Nature Singapore, Singapore, pp 243–269
41. Gholap AD, Hatvate NT, Sawant SD et al (2023e) Unlocking the potential of nanobiohybrids to combat environmental pollution. In: Nanobiohybrids for advanced wastewater treatment and energy recovery. IWA Publishing, pp 169–190
42. Gholap AD, Said RP, Pawar RD et al (2023f) Importance of carbohydrate-drug conjugates in vaccine development: a detailed review. In: Bioconjugated materials part 2 applications in drug delivery, vaccine formulations and important conjugates for cancer therapy. Elsevier, pp 191–256
43. Gholap AD, Sawant SD, Sayyad SF et al (2023g) Food biopackaging for human benefits: status and perspectives. In: Biobased packaging materials. Springer Nature Singapore, Singapore, pp 1–35
44. Gholap AD, Sayyad SF, Hatvate NT et al (2023h) Drug delivery strategies for avobenzone: a case study of photostabilization. Pharmaceutics 15:1008. https://doi.org/10.3390/pharmaceutics15031008
45. Gholap AD, Gupta J, Kamandar P et al (2024a) Harnessing nanovaccines for effective immunization—a special concern on COVID-19: facts, fidelity, and future prospective. ACS Biomater Sci Eng 10:271–297. https://doi.org/10.1021/acsbiomaterials.3c01247
46. Gholap AD, Hatvate NT, Khuspe PR et al (2024b) Decoding of nanoneutraceuticals and their delivery. In: Sustainable green nanotechnology. CRC Press, New York, pp 284–309
47. Gholap AD, Kapare HS, Pagar S et al (2024c) Exploring modified chitosan-based gene delivery technologies for therapeutic advancements. Int J Biol Macromol 260:129581. https://doi.org/10.1016/j.ijbiomac.2024.129581
48. Gholap AD, Pardeshi SR, Hatvate NT et al (2024d) Environmental implications and nanotechnological advances in octocrylene-enriched sunscreen formulations: a comprehensive review. Chemosphere 358:142235. https://doi.org/10.1016/j.chemosphere.2024.142235
49. Gholap AD, Rojekar S, Kapare HS et al (2024e) Chitosan scaffolds: expanding horizons in biomedical applications. Carbohydr Polym 323:121394. https://doi.org/10.1016/j.carbpol.2023.121394
50. Gholap AD, Uddin MJ, Faiyazuddin M et al (2024f) Advances in artificial intelligence for drug delivery and development: a comprehensive review. Comput Biol Med 178:108702. https://doi.org/10.1016/j.compbiomed.2024.108702
51. Gogtay N, Ravi R, Thatte U (2017) Regulatory requirements for clinical trials in India: what academicians need to know. Indian J Anaesth 61:192. https://doi.org/10.4103/ija.IJA_143_17
52. Guo Z, Poot AA, Grijpma DW (2021) Advanced polymer-based composites and structures for biomedical applications. Eur Polym J 149:110388. https://doi.org/10.1016/j.eurpolymj.2021.110388
53. Gupta P, Sharma S, Jabin S, Jadoun S (2024) Chitosan nanocomposite for tissue engineering and regenerative medicine: a review. Int J Biol Macromol 254:127660. https://doi.org/10.1016/j.ijbiomac.2023.127660

54. Habibzadeh F, Sadraei SM, Mansoori R et al (2022) Nanomaterials supported by polymers for tissue engineering applications: a review. Heliyon 8:e12193. https://doi.org/10.1016/j.heliyon.2022.e12193
55. Hasan A, Morshed M, Memic A et al (2018) Nanoparticles in tissue engineering: applications, challenges and prospects. Int J Nanomedicine 13:5637–5655. https://doi.org/10.2147/IJN.S153758
56. Heredia NS, Vizuete K, Flores-Calero M et al (2022) Comparative statistical analysis of the release kinetics models for nanoprecipitated drug delivery systems based on poly(lactic-co-glycolic acid). PLoS One 17:e0264825. https://doi.org/10.1371/journal.pone.0264825
57. Iftekar SF, Aabid A, Amir A, Baig M (2023) Advancements and limitations in 3D printing materials and technologies: a critical review. Polymers (Basel) 15:2519. https://doi.org/10.3390/polym15112519
58. Ikada Y (2006) Challenges in tissue engineering. J R Soc Interface 3:589–601. https://doi.org/10.1098/rsif.2006.0124
59. Intravaia JT, Graham T, Kim HS et al (2023) Smart orthopedic biomaterials and implants. Curr Opin Biomed Eng 25:100439. https://doi.org/10.1016/j.cobme.2022.100439
60. Jiang Z, Zheng Z, Yu S et al (2023) Nanofiber scaffolds as drug delivery systems promoting wound healing. Pharmaceutics 15:1829. https://doi.org/10.3390/pharmaceutics15071829
61. Kafle A, Luis E, Silwal R et al (2021) 3D/4D printing of polymers: fused deposition modelling (FDM), selective laser sintering (SLS), and stereolithography (SLA). Polymers (Basel) 13:3101. https://doi.org/10.3390/polym13183101
62. Kantaros A, Ganetsos T (2023) From static to dynamic: smart materials pioneering additive manufacturing in regenerative medicine. Int J Mol Sci 24:15748. https://doi.org/10.3390/ijms242115748
63. Kesireddy V, Kasper FK (2016) Approaches for building bioactive elements into synthetic scaffolds for bone tissue engineering. J Mater Chem B 4:6773–6786. https://doi.org/10.1039/C6TB00783J
64. Khan F, Tanaka M (2017) Designing smart biomaterials for tissue engineering. Int J Mol Sci 19:17. https://doi.org/10.3390/ijms19010017
65. Khan HM, Liao X, Sheikh BA et al (2022) Smart biomaterials and their potential applications in tissue engineering. J Mater Chem B 10:6859–6895. https://doi.org/10.1039/D2TB01106A
66. Khedkar A, Tupe D, Gholap AD (2012a) Pharmacosomes: an emerging vesicular system. Inven Rapid NDDS [ISSN 0976-3791] 1:1–3
67. Khedkar A, Tupe D, Gholap AD (2012b) Cubosomes: emerging drug delivery system. Inven Rapid NDDS [ISSN 0976-3791] 1:1–3
68. Kim Y-H, Vijayavenkataraman S, Cidonio G (2024) Biomaterials and scaffolds for tissue engineering and regenerative medicine. BMC Methods 1:2. https://doi.org/10.1186/s44330-024-00002-7
69. Kouhi M, de Souza Araújo IJ, Asa'ad F et al (2024) Recent advances in additive manufacturing of patient-specific devices for dental and maxillofacial rehabilitation. Dent Mater 40:700–715. https://doi.org/10.1016/j.dental.2024.02.006
70. Krishani M, Shin WY, Suhaimi H, Sambudi NS (2023) Development of scaffolds from bio-based natural materials for tissue regeneration applications: a review. Gels 9:100. https://doi.org/10.3390/gels9020100
71. Kumar KB, Rajitha A, Rao AK et al (2023) SMART materials for biomedical applications: advancements and challenges. E3S Web Conf 430:01133. https://doi.org/10.1051/e3sconf/202343001133
72. Lee EJ, Kasper FK, Mikos AG (2014) Biomaterials for tissue engineering. Ann Biomed Eng 42:323–337. https://doi.org/10.1007/s10439-013-0859-6
73. Li H, Yang Y, Hong W et al (2020) Applications of genome editing technology in the targeted therapy of human diseases: mechanisms, advances and prospects. Signal Transduct Target Ther 5:1. https://doi.org/10.1038/s41392-019-0089-y
74. Liu Y, Liu X, Lu J et al (2024) Post-treatment technologies for high-speed additive manufacturing: status, challenge and tendency. J Mater Res Technol 30:1057–1082. https://doi.org/10.1016/j.jmrt.2024.03.110

75. Loh QL, Choong C (2013) Three-dimensional scaffolds for tissue engineering applications: role of porosity and pore size. Tissue Eng Part B Rev 19:485–502. https://doi.org/10.1089/ten.teb.2012.0437
76. Madrid-Sánchez A, Duerr F, Nie Y et al (2022) Fabrication of large-scale scaffolds with microscale features using light sheet stereolithography. Int J Bioprinting 9:650. https://doi.org/10.18063/ijb.v9i2.650
77. Magar TV, Gholap AD, Musmade DS (2011) Elastic vesicles as drugs carriers through the skin. Inven Rapid NDDS [ISSN 0976-3791] 1:1–2
78. Mahajan AM, Gholap AD, Borude SS (2011a) recent updates on Q fever: diagnosis and treatment. Pharmacology 1:254–264
79. Mahajan AM, Gholap AD, Borude SS (2011b) POEMS syndrome: a review. Pharmacology 1:265–274
80. Mani MP, Sadia M, Jaganathan SK et al (2022) A review on 3D printing in tissue engineering applications. J Polym Eng 42:243–265. https://doi.org/10.1515/polyeng-2021-0059
81. Mantha S, Pillai S, Khayambashi P et al (2019) Smart hydrogels in tissue engineering and regenerative medicine. Materials (Basel) 12:3323. https://doi.org/10.3390/ma12203323
82. Mao AS, Mooney DJ (2015) Regenerative medicine: current therapies and future directions. Proc Natl Acad Sci 112:14452–14459. https://doi.org/10.1073/pnas.1508520112
83. Mao H, Yang L, Zhu H et al (2020) Recent advances and challenges in materials for 3D bioprinting. Prog Nat Sci Mater Int 30:618–634. https://doi.org/10.1016/j.pnsc.2020.09.015
84. Meng M, Wang J, Huang H et al (2023) 3D printing metal implants in orthopedic surgery: methods, applications and future prospects. J Orthop Transl 42:94–112. https://doi.org/10.1016/j.jot.2023.08.004
85. Mesut B, Al-Mohaya M, Gholap AD et al (2024) Demystifying the potential of lipid-based nanocarriers in targeting brain malignancies. Naunyn Schmiedeberg's Arch Pharmacol. https://doi.org/10.1007/s00210-024-03212-6
86. Mirzaali MJ, Moosabeiki V, Rajaai SM et al (2022) Additive manufacturing of biomaterials—design principles and their implementation. Materials (Basel) 15:5457. https://doi.org/10.3390/ma15155457
87. Mobarak MH, Islam MA, Hossain N et al (2023) Recent advances of additive manufacturing in implant fabrication—a review. Appl Surf Sci Adv 18:100462. https://doi.org/10.1016/j.apsadv.2023.100462
88. Montoya C, Du Y, Gianforcaro AL et al (2021) On the road to smart biomaterials for bone research: definitions, concepts, advances, and outlook. Bone Res 9:12. https://doi.org/10.1038/s41413-020-00131-z
89. Ni J, Ling H, Zhang S et al (2019) Three-dimensional printing of metals for biomedical applications. Mater Today Bio 3:100024. https://doi.org/10.1016/j.mtbio.2019.100024
90. Ntege EH, Sunami H, Shimizu Y (2020) Advances in regenerative therapy: a review of the literature and future directions. Regen Ther 14:136–153. https://doi.org/10.1016/j.reth.2020.01.004
91. Olson JL, Atala A, Yoo JJ (2011) Tissue engineering: current strategies and future directions. Chonnam Med J 47:1. https://doi.org/10.4068/cmj.2011.47.1.1
92. Pardeshi S, Gholap A, More M et al (2023) Dendrimers based antibacterial and antiviral materials. In: Antibacterial and antiviral functional materials. ACS Publications, pp 139–169
93. Parit S, Manchare A, Gholap AD et al (2024) Antibody-drug conjugates: a promising breakthrough in cancer therapy. Int J Pharm 659:124211. https://doi.org/10.1016/j.ijpharm.2024.124211
94. Park S, Shou W, Makatura L et al (2022) 3D printing of polymer composites: materials, processes, and applications. Matter 5:43–76. https://doi.org/10.1016/j.matt.2021.10.018
95. Pavan Kalyan B, Kumar L (2022) 3D printing: applications in tissue engineering, medical devices, and drug delivery. AAPS PharmSciTech 23:92. https://doi.org/10.1208/s12249-022-02242-8
96. Persaud A, Maus A, Strait L, Zhu D (2022) 3D bioprinting with live cells. Eng Regen 3:292–309. https://doi.org/10.1016/j.engreg.2022.07.002

97. Placone JK, Mahadik B, Fisher JP (2020) Addressing present pitfalls in 3D printing for tissue engineering to enhance future potential. APL Bioeng 4. https://doi.org/10.1063/1.5127860
98. Puri AV, Gholap AD (2018) Comparative evaluation of papaya and banana starch as disintegrant in diclofenac sodium tablet formulation. Int J Pharm Bio Sci 8:512–522
99. Qu M, Wang C, Zhou X et al (2021) Multi-dimensional printing for bone tissue engineering. Adv Healthc Mater 10. https://doi.org/10.1002/adhm.202001986
100. Rajurkar VG, Gholap AD (2012a) Potential applications of superparmagnetic nanoparticles technology. Inven Rapid NDDS [ISSN 0976-3791] 1:1–3
101. Rajurkar VG, Gholap AD (2012b) New threat of clinical toxicities evoked from nanocarrier systems. Inven Rapid NDDS [ISSN 0976-3791] 1:1–4
102. Rajurkar VG, Gholap AD (2012c) Lipid nanocapsules: a novel promising tool as drug delivery system. Inven Rapid NDDS [ISSN 0976-3791] 1:1–4
103. Rajurkar VG, Gholap AD (2012d) Potential applications of capillary electromigration methods for the improvement of separation efficiency. Inven Rapid Pharm Anal Qual Assur [ISSN 0976-3813] 1:1–3
104. Rajurkar VG, Gholap AD (2012e) Recent technologies in nanomedicine for treatment of cancer. Inven Rapid Pharm Tech 1:1–3
105. Rajurkar VG, Gholap AD (2012f) Recent technologies in nanomedicine for treatment of cancer. Inven Rapid Pharm Tech 4:1–3
106. Rambhia KJ, Ma PX (2015) Controlled drug release for tissue engineering. J Control Release 219:119–128. https://doi.org/10.1016/j.jconrel.2015.08.049
107. Rangel R, Swanson WB, Wu DT (2024) The future of cell-instructive biomaterials for tissue regeneration—a perspective from early career clinician-scientists. Front Mater 10. https://doi.org/10.3389/fmats.2023.1328904
108. Rojekar S, Gholap AD, Togre N et al (2024) Current status of mannose receptor-targeted drug delivery for improved anti-HIV therapy. J Control Release 372:494–521. https://doi.org/10.1016/j.jconrel.2024.06.002
109. Rouf S, Malik A, Singh N et al (2022) Additive manufacturing technologies: industrial and medical applications. Sustain Oper Comput 3:258–274. https://doi.org/10.1016/j.susoc.2022.05.001
110. Saberianpour S, Heidarzadeh M, Geranmayeh MH et al (2018) Tissue engineering strategies for the induction of angiogenesis using biomaterials. J Biol Eng 12:36. https://doi.org/10.1186/s13036-018-0133-4
111. Sen M (2020) Nanocomposite materials. In: Nanotechnology and the environment. IntechOpen
112. Shaikh MS, Faiyazuddin M, Khan MS et al (2024) Chikungunya virus vaccine: a decade of progress solving epidemiological dilemma, emerging concepts, and immunological interventions. Front Microbiol 15. https://doi.org/10.3389/fmicb.2024.1413250
113. Shirazi SFS, Gharehkhani S, Mehrali M et al (2015) A review on powder-based additive manufacturing for tissue engineering: selective laser sintering and inkjet 3D printing. Sci Technol Adv Mater 16:033502. https://doi.org/10.1088/1468-6996/16/3/033502
114. Simorgh S, Alasvand N, Khodadadi M et al (2022) Additive manufacturing of bioactive glass biomaterials. Methods 208:75–91. https://doi.org/10.1016/j.ymeth.2022.10.010
115. Singh T, Ravi Kumar V, Gholap AD, Sanap G et al (2016) In vitro antipyrial activity of psidium guajava leaf extract. Eur J Pharm Med Res 3:312–315
116. Singh D, Mathur A, Arora S et al (2022) Journey of organ on a chip technology and its role in future healthcare scenario. Appl Surf Sci Adv 9:100246. https://doi.org/10.1016/j.apsadv.2022.100246
117. Singh AK, Malviya R, Prajapati B, Singh S, Goyal P (2023) Utilization of stimuli-responsive biomaterials in the formulation of cancer vaccines. J Funct Biomater 14(5):247
118. Smagul S, Kim Y, Smagulova A et al (2020) Biomaterials loaded with growth factors/cytokines and stem cells for cardiac tissue regeneration. Int J Mol Sci 21:5952. https://doi.org/10.3390/ijms21175952

119. Solchaga LA, Tognana E, Penick K et al (2006) A rapid seeding technique for the assembly of large cell/scaffold composite constructs. Tissue Eng 12:1851–1863. https://doi.org/10.1089/ten.2006.12.1851
120. Stratton S, Shelke NB, Hoshino K et al (2016) Bioactive polymeric scaffolds for tissue engineering. Bioact Mater 1:93–108. https://doi.org/10.1016/j.bioactmat.2016.11.001
121. Sun L, Huang WM, Ding Z et al (2012) Stimulus-responsive shape memory materials: a review. Mater Des 33:577–640. https://doi.org/10.1016/j.matdes.2011.04.065
122. Szwed-Georgiou A, Płociński P, Kupikowska-Stobba B et al (2023) Bioactive materials for bone regeneration: biomolecules and delivery systems. ACS Biomater Sci Eng 9:5222–5254. https://doi.org/10.1021/acsbiomaterials.3c00609
123. Thomas DS, Gilbert SW (2014) Costs and cost effectiveness of additive manufacturing. National Institute of Standards and Technology, Gaithersburg
124. Tofail SAM, Koumoulos EP, Bandyopadhyay A et al (2018) Additive manufacturing: scientific and technological challenges, market uptake and opportunities. Mater Today 21:22–37. https://doi.org/10.1016/j.mattod.2017.07.001
125. Tollemar V, Collier ZJ, Mohammed MK et al (2016) Stem cells, growth factors and scaffolds in craniofacial regenerative medicine. Genes Dis 3:56–71. https://doi.org/10.1016/j.gendis.2015.09.004
126. Tom T, Sreenilayam SP, Brabazon D et al (2022) Additive manufacturing in the biomedical field-recent research developments. Result Eng 16:100661. https://doi.org/10.1016/j.rineng.2022.100661
127. Tupe D, Gholap AD (2012) The threat of photoaging: clinical conditions, evaluation and treatment. Inven Rapid Cosmeceuticals [ISSN 0976-3864] 1:1–3
128. Ul Haq I, Krukiewicz K (2023) Antimicrobial approaches for medical implants coating to prevent implants associated infections: insights to develop durable antimicrobial implants. Appl Surf Sci Adv 18:100532. https://doi.org/10.1016/j.apsadv.2023.100532
129. Velasco MA, Narváez-Tovar CA, Garzón-Alvarado DA (2015) Design, materials, and mechanobiology of biodegradable scaffolds for bone tissue engineering. Biomed Res Int 2015:1–21. https://doi.org/10.1155/2015/729076
130. Vesvoranan O, Anup A, Hixon KR (2022) Current concepts and methods in tissue interface scaffold fabrication. Biomimetics 7:151. https://doi.org/10.3390/biomimetics7040151
131. Vilas Bhagat AG (2012) Gold nanorods: a promising drug delivery vehicle. Inven Rapid NDDS [ISSN 0976-3864] 1:1–4
132. Vyas C, Pereira R, Huang B et al (2017) Engineering the vasculature with additive manufacturing. Curr Opin Biomed Eng 2:1–13. https://doi.org/10.1016/j.cobme.2017.05.008
133. Vyas J, Singh S, Shah I, Prajapati BG (2023a) Potential applications and additive manufacturing technology-based considerations of mesoporous silica: a review. AAPS PharmSciTech 25(1):6
134. Vyas J, Shah I, Singh S, Prajapati BG (2023b) Biomaterials-based additive manufacturing for customized bioengineering in management of otolaryngology: a comprehensive review. Front Bioeng Biotechnol 11:1234340
135. Vyas J, Raytthatha N, Singh S, Prajapati BG, Mohite P, Munde S (2024a) Sustainable sources of raw materials as substituting biomaterials for additive manufacturing of dental implants: a review. Peri Impl Res 8(1):3
136. Vyas J, Raytthatha N, Singh S, Prajapati B (2024b) Next-generation computational automation-based additive manufacturing of pharmaceuticals: an approach to fabricate precise medicine. In: Handbook of 3D printing in pharmaceutics. CRC Press, pp 153–175
137. Wan X, Xiao Z, Tian Y et al (2024) Recent advances in 4D printing of advanced materials and structures for functional applications. Adv Mater. https://doi.org/10.1002/adma.202312263
138. Wang Z, Wang Z, Lu WW et al (2017) Novel biomaterial strategies for controlled growth factor delivery for biomedical applications. NPG Asia Mater 9:e435–e435. https://doi.org/10.1038/am.2017.171
139. Wang M, Wu Y, Li G et al (2024) Articular cartilage repair biomaterials: strategies and applications. Mater Today Bio 24:100948. https://doi.org/10.1016/j.mtbio.2024.100948

140. Winarso R, Anggoro PW, Ismail R et al (2022) Application of fused deposition modeling (FDM) on bone scaffold manufacturing process: a review. Heliyon 8:e11701. https://doi.org/10.1016/j.heliyon.2022.e11701
141. Xing F, Xu J, Yu P et al (2023a) Recent advances in biofabrication strategies based on bioprinting for vascularized tissue repair and regeneration. Mater Des 229:111885. https://doi.org/10.1016/j.matdes.2023.111885
142. Xing Y, Qiu L, Liu D et al (2023b) The role of smart polymeric biomaterials in bone regeneration: a review. Front Bioeng Biotechnol 11:1240861. https://doi.org/10.3389/fbioe.2023.1240861
143. Yang X, Li S, Ren Y et al (2022) 3D printed hydrogel for articular cartilage regeneration. Compos Part B Eng 237:109863. https://doi.org/10.1016/j.compositesb.2022.109863
144. Yeo M, Sarkar A, Singh YP et al (2024) Synergistic coupling between 3D bioprinting and vascularization strategies. Biofabrication 16:012003. https://doi.org/10.1088/1758-5090/ad0b3f
145. Yu J, Park SA, Kim WD et al (2020) Current advances in 3D bioprinting technology and its applications for tissue engineering. Polymers (Basel) 12:2958. https://doi.org/10.3390/polym12122958
146. Zakrzewski W, Dobrzyński M, Szymonowicz M, Rybak Z (2019) Stem cells: past, present, and future. Stem Cell Res Ther 10:68. https://doi.org/10.1186/s13287-019-1165-5
147. Zaszczyńska A, Moczulska-Heljak M, Gradys A, Sajkiewicz P (2021) Advances in 3D printing for tissue engineering. Materials (Basel) 14:3149. https://doi.org/10.3390/ma14123149
148. Zeng C-W (2023) Advancing spinal cord injury treatment through stem cell therapy: a comprehensive review of cell types, challenges, and emerging technologies in regenerative medicine. Int J Mol Sci 24:14349. https://doi.org/10.3390/ijms241814349
149. Zhang K, Wang S, Zhou C et al (2018) Advanced smart biomaterials and constructs for hard tissue engineering and regeneration. Bone Res 6:31. https://doi.org/10.1038/s41413-018-0032-9
150. Zhao T, Li X, Li H et al (2023) Advancing drug delivery to articular cartilage: from single to multiple strategies. Acta Pharm Sin B 13:4127–4148. https://doi.org/10.1016/j.apsb.2022.11.021

6 Biomaterial-Additive Manufactured Scaffolds for Tissue Engineering and Regenerative Medicine

Viola Colaco, Deepanjan Datta, Sony Priyanka Bandi, Namdev Dhas, and Prabhanjan S. Giram

Abstract

Additive manufacturing (AM), referred to as 3D printing, is recognized as an innovative approach in current manufacturing and engineering design, especially for bio-based materials. This advancement has aided in the development of personalized biomaterials and medical devices, such as prosthetics, implants, and orthotics, with sophisticated internal microstructures and adjustable properties. It enables the fabrication of intricate structures, such as scaffolds used in tissue engineering (TE), which promotes tissue regeneration and ingrowth in soft tissue and hard bone applications. The selection of biomaterials for scaffold development has a significant influence on the architecture, characteristics, and cellular response. Thus, meticulous material selection is required to ensure optimal scaffold performance. The essential characteristics for scaffold fabrication in TE include biocompatibility, biodegradability, biomimicry, porous structure, and compatibility with manufacturing technologies. The material properties have a pivotal role in promoting cell proliferation, differentiation, and regeneration, and hence, the chosen material must align with the scaffold requirements. Materials

V. Colaco · D. Datta (✉) · N. Dhas
Department of Pharmaceutics, Manipal College of Pharmaceutical Sciences, Manipal Academy of Higher Education, Manipal, Karnataka, India
e-mail: deepanjan.datta@manipal.edu

S. P. Bandi
Loka Laboratories Private Limited, Technology Business Incubator, BITS Pilani Hyderabad Campus, Medchal, Telangana, India

P. S. Giram
Department of Pharmaceutics, Dr. D.Y. Patil Institute of Pharmaceutical Sciences and Research, Pune, Maharashtra, India

Department of Pharmaceutical Sciences, University at Buffalo, State University of New York, Buffalo, NY, USA

S. Singh et al. (eds.), *Biomaterial-based Additive Manufacturing in Tissue Engineering and Regeneration*, Biomaterials, Bioengineering and Sustainability 7,
https://doi.org/10.1007/978-3-031-96070-3_6

utilized in scaffold preparation include composites, metals, polymers, ceramics, and nanomaterials. Stainless steel was initially utilized, but its non-biocompatible nature led to the exploration of biocompatible metal alternatives. Polymers like alginate, chitosan, collagen, hyaluronic acid (HA), polycaprolactone, polyglycolic acid, and polylactic acid have found widespread use in scaffold fabrication for TE and regeneration. Furthermore, combinations of materials are frequently used to enhance scaffold qualities under optimal additive manufacturing conditions, aiming to achieve superior performance and functionality. To summarize, this chapter will discuss the principles of AM, the types of biomaterials used, the fabrication and characterization techniques of the biomaterial scaffold, and its applications in tissue engineering and regenerative medicine.

Keywords

Additive manufacturing · Biomaterials · Scaffolds · 3D printing · Tissue engineering · Regenerative medicine

Abbreviations

ADSC	Adipose-derived stem cells
Al_2O_3	Alumina
AM	Additive manufacturing
BJ	Binder jetting
BTE	Bone Tissue Engineering
CAD	Computer-aided design
dECM	Decellularized ECM
DED	Direct energy deposition
DLP	Digital light processing
DM	Decellularized matrix
EBM	Electron beam melting
ECM	Extracellular matrix
Ecs	Endothelial cells
FDA	Food and Drug Administration
FDM	Fused deposition modelling
GF	Growth factor
HA	Hyaluronic acid
Hap	Hydroxyapatite
HDFs	Human dermal fibroblasts
hEK	Human epidermal keratinocytes
HMVEC	Human microvascular endothelial cells
HPMC	Hydroxypropyl methylcellulose
LBL	Layer-by-layer
LOM	Laminated object manufacturing
MJ	Material jetting

OCIs	Osteochondral injuries
PBP	Powder bed process
PCL	Poly (ε-caprolactone)
PEGDA	Polyethylene-Glycol Diacrylate
PLA	Poly(lactic acid)
PLGA	Poly-Lactide-Glycolic Acid
RM	Regenerative medicine
SF	Silk fibroin
SL	Sheet lamination
SLA	Stereolithography
SLM	Selective laser melting
SLS	Selective laser sintering
TCP	Tricalcium phosphates
TE	Tissue engineering
TERM	Tissue engineering and regenerative medicine
TGFβ	Transforming growth factor-β
UAM	Ultrasonic additive manufacturing
VS	Vascular system
WE	Wire extrusion
ZrO_2	Zirconia
3D	Three-dimensional

6.1 Introduction

The convergence of biomaterials and additive manufacturing (AM) is establishing an innovative trend for scaffold design for tissue engineering (TE), facilitating the development of highly effective and personalized regenerative medicine (RM) therapies. AM, commonly referred to as three-dimensional (3D) printing, has transformed the fabrication of intricate structures, providing precise control over biomaterial architecture and composition. This advanced technique facilitates designing and fabricating customized, personalized biomaterials and biomedical devices, such as prosthetics, implants, and orthotics, with intricate microstructures and tunable properties [1, 2].

The TE, combined with AM, arose as an alternate strategy for repairing impaired tissues and organs by producing personalized alternatives that restore, enhance or preserve tissue function [3]. This multidisciplinary field integrates cell regeneration, scaffolds and growth factors (GFs) to generate biological alternatives that restore, maintain or enhance functioning, enhancing or substituting biological functions through clinical, biological, chemical, material science and engineering principles [4–6]. RM extends TE by incorporating additional strategies to promote *in vivo* tissue or organ regeneration. Tissue engineering and regenerative medicine (TERM) amalgamates basic sciences to achieve functional tissue or organ repair and construction [7]. The primary objective of TERM is to fabricate 3D biomaterial complexes that mimic the function of living tissue and can be utilized for repairing or regenerating injured tissue. These complexes must facilitate cell growth, nutrient and metabolic waste transport and gas exchange [4, 7]. TE is classified as *ex vivo* and *in situ* TE. In *ex vivo*, the stem cells are extracted and seeded onto an external scaffold within a appropriate environment for cell development and differentiation into the target tissue. After being implanted, the designed tissue that matches the size and shape of the defect gradually integrates with the surrounding tissue as the scaffold deteriorates. *In situ* TE involves implanting a pre-fabricated scaffold made from biocompatible materials directly into the tissue without pre-seeding, attracting surrounding cells to facilitate tissue regeneration, and providing an immunologically compatible alternative to *ex vivo* technique [8].

Biomaterials play a crucial role in TE, functioning as scaffolds that facilitate cellular growth, differentiation, and regeneration of tissues [3]. The choice of biomaterials depends on the specific requirements of the targeted tissue and must have properties that include biocompatibility, bioactivity, biodegradability and sterilizability [9]. These materials, derived from either natural or synthetic sources, must be biocompatible to avoid immunological response, sterilizable for safe incorporation into host tissue, biodegradable to disappear after achieving their function, and bioactive to promote tissue response [8]. Biomaterials facilitate tissue regeneration, induce cellular responses, and block natural responses such as immunological rejection of cell transplants [5, 8]. The choice of biomaterial for scaffold fabrication is crucial as it determines the scaffolds application. Scaffolds are temporary or permanent artificial extracellular matrix (ECM) that accommodate cells and assist 3D tissue regeneration. They act as cellular systems or delivery vehicles for tissue regeneration, requiring adequate colonization by host cells for successful regeneration and repair. The fundamental notion underlying TE involves the incorporation of scaffold with living cells or bioactive molecules to generate a "TE construct" that tissues repair and regenerate in the presence of an adequate blood supply [10]. Scaffold properties depend on their composition and must be tested both *in vitro* and *in vivo*. The ECM within scaffolds is crucial for enhancing cell proliferation and adhesion, providing biochemical signals and 3D mechanical support and influencing cell behaviour, shape, differentiation, proliferation and migration [5, 11]. The ideal scaffolds for TE should possess biocompatibility, biodegradability, porosity, appropriate pore size, mechanical properties, and surface properties (Fig. 6.1). As the scaffolds are fabricated from external materials, they must be biocompatible, safe and free from adverse effects, ensuring active and controllable interaction with tissue and the immune system to promote cellular activity and function [12].

Scaffolds temporarily support defective body structures, enabling stress transmission and degradation over time. They must degrade to allow cells to develop new ECM and, thus are not intended to be permanent implants. The biodegradability rate of scaffolds should align with tissue formation rates in the human body. Bioabsorbable and non-toxic material degrades over time or through metabolism. Biodegradable polymers are often preferred due to their flexible mechanical property and natural bioavailability [5].

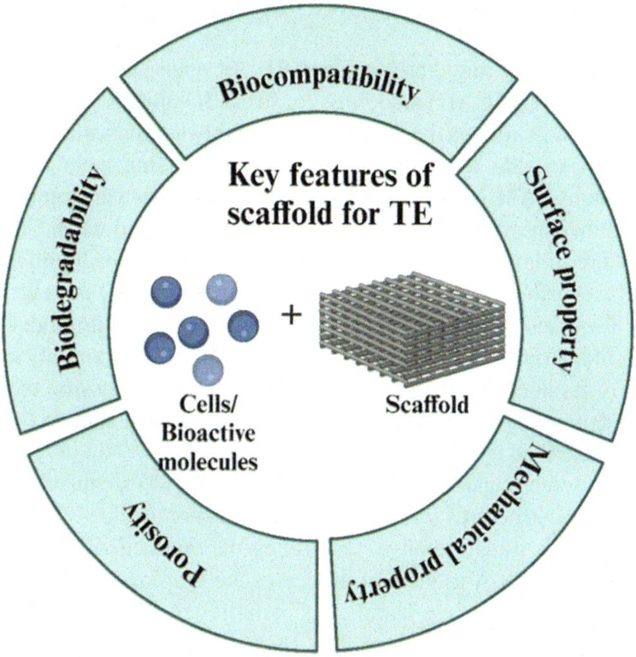

Fig. 6.1 Key features of scaffold for tissue engineering (TE). (Created with BioRender.com)

The materials used to fabricate scaffolds must mimic the tissues or organs they replace and provide mechanical integrity to the completion of remodeling. They should replicate the target tissue's composition, structure, and functions to transfer load efficiently and fully replace it with newly formed tissue after degradation. Biomaterials employed in scaffold fabrication must be selected according to the distinct requirements of individual tissues. This includes replicating the biochemical, mechanical, and topographical cues of the native ECM to enhance cellular adhesion, differentiation and proliferation [13, 14].

The fabricated scaffolds must have pores designed to facilitate cellular movement and tissue ingrowth, assist vasculature formation, and improve cell viability. The scaffold porous structure ensures that nutrients diffuse to cells within the construct. Ideally, the scaffold should have a substantial pore size, high porosity to cell motility, smaller pores to increase surface area and sufficient ligand density to optimal cell binding. Scaffolds for TE should have pore sizes between 50 and 900 μm and a porosity of at least 70% [5]. This chapter will explore the principle of AM, the types of biomaterials used, the fabrication and characterization techniques of the biomaterial scaffold, and its applications in TERM.

6.2 Fundamentals of Additive Manufacturing

The AM or 3D printing, rapid prototyping (RP), or open-ended fabrication, utilizes computer-aided design (CAD) models to construct 3D objects layer-by-layer (LBL) [15]. This method facilitates the fabrication of highly sophisticated and customized structures using a wide range of materials, encompassing metals, ceramics, and polymers. Notably, AM offers cost-efficiency and exclusive customization options [15, 16]. The invention of 3D printing dates to the year 1986 when Charles W Hull introduced stereolithography (SLA), a technique that solidifies liquid material LBL along three axes using ultraviolet light to generate a solid 3D structure [5]. AM is extensively used and researched in fields such as medical, automotive, aerospace, marine, and industrial spare parts manufacturing [16]. It is commonly known as RP, and free-form production is guided by 3D printing, encompassing techniques that build parts or products in an LBL approach, typically small size, in limited quantities, and characterized by intricate custom design [17].

AM has garnered considerable attention in TE due to its capability to precisely fabricate complex structures with reliability and customization. The application of AM in healthcare is transformative, facilitating the production of complex and tailored medical products. Various AM processes are employed to fabricate different materials, with critical process parameters including quality, manufacturing speed, resolution, cost, surface smoothness, build volume, and component strength [18]. These materials allow for the precise control of structural parameters like pore diameter, porosity and interconnectivity, which are vital for maintaining cell adhesion, growth, and differentiation [19].

AM offers numerous advantages in TE, making it a promising approaches for the fabrication of intricate structures and devices. It facilitates the fabrication of customizable structures that satisfy the mechanical properties of host tissue, thereby reducing stress concentration and improving device-tissue compatibility [20]. AM facilitates the production of complex structures using computer-assisted designs that are challenging or infeasible to fabricate using traditional manufacturing techniques [19]. It provides substantial control over the final product's structure and design, allowing the formation of structures with specific properties and functions [21]. Additionally, AM can fabricate biomimetic structures that emulate the natural structure and functionality of biological tissues, enhancing tissue integration and functionality [22].

Despite its advancements, AM faces several challenges, including that the limited range of materials suitable for AM technologies. Although advancement in the development of novel materials, there remains a need for more biocompatible and bioactive materials. Furthermore, the current state of AM technology often lacks the necessary standards for characterization and validation, impacting the scalability and repeatability of the scaffold. There is also a lack of standards and guidelines for clinical applications of AM [23].

To overcome these challenges, AM efforts for TE should focus on developing biocompatible and bioactive material, establishing standardization and validation protocol, and creating a regulatory framework, to ensure consistent and reliable use in clinical applications.

6.3 Types of Biomaterials in Additive Manufacturing

A variety of biomaterials have been used in the AM of scaffolds. Biobased natural materials possess biocompatibility, ensuring their safety and the absence of harmful chemical release during biodegradation. Thus, it is particularly advisable to construct scaffolds utilizing these materials. Thus far, there have been no individual materials that possess all the characteristics required for the fabrication of scaffolds. Thus, it is suggested to incorporate multiple materials in order to achieve the desired qualities. To fabricate a dependable scaffold through the integration of various materials, it is imperative to carefully select an effective manufacturing approach. To this end, various biomaterials, including ceramic, synthetic, natural, and composite hybrids, have been discussed and manufactured using additive techniques (SLA, selective laser sintering (SLS), and fused deposition modelling (FDM)).

6.3.1 Ceramic Biomaterials

Ceramics are a type of biomaterial that consists of inorganic salts, specifically calcium and phosphate. Owing to their capacity to stimulate osteogenesis and promote tissue regeneration, these biomaterials show great promise for applications in bone and dental TE. The bone tissue replicates the inorganic composition of calcium and phosphate salt [24]. Bio-ceramics are sourced from natural sources and have found extensive applications in bone tissue engineering (BTE). Bio-ceramics are non-metallic solids that possess favourable characteristics such as bio-inertness, compatibility, bioactivity, mechanical strength and osteo-conductivity [25, 26]. Furthermore, bio-ceramics have the ability to induce osteogenesis and enhance the osteogenic properties of scaffolds. Out of all the biomaterials, bio-ceramic scaffolds have demonstrated greater efficacy in the treatment of small bone defects, including applications in bone-filling and orthopaedic implants. Three primary categories of ceramics exist: bioactive, bioinert, and bioresorbable. The bioinert ceramics encompass pyrolytic carbon, alumina (Al_2O_3), and zirconia (ZrO_2). Bioactive ceramics consist of glass ceramics and bio-glasses, whereas bioresorbable ceramics comprise calcium phosphates. The ceramics most frequently utilized in BTE applications are tricalcium phosphates (TCP), hydroxyapatite (HAp), and their mixtures [27].

Composites encompass ceramic and polymer-based biomaterials. Poly-Lactide-Glycolic Acid (PLGA), Chitosan, and Polyethylene-Glycol Diacrylate (PEGDA) are frequently used polymers. The composition of the skin was enhanced by incorporating materials like graphene, silica, ZrO_2, and bioglass. This was done to create composites with mechanical properties similar to bones [28]. 3D printing allows for the production of grafts that are tailored to the unique needs of the patient, considering factors such as histocompatibility, graft size, and bone formation rates. Studies have shown the development of 3D-porous materials with the aim of improving vascularization within the scaffolds. The 3D-printed scaffolds demonstrated compressive strength which was determined to be 110 MPa, with a porosity of 34%. These scaffolds were made using gahnite, hardystonite, hydroxypropyl methylcellulose (HPMC), strontium, and sodium polyacrylate. Due to its exceptional mechanical properties and ability to promote blood vessel growth, BTE has significant promise [29, 30].

6.3.2 Natural Polymeric Biomaterials

Natural polymeric biomaterials in TE have shown promising alternatives to conventional treatment strategies [31]. A variety of natural polymers (cellulose, chitosan, alginate, starch, collagen, keratin, gelatin, HA, guar gum, pullulan, fibroin and elastin) among others, have been documented in the preparation of scaffolds for TERM [32]. Natural polymers display comparable traits to soft tissues, demonstrating bioactivity, exceptional cell adhesion and proliferation, while fulfilling the criteria of biodegradability and biocompatibility. Furthermore, they are well-known for their extensive accessibility, environmental friendliness, and versatility in diverse applications. Furthermore, natural polymers commonly exhibit inadequate mechanical and physical stability, which restricts their usage in load-bearing orthopaedic applications [33, 34].

The main hydrolyzed collagen constituent, gelatin, spontaneously forms in the ECM and suspends gel cells at low temperatures. Natural starch polymers with water-based binders have been successfully processed employing AM. The combination of synthetic and natural polymers has been shown to improve the mechanical properties of biomaterials [35]. Starch-based polymers have an extended breakdown time, larger porosity, and greater cellular integration, that makes them excellent for BTE [36]. 3D-printed material combinations can produce 3D pore scaffolds. Starch-cellulose-based polymeric composites are biocompatible in the delivery of bioactive compounds [37]. The findings demonstrated that by optimizing the process factors such as scanning speed and laser power, it was possible to formulate biopolymer scaffolds using the combination of starch and cellulose acetate. The utilization of AM in TE has enabled the creation of tissue-like structures through the use of cutting-edge technology for printing cells and matrix materials. Research has demonstrated that utilizing AM for producing cell charging structures can effectively locate the cells, therefore achieving a high level of cell survival in the fabricated constructs [38, 39].

6.3.3 Synthetic Polymeric Biomaterials

Synthetic polymers possess superior mechanical qualities compared to natural polymers. Nevertheless, they also pose a significant threat of immunological rejection, and the process of breakdown releases carbon dioxide, which can cause harm to cells [40]. Synthetic polymers function as biomaterials with predictable qualities, including a diverse spectrum of mechanical and physical characteristics such as controllable degradation rates. If they are produced through controlled synthesis, they pose minimal immunological hazards. Aliphatic polyesters, such as PLGA, poly (ε-caprolactone) (PCL), and poly(lactic acid) (PLA), are often utilized as synthetic polymers in BTE applications. PLA demonstrates excellent compatibility with living organisms, although it undergoes deterioration at a rather sluggish pace, taking approximately 3–5 years. Therefore, the combination of PLA with HAp is utilized to enhance its mechanical and physical strength [25, 41]. PLGA polymer, similar to poly(lactic acid), shows promise for use in the biomedical area and has received approval from the FDA. This is derived by the process of copolymerization involving glycolic acid and lactic acid. PLGA polymer is highly hydrophilic and degradable compared to PLA polymer. Modifying the lactide-to-glycolide monomer ratio enables the regulation of poly(lactic-co-glycolide) (PLGA) degradation [42]. PCL is a type of polyester that is partially crystalline, capable of being broken down by natural processes, and non-toxic. It exhibits water resistance and breaks down slowly, taking over 24 months to degrade. These issues can be resolved by including different polymers or creating composites [43].

6.3.4 Composite Biomaterials and Hybrid Constructs

Composite biomaterials are composed of multiple distinct material types. Composite biomaterials leverage the benefits of each constituent phase in the composite. For instance, composites that combine bioactive glasses with biodegradable polymers leverage the advantages of both components. The bioactive glass particles contribute to the bioactivity feature, while the polymer provides the desired flexibility. This novel material family has expanded the potential applications for diverse tissue regeneration. Recently, there has been a concept of green bio-composite, which is the combination of natural components and degradable inorganic fillers, including titanium, silica and Al_2O_3 [44]. PCL/HA composites have generated significant interest in the field of BTE. One of the study findings revealed that the mechanical properties of PCL/HA bio-composites were enhanced, in addition to the pure PCL [45, 46]. Farag et al. created composite scaffolds using the AM approach, incorporating Mg-phosphate and gelatin [47]. The study demonstrated that the use of gelatin significantly enhanced the scaffold's mechanical strength, with a compressive strength of 16.7 ± 1.9 MPa. Additionally, gelatin increased cell survival and drug release characteristics. Es-saddik et al. examined how the sintering temperature affects the microstructure, mechanical characteristics, and compaction of a composite material made of 10% zirconia and HAp [48]. The findings indicated that the composite's mechanical characteristics were highly responsive to both the sintering temperature and densification procedure. Nevertheless, despite various advantages, composite biomaterials have some limitations that include a slow rate of decay, less cellular contact, and typically restricted to rigid tissue, among others [49].

6.4 Manufacturing Techniques for Biomaterial Scaffolds

AM technologies in TE encompass seven processes: liquid polymerization, material deposition, powder processes, sheet lamination, material jetting,, binder jetting, and direct energy deposition (Fig. 6.2).

6.4.1 Liquid Polymerization

Liquid polymerization, also known as vat polymerization, was the first AM technology developed. It creates parts by repeatedly scanning successive layers derived from CAD files (.stl format). Components are fabricated from a liquid resin through light-initiated polymerization on a build framework, selectively curved by laser or ultraviolet light. This process is favoured in TE due to the flexibility of mixing different materials [5]. Materials used include photopolymers, bio-resins, and ceramic resins. Advantages include the capability to manufacture large parts with high accuracy and a good surface finish at a high build speed. Disadvantages involve low shelf life, reliance on expensive photopolymer precursors, and poor mechanical properties of photopolymers. The resolution ranges from 0.1 to 100 μm [17]. Liquid polymerization is further classified into two processes: stereolithography (SLA) process and digital light processing (DLP).

Stereolithography Process (SLA) The stereolithography process (SLA), a laser is utilized to scan and solidify each layer, creating 3D structures. After each layer is cured, the platform is reduced by one layer thickness. This method requires a photoinitiator for photopolymerization. SLA-produced scaffolds show biocompatibility and excellent cell proliferation [50].

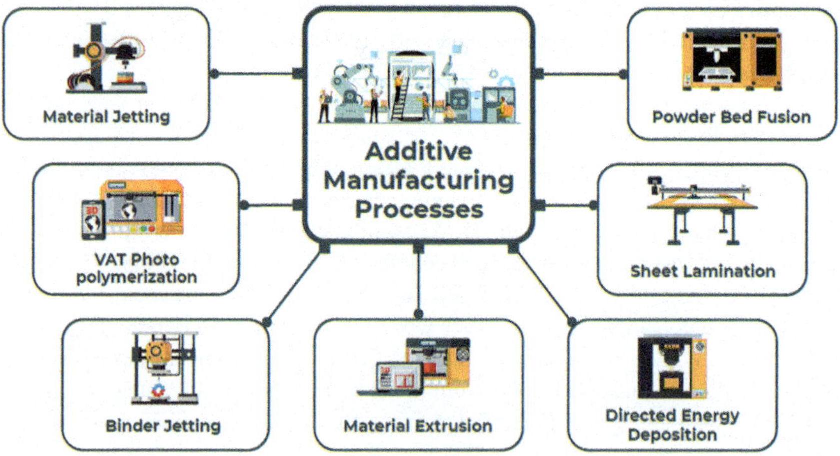

Fig. 6.2 Different fabrication techniques using AM for biomaterial scaffold. (Reprinted with permission from [55])

Digital Light Processing (DLP) The digital light processing (DLP) projects light from the bottom of the platform, curing entire layers simultaneously, leading in a more efficient process relative to SLA. DLP can fabricate bone scaffolds with high accuracy and fine internal geometries as thin as 60 μm, supporting increased cell density, proliferation, and distribution. DLP's application in TE is less explored due to its novelty [50].

6.4.2 Material Deposition Process

Material deposition processes employ extrusion-based techniques, melting thermoplastic filaments or metal wires via a nozzle to fabricate objects layer by layer. Filaments, typically with diameters of 0.1–0.4 mm, are heated and extruded based on CAD data file (.stl format) [50]. This cost-effective method offers open-source availability and is widely used. Typical resolution ranges from 100 μm to 1 cm [17]. Materials used include polymers, metals, ceramics and hybrids. Advantages include the ability to fabricate multi-material, multicoloured parts, scalability and low cost. However, drawbacks include vertical anisotropy, suboptimal surface finish, medium to high temperature requirements and relatively low resolution [17]. Two main processes exist within material deposition: fused deposition modelling (FDM) and wire extrusion (WE) process.

Fused Deposition Modeling (FDM) In fused deposition modeling (FDM), filament continuously feeds through the extruder and nozzle via rotating rollers. Material is deposited layer-by-layer onto a construction plate until the required product is formed and dimensions are achieved. The printer nozzle moves according to spatial coordinates from CAD models in G-code files, ensuring precise fabrication of the components [17, 51]. Polymers are commonly used in this process.

Wire Extrusion (WE) Wire extrusion (WE) is akin to FDM but utilizes either polymer or metal wire. Challenges with metal-based wire extrusion include limitations in extrusion head motion control accuracy and nozzle diameter, making it less explored in TE [5].

6.4.3 Powder Bed Process

Powder bed process (PBP) involves spreading a fine layer of powder and then utilizing an energy source such as laser, heat or electron beam to melt and selectively fuse the material to form a 3D object for diverse applications. The energy must be sufficient to melt the powder at its melting point, leading to the formation of a solid structure. The build platform is reduced by the thickness of each layer, followed by the application of a new powder layer, repeating the process [5, 52]. Materials suitable for PBP include metals, ceramics, composites, polymers and hybrids. Advantages are its cost-effectiveness, high specific strength and stiffness, built-in support structure from the powder bed, high resolution and material versatility. However, it is a slow process, has limited structural integrity and scalability, requires high energy, and the finish quality that is dependent on powder particle size with poor reusability of un-sintered powder [17]. Typical resolution ranges from 10 to 25 μm. PBP technologies include electron beam melting (EBM), selective laser sintering (SLS), and selective laser melting (SLM).

Selective Laser Sintering (SLS) The selective laser sintering (SLS) utilizes a laser to fuse powder materials. It is ideal for fabricating complex bone scaffolds in TE with biomaterials that can be fused without decomposing under laser [5].

Electron Beam Melting (EBM) Electron beam melting (EBM) utilizes an electron beam with higher energy density than a laser, reducing build time and cost. This technology is used in producing biocompatible metal implants, such as hip and shoulder replacements [53]. It is effective in producing porous metallic scaffolds with varying pore design and mechanical strength [5].

Selective Laser Melting (SLM) The selective laser melting (SLM) employs a laser to fuse metal powder, building parts LBL directly [5]. SLM fabricates fully functional parts, tools or prototypes [52]. This process is costly and slow due to the high temperatures required.

6.4.4 Sheet Lamination

Sheet lamination (SL) involves bonding layers of material sheet to form an object. Materials used include metals, ceramics, polymers, and hybrids. Advantages of this technology include rapid processing speed, cost-effective and ease material handling. The drawbacks include the need for post-processing and part quality depending on the adhesive used [17]. Typical resolution is 200–300 μm. SL includes laminated object manufacturing (LOM) and ultrasonic additive manufacturing (UAM).

Laminated Object Manufacturing (LOM) The laminated object manufacturing (LOM) integrates both additive and subtractive techniques, wherein layers are bonded using bonding mechanisms between sheets such as thermal bonding, adhesive bonding, and clamping. Each layer is then laser cut into the desired shape [18, 54].

Ultrasonic Additive Manufacturing The ultrasonic additive manufacturing (UAM) can be considered a hybrid sheet lamination process utilizing ultrasound to bind metal sheets. It involves ultrasonic metal welding and computer numerical control (CNC) milling [18]. This method can bond different materials without melting them [5].

6.4.5 Binder Jetting

Binder jetting (BJ) employs powder materials and liquid binders with a typical resolution of 100 μm. A nozzle deposits the binder, which functions as an adhesive to facilitate the cohesion of the powder, while the platform lowers incrementally to add layers as per a CAD design [18]. Materials include polymers, ceramics, composites, metals and hybrids. Advantages include diverse material options, increased printing speed and low cost. Drawbacks include limited mechanical property, porosity, low-viscosity ink requirements and significant post-processing [17].

6.4.6 Material Jetting

Material jetting (MJ) is an analogue to inkjet printing, in which liquid droplets are deposited onto a build framework and cured by ultraviolet light, with a typical resolution of 10–25 μm. This process primarily uses photopolymers. Mostly, water jet or sodium hydroxide solution is used post-processing to remove the support material [5]. Materials used include polymers, ceramics, composites, hybrids and biologicals. The advantages of this technology include high accuracy, low waste, multi-material and multicoloured capabilities and good surface finish. Drawbacks include the need for support material, restriction to photopolymers and thermosets, the need for controlled ink viscosity and low-strength applications [17].

6.4.7 Direct Energy Deposition

Direct energy deposition (DED) is similar to the material extrusion process with a typical resolution of 100 μm–1 cm. This process involves depositing materials in multiple directions (four or five axes), melted by a laser or electron beam. A high-intensity laser is directed onto a constant flow of material powder to generate a thin layer [5]. It uses metals and metal hybrids. Advantages are high grain structure control, high-quality parts, and suitability for repairs. Drawbacks include limitations to metals and metal hybrids and the need to balance surface quality with print speed [17].

6.5 Characterization Methods for Additive Manufactured Scaffolds

The success of cell growing in a 3D scaffold depends on intrinsic variability (cell properties) and extrinsic variability (microenvironmental properties). While inherent variability cannot be controlled during culture, reproducibility through AM minimizes extrinsic variability by precisely controlling scaffold features. Characterization of scaffolds ensures reproducibility and allows optimization, crucial for clinical application and commercialization. The characterization process differs based on the scaffold type: ready-made scaffolds must be stable and consistent across batches, while scaffolds produced on-demand during surgery need to be checked as they are fabricated. The characterization of scaffolds through rheological, mechanical, thermal, microscopic, and biological analysis will help in understanding the behaviour, mechanical strength, structural characteristics and sterility of additive manufactured scaffolds, aiding in optimizing scaffold design and manufacturing processes for effective TE (Fig. 6.3) [56].

6.5.1 Geometrical Properties

The material composition and 3D structural arrangement intrinsically influence the geometrical characteristics of an AM scaffold. These properties encompass pore dimension, porosity, ration of surface area-to-volume, strut thickness, interconnectivity, and cross-sectional dimension. These parameters critically impact the scaffold's mechanical, chemical, physical and biological properties [56]. Scaffolds can be designed to have graded porosity and pore size distribution using AM techniques that impact cell infiltration, nutrient transport and tissue ingrowth [57]. Scaffold design methodologies include regular and irregular structures. Regular structures fabricated using CAD tools with periodic unit cell arrangements like cylinders, spheres and blocks allow easier modeling, stimulation, and manufacturing but are limited in mimicking natural bone. Irregular structures, derived from fractal curves or clinical images, closely resemble natural tissue and can be fabricated using various techniques. These irregular structures may promote tissue regeneration due to non-uniform stress distribution and reduced susceptibility to catastrophic failure [58]. The geometric properties of AM scaffolds help to ensure that the scaffolds provide the appropriate microenvironment for cellular growth and tissue regeneration.

6.5.2 Rheological Characterization

Rheological analysis is studied to understand the behaviour of the scaffold material. It determines the processability of the material by assessing viscosity, extrusion temperature and shear-thinning behaviour. This helps control the printing process and ensure the scaffold material's integrity [59]. This method helps to evaluate the material's viscoelastic properties, which exhibit both viscous and elastic behaviour. The complex shear modulus describes the viscoelastic response and is characterized by time and frequency dependence [60]. Rheological assessment involves preparing samples as pellets or fragments. Rheological measurements are considered, including rotational rheometry, oscillatory rheometry and capillary rheometry [61]. Rotational rheometry evaluates viscosity and viscoelastic properties under controlled shear rates or stresses. In oscillatory rheometry, small-amplitude oscillatory shear is utilized to determine the elastic or viscous moduli of material [21, 59]. Capillary rheometry measures viscosity under high shear rates, stimulating the

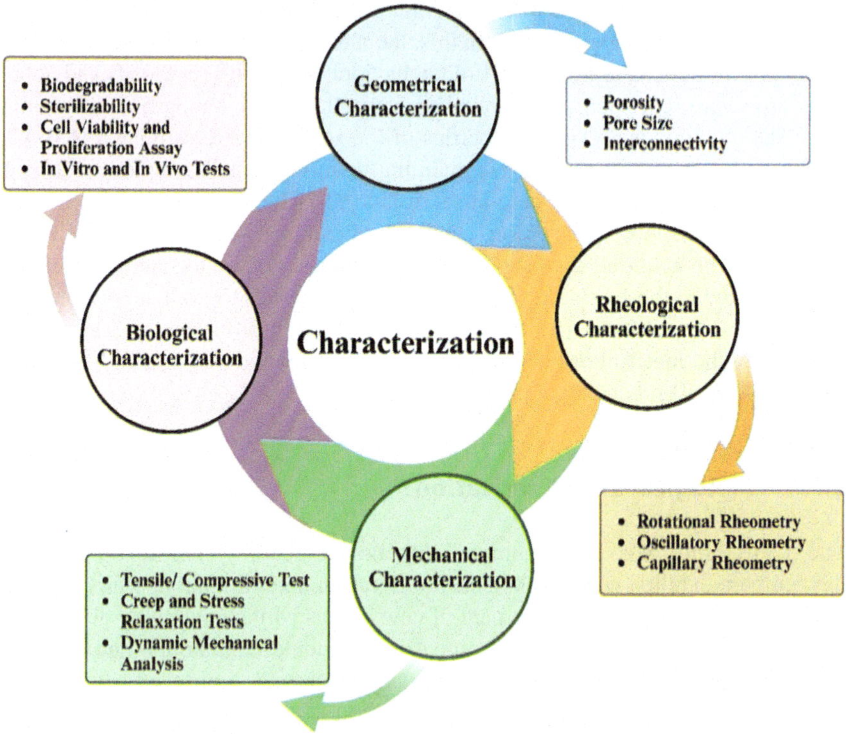

Fig. 6.3 Characterization methods for AM scaffolds. (Created with BioRender.com)

extrusion process in AM. The results of rheological analysis identify shear-thinning behaviour, determine the transition between viscous and elastic dominance and estimate the optimal processing temperature and shear rates based on the rheological properties [59, 62]. The rheological behaviour of the scaffold depends on the material composition, temperature, shear rate, cell shape, porosity and printing techniques [63]. Thermal analysis aids in understanding the thermal transition and melting point of the material, which are crucial for determining the optimal processing conditions for scaffolds [57]. The thermal property of the scaffold is not directly controlled by the AM process.

6.5.3 Mechanical Characterization

The mechanical properties of biomimetic scaffolds should closely resemble those of natural tissues to provide appropriate mechanical stimuli that support cell growth and differentiation, particularly in soft tissues. Additionally, the scaffold must withstand surgical procedures and maintain its qualities post-implantation for the desired period. This is especially crucial for load-bearing applications, stiffness and the ability to maintain the structure over time where degradation can significantly impact mechanical properties [56]. In this, the mechanical properties of the scaffold essential for understanding the scaffold behaviour under different loads and conditions are measured, such as compressive strength, tensile strength, and modulus [64]. The dynamic and elastic properties of the scaffold were estimated [65]. The testing scaffolds are prepared by segmenting them into small samples or utilising testing apparatus tailored for certain scaffold kinds. The scaffolds are subjected to various loads such as compression, tension, or bending to evaluate the mechanical properties. The acquired data is then assessed to assess the mechanical properties and compared to those of natural tissues or other materials to evaluate performance and potential applications [57, 65, 66]. The mechanical properties of scaffolds are affected by the material utilized, the type of stress applied, the scaffold geometry, and the physical and chemical parameters [56].

6.5.4 Biological Characterization

The scaffold must possess biocompatibility, biodegradability, non-toxic, water and oxygen permeability, cell attachment, proliferation and mitigation, which help the scaffold to stimulate the development of new tissues while undergoing degradation without interfering with other cell processes. It should emulate the natural ECM, aiding cell proliferation, adherence, and division through improved cell adhesion and signaling [67]. Biochemical characterization is conducted to analyze the biochemical structure of particular cellular activity on AM scaffolds [56]. It assesses the scaffold's capacity to support and promote cellular growth, proliferation and regeneration of tissues. Biodegradability and sterilization are the essential components of the scaffold. Biodegradability allows the body ECM to replace the scaffold over time without harming other tissues. The scaffolds should be sterilizable to prevent infections. Sterilization techniques include the use of chemical and gamma radiation. Antimicrobial solutions and argon plasma treatment offer alternative sterilization methods that maintain material properties and improve cytocompatibility [67]. This analyzes the sterility and biocompatibility, cell-scaffold interactions, tissue regeneration potential and *in vitro* and *in vivo* testing to ensure the scaffold's suitability and effectiveness.

6.6 Applications of Additive Manufactured Scaffolds in Tissue Engineering

6.6.1 Bone Tissue Engineering

The primary objective of Bone TE (BTE) is to restore and replace the functionalities of bone tissue by combining several technologies, such as cell biology, biomaterials, and construction techniques. Scaffolds are essential in aiding cell adherence, differentiation, proliferation, and bone tissue regulation in the fabrication process

[68]. A well-designed scaffold seeks to replicate the natural ECM of the specific tissue, providing a favourable condition for cell adhesion, proliferation, and specialization [69]. The scaffold creates an optimal environment for osteoblasts, and other important biological components necessary for bone repair (Fig. 6.4) [70].

It is noteworthy that the compressive resistance of human bones is 170–193 MPa for cortical bone and 7–10 MPa for cancellous bone [72, 73]. Thus, it is necessary to have a biomaterial that possesses mechanical properties comparable to those of bone to guarantee the durability of the implant. To this end, Castilho et al. evaluated scaffold biocompatibility *in vivo* [74]. In canine, 3D-printed TCP scaffolds were surgically placed into the stifle joint tibial crest. The reduced cage area observed between day 1 and 16 weeks verified the 3D-printed TCP scaffold's superior biological performance, biochemical response and superior mineralization of the 3D-printed TCP scaffolds within the osteotomy gap (Fig. 6.5a). In another study, Habibovic et al. examined the osteoinduction and osteoconduction of brushite-based 3D-printed scaffolds [75]. The implanted scaffolds in goats showed new bone growth in the pore channel after 12 weeks (Fig. 6.5b). Inzana et al. created porous scaffolds using a composite powder (HAp and TCP), with type-I collagen dissolved in a binder solution containing phosphoric acid [76]. The studies showed that the mice with femoral defects implanted with printed scaffolds experienced fresh bone development at the defect site (Fig. 6.5c).

Regarding porosity, closed pores are not involved in physiological processes since they are not accessible to cells and body fluids [78]. Pores with a diameter of 20–50 micrometres may facilitate efficient exchange of physiological fluids. Biomaterial structures with a diameter ranging from 100 to 350 micrometres are appropriate for cell colonisation and vascularization, which allows for bone penetration [79].

The design of 3D biomedical scaffolds in TE is vital for tissue regeneration process. It is important to fabricate a structure that closely resembles the complex and interconnected network observed in natural tissue [80]. A mixture of binders and post-processing parameters to create porous biomaterial constructs for BTE through AM has been reported by Kumar et al. [77]. The AM-based scaffolds should possess the unique characteristics of biocompatibility and biodegradability, integrated

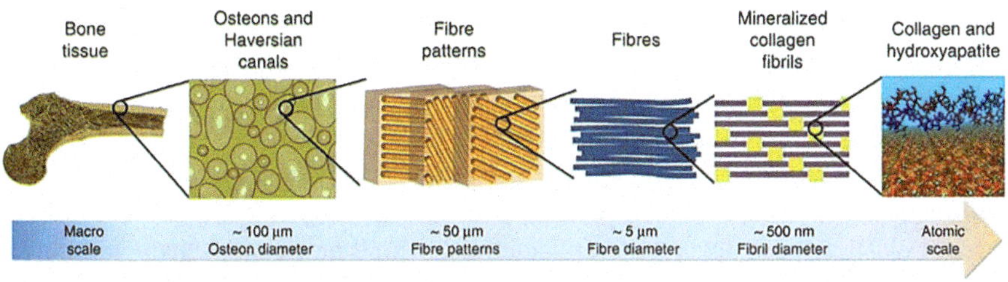

Fig. 6.4 Pictorial representation of bone tissue. (This image has been reprinted and licensed under a Creative Common Attributes CC-BY [71])

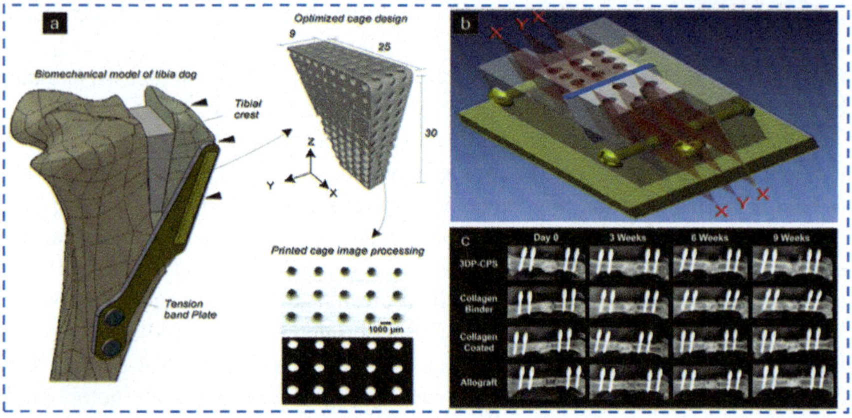

Fig. 6.5 A diagram depicting the procedure for developing an implant design utilizing a biochemical model obtained from a 3D scan of a canine's tibia (**a**). The design of the cage shows the positions of the peripheral sections in the X–X plane and the central sections in the Y–Y plane for histological analysis (**b**). Calcium phosphate scaffolds created using 3D printing technology promote bone formation in a severely sized lesion in the femur of mice. Chronological progression of bone healing as depicted by X-ray imaging (**c**). (Reprinted with permission from [75–77])

structural mechanical properties, and precise pore shape and size that decide their porous nature, surface roughness, and wettability [81].

6.6.2 Cartilage Regeneration

Cartilage is a tough connective tissue covering the ends of bones in joints, allowing for movement. It acts as a multi-layered material with certain mechanical qualities, such as being able to transmit loads across the joint [82]. The appealing approach for cartilage regeneration involves the use of pre-made biological scaffolds that not only mimic the natural environment for cells and promote their growth and specialization while also providing mechanical support and function within joints that bear weight, starting from the moment they are implanted (Fig. 6.6). The process of cartilage regeneration in *in-vivo* is expected to be time-consuming, spanning from months to years. Therefore, it is necessary to use implanted regenerative biomaterials that can offer mechanical support and maintain their structural integrity under the constant stress of joint movement. To this end, Wang et al. reported the development of porous alginate scaffolds, reinforced with 3D printed PLCL to mimic sulfated glycosaminoglycan [83]. The sulfated scaffolds were found to have the ability to promote chondrogenesis of MSCs and regulate the secretion of TGF-β3 in *in vitro* studies over an extended period. Overall, the 3D-printed scaffold exhibited greater elasticity, and accurately replicated the overall mechanical characteristics of natural cartilage tissue. RM aims to enhance, repair or regenerate impaired tissues or organs through the integration of cells, biomaterials and GFs. Both TE and

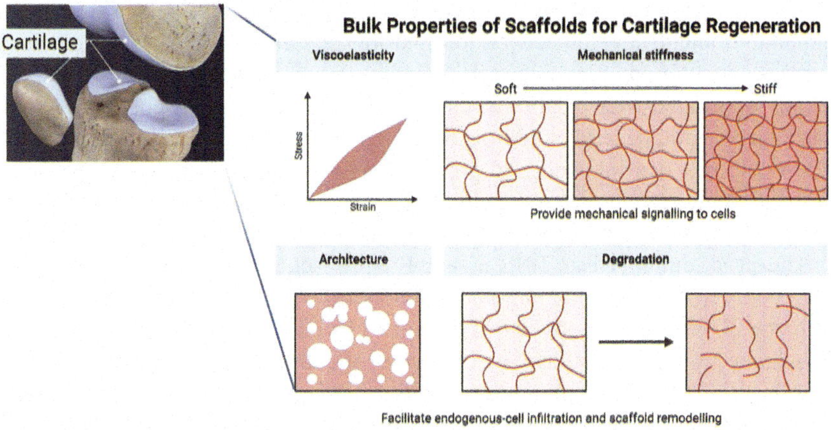

Fig. 6.6 Schematic illustration of desirable properties of additive manufactured scaffolds for cartilage regeneration. (Created with BioRender.com)

embryonic biology are presently focused on the mechanism of ECM deposition and tissue self-assembly. Kundu et al. fabricated 3D cell-printed scaffolds utilizing LBL deposition of PCL and alginate hydrogel encapsulating chondrocyte cells via AM and a multi-head deposition technique [84]. Scaffolds composed of PCL–alginate gels incorporated with transforming growth factor-β (TGFβ) demonstrated increased formation of ECM. The PCL–alginate gel 3D cell-printed scaffolds were implanted into female nude mice's dorsal subcutaneous spaces. After 4 weeks, histopathological and immunohistochemical analyses of the retrieved implants showed increased cartilage tissue formation and increased type II collagen fibril deposition in the PCL–alginate gel hybrid construct supplemented with TGFβ. Based on the same approach. Zare et al. printed PCL and loaded it with Alginate [85]. Alginate-sulfate hydrogel was prepared to mimic glycosaminoglycan structure. The hydrogel was also treated with varying decellularized ECM (dECM) concentrations to boost chondrogenesis. Overall, the scaffold demonstrated mechanical resemblance to endogenous nasal cartilage tissue and relevant biochemical characteristics, making it a promising choice for *in vivo* research.

Integrating AM and TE technologies may create scaffolds that provide a 3D environment ideal for the regeneration of cartilage. 3D TE scaffolds using silk fibroin (SF) were reported utilizing an indirect AM method [86]. The inherent qualities, like the chemical composition and thermal properties of SF, remained intact throughout the indirect AM method. The production of collagen Type II, increased on a per-cell basis when the culture duration reached 12 weeks. This suggests that the chondrocytes within the scaffold underwent re-differentiation. The implantation of a scaffold with chondrocyte constructions with subcutaneous tissue in nude mice was also verified to have resulted in the development of ectopic cartilage through histological investigation and immunostaining. Moroni et al. utilized the RP fabrication technique (3DF) and electrospinning to fabricate polymeric scaffolds consisting of both macro- and periodically interspersed micro-fibers [87]. The study aimed to evaluate how the network of microfibrillar affects cellular entrapment and subsequent differentiation into cartilaginous tissue. In a recent study by Gruber et al. 3D-printed PCL scaffolds integrated with PLA microspheres containing cartilage-specific decellularized matrix (DM) were fabricated (Fig. 6.7) [88]. The DM composite scaffolds elicited greater levels of biochemical signals associated with cartilage formation compared to the control group, therefore supporting their potential use in the treatment of osteochondral injuries (OCIs).

Cui et al. employed an LBL assembly approach and photo-polymerization of PEGDMA hydrogel containing human chondrocytes to patch flaws in OC plugs, utilizing 3D jetting methods [89]. The printed PEGDMA had a compressive modulus similar to human cartilage after 6 weeks of culture. The chondrocytes retained their initial deposited locations, and viability was improved by 26% owing to photo-polymerization. Schuurman et al. aimed to mimic native cartilage's zonal layout [90]. Expanded equine ACh cell pellets from the deep and superficial zone were pressed between 3FD PEGT/PBT layers. Cartilaginous tissue formation was observed across all groups, regardless of population or scaffold seeding method, however, the initial cell placement did not maintain or restore the original zonal

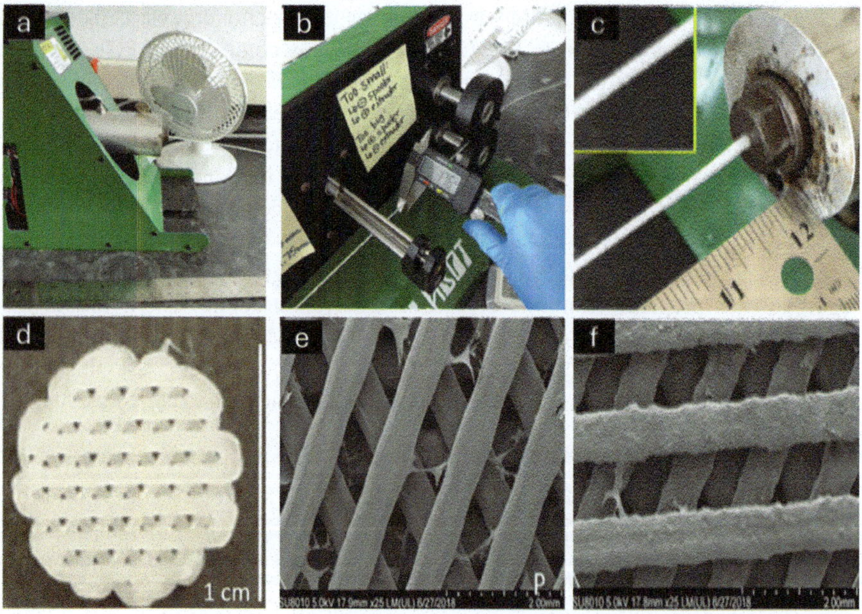

Fig. 6.7 Digital images captured for PCL filaments and their utilization in 3D printing: The DM-PLA microspheres were combined with PCL powder using an extruder. Combine assembly of components (**a** and **b**) to form filaments (**c**). The DMPLA microspheres were incorporated into PCL filaments, which were subsequently 3D printed with a FlashForge Dreamer 3D printer to create the scaffolds (**d**). The SEM analysis revealed that the PCL scaffolds had an even surface (**e**), whereas the scaffolds with embedded microspheres had an uneven surface (**f**). (Reprinted with permission from [88])

phenotype. Overall, AM-based 3D bioprinting is a progressive and effective method for the restoration of cartilage tissue. Advancements in printing techniques, the creation of innovative bioinks, and a deeper comprehension of the molecular processes that govern cartilage regeneration have speeded up the potential use of 3D-printed scaffolds as a highly promising therapeutic option in clinical settings in the near future.

6.6.3 Skin Tissue Engineering and Replacement

The human skin consists of three distinct layers: the epidermis, dermis, and hypodermis layers. The epidermis, the outermost layer of the skin, consists primarily of keratinocytes. These keratinocytes are arranged in a highly structured architecture that is characterised by both their physical appearance and their level of differentiation [91, 92]. The dermal layer has a lower number of cells compared to the epidermis and contains a diverse combination of ECM components, such as collagens, elastin, proteoglycans, and glycosaminoglycans.

The utilization of skin replacement technology presents a promising basis for enhancing the treatment of both chronic and acute skin injuries. However, for skin TE to effectively interact with the surrounding tissue, it is necessary to utilize cellular basis technology and simulated ECM, taking into consideration the mechanical and physiological features of active skin [77, 93]. There is currently no skin prototype that precisely replicates the composition, natural structure, consistency, or visual appearance of real skin. The skin alternatives may possess vital, user-friendly, and wound-specific attributes [44].

Bio-printing offers numerous advantages when compared to both hand seeding and other 3D production methods. Automated technologies, such as bioprinting, frequently result in enhanced reproducibility in comparison to manual methods. Obtaining a layered full-thickness skin-like structure, 3D skin models often takes at least 3–4 weeks. Bioprinting enables the simultaneous layering of several sheets of cells, facilitating the creation of complex structures from inception [94]. Recent research by Organovo, Inc. showcased a scaffold-free extrusion-based skin framework made with the NovoGen Bioprinter Platform (Fig. 6.8) [95]. The work used normal keratinocytes and human dermal fibroblasts (HDFs) to create tissue with discrete layers, stratified design, and collagen synthesis in the dermal layer. This multiwell model for screening applications showed damage response to Triton X or toxicants SDS, resulting in lower viability and increased IL1a production during exposure. Rimann et al. created a full-thickness model using an automated method of microvalve droplet-based technique [96]. The method involved fabricating dermal equivalents with seven layers of chemically specified UV cross-linked PEG-based bioink with suspended primary HDFs. The printing concept involved depositing seven concentric layers of acellular bioink onto a dermal structure, forming a cylindrical shape designed to manually collect a suspension of primary human keratinocytes. Dermal equivalents cultivated without keratinocytes remained viable for up to 7 weeks, demonstrating the potential of bioprinted tissues as long-term

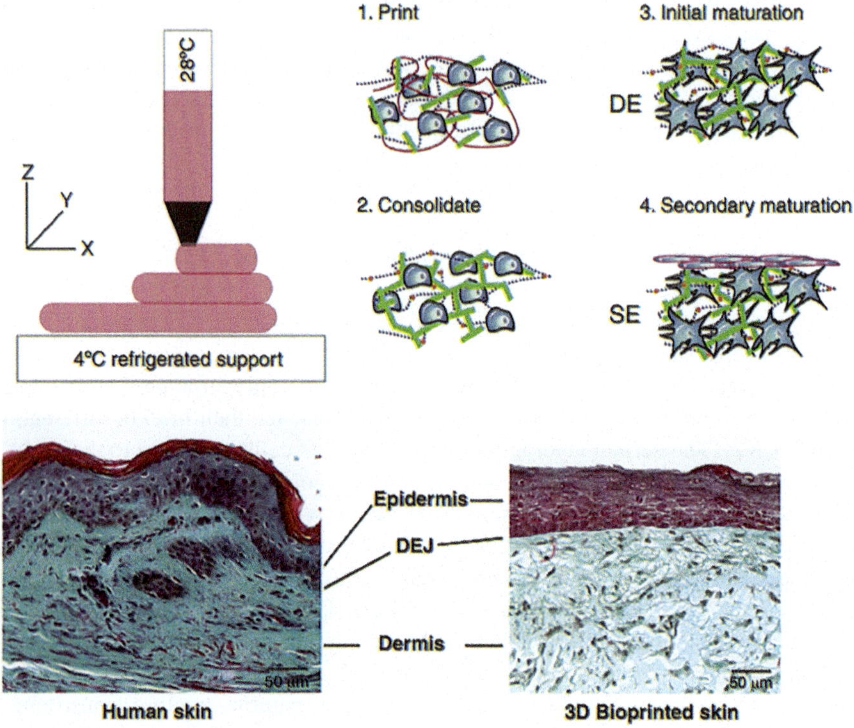

Fig. 6.8 The full-thickness skin tissue is created using bioprinting technology and a dermal bioink that may be adjusted after printing during the maturation process. The resultant skin tissue exhibits dermal and multilayered epidermal layers that closely resemble those found in natural skin. (Reprinted with permission from [95])

culture models. In a study conducted by Yanez et al., an adapted inkjet printer was used to apply human microvascular endothelial cells (HMVEC) in conjugation with a manually blended mixture of collagen, neonatal human epidermal keratinocytes (hEK) and neonatal human dermal fibroblast cells [97]. This process resulted in the creation of a complete skin graft with full thickness. Researchers have successfully shown laser-based skin tissue bioprinting [98]. The initial study used a donor slide with labeled murine NIH3T3 fibroblasts or human HaCaT keratinocytes, encased in an alginate and blood plasma hydrogel, and printed on a collector slide with Matrigel. The bioprinter developed a 2D checkerboard composed of interspersed green and blue cells, demonstrating its ability to manufacture precise patterns with great precision and minimal impact on cell viability and proliferation.

Lee et al. successfully utilized 3D bioprinting to manufacture human skin tissue [99]. AM assembly procedures were employed, incorporating collagen type I, fibroblasts, and keratinocytes. Another study described the effective bioprinting of collagen gel, including fibroblasts and keratinocytes (each consisting of 20 layers), onto a Matriderm sheet, which is a decellularized dermal matrix [100]. Following a period of 10 days in a laboratory culture, the skin constructions that were printed displayed the existence of cadherins and connexin 43 in the superficial layer of the skin, which is a clear indication of the development and unity of the tissue. Moakes et al. reported the utilization of suspended layer AM to fabricate a seamlessly integrated tri-layered implant that closely mimics the structure of human skin (Fig. 6.9) [101]. By meticulously regulating the composition of the bioink, both chemical and biological gradients were created across the printed structure. The model's culture revealed that within 3 weeks, the constituent cellular components significantly contributed to the reorganization of the supporting matrix to resemble the architecture of healthy skin.

Taken together, 3D bioprinting facilitates the accurate creation of intricate multicellular structures with controlled microenvironments, resulting in enhanced cellular communication and tissue development through the provision of native biophysical and biochemical cues. Recent advancements in bioprinting have demonstrated the capability of creating functional skin tissues with separate dermis and epidermis layers. The successful utilization of various procedures thus far demonstrates the adaptability of these automated computer-controlled design platforms as well as the skin's comprehensive compatibility with automated AM methods.

6.6.4 Vascular Tissue Engineering

The artificial advancement of a vascular system (VS), particularly the network of blood vessels, is much adored in the field of TE [102–104]. The VS, consisting mostly of blood vessels, is essential for the artificial generation of organs such as the heart, kidney, liver, pancreas, etc. [102, 105]. The VS functions as a transportation system that delivers oxygen, nutrients, hormones, immune cells and cytokines (proteins that facilitate cell signaling). It plays an important role in regulating pH balance, homeostasis, osmotic pressure, and blood pressure in the body [106]. Researchers have not yet achieved their goal of generating an accurate and comprehensive replica of the layers and structures that constitute a functional blood vessel [107]. Thus, while discussing artificial structures, blood vessels will be seen as a conduit similar to the VS with unique properties that should support blow flow without deformations; without triggering negative immunogenic responses; and should possess the capacity to undergo *in vivo* growth, remodeling, and self-repair; the necessary characteristics for implantation, including resistance to bending; and with an ability to be manufactured in large quantities, and cost-effective [108].

Over the last decade, researchers have relied on the strength of AM to develop vascular-like channels. On the contrary, Jia et al. 2016 created functional blood vessel structures utilizing a complex 3D bioprinting technology that involved many layers and coaxial extrusion [109]. They employed GelMA and sodium alginate, which were also utilized by Christensen et al., together with mouse fibroblast cells [110]. Xu et al. integrated the advantages of 3D-printed scaffolds and the benefits of acellular tissue matrices [111]. The authors initially printed a support framework and inserted a dECM. The sequence involved inserting cells in specified locations to

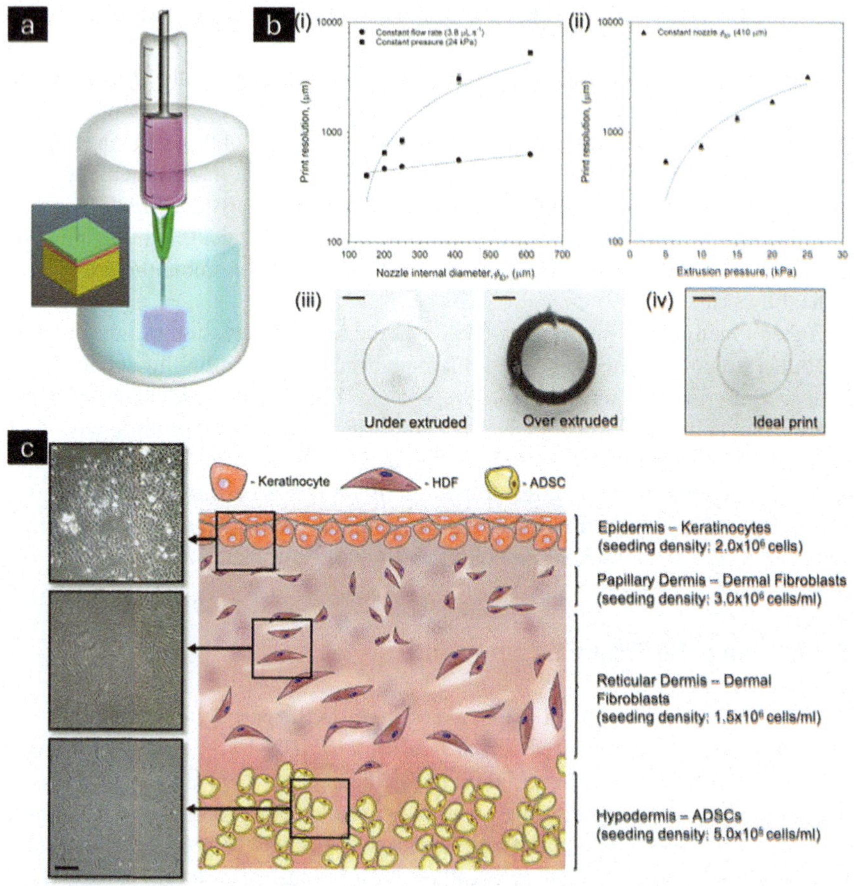

Fig. 6.9 The graphic illustrates the print being suspended in a print bath that contains a fluid gel. The inset of the diagram displays the CAD picture utilized to produce the three-layered construct (**a**). Obtained data on the print resolution for an individual layer by managing different printing parameters: (**b** i) maintain a consistent flow rate, (**b** ii) a constant nozzle internal diameter, (**b** iii) sample images showing the printed layers, and (**b** iv) a typical image representing the optimal printing settings. Diagram illustrating the structure of the skin transplant, which includes hypodermis, a dermis with two separate compartments, and an outermost layer called the epidermis. The included micrographs display adipocytes (ADSC), HDFs, and hEK that were cultured in a two-dimensional environment before being added to the structure (**c**). (This image has been reprinted and licensed under a Creative Common Attributes CC-BY [101])

replicate the three layers of blood arteries. The personalized printed scaffold showed optimized mechanical resistance and ensured great biocompatibility (Fig. 6.10). In another work, Suntornnond et al. created scaffold constructs resembling blood vessels using a 3D bioprinter that operates by extrusion [112]. The bioprinter used Pluronic 127 as the material. Direct printing enabled the production of complex branched structures that closely resemble real blood artery architecture. A cross-linked hollow alginate filament that was filled with fibroblasts and smooth muscle was prepared and reported by Gao et al. [113]. These filaments were extruded along a rotatable rod template, after which endothelial cells (ECs) were then seeded onto the inner wall. In continuation of these studies, various techniques of AM were used to construct vascular conduit (Fig. 6.11) [110, 114–120]. The utilization of thermal inkjet printing technique was employed for the simultaneous printing of HMVEC and N1H 3T3 mouse fibroblasts using fibrin and sodium alginate material, respectively [110, 121]. The extrusion technique was used to simultaneously print adipose-derived stromal cells that were stimulated to develop into endothelial-like cells (HUVECs), normal human lung fibroblasts (HUVECs and human mesenchymal stem cells (L929 mouse fibroblasts, mouse smooth muscle cells, and HUVECs),

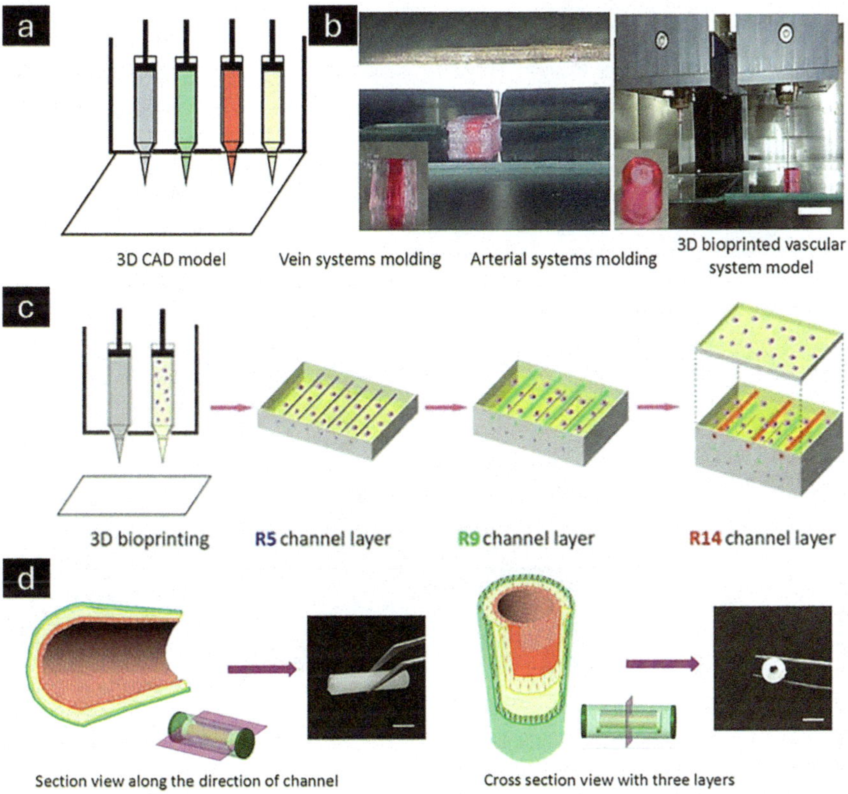

Fig. 6.10 Diagram illustrating four separate nozzles of the 3D bioprinter (**a**). Construction of tissue structures with several nozzles (**b**). Schematic demonstrating the process of biofabricating thick tissues with hollow channels of three distinct diameters: (wide radius: 1.4 mm), moderate (radius: 0.9 mm), and narrow (radius: 0.5 mm) channels (**c**). Schematic representation of the arteriovenous structure, consisting of three layers. The graphic shows both a transverse and longitudinal sectional view of the biofabricated arteriovenous structure (**d**). (This image has been reprinted and licensed under a Creative Common Attributes CC-BY [111])

L929 fibroblasts and HUVECs, human aortic vascular smooth muscle cells, and HDFs-neonatal cells, among others. This was achieved using various synthetic polymeric materials [109, 111–113, 122, 123].

Krishnamoorthy et al. developed and utilized an AM-based digital light processing method to fabricate 3D constructions resembling blood vessels [124]. The study primarily aimed to modify the composition to attain high cell viability. Shin et al. and Ware et al. employed the same technology to fabricate structures that resemble blood vessels [125, 126]. Kabirian et al. utilized FDA-approved PLA for the AM of vascular scaffolds. They successfully adjusted the polymer flow rate to solve the

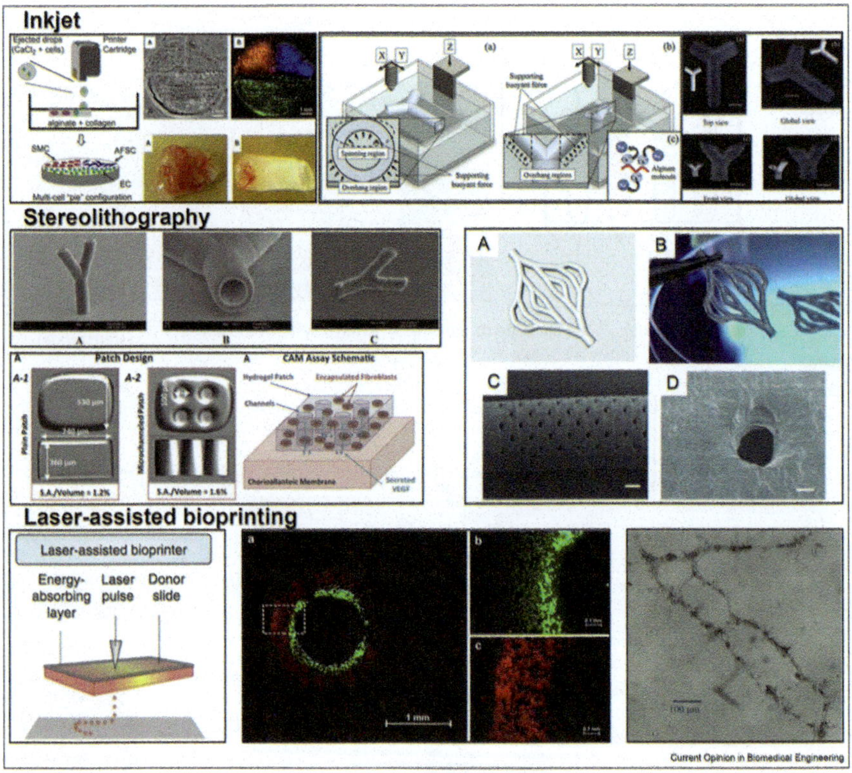

Fig. 6.11 Schematic illustration of Inkjet, SLA, and laser-assisted bioprinting process. Bioprinting multi-cell tissue constructions with light and fluorescence microscopy images assisted by inkjet (left). Gross images show significant vascularization after implantation compared to control (upper panel). Schematic of a printing technique that uses the surrounding solution as a buoyant support, producing structures with cantilevered and overhanging parts in both vertical and horizontal directions (right). Fabricated structures with SEM images of branched tubular structures made with 2 PP, hydrogel patch design (bottom-left) and schematic CAM assay and biomimetic microporous polyacrylate vascular network (middle panel). Laser-assisted bioprinting (left) methodology sequentially deposited ECs labeled with dual fluorescent probes in concentric circles, demonstrating accurate spatial placement (middle); and printed ECs' double branch structure showing complex vascular-like networks (right) (lower panel). (Reprinted with permission from [129])

issue of low-resolution [127]. Centola et al. utilized a blend of electrospinning and fused-deposition modeling techniques to create hybrid vascular grafts [128]. The biocompatible scaffold showed enhanced mechanical characteristics. Taken together, various approaches and technologies have emerged over time to facilitate the artificial generation of vesicular networks. Among these, AM technologies stand out. AM provides a high level of customization enabling the printing or production of individual cells or co-cultures. In addition, AM decreases processing durations and can utilize a diverse array of materials, such as bioinks and responsive materials. Additive manufacturing technologies provide a feasible and effective method for incorporating functional vascular networks with intricate organization into a three-dimensional structure. This has been achieved by depositing biomaterials, cells, and biomolecules in layers.

6.7 Conclusion

AM has evolved as a transformative technology in the domain of TERM. By facilitating the fabrication of complex, personalized scaffolds, AM has opened new avenues for the regeneration of damaged tissues and organs. The biomaterial scaffold acts as a framework for guiding tissue regeneration, providing the required structural and biochemical cues to guide cell growth and differentiation. This review provides an explanation of the different methods used to fabricate and evaluate biomaterial scaffolds using AM. It also covers the biomaterials that are used for fabrication, including natural and synthetic polymers, ceramics, and composite hybrids. The scaffolds that are developed using AM are used in TE, BTE, cartilage regeneration, skin replacement, and vascular TE. AM provides precise control over biomaterial architecture, enabling personalized RM therapies. Despite challenges, AM's potential in fabricating complex structures for TE is transformative. The materials used in the fabrication offer distinct advantages and limitations, requiring careful selection and combination to achieve desired properties. Using modern AM processes, various materials can be integrated, leading to the possibility of personalised scaffolds that are both effective and efficient for RM. After the fabrication of scaffolds, characterization is crucial to ensure reproducibility, optimization, and clinical application. Characteristic methods such as geometrical, rheological, mechanical, and biological analysis enable the design of effective TE scaffolds that provide the suitable microenvironment for cell growth and tissue regeneration. Overall, AM enables the precise creation of intricate, multicellular structures with controlled microenvironments, advancing the field of TE and paving the way for clinical applications.

CRediT authorship contribution statement **Viola Colaco:** Methodology, Investigation, Writing–original draft, Writing–review & editing. **Deepanjan Datta:** Conceptualization, Methodology, Investigation, Validation, Writing–original draft, Writing–review & editing. **Sony Priyanka Bandi:** Methodology, Validation, Investigation, Writing–original draft. **Namdev Dhas:** Validation, Review & editing. **Prabhanjan S. Giram:** Validation, Review & editing.

Declaration of Competing Interest The authors declare no competing interest.

References

1. Vyas J, Raytthatha N, Singh S, Prajapati B (2024) Next-generation computational automation-based additive manufacturing of pharmaceuticals: an approach to fabricate precise medicine. In: Handbook of 3D printing in pharmaceutics. CRC Press, pp 153–175
2. Vyas J, Singh S, Shah I, Prajapati BG (2023) Potential applications and additive manufacturing technology-based considerations of mesoporous silica: a review. AAPS PharmSciTech 25(1):6
3. Shah SR, Modi CD, Singh S, Mori DD, Soniwala MM, Prajapati BG (2024) Recent advances in additive manufacturing of polycaprolactone-based scaffolds for tissue engineering applications: a comprehensive review. Regen Eng Transl Med 11:1–20
4. Guo B, Ma PX (2018) Conducting polymers for tissue engineering. Biomacromolecules 19:1764–1782. https://doi.org/10.1021/acs.biomac.8b00276
5. Vyas J, Shah I, Singh S, Prajapati BG (2023) Biomaterials-based additive manufacturing for customized bioengineering in management of otolaryngology: a comprehensive review. Front Bioeng Biotechnol 11:1234340
6. Elitok MS, Gunduz E, Gurses HE, Gunduz M (2018) Chapter 20—tissue engineering: towards development of regenerative and transplant medicine. In: Barh D, Azevedo V (eds) Omics technologies and bio-engineering. Academic, pp 471–495
7. Han F, Wang J, Ding L et al (2020) Tissue engineering and regenerative medicine: achievements, future, and sustainability in Asia. Front Bioeng Biotechnol 8. https://doi.org/10.3389/fbioe.2020.00083
8. Kapoor DU, Gaur M, Kumar A, Singh S, Chidrawar VR, Prajapati BG (2024) Value-added lignocellulose from waste biomass as a potential source in fabrication of lignin-based composite for multifaceted applications. Regen Eng Transl Med 2024:1–30
9. Datta D, Prajapati B, Jethva H, Agrawal K, Singh S, Prajapati BG (2024) Value-added nanocellulose valorized from fruit peel waste for potential dermal wound healing and tissue regenerative applications. Regen Eng Transl Med 11:1–24
10. Sachdeva S, Mani A, Vora H et al (2022) Scaffolds in tissue engineering. J Cell Biotechnol 8:1–9. https://doi.org/10.3233/JCB-220070
11. Yi S, Ding F, Gong L, Gu X (2017) Extracellular matrix scaffolds for tissue engineering and regenerative medicine. Curr Stem Cell Res Ther 12:233–246
12. Ngadiman NHA, Noordin M, Idris A, Kurniawan D (2017) A review of evolution of electrospun tissue engineering scaffold: from two dimensions to three dimensions. Proc Inst Mech Eng H 231:597–616. https://doi.org/10.1177/0954411917699021
13. De Santis R, D'Amora U, Russo T et al (2015) 3D fibre deposition and stereolithography techniques for the design of multifunctional nanocomposite magnetic scaffolds. J Mater Sci Mater Med 26:250. https://doi.org/10.1007/s10856-015-5582-4
14. Chung JJ, Im H, Kim SH et al (2020) Toward biomimetic scaffolds for tissue engineering: 3D printing techniques in regenerative medicine. Front Bioeng Biotechnol 8. https://doi.org/10.3389/fbioe.2020.586406
15. Rezvani Ghomi E, Khosravi F, Neisiany RE et al (2021) Future of additive manufacturing in healthcare. Curr Opin Biomed Eng 17:100255. https://doi.org/10.1016/j.cobme.2020.100255
16. Salmi M (2021) Additive manufacturing processes in medical applications. Materials 14:191. https://doi.org/10.3390/ma14010191
17. Daminabo SC, Goel S, Grammatikos SA et al (2020) Fused deposition modeling-based additive manufacturing (3D printing): techniques for polymer material systems. Mater Today Chem 16:100248. https://doi.org/10.1016/j.mtchem.2020.100248
18. Lee J-Y, An J, Chua CK (2017) Fundamentals and applications of 3D printing for novel materials. Appl Mater Today 7:120–133. https://doi.org/10.1016/j.apmt.2017.02.004

19. Miar S, Shafiee A, Guda T, Narayan R (2018) Additive manufacturing for tissue engineering. In: Ovsianikov A, Yoo J, Mironov V (eds) 3D printing and biofabrication. Springer, Cham, pp 1–52
20. Giubilini A, Bondioli F, Messori M et al (2021) Advantages of additive manufacturing for biomedical applications of polyhydroxyalkanoates. Bioengineering 8:29. https://doi.org/10.3390/bioengineering8020029
21. Germaini M-M, Belhabib S, Guessasma S et al (2022) Additive manufacturing of biomaterials for bone tissue engineering—a critical review of the state of the art and new concepts. Prog Mater Sci 130:100963. https://doi.org/10.1016/j.pmatsci.2022.100963
22. Wu Y, Lu Y, Zhao M et al (2022) A critical review of additive manufacturing techniques and associated biomaterials used in bone tissue engineering. Polymers 14:2117. https://doi.org/10.3390/polym14102117
23. Placone JK, Mahadik B, Fisher JP (2020) Addressing present pitfalls in 3D printing for tissue engineering to enhance future potential. APL Bioeng 4:010901. https://doi.org/10.1063/1.5127860
24. Seitz H, Rieder W, Irsen S et al (2005) Three-dimensional printing of porous ceramic scaffolds for bone tissue engineering. J Biomed Mater Res B Appl Biomater 74B:782–788. https://doi.org/10.1002/jbm.b.30291
25. Perić Kačarević Ž, Rider P, Alkildani S et al (2020) An introduction to bone tissue engineering. Int J Artif Organs 43:69–86. https://doi.org/10.1177/0391398819876286
26. Sultana N, Hassan MI, Lim MM (2015) Composite synthetic scaffolds for tissue engineering and regenerative medicine. Springer, Cham
27. Chocholata P, Kulda V, Babuska V (2019) Fabrication of scaffolds for bone-tissue regeneration. Materials 12:568. https://doi.org/10.3390/ma12040568
28. Dhinakaran V, Stalin B, Sai MS et al (2021) Recent developments of graphene composites for energy storage devices. Mater Today Proc 45:1779–1782. https://doi.org/10.1016/j.matpr.2020.08.631
29. Du X, Fu S, Zhu Y (2018) 3D printing of ceramic-based scaffolds for bone tissue engineering: an overview. J Mater Chem B 6:4397–4412. https://doi.org/10.1039/C8TB00677F
30. Lee J-B, Maeng W-Y, Koh Y-H, Kim H-E (2018) Porous calcium phosphate ceramic scaffolds with tailored pore orientations and mechanical properties using lithography-based ceramic 3D printing technique. Materials 11:1711. https://doi.org/10.3390/ma11091711
31. Tappa K, Jammalamadaka U (2018) Novel biomaterials used in medical 3D printing techniques. J Funct Biomater 9:17. https://doi.org/10.3390/jfb9010017
32. Adel IM, ElMeligy MF, Elkasabgy NA (2022) Conventional and recent trends of scaffolds fabrication: a superior mode for tissue engineering. Pharmaceutics 14:306. https://doi.org/10.3390/pharmaceutics14020306
33. Alaribe FN, Manoto SL, Motaung SCKM (2016) Scaffolds from biomaterials: advantages and limitations in bone and tissue engineering. Biologia (Bratisl) 71:353–366. https://doi.org/10.1515/biolog-2016-0056
34. Krishani M, Shin WY, Suhaimi H, Sambudi NS (2023) Development of scaffolds from bio-based natural materials for tissue regeneration applications: a review. Gels 9:100. https://doi.org/10.3390/gels9020100
35. Richards DJ, Tan Y, Jia J et al (2013) 3D printing for tissue engineering. Isr J Chem 53:805–814. https://doi.org/10.1002/ijch.201300086
36. Guvendiren M, Molde J, Soares RMD, Kohn J (2016) Designing biomaterials for 3D printing. ACS Biomater Sci Eng 2:1679–1693. https://doi.org/10.1021/acsbiomaterials.6b00121
37. Yan Q, Dong H, Su J et al (2018) A review of 3D printing technology for medical applications. Engineering 4:729–742. https://doi.org/10.1016/j.eng.2018.07.021
38. Kirchmajer MD, Iii RG, Panhuis M (2015) An overview of the suitability of hydrogel-forming polymers for extrusion-based 3D-printing. J Mater Chem B 3:4105–4117. https://doi.org/10.1039/C5TB00393H
39. Skardal A, Atala A (2015) Biomaterials for integration with 3-D bioprinting. Ann Biomed Eng 43:730–746. https://doi.org/10.1007/s10439-014-1207-1

40. Wang K, Amin K, An Z et al (2020) Advanced functional polymer materials. Mater Chem Front 4:1803–1915. https://doi.org/10.1039/D0QM00025F
41. Custodio CL, Broñola PJM, Cayabyab SR et al (2021) Powder loading effects on the physicochemical and mechanical properties of 3D printed poly lactic acid/hydroxyapatite biocomposites. Int J Bioprinting 7:326. https://doi.org/10.18063/ijb.v7i1.326
42. Makadia HK, Siegel SJ (2011) Poly lactic-co-glycolic acid (plga) as biodegradable controlled drug delivery carrier. Polymers 3:1377–1397. https://doi.org/10.3390/polym3031377
43. Bartnikowski M, Dargaville TR, Ivanovski S, Hutmacher DW (2019) Degradation mechanisms of polycaprolactone in the context of chemistry, geometry and environment. Prog Polym Sci 96:1–20. https://doi.org/10.1016/j.progpolymsci.2019.05.004
44. Veeman D, Sai MS, Sureshkumar P et al (2021) Additive manufacturing of biopolymers for tissue engineering and regenerative medicine: an overview, potential applications, advancements, and trends. Int J Polym Sci 2021:4907027. https://doi.org/10.1155/2021/4907027
45. Liu C (2015) 10—collagen–hydroxyapatite composite scaffolds for tissue engineering. In: Mucalo M (ed) Hydroxyapatite (Hap) for biomedical applications. Woodhead Publishing, pp 211–234
46. Mondal S, Pal U (2019) 3D hydroxyapatite scaffold for bone regeneration and local drug delivery applications. J Drug Deliv Sci Technol 53:101131. https://doi.org/10.1016/j.jddst.2019.101131
47. Farag MM, Yun H (2014) Effect of gelatin addition on fabrication of magnesium phosphate-based scaffolds prepared by additive manufacturing system. Mater Lett 132:111–115. https://doi.org/10.1016/j.matlet.2014.06.055
48. Es-saddik M, Laasri S, Taha M et al (2021) Effect of the surface chemistry on the stability and mechanical properties of the Zirconia-Hydroxyapatite bioceramic. Surf Interfaces 23:100980. https://doi.org/10.1016/j.surfin.2021.100980
49. Masek A, Kosmalska A (2022) Technological limitations in obtaining and using cellulose biocomposites. Front Bioeng Biotechnol 10. https://doi.org/10.3389/fbioe.2022.912052
50. Melchels FPW, Feijen J, Grijpma DW (2010) A review on stereolithography and its applications in biomedical engineering. Biomaterials 31:6121–6130. https://doi.org/10.1016/j.biomaterials.2010.04.050
51. Mwema FM, Akinlabi ET (2020) Basics of fused deposition modelling (FDM). Fused Depos Model 1–15. https://doi.org/10.1007/978-3-030-48259-6_1
52. Yasa E (2021) Chapter 3—selective laser melting: principles and surface quality. In: Pou J, Riveiro A, Davim JP (eds) Additive manufacturing. Elsevier, pp 77–120
53. Ligon SC, Liska R, Stampfl J et al (2017) Polymers for 3D printing and customized additive manufacturing. Chem Rev 117:10212–10290. https://doi.org/10.1021/acs.chemrev.7b00074
54. Zhang Y, Wang J (2017) Fabrication of functionally graded porous polymer structures using thermal bonding lamination techniques. Procedia Manuf 10:866–875. https://doi.org/10.1016/j.promfg.2017.07.073
55. Mobarak MH, Islam MDA, Hossain N et al (2023) Recent advances of additive manufacturing in implant fabrication—a review. Appl Surf Sci Adv 18:100462. https://doi.org/10.1016/j.apsadv.2023.100462
56. Criscenti G, De Maria C, Vozzi G, Moroni L (2017) Characterization of additive manufactured scaffolds. In: Ovsianikov A, Yoo J, Mironov V (eds) 3D printing and biofabrication. Springer, Cham, pp 1–25
57. Sinha R, Sanchez A, Camara-Torres M et al (2021) Additive manufactured scaffolds for bone tissue engineering: physical characterization of thermoplastic composites with functional fillers. ACS Appl Polym Mater 3:3788–3799. https://doi.org/10.1021/acsapm.1c00363
58. Velasco MA, Lancheros Y, Garzón-Alvarado DA (2016) Geometric and mechanical properties evaluation of scaffolds for bone tissue applications designing by a reaction-diffusion models and manufactured with a material jetting system. J Comput Des Eng 3:385–397. https://doi.org/10.1016/j.jcde.2016.06.006

59. Daskalakis E, Hassan MH, Omar AM et al (2023) Rheological behaviour of different composite materials for additive manufacturing of 3D bone scaffolds. J Mater Res Technol 24:3670–3682. https://doi.org/10.1016/j.jmrt.2023.03.231
60. Jang J-W, Min K-E, Kim C et al (2024) Rheological properties and 3D printing behavior of PCL and DMSO2 composites for bio-scaffold. Materials 17:2459. https://doi.org/10.3390/ma17102459
61. Huang B, Bártolo PJ (2018) Rheological characterization of polymer/ceramic blends for 3D printing of bone scaffolds. Polym Test 68:365–378. https://doi.org/10.1016/j.polymertesting.2018.04.033
62. Sánchez-Cid P, Jiménez-Rosado M, Perez-Puyana V et al (2021) Rheological and microstructural evaluation of collagen-based scaffolds crosslinked with fructose. Polymers 13:632. https://doi.org/10.3390/polym13040632
63. Barrulas RV, Corvo MC (2023) Rheology in product development: an insight into 3D printing of hydrogels and aerogels. Gels 9:986. https://doi.org/10.3390/gels9120986
64. Pecorini G, Chiellini F, Puppi D (2022) Mechanical characterization of additive manufactured polymeric scaffolds for tissue engineering. In: Dash M (ed) Biomimetic biomaterials for tissue regeneration and drug delivery. Springer, Singapore, pp 99–148
65. Rojas-Rojas L, Tozzi G, Guillén-Girón T (2023) A Comprehensive mechanical characterization of subject-specific 3D printed scaffolds mimicking trabecular bone architecture biomechanics. Life 13:2141. https://doi.org/10.3390/life13112141
66. Dizon JRC, Espera AH, Chen Q, Advincula RC (2018) Mechanical characterization of 3D-printed polymers. Addit Manuf 20:44–67. https://doi.org/10.1016/j.addma.2017.12.002
67. Suamte L, Tirkey A, Barman J, Jayasekhar Babu P (2023) Various manufacturing methods and ideal properties of scaffolds for tissue engineering applications. Smart Mater Manuf 1:100011. https://doi.org/10.1016/j.smmf.2022.100011
68. Shafiee A, Atala A (2017) Tissue engineering: toward a new era of medicine. Annu Rev Med 68:29–40. https://doi.org/10.1146/annurev-med-102715-092331
69. Fattahi R, Mohebichamkhorami F, Taghipour N, Keshel SH (2022) The effect of extracellular matrix remodeling on material-based strategies for bone regeneration: review article. Tissue Cell 76:101748. https://doi.org/10.1016/j.tice.2022.101748
70. Alonzo M, Alvarez Primo F, Anil Kumar S et al (2021) Bone tissue engineering techniques, advances, and scaffolds for treatment of bone defects. Curr Opin Biomed Eng 17:100248. https://doi.org/10.1016/j.cobme.2020.100248
71. Nair AK, Gautieri A, Chang S-W, Buehler MJ (2013) Molecular mechanics of mineralized collagen fibrils in bone. Nat Commun 4:1724. https://doi.org/10.1038/ncomms2720
72. Hench LL, Wilson J (1993) An introduction to bioceramics. World Scientific
73. Murugan R, Ramakrishna S (2005) Development of nanocomposites for bone grafting. Compos Sci Technol 65:2385–2406. https://doi.org/10.1016/j.compscitech.2005.07.022
74. Castilho M, Dias M, Vorndran E et al (2014) Application of a 3D printed customized implant for canine cruciate ligament treatment by tibial tuberosity advancement. Biofabrication 6:025005. https://doi.org/10.1088/1758-5082/6/2/025005
75. Habibovic P, Gbureck U, Doillon CJ et al (2008) Osteoconduction and osteoinduction of low-temperature 3D printed bioceramic implants. Biomaterials 29:944–953. https://doi.org/10.1016/j.biomaterials.2007.10.023
76. Inzana JA, Olvera D, Fuller SM et al (2014) 3D printing of composite calcium phosphate and collagen scaffolds for bone regeneration. Biomaterials 35:4026–4034. https://doi.org/10.1016/j.biomaterials.2014.01.064
77. Kumar A, Mandal S, Barui S et al (2016) Low temperature additive manufacturing of three dimensional scaffolds for bone-tissue engineering applications: processing related challenges and property assessment. Mater Sci Eng R Rep 103:1–39. https://doi.org/10.1016/j.mser.2016.01.001
78. Ravaglioli A, Krajewski A (1997) Implantable porous bioceramics. Mater Sci Forum 250:221–230. https://doi.org/10.4028/www.scientific.net/MSF.250.221

79. Simske SJ, Ayers RA, Bateman TA (1997) Porous materials for bone engineering. Mater Sci Forum 250:151–182. https://doi.org/10.4028/www.scientific.net/MSF.250.151
80. Zhu Y, Goh C, Shrestha A (2021) Biomaterial properties modulating bone regeneration. Macromol Biosci 21:2000365. https://doi.org/10.1002/mabi.202000365
81. Selim M, Mousa HM, Abdel-Jaber GT et al (2024) Innovative designs of 3D scaffolds for bone tissue regeneration: understanding principles and addressing challenges. Eur Polym J 215:113251. https://doi.org/10.1016/j.eurpolymj.2024.113251
82. Temenoff JS, Mikos AG (2000) Review: tissue engineering for regeneration of articular cartilage. Biomaterials 21:431–440. https://doi.org/10.1016/s0142-9612(99)00213-6
83. Wang B, Chariyev-Prinz F, Burdis R et al (2022) Additive manufacturing of cartilage-mimetic scaffolds as off-the-shelf implants for joint regeneration. Biofabrication 14:024101. https://doi.org/10.1088/1758-5090/ac41a0
84. Kundu J, Shim J-H, Jang J et al (2015) An additive manufacturing-based PCL–alginate–chondrocyte bioprinted scaffold for cartilage tissue engineering. J Tissue Eng Regen Med 9:1286–1297. https://doi.org/10.1002/term.1682
85. Zare P, Pezeshki-Modaress M, Davachi SM et al (2022) An additive manufacturing-based 3D printed poly ε-caprolactone/alginate sulfate/extracellular matrix construct for nasal cartilage regeneration. J Biomed Mater Res A 110:1199–1209. https://doi.org/10.1002/jbm.a.37363
86. Chen C-H, Liu JM-J, Chua C-K et al (2014) Cartilage tissue engineering with silk fibroin scaffolds fabricated by indirect additive manufacturing technology. Materials 7:2104–2119. https://doi.org/10.3390/ma7032104
87. Moroni L, Schotel R, Hamann D et al (2008) 3D fiber-deposited electrospun integrated scaffolds enhance cartilage tissue formation. Adv Funct Mater 18:53–60. https://doi.org/10.1002/adfm.200601158
88. Gruber SMS, Murab S, Ghosh P et al (2022) Direct 3D printing of decellularized matrix embedded composite polycaprolactone scaffolds for cartilage regeneration. Biomater Adv 140:213052. https://doi.org/10.1016/j.bioadv.2022.213052
89. Cui X, Breitenkamp K, Finn MG et al (2012) Direct human cartilage repair using three-dimensional bioprinting technology. Tissue Eng Part A 18:1304–1312. https://doi.org/10.1089/ten.tea.2011.0543
90. Schuurman W, Harimulyo EB, Gawlitta D et al (2016) Three-dimensional assembly of tissue-engineered cartilage constructs results in cartilaginous tissue formation without retainment of zonal characteristics. J Tissue Eng Regen Med 10:315–324. https://doi.org/10.1002/term.1726
91. Raut S, Azheruddin M, Kumar R et al (2024) Lecithin organogel: a promising carrier for the treatment of skin diseases. ACS Omega 9:9865–9885. https://doi.org/10.1021/acsomega.3c05563
92. Datta D, Noor A, Rathee A et al (2024) Hypothesizing the oleic acid-mediated enhanced and sustained transdermal codelivery of pregabalin and diclofenac adhesive nanogel: a proof of concept. Curr Mol Med. https://doi.org/10.2174/0115665240291343240306054318
93. Gomes S, Leonor IB, Mano JF et al (2012) Natural and genetically engineered proteins for tissue engineering. Prog Polym Sci 37:1–17. https://doi.org/10.1016/j.progpolymsci.2011.07.003
94. Carlson MW, Alt-Holland A, Egles C, Garlick JA (2008) Three-dimensional tissue models of normal and diseased skin. Curr Protoc Cell Biol 41:19.9.1–19.9.17. https://doi.org/10.1002/0471143030.cb1909s41
95. Nguyen DG, Pentoney SL (2017) Bioprinted three dimensional human tissues for toxicology and disease modeling. Drug Discov Today Technol 23:37–44. https://doi.org/10.1016/j.ddtec.2017.03.001
96. Rimann M, Bono E, Annaheim H et al (2016) Standardized 3D bioprinting of soft tissue models with human primary cells. J Lab Autom 21:496–509. https://doi.org/10.1177/2211068214567146
97. Yanez M, Rincon J, Dones A et al (2015) In vivo assessment of printed microvasculature in a bilayer skin graft to treat full-thickness wounds. Tissue Eng Part A 21:224–233. https://doi.org/10.1089/ten.tea.2013.0561

98. Koch L, Kuhn S, Sorg H et al (2010) Laser printing of skin cells and human stem cells. Tissue Eng Part C Methods 16:847–854. https://doi.org/10.1089/ten.tec.2009.0397
99. Lee V, Singh G, Trasatti JP et al (2014) Design and fabrication of human skin by three-dimensional bioprinting. Tissue Eng Part C Methods 20:473–484. https://doi.org/10.1089/ten.tec.2013.0335
100. Koch L, Deiwick A, Schlie S et al (2012) Skin tissue generation by laser cell printing. Biotechnol Bioeng 109:1855–1863. https://doi.org/10.1002/bit.24455
101. Moakes RJA, Senior JJ, Robinson TE et al (2021) A suspended layer additive manufacturing approach to the bioprinting of tri-layered skin equivalents. APL Bioeng 5:046103. https://doi.org/10.1063/5.0061361
102. Sapoznik E, Niu G, Nomi M et al (2015) Chapter 27—regeneration of the vascular system. In: Atala A, Allickson JG (eds) Translational regenerative medicine. Academic, Boston, pp 357–373
103. Gao B, Yang Q, Zhao X et al (2016) 4D bioprinting for biomedical applications. Trends Biotechnol 34:746–756. https://doi.org/10.1016/j.tibtech.2016.03.004
104. Kakisis JD, Liapis CD, Breuer C, Sumpio BE (2005) Artificial blood vessel: the Holy Grail of peripheral vascular surgery. J Vasc Surg 41:349–354. https://doi.org/10.1016/j.jvs.2004.12.026
105. Kang H-W, Lee SJ, Ko IK et al (2016) A 3D bioprinting system to produce human-scale tissue constructs with structural integrity. Nat Biotechnol 34:312–319. https://doi.org/10.1038/nbt.3413
106. Hantusch B (2019) Morphological and functional characteristics of blood and lymphatic vessels. In: Fundamentals of vascular biology. Springer, Cham, pp 1–43
107. Cao X, Maharjan S, Ashfaq R et al (2021) Bioprinting of small-diameter blood vessels. Engineering 7:832–844. https://doi.org/10.1016/j.eng.2020.03.019
108. Solis DM, Czekanski A (2022) 3D and 4D additive manufacturing techniques for vascular-like structures—a review. Bioprinting 25:e00182. https://doi.org/10.1016/j.bprint.2021.e00182
109. Jia W, Gungor-Ozkerim PS, Zhang YS et al (2016) Direct 3D bioprinting of perfusable vascular constructs using a blend bioink. Biomaterials 106:58–68. https://doi.org/10.1016/j.biomaterials.2016.07.038
110. Christensen K, Xu C, Chai W et al (2015) Freeform inkjet printing of cellular structures with bifurcations. Biotechnol Bioeng 112:1047–1055. https://doi.org/10.1002/bit.25501
111. Xu Y, Hu Y, Liu C et al (2018) A novel strategy for creating tissue-engineered biomimetic blood vessels using 3D bioprinting technology. Materials 11:1581. https://doi.org/10.3390/ma11091581
112. Suntornnond R, Tan EYS, An J, Chua CK (2017) A highly printable and biocompatible hydrogel composite for direct printing of soft and perfusable vasculature-like structures. Sci Rep 7:16902. https://doi.org/10.1038/s41598-017-17198-0
113. Gao Q, Liu Z, Lin Z et al (2017) 3D bioprinting of vessel-like structures with multilevel fluidic channels. ACS Biomater Sci Eng 3:399–408. https://doi.org/10.1021/acsbiomaterials.6b00643
114. Wu PK, Ringeisen BR (2010) Development of human umbilical vein endothelial cell (HUVEC) and human umbilical vein smooth muscle cell (HUVSMC) branch/stem structures on hydrogel layers via biological laser printing (BioLP). Biofabrication 2:014111. https://doi.org/10.1088/1758-5082/2/1/014111
115. Meyer W, Engelhardt S, Novosel E et al (2012) Soft polymers for building up small and smallest blood supplying systems by stereolithography. J Funct Biomater 3:257–268. https://doi.org/10.3390/jfb3020257
116. Huber B, Engelhardt S, Meyer W et al (2016) Blood-vessel mimicking structures by stereolithographic fabrication of small porous tubes using cytocompatible polyacrylate elastomers, biofunctionalization and endothelialization. J Funct Biomater 7:11. https://doi.org/10.3390/jfb7020011

117. Raman R, Bhaduri B, Mir M et al (2016) High-resolution projection microstereolithography for patterning of neovasculature. Adv Healthc Mater 5:610–619. https://doi.org/10.1002/adhm.201500721
118. Murphy SV, Atala A (2014) 3D bioprinting of tissues and organs. Nat Biotechnol 32:773–785. https://doi.org/10.1038/nbt.2958
119. Guillotin B, Souquet A, Catros S et al (2010) Laser assisted bioprinting of engineered tissue with high cell density and microscale organization. Biomaterials 31:7250–7256. https://doi.org/10.1016/j.biomaterials.2010.05.055
120. Xu T, Zhao W, Zhu J-M et al (2013) Complex heterogeneous tissue constructs containing multiple cell types prepared by inkjet printing technology. Biomaterials 34:130–139. https://doi.org/10.1016/j.biomaterials.2012.09.035
121. Cui X, Boland T (2009) Human microvasculature fabrication using thermal inkjet printing technology. Biomaterials 30:6221–6227. https://doi.org/10.1016/j.biomaterials.2009.07.056
122. Li S, Xiong Z, Wang X et al (2009) Direct fabrication of a hybrid cell/hydrogel construct by a double-nozzle assembling technology. J Bioact Compat Polym 24:249–265. https://doi.org/10.1177/0883911509104094
123. Lee VK, Lanzi AM, Ngo H et al (2014) Generation of multi-scale vascular network system within 3D hydrogel using 3D bio-printing technology. Cell Mol Bioeng 7:460–472. https://doi.org/10.1007/s12195-014-0340-0
124. Krishnamoorthy S, Wadnap S, Noorani B et al (2020) Investigation of gelatin methacrylate working curves in dynamic optical projection stereolithography of vascular-like constructs. Eur Polym J 124:109487. https://doi.org/10.1016/j.eurpolymj.2020.109487
125. Shin S, Kwak H, Hyun J (2018) Melanin nanoparticle-incorporated silk fibroin hydrogels for the enhancement of printing resolution in 3D-projection stereolithography of poly(ethylene glycol)-tetraacrylate bio-ink. ACS Appl Mater Interfaces 10:23573–23582. https://doi.org/10.1021/acsami.8b05963
126. Ware HOT, Farsheed AC, Akar B et al (2018) High-speed on-demand 3D printed bioresorbable vascular scaffolds. Mater Today Chem 7:25–34. https://doi.org/10.1016/j.mtchem.2017.10.002
127. Kabirian F, Ditkowski B, Zamanian A et al (2018) An innovative approach towards 3D-printed scaffolds for the next generation of tissue-engineered vascular grafts. Mater Today Proc 5:15586–15594. https://doi.org/10.1016/j.matpr.2018.04.167
128. Centola M, Rainer A, Spadaccio C et al (2010) Combining electrospinning and fused deposition modeling for the fabrication of a hybrid vascular graft. Biofabrication 2:014102. https://doi.org/10.1088/1758-5082/2/1/014102
129. Vyas C, Pereira R, Huang B et al (2017) Engineering the vasculature with additive manufacturing. Curr Opin Biomed Eng 2:1–13. https://doi.org/10.1016/j.cobme.2017.05.008

Conductive Biomaterials-Based Additive Manufactured Products for Wound Healing and Skin Tissue Engineering

7

Sony Priyanka Bandi, Bahareh Alizadeh, Shreya Shashank Chauhan, and Deepanjan Datta

Abstract

Conductive biomaterials integrated into additive manufacturing processes have emerged as promising avenues for developing innovative products with enhanced functionalities. This chapter provides a comprehensive overview of the design, fabrication, and application of conductive biomaterial-based additive manufactured products for wound healing and skin tissue engineering. The chapter begins with an introduction to the challenges associated with conventional wound healing methods and the growing demand for novel approaches to address these challenges. The following section explores the fundamentals of additive manufacturing, highlighting its adaptability and appropriateness for the fabrication of complicated structures with exact control over the design and composition of materials. A notable comprehensive analysis of the properties and functions of conductive biomaterials, emphasizing their distinct electrical, mechanical, and biological traits that make them suitable for tissue engineering applications, has been discussed. The incorporation of conductive biomaterials,

S. P. Bandi (✉)
Loka Laboratories Private Limited, Technology Business Incubator, BITS Pilani Hyderabad Campus, Hyderabad, Telangana, India

B. Alizadeh
Protein Technology Research Centre, Shahid Beheshti University of Medical Sciences, Tehran, Iran

S. S. Chauhan
Department of Pharmacy, Birla Institute of Technology and Science (BITS) Pilani, Hyderabad Campus, Hyderabad, Telangana, India

D. Datta (✉)
Department of Pharmaceutics, Manipal College of Pharmaceutical Sciences, Manipal Academy of Higher Education, Manipal, Karnataka, India
e-mail: deepanjan.datta@manipal.edu

S. Singh et al. (eds.), *Biomaterial-based Additive Manufacturing in Tissue Engineering and Regeneration*, Biomaterials, Bioengineering and Sustainability 7,
https://doi.org/10.1007/978-3-031-96070-3_7

such as graphene, carbon nanotubes, and conductive polymers, among others into additive manufacturing processes enables the development of scaffolds and dressings with enhanced conductivity, which can modulate cellular behaviour and promote tissue regeneration. Furthermore, the chapter explores various strategies for optimizing the fabrication process to ensure the biocompatibility, mechanical integrity, and electrical conductivity of the resulting products. It also examines the influence of scaffold architecture, surface topography, and electrical stimulation parameters on cell proliferation, differentiation, and tissue regeneration. Taken together, this chapter underscores the potential of conductive biomaterial-based additive manufactured products as innovative solutions for addressing the unmet needs in wound healing and skin tissue engineering. It highlights the significance of materials science, and biomedical engineering, working together to progress this fascinating area and put research results into clinical practice.

Keywords

Conductive biomaterials · Additive manufacturing · 3D printing · Fabrication techniques · Wound healing · Skin tissue engineering

Abbreviations

Al	Aluminium
AM	Additive manufacturing
AMF	Additive manufacturing file
CAD	Computer-aided design
$CaSO_4$	Calcium sulphate
CLIP	Continuous liquid interface
CNTs	Carbon nanotubes
CPs	Conducting polymers
DED	Direct energy deposition
DIW	Direct ink writing
DLP	Direct light processing
DPP	Daylight polymer printing
EBM	Electron beam melting
ECM	Extracellular matrix
FDM	Fused deposition modelling
FFF	Fused filament fabrication
LCD	Liquid crystal display
LOM	Laminated object manufacturing
Mg	Magnesium
PANI	Polyaniline
PBF	Powder bed fusion
PCL	Polycaprolactone
PEDOT	Poly(3,4-ethylenedioxythiophene)

PLA	Polylactic acid
PLLA	Poly-l-lactide
PPy	Polypyrrole
PZT	Lead zirconate titanate
Si	Silica
SLA	Stereolithography
SLM	Selective laser melting
SLS	Selective laser sintering
TE	Tissue engineering
UAM	Ultrasonic additive manufacturing
UV	Ultraviolet
3D	Three-dimensional
3MF	3D manufacturing format

7.1 Introduction

Tissue engineering has become a valuable method for regenerating functional tissues to repair or replace damaged tissues. Biomaterials have essential functions as they provide pre-transplanted cells with mechanical and structural support [1–3]. It is commonly known that biomaterials ought to replicate the biochemical and biophysical characteristics of the cell microenvironment found in native tissues [4–6]. This allows cells to function in a manner consistent with their natural condition. Numerous types of biomaterials have been reported to be used in the creation of various scaffolds for tissue engineering. Three main types of biomaterials are widely employed to create various scaffold architectures: ceramics, biopolymers, and synthetic polymers. Nevertheless, the preparation of biomimicking biomaterials that specifically target certain tissues is a complex task that requires understanding and expertise [7–9]. Therefore, significant endeavours have been directed towards the development of biomaterials that mimic one or more microenvironmental signals, such as biochemical, mechanical, electrical, and structural cues [10, 11]. Electrical cues, such as electrical signals and conductive networks in tissues, have significant impacts on skin wounds [12]. Skin dysfunctions may result in a typical electrical impulse and interrupted conductive networks that have been shown to have a detrimental effect on the electrical integration of implanted biomaterials with the surrounding tissue matrices [13].

Conductive biomaterials derived from carbon or inorganic nanomaterials have shown potential for wound healing and skin tissue engineering (Fig. 7.1). Scientists have created many kinds of conductive biomaterials including films [14, 15], nanofibers [16, 17], hydrogels [18–20], and three-dimensional (3D) porous scaffolds [21, 22]. These conductive biomaterials have shown significant desirable outcomes in skin replacement or in wound management either *ex vivo* or *in vivo*. These outcomes were attributed to their similar conductivity to human skin, strong antioxidant and antibacterial properties, and their ability to deliver drugs through electrical stimulation, and to generate heat through light absorption, respectively. The skin, the body's biggest organ, is crucial for protecting individuals from external elements. The skin is

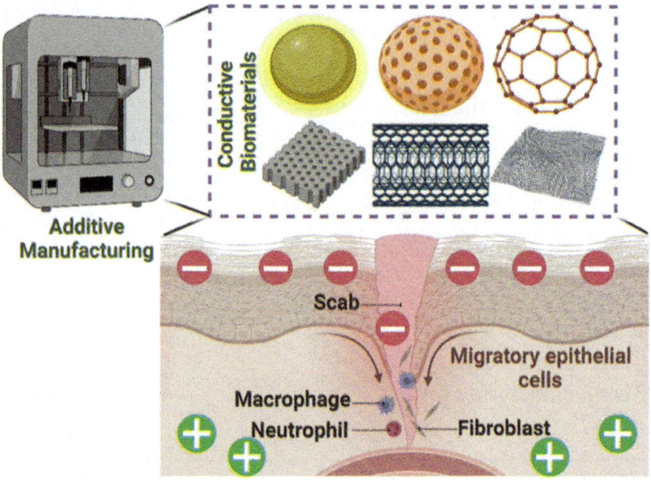

Fig. 7.1 Pictorial representation of additive manufacturing-based conductive biomaterials and their applications in skin wounds. (Created with BioRender.com)

composed of three layers: the dermis, which is the innermost layer, the epidermis, which is the outermost layer, and the hypodermis, which is the subcutaneous fat tissue. Their varying thicknesses primarily influence how these layers can be distinguished from one another [23]. Loss in the integrity in these layers results in the loss of barrier integrity which may cause a wound that takes a significant amount of time to heal in an individual [24]. The healing process consists of four phases that occur simultaneously: haemostasis, inflammation, proliferation, and remodeling. These phases involve many growth factors, enzymes, and cytokines that work together to regulate important cellular activities [25–28]. Thus, conductive scaffolds that imitate the extracellular matrix (ECM) and include bioactive chemicals or cells have been carefully created for treating severe wounds. These scaffolds have been shown to offer mechanical support and regulate cell activities [3, 29]. It is worth mentioning that besides other important parameters (inherent molecular structure, doping, conductivity, reversible oxidation-reduction reactions, pseudo capacitance, swelling/deswelling, electrochromism), among others, the charge (positive, negative or neutral) of the conductive biomaterials with that of the skin (negatively charge), also plays a significant role in determining the overall fate of the fabricated system [30]. So far, there have been no explicative studies performed or reported citing the interplay of charge activities. However, significant reports have been published over the decades explaining this concept concerning cell line studies and drug-polymeric composite interactions [31–34]. On this note, various conductive biopolymers including polyaniline (PANi), polypyrrole (PPy), and poly(3,4-ethylenedioxythiophene) (PEDOT) have shown enormous potential as conducting soft-biomaterials compared to inorganic materials based on graphite [35].

In today's manufacturing and engineering design, additive manufacturing (AM) has become a game-changing method for producing conductive bio-based materials. It makes it possible to construct intricate structures, such as the scaffolds used in tissue engineering (TE). AM or 3D printing develops 3D objects layer by layer using CAD models [36]. This process enables the construction of extremely intricate and personalized structures. To do this, He et al. used 3D printing with digital light processing to produce a variety of conductive hydrogels with remarkable qualities [37]. These hydrogels showed high strength, reaching up to 22.9 MPa, as well as excellent ionic conductivity, with values of up to 9.64 Sm^{-1}. Additionally, they displayed remarkable elasticity, stretching up to 583%. Both the achievement of hydrogels with complex structures and high resolution, as well as the attainment of high ionic conductivity for hydrogels, were accomplished by the ion coordination effect and chemical cross-linking.

Taken together, conductive biomaterials are typically produced by adding a minor quantity of conductive material to nonconducting polymers. Studies have shown the effectiveness of combining conductive biomaterials with other bioactive substances and cells to expedite the wound healing process in various pathways. Conductive materials also can guide and orient the movement of cells from the periphery into the centre of a wound, hence expediting the process of wound healing and regeneration. Briefly, this chapter explicitly reviews the properties and fabrication methods involved for conductive biopolymers, carbon nanomaterials, and conductive biomaterials with their influence on cellular behaviour. This chapter also delves into AM techniques with a specific focus on 3D bioprinting for the enhanced products developed, utilized and reported in wound healing and skin tissue engineering.

7.2 Fundamentals of Additive Manufacturing

AM has developed into a manufacturing technique over the past 30 years. Charles Hull developed AM technology in 1986 when he discovered how to make three-dimensional objects using ultraviolet light and UV-sensitive polymers [38]. The chronology of the most important findings in AM field from the 1980s to the present is shown in Fig. 7.2. This innovative method focuses on layer-by-layer material addition to producing complicated products quickly and efficiently. AM technology enables the direct fabrication of parts or the transformation of design concepts into functional prototypes [39]. Other names for this technology include layer manufacturing, solid free-form fabrication, quick prototyping, rapid manufacturing, direct digital manufacturing, additive fabrication, and 3D printing. The AM technology that utilizes 3D printing models is more helpful than conventional manufacturing processes, which utilize 2D models.

The combination of novel materials and digital information technology has been termed "the third industrial revolution" [41, 42]. Biomedical uses of AM include TE, personalized implants, hearing aids, drug delivery devices, research testing, and the production of surgical equipment [43, 44].

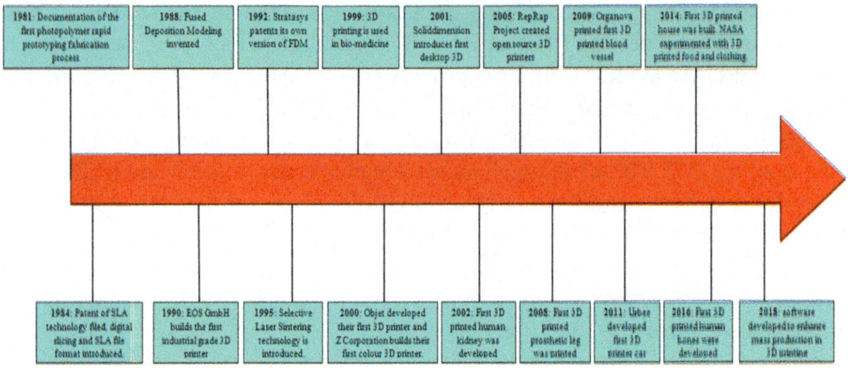

Fig. 7.2 The timeline of AM technology history and landmark accomplishments from the 1980s to the present. (This image has been adapted from [40] and licensed under a Creative Common Attribution CC-BY)

7.3 The Process Chain for Additive Manufacturing

To commence an additive manufacturing process, a virtual 3D CAD model must be created. It is possible to use almost any 3D CAD program. The CAD file is then transformed into a format that the AM machine can understand. The most used file format now is STL (standard triangle language). These files are triangulated versions of the original CAD files. When compared to the antiquated STL format, additive manufacturing file format (AMF) and 3D manufacturing format (3MF) offer improved features like colour, material, and curved triangles for higher model quality.

The model is positioned on the virtual build platform in the best orientation for printing by the AM machine software after it has opened the STL file from the CAD program. Print orientation adds strength and affects surface quality. Some techniques can create parts with weak layers or vertical print directions, resulting in extremely anisotropic parts. Other printing methods use support material for overhanging areas. The STL files are then divided into thin layers by the AM machine's software. Users can also set up other print parameters, like speed, material, fill patterns, and print resolution (layer thickness), using certain software. The machine starts layer-by-layer construction as soon as the software gives it part build instructions. Layer construction and material selection depend on the additive manufacturing technology (Fig. 7.3). When printing is finished, the part is taken out and post-processed. In many cases, post-processing involves additional processing such as machining if the part needs a finer finish, infiltration to strengthen the part, heat treatment for metal parts, or colouring and painting if the part needs to be a different colour from what the AM material can provide. Post-processing also involves removing support material and cleaning the leftover powder.

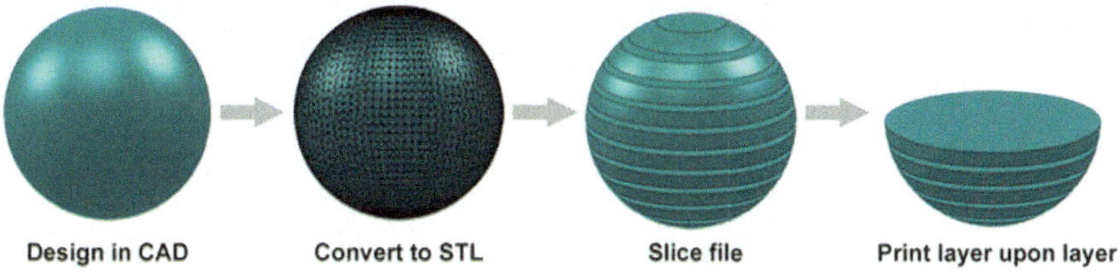

Fig. 7.3 The production chain of additive manufacturing. (Reprinted with permission from [45])

7.4 Categorization of AM Procedures

The AM processes can be categorized based on the materials used and the sort of energy used. According to the International Organization of Standardization, there are seven AM categories: vat-polymerization, material extrusion, material jetting, binder jetting, powder-based fusion, sheet lamination, and direct energy deposition [46]. Vat polymerization technologies can be classified according to the light source used by the 3D printer. Included in this category are liquid crystal display (LCD) printers, also known as daylight polymer printing (DPP), direct light processing (DLP), and stereolithography laser-based (SLA) [46]. Additionally, continuous liquid interface (CLIP) is a vat-polymerization technology that is a variation of DLP. Electron beam melting (EBM), selective laser sintering (SLS), and selective laser melting (SLM) are all powder bed fusion technologies (Fig. 7.4). Kruth has classified AM processes based on the material state before fabrication, which can be divided into four categories: (a) liquid, (b) filament/paste, (c) powder, and (d) solid sheet [47, 48].

Liquid-based technologies involve the solidification of a liquid resin through contact with a UV or laser light or the fusion process. Powder-based processes utilize energy or binding agents to create 3D structures. In solid-based techniques, solid plates are employed and can be joined with either laser bonding or adhesive bonding [47]. These techniques that are based on solid materials are more applicable to material removal and assembly procedures. AM may be further classified according to the medium employed to process the base material [49] (Fig. 7.5).

7.5 Types of Additive Manufacturing

As discussed earlier, AM is a sophisticated manufacturing technique utilized for creating prototypes, tools, and functional items. Each of the AM techniques constructs 3D structures by sequentially depositing layers of materials. This section will provide a comprehensive assessment of various AM technologies, to provide detailed explanations of their operational systems, advantages, limits, and applications.

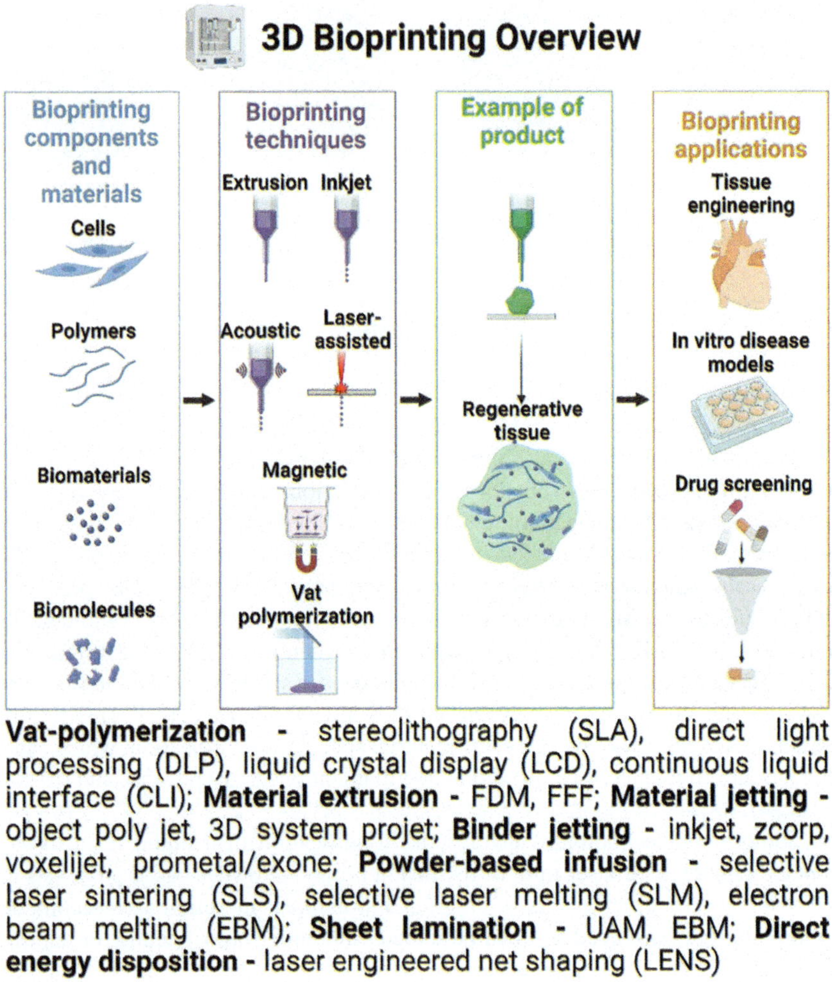

Fig. 7.4 Schematic representation of 3D bioprinting applications and approaches for the classified AM-based technologies. (Created with BioRender.com)

7.5.1 Vat Photopolymerization

This AM technology process solely uses light-activated polymerization to cure a liquid photopolymer. Lithographic techniques including stereolithography, digital light processing (DLP), and multiphoton polymerization fall within this category [50]. The main vat polymerization parameters are wavelength, exposure time, and power supply intensity. The materials employed in this study are initially liquid and hardened upon exposure to UV radiation [51].

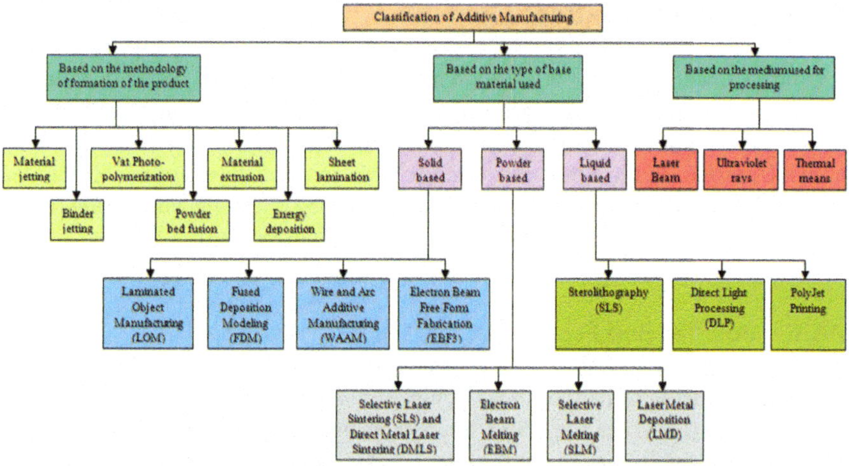

Fig. 7.5 Schematic illustration for the classified AM process based on product formation, base material and medium processing. (This image has been adapted from [40] and licensed under a Creative Common Attribution CC-BY)

7.5.2 Material Extrusion

It involves nozzle-sprayed material. Polymer is used to build parts of layer-by-layer by heating and extruding thermoplastic. Fused filament modeling (FFF), fused deposition modeling (FDM), 3D bio-plotting, and 3D dispensing are sub-classifications of material extrusion.

7.5.3 Material Jetting

Print head droplets of photosensitive materials are selectively placed and hardened under UV light. These include photopolymers, thermoplastics, composites, biological or hybrid materials, and ceramics. Notably, inkjet printing is included in this category [50, 52].

7.5.4 Binder Jetting

Fusing powdered build material requires a careful deposit of a liquid solution. It spreads powder to form a layer with a chemical binder jet. Printing big objects with metals, polymers, sands, ceramics, and hybrid materials is possible [50, 51, 53].

7.5.5 Sheet Lamination

A building material sheet is attached to make an object. Layered object manufacture and ultrasound additive manufacturing are examples. According to Mostafaei et al. (2021), it is affordable, recyclable, easy to handle, and produces full-color prints. Ultasonic additive manufacturing (UAM) connects layers of metal from featureless foil stock using sound waves, while laminated object manufacturing (LOM) produces complicated geometries quickly and cheaply [51–53].

7.5.6 Powder Bed Fusion

This process fuses powder bed areas using laser or electron beam heat [50]. Electron beam melting (EBM) and selective laser sintering are powder bed fusion methods. SLS employs a high-power laser to sinter polymer powders and produce 3D components at great speed, precision, and surface polish. Materials include metals, polymers, ceramics, hybrids, and composites [51, 52].

7.5.7 Direct Energy Deposition

Materials are melted and fused via a complex process that employs a laser or plasma. This method, which significantly controls the metal grain structure, is exclusively used for metals and metal-based composites. This raises the quality of the object. The nozzle can travel in any direction, making this operation comparable to material extrusion.

7.6 Materials Used in Additive Manufacturing

For the creation of parts with complicated geometries, AM has become more popular in recent decades as opposed to conventional manufacturing techniques. This eliminates a lot of post-processing processes [53]. In contrast to other methods that rely on material subtraction, AM utilizes filaments, powder, inks, tissues, and sheets, among others to fabricate tangible 3D-designed products [53]. To achieve consistent and desired results, 3D printing necessitates basic materials that are both high-quality and robust [51].

7.6.1 Metals and Metallic Alloys

Metal AM exhibits exceptional growth potential. For example, Boeing has employed it to manufacture the F-15 Pylon Rib. It is also employed in automotive, defense, and biomedical applications, e.g., human implants [54, 55].

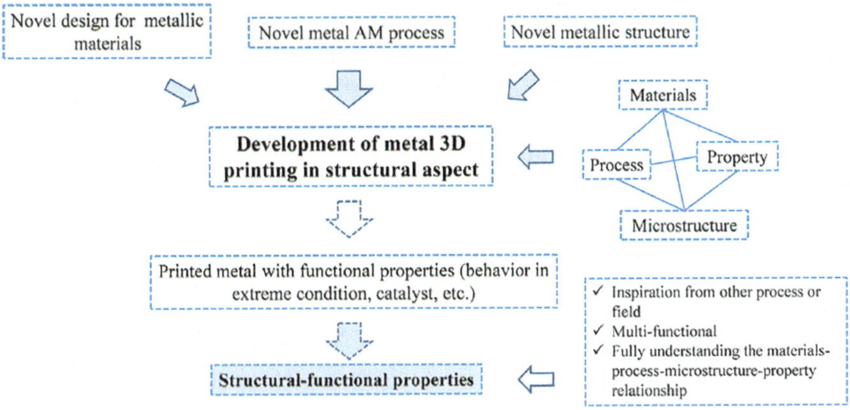

Fig. 7.6 Schematic representation for the development of metal 3D printing. (Reprinted with permission from [56])

AM of metals and alloys facilitates the production of complicated shapes with special connections. Multifunctional components can be created to simultaneously address structural, protective engineering, and insulation issues. Consequently, AM can reduce production costs and the number of necessary structural components, shortening the manufacturing timescale, and forming complex structures. Therefore, AM can modify the existing lifecycles of metal components. Figure 7.6 illustrates the outline of metal AM.

Typically, 3D printing of metals entails melting powder or wire, whereupon the liquid material solidifies layer by layer to form a solid component. The metal printing processes include diode-based procedures, direct energy deposition (DED), cold spraying, Powder bed fusion (PBF), and friction stir welding, among others [57–60]. These methods can improve speed or accuracy. In AM, many different metals and alloys are used. Murr et al. have reported that the components are manufactured with high precision and possess excellent mechanical properties, including complex shapes (0.02 mm) [61]. Many metallic materials are used in powder bed fusion (PBF). This process produces the components with high mechanical characteristics and with ±0.02 mm accuracy. Due to various limitations in AM, only a limited number of aluminium alloys are presently employed. They are less expensive than Ti alloys and simple to machine [62]. Additionally, the molten Al's low viscosity prevents the formation of a substantial melting pool. Al's high thermal conductivity enables faster AM processes and reduces internal thermal stresses. At present, the most frequently employed alloys are AlSi10 Mg and AlSi12 [63, 64].

7.6.2 Polymeric Materials

Polymers are the most popular and fundamental material for 3D printing due to their accessibility, affordability, superior mechanical qualities, and simplicity of

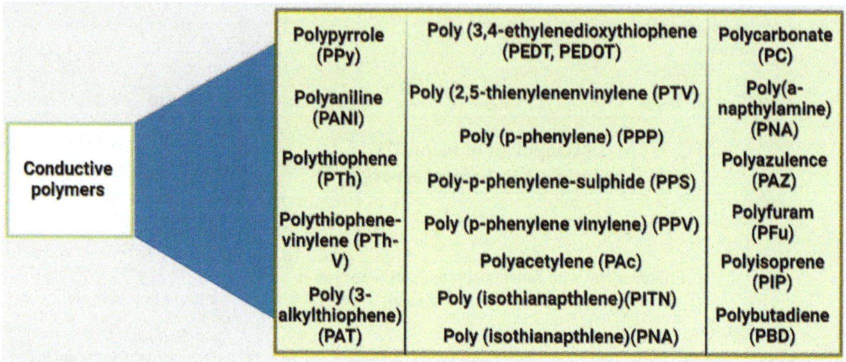

Fig. 7.7 Schematic representation of various types of conductive polymers explored. (Created with BioRender.com)

integration with various 3D printing techniques [65]. Polymers can be used in material extrusion and PBF, two more first-generation AM processes, in addition to vat polymerization. Because of their flexibility in changing their chemical structures and low melting temperatures, polymers are easier to manufacture than metals and ceramics. There are other qualities that polymers must have to be used in biological applications, in addition to their advantages in AM. For implants to be compatible with host tissues, they must be biocompatible. Moreover, during tissue self-repair, biodegradability is crucial for the breakdown of polymers.

Biocompatible polymers are suitable for both internal and external applications in the biomedical sector. The hearing aid industry is one example of an exterior application due to the robustness and durability of the polymeric material employed (the EnvisionTEC e-Shell 200 series). Conversely, coronary bypass implants with exceedingly tiny internal channels and lumbar implants composed of rigid tissue serve as illustrations of internal utilization. Biodegradable polymers are effective in the construction of human organs and actual tissue engineering [66]. Figure 7.7 shows the most explored polymers for the fabrication of conductive biomaterials processed using AM.

7.6.3 Ceramic Materials

Ceramic materials withstand heat, corrosive agents, brittleness, and rigidity. The creation of sophisticated ceramics for tissue engineering and biomaterials has made AM an indispensable approach. Slurry-based, powder-based, or mass solid-based feedstock, ceramics are employed in AM. Slurry-based ceramic AM uses finely ceramic particles dispersed across liquid or semi-liquid systems as the feedstock. Powder-based ceramic AM uses dispersed ceramic particles as its main feedstock, which are processed in powder beds. Tape-cast alumina, zirconia green sheets, silicon carbide, silicon-silicon carbide composite, glass-ceramic composite, and lead zirconate titanate (PZT) are among the ceramic materials used. These can be printed using AM techniques such as stereolithography, two-photon polymerisation, inkjet printing, 3D printing, DLP, SLS, SLM, laminated object manufacturing, and FDM. Furthermore, ceramics can be printed using a variety of other 3D printing techniques [52, 65, 67]. However, the process of refining sintered ceramic pieces to attain the intended form is both laborious and costly [68]. Consequently, the production of ceramics with complicated shapes through 3D printing and sintering has become increasingly appealing.

Ceramic scaffolds utilized in tissue engineering have demonstrated enhanced efficiency and accelerated production in comparison to conventional casting and sintering techniques [69]. A porous alumina ceramic that exhibited an extremely high flexural strength was created by incorporating dextrin and $CaSO_4$ [70]. Conversely, Maurath and Willenbacher demonstrated that a honeycomb structure without cracking and improved dimensional stability can be achieved by optimizing the sintering process and improving the rheology and homogeneity of the ceramic suspension during ink printing [71]. A hierarchical porous ceramic material with a high strength-to-weight ratio was investigated [72]. This technology achieved bigger pores by utilizing 3D printing to create a lattice structure, which refers to the space between filaments during the printing process. The foam created micro-pores by the presence of air bubbles.

7.6.4 Smart Materials

Smart materials are objects that can alter their significant properties, such as shape, color, size, and geometry, in response to changes in external conditions, like heat, humidity, pressure, temperature, light, and precipitation over time. Smart materials exhibit various characteristics that can be described as smart, such as shape memory, self-evolving, self-sensing, decision-making, self-assembly, self-adaptability, self-actuation, water absorption, self-degradation, and numerous functionalities of the material or composite. Self-evolving structures and flexible robotics employ smart materials generated through 3D printing [73, 74]. These are utilized in biomedical implants or microelectromechanical devices. Shape memory materials can be accurately reproduced and designed using 3D printing technology. Evaluating these materials involves assessing their dimensional correctness, density, and surface roughness [51].

7.6.5 Biomaterials

Biomaterials are the most frequently employed materials in the healthcare sector to produce products that meet specific requirements. The utilization of biomaterials in AM has been on the rise in recent studies due to advancements in 3D printing, including the ability to print complex structures, increased printing precision, printing speed, and reduced cost and material waste. These developments in AM technology have enabled considerable progress in the medical field, including advancements in implant materials, cell printing, and medical equipment. A significant obstacle in biomedical applications, such as the advancement of 3D models, organoids, personalized implants, implant-tissue interaction, and engineered tissues and organs, among others is the considerable amount of complexity that is necessary [75]. The biomedical industry is characterized by its patient-specific nature, which results in small production quantities [76]. In producing small, ordered quantities of customized implants, conventional manufacturing methods are prohibitively expensive due to the numerous pre-processing stages, including tooling. However, AM techniques can produce small quantities of customized parts at a faster pace and a lower cost than traditional manufacturing methods [77]. Additionally, researchers may access the CAD data to conduct research and replicate the results. AM techniques have the potential to fabricate intricate internal microstructures that possess the necessary mechanical, physical, and biological properties [78]. Additionally, the printing of porous biodegradable materials is a critical advancement in AM techniques, as it aids in the regeneration of tissues [79].

7.6.6 Hybrid Materials

3D printing of composites is a prominent focus in the printing industry, where different ink ingredients and printing procedures are combined to improve the characteristics of the produced items [80]. For instance, polymeric scaffolds are an example of this approach [81]. According to Wang and Liu, the present AM technology is incapable of concurrently processing with metallic materials [82]. Wang and Liu conducted a study in which they developed a hybrid 3D printed product that solidifies at ambient temperature after being left for a while. This product is composed of a low-melting alloy and silicone rubber. Therefore, accurately determining the melting temperatures of the input material is essential for the smooth functioning of hybrid materials 3D printing [82].

Gallium-indium alloy soft lithography created a liquid-phase electrical circuit pattern on an elastomer substrate [83]. A hybrid inkjet printer and a layer of pinhole-free polymer are used to make pixels that emit light from two different colours of polymer [84]. It has also created viable cartilage tissues [85]. Mannoor et al. developed a hybrid bionic ear by using a gel matrix that was seeded with cells and filled with silicone and silver nanoparticles [86]. For hybrid 3D printing, metal ink must always be liquid [82]. Hybrid products are produced by prospectively combining organic and inorganic components to attain superior thermal stability and toughness [87]. Organically modified silanes are the main precursors for hybrid materials.

7.7 Conductive Biomaterials for Tissue Engineering

Conducting polymers (CPs) have been shown to possess both electrical conductivity and compatibility. CPs are extremely popular in biological and medical applications as several tissues show responsivity towards electrical fields and stimuli. The field of tissue engineering has particularly benefitted from the advent of CPs due to their unique ability to support the electrical activity of cells, particularly in tissues that are naturally electrically active, such as neural, muscular, and cardiac tissues. Cells, scaffolds and growth-stimulating signals form the key components of bioengineered tissues.

7.7.1 Properties and Characteristics of Conductive Biomaterials

7.7.1.1 Electrical Conductivity

One essential characteristic of conductive biomaterials is their electrical conductivity, which makes it possible to use them in biomedical applications where electrical signal transmission is necessary. The electrical functionality of biological tissues is supported and improved by conductive biomaterials [35, 88]. This characteristic is essential for tissues like the brain and cardiac tissues that naturally show electrical activity. Electrically conductive scaffolds in cardiac tissue engineering can better integrate synthetic tissues with native heart tissue by synchronizing cardiomyocyte contractions [89]. Materials like carbon nanotubes and conductive hydrogels enable neurite outgrowth and signal transmission in neural tissue engineering, promoting neural regeneration and interacting with neural prostheses. These biomaterials' capacity to provide electrical stimulation promotes cell proliferation, differentiation, and overall tissue functionality, making electrical conductivity a key attribute for developing effective and integrative tissue engineering solutions [90, 91]. As research advances, optimizing the balance between electrical conductivity, biocompatibility, and mechanical properties remains crucial for the successful application of these materials in clinical settings.

7.7.1.2 Biocompatibility

One essential characteristic of conductive biomaterials is biocompatibility, which guarantees that the materials can interact with biological tissues without triggering unfavourable immunological reactions. To be considered biocompatible, a substance needs to be free of toxins, promote cell adhesion, proliferation, and differentiation, and blend in perfectly with the host tissues. The biocompatibility of conductive polymers, such as polypyrrole (PPy) and poly(3,4-ethylenedioxythiophene) (PEDOT), and carbon-based materials, such as graphene and carbon nanotubes, have been thoroughly investigated. For example, PEDOT is perfect for brain interfaces because of its high biocompatibility, which promotes the proliferation of neural cells and lowers inflammatory reactions. Analogous to this, the surface properties of graphene can be modified to enhance cell adhesion and proliferation, which is crucial for applications in tissue engineering. Biocompatible conductive materials guarantee that designed heart tissues can integrate with natural tissue in cardiac tissue engineering without producing toxicity or rejection. The obstacle lies in maximizing these materials' conductive qualities while improving their biocompatibility, frequently accomplished by modifying their surfaces or forming composites. Ongoing research in this area aims to create electrically conductive and completely biocompatible conductive biomaterials, guaranteeing their safe and efficient usage in medical applications [35, 88, 91].

7.7.1.3 Mechanical Properties

To guarantee appropriate performance and integration, conductive biomaterials must closely resemble the mechanical properties of the target biological tissues. Tensile strength, elasticity, and stiffness are among the mechanical qualities that are necessary to preserve the engineered tissue's structural integrity and functionality. Scaffolds employed in heart tissue engineering must endure repetitive mechanical pressures while facilitating electrical signal transfer. It is possible to develop conductive polymers like PEDOT and PPy to provide the required mechanical flexibility and durability for these kinds of applications. Similar to this, conductive hydrogels, which imitate the natural extracellular matrix by combining the characteristics of hydrogels with electrical conductivity, provide the elasticity and softness required for brain and muscle tissue engineering. Carbon-based materials are essential for applications like bone tissue engineering where both stiffness and electrical activity are required. They improve the mechanical strength of biomaterials while retaining flexibility and conductivity. For conductive biomaterials to be effectively used in tissue engineering, their mechanical qualities must be optimized while maintaining their biocompatibility and conductivity. This will guarantee that the biomaterials can withstand physiological conditions and promote tissue regeneration [88, 90, 92].

7.7.1.4 Degradability

In tissue engineering applications, where temporary scaffolds are necessary to promote tissue regeneration and later degrade without causing harm, degradability is a crucial attribute of conductive biomaterials. The advantages of electrical conductivity are combined with the capacity of degradable conductive biomaterials to be absorbed or removed by the body once they have served their intended purpose. In applications like cardiac and neurological tissue engineering, where scaffolds must initially give structural support and electrical stimulation but should not stay as permanent foreign entities, this feature is essential. It is possible to engineer polymers such as PPy and polyaniline (PANI) to break down gradually and release easily digested or eliminated non-toxic byproducts. Conductive hydrogels offer an adjustable disintegration rate while retaining essential mechanical and electrical properties. They frequently contain biodegradable polymers like polylactic acid (PLA) or polycaprolactone (PCL). To guarantee that the scaffold maintains the tissue until it is completely regenerated and then safely degrades, the degradation rate must be carefully regulated to match tissue healing and regeneration rates. The development of safe and efficient conductive biomaterials for therapeutic applications requires striking a balance between functional lifespan and degradation [90, 93, 94].

7.7.1.5 Surface Properties

For conductive biomaterials, surface characteristics are essential because they affect protein adsorption, cell-material interactions, and overall biocompatibility. These characteristics include charge, wettability, roughness, and surface chemistry; all of these have a significant impact on how cells react. For tissue engineering applications, for example, surface roughness might improve cell adherence by offering more anchoring places. Surface modifications are frequently applied to conductive polymers to enhance their biocompatibility and encourage cell adhesion and proliferation. Protein adsorption can be regulated by surface chemistry, including functional groups and coatings, which in turn influences cell behaviour and bio-integration. Another factor is wettability or surface hydrophilicity; too hydrophobic or hydrophilic of a surface might inhibit cell attachment, therefore wettability optimization is crucial. Because of their adjustable surface qualities, graphene and carbon nanotubes are widely utilized to produce conditions that promote cell division and growth. Surface charge plays a crucial role in the design of scaffolds for tissue regeneration since it can affect cellular processes like migration and proliferation. Therefore, optimizing conductive biomaterials' surface qualities is essential to boosting their functionality and guaranteeing their efficacy in biomedical applications [35, 92, 93].

7.8 Conductive-Based Biomaterials

Conductive biomaterials that are biocompatible have been used to create *in vitro* platforms for differentiating progenitor cell populations and constructing skin engineering scaffolds. The progress of innovative biomaterials and the creation of functional scaffolds are key to achieving successful skin engineering and regenerative therapeutics. The biocompatibility of new materials and their composites is a crucial factor for skin engineering applications. Replicating the inherent structure and chemical characteristics of the specific extracellular matrix is crucial for facilitating cellular functions. Numerous studies have shown that the interaction between cells and biomaterials in skin engineering can be controlled by certain features of the biomaterials, including mechanical stiffness, elasticity, surface topography, and wettability, which have been discussed earlier. However, these features are dependent and may vary based on the types and preparation of conductive biomaterials, which are discussed below.

7.8.1 Graphene-Based Biomaterials

Graphene-based conductive biomaterials with outstanding mechanical, electrical, and biocompatibility characteristics have made them extremely promising materials for tissue engineering. Graphene-based scaffolds have the potential to improve the functional integration of created cardiac tissues by improving the synchronization of cardiomyocyte contractions in cardiac tissue engineering. The tensile strength and flexibility impart graphene as a good material to use in the construction of scaffolds that can endure physiological loads. These characteristics guarantee that the scaffolds support the structure while allowing for the dynamic movements of tissues like muscles and nerves [95, 96].

Adding bioactive compounds to graphene or using biopolymers to create graphene composites can improve graphene's ability to interact with cells. Graphene-based materials in neural tissue engineering facilitate neurite outgrowth and improve neuronal electrical connection, both of which are essential for regaining brain function. Because of its conductive qualities, graphene is a great material for neural prosthetics and interfaces with neural tissues [88, 95, 96].

Scaffolds based on graphene offer the electrical stimulation required to promote myoblast differentiation and proliferation for the regeneration of skeletal muscle. Better muscle fibre development and functional recovery can be obtained.

According to a study, graphene-based cardiac patches have the potential to repair and regenerate heart since they can sustain cardiomyocyte development and electrical signal transmission. Graphene substrates may be useful in the treatment of brain injuries and illnesses, as evidenced by research demonstrating their ability to stimulate the proliferation and differentiation of neural stem cells. Methods such as incorporating hydrogels into graphene or covering it with proteins have been investigated to enhance its functionality and biocompatibility in a range of tissue engineering applications [93].

7.8.2 Carbon Nanotubes

Carbon nanotubes (CNTs) have become a promising conductive biomaterial for tissue engineering due to their extraordinary electrical, mechanical, and biocompatibility qualities. CNTs are cylindrical structures composed of rolled-up graphene sheets. They provide special benefits for a range of tissue engineering applications, especially those involving electrically active tissues such as neurological, muscular, and cardiac (Fig. 7.8).

Because of their superior electrical conductivity, CNTs are crucial for the growth of tissues like the heart and cerebral tissues that depend on electrical signals. Enhancing the electrical connection of cells by the incorporation of CNTs into scaffolds can result in increased tissue functionality. Mechanical properties, including high tensile strength and flexibility, are some of the notable outstanding properties of CNTs. These properties facilitate the formation of resilient and enduring scaffolds capable of enduring the mechanical stresses of biological tissues, hence assuring structural integrity while preserving the requisite flexibility for dynamic tissues such as muscles and nerves [90, 91, 95].

Improved biocompatibility of CNTs by functionalization with diverse biomolecules leads to improved cell adhesion, proliferation, and differentiation. Additional surface changes that regulate protein adsorption can improve biocompatibility even more. In neural tissue engineering, CNTs support neurite outgrowth and enhance the electrical activity of neurons, which is crucial for restoring neural functions.

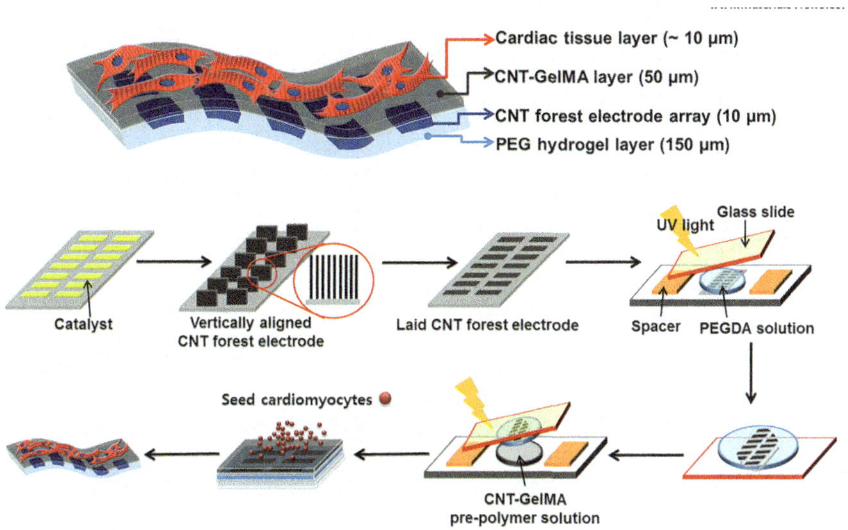

Fig. 7.8 A diagram depicting the process of creating 3D biohybrid actuators, which consist of cardiac tissue layered on a hydrogel sheet with several layers and embedded with aligned CNT microelectrodes. (Reprinted with permission from [97])

Their conductive properties facilitate the creation of neural interfaces and prosthetics that can effectively integrate with neural tissues. CNT-based scaffolds provide the necessary electrical stimulation to enhance myoblast differentiation and proliferation, facilitating skeletal muscle regeneration. This leads to improved muscle fiber formation and functional recovery.

Techniques such as coating CNTs with biocompatible polymers or integrating them into hydrogels have been explored to improve their biocompatibility and functionality (Fig. 7.9). These modifications ensure that CNT-based scaffolds are safe and effective for clinical use [88, 92, 93].

7.8.3 Other Conductive Biopolymers for Tissue Engineering

7.8.3.1 Polypyrrole

The conductive polymer known as polypyrrole (PPy) has attracted a lot of interest in the field of tissue engineering because of its exceptional electrical conductivity, biocompatibility, and simplicity of manufacture (Fig. 7.10). Due to its special qualities, it is ideal for use in regenerative medicine, especially when it comes to the engineering of electrically active tissues like cardiac, neurological, and muscular tissues [93].

Because PPy is intrinsically conductive, electrical signals can be transmitted through it more easily. This characteristic is essential for the growth of tissues like muscles and neurons that depend on electrical impulses. Depending on the dopants

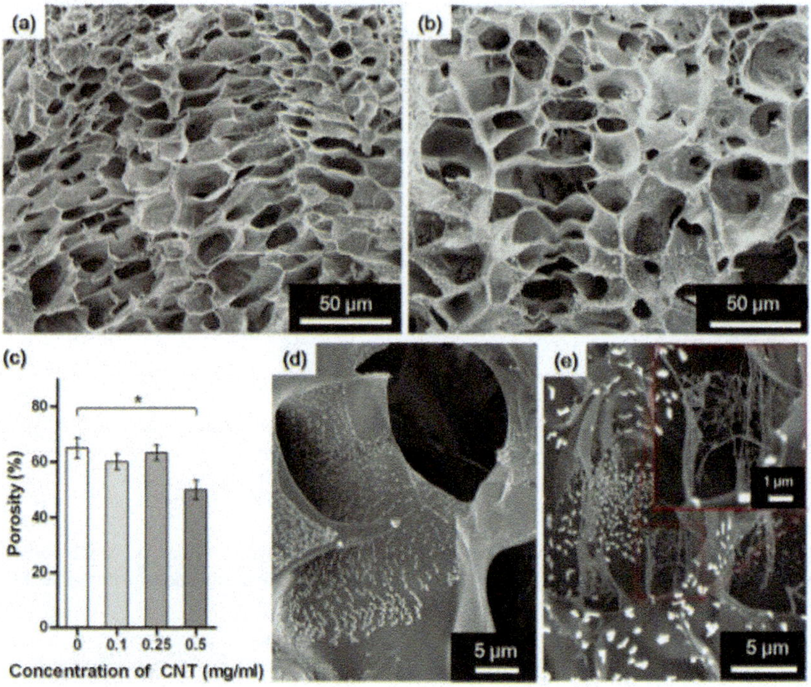

Fig. 7.9 SEM cross sections of GelMA (**a**) and CNT-GelMA (**b**) hydrogels with 0.5 mg/ml of CNTs. The porosity of CNT-GelMA hydrogels with different CNT concentrations was evaluated using ImageJ software based on SEM images (**c**). Cross-sections of GelMA hydrogel (**d**) and CNT-GelMA (**e**) hydrogel showing nanofiber junctions inside a porous structure. (Reprinted with permission from [98])

and synthesis techniques employed, various levels of conductivity can be achieved while creating PPy. Since PPy is biocompatible, it can promote cell attachment, expansion, and multiplication. Its biocompatibility can be further improved by surface changes and composites, making it appropriate for long-term implantation. PPy can be mixed with other materials to generate composites with customized mechanical properties, despite its relative stiffness when compared to other polymers. Because of their versatility, scaffolds built on PPy can replicate the mechanical properties of various tissues. Because PPy is chemically stable, its durability in physiological settings is guaranteed. In neural tissue engineering, PPy is frequently utilized to make coatings and scaffolds that promote neurite development and improve neuronal electrical connection [93, 95]. Research has demonstrated that PPy can enhance neural interface and prosthesis function as well as encourage the generation of neural stem cells.

In cardiac applications, PPy-based scaffolds can improve the electrical connectivity between cardiomyocytes, leading to better synchronization of contractions and enhanced integration of engineered cardiac tissues. PPy has been used to create conductive cardiac patches [94].

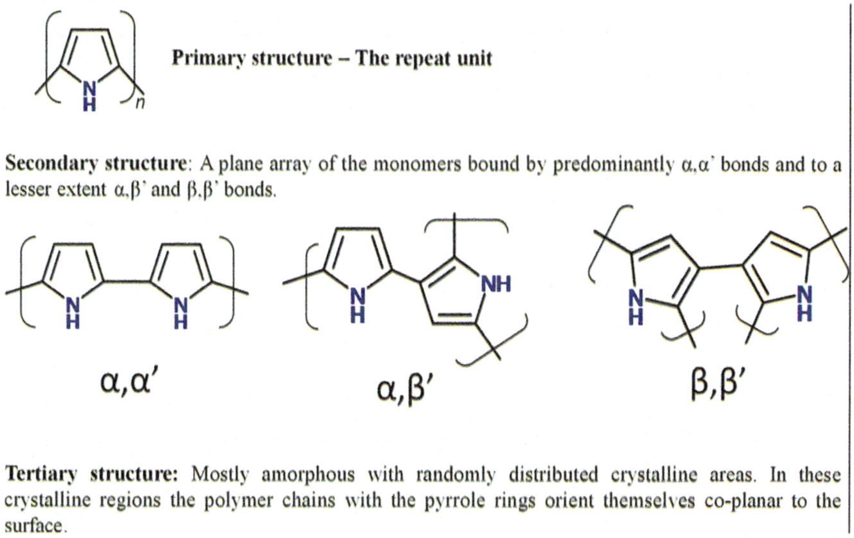

Fig. 7.10 The chemical structure of polypyrrole. (This image has been adapted from [92] and licensed under a Creative Common Attribution CC-BY)

PPy scaffolds can provide the necessary electrical stimulation to enhance myoblast proliferation and differentiation, which is critical for muscle tissue regeneration. PPy composites have been explored to create conductive and mechanically robust scaffolds for skeletal muscle engineering. PPy's surface modification can improve its interaction with cells and biocompatibility. Methods that enhance cell adhesion and proliferation include coating PPy with bioactive compounds or adding it to hydrogels. Composites with enhanced mechanical and electrical properties can be made by combining PPy with other polymers, hydrogels, or nanomaterials. These composites' overall performance in tissue engineering applications can be improved by customizing them to match the unique requirements of various tissues.

PPy is one of the multifunctional materials for the creation of tissue engineering scaffolds because of its electrical conductivity, biocompatibility, and flexibility. These highlight its versatility in enhancing the integration and performance of engineered tissues. Further research and development of PPy-based materials are expected to improve their properties and increase the application of the materials in regenerative medicine [92].

7.8.3.2 Polyaniline

Polyaniline (PANI) is a conductive polymer that has numerous applications in tissue engineering application (Fig. 7.11). Its conductivity, biocompatibility, and chemical inertness make it highly possible to use it in creating scaffold devices for the regeneration of electrically excitable tissues such as cardiac, neural, and muscular tissues [90, 91, 95].

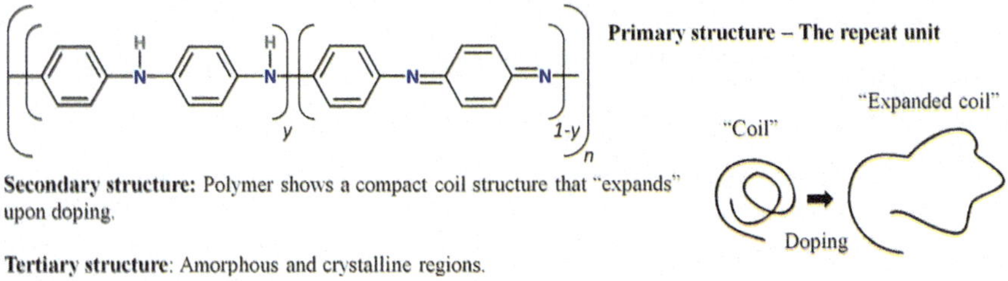

Fig. 7.11 The chemical structure of polyaniline. (This image has been adapted from [99] and licensed under a Common Creative Attribution CC-BY-NC-ND)

The electrical characteristics of PANI can be modified by varying its dopant acids, making it suitable for applications requiring conductivity control. This is important for tissues that depend on electrical impulses like the human heart and nerve tissues. PANI shows excellent biocompatibility which can be improved by exploiting the surface-modified morphology and accessibility to composite structures. For this purpose, micro and nano-structuring methods allow for the enhancement of cell attachment, proliferation, and differentiation as it is ideal for tissue engineering applications. The employ of PANI as a proximal sensor is advantageous because it has enhanced chemical stability at physiological conditions, hence having constant properties once implanted in the body. This stability is crucial to the viability of tissue-engineered constructs where their application is substantial [92, 93].

In cardiac applications, PANI has been utilized in the creation of conductive scaffolds capable of improving the coupling of cardiomyocytes. These help to enhance the electrical coupling of the heart muscles and bring about better coordination of contraction to the cardiac muscles. PANI is also used in neural tissue engineering, for the development of scaffolds and coatings that would allow nerve cells to grow and differentiate. These properties enable the development of neural interfaces and prosthetics that can engage with brain tissues and facilitate biopsy procedures. In skeletal muscle regeneration, the electrical stimulation from the pan base scaffold is also offered to support myoblast formation and differentiation. This results in enhanced formation of lean muscle fibres and functional aspects of the muscles [35, 92].

PANI's surface modification can greatly improve its biocompatibility and cell-interaction potential. Examples of surface modification techniques include grafting bioactive compounds or hydrogel incorporation. Composite materials with improved mechanical and electrical properties can be made by combining PANI with additional biocompatible polymers, nanoparticles, or hydrogels (Fig. 7.12). By customizing these composites for particular tissue engineering applications, their overall performance can be enhanced [92, 94].

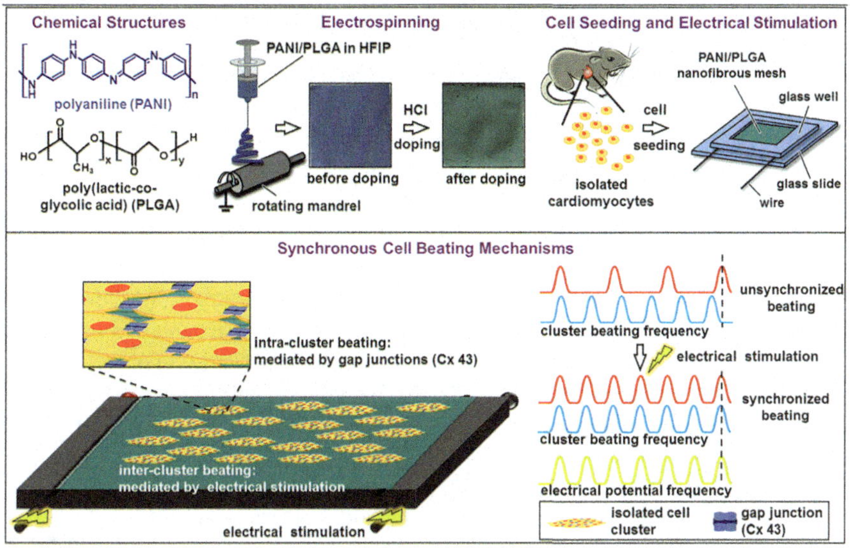

Fig. 7.12 Schematic illustration showing aligned conductive PANI/PLGA nanofibrous mesh manufacturing, cell seeding, electrical stimulation, and synchronised cell beating mechanisms. PANI- polyaniline; PLGA- poly-lactic-co-glycolic acid. (Reprinted with permission from [100])

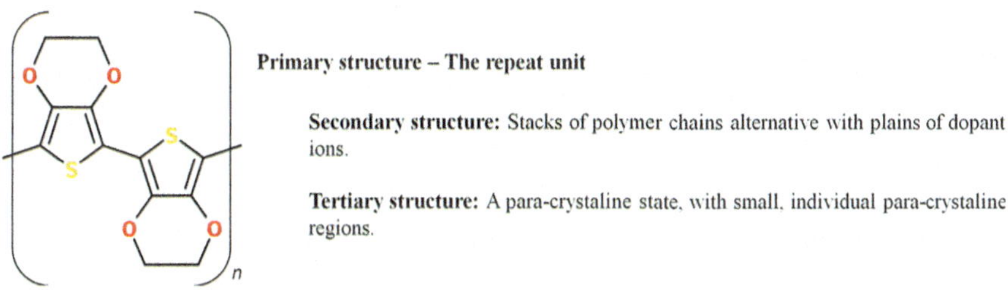

Fig. 7.13 The chemical structure of PEDOT. (Reproduced from [101])

7.8.3.3 Poly(3,4-ethyl-enedioxythiophene)

The conductive polymer, Poly(3,4-ethyl-enedioxythiophene) (PEDOT) has garnered considerable attention owing to its exceptional chemical stability, biocompatibility, and electrical conductivity (Fig. 7.13). Because developing tissues depend on electrical signals, PEDOT's high electrical conductivity is crucial. Its conductivity may be adjusted using synthesis and doping techniques, giving it great versatility for a range of biomedical applications [92, 93].

PEDOT supports cell attachment, growth, and proliferation due to its sturdy biocompatibility. This is essential for tissue engineering applications where it is necessary to have long-term contact with biological tissues [91]. When utilized *in vivo*, PEDOT's chemical stability under physiological settings guarantees its long-term durability and functioning. The integrity and functionality of tissue-engineered structures depend on this stability. PEDOT can be mixed with other materials to create composites to obtain the required mechanical qualities. Because of its versatility, scaffolds tailored to various tissues' mechanical properties can be made [35, 93]. In neural tissue engineering, PEDOT is widely utilized to make conductive scaffolds and coatings for prosthetics and brain interfaces. It improves neuronal electrical connection, promoting neurite development and neural stem cell differentiation.

PEDOT-based scaffolds have the potential to promote electrical communication among cardiomyocytes in cardiac tissue engineering, which could result in better integration of created cardiac tissues and improved contraction synchronization. Hydrogels and conductive cardiac patches that promote the growth and function of heart cells have been created using PEDOT. PEDOT scaffolds offer the electrical stimulation required to promote myoblast differentiation and proliferation, which is critical for the regeneration of muscle tissue. For skeletal muscle engineering, conductive and mechanically sturdy scaffolds made of PEDOT-based materials have been investigated [89, 94, 97]. PEDOT's surface modification can greatly improve its cell interaction and biocompatibility. Examples of these modifications include grafting bioactive compounds onto the material or encapsulating it in hydrogels. Composite materials with improved mechanical and electrical properties can be made by combining PEDOT with other biocompatible polymers, nanoparticles, or hydrogels. By customizing these composites for particular tissue engineering applications, their overall performance can be enhanced [90, 94].

7.9 Advantages and Challenges of Conductive Biopolymers

Conductive biopolymers can be considered a major improvement in the field of tissue engineering because these materials are electrically conductive yet biocompatible. These materials provide features that render them very suitable for applications involving electrically responsive tissues, such as neurological, cardiac, and muscular tissues. The former can be effectively used in electrical signalling, the importance of which cannot be overestimated when it comes to the functioning and development of these tissues. Through the conductive path, biopolymers can facilitate cell signalling and connections, thus increasing cell development and the ability of the tissue to restore its function [90, 92, 93].

The last and yet very important feature is their biocompatibility as far as the human body is concerned. By designing different forms of conductive biopolymers, the substrate surface can be optimized for cell attachment, growth, and differentiation and at the same time, there will not be retroactive immune reactions in the surrounding tissues. Further, chemical changes can also be made on the surface of these biopolymers. Finally, they derive their chemical stability under physiological conditions which makes it possible for them to offer their applications for extended periods as deemed useful in the long-term applications [35, 92, 93].

However, conductive biopolymers have some limitations which hamper them from being widely used in most applications. The first constraint area for the invention of this knowledge is the trade-off between biodegradability and electrical conductivity. People synthesize many conductive biopolymers with the threat of rapid degradation, which weakens the structure of the implant, or with delayed biodegradation, which entails the hazards of implantation. Also, to attain the necessary mechanical properties it is necessary to incorporate the above biopolymers with other materials which makes the processing and manufacturing a complex task and expensive. Cytotoxicity is also an issue as some of the conductive materials or their derivatives can be toxic to cells and therefore the choice of materials and their degradation products need to be rigorously evaluated [88, 94].

Other more critical issues are added in manufacturing, and these include manufacturing complications. Another factor in the cost of conductive biopolymers is that both the synthesis process and subsequent processing can be challenging, which renders it hard to attain a standard physical and electrical characteristics of the biopolymers. Finally, the stability of the electrical conductivity and the mechanical structure of these materials after long-term implantation into the biological environment returns to the greatest challenge that still requires long-term studies and improvements. Mitigating these issues is rather important for the further development of conductive biopolymers to enhance tissue engineering and the overall usage of biopolymers in tissue regeneration [35, 88, 92, 93].

7.10 Integration of Conductive Biomaterials into Additive Manufacturing

Precision manufacturing and functional materials are interestingly combined in additive manufacturing of conductive biomaterials, which can be used in innovative applications. To achieve good printability, each AM process has its own set of criteria, which may include rheological and chemical qualities, as well as particular transitional temperatures [102]. Because of this, conductive biomaterials are frequently combined with other polymers to allow AM without compromising electrical functionality [103].

Nevertheless, the presence of a secondary insulating polymer generally leads to a reduction in the total electrical conductivity of the material [104]. Incorporating conductive particles increases the polymer's total conductivity [105].

Conductive fillers can be dispersed into a polymer matrix as an alternative to employing conjugated polymers for creating electrical pathways. The incorporation of these fillers can have a substantial influence on the optical, mechanical, thermal stability, and electrical characteristics of the composite. Because conductive fillers allow a polymer to be utilized as the matrix material, a variety of polymers, including biopolymers and polymers with particular mechanical, chemical, or optical properties, can be employed. Additionally, the polymers can be customized for particular AM processes; for example, photocurable polymers can be utilized for stereolithography (SLA), low-viscosity resins can be employed for inkjet printing, and high-viscosity resins can be utilized for direct ink writing (DIW). This gives AM conductive polymer composites made with diverse processes a degree of freedom [106].

When creating conductive inks for AM, the conductive filler's percolation threshold in the polymer matrix is crucial. The point at which polymers go from being electrically insulators to being electrically conductive is known as the percolation threshold. Conductive fillers are often distributed randomly in polymer composites. As a result, the composite can exhibit either electrical conductivity or electrical insulation properties, depending on whether the concentration of fillers is above or below the percolation threshold [107]. The critical amount is influenced by various factors such as the filler material type, geometrical elements, filler alignment, composite processing, and the polarity of the polymer matrix [108–110].

Ma et al. investigated the impact of polypyrrole nanoparticle shape and concentration on scaffolds used in tissue engineering [22]. Both spherical (50 nm diameter) and tubular (10 μm length) particles were created using DIW and distributed in a poly-l-lactide (PLLA) matrix. According to Ma et al., tubular PPy had a greater conductivity for the AM portions (about 2×10^{-5} S cm^{-1}) at the same concentration than spherical PPy (approximately 7.5×10^{-5} S cm^{-1}). The composites were shown to be cytocompatible, making them suitable for use as scaffolds in tissue engineering. Weng et al. have investigated the use of CPs in tissue engineering with an inkjet-printed polypyrrole/collagen composite [111]. When printed on polyacrylate, the composite material yielded an electrical conductivity of 1.1 S cm^{-1} for cell patterning. *In vitro* cell research trials demonstrated the material's suitability for tissue engineering applications, as over 90% of the cells adhered to the printed collagen and polypyrrole pattern.

7.11 Influence of Conductive Biomaterials on Cellular Behaviour

Scaffold-based 3D cell cultures are the subject of extensive study in pharmacology, drug development, tissue engineering, and regenerative medicine, among other areas. Their primary usefulness comes from their capacity to replicate an environment that closely resembles the *in vivo* circumstances that cells experience. In terms of cell response to medications, nanomaterials, and biomolecules, this method enhances the validity and use of the data from *in vitro* investigations, addressing the transfer to *in vivo* applications [112]. Beginning in 1970, numerous attempts have been made to put into practice non-invasive techniques for keeping an eye on cells and communicating with them by various physical cues (such as optical, electrical, or magnetic) [113]. Of all the methods, some that involve applying an electrical field are appealing due to their non-invasiveness, precision, sensitivity, affordability, and time efficiency. In light of this, scientists began looking into several ways to record or stimulate cell development using certain electrical signals [114]. Conductivity has been extensively studied as a potential cue for enhancing the functions of synthetic tissues as well as for favourably influencing cell proliferation and differentiation when paired with appropriate biocompatibility and mechanical strength [92].

Research has shown that conductive polymers are widely used in biological contexts and can control and stimulate the development of many cell types [115]. Additionally, conductive polymers have demonstrated their value in regenerative medicine by enabling the creation of polymeric composites with enhanced mechanical, electrical, and structural qualities [116]. In applications requiring non-electroactive cells (such as bone, cartilage, and skin), it is essential to provide the proper tensile strength, compressive strength, and elastic modulus to support the physiological loads. Proper biomaterials should be able to replicate the host tissue's microstructure in addition to these mechanical characteristics, guaranteeing the structure's integration into the host environment. It has been demonstrated that PPy and PANI, two conductive polymers, assist and promote cell adherence and proliferation when combined with other materials like PLA and gelatine [117]. Enzymes in the body have the potential to break down the ester connections, producing conductive polymers with the special ability to degrade naturally [118]. Zhang et al. assessed the polymer's biological compatibility using Schwann cells after designing a biodegradable and electrically conductive polymer employing glycine ethyl ester side chains and aniline pentamer [115]. Their usefulness for scaffold production and other biological systems that require electroactivity was indicated by the lack of cytotoxicity or symptoms of inflammation, which validated their biocompatibility. Additionally, a distinct biodegradable conductive scaffold was created through the blending of conductive polypyrrole (PPy, 2.5%) and biodegradable chitosan (97.5%) [119].

7.12 Conclusion

An explicative review based on the design of conductive biomaterials in wound healing and skin tissue engineering has been portrayed in this chapter. With an emphasis on their conductivity, the characteristics of biomaterials and conductive polymers were discussed and examined. These biomaterials have an advantage over other non-conductive materials for scaffold construction because of the described features. They can measure cellular functions such as cell adhesion and proliferation using potential- or impedance-based strategies. Additionally, they can guide cellular growth through electric stimuli. Nevertheless, there are practical challenges that arise when utilizing conductive biomaterials in the field of tissue engineering. The limitations of the existing systems include inadequate polymer-cell interactions, absence of cell interface sites, hydrophobicity, and limited solubility, among others. Currently, researchers are primarily focusing on converting a conductive scaffold into a biosensor, eliminating the requirement for an external sensing system [120]. By creating a direct conducting pathway, this technique will allow for a more effective and responsive interaction between the scaffold and the cell. Even slight changes in voltage or impedance will precisely correspond to particular cellular activity. Moreover, the cytotoxicity of carbon nanomaterials, metals, and metal oxides is also a limiting factor that restricts their application, because most of them are modified with conductive biopolymeric materials.

The latest developments in 3D printing and bioprinting of conductive biomaterials have demonstrated significant potential in the field of wound and skin tissue engineering. The performance of regeneration has been demonstrated to experience a substantial enhancement, both *in vitro* and *in vivo*, due to the benefits of biomimetic potential transfer, enhanced cell functions, and comparable mechanical qualities to those of the original skin tissue. Studies have shown that conductive biomaterials can be shaped to suit different wounds. 2D biomaterials heal acute wounds with fewer exudates, while 3D biomaterials are utilized in complex wounds and skin substitutes. Due to their softness, flexibility, and mechanical qualities, conductive thin films and hydrogel membranes can be used as a bioelectronic substrate. Notably, conductive biomaterials have frequently been used to improve wound healing by including bioactive molecules such as biotherapeutics, proteins, and growth factors using various additive techniques. On a different note, to measure conductivity in wet conditions, electrochemical spectroscopy is recommended. Understanding the distinct phases of electrically conductive biomaterials and their relative contributions to total conductivity is crucial due to their complexity. This spectroscopic technique is hypothesized to enable better comparisons to identify future printing and functionalization of conductive 3D-printed biomaterials. In addition, chemical functionalization and composite methods for cell-friendly, printable, and electrically conductive bio-fabrication must be studied.

CRediT Authorship Contribution Statement **Sony Priyanka Bandi:** Conceptualization, Methodology, Validation, Investigation, Writing–original draft, Writing–review & editing. **Bahareh Alizadeh:** Investigation, Writing–original draft. **Shreya Shashank Chauhan:** Investigation, Writing–original draft. **Deepanjan Datta:** Conceptualization, Validation, Investigation, Writing–original draft, Writing–review & editing.

Declaration of Competing Interest The authors declare no competing interest.

References

1. Singh S, Chunglok W (2022) Biopolymers towards green and sustainable development. Bentham Science Publishers. https://doi.org/10.2174/97898150793021220101
2. Kapoor DU, Gaur M, Kumar A, Singh S, Chidrawar VR, Prajapati BG (2024) Value-added lignocellulose from waste biomass as a potential source in fabrication of lignin-based composite for multifaceted applications. Regen Eng Transl Med:1–30. https://doi.org/10.1007/s40883-024-00357-x
3. Shah SR, Modi CD, Singh S, Mori DD, Soniwala MM, Prajapati BG (2024) Recent advances in additive manufacturing of polycaprolactone-based scaffolds for tissue engineering applications: a comprehensive review. Regen Eng Transl Med 11(1):1–20
4. Datta D, Prajapati B, Jethva H, Agrawal K, Singh S, Prajapati BG (2024) Value-added nanocellulose valorized from fruit peel waste for potential dermal wound healing and tissue regenerative applications. Regen Eng Transl Med 11(3):1–24

5. Kiran NS, Vaishnavi G, Singh S, Yashaswini C, Parihar A, Pal S et al (2024) Biomaterials comprising implantable and dermal drug delivery targeting brain in management of Alzheimer's disease: a review. Regen Eng Transl Med 11(2):1–24
6. Mohite P, Asane G, Rebello N, Munde S, Ade N, Boban T et al (2024) Polymeric hydrogel sponges for wound healing applications: a comprehensive review. Regen Eng Transl Med 10:1–22
7. Singh S, Chittasupho C, Prajapati BG, Chandel AS (2023) Biodegradable polymeric materials in tissue engineering and their application in drug delivery. Front Bioeng Biotechnol 11:1296119
8. Sharma R, Malviya R, Singh S, Prajapati B (2023) A critical review on classified excipient sodium-alginate-based hydrogels: modification, characterization, and application in soft tissue engineering. Gels 9(5):430
9. Vyas J, Shah I, Singh S, Prajapati BG (2023) Biomaterials-based additive manufacturing for customized bioengineering in management of otolaryngology: a comprehensive review. Front Bioeng Biotechnol 11:1234340
10. Li K, Fan L, Lin J et al (2022) Nanosecond pulsed electric fields prime mesenchymal stem cells to peptide ghrelin and enhance chondrogenesis and osteochondral defect repair in vivo. Sci China Life Sci 65:927–939. https://doi.org/10.1007/s11427-021-1983-y
11. Dong R, Ma PX, Guo B (2020) Conductive biomaterials for muscle tissue engineering. Biomaterials 229:119584. https://doi.org/10.1016/j.biomaterials.2019.119584
12. Reid B, Zhao M (2014) The electrical response to injury: molecular mechanisms and wound healing. Adv Wound Care (New Rochelle) 3:184–201. https://doi.org/10.1089/wound.2013.0442
13. Post AD, Buchan S, John M et al (2021) Reconstituting electrical conduction in soft tissue: the path to replace the ablationist. Europace 23:1892–1902. https://doi.org/10.1093/europace/euab187
14. Aycan D, Selmi B, Kelel E et al (2019) Conductive polymeric film loaded with ibuprofen as a wound dressing material. Eur Polym J 121:109308. https://doi.org/10.1016/j.eurpolymj.2019.109308
15. Heald R, Bennett M, Subramanian V et al (2020) Printed electroceutical dressings for the inhibition of biofilms and treatment of chronic wounds. J Microelectromech Syst 10:1424–1431
16. Gizdavic-Nikolaidis M, Ray S, Bennett JR et al (2010) Electrospun functionalized polyaniline copolymer-based nanofibers with potential application in tissue engineering. Macromol Biosci 10:1424–1431. https://doi.org/10.1002/mabi.201000237
17. Moutsatsou P, Coopman K, Georgiadou S (2017) Biocompatibility assessment of conducting PANI/chitosan nanofibers for wound healing applications. Polymers (Basel) 9. https://doi.org/10.3390/polym9120687
18. Wang C, Jiang X, Kim HJ et al (2022) Flexible patch with printable and antibacterial conductive hydrogel electrodes for accelerated wound healing. Biomaterials 285:121479. https://doi.org/10.1016/j.biomaterials.2022.121479
19. Qiao L, Liang Y, Chen J et al (2023) Antibacterial conductive self-healing hydrogel wound dressing with dual dynamic bonds promotes infected wound healing. Bioact Mater 30:129–141. https://doi.org/10.1016/j.bioactmat.2023.07.015
20. Nie L, Wei Q, Li J et al (2023) Fabrication and desired properties of conductive hydrogel dressings for wound healing. RSC Adv 13:8502–8522. https://doi.org/10.1039/d2ra07195a
21. Lee JY, Jeong ED, Ahn CW, Lee JW (2013) Bioactive conducting scaffolds: active ester-functionalized polyterthiophene. Synth Met 185–186:66–70. https://doi.org/10.1016/j.synthmet.2013.09.044
22. Ma C, Jiang L, Wang Y et al (2019) 3D printing of conductive tissue engineering scaffolds containing polypyrrole nanoparticles with different morphologies and concentrations. Materials 12:2491–2507
23. Raut S, Azheruddin M, Kumar R et al (2024) Lecithin Organogel: a promising carrier for the treatment of skin diseases. ACS Omega 9:9865–9885. https://doi.org/10.1021/acsomega.3c05563

24. Singh S, Young A, McNaught CE (2017) The physiology of wound healing. Surgery (United Kingdom) 35:473–477. https://doi.org/10.1016/j.mpsur.2017.06.004
25. Nethi S, Das S, Patra C, Mukherjee S (2019) Recent advances in inorganic nanomaterials for wound-healing applications. Biomater Sci 7:2652–2674
26. Mandla S, Davenport Huyer L, Radisic M (2018) Review: multimodal bioactive material approaches for wound healing. APL Bioeng 2. https://doi.org/10.1063/1.5026773
27. Schultz GS, Davidson JM, Kirsner RS et al (2011) Dynamic reciprocity in the wound microenvironment. Wound Repair Regen 19:134–148. https://doi.org/10.1111/j.1524-475X.2011.00673.x
28. Li J, Zhang YP, Kirsner RS (2003) Angiogenesis in wound repair: angiogenic growth factors and the extracellular matrix. Microsc Res Tech 60:107–114. https://doi.org/10.1002/jemt.10249
29. Schultz GS, Wysocki A (2009) Interactions between extracellular matrix and growth factors in wound healing. Wound Repair Regen 17:153–162. https://doi.org/10.1111/j.1524-475X.2009.00466.x
30. Le TH, Kim Y, Yoon H (2017) Electrical and electrochemical properties of conducting polymers. Polymers (Basel) 9. https://doi.org/10.3390/polym9040150
31. Yeo WS, Mrksich M (2006) Electroactive self-assembled monolayers that permit orthogonal control over the adhesion of cells to patterned substrates. Langmuir 22:10816–10820. https://doi.org/10.1021/la061212y
32. Liu X, Miller AL, Park S et al (2017) Functionalized carbon nanotube and graphene oxide embedded electrically conductive hydrogel synergistically stimulates nerve cell differentiation. ACS Appl Mater Interfaces 9:14677–14690. https://doi.org/10.1021/acsami.7b02072
33. Gan D, Han L, Wang M et al (2018) Conductive and tough hydrogels based on biopolymer molecular templates for controlling in situ formation of polypyrrole nanorods. ACS Appl Mater Interfaces 10:36218–36228. https://doi.org/10.1021/acsami.8b10280
34. Zhang Z, Rouabhia M, Wang Z et al (2007) Electrically conductive biodegradable polymer composite for nerve regeneration: electricity-stimulated neurite outgrowth and axon regeneration. Artif Organs 31:13–22. https://doi.org/10.1111/j.1525-1594.2007.00335.x
35. Gajendiran M, Choi J, Kim SJ et al (2017) Conductive biomaterials for tissue engineering applications. J Ind Eng Chem 51:12–26. https://doi.org/10.1016/j.jiec.2017.02.031
36. Rezvani Ghomi E, Khosravi F, Neisiany RE et al (2021) Future of additive manufacturing in healthcare. Curr Opin Biomed Eng 17:100255. https://doi.org/10.1016/j.cobme.2020.100255
37. He Y, Yu R, Li X et al (2021) Digital light processing 4D printing of transparent, strong, highly conductive hydrogels. ACS Appl Mater Interfaces 13:36286–36294. https://doi.org/10.1021/acsami.1c08438
38. Prasad LK, Smyth H (2016) 3D printing technologies for drug delivery: a review. Drug Dev Ind Pharm 42:1019–1031. https://doi.org/10.3109/03639045.2015.1120743
39. Wang L, Liu J (2014) Liquid phase 3D printing for quickly manufacturing conductive metal objects with low melting point alloy ink. Sci China Technol Sci 57:1721–1728. https://doi.org/10.1007/s11431-014-5583-4
40. Alghamdi SS, John S, Choudhury NR, Dutta NK (2021) Additive manufacturing of polymer materials: progress, promise and challenges. Polymers (Basel) 13:1–39. https://doi.org/10.3390/polym13050753
41. Peter W, Katie D (2010) 3D printing for artists: research and creative practice. Rapport J Norwegian Print Assoc 1:12–15
42. Lu B, Li D (2013) Development of the additive manufacturing (3D printing) technology. Mach Building Automat 42:1–4
43. Shafiee A, Atala A (2016) Printing technologies for medical applications. Trends Mol Med 22:254–265. https://doi.org/10.1016/j.molmed.2016.01.003
44. Phan K, Sgro A, Maharaj MM et al (2016) Application of a 3D custom printed patient specific spinal implant for C1/2 arthrodesis. J Spine Surg 2:314–318. https://doi.org/10.21037/jss.2016.12.06

45. Diegel O, Nordin A, Motte D (2019) Additive manufacturing technologies BT—a practical guide to design for additive manufacturing. Springer
46. ISO/ASTM (2021) Additive manufacturing, general principles, fundamentals and vocabulary. Iso/Astm 52900:1–28
47. Kruth JP (1991) Material incress manufacturing by rapid prototyping techniques. CIRP Ann Manuf Technol 40:603–614. https://doi.org/10.1016/S0007-8506(07)61136-6
48. Kruth JP, Leu MC, Nakagawa T (1998) Progress in additive manufacturing and rapid prototyping. CIRP Ann Manuf Technol 47:525–540. https://doi.org/10.1016/S0007-8506(07)63240-5
49. Dutta Majumdar J, Manna I (2011) Laser material processing. Int Mater Rev 56:341–388
50. Ligon SC, Liska R, Stampfl J et al (2017) Polymers for 3D printing and customized additive manufacturing. Chem Rev 117:10212–10290. https://doi.org/10.1021/acs.chemrev.7b00074
51. Shahrubudin N, Lee TC, Ramlan R (2019) An overview on 3D printing technology: technological, materials, and applications. Procedia Manuf 35:1286–1296. https://doi.org/10.1016/j.promfg.2019.06.089
52. Chen Z, Li Z, Li J et al (2019) 3D printing of ceramics: a review. J Eur Ceram Soc 39:661–687. https://doi.org/10.1016/j.jeurceramsoc.2018.11.013
53. Mostafaei A, Elliott AM, Barnes JE et al (2021) Binder jet 3D printing—process parameters, materials, properties, modeling, and challenges. Prog Mater Sci 119:100707. https://doi.org/10.1016/j.pmatsci.2020.100707
54. Apelian D, Materials N, Board A et al (2004) Accelerating technology transition: bridging the valley of death for materials and processes in defense systems. The National Academies Press
55. Wohlers T (2017) 3D printing and additive manufacturing state of the industry, annual worldwide progress report Wohlers report: chapters titles: the Middle East, and other countries. Wohlers Associates Inc, Fort Collins
56. Liu G, Zhang X, Chen X et al (2021) Additive manufacturing of structural materials. Mater Sci Eng R Rep 145:100596. https://doi.org/10.1016/j.mser.2020.100596
57. Bai Y, Williams CB (2015) An exploration of binder jetting of copper. Rapid Prototyp J 21:177–185. https://doi.org/10.1108/RPJ-12-2014-0180
58. Sova A, Grigoriev S, Okunkova A, Smurov I (2013) Potential of cold gas dynamic spray as additive manufacturing technology. Int J Adv Manuf Technol 69:2269–2278. https://doi.org/10.1007/s00170-013-5166-8
59. Sharma A, Bandari V, Ito K et al (2017) A new process for design and manufacture of tailor-made functionally graded composites through friction stir additive manufacturing. J Manuf Process 26:122–130. https://doi.org/10.1016/j.jmapro.2017.02.007
60. Matthews MJ, Guss G, Drachenberg DR et al (2017) Diode-based additive manufacturing of metals using an optically-addressable light valve. Opt Express 25:11788. https://doi.org/10.1364/oe.25.011788
61. Murr LE, Martinez E, Hernandez J et al (2012) Microstructures and properties of 17-4 PH stainless steel fabricated by selective laser melting. J Mater Res Technol 1:167–177. https://doi.org/10.1016/S2238-7854(12)70029-7
62. Brice C, Shenoy R, Kral M, Buchannan K (2015) Precipitation behavior of aluminum alloy 2139 fabricated using additive manufacturing. Mater Sci Eng A 648:9–14. https://doi.org/10.1016/j.msea.2015.08.088
63. Rosenthal I, Stern A, Frage N (2017) Strain rate sensitivity and fracture mechanism of AlSi10Mg parts produced by selective laser melting. Mater Sci Eng A 682:509–517. https://doi.org/10.1016/j.msea.2016.11.070
64. Vora P, Mumtaz K, Todd I, Hopkinson N (2015) AlSi12 in-situ alloy formation and residual stress reduction using anchorless selective laser melting. Addit Manuf 7:12–19. https://doi.org/10.1016/j.addma.2015.06.003
65. Ngo TD, Kashani A, Imbalzano G et al (2018) Additive manufacturing (3D printing): a review of materials, methods, applications and challenges. Compos Part B Eng 143:172–196. https://doi.org/10.1016/j.compositesb.2018.02.012
66. Guo N, Leu MC (2013) Additive manufacturing: technology, applications and research needs. Front Mech Eng 8:215–243. https://doi.org/10.1007/s11465-013-0248-8

67. Hwa LC, Rajoo S, Noor AM et al (2017) Recent advances in 3D printing of porous ceramics: a review. Curr Opin Solid State Mater Sci 21:323–347. https://doi.org/10.1016/j.cossms.2017.08.002
68. Travitzky N, Bonet A, Dermeik B et al (2014) Additive manufacturing of ceramic-based materials. Adv Eng Mater 16:729–754. https://doi.org/10.1002/adem.201400097
69. Wen Y, Xun S, Haoye M et al (2017) 3D printed porous ceramic scaffolds for bone tissue engineering: a review. Biomater Sci 5:1690–1698. https://doi.org/10.1039/c7bm00315c
70. Li X, Gao M, Jiang Y (2016) Microstructure and mechanical properties of porous alumina ceramic prepared by a combination of 3–D printing and sintering. Ceram Int 42:12531–12535. https://doi.org/10.1016/j.ceramint.2016.05.027
71. Maurath J, Willenbacher N (2017) 3D printing of open-porous cellular ceramics with high specific strength. J Eur Ceram Soc 37:4833–4842. https://doi.org/10.1016/j.jeurceramsoc.2017.06.001
72. Minas C, Carnelli D, Tervoort E, Studart AR (2016) 3D printing of emulsions and foams into hierarchical porous ceramics. Adv Mater 28:9993–9999
73. Van Humbeeck J (2018) Additive manufacturing of shape memory alloys. Shape Mem Superelasticity 4:309–312. https://doi.org/10.1007/s40830-018-0174-z
74. Yang Y, Chen Y, Wei Y, Li Y (2016) 3D printing of shape memory polymer for functional part fabrication. Int J Adv Manuf Technol 84:2079–2095. https://doi.org/10.1007/s00170-015-7843-2
75. Ahangar P, Cooke ME, Weber MH, Rosenzweig DH (2019) Current biomedical applications of 3D printing and additive manufacturing. Appl Sci (Switzerland) 9. https://doi.org/10.3390/app9081713
76. Ackland DC, Robinson D, Redhead M et al (2017) A personalized 3D-printed prosthetic joint replacement for the human temporomandibular joint: from implant design to implantation. J Mech Behav Biomed Mater 69:404–411. https://doi.org/10.1016/j.jmbbm.2017.01.048
77. Banks J (2013) Adding value in additive manufacturing: researchers in the United Kingdom and Europe look to 3D printing for customization. IEEE Pulse 4:22–26. https://doi.org/10.1109/MPUL.2013.2279617
78. Singh S, Ramakrishna S, Singh R (2017) Material issues in additive manufacturing: a review. J Manuf Process 25:185–200. https://doi.org/10.1016/j.jmapro.2016.11.006
79. Krishna BV, Xue W, Bose S, Bandyopadhyay A (2008) Engineered porous metals for implants. JOM 60:45–48. https://doi.org/10.1007/s11837-008-0059-2
80. Fafenrot S, Grimmelsmann N, Wortmann M, Ehrmann A (2017) Three-dimensional (3D) printing of polymer-metal hybrid materials by fused deposition modeling. Materials 10. https://doi.org/10.3390/ma10101199
81. Kim G, Son J-G, Park S, Kim W (2008) Hybrid process for fabricating 3D hierarchical scaffolds combining rapid prototyping and electrospinning. Macromol Rapid Commun 29:1577–1581
82. Wang L, Liu J (2014) Compatible hybrid 3D printing of metal and nonmetal inks for direct manufacture of end functional devices. Sci China Technol Sci 57:2089–2095. https://doi.org/10.1007/s11431-014-5657-3
83. Tabatabai A, Fassler A, Usiak C, Majidi C (2013) Liquid-phase gallium-indium alloy electronics with microcontact printing. Langmuir 29:6194–6200. https://doi.org/10.1021/la401245d
84. Chang SC, Bharathan J, Yang Y et al (1998) Dual-color polymer light-emitting pixels processed by hybrid inkjet printing. Appl Phys Lett 73:2561–2563. https://doi.org/10.1063/1.122533
85. Xu T, Binder KW, Albanna MZ et al (2013) Hybrid printing of mechanically and biologically improved constructs for cartilage tissue engineering applications. Biofabrication 5. https://doi.org/10.1088/1758-5082/5/1/015001
86. Mannoor MS, Jiang Z, James T et al (2013) 3D printed bionic ears. Nano Lett 13:2634–2639. https://doi.org/10.1021/nl4007744
87. Shukrun E, Cooperstein I, Magdassi S (2018) 3D-printed organic–ceramic complex hybrid structures with high silica content. Adv Sci 5:1–7. https://doi.org/10.1002/advs.201800061

88. Saberi A, Jabbari F, Zarrintaj P et al (2019) Electrically conductive materials: opportunities and challenges in tissue engineering. Biomolecules 9(9):448
89. Dvir T, Timko BP, Brigham MD et al (2011) Nanowired three-dimensional cardiac patches. Nat Nanotechnol 6:720–725. https://doi.org/10.1038/nnano.2011.160
90. Kai D, Prabhakaran MP, Stahl B et al (2012) Mechanical properties and in vitro behavior of nanofiberhydrogel composites for tissue engineering applications. Nanotechnology 23. https://doi.org/10.1088/0957-4484/23/9/095705
91. Cezar CA, Roche ET, Vandenburgh HH et al (2016) Biologic-free mechanically induced muscle regeneration. Proc Natl Acad Sci USA 113:1534–1539. https://doi.org/10.1073/pnas.1517517113
92. Balint R, Cassidy NJ, Cartmell SH (2014) Conductive polymers: towards a smart biomaterial for tissue engineering. Acta Biomater 10:2341–2353. https://doi.org/10.1016/j.actbio.2014.02.015
93. Guimard NK, Gomez N, Schmidt CE (2007) Conducting polymers in biomedical engineering. Prog Polym Sci (Oxford) 32:876–921. https://doi.org/10.1016/j.progpolymsci.2007.05.012
94. Mousavi A, Vahdat S, Baheiraei N et al (2021) Multifunctional conductive biomaterials as promising platforms for cardiac tissue engineering. ACS Biomater Sci Eng 7:55–82. https://doi.org/10.1021/acsbiomaterials.0c01422
95. Bendali A, Hess LH, Seifert M et al (2013) Purified neurons can survive on peptide-free graphene layers. Adv Healthc Mater 2:929–933. https://doi.org/10.1002/adhm.201200347
96. Ghosal K, Sarkar K (2018) Biomedical applications of graphene nanomaterials and beyond. ACS Biomater Sci Eng 4:2653–2703. https://doi.org/10.1021/acsbiomaterials.8b00376
97. Shin SR, Shin C, Memic A et al (2015) Aligned carbon nanotube-based flexible gel substrates for engineering biohybrid tissue actuators. Adv Funct Mater 25:4486–4495. https://doi.org/10.1002/adfm.201501379
98. Shin SR, Bae H, Cha JM et al (2012) Carbon nanotube reinforced hybrid microgels as scaffold materials for cell encapsulation. ACS Nano 6:362–372. https://doi.org/10.1021/nn203711s
99. MacDiarmid AG, Epstein AJ (1995) Secondary doping in polyaniline. Synth Met 69:85–92. https://doi.org/10.1016/0379-6779(94)02374-8
100. Hsiao CW, Bai MY, Chang Y et al (2013) Electrical coupling of isolated cardiomyocyte clusters grown on aligned conductive nanofibrous meshes for their synchronized beating. Biomaterials 34:1063–1072. https://doi.org/10.1016/j.biomaterials.2012.10.065
101. Aasmundtveit KE, Samuelsen EJ, Pettersson LAA et al (1999) Structure of thin films of poly(3,4-ethylenedioxythiophene). Synth Met 101:561–564. https://doi.org/10.1016/S0379-6779(98)00315-4
102. Holness FB, Price AD (2017) Direct ink writing of 3D conductive polyaniline structures and rheological modelling. Smart Mater Struct 27(1):1–13
103. Cullen AT, Price AD (2018) Digital light processing for the fabrication of 3D intrinsically conductive polymer structures. Synth Met 235:34–41. https://doi.org/10.1016/j.synthmet.2017.11.003
104. Kurselis K, Kiyan R, Bagratashvili VN et al (2013) 3D fabrication of all-polymer conductive microstructures by two photon polymerization. Opt Express 21:31029. https://doi.org/10.1364/oe.21.031029
105. Salaoru I, Maswoud S, Paul S (2019) Inkjet printing of functional electronic memory cells: a step forward to green electronics. Micromachines (Basel) 10:1–10. https://doi.org/10.3390/mi10060417
106. Hamzah KA, Yeoh CK, Noor MM et al (2022) Mechanical properties and thermal and electrical conductivity of 3D printed ABS-copper ferrite composites via 3D printing technique. J Thermoplast Compos Mater 35:3–16. https://doi.org/10.1177/0892705719869405
107. Motaghi A, Hrymak A, Motlagh GH (2015) Electrical conductivity and percolation threshold of hybrid carbon/polymer composites. J Appl Polym Sci 132:1–9. https://doi.org/10.1002/app.41744

108. Choi HJ, Kim MS, Ahn D et al (2019) Electrical percolation threshold of carbon black in a polymer matrix and its application to antistatic fibre. Sci Rep 9:1–12. https://doi.org/10.1038/s41598-019-42495-1
109. Marsden AJ, Papageorgiou DG, Vallés C et al (2018) Electrical percolation in graphene-polymer composites. 2d Mater 5. https://doi.org/10.1088/2053-1583/aac055
110. Coelho PH, Morales AR (2014) Electrical conductivity, percolation threshold and dispersion properties of PMMA nanocomposites of hybrid conducting fillers. In: Proceedings of the IEEE conference on nanotechnology, pp 706–710. https://doi.org/10.1109/NANO.2014.6968023
111. Weng B, Liu X, Shepherd R, Wallace GG (2012) Inkjet printed polypyrrole/collagen scaffold: a combination of spatial control and electrical stimulation of PC12 cells. Synth Met 162:1375–1380. https://doi.org/10.1016/j.synthmet.2012.05.022
112. Edmondson R, Broglie JJ, Adcock AF, Yang L (2014) Three-dimensional cell culture systems and their applications in drug discovery and cell-based biosensors. Assay Drug Dev Technol 12:207–218. https://doi.org/10.1089/adt.2014.573
113. Rangarajan S, Madden L, Bursac N (2014) Use of flow, electrical, and mechanical stimulation to promote engineering of striated muscles. Ann Biomed Eng 42:1391–1405. https://doi.org/10.1007/s10439-013-0966-4.Use
114. Gu W, Zhao Y (2010) Cellular electrical impedance spectroscopy: an emerging technology of microscale biosensors. Expert Rev Med Devices 7:767–779. https://doi.org/10.1586/erd.10.47
115. Zhang Q, Yan Y, Li S, Feng T (2010) The synthesis and characterization of a novel biodegradable and electroactive polyphosphazene for nerve regeneration. Mater Sci Eng C 30:160–166. https://doi.org/10.1016/j.msec.2009.09.013
116. Bettinger CJ, Bruggeman JP, Misra A et al (2009) Biocompatibility of biodegradable semiconducting melanin films for nerve tissue engineering. Biomaterials 30:3050–3057. https://doi.org/10.1016/j.biomaterials.2009.02.018.Biocompatibility
117. Pelto J, Björninen M, Pälli A et al (2013) Novel polypyrrole-coated polylactide scaffolds enhance adipose stem cell proliferation and early osteogenic differentiation. Tissue Eng Part A 19:882–892. https://doi.org/10.1089/ten.tea.2012.0111
118. Shi G, Rouabhia M, Wang Z et al (2004) A novel electrically conductive and biodegradable composite made of polypyrrole nanoparticles and polylactide. Biomaterials 25:2477–2488. https://doi.org/10.1016/j.biomaterials.2003.09.032
119. Huang J, Hu X, Lu L et al (2010) Electrical regulation of Schwann cells using conductive polypyrrole/chitosan polymers. J Biomed Mater Res A 93:164–174. https://doi.org/10.1002/jbm.a.32511
120. Distler T, Boccaccini AR (2020) 3D printing of electrically conductive hydrogels for tissue engineering and biosensors—a review. Acta Biomater 101:1–13. https://doi.org/10.1016/j.actbio.2019.08.044

Synthetic Polymer-Based Additive Manufactured Goods for Regenerative Medicine and Tissue Engineering

8

Sachin Kothawade, Pooja Bhane, Rachana Hajare, and Vishal Pande

Abstract

Synthetic polymer-based additive manufacturing holds significant promise for regenerative medicine and tissue engineering, enabling the creation of customized scaffolds and implants. These scaffolds, essential for tissue regeneration, provide structural support and promote cellular activity. Factors such as chemical composition, biodegradability, and physical properties are crucial in scaffold design, impacting cell adhesion, proliferation, and differentiation. Polymers, both natural (e.g., collagen, hyaluronic acid) and synthetic (e.g., PLA, PCL), are widely used due to their favourable characteristics. Additive manufacturing techniques, including fused deposition modelling (FDM), stereolithography (SLA), and selective laser sintering (SLS), allow precise control over scaffold architecture, enhancing their suitability for specific tissue types. However, challenges remain, such as achieving the optimal balance between scaffold biodegradability and mechanical stability. The chapter explores the potential of biopolymers and biodegradable synthetic polymers in creating lightweight, sustainable materials for biomedical applications. It also addresses the need for advancements in material properties and manufacturing techniques to meet the complex requirements of tissue engineering. Future research directions include improving the biocompatibility and functional integration of these scaffolds to enhance their clinical applicability.

S. Kothawade (✉)
SCSSS's Sitabai Thite College of Pharmacy, Shirur, Maharashtra, India

P. Bhane · R. Hajare · V. Pande
RSM's N. N. Sattha College of Pharmacy, Ahmednagar, Maharashtra, India

S. Singh et al. (eds.), *Biomaterial-based Additive Manufacturing in Tissue Engineering and Regeneration*, Biomaterials, Bioengineering and Sustainability 7,
https://doi.org/10.1007/978-3-031-96070-3_8

Keywords

Additive manufacturing · Regenerative medicine · Tissue engineering · Biodegradable polymers · 3D printing · Scaffolds

8.1 Introduction

The body can repair astonishingly, therefore the many factors, including the type of tissue and the hormones required for the discrepancy and physical size, are limited to this regeneration (critical flaw). Any tissue damage that exceeds this critical dimension necessitates the use of outside aid techniques like regenerative medicine (RM) and tissue engineering (TE), where the outside hollows are referred to as yardsticks. These tissues provide a substrate for the creation of new tissue and cellular activity [1, 2]. In both TE and RM, the scaffolds are essential.

Growth agents are often used in these tissues to promote the formation of new tissue by quickening the process by which cells and certain lines differentiate from one another. Regarding the survival and proliferation of cells, the chemical and physical composition is essential. Numerous factors, including as origin, generations of modification, chemical and physical composition, and biodegradability, are considered when classifying biomaterials. The target tissue determines which biomaterial is best. Based on their chemical makeup, biomaterials are separated into ceramics, polymers, and composites. Calcium salts and significant inorganic metal components are included in the biomaterial class of ceramics. Most of these biomaterials have been used primarily in orthodontic applications. Soft tissue engineering uses polymers because they are similar to tissues that link together. Biomaterials in the composite class are mixtures of polymers and ceramics. The composites are used in dentistry and orthopaedics. Both synthetic and natural TE as well as RM polymers are commonly utilized [3].

Most these biomaterials have been used primarily in orthodontic applications. Soft tissue engineering uses polymers because they are similar to tissues that link together. Biomaterials in the composite class are mixtures of polymers and ceramics. The composites are used in dentistry and orthopaedics. Both synthetic and natural TE as well as RM polymers are commonly utilized [4]. Natural biomaterials with biodegradability and biocompatibility are widely available. Examples of these materials consist of collagen, hyaluronic acid, alginate, as well as chitosan. One of the primary features of natural polymers is the degradation of biomaterials [5].

The matrix outside of cells (ECM) contains various biomatters, which make cells highly compatible and growth responsive. One of the most widely utilized natural biomaterials in a variety of scaffold applications is collagen. In order to create biocomposites with multiple uses, high efficiency, little environmental impact, exclusive accessibility, renewable resources, environmental friendliness, and lightweight characteristics, biopolymers have recently attracted a lot of attention [6].

Biopolymeric composites ought to take the place of synthetic materials in the various applications they find in biochemistry, biomedical engineering, and optics [7].

The extracellular matrix (ECM) contains various biomatters, which make cells highly compatible and growth responsive. One of the most widely utilized natural biomaterials in a variety of scaffold applications is collagen. In order to develop bio-composites with multiple uses, high efficiency, little environmental impact, exclusive accessibility, renewable resources, environmental friendliness, and lightweight characteristics, biopolymers have recently attracted a lot of attention [8]. Biopolymeric composites ought to take the place of synthetic materials in the various applications they find in biochemistry, biomedical engineering, and optics [9]. There are two dimensions to the product data. As demonstrated in Fig. 8.1, commencing at the lowest point of the product, substance is lowered layer by layer as the data are sent via the machine. This process adds to the material by fusing the most recent layer with the oldest one. There has been a huge rise in interest in recent years for the researchers due to biopolymers and biodegradable synthetic polymers. The development of biopolymeric materials that are lightweight, robust, and sustainable is necessary for biomedical applications [11]. However, the development or selection of approaches to address architectural design difficulties requires a compromise between goals and visions, which typically clashes with novel biomaterials.

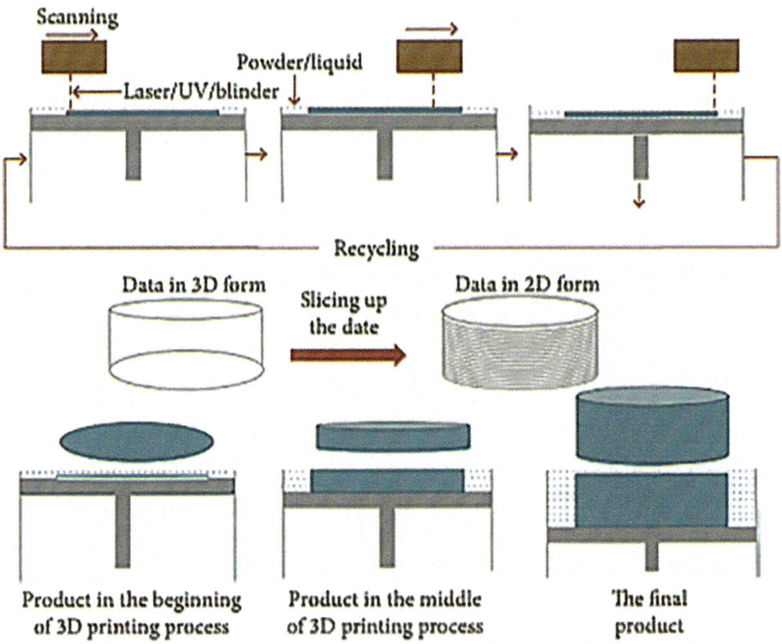

Fig. 8.1 Schematic Representation of 3D printing. (Reproduce with permission from [10] under CCBY 4.0)

Degenerative diseases of the kidneys, liver, pancreas, following heart can only be effectively treated by failing organs and transplants from another individual. The production of scaffolds and nanomaterials is the first step in the TE process. During the production process, certain characteristics such size, shape, porosity, biodegradability, and bioactivity must be met. These text tiles undergo chemical and physical modifications [12]. These specifications may vary depending on the target tissue, production process, and type of biomaterial. To allow Cells may seeds the tissue and produce the tissue that you want in the human organism or in vitro, allowing host cells to enter and replace the damaged tissue. Growth-promoting hormones, and chemical makers are used in these two methods essential because they specify tissue function and cell differentiation [13].

In addition to enabling physical cell attachment, biomaterials offer therapeutic substances such drugs, proteins, growth factors, and chemical indicators. Anchor feasibility is essential to most mammalian cells [14]. Cell death is frequently brought on by the absence of a substrate for cellular adhesion. For scrub materials, cell survival and function are therefore crucial for surface chemical and structural characteristics. Three techniques—chemical modification, physical modification, and surface coating—improve cell adherence to biomaterial and scaffold surfaces; Table 8.1 lists various Three-dimensional porous scaffolding methods for tissue engineering uses. The objective is to promote scaffolds and biomaterial cell growth and activity. It opens the door to the following phase, which involves adding cells to the yards. Conventionally, methods for introducing cells to scaffolds are included in the locations where the required cells are seeded [22]. During the complex process of creating an organ, the main cells develop a supportive matrices, vasculature, and structural framework, while the supporting cells carry out the specialized functions of the organ. To differentiate these organizations primary cells of different genotypes and phenotypes among the required cell lines, or pluripotent cells, can serve as implanted [23].

Table 8.1 3D porous scaffold approaches for tissue engineering applications

3D technology	Materials	Growth factors	Outcomes	References
Robocasting	PCL and bioactive borate glass	Hasc, laden	Controlled release of bioactive glass	[15]
3D bioprinting	PCL	Saos 2 cells seeding		[16]
Laser sintering technique	PCL and HAp	Not defined	Subchondral bone regeneration	[17]
Sol-gel method combined with 3D plotting	Hap/ chitosan/silica	Mouse BMSC seeding	Compressive strengtheq.to human trabecular bone	[18]
Electrospinning combined with 3D bioprinting	PCL	Laden with I.929mouse fibroblast	Multi-layered structure—3Dscaffolds	[19]
Phase separation process	Cartilage ECM derived/PLGA	BMSC seeding		[20, 21]

8.2 Need of Bio Printed Scaffolds and Its Fabrication

Natural polymers are defined as those that come from living things or biological resources, such as biological systems and microorganisms of plants or animals. Biopolymers (starch, chitosan, arabinoxylan, and proteins like keratin and gelatine, as well as polyhydroxyalkanoates (PHAs) and poly(3-hydroxybutyrate) [P(3HB)]) are other names for carbohydrates. One or more biopolymers are used in the synthesis of biopolymer composites in order to enhance the structural and functional properties of the final composites [24]. A biopolymer's functionality is determined by its composition, whereas its functional potential is mostly determined by the behaviour of its crystalline or amorphous constituents [25].

It is easily made from marine resources (shrimp, crab, and lobster) and used in a variety of biopolymer composites. Prior to the widespread use of bio-sustainable polymers, several issues, including production and technological difficulties, must be considered. Their functional characteristics seem inferior to well-recognized polymers derived from petroleum and provide a serious barrier to the wider application of biopolymer in many different contexts [26, 27].

For instance, a biopolymer component with a single bond performs worse than a plastic material because of its inferior mechanical property. Numerous techniques, such as lubricating, merging, mixing, and reinforcing with additional appropriate ceramics and polymers, can be used to address these problems [28]. Polymers, which are essential for the structure and operation of biological systems, are derived from a range of sources, including naturally occurring biopolymers including cellulose, protein, and microbial polyesters and the widely manufactured polystyrene [29].

Depending on where they come from, all three fundamental kinds are classified as biopolymers or composites of biobased polymers. (1) Biomass is used to extract or separate polymers like cellulose, starch, arabinoxylan, and keratin; (2) Ordinary chemical processing is used to create renewable biopolymers by fermenting carbohydrates with monomers like cellulose acetate (CA) and polylactic acid (PLA). (3) Although bacterial cellulose is produced by the digesting of microorganisms, polymers are mostly derived from PHAs. Materials made using bioderived or naturally occurring polymers, such PLA, PHAs, arabinoxylan, and chitosan, are created using bio-composite [30, 31].

Efficacious and sustainable bio-composites are green composites made of natural biopolymer constituents and biodegradable inorganic filler. Environmental issues and regulations have drawn attention to them [32, 33]. Biopolymers are extremely robust molecules generated from natural resources. These goods are known as bio-sustained items because they are readily recyclable, biodegradable, and sustainable in the marketplace and environment [34]. Chemically modified biopolymers, such as Both cellulose acetate and thiolated arabinoxylan have been utilized in several possible medicinal applications as well as other uses, as illustrated in Fig. 8.2.

An environment that is favourable for their destruction is created by microbial breakdown and environmental variables. "Green organic composites" are

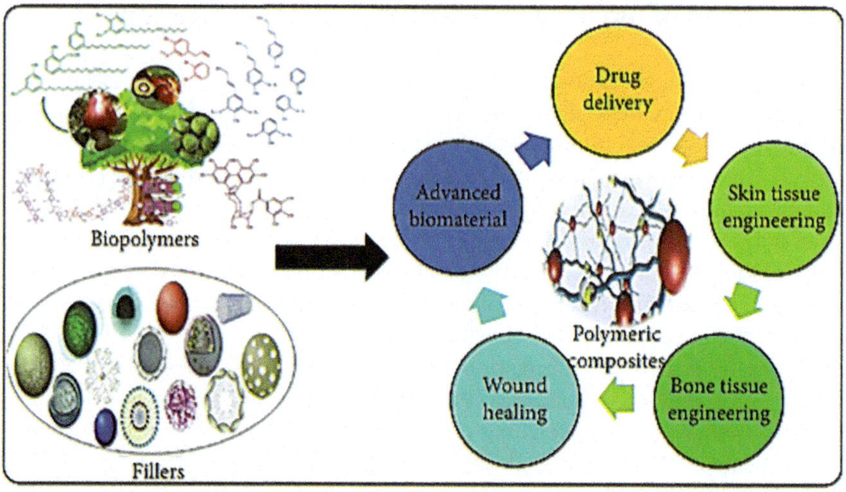

Fig. 8.2 Potential applications of biopolymers in different areas. (Reproduce with permission from [10] under CCBY 4.0)

biodegradable biopolymer composites. Many green bio compounds have several in-organic fillers in common, such as alumina, silica, and titanium. Bioplastic advancements have already brought several practical and environmentally beneficial products with synthetic oil to market [35].

8.3 Additive Manufacturing

Additive manufacturing (AM) is innovative technology that allows for the quick production a complicated one and novel components for the market [36]. In the scientific world, "an Additive Processing" is the term used to describe 3D printing (AM). It belongs among the newly technologies used in large manufacturing companies. At first, 3D printing was only used in the aerospace and prototype sectors. Its applicability was long limited by costly machinery and a lack of understanding. However, the invention of specific polymers for 3D printing makes it possible to apply the AM process in every industry. Facilities for 3D printing are being developed in the fields of biomedical, automotive, electronic, and information engineering. With subtractive manufacture, the material's intended geometric shape is achieved by lasers, drilling, cutting, or any other technique that eliminates certain material locations. Fortunately, additive manufacturing avoids this. On the other hand, layer-by-layer material shaping is known as additive manufacturing. Several additive manufacturing (AM) techniques, including following are utilized to create scaffolds and implants as shown in Fig. 8.3.

The flow and temperature behaviour of the filament passing through the nozzle are influenced by its design. This section is devoted to examining how the liquid

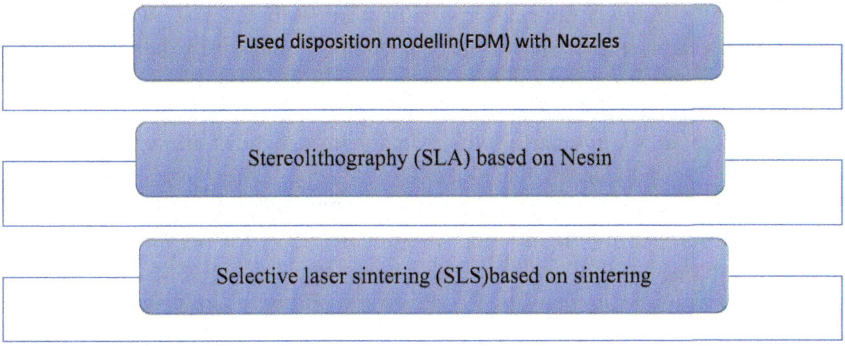

Fig. 8.3 Nozzle-based fused deposition modelling (FDM)

CF-PLA behaviour is affected by the geometries of the exit nozzle design [37]. A major challenge has been achieving a high standard for the completed components. The current investigation focuses on how nozzle diameter affects extrusion time, geometric error, and pressure drop. In the study, polylactic acid (PLA) was employed as the material, and the nozzle diameter was adjusted using finite element analysis (FEA). The Pressure started to drop along the liquefier then monitored. With the advancement of FDM printing methods, particularly Arburg Plastic Free forming, there is now a greater selection of polymeric materials available for use with FDM printers. Comparatively low melting point (Tm) as well as suitably elevated glass transition temperature (Tg) are minimum requirements for these materials. This is since low Tm makes 3D printing more inexpensive, but high Tg allows the polymer to harden quickly enough following extrusion to maintain the printed shape [38]. It is important to note that Tg is a feature of semi-crystalline materials amorphous regions or of amorphous materials themselves [39].

The amorphous phases or materials' molecular chains become locked put in position and behave similar to solid glass when the surrounding temperature falls below Tg. Materials used for FDM are:

- Polycarbonate: One carbonate-bound polymer is polycarbonate (PC). Polycarbonates are categorized as either aromatic or aliphatic PCs based on the structure of the R group. PC has excellent heat resistance, dimensional stability, toughness, and rigidity. It is used for water bottles and has great clarity [40]. Compared to ABS and other polymers, it has better qualities. It is primarily utilized in safety eyewear, CDs & DVDs, bottles, and greenhouse sheets [41].
- High-density polyethylene: A high-grade polyethylene polymer for engineering is HDPE. Material with the PE4710 and PE100 designations. In the oil and gas sector, these grades are frequently utilized for linings and pressure pipes. In low-pressure manifolds, the HDPE pressure pipes with thick walls include frequently the favoured option [42].
- Styrene butadiene acrylonitrile: The thermoplastic Styrene butadiene acrylonitrile (ABS) finds widespread application in many sectors owing to its exceptional qualities such as flexibility, durability, toughness, strong chemical resilience, as

well as comparatively poor absorption of water. Moreover, it guarantees excellent antioxidants and colour fastness enhance stability of oxidation [43]. Thus, ABS represents a material that is often utilized in FDM. About 200 °C is the melting temperature of ABS.

- Polylactic acid: Corn starch is the primary ingredient of Biodegradable polymeric polymer polylactic acid (PLA) biocompatible, and replenishable. PLA has strong mechanical qualities and is process able. It can be utilized in place of polymers derived from petroleum. Additionally, PLA-ISO as well as PLA-ABS were developed to utilized as FDM threads to fabricate parts. But PLA is a hydrophobic polymer that has limited heat stability, a sluggish rate of breakdown, and poor toughness [44]. The melting point of PLA, a homopolymer, is between 170 °C and 180 °C. Plastic films, bottles, biodegradable medical equipment, and scaffolds are among the products made with PLA.
- Nylon: In 1939, nylon—the first entirely synthetic textile fibre—was created. The most common and extensively used plastic is nylon. The polyamide group includes the several grades of nylon polymers that are available, including nylon 6, nylon 66, and nylon 12. Due of its exceptional qualities, the industry has been paying more attention to Nylon 12 in recent decades [45]. The melting point of nylon 12 is 224 °C.

8.4 Nesin-Based Stereo Lithography

Stereo lithography has emerged as a useful technique for creating biocompatible scaffolds for tissue engineering because of its capacity to create exact exterior and interior geometries that resemble human tissue. Digital light projection stereo lithography and laser-based are two distinct irradiation techniques that can be used in stereo lithography. The approach using a laser is a direct write technique that creates structures vector by vector using a computer-manipulated laser beam.

Scaffolds for the design of both soft and rigid tissues have been created by stereo lithography, offering a broad variety of mechanical attributes and polymeric resin materials. Trimethylene carbonate (TMC)/tri methylol propane (TMP) scaffolds are more pliable than the mechanically stiff PPF/DEF scaffolds. Studies conducted in vitro have shown the importance of intrinsic scaffold design and adherence of chondrocytes, and TMC/TMP the use of scaffolds in cartilage tissue engineering has been investigated [46]. When in contrast to scaffolds that have a triangle design, structure of a lattice that is staggered demonstrated increased tissue adherence as well as survival. In a later investigation, a hybrid scaffold was created by injecting chondrocytes encased in an agarose polymer placed within a TMC/TMP matrix. The encapsulated chondrocytes phenotypic was well preserved by the scaffolds, according to in vitro examination; normal lacunae development was seen. Recently, the stereolithographic manufacture of TMC scaffolds was optimized by altering the concentration of diluent [47].

The polymers including Propylene fumarate or poly (PPF), poly (trim ethylene carbonate) (TMC), Poly (D, L-lactide) as well as Polymer(e-caprolactone) (PCL) (PDLLA) are often investigated for stereo lithography of biodegradable architectures. Most resins produce stiff networks that are suitable for hard tissues like tissue from the bone engineering. It ought to be mentioned that lately studies worked on developing elastic networks for use in softer tissue engineering projects.

One of the most researched polymers for stereo lithography, poly (propylene fumarate) demonstrates photo cross linking and biodegradability. PPF, or polypropylene fibre, has an unsaturated backbone that makes it easier to crosslink it with other substances, such methyl methacrylate, N-vinyl pyrrolidone (NVP) as well as diethyl fumarate (DEF), to adjust viscosity and other material qualities. Propylene glycol is a substance that and fumaric acid sections have been produced during PPF degradation, making it a friendly material for tissue engineering. Since Fumaric acid, which is a chemical, is a naturally occurring intermediate by product of the metabolism of glucose, neither of breakdown products is harmful. Moreover, photo cross linked PPF's mechanical characteristics are like those of trabecular bone, making it an excellent biologic material for skeletal tissue engineering.

8.5 Selective Laser Sintering-Based on Sintering

Selective laser sintering (SLS), which stands for selective laser sintering, is one of the additive manufacturing methods that belong to the powder beds fusion family. A laser therapy selectively sinters the powdered polymer particles in SLS 3D printing, fusing them together and creating a component layer by layer. Thermoplastic polymers in granular form are the materials utilized in SLS. Materials having excellent structure-property correlations are known as engineering polymers [48]. They fall into three categories: elastomers, thermosetting polymers, and thermoplastics [49]. The mechanical and material qualities of thermoplastic parts made with SLS are comparable to those produced using traditional techniques [50]. Only a few various kinds of polymers are utilized in SLS, such as polyamide (PA11 and PA12), certain PC, PS, and their derivatives [51].

The most common material used in SLS is nylon [52]. An increasing amount of focus is being directed on some semi-crystalline polymers, including polyethylene [53], polycaprolactone (PCL), and polyetheretherketone (PEEK).

8.6 Advantages and Disadvantages

SLS printing lower's part costs by doing away with the need for molds, CNC bandwidth, and any additional high-fidelity component production method. Fast iteration is made possible by quick on-demand part printing. Moreover, SLS elements are isotropic, meaning that, unlike multilayered 3D printed objects, which frequently are not the case, they are of an identical strength in all directions. Although SLS is a great printing technology, entrance hurdles are significant. Machine cost and maintenance are the highest as contrasted with plastic printers, and they require highly skilled operators. Additionally, they can only make pieces of plastic strength.

8.7 Criteria for Selecting Synthetic Polymers for Additive Manufacturing in Tissue Engineering

Material Utilization

- Any material, natural or synthetic, can be used wholly or partially for tissue, organ, or bodily function enhancement.
- The goal is to preserve or improve an individual's quality of life at any given time.
- Biomaterial evaluation
- Replacements are evaluated as biomaterials [54].

Role of Saffold's

- Scaffolds are essential in various biomedical applications. They are employed in immunomodulation, controlled medication administration and bioengineering.
- Composites consist of considered indispensable tools these applications [55].

Regenerative Engineering

- A multidisciplinary field combining developmental biology, advanced materials science, physics, and clinical translation.
- Aimed at regenerating injured tissue [56].

Commercialization and Adoption

- Medicinal application of results has led to increased commercialization.
- Despite multiple discoveries in tissue engineering (TE), adoption in clinical applications requires consideration of biological, clinical, and mechanical aspects.

Scaffold Requirements

- Adequate scaffold conditions are essential for the growth, vascularization, multiplication, and integration of the host cell.
- Materials must deteriorate organically either during or following the healing process.

Unique Framework Properties

- Scaffold effectiveness is influenced by its unique properties, including chemical makeup, structure, and biological aspects [57].

8.8 Biological Characteristics

Among the biological characteristics of scaffolds are their nontoxicity and biocompatibility. Cells developing on scaffolds must be allowed to divide and distinguish without obstruction to produce a new matrix. For TE applications, a scaffold is therefore deemed optimal if it can imitate the properties of tissues' extracellular matrix (ECM) to ensure flawless and whole regeneration. But as was previously shown, several of the selected cell line, the underpinning material, the surface features, and the framework structure are some of the variables that influence the supporting cell's functionality.

Breakdown the matrix as well as the creation of new tissue can occur at the same time when a material is biocompatible. To ensure that the system can get rid It should not be dangerous if it stays out of the grid without disturbing other members. Challenge since they affect how scaffold connect to tissues and organs. Because biological material is not particularly good at communicating with its environment, efforts have been made to apply bioactive scaffolds to promote right cellular interactions, migration, or differentiation, thereby tissue information, and incorporating into the host, and to stop undesirable processes like scarring. The structure also needs to shield against host immunological responses. The immunological-inert biomaterials concept, which has been used recently, regulates the immune system. This includes T as well as B cell-mediated immunity as well as a decrease in NK cell expression [58, 59].

8.9 Fundamental Characteristics

The complex mechanical properties of biological tissue are linked to characteristics of mass transfer, making it a very sophisticated three-dimensional structure. Tissue engineering (TE) must employ biological scaffolds that provide the living cells, amino acids, and genes required for tissue reconstruction in order to get around this structural and functional complexity. Complex connective tissue environments, which consist of various cell types interacting with distinct inflammatory mediators to form outside of cells matrices within cells alongside hierarchical characteristics suggesting mechanical function that have elevated nonlinearity and two-phase behaviour, are obviously impossible for biological structures and substances to replicate [60].

An important barrier to the development of vascularized engineering substrates is the lack of vascular insufficient blood flow, which makes osseointegration inefficient, indicating that material selection influences the scaffold's final physical characteristics. In order to encourage cell development, it is frequently desirable for the scaffold's porosity to increase its mechanical qualities. Furthermore, a scaffold allows for high mass transport of vital nutrients across the scaffold when the pore size is appropriates well as enhances cell motility and water absorption [61].

8.10 Chemical Composition

Bio-ceramics, polymers, and hybrid materials—natural or artificial usually make up the majority of scaffolds [62]. There are issues with these matrices' composition, breakdown products, and biocompatibility depending on where the components used to make the scaffold came from. Despite the fact that many varied materials have been tested for use as scaffolds, some have reportedly been shown to not promote cell development in scaffolds.

There are two kinds of polymers: synthetic and natural. Natural polymers such as collagen, which is hyaluronic acid, fibrin, and chitosan have favourable biological compatibility, low immunogenicity, and osteoconductivity. But they have poor mechanical stability and high rates of free deterioration [63]. Controlled degradation rates are exhibited by synthetic polymers such as propylene fumarate both (PPF), polyanhydride, polycaprolactone (the material), polyphosphazene, polyether ether ketone (PEEK), poly (glycolic acid) (PGA). Additionally, they may be formed into intricate shapes and have superior cell adhesion (chemical groups that are negatively charged), and can transfer soluble substances. Moreover, synthetic polymers have a longer shelf life and may be made in big quantities at a low cost. In contrast to genuine tissue matrices, certain Studies conducted in vitro have demonstrated that the material can counteract the effects of naturally occurring tissue formation. Furthermore, in vivo environment, immunogenicity of the material, unanticipated degradation duration, as well as adverse impacts brought on by degradation product can all have a significant impact on poor regeneration. The matrices that are most promising in TE are those that are closest to the natural extracellular matrix based on these factors [64]. Consequently, nonbiodegradable scaffolds that reabsorb at a period distinct from skeletal tissue regeneration are to be avoided using the recently discovered techniques in the extracorporeal tissue engineering process. As a result, innovative techniques have been created to do away with scaffolds and solve these issues.

8.11 Synthetic Polymer-Based Additive Manufacturing in Regenerative Medicine

Synthetic polymers such as polylactic acid (PLA), polycaprolactone (PCL), and poly (lactic-co-glycolic acid) (PLGA) have been extensively used in tissue engineering due to their biocompatibility, biodegradability, and mechanical properties that can be tailored to match the requirements of different tissues [65]. These materials offer an excellent platform for creating scaffolds through additive manufacturing techniques like fused deposition modelling (FDM) and selective laser sintering (SLS). They allow precise control over scaffold architecture, which is crucial for promoting cellular activities like adhesion, proliferation, and differentiation.

Fused Deposition Modelling FDM has been widely studied for its ability to fabricate scaffolds from thermoplastic polymers such as PLA and PCL. Several research demonstrated that scaffolds created using FDM with PCL showed excellent support for osteogenic differentiation, making them suitable for bone tissue engineering applications. Additionally, these scaffolds have been used to support the growth of fibroblasts and endothelial cells, which are critical for vascularized tissue engineering [66].

Selective Laser Sintering SLS has been used to create scaffolds from various synthetic polymers like polyamide (PA) and polyetheretherketone (PEEK).

Stereolithography SLA has emerged as a technique for creating scaffolds with high-resolution features using photo-crosslink able polymers like poly (propylene fumarate) (PPF).

3D Bioprinting this technique has allowed the integration of cells and growth factors directly into scaffolds during fabrication.

8.11.1 Comparative Studies of Synthetic and Natural Polymers

Studies comparing synthetic polymers with natural ones (like collagen or hyaluronic acid) have shown that while natural polymers offer excellent biocompatibility and bioactivity, synthetic polymers provide better mechanical strength and durability. For example, PCL and PLGA have been found to degrade at controlled rates suitable for long-term applications, unlike some natural polymers that degrade too rapidly, limiting their use in load-bearing applications.

Emerging Trends and Innovations Recent innovations in synthetic polymer synthesis, such as the development of copolymers and polymer blends, have allowed for better control over mechanical properties and degradation rates. Research focused on developing polyhydroxyalkanoates (PHAs), which combine the favourable properties of both natural and synthetic polymers, offering a new direction for scaffold development.

8.12 Future Prospects in Synthetic Polymer-Based Additive Manufacturing for Regenerative Medicine

Future research should focus on enhancing the mechanical properties and biocompatibility of synthetic polymers through molecular engineering and the incorporation of nanomaterials. For instance, incorporating graphene or carbon nanotubes into synthetic polymers could significantly improve their mechanical strength and electrical conductivity, making them more suitable for nerve regeneration applications.

Integration of Bioactive Molecules and Cells Developing techniques for the controlled release of growth factors and other bioactive molecules from synthetic scaffolds is a promising area for future research. This could involve creating multi-layered scaffolds where different layers contain varying concentrations of growth factors, allowing for a staged release that mimics natural tissue healing processes.

Hybrid Scaffolds and Composite Materials The use of hybrid scaffolds that combine synthetic and natural polymers offers a promising strategy for tissue engineering. Such scaffolds could leverage the mechanical strength of synthetic polymers while benefiting from the bioactivity of natural ones. Additionally, incorporating ceramic materials could enhance scaffold stiffness, which is particularly useful in bone tissue engineering.

Personalized Medicine and Patient-Specific Scaffolds Additive manufacturing allows for the creation of patient-specific scaffolds tailored to the unique anatomical and pathological needs of everyone. Advances in imaging techniques and computer-aided design will further enhance the ability to create personalized implants, reducing the risk of rejection and improving patient outcomes.

Regulatory and Clinical Translation There is a need for robust regulatory frameworks to ensure the safety and efficacy of synthetic polymer-based scaffolds. Collaborative efforts between researchers, clinicians, and regulatory bodies are essential to accelerate the clinical translation of these technologies. Additionally, long-term studies on the biocompatibility and degradation of synthetic polymers in vivo are needed to address safety concerns.

8.13 Conclusions

Synthetic polymer-based additive manufacturing presents a transformative approach in regenerative medicine and tissue engineering. The ability to create scaffolds with precise control over their architecture and composition has opened new avenues for developing more effective and personalized medical treatments. However, challenges such as achieving the optimal balance between biodegradability and mechanical stability, enhancing biocompatibility, and integrating bioactive molecules remain. Future research should focus on addressing these challenges through material innovation, integration of advanced manufacturing techniques, and rigorous clinical evaluation. The field of synthetic polymer-based additive manufacturing is poised for significant growth, driven by technological advancements and an increasing understanding of the complex requirements of tissue engineering. The continued development of new materials, coupled with advancements in additive manufacturing technologies, will likely lead to more effective, customized solutions for a wide range of medical applications.

This chapter is incomplete no study presented which is previously published on the theme of chapter, no future prospectus, no conclusions.

References

1. Veeman D, Sai MS, Sureshkumar P, Jagadeesha T, Natrayan L, Ravichandran M, Mammo WD (2021) Additive manufacturing of biopolymers for tissue engineering and regenerative medicine: an outline, possible uses, developments, and patterns. Glob J Plast Sci 2021:1–20. https://doi.org/10.1155/2021/4907027
2. Koski C, Bose S, Vu AA (2020) Organic biopolymers and composites are manufactured additively for use in bone tissue engineering. Horiz Mater 7(8):2011–2027. https://doi.org/10.1039/D0MH00277A
3. Kuhnt T, Camarero-Espinosa S (2021) Additive manufacturing of nanocellulose based scaffolds for tissue engineering: beyond a reinforcement filler. Carbohydr Polym 252:117159. https://doi.org/10.1016/j.carbpol.2020.117159
4. Pina S, Oliveira JM, Reis RL (2015) Natural-based nanocomposites for bone tissue engineering and regenerative medicine: a review. Adv Mater 27(7):1143–1169. https://doi.org/10.1002/adma.201403354
5. Vyas J, Singh S, Shah I, Prajapati BG (2023) Potential applications and additive manufacturing technology-based considerations of mesoporous silica: a review. AAPS PharmSciTech 25(1):6
6. Vyas J, Shah I, Singh S, Prajapati BG (2023) Biomaterials-based additive manufacturing for customized bioengineering in management of otolaryngology: a comprehensive review. Front Bioeng Biotechnol 11:1234340
7. Di Marzio N, Eglin D, Serra T, Moroni L (2020) Bio-fabrication: convergence of 3D bioprinting and nano-biomaterials in tissue engineering and regenerative medicine. Front Bioeng Biotechnol 8:326. https://doi.org/10.3389/fbioe.2020.00326
8. Kumar P, Choonara YE, Khan RA, Pillay V (2017) The chemo-biological outreach of nanobiomaterials: implications for tissue engineering and regenerative medicine. Curr Pharm Des 23(24):3538–3549
9. Singh S, Nwabor OF, Syukri DM, Voravuthikunchai SP (2021) Chitosan-poly (vinyl alcohol) intelligent films fortified with anthocyanins isolated from Clitoria ternatea and Carissa carandas for monitoring beverage freshness. Int J Biol Macromol 182:1015–1025
10. Veeman D, Sai MS, Sureshkumar P, Jagadeesha T, Natrayan L, Ravichandran M, Mammo WD (2021) Additive manufacturing of biopolymers for tissue engineering and regenerative medicine: an overview, potential applications, advancements, and trends. Int J Polym Sci 2021(1):4907027
11. Vyas C, Pereira R, Huang B, Liu F, Wang W, Bartolo P (2017) Engineering the vasculature with additive manufacturing. Curr Opin Biomed Eng 2:1–3. https://doi.org/10.1016/j.cobme.2017.05.008
12. Li X, Cui R, Sun L, Aifantis KE, Fan Y, Feng Q, Cui F, Watari F (2014) 3D-printed biopolymers for tissue engineering application. Int J Polym Sci 2014:1–11. https://doi.org/10.1155/2014/829145
13. Mantha S, Pillai S, Khayambashi P, Upadhyay A, Zhang Y, Tao O, Pham HM, Tran SD (2019) Smart hydrogels in tissue engineering and regenerative medicine. Materials 12(20):3323. https://doi.org/10.3390/ma12203323
14. Kim JE, Kim SH, Jung Y (2016) Current status of three-dimensional printing inks for soft tissue regeneration. Tissue Eng Regen Med 13:636–646. https://doi.org/10.1007/s13770-016-0125-8
15. Murphy C, Kolan K, Li W, Semon J, Day D, Leu M (2017) 3D bioprinting of stem cells and polymer/bioactive glass composite scaffolds for bone tissue engineering. Int J Bioprint 3(1):1–12. https://doi.org/10.18063/IJB.2017.01.005
16. Fonseca DR, Sobreiro-Almeida R, Sol PC, Neves NM (2018) Development of non-orthogonal 3D-printed scaffolds to enhance their osteogenic performance. Biomater Sci 6(6):1569–1579. https://doi.org/10.1039/C8BM00073E
17. Oliveira AL, Sampaio SC, Sousa RA, Reis RL (2006) Controlled mineralization of nature-inspired silk fibroin/hydroxyapatite hybrid bioactive scaffolds for bone tissue engineering

applications. In: Proceedings of the 20th European conference on biomaterials, Nantes, France, September 2006, vol 27, pp 111–112
18. Du Y, Liu H, Yang Q, Wang S, Wang J, Ma J, Noh I, Mikos AG, Zhang S (2017) Selective laser sintering scaffold with hierarchical architecture and gradient composition for osteochondral repair in rabbits. Biomaterials 137:37–48. https://doi.org/10.1016/j.biomaterials.2017.05.021
19. Dong Y, Liang J, Cui Y, Xu S, Zhao N (2018) Fabrication of novel bioactive hydroxyapatite-chitosan-silica hybrid scaffolds: combined the sol-gel method with 3D plotting technique. Carbohydr Polym 197:183–193. https://doi.org/10.1016/j.carbpol.2018.05.086
20. Gao Q, Gu H, Zhao P, Zhang C, Cao M, Fu J, He Y (2018) Fabrication of electrospun nanofibrous scaffolds with 3D controllable geometric shapes. Mater Des 157:159–169. https://doi.org/10.1016/j.matdes.2018.07.042
21. Da H, Jia SJ, Meng GL, Cheng JH, Zhou W, Xiong Z, Mu YJ, Liu J (2013) The impact of compact layer in biphasic scaffold on osteochondral tissue engineering. PLoS One 8(1):e54838. https://doi.org/10.1371/journal.pone.0054838
22. Lynch CR, Kondiah PP, Choonara YE (2021) Advanced strategies for tissue engineering in regenerative medicine: a bio fabrication and biopolymer perspective. Molecules 26(9):2518. https://doi.org/10.3390/molecules26092518
23. Safinsha S, Ali MM (2020) Composite scaffolds in tissue engineering. Mater Today Proc 24:2318–2329. https://doi.org/10.1016/j.matpr.2020.03.761
24. Mehrpouya M, Vahabi H, Barletta M, Laheurte P, Langlois V (2021) Additive manufacturing of polyhydroxyalkanoates (PHAs) biopolymers: Materials, printing techniques, and applications. Mater Sci Eng C 127:112216. https://doi.org/10.1016/j.msec.2021.112216
25. Chaunier L, Guessasma S, Belhabib S, Della Valle G, Lourdin D, Leroy E (2018) Material extrusion of plant biopolymers: opportunities & challenges for 3D printing. Addit Manuf 21:220–233. https://doi.org/10.1016/j.addma.2018.03.016
26. Velu R, Jayashankar DK, Subburaj K (2021) Additive processing of biopolymers for medical applications. In: Additive manufacturing. Elsevier, pp 635–659. https://doi.org/10.1016/B978-0-12-818411-0.00019-7
27. Cruz F (2010) Fabrication of HA/PLLA composite scaffolds for bone tissue engineering using additive manufacturing technologies. Biopolymers 11:227–242. https://doi.org/10.5772/10264
28. Castro-Ladino JR, Cuy-Hoyos CA, Prías-Barragán JJ (2023) Basic physical properties and potential application of graphene oxide fibers synthesized from rice husk. Sci Rep 13:17967. https://doi.org/10.1038/s41598-023-45251-8
29. Wu CS (2018) Characterization, functionality and application of siliceous sponge spicules additive-based manufacturing biopolymer composites. Addit Manuf 22:13–20. https://doi.org/10.1016/j.addma.2018.04.034
30. Ecker JV, Kracalik M, Hild S, Haider A (2017) 3D-material extrusion-printing with biopolymers: a review. Chem Mater Eng 5:83–96. https://doi.org/10.13189/cme.2017.050402
31. Chaunier L, Della Valle G, Lourdin D, Réguerre AL, Cochet K, Leroy E (2019) Viscous sintering kinetics of biopolymer filaments extruded for 3D printing. Polym Test 77:105873. https://doi.org/10.1016/j.polymertesting.2019.04.020
32. Ilyas RA, Sapuan SM (2020) Biopolymers and biocomposites: chemistry and technology. Curr Anal Chem 16:500–503. https://doi.org/10.2174/157341101605200603095311
33. Singh A, Maheshwari S, Verma A, Ansari VA, Prajapati BG, Singh S (2024) Biopolymer composites for sustainable and green technology. In: Advances in sustainable biomaterials. CRC Press, Boca Raton, pp 1–16
34. Sarker MD, Naghieh S, Sharma NK, Ning L, Chen X (2019) Bioprinting of vascularized tissue scaffolds: influence of biopolymer, cells, growth factors, and gene delivery. J Healthc Eng 2019:1–16. https://doi.org/10.1155/2019/9156921
35. Dhinakaran V, Ajith J, Fahmidha AF, Jagadeesha T, Sathish T, Stalin B (2020) Wire arc additive manufacturing (WAAM) process of nickel based superalloys–a review. Mater Today Proc 21:920–925. https://doi.org/10.1088/1757-899X/988/1/012102

36. Sukindar NA, Ariffin MKA, Baharudin BTHT, Jaafar CNA, Ismail MIS (2016) Analyzing the effect of nozzle diameter in fused deposition modeling for extruding polylactic acid using open source 3D printing. J Teknol 78:71–76. https://doi.org/10.11113/jt.v78.6265
37. Xu W, Jambhulkar S, Zhu Y, Ravichandran D, Kakarla M, Vernon B, Lott DG, Cornella JL, Shefi O, Miquelard-Garnier G, Yang Y, Song K (2021) 3D printing for polymer/particle-based processing: a review. Compos Part B Eng 223:109102. https://doi.org/10.1002/smll.202100817
38. Lu H, Liu Y, Leng J, Du S (2010) Qualitative separation of the physical swelling effect on the recovery behavior of shape memory polymer. Eur Polym J 46:1908–1914. https://doi.org/10.1016/j.eurpolymj.2010.06.013
39. Zhang P, Wang Z, Li J, Li X, Cheng L (2020) From materials to devices using fused deposition modeling: a state-of-art review. Nanotechnol Rev 9:1594–1609. https://doi.org/10.1515/ntrev-2020-0101
40. Shimpi NG et al (2018) Biodegradable and biocompatible polymer composites. Matthew Deans, Duxford. https://doi.org/10.1016/B978-0-08-100970-3.02001-0
41. Asif AKMAH, Rahman M, Sarker P, Hasan Z (2019) Hydrogel fibre: future material of interest for biomedical applications. J Text Sci Technol 5:92–107
42. Piland SG, Gould TE, Jesunathadas M, Wiggins JS, Mcnair O, Caswell SV (2019) Protective helmets in sports. In: Materials in sports equipment. Elsevier, pp 71–121. https://doi.org/10.1016/B978-0-08-102582-6.00003-4
43. Esaklul KA, Mason J, Oil O, Corporation G, States U, Materials M et al (2017) Nonmetallics applications in oil and gas production (pipes, liners, rehabilitations). In: Trends in oil and gas corrosion research and technologies. https://doi.org/10.1016/B978-0-08-101105-8.00027-9
44. Landrock AH (1995) Thermoplastic foams. In: Handbook of plastic foams: types, properties, manufacture and applications. William Andrew, Norwich, pp 221–245
45. Brebu M, Sakata Y, Uddin A (2004) The individual and cumulative effect of brominated flame retardant and polyvinylchloride (PVC) on thermal degradation of acrylonitrile-butadiene-styrene (ABS) copolymer. Chemosphere 56:433–440. https://doi.org/10.1016/j.chemosphere.2004.04.002
46. Deshmukh K, Ahamed MB, Deshmukh RR, Pasha SKK, Bhagat PR, Chidambaram K (2017) Biopolymer composites with high dielectric performance: interface engineering. In: Biopolymer composites in electronics. https://doi.org/10.1016/B978-0-12-809261-3.00003-6
47. MatWeb (2021) Overview of materials for Nylon 12. MatWeb:1–3. http://www.matweb.com/search/datasheetText.aspx?bassnum=O2640. Accessed 2 Sept 2024
48. Lee SJ, Kang HW, Park JK, Rhie JW, Hahn SK, Cho DW (2008) Application of microstereolithography in the development of three-dimensional cartilage regeneration scaffolds. Biomed Microdev 10:233–241
49. Mittal V (2015) Nanocomposites with engineering polymers: a review. In: Manufacturing of nanocomposites with engineering plastics. Woodhead Publishing, Cambridge, pp 15–29
50. Dobbelstein H, George EP, Gurevich EL, Kostka A, Ostendorf A, Laplanche G (2020) Laser metal deposition of refractory high-entropy alloys for high-throughput synthesis and structure-property characterization. Int J Extreme Manuf 3(1):015201
51. Shi Y, Li Z, Sun H, Huang S, Zeng F (2004) Effect of the properties of the polymer materials on the quality of selective laser sintering parts. Proc Inst Mech Eng L J Mater Des Appl 218(3):247–252
52. Lammens N, Kersemans M, De Baere I, Van Paepegem W (2017) On the visco-elasto-plastic response of additively manufactured polyamide-12 (PA-12) through selective laser sintering. Polym Test 57:149–155
53. Rimell JT, Marquis PM (2000) Selective laser sintering of ultra high molecular weight polyethylene for clinical applications. J Biomed Mater Res 53(4):414–420
54. Xue J, Wu T, Dai Y, Xia Y (2019) Electrospinning and electrospun nanofibers: methods, materials, and applications. Chem Rev 119(8):5298–5415. https://doi.org/10.1021/acs.chemrev.8b00593
55. Ogueri KS, Laurencin CT (2020) Nanofiber technology for regenerative engineering. ACS Nano 14(8):9347–9363

56. Ogueri KS, Jafari T, Escobar Ivirico JL, Laurencin CT (2019) Polymeric biomaterials for scaffold-based bone regenerative engineering. Regen Eng Transl Med 5:128–154
57. Roseti L, Parisi V, Petretta M et al (2017) Scaffolds for bone tissue engineering: state of the art and new perspectives. Mater Sci Eng C 78:1246–1262. https://doi.org/10.1016/j.msec.2017.05.017
58. Hama R, Ulziibayar A, Reinhardt JW, Watanabe T, Kelly J, Shinoka T (2023) Recent developments in biopolymer-based hydrogels for tissue engineering applications. Biomol Ther 13(2):280. https://doi.org/10.3390/biom13020280
59. Li Z, Xie MB, Li Y, Ma Y, Li JS, Dai FY (2016) Recent progress in tissue engineering and regenerative medicine. J Biomater Tissue Eng 6(10):755–766. https://doi.org/10.1166/jbt.2016.1510
60. Nasiri B, Mashayekhan S (2017) Fabrication of porous scaffolds with decellularized cartilage matrix for tissue engineering application. Biologicals 48:39–46. https://doi.org/10.1016/j.biologicals.2017.05.008
61. Yang Y, Ritchie AC, Everitt NM, Chen E (2017) Comparison of glutaraldehyde and procyanidin cross-linked scaffolds for soft tissue engineering. Mater Sci Eng C 80:263–273. https://doi.org/10.1016/j.msec.2017.05.141
62. Sari M, Hening P, Chotimah AID, Yusuf Y (2021) Bioceramic hydroxyapatite-based scaffold with a porous structure using honeycomb as a natural polymeric porogen for bone tissue engineering. Biomater Res 25(1):2. https://doi.org/10.1186/s40824-021-00203-z
63. Islam MM, Shahruzzaman M, Biswas S, Sakib MN, Rashid TU (2020) Chitosan-based bioactive materials in tissue engineering applications – a review. Bioact Mater 5(1):164–183. https://doi.org/10.1016/j.bioactmat.2020.01.012
64. Karamanos NK, Theocharis AD, Piperigkou Z, Manou D, Passi A, Skandalis SS, Vynios DH, Orian-Rousseau V, Ricard-Blum S, Schmelzer CE, Duca L (2021) A guide to the composition and functions of the extracellular matrix. FEBS J 288(24):6850–6912. https://doi.org/10.1111/febs.15776
65. Shah SR, Modi CD, Singh S, Mori DD, Soniwala MM, Prajapati BG (2024) Recent advances in additive manufacturing of polycaprolactone-based scaffolds for tissue engineering applications: a comprehensive review. Regen Eng Transl Med 11:1–20
66. Vyas J, Raytthatha N, Singh S, Prajapati BG, Mohite P, Munde S (2024) Sustainable sources of raw materials as substituting biomaterials for additive manufacturing of dental implants: a review. Periodontal Implant Res 8(1):3

9 Biomaterials-Based Additive Manufactured Products for Chronic Wound Healing

Popat Mohite, Abhijeet Puri, Ramdas Pandhare, Sudarshan Singh, Manisha Sonawane, and Anil Pawar

Abstract

Chronic wounds pose a significant healthcare challenge globally, necessitating innovative solutions to improve healing outcomes. Biomaterials and additive manufacturing (AM) have emerged as promising technologies in this context. This chapter explores the synergistic potential of biomaterials and AM in the development of advanced wound care products. Firstly, it discusses the distinctive properties of biomaterials that make them appropriate for wound healing applications, such as biocompatibility, bioactivity, and tunable degradation profiles. Secondly, it delves into the capabilities of AM techniques, including 3D printing and bioprinting, in fabricating customized wound dressings and scaffolds with precise geometries and controlled porosities. Moreover, the chapter reviews recent advancements in biomaterial design tailored for chronic wound healing, highlighting case studies where applicable. It also addresses challenges such as scalability, regulatory considerations, and clinical translation of biomaterial-based AM products. In conclusion, this chapter underscores the transformative potential of integrating biomaterials and AM technologies to revolutionize chronic wound care, paving the way for personalized, efficient, and effective treatment strategies.

P. Mohite · A. Puri
AETs St. John Institute of Pharmacy and Research, Palghar, Maharashtra, India

R. Pandhare · A. Pawar (✉)
MESs College of Pharmacy, Sonai Tal-Newasa, Maharashtra, India

S. Singh
Faculty of Pharmacy, Chiang Mai University, Chiang Mai, Thailand

Office of Research Administration, Chiang Mai University, Mueang Chiang Mai, Thailand

M. Sonawane
Pravara Rural College of Pharmacy, Loni, Maharashtra, India

S. Singh et al. (eds.), *Biomaterial-based Additive Manufacturing in Tissue Engineering and Regeneration*, Biomaterials, Bioengineering and Sustainability 7,
https://doi.org/10.1007/978-3-031-96070-3_9

Keywords
Additive manufacturing · Biomaterial · Computer-aided design · Polycaprolactone · Three-dimensional printing

Abbreviations

2D	Two Dimensional
3D	Three Dimensional
AM	Additive Manufacturing
CAD	Computer-Aided Design
CIJ	Continuous Inkjet
ECM	Extracellular Matrix
FDM	Fused Deposition Modeling
FFF	Fused Filament Fabrication
HA	Hyaluronan
IJP	Inkjet Printing
PCL	Polycaprolactone
PEG	Polyethylene glycol
PU_S	Pressure Ulcers
SLA	Stereolithography
SLS	Selective Laser Sintering
TE	Tissue Engineering

9.1 Introduction

The phrase "chronic wound" was initially used in literature in the 1950s to describe wounds that did not heal normally or were difficult to heal [1]. The phrase has been criticized, though, because it is unclear how long anything is chronic [2]. Other terms including non-healing wounds, complicated wounds, and difficult-to-heal wounds have been proposed [3]. Chronic wounds do not heal in a predictable, timely, or well-organized manner. They are common and often receive inappropriate care. The morbidity and related costs of chronic wounds highlight the need for wound prevention and management guidelines to be implemented.

A wound is considered chronic when it exhibits abnormalities in the phases of hemostasis, inflammation, proliferation, and remodeling in the healing process. This kind of wound typically has an underlying pathology, which causes the healing process to be delayed. There is a lack of randomized controlled trial data for chronic wounds, hence the impact of occlusion is not as well established in these wound types [4]. According to standard definitions, chronic wounds are those that, after 3 months, have not undergone prompt, systematic healing that would have produced anatomic and functional integrity [5]. On the other hand, Leaper & Durani classified a chronic wound as one that does not heal completely after 6 weeks or does not reduce in size by 20–40% after 2–4 weeks of optimal treatment [6]. A chronic wound is defined by Martin & Nunan as a barrier defect that does not heal after 3 months. The lack of agreement about what constitutes a chronic wound has also been brought to light by recent studies, underscoring the need for more study in this field [7]. Chronic sores include diabetic foot ulcers, pressure ulcers, and venous leg ulcers. They drain a substantial amount of healthcare resources and present a challenge to wound care providers globally. Treatment effectiveness is achieved by comprehending the underlying pathophysiology and treating these perturbations with special attention, even though these illnesses are often challenging to treat.

There are three types of chronic wounds: pressure ulcers (PUs), diabetic ulcers, and vascular ulcers (venous and arterial). Some features are shared by all these wounds, including persistent or severe inflammation, continuous infections, the growth of drug-resistant microbial biofilms, and the inability of dermal and/or epidermal cells to respond to stimuli that encourage healing. Together, these pathophysiologic processes result in the nonhealing of these wounds. Diverse types of chronic wounds, however, have different underlying diseases [8].

The fiscal impact of chronic wounds on the world economy has led to an increase in patient morbidity and death. The most recent developments in imaging, mechanics, and computer modeling of wound biology have all contributed to the identification of chronic wounds. Tissue engineering advances of late may provide solutions for the management of chronic wounds. It is still possible to create specialized drug regimens using the tools available today to treat chronic wounds in particular patient populations [9]. At $20 billion yearly, chronic wounds affect 5.7 million patients and pose a considerable burden to patients, healthcare providers, and the US health care system. Understanding the typical healing process and creating a healthy physical and biochemical environment are essential for managing these issues [8].

9.2 Challenges in Treating Chronic Wounds

Preclinical and clinical wound research encounters several obstacles that have impeded advancement and the creation of effective treatments. Overcoming these obstacles will open new possibilities for future research design improvements [10]. The Standard treatment approaches and Challenges of treating chronic wounds are given in Table 9.1.

A comprehensive approach to wound treatment necessitates addressing the causes that cause delayed healing, as they are the primary barriers to treating chronic

Table 9.1 Challenges of treating chronic wounds

Type of wound	Standard treatment approaches	Challenges to treating/healing
Chronic	Debridement Moisture balance Infection control Optimization of comorbidities	Excessive and painful debridement Excess moisture Inadequate control of inflammation Heavy bioburden Lack of control of comorbidities Adherence to treatment

wounds. If wounds do not heal enough after 4 weeks of standard therapy, it is time to reassess the underlying pathology and determine whether more advanced therapeutic medications are required. However, there are often no facts to support the choice of an appropriate therapy.

Chronic wounds have a catastrophic effect on morbidity and constitute a significant financial and medical burden. Over 80% of medical expenses in India are covered by out-of-pocket expenditures, with public healthcare funding estimated to be 5% of GDP annually. These do not adequately capture the economic loss, frustration, and reduced quality of life that people with chronic wounds go through. They continue to be a significant clinical problem in long-term care, affecting both patient quality of life and medical expenses [11].

A chronic wound infection is the term used to describe a wound that heals slowly or shows signs of delayed, halted, or stalled healing as a result of both internal and external factors impacting the patient and their wound [12].

Chronic wounds are more challenging to treat because of their delayed rate of wound healing. Individuals suffering from chronic mechanical stress, obesity, diabetes, vascular disease, and malnutrition, and the elderly are more likely to suffer from chronic wounds, or a combination of these conditions. Polymicrobial communities that can create biofilms inhabit most chronic wounds, causing excessive inflammation and increased susceptibility to infections that greatly impede wound healing. One of the main obstacles to treating chronic wound infections is biofilm. Just 6% of wounds are acute and more than 60% of chronic wound infections have it. Encased within the complex microbial communities of bacteria and fungi that exist as unicellular, planktonic, or multicellular aggregates and communities is a polymeric matrix composed of polysaccharides, lipids, proteins, and nucleic acids. These communities are known as biofilms. In addition to serving as a temporary covering for wounds, wound dressings can help control or prevent infections. On the other hand, a wound dressing might attract microbes and biofilm formation, which would raise the microbial burden and slow down the healing process [13].

9.3 Role of Biomaterials in Wound Healing

Biomaterials are employed to assist the body during the healing process of wounds, encouraging migration and growth of cells and the creation of new tissue [14]. The mechanical characteristics of biomaterials employed in wound healing are critical for their ability to support these processes. Biomaterials for skin regeneration should have the proper porosity, mechanical strength, biocompatibility, structure, flexibility, and biodegradability [15].

9.4 Importance of Biomaterials in Medical Applications

Because they support a patient's recovery from diseases and accidents and help them regain function, biomaterials are crucial to modern medicine. Synthetic or natural biomaterials are used to replace, enhance, or maintain biological functions or damaged tissue. Materials derived from live organisms, or their byproducts, are known as natural biomaterials. Because of the numerous benefits these materials offer including their non-toxicity, biocompatibility, and biodegradability and the growing need for medical applications, the use of natural biomaterials has been rapidly rising. Collagen, chitosan, and alginate are among the materials effectively employed in tissue engineering, Nanotechnology as matrices, hydrogels, scaffolds, and implants; wound care; and medication delivery. Because natural biomaterials are biocompatible and biodegradable, the body may safely absorb them without the need for further operations, which could lead to savings in healthcare costs [16].

9.5 Biomaterials Are Used by Medical Professionals, Scientists, and Bioengineers for the Following Purposes

- Dental implants; tendons; ligaments, and Artificial joints, implants for hearing loss; stents, grafts; and heart valves, and nerve-stimulation devices.
- Techniques for encouraging the healing of human tissues, such as dissolvable dressings and the use of staples, clips, and sutures to close wounds.
- Regenerated human tissues were created using cells, bioactive substances, and biomaterial scaffolds. A human bladder grown in a lab and a hydrogel that repairs bones are two examples.
- Breakthrough molecular probes and nanoparticles that support molecular cancer imaging and treatment beyond biological barriers.
- Biosensors that can send data about the presence and concentration of chemicals. Brain activity sensors and blood glucose monitors are two examples.
- Equipment for delivering drugs to a disease target or for administering them. Two examples are implantable chemotherapy wafers for cancer patients and drug-coated vascular stents [16, 17].

9.6 Advantages of Using Biomaterials for Chronic Wounds

Biomaterials are designed to encourage or accelerate the wound-healing processes. Animal byproducts such as eggshell membrane, bovine serum albumin, and keratin can also be found in natural biomaterials. Biomaterials originating from animals, such keratin, have been shown to release growth factors and are useful in wound curative applications [17]. When more than two unique materials are combined with a matrix to attain specific desirable features for use in biomedical applications, the resultant material is referred to as a composite biomaterial [18].

To aid in wound healing, cellulose and chitosan nanoparticles are mixed with cellulose poly (methyl methacrylate) fibers to create a composite biomaterial [17]. Antimicrobial agents, such as antibiotics, antimicrobial nanoparticles, cationic chemical compounds, and others, can be put into biomaterials in order to treat infection and stop wound colonization [19]. Highly absorbent materials having the ability to offer a multimodal curative approach on a sole platform are hydrogel-based biomaterials [20].

Metal-based biomaterials can actively stimulate wound repair and prevent infection, contingent on the specifics of the several stages of wound healing. Metal-based biomaterials show enhanced functionality in the treatment of wound inflammation, angiogenesis promotion, re-epithelialization, antimicrobial activity, and even multistage combination therapy, especially after wound hemostasis [21].

Synthetic wound dressings have shown success in several animal studies and may promote wound regeneration. The use of biomaterials in skin tissue engineering holds enormous potential for reducing the burden associated with chronic wounds [17]. Conductive biomaterials can be shaped into a variety of shapes to fit the needs of various kinds of wounds. 3D hydrogels and scaffolds with ECM-like structures are frequently employed in more complex wounds and skin substitutes, while 2D biomaterials such as micro and nanofibers, membranes, and films can heal acute wounds with little exudates [22].

9.7 Additive Manufacturing Techniques in Biomaterial Production

A new method of producing components is called additive manufacturing (AM), it entails layer-by-layer material deposit in line with a digital model [23]. The basic idea behind additive manufacturing (AM) is that a 3D model or design is initially produced using 3D CAD. This allows for fabrication without the need for direct process planning, which can save a significant amount of cycle time [24].

9.8 Comparison with Traditional Manufacturing Methods

The fact that additive manufacturing requires a significant reduction in the amount of waste material is another functional distinction between it and traditional manufacturing. Machining by removal causes a specific amount of waste to be produced in a subtractive manufacturing process. Additive manufacturing ensures little waste by using only the material needed to make an item. One other use for additive manufacturing is the printing of a set of parts. Conventional methods require that every item be produced independently before it can be assembled. The production and assembly stages can be integrated into a single production process using next-generation 3D printing. The ability of additive manufacturing to produce customized goods quickly and on a wide scale (mass customization) sets it apart from subtractive technologies in another crucial practical way [25, 26].

9.9 Types of Additive Manufacturing Technologies

9.9.1 Stereolithography

In the 1980s, Chuck Hull created stereolithography (SLA), which is the first AM processes. The fundamental idea behind SLA and related processes is vat photopolymerization. On a resin layer, the chain reaction is started by applying light from a light source typically ultraviolet or electron beams. Radicalization is the process by which the monomers instantly become polymer chains following activation. In order to secure the subsequent layers in place, the resin pattern layer is hardened after polymerization, and process parameters are adjusted based on the intended use [27, 28]. The laser thickness, temperature of the printing bed, laser power, angle of stratification, duration of post-curing, and orientation of manufacturing are a few of the process parameters that SLA technology researchers use [28, 29]. The energy and exposure of the light source are the primary factors that determine each layer's thickness [30]. SLA may be used to create intricate nanocomposites. It offers the best accuracy and resolution together with a wide range of material options. Functional prototypes, patterns, molds, and tooling are among the many applications for SLA. Despite having a greater resolution, SLA has several drawbacks, such as limited materials and slow printing. SLA has a resolution of 10 μm [31].

9.9.2 Fused Deposition Modeling

Often called fused filament fabrication (FFF), fused deposition modeling (FDM) is one of the most widely utilized extrusion-based additive manufacturing techniques [32]. The material is added in the form of filaments into the 3D printer when using the FFF technique. Any FFF-based 3D printer's extruder typically consists of a feed material control system, nozzle, and heating chamber. The control system, which gets the material in filament form, regulates the feed rate. After going through the heating chamber and becoming semi-solid, the filament exits the nozzle and lands on the printing bed. The extruder's mechanical and thermal controls, which depend on the feeding material properties, govern the printing process. The FFF technique has been used to create thermoplastics, hydrogels, and particles-reinforced polymers [33].

Nylon, acrylonitrile butadiene styrene (ABS), and polylactic acid (PLA) are the three polymers most frequently used in FFF. The main reasons this technology is preferred are its affordability, ease of application, and the capacity to use a wider range of materials [34].

9.9.3 Selective Laser Sintering (SLS)

SLS printing may be more advantageous than other 3D printing methods for creating greater quantities of dose forms (e.g., 200 tablets/print) because of its high packing density and high print volume. The process melts the surface of a powder bed covered in powder particles selectively using a laser beam of a certain wavelength as the energy source. A laser could be anything from a laser diode to a CO_2-laser, based on the initial powder formulation's optical properties and the required power. Amorphous solid dispersions (ASDs) can be produced by a variety of polymers and active pharmaceutical ingredients (APIs), as demonstrated by SLS printing. The production of delivery system comprising BCS class pharmaceuticals may benefit from this strategy [35].

The SLS technique has demonstrated potential for producing medical products, such as knee trays, hip cups, dental crowns, and hearing aids [36]. A well-liked method for producing huge quantities of powder, SLS finds applications in pharmaceutical and biomedical sectors. It was invented by Carl Deckard at the University of Texas in 1984, and in 1990 a patent was issued for it [37].

9.9.4 Inkjet Printing and Bioprinting

The term "inkjet printing" describes a wide variety of digitally controlled techniques for producing and guiding microscopic liquid droplets. The printing methods of continuous inkjet (CIJ) and on-demand are the two broad categories for inkjet technology, as was previously established for 3D printing methods. Depending on the printing type, different techniques are used to apply drops to the printed material. As per CIJ, the process of printing entails the constant transfer of a liquid through a nozzle, wherein the liquid condenses into a stream of drops due to surface tension forces. Only single drops from an endless solutions flow need to be heading for to a particular spot on the substrate in order to produce a printed pattern. This is typically accomplished by giving some of the droplets an electrical charge, which causes them to wander starting with the flow's primary axis as they pass through an electrostatic field. The liquid then circulates throughout the system after some of the droplet's land in a designated chute [38].

Much research has been conducted in the last few years on the use of inkjet printing to create customized pharmaceuticals due to their affordability and versatility. However, formulation and optimization of printing parameters are lengthy procedures due to the multi-factorial nature of inkjet printing [39].

Since inkjet printing (IJP) technology provides excellent precision, reproducibility, robustness, and throughput at an affordable price, it is widely considered to be the best option. Because of its adaptability, inkjet printing is used comprehensively in the medical field and tissue engineering [40].

"Inkjet printing" is a catch-all term for systems that can use a pattern-generating device to create digitally prohibited formulation and position of microscopic fluid droplets on a substrate [41].

A possible fabrication method for creating digitally controlled, highly replicated tissue is bioprinting technology. Pre-processing, processing, and post-processing stages are commonly included in bioprinting techniques. Initially, computer-aided design (CAD) is used to construct the tissue blueprint during the pre-processing step. Imaging techniques can be used to retrieve the crucial data for the design regarding human organ architecture, anatomy, and histological structure and composition. Additionally, at this point, biomaterials parameters are determined. During the processing step, an appropriate bioprinter prints the desired structure. Delivering the intended scaffold depends critically on the bio-ink used in the bioprinter. The resulting scaffold is matured by post-processing processes prior to host implantation. This also includes scaling up the process using an optimal bioreactor. In all three stages, two crucial tools are computer-aided design and computer-aided manufacturing [42].

9.10 Biomaterials Used in Chronic Wound Healing

9.10.1 Natural Biomaterials

In general, natural polymers are biocompatible, biodegradable, and bioabsorbable. Certain substances, including keratin and chitosan, also exhibit antibacterial or anti-inflammatory characteristics. Because of their resemblance to macromolecules that the human body recognizes, natural proteins and polysaccharides have drawn more attention in the creation of wound dressings. This is because they achieve ECM biomimicry.

9.10.1.1 Collagen

One of the key ingredients of the extracellular matrix (ECM) is collagen, which is key for the various stages of wound healing. The immune response is triggered by the collagen fragments through neutrophil mediation [43]. Animal sources provide most of the collagen used as biomaterials. Studies have indicated that using bovine collagen as a skin substitute can enhance wound healing. Increased tensile strength and biochemical indicators of wound healing, such as albumin, prealbumin, transferrin, and hydroxyproline, were observed in the animals given collagen oligopeptides [44]. Although collagen derived from animals has demonstrated promise in promoting wound healing, using it may also raise the risk of infection [45].

Although collagen derived from animals have demonstrated promise in promoting wound healing, using it may also raise the risk of infection [43]. The collagen that has been removed can be used to make sponges and provide an atmosphere that is conducive to the regeneration of skin tissue. Burns, pressure sores, and ulcers are frequently treated using these collagen sponges. Collagen remains a preferred biomaterial in tissue engineering because of its excellent biocompatibility, low immunological activation, and ease of manipulation [17, 46],

9.10.1.2 Chitosan

One of the most popular ligands is chitosan. Its positive charge has been documented to enhance the absorption of metal anion medications [47]. Chitosan has been used as a coating layer by electrostatic adsorption on negatively charged particle surfaces due to its cationic charge [48]. Additionally, chitosan has demonstrated its ability to modify drug release behavior [49]. The negatively charged cell membrane is drawn to the positively charged chitosan-coated nanoparticles, increasing the penetration of weakly water-soluble medications. Because of its many uses, chitosan is commonly utilized in hydrogels, membranes, and electro spun nanofibers for wound recovery [50]. Because of their antibacterial qualities, wound tissues are being treated with chitosan dressings [51].

Hesperidin was added to chitosan/alginate hydrogels at varying concentrations [52]. In mouse model, these hydrogel products were evaluated on a cutaneous wound that was full thickness. The results demonstrated that the hydrogels showed a higher wound closure rate than the gauze-treated group [52]. A nanofiber mat containing zinc oxide (ZnO), PVA, and chitosan showed strong antibacterial and antioxidant properties in a different study, pointing to the mat's prospective for accelerating the recovery of diabetic injuries [53].

9.10.1.3 Alginate

Alginate is found naturally in brown seaweed, a naturally occurring anionic biopolymer. It is non-toxic, easily accessible, and biocompatible. Alginate-based wound dressings are known for their moist environment and low bacterial infection rates, both of which are critical elements in the healing process. Alginate is formed from brown algae (Phaeophyceae) such as Macrocystis pyrifera, Laminaria japonica, Laminaria hyperborea, Laminaria digitate, and Ascophyllum nodosum [54]. It is made up of guluronate and mannuronate in varying ratios. It is composed of a block of residues called β-L-guluronate (G) and β-D-mannuronate (M) joined by a [1, 4] bond. The blocks are made up of M and G residues that alternate (GMGMGM) and G residues that are consecutive (GGGGGG) and M residues (MMMMMM). Alginates differ in the ratios and residue orders of M and G, which dictate their molecular weight and physical characteristics. Even in the absence of temperature, an alginate-based gel can form. Alginate gels can be produced via acid precipitation or ionic crosslinking with cations. Alginate gels because of the cooperative divalent cation binding and the dimerization of G residues in the polymer's G-block regions. When divalent cations, such Ca ions, are added to alginate, the binding of G chains on opposing sides creates a hole with a diamond shape and a hydrophilic cavity that uses the oxygen atoms from the carboxyl groups to attach the Ca ions by multicoordination. Other kinds of divalent cations can also be used to generate this arrangement, which results in an egg-box junction zone shape. The cations in alginate have the ability to bind to block sequences other than G-blocks [55]. Trivalent ions, including Al and Fe, can also cause alginate gels to develop. In contrast to divalent cations, trivalent cations have a more robust binding to alginate, creating a more compact gel network. Alginate gel can be created via internal or external gelation. A homogenous delivery of alginate in hydrogel is the outcome of regulated alginate exposure to cations during internal gelation. It occurs simultaneously in various locations within the polymer. Alginate is distributed unevenly across the gel matrix network as a result of the external gelation process, which involves the migration of cations from a region of higher concentration into the inner region of alginate particles [56].

9.10.2 Synthetic Biomaterials

9.10.2.1 Poly (Lactic Acid) (PLA)

Because PLA has strong biocompatibility, biodegradability, and moisture management qualities, it may aid in wound healing through wound area repair, collagen deposition, angiogenesis, and cell activity [57].

PLA is a biodegradable polyester made from renewable resources like corn flour or sugar cane [58]. This polymer is useful for medical applications due to its many qualities, including its low cost, hydrophobicity, structural stability, biocompatibility, and mechanical integrity. Since PLA is an aliphatic polyester, it can be hydrolyzed under physiological conditions or broken down by enzymes to produce compounds that the body can readily absorb [59]. A new dressing scaffold made of Polyvinylpyrrolidone/PLA-Polyethylene oxide fibers with encapsulated collagen and cefazolin was assessed as a potential wound-healing material for tissue engineering applications. By changing the concentration of collagen, several scaffolds were created. According to the study, as collagen levels rise, scaffold adhesion increases, enhancing the healing process. Additionally, the synthetic material had antibacterial activity, and in vivo histology examinations showed that the best healing rates were observed with collagen dosages of 10% and 20%. In a different study, the encapsulation of mono, di, or tri-phase graphene oxide, zirconia, and hydroxyapatite nanosheets into PLA nanofibrous scaffolds was described. It has been shown that the scaffolds increased cell viability and showed improved antibacterial activity. Furthermore, human fibroblasts multiply within the nanofibrous scaffold's pores and on its surface [60].

9.10.2.2 Polycaprolactone

Biodegradable and synthetic, polycaprolactone is a member of the aliphatic polyester family. It is made up of several units of hexanoate [61]. It is used in the creation of scaffolds for bone tissue regeneration and wound healing because of its biodegradability. The research was done on polycaprolactone (PCL) because of its unique mechanical characteristics, biodegradability, and miscibility with various polymers [62]. The primary factors that determine PCL's mechanical, thermal, and physical qualities are its degree of crystallinity and molecular weight [63].Semi-crystalline PCL has a strong hydrophobic character, is readily processed, and is highly soluble at room temperature. Because of its high mix compatibility and low melting temperature, researchers from all over the world are tremendously provoked in the biomedical arena. By adjusting a few factors, such as the breakdown rate and molecular weight, PCL-based materials can be applied to any type of tissue [64]. Methods like adding nanoparticles could be used to track how quickly PCL scaffolds made by electrospinning deteriorate. The hydrophobic property of PCL limits the application potential of electro spun PCL scaffolds for wound repair, which is another significant disadvantage of the material. Nonetheless, several investigations revealed that quick wound healing was exhibited by composite PCL scaffolds. Moreover, the cellular adhesion characteristics of electro spun PCL scaffolds can be improved through the use of methods like plasma treatment and gamma irradiation [65].

9.10.2.3 Polyethylene Glycol (PEG)

Hydrophilic, biocompatible, and non-immunogenic, polyethylene glycol is a flexible ether-based polymer. These characteristics make it a popular synthetic material for healing wounds. Growth factors such as EGF and PEG macromers are highly attracted to one other and can form connections. These can be applied directly to the wound [66]. Combining PLGA with chitosan with PEG stabilizes its mechanical, thermal, and crystallinity properties. Diabetic sores have been treated using these PEG-based bandages. By encouraging the production of collagen and the development and multiplication of skin cells, these bandages promote wound healing. Such dressing also lessens the production of scars [66].

9.11 Composite Biomaterials

To aid in wound healing, cellulose and chitosan nanoparticles are mixed with cellulose poly (methyl methacrylate) fibers to create a composite biomaterial. This biomaterial has demonstrated efficacy against Staphylococcus aureus and is an effective anti-infective bandage [67]. But frequently, one element cannot provide all the elements required to make a suitable wound dressing. For instance, synthetic polymers showed good mechanical qualities but low cell affinity. Conversely, natural polymers have a short shelf life and typically poor mechanical strength despite being biocompatible and biodegradable. An overview of bio-composite applications in medicine is summarized in Table 9.2.

A collagen, alginate, and silver nanoparticle blend was created by dissolving sodium alginate powder in a silver nanoparticle solution and then adding it to a collagen solution. After that, the mixture was put in rectangular molds. Mouse embryonic fibroblasts were grafted onto the scaffolds in order to assess the bio-composite's biocompatibility as the quantity of silver was less than 50 μM, the cells showed a relative vitality of more than 72%. The antimicrobial properties of S. aureus and E. coli were also assessed in the bio-composites. The bio-composite wound dressing was revealed by the silver nanoparticles' capacity to stop the growth of both bacteria, even as composites lacking in these particles showed no antibacterial action [73].

Table 9.2 An overview of bio-composite applications in medicine

Application	Bio-composite composition	References
Wound recovery	Gelatin and Polycaprolactone	[68]
	Fibrinogen, polyethylene oxide and chitosan	[69]
	Silver nanoparticles collagen, and alginate	[70]
	Silver nanoparticles, keratin, and polyurethane	[71]
	Dextran and collagen	[72]

9.11.1 Advantages of Composites for Wound Healing

- The benefits of composite hydrogel dressing stem from its ability to effectively treat three distinct kinds of skin wounds: burns, diabetic wounds, and full-layer skin defects. The reason hydrogels are so popular in wound treatments is because of their high-water content and biocompatibility, which help to keep the surrounding area wet [73].
- Composite biodegradable materials have exceptional biocompatibility [74].
- In situ produced wound dressings have several benefits, including as easy application, improved therapeutic effects, patient compliance, and appropriate adaptability for the microstructure and architecture of the wound bed. Biomaterials make them excellent choices for accelerated wound healing [75].
- Composites are essential to the production of several components that can replace these limitations. Composites are hybrid materials that incorporate polymer resin for reinforcement together with the superior mechanical and physical qualities of fibers.
- Polymer composites possessing outstanding stiffness, endurance, and light weight also exhibit superior resistance to corrosion. Because they require little maintenance and employ inexpensive raw materials, these composites are inexpensive [76].
- Hydrogels have been extensively employed in wound treatment because of their extracellular matrix resemblance and biocompatibility [77].

9.11.2 Design and Fabrication of AM-Based Products for Wound Healing

The physiological mechanism via which wounds heal is intricate and dynamic, including a sequence of cellular and molecular activities that commence promptly after a tissue injury to restore the skin layer. Patients with chronic wounds, wounds that are highly infected, or any metabolic disease affecting the wound microenvironment undoubtedly experience excruciating agony and discomfort that negatively impacts their quality of life. Chronic wounds must be treated in order to preserve the patients' physical and emotional health and to aid in their recovery and quality of life. Choosing the appropriate wound dressing is crucial for promoting and enhancing the healing process.

9.12 Customizable Wound Dressings/Tailored Design for Specific Wound Types/Super-Wettable Bio-interfaces for Exudate Management

In addition to controlling bleeding in the initial stage of wound care, exudate management is essential to wound healing. Wound exudate is the fluid that surrounds a wound and is produced during the inflammatory and proliferative phases following hemostasis. Neutrophils, macrophages, inflammatory mediators, water electrolytes, and enzymes that break down proteins make up most of the exudate from acute wounds. These components allow the injured tissue to autolyze and supply nutrients for cell metabolism. In acute wounds, the exudate is considered positive, and as time goes on, less of it is produced. In this instance, the dressings should handle both moisture control and protection. Chronic wound exudate, on the other hand, differs significantly in its composition. It has lower levels of growth factors and increased levels of matrix metalloproteinases (MMP) and pro-inflammatory cytokines. Furthermore, exudate affects the skin around wounds, and because of inflammation, it may produce copious amounts of it continuously in chronic wounds. Furthermore, because the saturated dressing serves as a conduit for bacterial invasion, high-volume exudate may raise the possibility of infection. Considering this, it is crucial to keep the wound healing environment moist while eliminating excessive exudate. Water-absorbing dressings composed of hydrophilic materials are excellent options for exudate control because they provide many routes for exudate drainage. Generally speaking, hydrophilic dressings also possess electroactivity and antibacterial properties [78].

9.13 Advancements in Nanofibers for Wound Dressing

Chronic wounds require a lengthy, intricate healing process that often takes 12 weeks (about 3 months) to complete. Additionally, bacteria that might worsen immune system function and encourage wound infection can cause it to become sensitive to external biological agents. As a result, using the proper barrier is crucial to establishing a regulated environment for wound healing. Numerous topical wound barrier compositions, such as films, hydrogels, emulsions, and nano/micro-fibers, have been researched during the past few decades. The pH of the wound bed is 7.4, which is normal for blood, and its temperature is 37 °C. These environments cause the medication to release and the nano- and micro-fibers to break down at the same time. A drug with superior antibiotic and antimicrobial properties, along with a matrix with good biocompatibility and a notable biodegradation rate (20% degradation in 15 days (about 2 weeks) for chitin nanofiber, for example), makes up drug-embedded nano/microfiber. A few medications, such as captopril, can cause vasodilation. This enhances vascular permeability and improves the healing process by enabling leukocytes to penetrate the injured tissue more quickly [79].

9.14 Layered and Structured Dressings

Multilayer dressings are better because they can be made with more composite functionality. He et al. reports that 3-layer dressing consisting of PCL/polystyrene microspheres, collagen/quaternized chitosan sponge, and PCL/gelatin nanofiber [80]. Exogenic bacterial adhesion was inhibited by hydrophobic microspheres, exudates were absorbed by sponges, and directed cell growth and migration was promoted by aligned PCL/gelatin mixed with curcumin. Both the inner and outer sides' antibacterial functions are made possible by this tactic. Following the same principle, four-layer hybrid substrates that were super-hydrophilic, hydrophilic, hydrophobic, and superhydrophobic were created by integrating astaxanthin and ciprofloxacin into the hydrophilic layer [81]. This membrane has potential uses to absorb fluids, be antibacterial, antioxidant, and self-cleaning.

To make the most of exudates, Chang's group developed a 4-layer dressing for ion backflow [82]. The top layers consisted of hydrophilic deacetylated cellulose acetate, superabsorbent particles, and sequentially ordered bioactive silicate bioglass particles. The bottom layer was a PU membrane featuring micropore arrays. This technique allows for the unidirectional transmission of large volumes of wound bed exudates with the concomitant backflow. This new system aided in the healing of diabetic lesions. Additionally, this is made simpler by its sandwich construction, which consists of hydrophobic PLA on one side and hydrophilic zinc silicate bioceramics on the other [83]. Exudate absorption and the ensuing release of Zn_2+ and $SiO_3$2-ions promoted the recovery of the burn site.

Xu and colleagues cleverly created a three-layer patch combining filter paper, medical adhesive tape, and Janus polydimethylsiloxane layer to overcome the poor stretching performance of conventional band-aids [84]. A femtosecond laser was used to penetrate a layer of polydimethylsiloxane, which was subsequently single-sided modified for the Janus property. In both the stretched and relaxed phases, this patch exhibited unidirectional liquid transfer, as indicated.

Hydro-conductive dressings are a novel category of bandages that have recently been presented. These dressings have a unique layered structure that allows them to draw out waste items, absorb exudate, and transport them to the dressing's core. The first commercially produced hydro-conductive dressing was called Drawtex® (SteadMed Medical, USA), and it was distinguished by its high absorption capacity (30–50 folds), non-traumatic removal, and ability to maintain its structure for 7 days. Many efforts have been made to investigate the qualities of Drawtex® dressing because of its possible better effectiveness [85]. For example, the first in vitro assessment of Drawtex®'s ability to lower bacterial colonies in the medium it meets was conducted by Ortiz et al. The author specifically saw a simultaneous drop in bacteria in the media and an increase in the dressing itself, demonstrating the dressing's potential to trap pathogens inside its structure.

9.15 Scaffolds for Tissue Engineering

For almost 30 years, scaffold-based tissue engineering approaches have been studied, and numerous methods have been created to engineer diverse bodily tissues [86]. Artificial tissue scaffolds are made to resemble, at least in part, the extracellular matrix seen in real tissues. Consequently, their actions ought to mimic the extracellular matrix of the target tissue. The last several decades have seen the development of four main scaffolding techniques: the injection of cell-encapsulated self-assembled hydrogels; the implantation of laminated cell sheets containing secreted extracellular matrix; and the insertion of prefabricated porous scaffolds containing cell-seeded, decellularized allograft or xenograft extracellular matrix. Every strategy has advantages and disadvantages as well as favored tissue engineering uses. These scaffolding approaches are useful guides for planning tissue engineering procedures for complicated tissues like the intervertebral disc. They can also be combined. Furthermore, particular awareness to tissue-specific factors like the degree of injury, the distinct structural-functional connection, the various tissue compositions, and the interfaces within the intervertebral disc [87]. For tissue engineering applications, scaffold fabrication efficiency is crucial. Specific characteristics are required of scaffolds. The mechanical and biological characteristics of tissue engineering scaffolds are among their most crucial attributes [42].

In tissue engineering applications, 3D scaffolds are widely employed to produce a proper milieu to heal damaged tissues. These scaffolds aid in our comprehension of how cells interact and react to mechanical cues from their three-dimensional (3D) surrounds by mimicking the actual in vivo milieu. Therefore, cellular response and fate are determined by the scaffolds' material qualities. These three-dimensional scaffolds are often quite porous, with linked pore networks that help with waste disposal and the transport of nutrients and oxygen [88].

In order to create a three-dimensional environment that will induce tissue development, biodegradable scaffolds are essential. Tissue regeneration is seen to have enormous potential when scaffolding materials and stem cell technologies are used together [89]. Thousands of surgical procedures are carried out daily to replace or repair tissue that has been harmed by trauma or illness. Tissue Engineering (TE) is a rapidly emerging field that combines human cells with highly porous scaffold biomaterials to try to repair damaged tissues. These scaffolds serve as templates for tissue regeneration and direct the creation of new tissue [90]. Vascular disorders and fractures related to the bones create serious socioeconomic issues. To repair injured bone, conventional bone grafts such autograft and allograft have been utilized. But despite their shortcomings like scarce donor supply and elevated infection risk great advancements have been achieved in biomaterials and scaffolds, which offer a viable alternative for bone-tissue engineering [91].

Cancer, birth defects, and trauma-related tissue damage require novel, effective therapies that promote tissue regeneration. In this regard, tissue engineering exhibits significant promise in regaining the original structure and functionality of injured tissues by the mingling of cells with particular scaffolds. A vital function in directing cell growth and the generation of new tissues is played by scaffolds composed of natural and/or synthetic polymers and occasionally ceramic materials. It has been revealed that monolayered scaffolds with a homogenous material structure are insufficient to replicate the intricate biological milieu found in tissues. Since many tissues, including cutaneous, vascular, and osteochondral, have multiple structures, multilayered scaffolds are more beneficial for regenerating these tissues [92].

9.16 Functionalization with Growth Factors and Bioactive Molecules

The goal of tissue engineering is to enhance or repair tissue function, which will significantly improve human life. In tissue repair, scaffolds should have a physical and chemical makeup like that of native extracellular matrix. Substrates made of conducting polymers can stimulate electrically on scaffolds, which can significantly improve the regeneration of electroactive tissues like muscles and nerves in the body. Solution casting/particle leaching, conducting polymer deposition on template scaffolds, and electrospinning of conducting polymer blends with other polymers have all been used to create electrical conducting scaffolds. Conducting scaffolds can immobilize biologically active compounds, making them multifunctional indicators for tissue engineering. Additionally, research has been done on the interactions between conducting scaffolds and brain and cardiac cells. Therefore, tissue engineering and other bio-related fields will be impacted by the creation and use of scaffolds in tissue engineering [93].

Tissue engineering is a potentially effective solution because conventional approaches are not suitable for treating myocardial infarction. In a typical tissue engineering approach, a scaffold, or patch, is needed to harbor the cells and bolster the infarcted heart mechanically. A perfect CTE scaffold should encourage cell adhesion and mechanically and physiologically resemble the original tissue. Although many synthetic polymers meet the necessary mechanical requirements, they are not successful at interacting between cells and materials. Therefore, a great deal of research has been done on bulk and surface functionalization techniques [94].

For tissue engineering, a range of designed scaffolds made of ceramics, polymers, and their composites have been developed. For better tissue regeneration, biomimicry has been incorporated into most 3D scaffold designs, both concerning bioactivity and physicochemical characteristics. It has been demonstrated that scaffolds made using hydrogels, lithography, salt leaching, and particle sintering effectively encourage tissue regeneration and cell development. The decellularization of entire part or tissues produces scaffold systems that are widely used due to guaranteed biocompatibility. Conventional scaffold construction methods required numerous phases, were not repeatable, and frequently failed to produce complex structures with higher resolution. Several of the drawbacks of conventional methods were addressed by 3D printing technology, using a variety of thermoplastics and hydrogels, this also made the synthesis of micro-nanostructured scaffolds and devices for tissue engineering and drug delivery easier [95].

The PCL-PEG-PCL (PCEC) copolymer is involved in the field of bone tissue engineering due to its ability to promote osteogenesis and its biocompatibility [96]. Considerable work has gone into investigating methods for creating bioactive scaffolds. Electro spun scaffolds, which are ideal for delivering biomolecules due to their huge surface-volume ratio and ultrathin fiber diameter, have been increasingly popular in this field during the last 5 years [97].

9.17 Drug Delivery Systems

In order to improve medication dispersion and provide sustained release, simple polymer layers were added and the thickness of the barrier layers surrounding the dressing was changed [98]. By utilizing an iron-containing metal-organic framework (Fe-MOF), Zhang et al. were able to control the release of glucose oxidase (Gox). They accomplished this by linking Gox by amide bonding and generating G@Fe/PGD fiber membranes utilizing PCL, gelatin, and d-glucose (PGD) fiber networks [99].

A medication ingredient (triclosan or levofloxacin) was electro spun with PLA to create type I collagen. Against this chemical, *S. aureus* and *E. coli* both showed good antibacterial action. Moreover, because the drug was close to the surface, there was an initial burst release that was noticed. Moreover, the collagen incorporated into the fibers changed the behavior of the drug release and resulted in a regulated drug dissolution profile after the burst release. A common method for creating nanofibers with controlled-release growth factors (GFs) is electrospinning [100].

In 2014, Kulkarni et al. introduced a revolutionary technique https://www.ncbi.nlm.nih.gov/pmc/articles/PMC9544096/. Layer by layer, they created poly (acrylic acid)-modified polyurethane nanofibers containing EGF. EGF release was successfully prolonged by changing the quantity of layers applied to the surface [101].

Using commercially available sutures, Centeno-Cerdas et al. (2018) inserted photosynthetic genetically modified microalgae to obtain regulated release of growth factors (VEGF, PDGF-BB, or SDF-1α) at the wound site [102]. Pacheco et al. (2020) reported on a multilayer membrane with regulated release of drug capabilities in another investigation [103].

The strength and gradual degradation of silk fibroin are combined with the hydrophilicity of HA in (Hyaluronan) HA-silk interpenetrating network-based hydrogels to create a single material [104]. In contrast, a hydrogel of carboxymethyl chitosan and Schiff-base crosslinked oxidized HA permitted the prolonged release of curcumin and EGF to encourage wound recovery [104].

Biomaterials based on gelatin facilitate the healing of chronic wounds. Gelatin-based nanocomposites support recovery of chronic wounds and improve the antibacterial qualities surrounding the wound. Because they are porous, absorbent, and expandable, gelatin-based sponges help chronic wound healing by absorbing water around the wound [105].

Numerous wound dressings have built-in antimicrobial qualities that prevent pathogen invasion; incorporating anti-infectious targeted drugs into the dressing could boost its effectiveness. Antimicrobial biomaterials could potentially serve as an alternative to systemic therapy for persistent wound infections [19].

By preventing bacterial attachment, colonization, and growth, antimicrobial biomaterials have emerged as a viable and effective means of preventing implant-related illnesses. Thus, it has been shown that antimicrobial therapy reduces infection-related morbidity and death in patients in critical condition. Antimicrobial scaffolds and antimicrobial coatings for medical devices are produced using a variety of antimicrobial chemicals, such include peptides, antibacterial marine extracts, inorganic nanoparticles, and antimicrobial polymers [106].

Human health is seriously threatened by bacterial infections and the rising resistance of microorganisms. Furthermore, bacterial infiltration poses a serious threat to the process of wound recovery. Taking into consideration the complex wound healing process, novel approaches to skin tissue engineering focus on the integration of bioactive substances, such as antibacterial agents, into biomaterials with different morphologies to improve cell behaviors and promote wound healing [107].

Chitosan microspheres encapsulated in PEG fibrin gels and loaded with silver sulfadiazine showed good antibacterial activity [108].Collagen-coated AgNPs encased in collagen hydrogels have been demonstrated to be safe and effective against S. aureus, *Staphylococcus epidermidis, E. coli,* and *P. aeruginosa,* as well as possess antimicrobial qualities in primary human skin fibroblasts and keratinocytes [109]. For stronger antibacterial activity and better wound healing, genipin cross-linked with chitosan, PEG, and silver nanoparticles is mixed in a nanocomposite [110]. It has been demonstrated that titanium dioxide nanoparticle-loaded wound dressings with chitosan and pectin coatings have excellent wound-healing capabilities and are antibacterial [111].

9.18 Advantages and Limitations

The primary benefits of hydrogels are their high-water content, which promotes better encapsulation, growth, and protection of fragile drugs and cells. It also allows for easier patterning using 3D printing and allows for controllable drug release. Additionally, hydrogels' mechanical properties can be modified due to crosslinking. Due to their poor mechanical qualities, they are not recommended for use in load-bearing applications; they are also time-consuming to optimize for printing; and they are difficult to sterilize [112].

A patient's customized implants, prosthetics, and even living tissues can be created via additive manufacturing technology. By employing medical imaging data, 3D printers may produce implants and prosthesis that exactly fit a person's unique anatomy, increasing comfort, and reducing issues [113]. The benefits of AM include cheap cost and complicated design small-scale part manufacturing, which is advantageous for individual patient demands such as replacing teeth and bone [114].

Adaptability, scalability, and high resolution and precision are just a few benefits of droplet 3D printing. In the end, post-curing procedures are usually needed to enhance the created object's mechanical properties and stability. The advantages of high-resolution processing, nozzle-free operation, reduced heat stress, and improved layer connectivity outweigh these drawbacks, making additive manufacturing a feasible choice for the production of pharmaceuticals [115].

9.19 Conclusion

This review's main goal was to highlight the role that bioactive compounds, both natural and manufactured, play in the control of wound recovery. Scaffolds made of a variety of Polymers can be altered to resemble the texture of skin and properties, offering various kinds of wounds suitable environments and configurations. Furthermore, wound care can be enhanced and the healing process accelerated by the inclusion of GF, pharmaceuticals, stem cells, active agents, and silver. Wound healing is a complicated, organized physiological process that involves the deposition of collagen fibers, cell proliferation, re-epithelialization, and tissue regeneration. As a result, selecting suitable polymers, bioactive compounds, and dressings for wounds is crucial to facilitate and accelerate the healing process. In an effort to speed up wound healing and avoid infection, innovative wound dressings based on synthetic or natural biomaterials have been developed as a result of scientific and technological advancements in skin tissue engineering and medicine. These dressings are far more durable, hydrophilic, low toxicity, biocompatible, biodegradable, and have antibacterial and antimicrobial effects as compared to standard dressings. It has been shown that tissue scaffolds work well for skin tissue engineering and wound dressing in a range of wound situations. In addition to covering the wound and acting as a physical barrier, tissue scaffolds can act as a cellular skin that contains excretive biological components that promote granulation tissue production and re-epithelialization.

References

1. Singh S, Syukri DM, Ushir YV, Mishra A, Ontong JC, Nwabor OF et al (2024) Post-operative wound healing efficacy of eucalyptus Camaldulensis phenolic-rich extracts incorporated hydrogel with enhanced antioxidant, antibacterial, and anti-inflammatory activities. J Polym Environ 33:1–26
2. Israni DK, Raghani NR, Soni J, Shah M, Prajapati BG, Chorawala MR et al (2024) Harnessing Cannabis sativa oil for enhanced skin wound healing: the role of reactive oxygen species regulation. Pharmaceutics 16(10):1277
3. Datta D, Prajapati B, Jethva H, Agrawal K, Singh S, Prajapati BG (2024) Value-added nanocellulose valorized from fruit peel waste for potential dermal wound healing and tissue regenerative applications. Regen Eng Transl Med 11:1–24
4. Mohite P, Shah SR, Munde S, Ade N, Boban T, Singh S, Prajapati B (2024) Biopolymers in textile-based scaffolding and wound healing. In: Biopolymers in the textile industry: opportunities and limitations. Springer Nature, Singapore, pp 175–202
5. Singh SA, Vellapandian C, Shah DD, Jayeoye TJ, Chorawala MR, Singh S, Prajapati BG (2024) Valorised calcium-rich biomass from fish waste and eggshells in the fabrication of antibacterial scaffold for wound healing applications: a review. Waste and Biomass Valor 15(4):1917–1941
6. Mohite P, Asane G, Rebello N, Munde S, Ade N, Boban T et al (2024) Polymeric hydrogel sponges for wound healing applications: a comprehensive review. Regen Eng Transl Med 10:1–22
7. Martin P, Nunan R (2015) Cellular and molecular mechanisms of repair in acute and chronic wound healing. Br J Dermatol 173(2):370–378

8. Frykberg RG, Banks J (2015) Challenges in the treatment of chronic wounds. Adv Wound Care (New Rochelle) 4(9):560–582
9. Singh S, Dodiya TR, Singh S, Dodiya R (2021) Topical wound healing, antimicrobial and antioxidant potential of Mimosa pudica Linn root extracted using n-hexane followed by methanol, fortified in ointment base. Int J Pharm Sci Nanotechnol 14(3):5472–5480
10. Singh S, Nwabor OF, Sukri DM, Wunnoo S, Dumjun K, Lethongkam S et al (2022) Poly (vinyl alcohol) copolymerized with xanthan gum/hypromellose/sodium carboxymethyl cellulose dermal dressings functionalized with biogenic nanostructured materials for antibacterial and wound healing application. Int J Biol Macromol 216:235–250
11. Monika P, Chandraprabha MN, Rangarajan A, Waiker PV, Chidambara Murthy KN (2022) Challenges in healing wound: role of complementary and alternative medicine. Front Nutr 8:791899
12. Falcone M, De Angelis B, Pea F, Scalise A, Stefani S, Tasinato R et al (2021) Challenges in the management of chronic wound infections. J Glob Antimicrob Resist 26:140–147
13. Ding X, Tang Q, Xu Z, Xu Y, Zhang H, Zheng D et al (2022) Challenges and innovations in treating chronic and acute wound infections: from basic science to clinical practice. Burns Trauma 10:tkac014
14. Ma PX, Choi JW (2001) Biodegradable polymer scaffolds with well-defined interconnected spherical pore network. Tissue Eng 7(1):23–33
15. Niculescu AG, Grumezescu AM (2022) An up-to-date review of biomaterials application in wound management. Polymers (Basel) 14(3):421
16. Wang H (2023) Biomaterials in medical applications. Polymers (Basel) 15(4):847
17. Downer M, Berry CE, Parker JB, Kameni L, Griffin M (2023) Current biomaterials for wound healing. Bioengineering [Internet] 10(12):1378
18. Chaudhari AA, Vig K, Baganizi DR, Sahu R, Dixit S, Dennis V et al (2016) Future prospects for scaffolding methods and biomaterials in skin tissue engineering: a review. Int J Mol Sci 17(12):1974
19. Miron A, Giurcaneanu C, Mihai MM, Beiu C, Voiculescu VM, Popescu MN et al (2023) Antimicrobial biomaterials for chronic wound care. Pharmaceutics [Internet] 15(6):1606
20. Priddy-Arrington TR, Ward MS, Edwards RE, Caldorera-Moore ME (2021) Proactive biomaterials for chronic wound management and treatment. Curr Opin Biomed Eng 20:100327
21. Wang H, Xu Z, Li Q, Wu J (2021) Application of metal-based biomaterials in wound repair. Eng Regen 2:137–153
22. Yu R, Zhang H, Guo B (2021) Conductive biomaterials as bioactive wound dressing for wound healing and skin tissue engineering. Nanomicro Lett 14(1):1
23. Jemghili R, Aittaleb A, Mansouri K (2022) Additive manufacturing technology, process parameters influencing product quality and its applications. Int J Adv Technol Eng Explor 9:1678–1703
24. Vyas J, Singh S, Shah I, Prajapati BG (2023) Potential applications and additive manufacturing technology-based considerations of mesoporous silica: a review. AAPS PharmSciTech 25(1):6
25. Vyas J, Raytthatha N, Singh S, Prajapati BG, Mohite P, Munde S (2024) Sustainable sources of raw materials as substituting biomaterials for additive manufacturing of dental implants: a review. Periodontal Implant Res 8(1):3
26. Mecheter A, Tarlochan F, Kucukvar M (2023) A review of conventional versus additive manufacturing for metals: life-cycle environmental and economic analysis. Sustainability [Internet] 15(16):12299
27. Kausar Z, Shah F, Masood Z, Rehman H, Khaydarov S, Saeed MT et al (2021) Energy efficient parallel configuration based six degree of freedom machining bed. Energies 14:2642
28. Yang Y, Li L, Zhao J (2019) Mechanical property modeling of photosensitive liquid resin in stereolithography additive manufacturing: bridging degree of cure with tensile strength and hardness. Mater Des 162:418–428
29. Kazemi M, Rahimi A (2018) Stereolithography process optimization for tensile strength improvement of products. Rapid Prototyp J 24:688–697

30. Melchels FPW, Feijen J, Grijpma DW (2010) A review on stereolithography and its applications in biomedical engineering. Biomaterials 31(24):6121–6130
31. Wang X, Jiang M, Zhou Z, Gou J, Hui D (2017) 3D printing of polymer matrix composites: a review and prospective. Compos Part B 110:442–458
32. Mengesha Medibew T (2022) A comprehensive review on the optimization of the fused deposition modeling process parameter for better tensile strength of PLA-printed parts. Adv Mater Sci Eng 2022(1):5490831
33. Rashid AA, Koç M (2021) Fused filament fabrication process: a review of numerical simulation techniques. Polym Int 13(20):3534
34. Durga Prasad Reddy R, Sharma V (2020) Additive manufacturing in drug delivery applications: a review. Int J Pharm 589:119820
35. Tikhomirov E, Åhlén M, Di Gallo N, Strømme M, Kipping T, Quodbach J et al (2023) Selective laser sintering additive manufacturing of dosage forms: effect of powder formulation and process parameters on the physical properties of printed tablets. Int J Pharm 635:122780
36. Balasankar A, Anbazhakan K, Arul V, Mutharaian VN, Sriram G, Aruchamy K et al (2023) Recent advances in the production of pharmaceuticals using selective laser sintering. Biomimetics [Internet] 8(4):330
37. Tabriz AG, Kuofie H, Scoble J, Boulton S, Douroumis D (2023) Selective laser sintering for printing pharmaceutical dosage forms. J Drug Deliv Sci Technol 86:104699
38. Blynskaya EV, Tishkov SV, Alekseev KV, Vetcher AA, Marakhova AI, Rejepov DT (2022) Polymers in technologies of additive and inkjet printing of dosage formulations. Polym Int 14(13):2543
39. Carou-Senra P, Ong JJ, Castro BM, Seoane-Viaño I, Rodríguez-Pombo L, Cabalar P et al (2023) Predicting pharmaceutical inkjet printing outcomes using machine learning. Int J Pharm X 5:100181
40. Uddin MJ, Hassan J, Douroumis D (2022) Thermal inkjet printing: prospects and applications in the development of medicine. Technologies [Internet] 10(5):108
41. Zhou LY, Fu J, He Y (2020) A review of 3D printing technologies for soft polymer materials. Adv Funct Mater 30(28):2000187
42. Krishani M, Shin WY, Suhaimi H, Sambudi NS (2023) Development of scaffolds from bio-based natural materials for tissue regeneration applications: a review. Gels [Internet] 9(2):100
43. Mathew-Steiner SS, Roy S, Sen CK (2021) Collagen in wound healing. Bioengineering [Internet]. 8(5):63
44. Li D, Ren J-w, Xu T, Li L, Liu P, Li Y (2021) Effect of bovine bone collagen oligopeptides on wound healing in mice. Aging (Albany NY) 13(6):9028
45. Davison-Kotler E, Marshall WS, García-Gareta E (2019) Sources of collagen for biomaterials in skin wound healing. Bioengineering [Internet] 6(3):56
46. Chattopadhyay S, Raines RT (2014) Collagen-based biomaterials for wound healing. Biopolymers 101(8):821–833
47. Das U, Kapoor DU, Singh S, Prajapati BG (2024) Unveiling the potential of chitosan-coated lipid nanoparticles in drug delivery for management of critical illness: a review. Z Naturforsch C J Biosci 79(5-6):107–124
48. Begum RF, Singh S, Prajapati B, Sumithra M, Patel RJ (2024) Advanced targeted drug delivery of bioactive agents fortified with graft chitosan in management of cancer: a review. Curr Med Chem 32:3759–3789
49. Mohite P, Shah SR, Singh S, Rajput T, Munde S, Ade N et al (2023) Chitosan and chito-oligosaccharide: a versatile biopolymer with endless grafting possibilities for multifarious applications. Front Bioeng Biotechnol 11:1190879
50. Singh S, Nwabor OF, Syukri DM, Voravuthikunchai SP (2021) Chitosan-poly (vinyl alcohol) intelligent films fortified with anthocyanins isolated from Clitoria ternatea and Carissa carandas for monitoring beverage freshness. Int J Biol Macromol 182:1015–1025
51. Mohite P, Rahayu P, Munde S, Ade N, Chidrawar VR, Singh S et al (2023) Chitosan-based hydrogel in the management of dermal infections: a review. Gels 9(7):594

52. Bagher Z, Ehterami A, Safdel MH, Khastar H, Semiari H, Asefnejad A et al (2020) Wound healing with alginate/chitosan hydrogel containing hesperidin in rat model. J Drug Deliv Sci Technol 55:101379
53. Deng X, Gould M, Ali MA (2022) A review of current advancements for wound healing: biomaterial applications and medical devices. J Biomed Mater Res B Appl Biomater 110(11):2542–2573
54. Sharma R, Malviya R, Singh S, Prajapati B (2023) A critical review on classified excipient sodium-alginate-based hydrogels: modification, characterization, and application in soft tissue engineering. Gels 9(5):430
55. Ching SH, Bansal N, Bhandari B (2017) Alginate gel particles – a review of production techniques and physical properties. Crit Rev Food Sci Nutr 57(6):1133–1152
56. Topuz F, Henke A, Richtering W, Groll J (2012) Magnesium ions and alginate do form hydrogels: a rheological study. Soft Matter 8(18):4877–4881
57. Chen HL, Chung JWY, Yan VCM, Wong TKS (2023) Polylactic acid-based biomaterials in wound healing: a systematic review. Adv Skin Wound Care 36(9):1–8
58. Avinc O, Khoddami A (2009) Overview of poly(lactic acid) (PLA) fibre. Fibre Chem 41(6):391–401
59. Yoon SD, Kwon YS, Lee KS (2017) Biodegradation and biocompatibility of poly L-lactic acid implantable mesh. Int Neurourol J 21(Suppl 1):S48–S54
60. Hajikhani M, Emam-Djomeh Z, Askari G (2021) Fabrication and characterization of mucoadhesive bioplastic patch via coaxial polylactic acid (PLA) based electrospun nanofibers with antimicrobial and wound healing application. Int J Biol Macromol 172:143–153
61. Dwivedi R, Kumar S, Pandey R, Mahajan A, Nandana D, Katti DS et al (2020) Polycaprolactone as biomaterial for bone scaffolds: review of literature. J Oral Biol Craniofac Res 10(1):381–388
62. Shah SR, Modi CD, Singh S, Mori DD, Soniwala MM, Prajapati BG (2024) Recent advances in additive manufacturing of polycaprolactone-based scaffolds for tissue engineering applications: a comprehensive review. Regen Eng Transl Med 11:1–20
63. Jenkins MJ, Harrison KL (2006) The effect of molecular weight on the crystallization kinetics of polycaprolactone. Polym Adv Technol 17(6):474–478
64. Augustine R, Kalarikkal N, Thomas S (2016) Effect of zinc oxide nanoparticles on the in vitro degradation of electrospun polycaprolactone membranes in simulated body fluid. Int J Polym Mater Polym Biomater 65(1):28–37
65. Augustine R, Saha A, Jayachandran VP, Thomas S, Kalarikkal N (2015) Dose-dependent effects of gamma irradiation on the materials properties and cell proliferation of electrospun polycaprolactone tissue engineering scaffolds. Int J Polym Mater Polym Biomater 64(10):526–533
66. Shahverdi S, Hajimiri M, Esfandiari MA, Larijani B, Atyabi F, Rajabiani A et al (2014) Fabrication and structure analysis of poly(lactide-co-glycolic acid)/silk fibroin hybrid scaffold for wound dressing applications. Int J Pharm 473(1):345–355
67. Chaudhari AA, Vig K, Baganizi DR, Sahu R, Dixit S, Dennis V et al (2016) Future prospects for scaffolding methods and biomaterials in skin tissue engineering: a review. Int J Mol Sci [Internet] 17(12):1974
68. Anjum F, Agabalyan NA, Sparks HD, Rosin NL, Kallos MS, Biernaskie J (2017) Biocomposite nanofiber matrices to support ECM remodeling by human dermal progenitors and enhanced wound closure. Sci Rep 7(1):10291
69. Yuan TT, DiGeorge Foushee AM, Johnson MC, Jockheck-Clark AR, Stahl JM (2018) Development of electrospun chitosan-polyethylene oxide/fibrinogen biocomposite for potential wound healing applications. Nanoscale Res Lett 13(1):88
70. Zhang H, Peng M, Cheng T, Zhao P, Lipeng Q, Zhou J et al (2018) Silver nanoparticles-doped collagen–alginate antimicrobial biocomposite as potential wound dressing. J Mater Sci 53:14944–14952

71. Wang Y, Li P, Xiang P, Lu J, Yuan J, Shen J (2016) Electrospun polyurethane/keratin/AgNP biocomposite mats for biocompatible and antibacterial wound dressings. J Mater Chem B 4(4):635–648
72. Roussakis E, Ortines RV, Pinsker BL, Mooers CT, Evans CL, Miller LS et al (2019) Theranostic biocomposite scaffold membrane. Biomaterials 212:17–27
73. Valente KP, Brolo A, Suleman A (2020) From dermal patch to implants—applications of biocomposites in living tissues. Molecules [Internet] 25(3):507
74. Ren S, Guo S, Yang L, Wang C (2022) Effect of composite biodegradable biomaterials on wound healing in diabetes. Front Bioeng Biotechnol 10:1060026
75. Negut I, Dorcioman G, Grumezescu V (2020) Scaffolds for wound healing applications. Polymers (Basel). 12(9):2010
76. Faheed NK (2024) Advantages of natural fiber composites for biomedical applications: a review of recent advances. Emerg Mater 7(1):63–75
77. Michalicha A, Belcarz A, Giannakoudakis DA, Staniszewska M, Barczak M (2024) Designing composite stimuli-responsive hydrogels for wound healing applications: the state-of-the-art and recent discoveries. Materials [Internet] 17(2):278
78. Zhong D, Zhang H, Ma Z, Xin Q, Lu Y, Shi P et al (2022) Recent advancements in wound management: tailoring superwettable bio-interfaces. Front Bioeng Biotechnol 10:1106267
79. Ambekar RS, Kandasubramanian B (2019) Advancements in nanofibers for wound dressing: a review. Eur Polym J 117:304–336
80. He C, Yu B, Lv Y, Huang Y, Guo J, Li L et al (2022) Biomimetic asymmetric composite dressing by electrospinning with aligned nanofibrous and micropatterned structures for severe burn wound healing. ACS Appl Mater Interfaces. https://doi.org/10.1021/acsami.2c04323
81. Zhang W, Guan X, Qiu X, Gao T, Yu W, Zhang M et al (2023) Bioactive composite Janus nanofibrous membranes loading Ciprofloxacin and Astaxanthin for enhanced healing of full-thickness skin defect wounds. Appl Surf Sci 610:155290
82. Bao F, Pei G, Wu Z, Zhuang H, Zhang Z, Huan Z et al (2020) Bioactive self-pumping composite wound dressings with micropore array modified Janus membrane for enhanced diabetic wound healing. Adv Funct Mater 30(49):2005422
83. Zhang Z, Li W, Liu Y, Yang Z, Ma L, Zhuang H et al (2021) Design of a biofluid-absorbing bioactive sandwich-structured Zn–Si bioceramic composite wound dressing for hair follicle regeneration and skin burn wound healing. Bioact Mater 6(7):1910–1920
84. Xu B, Li A, Wang R, Zhang J, Ding Y, Pan D et al (2021) Elastic Janus film for wound dressings: unidirectional biofluid transport and effectively promoting wound healing. Adv Funct Mater 31(41):2105265
85. Laurano R, Boffito M, Ciardelli G, Chiono V (2022) Wound dressing products: a translational investigation from the bench to the market. Eng Regen 3(2):182–200
86. Henkel J, Hutmacher D (2013) Design and fabrication of scaffold-based tissue engineering. BioNanoMaterials 14:171–193
87. Chan BP, Leong KW (2008) Scaffolding in tissue engineering: general approaches and tissue-specific considerations. Eur Spine J 17(Suppl 4):467–479
88. Loh QL, Choong C (2013) Three-dimensional scaffolds for tissue engineering applications: role of porosity and pore size. Tissue Eng Part B Rev 19(6):485–502
89. Hosseinkhani M, Mehrabani D, Karimfar MH, Bakhtiyari S, Manafi A, Shirazi R (2014) Tissue engineered scaffolds in regenerative medicine. World J Plast Surg 3(1):3
90. O'Brien FJ (2011) Biomaterials & scaffolds for tissue engineering. Mater Today 14(3):88–95
91. Liu X, Ma PX (2004) Polymeric scaffolds for bone tissue engineering. Ann Biomed Eng 32(3):477–486
92. Bertsch C, Maréchal H, Gribova V, Lévy B, Debry C, Lavalle P et al (2023) Biomimetic bilayered scaffolds for tissue engineering: from current design strategies to medical applications. Adv Healthc Mater 12(17):2203115
93. Guo B, Lei B, Li P, Ma PX (2015) Functionalized scaffolds to enhance tissue regeneration. Regen Biomater 2(1):47–57

94. Tallawi M, Rosellini E, Barbani N, Cascone MG, Rai R, Saint-Pierre G et al (2015) Strategies for the chemical and biological functionalization of scaffolds for cardiac tissue engineering: a review. J R Soc Interface 12(108):20150254
95. Stratton S, Shelke NB, Hoshino K, Rudraiah S, Kumbar SG (2016) Bioactive polymeric scaffolds for tissue engineering. Bioact Mater 1(2):93–108
96. Abedi F, Moghaddam SV, Ghandforoushan P, Aghazadeh M, Ebadi H, Davaran S (2022) Synthesis and characterization of growth factor free nanoengineered bioactive scaffolds for bone tissue engineering. J Biol Eng 16(1):28
97. Ji W, Sun Y, Yang F, van den Beucken JJ, Fan M, Chen Z et al (2011) Bioactive electrospun scaffolds delivering growth factors and genes for tissue engineering applications. Pharm Res 28(6):1259–1272
98. Tabriz AG, Douroumis D (2022) Recent advances in 3D printing for wound healing: a systematic review. J Drug Deliv Sci Technol 74:103564
99. Arkan E, Karabiber A, Topçu MA, Kinas Z, Sarilmaz A, Ozel SS et al (2023) Nano-ceria based TENGs: effect of dopant structure on energy harvesting performance. Surf Interfaces 37:102683
100. Hall Barrientos IJ, Paladino E, Szabó P, Brozio S, Hall PJ, Oseghale CI et al (2017) Electrospun collagen-based nanofibres: a sustainable material for improved antibiotic utilisation in tissue engineering applications. Int J Pharm 531(1):67–79
101. Kulkarni A, Diehl-Jones W, Ghanbar S, Liu S (2014) Layer-by-layer assembly of epidermal growth factors on polyurethane films for wound closure. J Biomater Appl 29(2):278–290
102. Centeno-Cerdas C, Jarquín-Cordero M, Chávez MN, Hopfner U, Holmes C, Schmauss D et al (2018) Development of photosynthetic sutures for the local delivery of oxygen and recombinant growth factors in wounds. Acta Biomater 81:184–194
103. Pacheco MS, Kano GE, Paulo LA, Lopes PS, de Moraes MA (2020) Silk fibroin/chitosan/alginate multilayer membranes as a system for controlled drug release in wound healing. Int J Biol Macromol 152:803–811
104. Onder OC, Batool SR, Nazeer MA (2022) Self-assembled silk fibroin hydrogels: from preparation to biomedical applications. Mater Adv 3(18):6920–6949
105. Cao H, Wang J, Hao Z, Zhao D (2024) Gelatin-based biomaterials and gelatin as an additive for chronic wound repair. Front Pharmacol 15:1398939
106. Sam S, Joseph B, Thomas S (2023) Exploring the antimicrobial features of biomaterials for biomedical applications. Results Eng 17:100979
107. Liang Y, Liang Y, Zhang H, Guo B (2022) Antibacterial biomaterials for skin wound dressing. Asian J Pharm Sci 17(3):353–384
108. Seetharaman S, Natesan S, Stowers RS, Mullens C, Baer DG, Suggs LJ et al (2011) A PEGylated fibrin-based wound dressing with antimicrobial and angiogenic activity. Acta Biomater 7(7):2787–2796
109. Alarcon EI, Udekwu KI, Noel CW, Gagnon LBP, Taylor PK, Vulesevic B et al (2015) Safety and efficacy of composite collagen–silver nanoparticle hydrogels as tissue engineering scaffolds. Nanoscale 7(44):18789–18798
110. Shen J, Wang Y, Feng Y (2023) Tuning photoluminescence of polysiloxane functionalized carbon dots through changing polysiloxane compositions. Polym Adv Technol 34(9):2936–2945
111. Archana D, Dutta J, Dutta PK (2013) Evaluation of chitosan nano dressing for wound healing: characterization, in vitro and in vivo studies. Int J Biol Macromol 57:193–203
112. Mirzaali MJ, Moosabeiki V, Rajaai SM, Zhou J, Zadpoor AA (2022) Additive manufacturing of biomaterials—design principles and their implementation. Materials [Internet] 15(15):5457
113. Jayanth N, Venkata Roshan M, Sakthi Balaji S, Durga Karthik P, Barathwaj A, Rishiyadhav G (2023) Additive manufacturing of biomaterials: a review. Mater Today Proc 113:287–298

114. Bose S, Ke D, Sahasrabudhe H, Bandyopadhyay A (2018) Additive manufacturing of biomaterials. Prog Mater Sci 93:45–111
115. Vyas J, Shah I, Singh S, Prajapati BG (2023) Biomaterials-based additive manufacturing for customized bioengineering in management of otolaryngology: a comprehensive review. Front Bioeng Biotechnol 11:1234340

Biomaterials-Based Additive Manufactured Products for Diabetic Wound Healing

10

Abhijeet Puri, Popat Mohite, Sudarshan Singh, Yogesh Chaudhari, Manisha Chaudhari, Yunus Ansari, and Amol Gholap

Abstract

The detrimental effects of diabetes extend to global healthcare systems due to its long-term nature, including chronic ulcers and poor wound healing. Understanding the pathophysiological mechanisms behind this is challenging. Biodegradable biomaterials can help address this problem effectively. Integrating materials, cells, and growth factors creates a comprehensive wound repair system. Composite biodegradable biomaterials excel in wound healing due to their biocompatibility and ability to enhance the process. Films, foams, hydrogels, antimicrobials, and dressings made from biomaterials have been used in various wound treatment applications. Despite extensive research, an ideal treatment for non-healing diabetic ulcers has yet to be found. With a specific emphasis on biomaterials for diabetic wound care, this chapter delves into recent advancements and challenges. It explores the primary categories of biomaterials utilized in wound dressings and skin substitutes, their properties, processing methods, and emerging therapies incorporating biomaterials. The discussion explores

A. Puri (✉) · P. Mohite · A. Gholap
AETs St. John Institute of Pharmacy and Research, Palghar, India
e-mail: abhijeetp@sjipr.edu.in

S. Singh
Faculty of Pharmacy, Chiang Mai University, Chiang Mai, Thailand

Office of Research Administrations, Chiang Mai University, Mueang Chiang Mai, Thailand

Y. Chaudhari
Dr. L. H. Hiranandani College of Pharmacy, Ulhasnagar (M.S), India

M. Chaudhari
D. Y. Patil University School of Pharmacy, Navi Mumbai (M.S), India

Y. Ansari
SNJB's Shriman Suresh dada Jain College of Pharmacy, Chandwad (M.S), India

S. Singh et al. (eds.), *Biomaterial-based Additive Manufacturing in Tissue Engineering and Regeneration*, Biomaterials, Bioengineering and Sustainability 7,
https://doi.org/10.1007/978-3-031-96070-3_10

various treatments, such as alternative biophysical therapies, cell- and drug-delivery therapies, and the use of synthetic and natural polymers. Considering the multiple explored approaches and the rapid advancement of biomaterial technology, cautious optimism exists regarding a potential breakthrough for healing diabetic wounds and addressing chronic diabetic ulcers.

Keywords

Diabetes · Wound healing · Biomaterial · Smart biomaterials · 3D bio-printed scaffolds

Abbreviations

ASTM	American Society for Testing of Materials
AM	Additive manufacturing
ASCs	Adipose-derived stem cells
AGEs	Advanced glycation end products
BSA	Bovine serum albumin
CNTs	Carbon nanotubes
CAT	Catalase
CAD	Computer-Aided Design
DFO	Deferoxamine
DFUs	Diabetic foot ulcers
dECM	decellularized Extracellular Matrix
FDM	Fused deposition modelling
ECM	extracellular matrix
ECM	Extracellular matrix
Gox	Glucose Oxidase
Gpx	Glutathione Peroxidase
HIF-1α	Hypoxia-inducible factor-1 α
MMPs	Matrix Metalloproteinases
PRP	Platelet-rich plasma
PCL	Polycaprolactone
PLA	Polylactide
PPy	Poly pyrrole
PANI	Polyaniline
PTh	Polythiophene
P.C.	Panthenol Citrate
POD	Peroxidase
STZ	Streptozotocin
SIS	Small Intestinal Submucosa
SeNPs	Selenium nanoparticles
VEGF	vascular endothelial growth factor

10.1 Introduction

Biomedical engineering in the past used essential biomaterials for tissue replacement and restoration of lost function, as there was limited knowledge of biological mechanisms. As our understanding of tissues, disease, and trauma grew, we began to delve into the concept of tissue repair. Through cell-based tissue engineering, advancements in stem cell research have allowed medicine to see the potential of regenerating tissues [1]. The definition of a biomaterial has changed alongside the development of medical concepts. As defined by Williams in 1987, biomaterials are used in medical devices to interact with biological systems. Despite the unchanged definition, notable breakthroughs have occurred in understanding how biomaterials interact with the biological system, especially in humans. Since biomaterials are no longer expected to be inert in the body, they are anticipated to be bioactive and assist in regeneration [2]. By triggering a biological response once implanted, bioactive materials exhibit an incredible ability, including cell adhesion, proliferation, and the most exciting part: stem cell differentiation for tissue and organ regeneration. Due to the intricate reactions of cells and tissues to these materials, biomaterial design, selection, synthesis, and fabrication find valuable guidance.

Biomimetics, a subset of biomaterials, involves replicating the chemical composition, microstructure, and manufacturing process of natural materials or living tissues [3]. Integrating biomaterials and living systems is essential for restoring injured body parts, involving natural and synthetic materials. These materials are designed to replace a specific body part or aid physiological functions. The effective execution of their tasks is made possible by biomaterials through connections with human cells, tissues, and organs. Engineers and scientists have been interested in dissimilar materials, like metallic, ceramic, polymeric, and composite materials, in biomedical applications for the past five decades [4]. A restricted comprehension of the interactions between materials and living systems with biocompatibility existed during the initial phases. The primary focus of researchers is to uncover and utilize exceptional material properties that exhibit desired biological characteristics to advance medical treatments and enhance the overall human experience. The production of biomaterials is gaining importance in improving reliability and decreasing the chances of rejection by the human body [5]. Researchers are currently focused on exploring new manufacturing techniques to process biomaterials. The goal is to reduce expenses, improve inventory control, and maximize efficiency in living organisms. Apart from manufacturing individual components, there is a rising interest in delving into advanced manufacturing sciences to create intricate multi-part structures, which pose challenges in traditional manufacturing processes. Devices in the biomedical field with gradients in their structure or composition can be highly versatile and enhance their performance when used in living organisms [6].

Biomaterials are specifically engineered to boost and enhance the mechanisms responsible for wound healing. Using polymeric biomaterials with plant extracts, antimicrobial agents, and extracellular components moisturizes wounds and forms a protective barrier against pathogens, promoting wound healing [7]. The development of biomaterials for non-healing wounds progresses with implants derived from the extracellular matrix, enhancing biological and physical aspects to drive innovation. By employing tissue-processing techniques, it becomes possible to manufacture devices made from the extracellular matrix that can induce diverse responses when implanted. Potential consequences of these reactions include the formation of a foreign body capsule around implants, degradation and assimilation of the implant, or colonization of the extracellular matrix by local fibroblasts and blood vessels, producing active tissue [8]. Bio-based polymeric materials, including proteins, polysaccharides, and aliphatic polyesters, can be created through plant, animal, or microbial synthesis. These polymers stand out from other biopolymers because they can either biodegrade, mimicking starch characteristics or remain non-biodegradable, resembling bio-polyethylene. In biomedical science, Polycaprolactone (PCL) and polylactide (PLA) have been scrutinized for their potential in additive manufacturing (AM), specifically in the creation of biodegradable and biocompatible scaffolds. Over the last decade, there has been a noticeable shift towards adopting sustainable biomaterials, surpassing traditional materials. The decomposition process of sustainable materials can be accurately controlled by incorporating harmless components during implantation. By utilizing a precise blend of synthetic and natural biopolymers, the unique properties of these materials enable the construction of both hard and soft tissues. Moreover, these materials come at a lower cost and exhibit similar chemical, physical, and biological properties as specific living cells and tissues [9].

The additive manufacturing process (AM) involves constructing structures layer by layer using metallic, polymeric, ceramic, or composite materials based on a computer-generated design file [10]. AM fabricators meticulously build each layer, employing various traditional manufacturing methods to ensure precision [11]. Two broad categories of biomaterial additive manufacturing techniques exist: acellular and cellular. Utilizing substances without living cells is the crucial characteristic of printing materials in the acellular category. The cellular category highlights an intriguing combination of live cells and additional materials, creating a captivating display [12]. The American Society for Testing of Materials (ASTM) has developed a guideline to classify diverse additive manufacturing techniques to bring uniformity. Figure 10.1 highlights various schematic depictions of AM techniques employed in biomaterials. In adherence to the ISO/ASTM 52900 standard, seven specific categories of additive manufacturing have been identified and clearly defined [13]. The development of biomaterials is influenced by the design of their microarchitecture and the selection of additive manufacturing techniques and printing parameters. These techniques include binder jetting, directed energy deposition, material extrusion, material jetting, powder bed fusion, sheet lamination, and vat photopolymerization. Porous or lattice structures are frequently needed in biomedical applications [14]. To effectively transport nutrients and oxygen to cells, the pores in biomaterials must be open and interconnected, allowing for easy passage. With the advent of AM technologies, it is now possible to intricately manipulate pore sizes and internal architectures on a microscopic level while fabricating detailed organic structures. Unlike conventional fabrication methods like casting

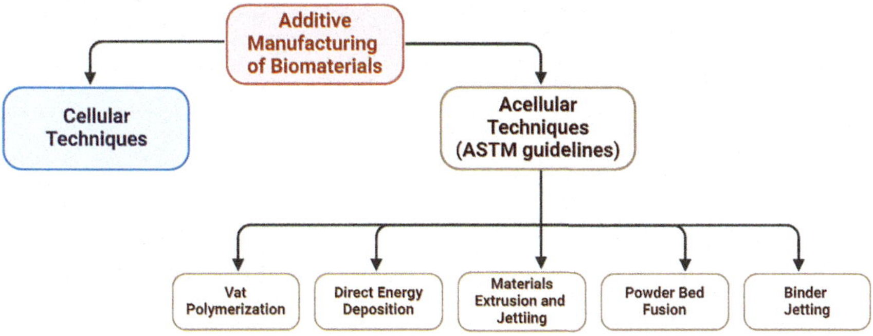

Fig. 10.1 Several types of additive manufacturing of biomaterials. (Created by using Biorender.com)

and moulding, AM offers designers a higher control over the internal geometries of porous materials [15].

Various biomaterials, including films, foams, hydrogels, antimicrobials, and multiple dressings, have been utilized in therapeutic applications to treat chronic wounds [16]. Despite numerous achievements, there is still no perfect solution available to directly tackle the intricate challenges linked to the impaired healing of diabetic ulcers. This chapter delves into the recent breakthroughs and ongoing obstacles encountered in biomaterials for treating diabetic wounds. The chapter examines the main categories of biomaterials (such as synthetic, natural, and hybrid) utilized in wound dressings and skin replacements. It covers essential considerations in manufacturing biomaterials and the latest advancements in diabetic wound treatment that incorporate biomaterials as a crucial element.

10.2 Diabetic Wound Healing and Related Influencing Factors

Diabetes mellitus is an intricate metabolic condition that has surged as one of the most rapidly growing health emergencies in present-day society. Various complications can arise in diabetic patients, including the development of diabetic foot [17]. The consequences include an increased likelihood of lower-extremity amputation and mortality, posing a severe threat to the life and well-being of patients. The management and treatment of this condition are incredibly challenging due to the impact of multiple factors on its healing process, as depicted in Fig. 10.2. Diabetes mellitus is a global health issue, affecting approximately 537 million people worldwide. In 2045, the population is estimated to have risen to 783 million [19]. Non-traumatic lower limb amputations are predominantly caused by diabetes mellitus worldwide. The increased prevalence of diabetic foot complications stemming from the growing incidence of DM is now a pressing global concern with far-reaching medical, social, and economic implications [20]. Foot ulcers in diabetic patients, called diabetic foot ulcers, harm the skin, soft tissues, and bones in the legs and feet.

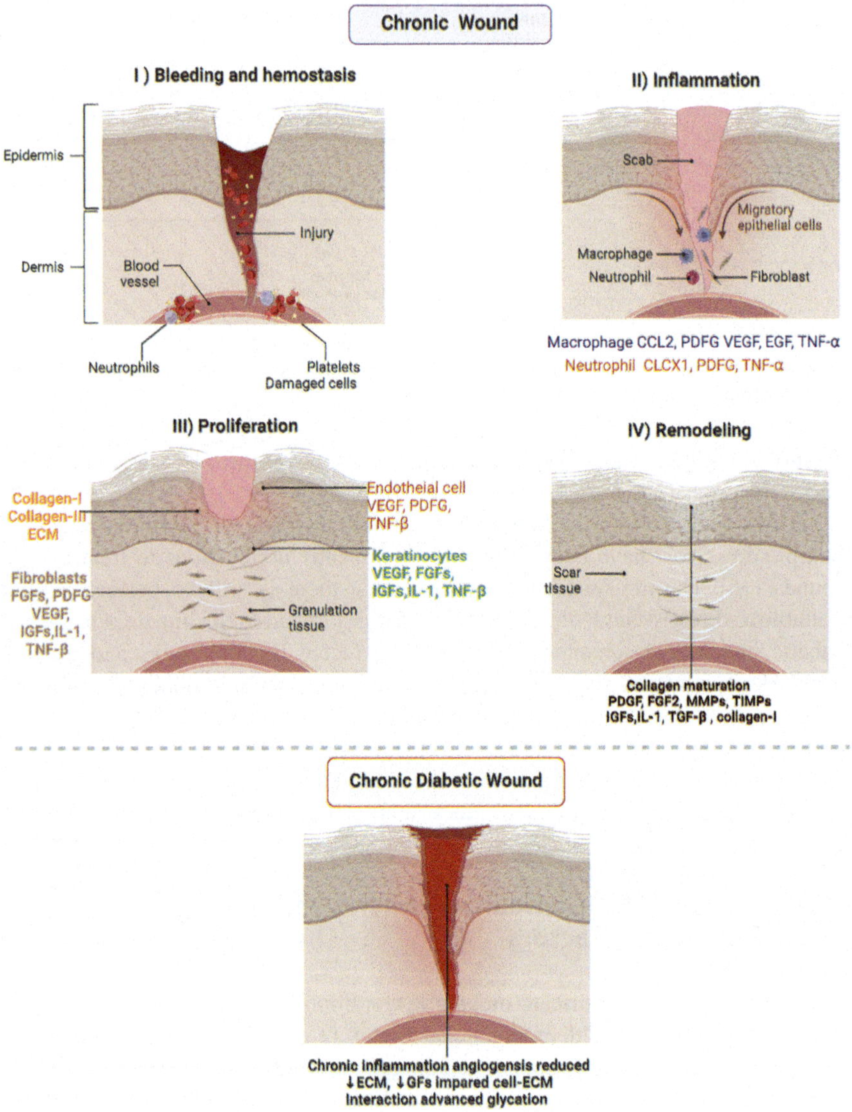

Fig. 10.2 Contrasting wound-healing mechanisms in normal and diabetes patients. (Adapted with modification using Biorender from [18] under Creative Commons Attribution (CC BY 4.0))

Additionally, they carry a significant risk of infection and amputations [21]. Diabetic wounds, known for being chronic and challenging to heal, are a prevalent complication among individuals with diabetes. Out of all diabetic patients, more than 20% experience the painful condition of diabetic foot ulcers. Astonishingly, about 28% of these individuals are forced to undergo the traumatic experience of having their lower limbs amputated simply because they are denied the necessary treatment [22].

There is a 15–25% likelihood that patients with diabetes will develop a chronic wound, causing discomfort and a constant need for wound care. Chronic wounds commonly seen in individuals with diabetes include ulcers that affect the foot and veins and those caused by pressure [23]. Classified as chronic wounds, diabetic foot ulcers (DFUs) create a frustratingly slow healing process, leaving a lasting disruption in the skin. Public health is heavily burdened by DFUs, which have a global prevalence of 6.3%. Diabetic ulceration often occurs due to peripheral neuropathy, a condition characterized by damage to the peripheral nerves in the limbs. Sensory loss makes it difficult to perceive the heightened pressure or pain resulting from minor injuries [24]. Over time, the absence of reaction to what might seem like insignificant events, along with the compromised blood flow and diminished ability to heal in individuals with diabetes, ultimately culminates in the formation of ulcers. The risk of lower extremity amputation is elevated for those affected by DFUs and other chronic wounds, as these wounds can become infected and lead to osteomyelitis and sepsis. Numerous studies have shown that ulcers are a common precursor to about 85% of amputations, and there is a significant risk of new ulcer development at a neighbouring wound site, with an incidence rate of up to 50% [25]. The formation of diabetic wounds is primarily attributed to secondary infection, necrosis, and compromised arterial blood flow due to peripheral vasculopathy or neuropathy, which are consequences of prolonged hyperglycaemia. Neuropathy and chronic wounds are expected consequences of poor blood circulation in specific areas like the feet and fingers [26]. Conventional clinical methods for treating diabetic wounds include managing blood sugar levels, using antibiotics to prevent infection, removing dead tissue through surgery, applying negative pressure to promote healing, using oxygen therapy, and dressing the wound with gauze, bandages, hydrocolloids, or films. However, these treatments have drawbacks, such as their inability to heal cellular damage on the outer edges of the wound and efficiently transport therapeutic agents to the affected tissue. Hence, the management of diabetic wounds proves to be a challenging task in clinical practice [27].

10.2.1 Factors Affecting the Healing Process of Diabetic Wounds

Chronic wounds indicate a disruption in the normal process of wound healing caused by numerous factors that prevent the wounds from progressing through the usual stages of closure. Chronic wounds, which are formed through a complex pathophysiological mechanism, are most associated with diabetes, making it the leading cause of this type of wound [28]. Diabetic wound healing is affected by various pathophysiological factors such as high blood sugar levels, nerve damage, complications in the small blood vessels (peripheral arterial disease, lack of oxygen, anaemia), skin barrier breakdown and infection, as illustrated in Fig. 10.3. Sustained hyperglycaemia in diabetic patients significantly impacts endothelial cells, altering their metabolism and function [29]. This disrupts the flow of blood in both small and large vessels, hindering the growth of new blood vessels. The function of keratinocytes and fibroblasts is compromised by the accumulation of advanced

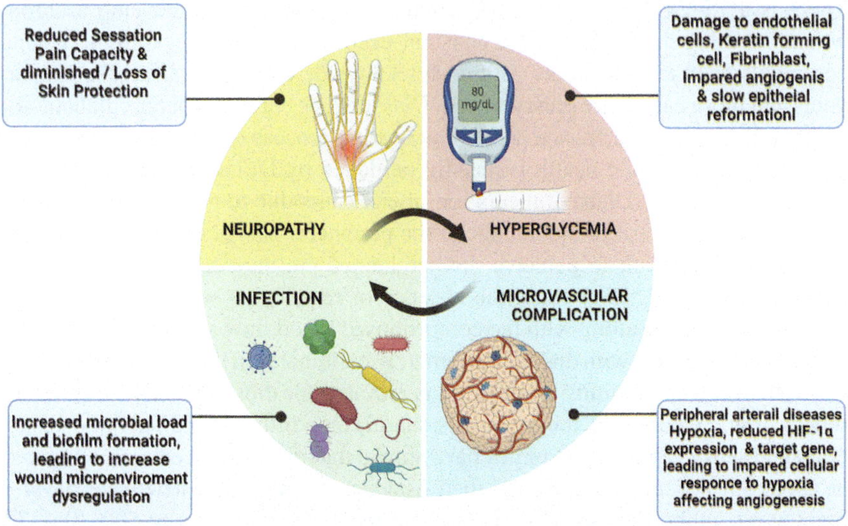

Fig. 10.3 Several factors, including hyperglycaemia, neuropathy, microvascular complications, and infection, influence diabetic wound healing. (Created by using Biorender.com)

glycation end products (AGEs) and endothelial cells. High blood sugar levels, known as hyperglycaemia, significantly contribute to oxidative stress, leading to detrimental consequences on the peripheral nerve's blood circulation, structure, and metabolism [30]. Diabetic patients with impaired nerve function may experience decreased sensation and pain perception from wounds. The most prevalent type of neuropathy is hand-foot neuropathy, which is also a leading cause of foot wound healing failure in diabetic individuals [31].

Additionally, diabetic wounds are prone to experiencing severe hypoxia following tissue injury. Hypoxia-inducible factor-1 α (HIF-1α) and its target genes are crucial in the cellular response to low blood oxygen levels. When the levels decrease, the expression of HIF-1α and its target genes decreases, leading to a diminished cellular response to hypoxia [32, 33]. The ability of blood vessels to develop is compromised by this impairment, causing a delay in the healing of wounds. Furthermore, the vulnerability of diabetic patients to infections adds to the already prolonged recovery time of diabetic foot ulcers.

10.3 Biomaterials as a Promising Therapeutic Platform in the Fabrication of Wound Dressings

As mentioned earlier, traditional methods of treating wounds involve using medication to manage pain, prevent or treat infections, and clean the wound. Creating a conducive environment for healing frequently involves using bandages and closure systems. The use of polymeric biomaterials can enhance conventional wound therapy, whether they are synthetic or natural. These specially designed polymeric materials retain moisture and warmth at the wound site while simultaneously sealing it from potential infections. The novel highlights smart biomaterials that stand apart from traditional wound dressings by promoting a moist environment, facilitating the exchange of oxygen and the removal of exudate [34].

Furthermore, they enable the controlled release of growth factors or medications, promoting tissue regeneration. Biocompatibility at the site of the application is a notable advantage of using biomaterials in wound care. It is worth noting that these materials are biodegradable as well. The biomaterial's ability to degrade after drug delivery means there is no need for surgical extraction. The construction of biomaterials aims to facilitate and encourage wound healing. In ulcer and chronic wound treatment, diverse dressing choices are accessible, encompassing smart biomaterials derived from natural and synthetic polymers that support re-epithelialization [25]. Biomaterials, including films, hydrogels, foams, natural polymers, nanoparticles, antimicrobials, and several types of dressings, have been utilized in various therapeutic uses for addressing persistent wounds, as shown in Fig. 10.4.

10.3.1 Polymeric Biomaterials

Biocompatibility and resemblance to the extracellular matrix (ECM) make polymers popular in pharmaceutical and biomedical applications. The generated substrate can recreate the biological environment, facilitating cell involvement in tissue growth, specialization, and healing. Furthermore, specific polymers possess the ability to biodegrade and be absorbed by the body, aiding in the regeneration of tissue without causing inflammation. Biocompatible materials, like chitosan, are essential for wound treatment to promote proper healing [35]. The properties of polymers are fascinating and varied. Their versatility can be modified through chemical processes, creating complex 3D structures that serve as scaffolds. Furthermore, the functionality of their surface can be personalized to suit specific demands. Polymers possess remarkable flexibility in their topology, dimensions, and chemistry, enabling them to act as an effective drug delivery system that enhances wound repair. Consequently, polymers are ideal for developing skin substitutes [36].

10.3.2 Natural Polymers

Wound healing often involves using natural polymers such as collagen, gelatine, chitosan, and hyaluronic acid. Chitosan is a remarkable substance for burns and wound healing due to its biocompatibility, ability to facilitate tissue repair, and lack of adverse effects. In addition, chitosan can transport large molecules such as proteins, antigens, and peptides [37]. Ahmad et al. investigated the potential of mupirocin-loaded chitosan-based hydrogel membranes in their study, which showed positive effects on wound healing, controlled release, and no skin irritation. Multiple

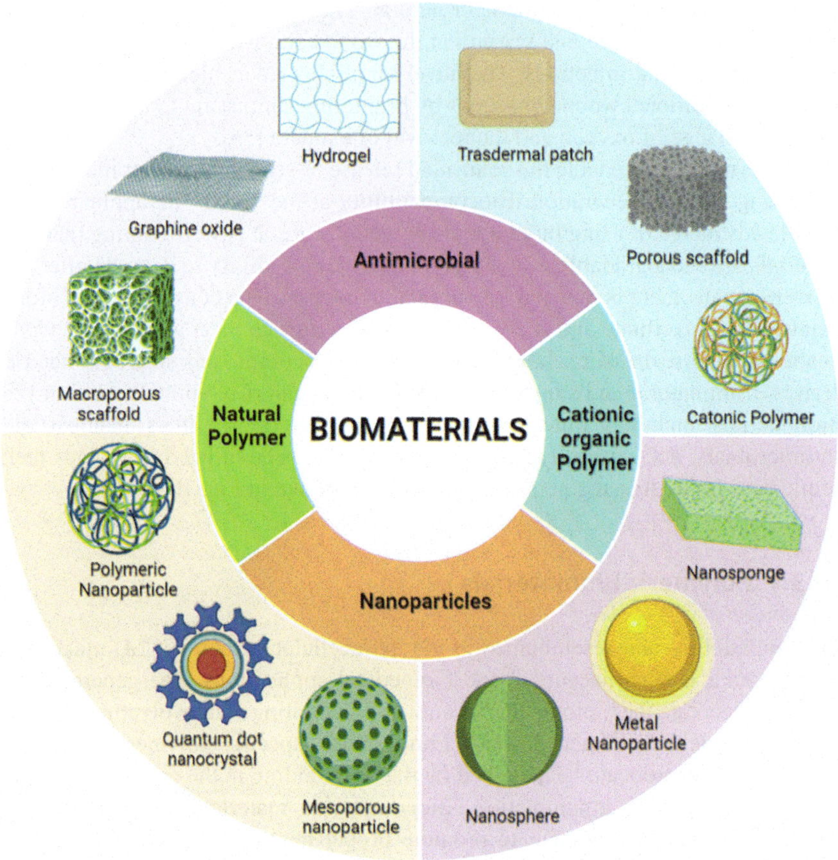

Fig. 10.4 Biomaterials employed in the treatment of diabetic wound healing. (Created by using Biorender.com)

applications are typically needed for traditional treatment with mupirocin ointment, but it is less favoured by patients who often complain about feeling soiled. Another study discovered that wounds treated with high molecular weight chitosan exhibited faster re-epithelialization and wound closure than those treated with fucidin ointment [38]. Nanofibrous scaffolds made of collagen and gelatine are employed in wound healing. In contrast, hyaluronic acid nanofibrous scaffolds imitate the crucial extracellular matrix to regulate cellular function in the regeneration of cartilaginous tissue [39]. A composite biomaterial for wound healing can be created by combining cellulose and chitosan nanoparticles with cellulose poly (methyl methacrylate) fibres. Not only does this exceptional material act as a bandage that prevents infections, but it also highlights antibacterial abilities [40]. The eco-friendly nature of natural polymers, obtained from microbial, vegetable, and animal biomass, aligns with their name. The remarkable suitability of these polymers for health-related applications stems from their inherent biodegradability, biocompatibility, and biological activity. Furthermore, natural-derived polymers present a feasible substitute for the structural elements of the natural ECM and emulate the cellular environment of the skin [41]. The biopolymers are valuable in creating versatile materials as they meet the requirements for tissue engineering applications. The breakdown of natural polymers through enzymatic degradation yields by-products that exhibit low toxicity and are generally well-received by living organisms. Nevertheless, controlling natural polymers' degradation rates and processes can be pretty challenging despite their numerous advantages [42]. The utilization of natural polymers in wound healing encompasses a range of applications, including enhancing regeneration, drug delivery, and promoting tissue regeneration via forming 3D scaffolds. In biomedical applications, selecting polymeric materials requires careful consideration of multiple factors. These include the structure, solubility, pore size, degradation rate, water absorbency, and electrical charge [43].

10.3.3 Synthetic Polymers

Synthetic biomaterials, specifically designed and engineered for biomedical and health applications, are artificial materials with diverse uses. Synthetic polymers in wound healing encompass various applications, including suture material, aneurysm coils, drug delivery systems, growth factor delivery systems, and scaffolds. The utilization of these applications significantly accelerates the wound-healing process [44]. The primary intention behind creating polymer films was to use them as surgical incision drapes. PEG polymers are employed as carriers to promote the precise delivery of growth factors, specifically EGF, to the site of the wound. In addition, the utilization of electro spun scaffolds derived from polycaprolactone, a polymer known for its biocompatibility and bio-resorb ability, creates an environment that closely resembles the ECM, thus making it a viable option for the treatment of both acute and chronic wounds [45]. The integration of polyvinyl alcohol and eudragit polymers can contribute to the advancement of tissue engineering. Research has demonstrated that incorporating polymers in dressings or medical devices holds promise for enhancing wound healing control. Nevertheless, they are currently called occlusive dressings in wound healing management. The field of wound dressing has seen notable progress in recent years, with synthetic polymers emerging as the top choice for managing wounds [46]. Synthetic polymer dressings are categorized into passive and interactive types. Passive synthetic polymer dressings serve the purpose of covering wounds and supporting the recovery process through the application of a polymer film. Passive synthetic polymers include gauze and tulle. Interactive synthetic polymer dressings can be occlusive or semi-occlusive to prevent bacteria from entering wounds. Viable options include films, foams, hydrogels, or hydrocolloids. Polyurethane and other synthetic polymers are commonly used to make semi-permeable dressings. These materials offer a dual function of allowing the passage of moisture and vapour while also serving as a protective barrier against bacteria [47]. Poly(lactide-co-glycolide) polymers have been observed in creating fibrous scaffolds for bone tissue regeneration. Moreover, it is possible to formulate these polymers into injectable scaffolds at the site. Synthetic ones like aliphatic polyesters are among the biomaterials utilized in tissue engineering due to their advantageous properties. Their high mechanical strength, non-toxic nature, and ease of processing make them desirable materials, and they also possess flexibility [48]. Polymers such as polylactic acid (PLA) and polyglycolic acid (PGA) fall under the aliphatic polyesters, composed of repeating units derived from nonaromatic hydrocarbon synthetic biomaterials. Utilizing these biomaterials has yielded positive outcomes in wound healing and skin regeneration by using them to deliver growth factors [49].

10.4 Biomaterials in Managing Diabetic Wounds: Successes and Challenges

In the realm of diabetic wound management, the utilization of biomaterials stands as a beacon of hope amidst the complexities and challenges posed by chronic ulcers associated with diabetes mellitus. These wounds, characterized by an imbalanced wound microenvironment and persistent inflammation at the molecular and cellular level, present a formidable obstacle to patients' well-being and healthcare systems worldwide. Amidst this backdrop, biomaterials emerge as a promising therapeutic platform, offering diverse solutions to address the multifaceted nature of diabetic wound healing. In this discussion, let's explore the successes and challenges of applying biomaterials to manage diabetic wounds [50]. By navigating through the landscape of achievements and obstacles, we aim to provide a comprehensive understanding of the current situation in this critical area of healthcare. From accelerating wound healing rates to mitigating infection risks and creating conducive wound microenvironments, biomaterials offer a spectrum of benefits for diabetic wound care. However, alongside these successes lie significant challenges that must be addressed to fully harness the potential of biomaterial-based interventions. Issues such as the lack of standardized protocols, immune responses, and regulatory hurdles present formidable barriers to widespread adoption and efficacy [51]. Nevertheless, amidst these challenges lie opportunities for innovation and advancement, paving the way for a future where biomaterial technologies revolutionize diabetic wound management. Through a balanced exploration of successes and challenges, this discussion aims to equip clinicians, researchers, and stakeholders with the insights to navigate the complexities of utilizing biomaterials for diabetic wound management. By fostering a deeper understanding of the triumphs and tribulations, we strive to catalyse collaborative efforts and drive progress toward more effective treatments for diabetic wounds [52].

The primary successes observed with the use of biomaterials in diabetic wound management include enhanced wound closure, improved tissue regeneration, reduced inflammation, oxidative stress, and infection rates, as well as promotion of angiogenesis [53]. Biomaterials have been proven to possess distinct characteristics, including biocompatibility, biodegradability, and the capacity to serve as a localized reservoir for bioactive molecules. These properties are crucial in promoting effective wound healing by offering anti-inflammatory and antimicrobial benefits. Moreover, intelligent biomaterial scaffolds can react to particular stimuli in the wound microenvironment, effectively speeding up the healing of diabetic wounds and enhancing the effectiveness of treatment. Studies have also highlighted the comparative efficacy of different biomaterials, with placenta-based tissue products showing the highest wound healing probability in diabetic foot ulcers [54]. Diabetic patients have experienced improved wound healing rates by applying diverse biomaterials, yielding promising results. One remarkable case is a composite hydrogel from bovine serum albumin (BSA) and *Aloe vera* (AV) gel. This unique combination has proven to be highly effective in promoting wound healing by increasing collagen deposition and elevating the expression of vascular endothelial growth factor (VEGF) [55].

Additionally, panthenol citrate (PC) has been developed as a compound with antioxidant, antibacterial, and proangiogenic properties, improving re-epithelialization and tissue quality in diabetic wounds [53]. Studies have indicated that biomaterials like placenta-based tissue products, living cell skin substitutes, and acellular skin substitutes offer higher probabilities of wound healing in diabetic foot ulcers, surpassing those of standard care. Nanofibrous scaffolds have gained attention for their ability to incorporate bioactive agents, providing a biomimetic environment for cell proliferation and enhancing diabetic wound healing [56].

The current challenges in developing biomaterials for diabetic wound treatment include the complexity of wound healing in diabetes, the need for effective wound dressings that can absorb exudates and prevent infections, and the demand for materials that support the synthesis of biochemical mediators crucial for healing [57]. Researchers are addressing these challenges by incorporating bioactive compounds into polymeric wound dressings, utilizing innovative biomaterials that respond to wound microenvironment stimuli, and exploring nanofibrous scaffolds for their flexibility and ability to load bioactive agents. These approaches seek to accelerate the healing of diabetic wounds, enhance treatment results, and establish a more productive and streamlined approach to managing diabetic foot ulcers [58]. Foreign body reactions can complicate the use of biomaterials in diabetic wound management. These reactions involve immune responses and encapsulation of biomaterials, often leading to inflammation and tissue damage. For instance, non-metallic foreign bodies like wood or glass left in wounds can trigger inflammatory processes and hinder healing.

Moreover, the extent and pattern of foreign body reactions influence biomaterial functionality, with polysaccharides showing varying immune responses based on their glycan structure. The involvement of macrophage subsets, including foreign body giant cells, in the early phases of the reaction is crucial, with different macrophage types linked to either fibrotic or regenerative outcomes. Additionally, adaptive immune cells may play a role in the foreign body's reaction to synthetic polymers, highlighting the complexity of immune responses to biomaterials in diabetic wound care [59].

10.5 Smart Biomaterials in Diabetic Wounds

Smart biomaterials play a crucial role in diabetic wound care by addressing the complexities of wound healing in diabetic patients. These biomaterials are designed to expedite wound closure, improve treatment outcomes, and manage the challenging healing process of diabetic foot ulcers. By responding to stimuli like pH, reactive oxygen species, and glucose concentration, these delivery systems can effectively transport drugs with anti-inflammatory, bioactive, and antibacterial properties in the wound microenvironment. Hydrogels and other innovative biomaterials are ideal for chronic wound treatment due to their versatility, biocompatibility, and ability to adjust therapeutic release according to external and internal changes. In terms of diabetic wound management, these advanced biomaterials offer a promising solution that can significantly enhance patient outcomes [60]. The purpose of discussing smart biomaterials and their applications is to highlight their innovative features and potential to revolutionize diabetic wound management. By exploring the dynamic capabilities of smart biomaterials, such as real-time monitoring and targeted interventions, this discussion aims to display their role in improving treatment outcomes for diabetic wounds. Additionally, by examining the challenges and opportunities associated with smart biomaterials, we can pave the way for further advancements in this critical area of healthcare.

10.6 Emerging Biomaterial Technologies to Treat Diabetic Wounds

Diabetic wounds are a major challenge in healthcare due to slow healing and complications. Recent advancements in biomaterial technologies offer improved wound-healing outcomes for diabetic wounds. Some emerging biomaterial technologies for diabetic wound treatment include hydrogels, nanotechnology-based therapies, scaffolds, bioactive dressings, biodegradable polymers, gene therapy vectors, collagen-based dressings, ceramic-based biomaterials, stem cell therapy, and phototherapy [61, 62].

Developing smart dressings for monitoring wounds and providing tailored treatment is a critical advancement in biomaterial technologies. Traditional wound care methods have limitations such as antibiotic resistance, inadequate moisture balance, and infection protection. Diabetic wounds, prevalent in approximately 15% of diabetic patients, pose significant health and economic burdens globally [63, 64].

The use of stem cell delivery and skin substitutes in research on diabetic wound healing is gaining attention. Stem cells can naturally regenerate through paracrine effects. Studies have shown that delivering adipose-derived stem cells (ASCs) using a poly (ethylene glycol)-gelatine-based injectable hydrogel system decreased inflammatory cells, enhanced blood vessel development, and accelerated wound healing in a mouse model [65]. However, the clinical application of stem cell therapies has been limited due to ethical concerns, reduced potency when cells are expanded outside the body and low survival rates after implantation into wounds

[66]. In a study by Ding et al., Deferoxamine (DFO) was used to treat diabetic foot ulcers. The study used silk nanofiber hydrogels containing DFO, which significantly improved healing diabetic wounds. The DFO-loaded hydrogels enhanced healing, reduced chronic inflammation, and increased collagen deposition in diabetic wounds. This approach holds potential for clinical applications in treating diabetic ulcers [67].

10.7 Conductive Biomaterials for Diabetic Wound Healing

10.7.1 Hydrogels

Hydrogels represent a new category of functional polymer materials garnering significant attention within the biomedical realm, owing to their distinctive characteristics such as high-water content, rapid swelling kinetics, and favourable biocompatibility. These hydrogels are classified as natural and synthetic types, with natural hydrogels further subdivided based on their source material, including chitosan, hyaluronic acid, alginate, gelatine, and cellulose [68]. The main feature of hydrogels is their three-dimensional network structure, formed via either physical or chemical crosslinking methods. Physical crosslinking is primarily based on intermolecular forces, such as electrostatic/ionic, hydrophobic, and hydrogen bonding interactions. Chemical crosslinking involves the formation of covalent bonds through specific chemical reactions. Chemically crosslinked hydrogels offer superior mechanical stability due to the strength of covalent interactions compared to noncovalent ones. Certain hydrogels, particularly those derived from chitosan and alginate, can degrade over time, minimizing the risk of secondary damage during dressing changes. Despite these advantages, in their swollen state, hydrogels often suffer from inadequate mechanical stability, limiting their efficacy in wound management. Some hydrogel formulations necessitate the application of a secondary dressing to prevent dehydration, adding complexity to their clinical use [69].

10.7.2 Nanoparticles

Nanoparticles hold considerable promise in diabetes wound healing due to their unique properties, small size, large surface area-to-volume ratio, and tunable surface chemistry. These characteristics enable nanoparticles to interact effectively with biological systems and deliver therapeutic agents precisely to target sites, as depicted in Table 10.1.

The wound healing effects of a unique combination therapy involving selenium nanoparticles and platelet-rich plasma (PRP) in diabetic mice. Selenium nanoparticles (SeNPs) with potent antioxidant and antimicrobial properties were paired with PRP, which is rich in growth factors and shows a synergistic effect, significantly increasing healing in diabetic mice [78]. Silver nanoparticles (AgNPs) have also enhanced wound healing in diabetic patients by penetrating bacterial cell

Table 10.1 Different nanoparticles used in diabetic wound healing

Type of nanoparticles	Biological activity	References
Selenium	Wound healing	[70]
Silver	Wound healing	[71]
Gold	Antimicrobial and proangiogenic	[72]
Zinc oxide	Diabetic foot ulcers healing	[73]
Titanium dioxide	Healing efficacy in diabetic wounds	[74]
Copper oxide	Enhanced wound angiogenesis	[75]
Iron oxide	Wound healing	[76]
Graphene oxide	Diabetic foot ulcer healing	[77]

membranes, disrupting their structural integrity, and inducing leakage of cellular contents, causing death of the bacterial cell. Silver nanoparticles generate reactive oxygen species [79] within bacterial cells, causing oxidative damage to proteins, lipids, and nucleic acids [80]. Gold nanoparticles exhibit anti-inflammatory properties and promote angiogenesis, the formation of new blood vessels, which is crucial for supplying oxygen and nutrients to the wound site. Gold nanoparticles can modulate immune responses, reducing excessive inflammation and supporting tissue repair processes. Zinc oxide nanoparticles have antimicrobial, anti-inflammatory, and wound-healing properties, and they have been used in wound dressings and topical formulations to accelerate the healing process, lower inflammation, and prevent infections in diabetic wounds [73]. Titanium dioxide nanoparticles also exhibit antimicrobial activity and have been investigated for their potential use in wound dressings and coatings to inhibit bacterial growth and promote wound healing in diabetic ulcers [74]. Calcium phosphate nanoparticles have osteoconductive properties, which support the formation of new bone tissue. In diabetic patients with foot ulcers that extend to bone, calcium phosphate nanoparticles could be used to regenerate bone and facilitate wound closure [81]. Copper oxide nanoparticles possess antimicrobial properties and have been explored for their potential against bacterial infections in diabetic wounds. They exhibit broad-spectrum antimicrobial activity and can help prevent pathogenic bacteria from colonizing the wound site, promoting wound healing [82]. Graphene nanoparticles and graphene-based materials have been used in wound healing due to their extraordinary mechanical strength, conductivity, and biocompatibility. They can be incorporated into wound dressings or scaffolds to enhance mechanical properties, provide electrical stimulation, and promote cellular adhesion and proliferation, accelerating wound closure in diabetic ulcers [76]. Carbon nanotubes (CNTs) have garnered significant interest in biomedical applications due to their exceptional physicochemical properties at the nanoscale. CNTs-based composite hydrogels exhibit superior antibacterial efficacy compared to pure polymer hydrogels, leading to accelerated wound healing [83]. Liposomes are spherical vesicles composed of lipid bilayers encapsulating hydrophilic and hydrophobic drugs. They offer biocompatibility and versatility in drug delivery and have been explored for their potential to deliver therapeutic agents to diabetic wounds. Liposomes can improve drug solubility, stability, and bioavailability, enhancing their efficacy in promoting wound healing [84].

10.7.3 Scaffolds

Diabetic ulcers require specialized care to prevent complications like sepsis, amputation, and deformities. Conventional wound dressings like hydrogels, gauze, films, and foams are commonly used, but researchers are increasingly turning to nanofibrous scaffolds. Scaffolds are three-dimensional (3D) structures designed to mimic natural tissues' extracellular matrix (ECM). These scaffolds provide a framework or support structure for cells to adhere, proliferate, and differentiate, ultimately forming new tissue. These scaffolds have good flexibility, carry various bioactive compounds, and possess a high surface area-to-volume ratio, mimicking natural tissue and promoting cell growth. Polymers like chitosan, polyvinyl alcohol, polycaprolactone, tragacanth gum, gelatine, cellulose, polylactide, and silk fibroin fabricate nanofibrous scaffold [85].

10.7.4 Bioactive Dressings

These dressings contain bioactive compounds or growth factors that stimulate wound healing processes, lower inflammation, and improve tissue regeneration in diabetic ulcers. Cellulose/collagen dressings have excellent biological properties and are used in non-healing wounds. The use of cellulose/collagen dressings shows a substantial improvement in wound closure rates compared to conventional dressings. Their efficacy in promoting angiogenesis and expediting re-epithelialization has been effectively demonstrated in clinical trials, indicating a strong potential for widespread adoption in diabetes wound treatment protocols [86].

10.7.5 Biodegradable Materials

Materials gradually degrade over time, release therapeutic agents, or promote tissue integration while minimizing the risk of foreign body reactions or complications. Repairing diabetic wounds has long posed a challenge in plastic surgery. So, employing biocompatible, biodegradable biomaterials as dressings loaded with active substances or cells has become crucial. These materials aim to integrate into a comprehensive wound repair system, combining materials, cells, and growth factors. Composite biodegradable biomaterials have shown strong biocompatibility and wound-healing promotion. Stem cells and their exosomes regulate immune responses, inflammation, angiogenesis, skin cell proliferation, re-epithelialization, and collagen remodelling to inhibit scar hyperplasia [87].

10.7.6 Collagen-Based Dressings

Dressings derived from animal or synthetic sources; collagen dressings provide a scaffold for cell attachment and proliferation, promoting wound closure and tissue regeneration in diabetic ulcers. The delayed healing of diabetic wounds is believed to stem from several critical factors, including a persistent inflammatory response, disruptions in extracellular matrix degradation, and heightened bacterial presence. Collagen and its constituents, like fibroblasts and keratinocytes, play an essential role in wound healing and skin regeneration. Dressings incorporating collagen materials establish a biological scaffold matrix, which regulates extracellular components and facilitates wound healing.

10.7.7 Ceramic-Based Biomaterials

Ceramics such as calcium phosphate or bioactive glasses can fill bone defects in diabetic foot ulcers, promoting bone regeneration and stability. Fibrin ceramic nanofibers are emerging as a promising frontier in biomedical research due to their unique characteristics, including distinctive morphology, nanoscale roughness, intricate nano topography, favourable wettability, inherent bioactivity, and potential for chemical functionalization [86].

10.7.8 Conductive Polymers

Conductive polymers represent a class of materials with unique electrical properties. These polymers can conduct electricity while retaining the advantageous properties of traditional polymers, like biocompatibility, flexibility, and ease of processing. In diabetes, impaired wound healing processes pose significant clinical challenges; conductive polymers offer promising solutions to enhance tissue regeneration and expedite wound closure. The following conductive polymers are used for the diabetic wound healing process [88].

Poly Pyrrole Poly pyrrole is one of the most extensively researched conductive polymers due to its high conductivity, biocompatibility, and ease of synthesis. It can be electrochemically deposited on substrates to form conductive coatings or incorporated into biomaterials such as hydrogels, scaffolds, and dressings. Poly pyrrole-based materials have been shown to promote cell adhesion, proliferation, and differentiation, making them promising biomaterials for enhancing diabetic wound healing [89].

Polyaniline Polyaniline is a versatile conducting polymer with a wide range of characteristics that make it a choice for diabetic wound healing applications [90]. Its electrical conductivity, biocompatibility, antioxidant properties, and ability to deliver bioactive molecules make polyaniline-based materials promising candidates for enhancing wound healing outcomes in diabetic patients [91].

Polythiophene Polythiophene is a conductive polymer with tuneable electrical conductivity and optical properties. Its characteristics can be customized by introducing different substituents and side chains during the polymerization process and enhancing its processability. A polythiophene scaffold or dressing provides electrical conductivity and maintains a wound gap, promoting tissue regeneration to treat diabetic ulcers [92].

10.8 Enzyme-Based Biomaterials Toward Diabetic Wound Healing

Enzyme-containing biomaterials play a crucial role in treating diabetic wounds by effectively addressing the challenges associated with delayed wound healing and bacterial infections in individuals with diabetes. These biomaterials, including enzyme-loaded microneedles, enzyme cascade inverse opal microparticles, and panthenol citrate (PC) hydrogel dressings, possess properties such as antibacterial, antioxidant, anti-inflammatory, and proangiogenic effects [93]. They improve tissue penetration, support the removal of biofilms, prevent reinfection, and enhance wound healing outcomes. Moreover, enzyme-based biomaterials can offer real-time monitoring of wound pH and healing progress, providing a comprehensive approach to managing diabetic wounds. In conclusion, these innovative biomaterials are promising to enhance the quality of healed skin, foster tissue regeneration, and improve wound care for diabetic patients [94].

Enzymes such as glucose oxidase (GOx), peroxidase (POD), glutathione peroxidase (GPx), and catalase (CAT) have been demonstrated to promote diabetic wound healing. These enzymes play crucial roles in addressing the complex microenvironment of diabetic wounds by enhancing oxygen supply, reducing reactive oxygen species [79] levels, facilitating cellular proliferation, and stimulating angiogenesis [95]. Specifically, the multifunctional glucose oxidase (GOx) and catalase (CAT) nano enzyme-chitosan (GCNC) hydrogel complex has been designed to enhance the wound microenvironment and ensure the continuous delivery of oxygen, thus optimizing the healing process. Additionally, the glucose/ROS cascade-responsive enzyme (CHA@GOx) has shown promising results in regulating the oxygen balance of wounds and promoting diabetic wound repair synergistically. Enzyme-based approaches provide innovative techniques to optimize the healing outcomes of diabetic wounds [96, 97].

10.8.1 Mechanisms of Action

In individuals with diabetes, Matrix Metalloproteinases (MMPs) break down the extracellular matrix components during the wound-healing process. During the inflammatory phase, MMPs break down the extracellular matrix, while in the proliferation phase, they facilitate angiogenesis and cell migration [98]. In the remodelling phase, MMPs assist in tissue remodelling. In diabetic foot ulcers (DFUs), overexpression of MMPs can lead to excessive tissue degradation and impaired wound healing, emphasizing the importance of regulating MMP levels to promote healing in DFUs. Imbalance in MMPs, seen in diabetic patients, can exacerbate chronic wounds, leading to complications like chronic foot ulcers and amputations [99]. Strategies targeting MMPs, such as natural or synthetic MMP inhibitors, are being explored to enhance wound healing in diabetic patients [100]. Proteases and serine proteases are critical players in extracellular matrix remodelling and growth factor activation, impacting processes like cell motility, invasion, and wound healing. Disturbances in proteolytic activity can hinder wound healing, especially in chronic wounds, emphasizing the importance of regulating proteases in the wound microenvironment. Enzymes like enzymes have been developed to regulate the complex microenvironment of diabetic wounds, effectively removing excess reactive oxygen species, [79] and promoting wound healing. Understanding and controlling the changing rules and characteristics of the wound microenvironment is essential for purposefully regulating the wound healing process and achieving optimal tissue repair [101].

10.8.2 Types of Enzyme-Based Biomaterials

The application of enzyme-loaded scaffolds has shown immense potential in diabetic wound healing. Utilizing these scaffolds can effectively address the numerous factors that hinder wound recovery in individuals with diabetes [102]. Incorporating enzyme-mimicking enzymes, including catalase-mimic enzymes, into hydrogels offers a potential solution for mitigating the adverse microenvironment in diabetic wounds. Through the integration of enzymes, these hydrogels manifest an impressive ability to capture heightened reactive oxygen species [79], producing oxygen and shielding skin cells from ROS-related damage while stimulating cell proliferation. Additionally, including scaffolds containing enzymes can facilitate a transition in macrophage polarization, assisting in the shift from a pro-inflammatory to an anti-inflammatory subtype, thereby accelerating the wound healing process. Using enzyme-loaded scaffolds offers a distinctive strategy for enhancing diabetic wound healing. This approach focuses on addressing inflammation, promoting cell growth, restoring epithelial layers, facilitating collagen formation, and promoting the development of new blood vessels [103]. Critical characteristics of enzyme-loaded scaffolds include precise control over structural parameters like pore size and geometry to prevent mass-transport limitations. Enzyme-loaded scaffolds exhibit lower activity than free enzymes due to diffusion constraints within the scaffold pores, but they offer long-term stability and increased thermal tolerance [104]. In addition, the self-assembly of small molecules in enzyme-manipulated hydrogenation provides region- and enantioselectivity and allows for gentle reaction conditions and the encapsulation of delicate bio-medications or cells [105]. Additionally, using genetically encodable scaffolds in synthetic biology enhances the properties of enzymes through the optimization of reaction conditions, improvement of metabolic flux, and enhancement of enzyme stability. The attributes mentioned above contribute to the high potential of enzyme-loaded scaffolds in biomedicine [106].

Several limitations and hurdles impede the progress and marketability of enzyme-based biomaterials for diabetic wound healing. Challenges encompass the emergence of drug-resistant bacteria caused by extended antibiotic usage, consequently reducing efficacy. Additionally, traditional management systems for diabetes are criticized for prohibitive costs and limited disposal options, hindering widespread adoption of new treatments. In addition, the effectiveness of biomaterials in wound healing fluctuates depending on the severity of the wound, and specific biomaterials encounter obstacles such as limited availability, costly transportation, and processing difficulties, thus rendering them unsuitable for broad medical utilization. The application of smart biomaterials holds immense potential in accelerating the closure of diabetic wounds. Nonetheless, successfully commercializing these biomaterials requires addressing obstacles related to the complex nature of wound healing processes and developing responsive systems that can adapt to the wound microenvironment [107].

10.9 3D Bio-printed Scaffolds for Diabetic Wound-Healing Applications

Since its first demonstration in the 1980s, 3D printing and bioprinting has experienced rapid growth. The credit for this can be given to the advancements in technology and the extensive knowledge we now have about how to utilize these technologies, particularly in tissue engineering, where scaffolds are vital. To create 3D (bio) printed scaffolds, the first step is for the user to utilize computer-aided design (CAD) software to design a scaffold, ensuring its precise specifications. Subsequently, dedicated software can segment the design before it is printed, giving the printer a detailed set of instructions. 3DP offers many benefits, such as the ability to create intricate (scaffold) designs quickly and affordably in assorted sizes. 3D-bio-printed scaffolds have demonstrated impressive cell loading capabilities and have successfully preserved cell viability, all while offering crucial structural support to the wound. The degree to which these benefits are realized depends on the specific bioprinting technology chosen. 3D bio-printing technology utilizes computer-controlled systems to deposit biomaterials, including cells and bio inks, layer by layer, as illustrated in Fig. 10.5.

This innovative approach enables the precise placement of cells, growth factors, and other bioactive substances, facilitating tissue regeneration and the creation of complex structures like organs and tissues. Various polymers and cells are employed

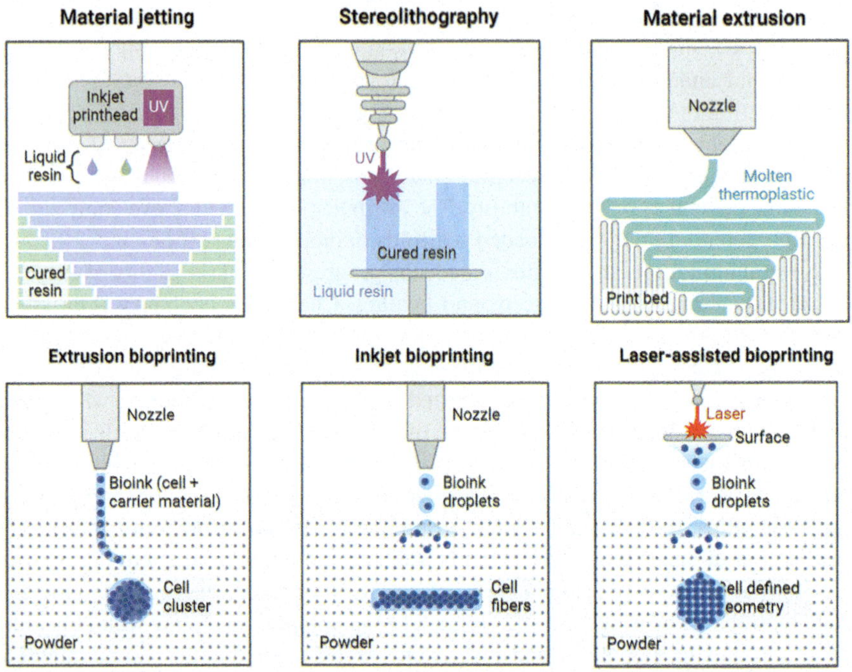

Fig. 10.5 3D bio-printing technology utilizes in Diabetic wound healing. (Created by using Biorender.com)

in bio-printing, offering advantages such as controlled cell deposition and enhanced tissue regeneration. Challenges in the field include finding suitable bio-inks, optimizing printing parameters, and detecting printing defects, which can be addressed by integrating machine learning algorithms with 3D bio-printing technology. Additionally, advancements like digital light processing (DLP) and novel techniques such as photopolymerization of orderly extruded multi-materials (POEM) are enhancing the capabilities of 3D bio-printing by enabling high-resolution, multi-material and cell-laden structure printing [108]. By using bioprinting, the limitations of traditional scaffold fabrication techniques, like using volatile solvents in electrospinning, can be overcome to produce scaffolds for diabetic wound healing. Bioprinters often come equipped with multiple print heads, allowing the simultaneous extrusion of varied materials during printing. The unique capabilities of bio-printing make it an ideal method for utilizing cells and other materials to facilitate wound healing.

Fused deposition modelling (FDM) printers, which typically operate at elevated temperatures, cannot support the printing of biologics. In addition, bioprinters offer a sterile manufacturing process for scaffolds, thanks to their enclosed environment and built-in UV sterilization capabilities. For successful DFU healing to occur, the scaffold should replicate the skin's physiological properties. The scaffold should provide structural support and durability to the wound while also displaying enough flexibility to seamlessly integrate with the skin's elastic properties during typical body movements [109]. The significance of this technology lies in its contribution to diabetic wound healing through the provision of innovative solutions for personalized treatment.

Researchers have developed functional wound dressings that facilitate the management of diabetic foot ulcers through the fusion of bioactive compounds and advanced manufacturing techniques. These bio-printed constructs can provide a controlled release of bioactive compounds, promote tissue repair, manage infection and inflammation, and accelerate wound healing. Additionally, incorporating DNA-induced biomineralization strategies in 3D printing inks has shown promising results in chronic wound repair, offering a functional platform for clinical applications. Using bioinks containing DNA and bio silica in 3D-printed hydrogels enhances biological activity, such as reactive oxygen species scavenging, angiogenesis, and anti-inflammation, thereby facilitating acute and diabetic wound healing [110, 111].

Fratini et al. employed two novel techniques to fabricate wound dressings, setting a new standard in manufacturing. The distinguishing factor of this research lies in its innovative approach of incorporating drug-loaded liposomes into a hydrogel, which is subsequently 3D printed into a scaffold for wound dressing. Through coaxial 3D bioprinting, researchers successfully fabricated a grid-shaped scaffold that harmoniously blends two separate bioinks in a single filament. The filament's core, produced through microbore extrusion, comprises a nanocomposite hydrogel of hydroxyethyl cellulose (HEC) and PEGylated liposomes housing thyme oil (TO). Conversely, the filament's external layer is comprised of a hybrid hydrogel consisting of a combination of sodium alginate and cellulose nanocrystals (SA/CNC), which is subsequently infused with thyme oil. As a result, a unique combination of release profiles for the active pharmaceutical ingredient (API) is achieved. The presence of thyme oil in the filament shell facilitates rapid release within the first 24 h, whereas the liposomes ensure sustained release over 10 days. Confocal Microscopy was employed to confirm the presence of liposomes within the scaffold. In addition, the assessment of the manufactured scaffolds using the zone of inhibition test confirmed their antibacterial activity against both Gram-positive and negative bacteria [112].

Ding et al. developed a hydrogel scaffold that expedites the gelling process of MoS2 through in situ three-dimensional (3D) bioprinting, intending to improve the healing of chronic diabetic wounds. A simple chemical conjugation process was used to prepare DEX-BA and DEX-CA, and dextran derivatives were modified with benzaldehyde and cyanoacetate groups. The DEX hydrogel was synthesized through the amalgamation of solutions of DEX-BA and DEX-CA under ambient conditions. The introduction of an alkaline MoS2 nanosheet, which possesses relatively weak properties, facilitated the development of a Decellularized Extracellular Matrix (DEM) hydrogel featuring a porous structure akin to the DEX hydrogel. The in-situ printing of the DEM hydrogel displayed shear-thinning behaviour, enabling seamless flow through needles, and displayed post-extrusion self-repairing ability. As the shear rate increased, a decrease in viscosity was observed. The hydrogel exhibited outstanding injectability, forming continuous fibres when introduced into a solution.

Additional notable factors are yield stress and breakage strain, which displayed commendable self-supporting capability. The hydrogel met the conditions for in situ 3D printing and displayed adhesive properties on wet surfaces of biological tissue. The wound healing potential of MoS2 hydrogel scaffolds printed in-situ was evaluated in an infected diabetic mouse model. Streptozotocin (STZ) was administered via intraperitoneal injection to induce diabetes in the mice. Different treatment methods were used to address chronic wounds infected with S. aureus, namely G1 and PBS control. In the case of the G3 and G4 groups, the DEM hydrogel scaffolds were directly printed into the wounds. The findings demonstrated that the in-situ printed DEM-3NIR hydrogel scaffolds exhibited the highest efficacy in promoting the healing of infected chronic diabetic wounds. The application of Masson's trichrome staining allowed for the identification of collagen deposition at the wound sites. Following a 15-day treatment period, it was observed that both the DEM-3 and DEM-3-NIR groups displayed a higher level of collagen deposition compared the control group treated with PBS. This finding provides evidence of the considerable enhancement of tissue remodelling and wound healing through collagen deposition facilitated by the in-situ printed DEM-3-NIR hydrogel scaffold. In wound healing and tissue regeneration, particularly in the case of infected chronic wounds, the inflammatory response assumes a critical role. The excessive buildup of reactive oxygen species (ROS) in the wounds can worsen the inflammatory response and increase the susceptibility of the wounds [113].

Through the implementation of 3D bioprinting and double crosslinking with UV light, Liao et al. successfully developed patches for treating diabetic wounds. The promotion of angiogenesis was facilitated by crosslinking the patches with vascular endothelial growth factor (VEGF). A precise match to different clinical trauma shapes and specific needs was achieved through 3D printing technology in creating customized patch structures. The patches were created using biomaterials such as alginate and chondroitin sulfate methacryloyl, which are biological polysaccharides. The process of achieving cross-linking involved the utilization of calcium ions and photo-crosslinking. Through the implementation of this approach, notable improvements were achieved in the mechanical properties and bioactivity of the patches. The process of chemically coupling growth factors and extending the release time of VEGF was simplified by easily and rapidly photo-crosslinking acrylate-modified VEGF (mVEGF) using UV irradiation. The functionality of the 3D-printed patches is based on harnessing their inherent activity and the presence of loaded VEGF to induce a healing response. Incorporating these features renders the 3D-bio printed double-crosslinked angiogenic patches highly favourable for treating diabetic wounds and various tissue engineering endeavours [114].

Hu et al. carried out an investigation in which they effectively produced Mesoporous bioglass (MBG) with distinctive attributes. Upon the establishment of MBG, scientists formulated four types of slurries by employing decellularized small intestinal submucosa (SIS). Among the components were pure SIS and multiple concentrations of MBG. After preparing the slurry, they utilized an advanced extrusion cryogenic 3D printing system to fabricate SIS/MBG hydrogel scaffolds. The slurry was carefully loaded into the material barrel with utmost precision, followed by the application of pneumatic force to propel it through the nozzle.

Subsequently, the slurry gracefully descended onto a glass slide carefully positioned on the cryogenic stage. An automatic cooling pump was connected to the cryogenic stage, and its gentle vibrations indicated the subzero temperature of −20 °C. After the completion of the printing process, a freeze-drying period of 24 h was applied to the 3D SIS/MBG hydrogel scaffolds. Next, they were immersed in a solution containing 95% ethanol and 1.0 wt% of 1-ethyl-3-(3-dimethylaminopropyl) carbodiimide (EDC) at room temperature for 30 min. The cross-linking of the SIS hydrogel occurred during this step. In order to guarantee the thorough elimination of any remaining residue, the fabricated hydrogel scaffolds went through a series of three extensive rinses utilizing distilled water. A team of researchers has developed a novel 3D scaffold dressing (SIS/MBG@Exos) by combining mesoporous bioactive glass (MBG) and exosomes. This integration enables the controlled and continuous release of bioactive exosomes. By possessing a meticulously organized 3D framework, optimal porosity, commendable biocompatibility, as well as remarkable haemostasis ability, the hydrogel scaffolds of SIS/MBG@Exos create a favourable milieu for the proliferation, migration, and angiogenesis of Human umbilical vein endothelial cells (HUVECs). The findings from in vivo experiments on diabetic wounds indicate that the SIS/MBG@Exos hydrogel scaffolds enhance the healing of diabetic wounds by increasing blood circulation and facilitating angiogenesis.

Additionally, the SIS/MBG@Exos hydrogel scaffolds contribute to the promotion of granulation tissue formation, the facilitation of organized collagen fibre deposition, and the facilitation of functional new blood vessel growth. Efficient wound healing relies heavily on the execution of these specific processes. In recapitulation, this study introduces an innovative and promising technique to tackle diabetic wounds, providing a potential avenue for improved treatment outcomes [115].

10.9.1 Bio-inks and Materials

Current trends in developing bio-inks for diabetic wound healing involve innovative approaches such as 3D bioprinting, functional 3D-printing inks, and portable handheld bioprinters. These technologies aim to create customized wound dressings that can effectively manage diabetic foot ulcers by promoting tissue repair, controlling infection and inflammation, and enhancing wound healing [112]. Bio-inks incorporating bioactive compounds like DNA-induced bio silica and extracellular vesicles have shown promising results in regulating the inflammatory microenvironment, promoting vascular regeneration, and accelerating wound healing [116]. Using bio-based materials in these bio-inks stimulates cellular and molecular processes that aid in wound healing, offering potential solutions to the challenges faced in diabetic wound care [116].

In 3D bio-printing for wound healing, researchers often turn to biocompatible and biodegradable materials. Included in the array of materials are Ca2+ ion-loaded oxidized starch hydrogels (CaOMS), astaxanthin-incorporated alginate dialdehyde-gelatine (ADA-GEL) hydrogels, borate bioactive glass (BBG) microparticles, and biodegradable polymers like Poly (lactic acid) (PLA) and Polyhydroxyalkanoates (PHA) [117]. These materials provide benefits such as high-water absorption capacity, improved blood clotting, increased density of network structure, and the release of biologically active ions and growth factors to facilitate the healing of wounds. By combining these materials in 3D printing technologies, the creation of smart wound dressings that effectively expedite the wound healing process becomes possible, positioning them as highly promising biomaterials for accelerating haemostasis and wound recovery [118].

Diverse techniques are employed to design patient-specific scaffolds in 3D bio-printing. A possible method encompasses the utilization of individualized implants derived from C.T. images, although its laborious nature and excessive cost characterize this approach. An additional approach encompasses multi-nozzle extrusion-based 3D bioprinting systems, enabling the construction of scaffolds that enclose living cells in predetermined sites [119]. Moreover, by utilizing FRESH 3D bio-printing and C.T. imaging, it becomes possible to produce custom-made dECM patches for soft tissue injuries, offering a solution for significant volume needs. Using the continuous extrusion technique, researchers can produce gradient scaffolds connecting tendons and muscles. These scaffolds incorporate tissue-specific growth factors from tendons and muscles' decellularized extracellular matrix (dECM) [120]. The layer-by-layer fabrication process in creating patient-specific scaffolds in 3D bio-printing involves the deposition of materials to build up the scaffold structure. This process allows precise control over the scaffold's internal geometry design and porosity percentage. A frequently employed technique for achieving this involves using a multi-nozzle extrusion-based 3D bioprinting system, which permits printing diverse biomaterials that enclose living cells in a predetermined spot [120]. Researchers can guarantee the accuracy and strength of the fabricated scaffolds by optimizing printing parameters and analysing natural polymer properties. This precise layering technique ensures that the scaffold remains structurally sound, maintains its shape accurately, and promotes the survival and growth of live cells [121].

The immunomodulation process involves carefully regulating the immune response to achieve the desired level. Immunomodulating biomaterials present a potential solution in addressing prevalent chronic diseases like type 1 diabetes, an autoimmune condition characterized by the destruction of insulin-producing cells in the pancreas by the body's immune system. A recent breakthrough by researchers involved the development of an injectable, synthetic biomaterial that successfully reversed type 1 diabetes in non-obese diabetic mice. This accomplishment signifies a notable milestone in developing a biodegradable platform designed to mitigate the effects of the disease. The utilization of injectable biomaterials for administering therapeutic agents, including medications, genetic materials, and proteins, is experiencing a steady rise. They provide the option to address various conditions through precise delivery while circumventing immune system uptake. Presently, research is being conducted that explores the application of synthetic and naturally derived injectable biomaterials in treating bone defects, cancer, and heart attacks. Supramolecular biomaterials, characterized by their molecular complexes that surpass the intrinsic capacities of individual molecules, hold promising potential as materials for injury or disease treatment due to their dual capacity for sensing and responding. Scientists are investigating the advancement of supramolecular biomaterials that can be activated or deactivated in response to physiological signals or imitate natural biological communication.

10.10 Conclusion

Wound healing in individuals with diabetes involves a complicated and ever-changing process of repairing cellular structures and tissue layers in injured tissues to return them to a normal state as much as possible. Due to the rising occurrence of diabetes and its associated complications, such as chronic nonhealing sores, it is imperative to create bioactive biomaterials that can provide effective treatment options for long-term diabetic ulcers. Biomaterial frameworks with wound healing properties can assist in re-epithelization and tissue regeneration, thereby expediting the healing process. Utilizing suitable biomaterials as carriers can improve the efficacy of plant extracts in stimulating the inflammatory, proliferative, and remodelling phases of chronic wounds without the disadvantages of antibiotic resistance and hypersensitivity linked to a limited range of medications. As discussed in this section, the recent advancements in biomaterials research demonstrate the significant potential of various methods currently being developed and tested in laboratory settings. Notably, approaches involving bioactive compounds or encapsulated cell therapies show promise in addressing the persistent problem of nonhealing diabetic ulcers. Expanded use of traditional plant extracts with established wound-healing properties will increase the options for managing chronic wounds. Despite the challenges associated with utilizing extracellular matrix-loaded biomaterials and cellular and extracellular treatment options, they can deliver multiple molecules to the wound site without degradation. It is crucial to ensure these technologies are cost-effective to encourage widespread use. Nevertheless, the field of biomaterial technology is advancing quickly due to the various methods being explored, including polymers and composites, self-assembling peptides, hydrogels and nanoparticles, microRNAs, alternative biophysical therapies, as well as cell- and drug-delivery therapies. The review presents a cautiously optimistic perspective, suggesting a potential breakthrough that could transform the treatment of diabetic wounds and provide a solution for chronic diabetic ulcers.

References

1. Todros S, Todesco M, Bagno A (2021) Biomaterials and their biomedical applications: from replacement to regeneration. Processes. https://doi.org/10.3390/pr9111949
2. Bonferoni MC, Caramella C, Catenacci L, Conti B, Dorati R, Ferrari F et al (2021) Biomaterials for soft tissue repair and regeneration: a focus on Italian research in the field. Pharmaceutics. https://doi.org/10.3390/pharmaceutics13091341
3. Kuhn LT (2005) 6—Biomaterials. In: Enderle JD, Blanchard SM, Bronzino JD (eds) Introduction to biomedical engineering, 2nd edn. Academic, Boston, pp 255–312
4. Spoială A, Ilie C-I, Ficai D, Ficai A (2023) Biomaterials. In: Gunduz O, Egles C, Pérez RA, Ficai D, Ustundag CB (eds) Biomaterials and tissue engineering. Springer International Publishing, Cham, pp 97–129
5. Al-Zyoud W, Haddadin D, Hasan SA, Jaradat H, Kanoun O (2023) Biocompatibility testing for implants: a novel tool for selection and characterization. Materials. https://doi.org/10.3390/ma16216881
6. Bose S, Ke D, Sahasrabudhe H, Bandyopadhyay A (2018) Additive manufacturing of biomaterials. Prog Mater Sci 93:45–111. https://doi.org/10.1016/j.pmatsci.2017.08.003
7. Singh S, Nwabor OF, Sukri DM, Wunnoo S, Dumjun K, Lethongkam S et al (2022) Poly (vinyl alcohol) copolymerized with xanthan gum/hypromellose/sodium carboxymethyl cellulose dermal dressings functionalized with biogenic nanostructured materials for antibacterial and wound healing application. Int J Biol Macromol 216:235–250
8. Oliveira C, Sousa D, Teixeira JA, Ferreira-Santos P, Botelho CM (2023) Polymeric biomaterials for wound healing. Front Bioeng Biotechnol:11. https://doi.org/10.3389/fbioe.2023.1136077
9. Arif ZU, Khalid MY, Noroozi R, Hossain M, Shi HH, Tariq A et al (2023) Additive manufacturing of sustainable biomaterials for biomedical applications. Asian J Pharm Sci 18(3):100812. https://doi.org/10.1016/j.ajps.2023.100812
10. Zhou L, Miller J, Vezza J, Mayster M, Raffay M, Justice Q et al (2024) Additive manufacturing: a comprehensive review. Sensors. https://doi.org/10.3390/s24092668
11. Liu G, Zhang X, Chen X, He Y, Cheng L, Huo M et al (2021) Additive manufacturing of structural materials. Mater Sci Eng R Rep 145:100596. https://doi.org/10.1016/j.mser.2020.100596
12. Gonzalez-Gutierrez J, Cano S, Schuschnigg S, Kukla C, Sapkota J, Holzer C (2018) Additive manufacturing of metallic and ceramic components by the material extrusion of highly-filled polymers: a review and future perspectives. Materials. https://doi.org/10.3390/ma11050840
13. Rahman M, Islam KS, Dip TM, Chowdhury MFM, Debnath SR, Hasan SMM et al (2023) A review on nanomaterial-based additive manufacturing: dynamics in properties, prospects, and challenges. Prog Addit Manuf. https://doi.org/10.1007/s40964-023-00514-8
14. Alexander AE, Wake N, Chepelev L, Brantner P, Ryan J, Wang KC (2021) A guideline for 3D printing terminology in biomedical research utilizing ISO/ASTM standards. 3D Print Med 7(1):8. https://doi.org/10.1186/s41205-021-00098-5
15. Mirzaali MJ, Moosabeiki V, Rajaai SM, Zhou J, Zadpoor AA (2022) Additive manufacturing of biomaterials—design principles and their implementation. Materials. https://doi.org/10.3390/ma15155457
16. Singh S, Syukri DM, Ushir YV, Mishra A, Ontong JC, Nwabor OF et al (2024) Post-operative wound healing efficacy of Eucalyptus camaldulensis phenolic-rich extracts incorporated hydrogel with enhanced antioxidant, antibacterial, and anti-inflammatory activities. J Polym Environ 33:814–839
17. Mishra T, Verma S, Verma VS, Mishra A, Jeswani G, Sahu D et al (2024) In-silico molecular docking and screening of Costus specious bioactive compounds as potential herb in management of diabetes mellitus. Vegetos, pp 1–14
18. Mohite P, Rahayu P, Munde S, Ade N, Chidrawar VR, Singh S et al (2023) Chitosan-based hydrogel in the management of dermal infections: a review. Gels 9(7):594

19. Ansari P, Akther S, Khan JT, Islam SS, Masud MSR, Rahman A et al (2022) Hyperglycaemia-linked diabetic foot complications and their management using conventional and alternative therapies. Appl Sci. https://doi.org/10.3390/app122211777
20. Ong KL, Stafford LK, McLaughlin SA, Boyko EJ, Vollset SE, Smith AE et al (2023) Global, regional, and national burden of diabetes from 1990 to 2021, with projections of prevalence to 2050: a systematic analysis for the Global Burden of Disease Study 2021. Lancet 402(10397):203–234. https://doi.org/10.1016/S0140-6736(23)01301-6
21. Edmonds M, Manu C, Vas P (2021) The current burden of diabetic foot disease. J Clin Orthop Trauma 17:88–93. https://doi.org/10.1016/j.jcot.2021.01.017
22. Worsley AL, Lui DH, Ntow-Boahene W, Song W, Good L, Tsui J (2023) The importance of inflammation control for the treatment of chronic diabetic wounds. Int Wound J 20(6):2346–2359. https://doi.org/10.1111/iwj.14048
23. Israni DK, Raghani NR, Soni J, Shah M, Prajapati BG, Chorawala MR et al (2024) Harnessing Cannabis sativa oil for enhanced skin wound healing: the role of reactive oxygen species regulation. Pharmaceutics 16(10):1277
24. Datta D, Prajapati B, Jethva H, Agrawal K, Singh S, Prajapati BG (2024) Value-added nanocellulose valorized from fruit peel waste for potential dermal wound healing and tissue regenerative applications. Regen Eng Transl Med:1–24
25. Gianino E, Miller C, Gilmore J (2018) Smart wound dressings for diabetic chronic wounds. Bioengineering. https://doi.org/10.3390/bioengineering5030051
26. Mohite P, Shah SR, Munde S, Ade N, Boban T, Singh S, Prajapati B (2024) Biopolymers in textile-based scaffolding and wound healing. In: Biopolymers in the textile industry: opportunities and limitations. Singapore, Springer Nature Singapore, pp 175–202
27. Singh SA, Vellapandian C, Shah DD, Jayeoye TJ, Chorawala MR, Singh S, Prajapati BG (2024) Valorised calcium-rich biomass from fish waste and eggshells in the fabrication of antibacterial scaffold for wound healing applications: a review. Waste Biomass Valor 15(4):1917–1941
28. Mudigonda J, Chenicheri S, Ramachandran R (2022) Chapter 6—Chronic wounds and tissue engineering: prospective and promise. In: Sharma CP, Chandy T, Thomas V, Thankam FG (eds) Tissue engineering. Academic, pp 101–122
29. Zheng SY, Wan XX, Kambey PA, Luo Y, Hu XM, Liu YF et al (2023) Therapeutic role of growth factors in treating diabetic wound. World J Diabetes 14(4):364–395. https://doi.org/10.4239/wjd.v14.i4.364
30. Lee SH, Kim SH, Kim KB, Kim HS, Lee YK (2024) Factors influencing wound healing in diabetic foot patients. Medicina. https://doi.org/10.3390/medicina60050723
31. Bansal V, Kalita J, Misra UK (2006) Diabetic neuropathy. Postgrad Med J 82(964):95–100. https://doi.org/10.1136/pgmj.2005.036137
32. Catrina S-B, Zheng X (2021) Hypoxia and hypoxia-inducible factors in diabetes and its complications. Diabetologia 64(4):709–716. https://doi.org/10.1007/s00125-021-05380-z
33. Zhu D, Wei W, Zhang J, Zhao B, Li Q, Jin P (2024) Mechanism of damage of HIF-1 signaling in chronic diabetic foot ulcers and its related therapeutic perspectives. Heliyon 10(3):e24656. https://doi.org/10.1016/j.heliyon.2024.e24656
34. Mir M, Ali MN, Barakullah A, Gulzar A, Arshad M, Fatima S et al (2018) Synthetic polymeric biomaterials for wound healing: a review. Prog Biomater 7(1):1–21. https://doi.org/10.1007/s40204-018-0083-4
35. Mohite P, Shah SR, Singh S, Rajput T, Munde S, Ade N et al (2023) Chitosan and chito-oligosaccharide: a versatile biopolymer with endless grafting possibilities for multifarious applications. Front Bioeng Biotechnol 11:1190879
36. Nguyen HM, Ngoc Le TT, Nguyen AT, Thien Le HN, Pham TT (2023) Biomedical materials for wound dressing: recent advances and applications. RSC Adv 13(8):5509–5528. https://doi.org/10.1039/D2RA07673J
37. Begum RF, Singh S, Prajapati B, Sumithra M, Patel RJ (2024) Advanced targeted drug delivery of bioactive agents fortified with graft chitosan in management of cancer: a review. Curr Med Chem

38. Ahmad S, Minhas MU, Ahmad M, Sohail M, Abdullah O, Badshah SF (2018) Preparation and evaluation of skin wound healing chitosan-based hydrogel membranes. AAPS PharmSciTech 19(7):3199–3209. https://doi.org/10.1208/s12249-018-1131-z
39. Ahmadian E, Eftekhari A, Janas D, Vahedi P (2023) Nanofiber scaffolds based on extracellular matrix for articular cartilage engineering: a perspective. Nanotheranostics 7(1):61–69. https://doi.org/10.7150/ntno.78611
40. Rajinikanth BS, Rajkumar DSR, K K, Vijayaragavan V. (2024) Chitosan-based biomaterial in wound healing: a review. Cureus 16(2):e55193. https://doi.org/10.7759/cureus.55193
41. Bejenaru C, Radu A, Segneanu A-E, Biţă A, Ciocîlteu MV, Mogoşanu GD et al (2024) Pharmaceutical applications of biomass polymers: review of current research and perspectives. Polymers. https://doi.org/10.3390/polym16091182
42. Baranwal J, Barse B, Fais A, Delogu GL, Kumar A (2022) Biopolymer: a sustainable material for food and medical applications. Polymers. https://doi.org/10.3390/polym14050983
43. Das A, Ringu T, Ghosh S, Pramanik N (2023) A comprehensive review on recent advances in preparation, physicochemical characterization, and bioengineering applications of biopolymers. Polym Bull 80(7):7247–7312. https://doi.org/10.1007/s00289-022-04443-4
44. Prete S, Dattilo M, Patitucci F, Pezzi G, Parisi OI, Puoci F (2023) Natural and synthetic polymeric biomaterials for application in wound management. J Funct Biomater. https://doi.org/10.3390/jfb14090455
45. Ujjwal RR, Yadav A, Tripathi S, Krishna SSTV (2022) Polymer-based nanotherapeutics for burn wounds. Curr Pharm Biotechnol 23(12):1460–1482. https://doi.org/10.2174/1389201022666210927103755
46. Nathan KG, Genasan K, Kamarul T (2023) Polyvinyl alcohol-chitosan scaffold for tissue engineering and regenerative medicine application: a review. Mar Drugs. https://doi.org/10.3390/md21050304
47. Alven S, Peter S, Mbese Z, Aderibigbe BA (2022) Polymer-based wound dressing materials loaded with bioactive agents: potential materials for the treatment of diabetic wounds. Polymers. https://doi.org/10.3390/polym14040724
48. Zarrintaj P, Seidi F, Youssefi Azarfam M, Khodadadi Yazdi M, Erfani A, Barani M et al (2023) Biopolymer-based composites for tissue engineering applications: a basis for future opportunities. Compos Part B 258:110701. https://doi.org/10.1016/j.compositesb.2023.110701
49. Downer M, Berry CE, Parker JB, Kameni L, Griffin M (2023) Current biomaterials for wound healing. Bioengineering. https://doi.org/10.3390/bioengineering10121378
50. Da Silva J, Leal EC, Carvalho E, Silva EA (2023) Innovative functional biomaterials as therapeutic wound dressings for chronic diabetic foot ulcers. Int J Mol Sci. https://doi.org/10.3390/ijms24129900
51. Deng X, Gould M, Ali MA (2022) A review of current advancements for wound healing: biomaterial applications and medical devices. J Biomed Mater Res B Appl Biomater 110(11):2542–2573. https://doi.org/10.1002/jbm.b.35086
52. Jiang X, Zeng Y-E, Li C, Wang K, Yu D-G (2024) Enhancing diabetic wound healing: advances in electrospun scaffolds from pathogenesis to therapeutic applications. Front Bioeng Biotechnol:12. https://doi.org/10.3389/fbioe.2024.1354286
53. Wang H, Duan C, Keate RL, Ameer GA (2023) Panthenol citrate biomaterials accelerate wound healing and restore tissue integrity. Adv Healthc Mater 12(31):2301683. https://doi.org/10.1002/adhm.202301683
54. Renuka RR, Julius A, Yoganandham ST, Umapathy D, Ramadoss R, Samrot AV et al (2023) Diverse nanocomposites as a potential dressing for diabetic wound healing. Front Endocrinol:13. https://doi.org/10.3389/fendo.2022.1074568
55. Naik K, Singh P, Yadav M, Srivastava SK, Tripathi S, Ranjan R et al (2023) 3D printable, injectable amyloid-based composite hydrogel of bovine serum albumin and aloe vera for rapid diabetic wound healing. J Mater Chem B 11(34):8142–8158. https://doi.org/10.1039/D3TB01151H

56. Chen AC-Y, Lu Y, Hsieh C-Y, Chen Y-S, Chang K-C, Chang D-H (2023) Advanced biomaterials and topical medications for treating diabetic foot ulcers: a systematic review and network meta-analysis. Adv Wound Care 13(2):97–113. https://doi.org/10.1089/wound.2023.0024
57. Sathyaraj WV, Prabakaran L, Bhoopathy J, Dharmalingam S, Karthikeyan R, Atchudan R (2023) Therapeutic efficacy of polymeric biomaterials in treating diabetic wounds—an upcoming wound healing technology. Polymers. https://doi.org/10.3390/polym15051205
58. Yusuf Aliyu A, Adeleke OA (2023) Nanofibrous scaffolds for diabetic wound healing. Pharmaceutics. https://doi.org/10.3390/pharmaceutics15030986
59. Kyriakides TR, Kim H-J, Zheng C, Harkins L, Tao W, Deschenes E (2022) Foreign body response to synthetic polymer biomaterials and the role of adaptive immunity. Biomed Mater 17(2):022007. https://doi.org/10.1088/1748-605X/ac5574
60. Singh SS, Behera SK, Rai S, Tripathy SK, Chakrabortty S, Mishra A A critical review on nanomaterial based therapeutics for diabetic wound healing. Biotechnol Genet Eng Rev 1–35. https://doi.org/10.1080/02648725.2022.2161732
61. Ataide JA, Zanchetta B, Santos ÉM, Fava ALM, Alves TFR, Cefali LC et al (2022) Nanotechnology-based dressings for wound management. Pharmaceuticals. https://doi.org/10.3390/ph15101286
62. Kushwaha A, Goswami L, Kim BS (2022) Nanomaterial-based therapy for wound healing. Nanomaterials. https://doi.org/10.3390/nano12040618
63. Monika P, Chandraprabha MN, Rangarajan A, Waiker PV, Chidambara Murthy KN (2022) Challenges in healing wound: role of complementary and alternative medicine. Frontiers. Nutrition:8. https://doi.org/10.3389/fnut.2021.791899
64. Metwally WM, El-Habashy SE, El-Hosseiny LS, Essawy MM, Eltaher HM, El-Khordagui LK (2024) Bioinspired 3D-printed scaffold embedding DDAB-nano ZnO/nanofibrous microspheres for regenerative diabetic wound healing. Biofabrication 16(1):015001. https://doi.org/10.1088/1758-5090/acfd60
65. Dong Y, Rodrigues M, Kwon SH, Li X, Sigen A, Brett EA et al (2018) Acceleration of diabetic wound regeneration using an in situ–formed stem-cell-based skin substitute. Adv Healthc Mater 7(17):1800432. https://doi.org/10.1002/adhm.201800432
66. Tsou Y-H, Khoneisser J, Huang P-C, Xu X (2016) Hydrogel as a bioactive material to regulate stem cell fate. Bioact Mater 1(1):39–55. https://doi.org/10.1016/j.bioactmat.2016.05.001
67. Ding Z, Zhang Y, Guo P, Duan T, Cheng W, Guo Y et al (2021) Injectable desferrioxamine-laden silk nanofiber hydrogels for accelerating diabetic wound healing. ACS Biomater Sci Eng 7(3):1147–1158. https://doi.org/10.1021/acsbiomaterials.0c01502
68. Mohite P, Asane G, Rebello N, Munde S, Ade N, Boban T et al (2024) Polymeric hydrogel sponges for wound healing applications: a comprehensive review. Regen Eng Transl Med 10:416–437
69. Zhang S, Ge G, Qin Y, Li W, Dong J, Mei J et al (2023) Recent advances in responsive hydrogels for diabetic wound healing. Mater Today Bio 18:100508. https://doi.org/10.1016/j.mtbio.2022.100508
70. Puri A, Mohite P, Patil S, Chidrawar VR, Ushir YV, Dodiya R, Singh S (2023) Facile green synthesis and characterization of Terminalia arjuna bark phenolic–selenium nanogel: a biocompatible and green nano-biomaterial for multifaceted biological applications. Front Chem 11:1273360
71. Syukri DM, Singh S, Nwabor OF, Ontong JC, Dejyong K, Sunghan J et al (2024) Prevention of post-operative bacterial colonization on mice buccal mucosa using biogenic silver nanoparticles-coated nylon sutures. Regen Eng Transl Med 10:294–308
72. Jayeoye TJ, Eze FN, Singh S, Olatunde OO, Benjakul S, Rujiralai T (2021) Synthesis of gold nanoparticles/polyaniline boronic acid/sodium alginate aqueous nanocomposite based on chemical oxidative polymerization for biological applications. Int J Biol Macromol 179:196–205
73. Loera-Valencia R, Neira RE, Urbina BP, Camacho A, Galindo RB (2022) Evaluation of the therapeutic efficacy of dressings with ZnO nanoparticles in the treatment of diabetic foot ulcers. Biomed Pharmacother 155:113708. https://doi.org/10.1016/j.biopha.2022.113708

74. Ahmad MZ, Alasiri AS, Ahmad J, Alqahtani AA, Abdullah MM, Abdel-Wahab BA et al (2022) Green synthesis of titanium dioxide nanoparticles using *Ocimum sanctum* leaf extract: in vitro characterization and its healing efficacy in diabetic wounds. Molecules. https://doi.org/10.3390/molecules27227712
75. Salvo J, Sandoval C (2022) Role of copper nanoparticles in wound healing for chronic wounds: literature review. Burns Trauma 10:tkab047. https://doi.org/10.1093/burnst/tkab047
76. Lu Z, Yu D, Nie F, Wang Y, Chong Y (2023) Iron nanoparticles open up new directions for promoting healing in chronic wounds in the context of bacterial infection. Pharmaceutics. https://doi.org/10.3390/pharmaceutics15092327
77. Hao P-C, Burnouf T, Chiang C-W, Jheng P-R, Szunerits S, Yang J-C et al (2023) Enhanced diabetic wound healing using platelet-derived extracellular vesicles and reduced graphene oxide in polymer-coordinated hydrogels. J Nanobiotechnol 21(1):318. https://doi.org/10.1186/s12951-023-02068-x
78. Karas RA, Alexeree S, Elsayed H, Attia YA (2024) Assessment of wound healing activity in diabetic mice treated with a novel therapeutic combination of selenium nanoparticles and platelets rich plasma. Sci Rep 14(1):5346. https://doi.org/10.1038/s41598-024-54064-2
79. Singh S, Chunglok W, Nwabor OF, Ushir YV, Singh S, Panpipat W (2022) Hydrophilic biopolymer matrix antibacterial peel-off facial mask functionalized with biogenic nanostructured material for cosmeceutical applications. J Polym Environ 30(3):938–953
80. Nagime PV, Syukri DM, Sjahriani T, Hermawan D, Shaikh NM, Shafi S et al (2024) Phyto-mediated biosynthesis of silver nanoparticles using Aloe barbadensis Miller leaves gel with improved antibacterial, anti-fungal, antioxidant, anti-inflammatory, anti-diabetic, and anti-cancer activities. Nano-Struct Nano-Objects 40:101368
81. Rahman MA, Islam MS, Haque P, Khan MN, Takafuji M, Begum M et al (2020) Calcium ion mediated rapid wound healing by nano-ZnO doped calcium phosphate-chitosan-alginate biocomposites. Materialia 13:100839. https://doi.org/10.1016/j.mtla.2020.100839
82. Huang W, Xu P, Fu X, Yang J, Jing W, Cai Y et al (2023) Functional molecule-mediated assembled copper nanozymes for diabetic wound healing. J Nanobiotechnol 21(1):294. https://doi.org/10.1186/s12951-023-02048-1
83. Shariati A, Hosseini SM, Chegini Z, Seifalian A, Arabestani MR (2023) Graphene-based materials for inhibition of wound infection and accelerating wound healing. Biomed Pharmacother 158:114184. https://doi.org/10.1016/j.biopha.2022.114184
84. Eid HM, Ali AA, Ali AMA, Eissa EM, Hassan RM, Abo El-Ela FI et al (2022) Potential use of tailored citicoline chitosan-coated liposomes for effective wound healing in diabetic rat model. Int J Nanomedicine 17:555–575. https://doi.org/10.2147/ijn.S342504
85. Park D, Lee SJ, Choi DK, Park J-W (2023) Therapeutic agent-loaded fibrous scaffolds for biomedical applications. Pharmaceutics. https://doi.org/10.3390/pharmaceutics15051522
86. Naomi R, Fauzi MB (2020) Cellulose/collagen dressings for diabetic foot ulcer: a review. Pharmaceutics. https://doi.org/10.3390/pharmaceutics12090881
87. Ren S, Guo S, Yang L, Wang C (2022) Effect of composite biodegradable biomaterials on wound healing in diabetes. Front Bioeng Biotechnol:10. https://doi.org/10.3389/fbioe.2022.1060026
88. Nezakati T, Seifalian A, Tan A, Seifalian AM (2018) Conductive polymers: opportunities and challenges in biomedical applications. Chem Rev 118(14):6766–6843. https://doi.org/10.1021/acs.chemrev.6b00275
89. Lin Z, Fan D, Li G, He L, Qin X, Zhao B et al (2023) Antibacterial, adhesive, and conductive hydrogel for diabetic wound healing. Macromol Biosci 23(2):2200349. https://doi.org/10.1002/mabi.202200349
90. Beygisangchin M, Abdul Rashid S, Shafie S, Sadrolhosseini AR, Lim HN (2021) Preparations, properties, and applications of polyaniline and polyaniline thin films—a review. Polymers. https://doi.org/10.3390/polym13122003
91. Rai R, Roether JA, Boccaccini AR (2022) Polyaniline based polymers in tissue engineering applications: a review. Prog Biomed Eng 4(4):042004. https://doi.org/10.1088/2516-1091/ac93d3

92. Talikowska M, Fu X, Lisak G (2019) Application of conducting polymers to wound care and skin tissue engineering: a review. Biosens Bioelectron 135:50–63. https://doi.org/10.1016/j.bios.2019.04.001
93. Sun C, Zhou X, Liu C, Deng S, Song Y, Yang J et al (2023) An integrated therapeutic and preventive nanozyme-based microneedle for biofilm-infected diabetic wound healing. Adv Healthc Mater 12(30):2301474. https://doi.org/10.1002/adhm.202301474
94. Wang L, Chen G, Fan L, Chen H, Zhao Y, Lu L et al (2023) Biomimetic enzyme Cascade structural color hydrogel microparticles for diabetic wound healing management. Adv Sci 10(14):2206900. https://doi.org/10.1002/advs.202206900
95. He M, Wang Z, Yang H, Wang Q, Xiang D, Pang X et al (2023) Multi-functional bio-HJzyme: revolutionizing diabetic skin regeneration with its glucose-unlocked sterilization and programmed anti-inflammatory effects. Adv Sci 10(21):2300986. https://doi.org/10.1002/advs.202300986
96. Jiang J, Li X, Li H, Lv X, Xu Y, Hu Y et al (2023) Recent progress in nanozymes for the treatment of diabetic wounds. J Mater Chem B 11(29):6746–6761. https://doi.org/10.1039/D3TB00803G
97. Li Z, Fan X, Luo Z, Loh XJ, Ma Y, Ye E et al (2022) Nanoenzyme–chitosan hydrogel complex with cascade catalytic and self-reinforced antibacterial performance for accelerated healing of diabetic wounds. Nanoscale 14(40):14970–14983. https://doi.org/10.1039/D2NR04171E
98. Chen J, Qin S, Liu S, Zhong K, Jing Y, Wu X et al (2023) Targeting matrix metalloproteases in diabetic wound healing. Front Immunol:14. https://doi.org/10.3389/fimmu.2023.1089001
99. Fu K, Zheng X, Chen Y, Wu L, Yang Z, Chen X et al (2022) Role of matrix metalloproteinases in diabetic foot ulcers: potential therapeutic targets. Front Pharmacol:13. https://doi.org/10.3389/fphar.2022.1050630
100. Ayuk SM, Abrahamse H, Houreld NN (2016) The role of matrix metalloproteinases in diabetic wound healing in relation to photobiomodulation. J Diabetes Res 2016(1):2897656. https://doi.org/10.1155/2016/2897656
101. Hermes O, Schlage P, dem Keller Ua (2011) Wound degradomics—current status and future perspectives. Biol Chem 392(11):949–954. https://doi.org/10.1515/BC.2011.092
102. Chappidi S, Buddolla V, Ankireddy SR, Lakshmi BA, Kim Y-J (2023) Recent trends in diabetic wound healing with nanofibrous scaffolds. Eur J Pharmacol 945:175617. https://doi.org/10.1016/j.ejphar.2023.175617
103. Kaparekar PS, Anandasadagopan SK (2023) In vitro and in vivo effect of novel GA-CSNPs loaded col-fibrin nanocomposite scaffold on diabetic wound healing. J Biomed Mater Res B Appl Biomater 111(5):1074–1088. https://doi.org/10.1002/jbm.b.35215
104. Negut I, Dorcioman G, Grumezescu V (2020) Scaffolds for wound healing applications. Polymers. https://doi.org/10.3390/polym12092010
105. Cheng C, Sun Q, Wang X, He B, Jiang T (2022) Enzyme-manipulated hydrogelation of small molecules for biomedical applications. Acta Biomater 151:88–105. https://doi.org/10.1016/j.actbio.2022.08.016
106. Steier A, Schmieg B, Irtel von Brenndorff Y, Meier M, Nirschl H, Franzreb M et al (2020) Enzyme scaffolds with hierarchically defined properties via 3D jet writing. Macromol Biosci 20(9):2000154. https://doi.org/10.1002/mabi.202000154
107. Ramirez-Acuña JM, Cardenas-Cadena SA, Marquez-Salas PA, Garza-Veloz I, Perez-Favila A, Cid-Baez MA et al (2019) Diabetic foot ulcers: current advances in antimicrobial therapies and emerging treatments. Antibiotics. https://doi.org/10.3390/antibiotics8040193
108. Teixeira MC, Singh KK, de Melo BAG, Severino P, Cardoso JC, Souto EB (2023) Chapter 9—3D bioprinting: An innovative technique for biofabrication applied to regenerative medicine and tissue engineering. In: Santana MH, Souto EB, Shegokar R (eds) Nanotechnology and regenerative medicine. Academic, pp 195–232
109. Glover K, Mathew E, Pitzanti G, Magee E, Lamprou DA (2023) 3D bioprinted scaffolds for diabetic wound-healing applications. Drug Deliv Transl Res 13(8):2096–2109. https://doi.org/10.1007/s13346-022-01115-8

110. Kim N, Lee H, Han G, Kang M, Park S, Kim DE et al (2023) 3D-printed functional hydrogel by DNA-induced biomineralization for accelerated diabetic wound healing (Adv. Sci. 17/2023). Adv Sci 10(17):2370104. https://doi.org/10.1002/advs.202370104
111. Choudary R, Saini N, Chopra DS, Singh D, Singh N (2023) A comprehensive review of 3D bioprinting biomaterials: properties, strategies and wound healing application. J Mater Res 38(13):3264–3300. https://doi.org/10.1557/s43578-023-01078-7
112. Fratini C, Weaver E, Moroni S, Irwin R, Dallal Bashi YH, Uddin S et al (2023) Combining microfluidics and coaxial 3D-bioprinting for the manufacturing of diabetic wound healing dressings. Biomater Adv 153:213557. https://doi.org/10.1016/j.bioadv.2023.213557
113. Ding X, Yu Y, Li W, Zhao Y (2023) In situ 3D-bioprinting MoS2 accelerated gelling hydrogel scaffold for promoting chronic diabetic wound healing. Matter 6(3):1000–1014. https://doi.org/10.1016/j.matt.2023.01.001
114. Liao W, Duan X, Xie F, Zheng D, Yang P, Wang X et al (2023) 3D-bioprinted double-crosslinked angiogenic alginate/chondroitin sulfate patch for diabetic wound healing. Int J Biol Macromol 236:123952. https://doi.org/10.1016/j.ijbiomac.2023.123952
115. Hu Y, Wu B, Xiong Y, Tao R, Panayi AC, Chen L et al (2021) Cryogenic 3D printed hydrogel scaffolds loading exosomes accelerate diabetic wound healing. Chem Eng J 426:130634. https://doi.org/10.1016/j.cej.2021.130634
116. Zhou F, Xin L, Wang S, Chen K, Li D, Wang S et al (2023) Portable handheld “SkinPen” loaded with biomaterial ink for in situ wound healing. ACS Appl Mater Interfaces 15(23):27568–27585. https://doi.org/10.1021/acsami.3c02825
117. Teo YC, Abbas A, Park EJ, Barbut C, Guo J, Goh D et al (2023) 3D printed bioactive PLGA dermal scaffold for burn wound treatment. ACS Materials Au 3(3):265–272. https://doi.org/10.1021/acsmaterialsau.2c00079
118. Faidallah RF, Hanon MM, Szakál Z, Oldal I (2022) Biodegradable materials used in FDM 3D printing technology: a critical review. J Mod Mech Eng Technol 9:90–105. https://doi.org/10.31875/2409-9848.2022.09.11
119. Lee SS, Du X, Smit T, Bissacco EG, Seiler D, de Wild M et al (2023) 3D-Printed LEGO®-inspired titanium scaffolds for patient-specific regenerative medicine. bioRxiv. 2023.03.30.534953. https://doi.org/10.1101/2023.03.30.534953
120. Liu N, Zhang X, Guo Q, Wu T, Wang Y. (2022). 3D bioprinted scaffolds for tissue repair and regeneration. Frontiers in Materials, 9, 925321. https://doi.org/10.3389/fmats.2022.925321
121. Akilbekova D, Turlybekuly A (2023) 6—Patient-specific 3D bioprinting for in situ tissue engineering and regenerative medicine. In: Kalaskar DM (ed) 3D printing in medicine, 2nd edn. Woodhead Publishing, pp 149–178
122. Quigley C, Tuladhar S, Habib MA (2022). A bio-printing strategy to fabricate geometrically accurate 3d scaffolds. In: ASME 2022 international mechanical engineering congress and exposition 2022

Biomaterials-Based Additive Manufactured Products for *In Situ* Tissue Engineering

11

Yash Tarwate, Akshay Khandke, Madhur Kulkarni, Kunal Banode, Mital Patel, Krutika Pardeshi, and Amarjitsing Rajput

Abstract

In situ, tissue engineering harnesses the body's innate healing capabilities to repair damaged tissues and organs, a technique increasingly utilized for tissue regeneration. Customized scaffolds are vital in controlling the injury site microenvironment, guiding host stem/progenitor cell migration, and supporting tissue regrowth. This method proves effective by transforming the living body into a bioreactor, were natural biochemical and biophysical signals spur successful tissue and organ restoration. Crafting tissue-specific scaffolds accurately is essential for prompt and efficient tissue regeneration. Prominent additive manufacturing methods like stereolithography, fused deposition modelling, selective laser sintering, inkjet bioprinting, and laser-assisted bioprinting are pivotal in on-site tissue engineering for cartilage, bone, heart, liver, and skin tissues. Key processing parameters in 3D bioprinting utilizing hydrogel-based bioinks—such as viscosity, flow behaviour, flow rate, shear stress, viscoelasticity, nozzle size, and flow rate—affect print quality, structure stability, and cell vitality significantly. The

Y. Tarwate · A. Khandke · A. Rajput (✉)
Department of Pharmaceutics, Bharti Vidyapeeth Deemed to be University, Poona College of Pharmacy, Erandwane, Pune, Maharashtra, India
e-mail: amarjit.rajput@bharatividyapeeth.edu

M. Kulkarni · K. Banode
Department of Pharmaceutics, SCES's Indira College of Pharmacy, Pune, Maharashtra, India

M. Patel
Shobhaben Pratapbhai Patel School of Pharmacy & Technology Management, SVKM's Narsee Monjee Institute of Management Studies, Deemed to be University, Vile Parle (W), Mumbai, Maharashtra, India

K. Pardeshi
Department of Pharmaceutics, School of Pharmaceutical Sciences, Sandip University, Nashik, Maharashtra, India

S. Singh et al. (eds.), *Biomaterial-based Additive Manufacturing in Tissue Engineering and Regeneration*, Biomaterials, Bioengineering and Sustainability 7,
https://doi.org/10.1007/978-3-031-96070-3_11

potential of both natural and synthetic biopolymers like alginate, collagen, chitosan, polycaprolactone, polyvinyl alcohol, and polyurethane in scaffold production is studied extensively. These biopolymers offer suitable biocompatibility, degradability, and mechanical properties for tissue engineering. Challenges such as variations between batches, standardizing parameters, and preserving cell vitality against shear stress are analysed rigorously. Future research focuses on boosting reproducibility and efficiency through enhanced material formulations and advanced bioprinting techniques, advancing personalized medicine and regenerative therapies.

Keywords

Biomaterials · Additive manufacturing · Tissue engineering · Biopolymer

Abbreviations

AA	L-ascorbic acid
ABS	Acrylonitrile butadiene styrene
ACP	Acetophenone
$AgNO_3$	Silver nitrate
AM	Additive manufacturing
AMF	Additive Manufacturing File Format
ASTM	American Society for Testing and Materials
BPADMA	Bisphenol A ethoxylate Di methacrylate
CAD	Computer-aided design
CJIP	Continuous inkjet printing
CLIP	Continuous Liquid Interface Production
CMC	Carboxy methyl cellulose
CNC	Cellulose nano crystals
CNPs	Carbon nanotubes
Co-Cr	Cobalt-Chromium
CQ-EDAB	Camphor quinone ethyl-4-dimethylaminobenzoate
CTAB	Cetyltrimethylammonium bromide
DLP	Digital Light Processing
DoD	Drop-on-demand
DoDP	Drop-on-demand printing
ECM	Extracellular matrix
FDA	Food and Drug Administration
FDM	Fused deposition modelling
FFF	Fused filament fabrication
G-Code	Geometric Code
GelAGE	allylated gelatin
GelMA	Gelatin methacryloyl
GNPs	Graphene nanoplatelets

HAMA	hyaluronic acid methacrylate
$HAuCl_4{\cdot}3H_2O$	Hydrogen tetrachloroauratetrihydrate
HCl	Hydrochloric acid
HEMA	2-hydroxyethyl methacrylate
HHEMPP	2-hydroxy-4-(2-hydroxy ethoxy)-2-methyl propiophenone
HIPS	High-impact polystyrene
HME	Hot-melt extrusion
HMPP	2- hydroxy-2-methylpropiophenone
HNO_3	Nitric acid
HPMC	Hydroxypropyl methylcellulose
HPMCAS	Hydroxypropyl methylcellulose acetate succinate
IJP	Inkjet printing
IND	Investigational new drug
IR	Infrared
ISO	International Organization for Standardization
LAB	Laser-assisted bioprinting
LIFT	Laser-induced forward transfer
MAPLE-DW	Matrix-assisted pulsed laser evaporation direct writing
MNPs	Metal nanoparticles
MNs	Microneedles
$NaBH_4$	Sodium borohydride
NaCl	Sodium chloride
NaOL	Sodium oleate
PA	Polyamide
PAEK	Polyaryletherketone
PANI	Polyaniline
PBF	Powder bed fusion
PC	Polycarbonate
PCL	Polycaprolactone
PDLLA	Poly-DL-lactide
PDLLA	Poly (D, L-lactide)
PE	Polyethylene
PEDOT: PSS	Polymer blend of poly (3,4-ethylene dioxythiophene) and polystyrene sulfonate
PEEK	Polyetheretherketone
PEGDA	Polyethylene glycol diacrylate
PEI	Polyetherimide
PEKK	Polyetherketoneketone
PETG	Polyethylene Terephthalate Glycol
PG	Propylene glycol
PGA	Polyglycolide / Polyglycolic acid
PGLA/PLGA	Polylactic co-glycolic acid
PGSM	Poly(glycerol-co-sebacate) (PGS)-methacrylate
PHBA	p-hydroxybenzoic acid
PIJ	Piezoelectric inkjet printing

PLA	Polylactide
PLCL	Poly(L-lactide-co-ε-caprolactone)
PLLA	Poly-L-lactide
PMMA	Polymethyl methylacrylate
PP	Polypropylene
PPF	Polypropylene fumarate
PVA-MA	Poly (vinyl alcohol) methacrylate
PVP	Polyvinylpyrrolidone
RF	Riboflavin
Sil-MA	Methacryloyl-modified silk
SLA	Stereolithography
SLS	Selective Laser Sintering
SPS	Sodium persulfate
STL	Standard Tessellation Language
TCP	Nano β-tricalcium phosphate
TIJ	Thermal inkjet printing
TPO	diphenyl (2,4,6-tri methyl benzoyl) phosphine oxide
TPU	Thermoplastic polyurethane
TTCP	Tetracalcium Phosphate
UV	Ultra-violet
VP	Vat-photopolymerization
°C	Degree Celsius
2D	Two-dimensional
2PP	Two-photon polymerization
3D	Three-dimensional
4D	Four-dimensional
5D	Five-dimensional

11.1 Introduction

Three-dimensional printing, often called additive manufacturing, creates objects by gradually adding layers of material reproducing computer-aided design (CAD) models. Amidst the dominance of meticulousness, this technology can produce tailored medications and medical devices with high drug loading and complicated architectures, overcoming the constraints of traditional dosage forms. The commercially available formulation is often manually pulverized or fragmented to modify the dosage while administered by a paediatric and geriatric patient. 3D printing technology can personalize patient-specific doses, reducing improper dosing and improving patient compliance.

Chuck Hull, often credited as the pioneer of 3D printing, filed an initial patent regarding a 3D printing technique in the early 1980s. Three-dimensional (3D) printing is a process used to fabricate three-dimensional objects [1]. The additive manufacturing (AM) technique offers many benefits to humanity. Still, its industrial application is often limited by its slow manufacturing speed compared to traditional manufacturing processes, the scarcity of pharmaceutical-grade approved printing materials, and its suboptimality in large-scale production [2].

Material extrusion, VAT photopolymerization, Powder Bed Fusion, Material Jetting, Binder Jetting, Direct Energy Deposition, and Sheet lamination are among the several processes that make up 3D printing, according to the ISO and ASTM [3]. 3D printing allows for quick prototyping of pharmaceutical items, allowing scientists to evaluate different formulas and choose the best option swiftly. In biofabrication and organ regeneration, 3D printing has been instrumental in developing biocompatible and biodegradable materials. These materials can be scaffolds to support cell growth and tissue regeneration. By printing with these specialized materials, researchers can create complex structures that mimic the natural extracellular matrix, providing a suitable environment for cells to proliferate and differentiate [4].

The concept of three-dimensional biological printing, also known as bioprinting, was first introduced by a team of researchers led by Vladimir Mironov in 2003. This innovative technology represents a significant advancement in additive manufacturing, building upon the foundations of traditional 3D printing. 3D bioprinting is an innovative technology with immense potential for transforming the realms of biofabrication and organ regeneration. This layered manufacturing method enables the creation of intricate composite tissue constructs by precisely depositing cellular units and biological materials in a sequential layering manner. This technology integrates biological materials, such as cells, growth factors, and other biomaterials, into a specialized "bioink." Utilizing a person's cells, the engineered tissues created through this process can bypass the risks associated with traditional transplant surgeries, such as infection and immunocompromise. 3D bioprinting possesses the capacity to revolutionize the domain of organ transplantation by imparting a reliable and safe alternative to donor-reliant transplants. Additionally, this technology can be used to create in vitro models for drug testing and disease research, reducing the need for animal testing and providing more accurate representations of human physiology [5–7].

The materials employed within the methodology of bioprinting are denoted as biomaterials, necessitating adherence to the ensuing criteria: biocompatibility, denoting compatibility with the human body; bio-resorbability; biodegradability; and requisite mechanical attributes, contingent upon the site of implantation. Three-dimensional bioprinting is crucial in tissue engineering endeavours, focused on developing functional tissues applicable in regenerative medicine and pharmaceutical testing. Bioprinting facilitates the creation of tissue engineering scaffolds with patient-specific spatial geometry, regulated microstructures, and precise positioning of various cell types. The potential of tissue regeneration and reconstruction holds promise for repairing or substituting damaged tissues and organs [8].

11.2 General Process Involved in the 3D Printing

11.2.1 3D Modelling

The first crucial step is to use computer-aided design (CAD) software to produce a 3D model of the desired object. The physical thing that will be printed is designed using this digital model as a blueprint. The model is saved in a standard file format, like STL (Standard Tessellation Language) or AMF (Additive Manufacturing File Format), that can be read by 3D printing software.

11.2.2 Slicing and Geometric Code (G-Code) Generation

Slicing software imports the 3D model file and "slices" the model into thin horizontal layers. This software instructs the 3D printer to deposit material layer by layer to replicate the object precisely. It does this by analysing the model and producing a set of instructions known as G-code. The G-code specifies the precise movements of the print head and other settings like temperature and extrusion rate.

11.2.3 Printing

The G-code instructions are sent to the 3D printer. The printer deposits the build material, typically a thermoplastic filament, and extrudes it through a heated nozzle. The print head moves precisely in the X and Y axes while the build plate moves down slightly after each layer is printed, building the object up from the bottom. The printer follows the G-code to deposit material in the correct locations for each layer.

11.2.4 Post-processing

After printing, a few post-processing activities could be needed for the printed item. This can involve smoothing the surface, applying a finish, or coating, or removing any printed support structures to hold the model's overhanging portions.

The specific steps and technologies used can vary between different 3D printing methods, like Fused deposition modelling (FDM), Stereolithography (SLA), or Selective Laser Sintering (SLS). Generally, the workflow consists of creating a 3D model, converting it to instructions the printer can follow, printing the layers, and finishing the final part.

11.3 Process Involved in Bioprinting

The bioprinting process can be divided into three main stages: pre- and post-bioprinting [9].

11.3.1 Pre-bioprinting

- The selection and preparation of bioinks are materials composed of cells and biomaterials such as hydrogels or polymers.
- Design and create digital models for the desired tissues or organs using CAD software.
- Optimization of bioink properties, such as viscosity and crosslinking, to ensure printability and cell viability.

11.3.2 Bioprinting

- Transfer of the bioink into the bioprinter's cartridges or reservoirs.
- Precise deposition of the bioink layer by layer using various bioprinting techniques, such as extrusion-based, inkjet, stereolithography, or laser-assisted bioprinting.
- Incorporation of supporting structures such as scaffolds to maintain the desired shape and structure during the printing process.

11.3.3 Post-bioprinting

- Crosslinking or solidification of the printed construct to stabilize the structure and maintain its integrity.
- Removal of any supporting materials or scaffolds, if applicable.
- Incubation of the printed construct in a suitable culture medium to foster cell growth, lineage, and senescence.
- Evaluation of the printed construct's properties, such as mechanical strength, cell viability, and tissue-specific functionality.

11.4 Additive Manufacturing Techniques

11.4.1 Stereolithography

Charles W. Hull invented Stereolithography (SLA) during the 1980s to fabricate 3D objects from digital data, marking the inception of 3D printing and making it the first commercially available 3D printing technique [10, 11]. SLA bioprinting emerges as a cutting-edge technology in the bioprinting realm, offering vast potential and clinical utility owing to its customizable architecture and multi-functionality. Leveraging light-based mechanisms akin to stereolithography (SLA), this method initiates chemical processes to solidify or cure bioinks exclusively within illuminated regions, facilitating precise construction. Vat-photopolymerization (VP) technology employs a laser beam to create intricate 3D objects with excellent resolution, precision, accuracy, and speed. It is also known as the reductive polymerization technique. Vat photopolymerization outperforms other 3D printing processes in generating intricate structures at a rapid pace with high resolution and precision. It utilizes a light source to induce photo-curing in liquid resin, resulting in composed structures with superior spatial resolution and thickness as small as 1μm [2, 8, 12].

SLA, a pioneering commercial AM technology, was first applied in bioprinting in 2004. Researchers at Clemens University, led by Boland, utilized SLA to create porous tissue engineering scaffolds made of cell-encapsulated polyethylene glycol-diacrylate. This pioneering work heralded the beginning of applying the SLA technique in bio-printing [10, 13]. Using a light source in SLA printing affords distinct advantages, including producing smaller prints with heightened resolution compared to the FDM printing method. The horizontal resolution of an SLA scanner is contingent upon the dimensions of the light source spot, typically ranging from 30 to 140 microns. Likewise, the vertical resolution spans from 25 to 200 microns. Ink materials conducive to this process should exhibit optical transparency, low viscosity, biocompatibility, and UV stability. These traits are essential for optimal performance and accuracy in fabricating intricate structures via SLA technology [8].

A significant advantage of this technique is that the printing resolution can be improved by reducing the laser's spot size, enabling a resolution of up to 10μm. This resolution is influenced by several factors, including printing speed, direction, and volume, but the laser's spot size is a critical determinant. The ability to achieve high-resolution prints is a crucial benefit of SLA technology, making it a popular choice for applications requiring fine details and precision. In bioprinting, SLA techniques that use water-insoluble photo-initiators are not ideal, and Ultraviolet (UV) light is not the most favourable choice for cells. Nevertheless, researchers successfully have bio-printed cells by employing customized bioinks. Bioinks, a combination of cells and biocompatible materials, are crucial to 3D bioprinting as they enable precise placement of cells in a defined 3D microenvironment, forming tissue-like structures [14]. Several innovative SLA-based techniques, including two-photon polymerization (2PP), digital light processing (DLP), and multi-photon polymerization (MPP), were created. Continuous Liquid Interface Production (CLIP) is an unconventional stereolithography (SLA) variant with an inverted configuration. In this approach, a curing light source is strategically positioned beneath the resin bath, illuminating the resin through a transparent window at the base of the container. The fabrication platform on which the desired structure is built is lowered into the resin bath from above. The resin solidifies on the platform by selectively exposing the resin to the curing light based on the 3D CAD model. As the platform is continuously lifted, fresh resin fills the void, allowing the structure to grow in size [15].

11.4.2 Fused Deposition Modelling

A continuous filament of thermoplastic material is fed through a heated printer extruder head in the fused deposition modelling (FDM) or fused filament fabrication (FFF) 3D printing technique. The material is subsequently deposited onto the workpiece layer by layer to construct the desired shape under computer control. This innovative technology, created by S. Scott Crump in 1988, is known for its cost-effectiveness and flexibility in producing a wide range of objects.

Fused Deposition Modelling (FDM) is a prevalent and widely studied method of 3D printing that is popular due to its versatility and ease of use. This technique is responsible for 80% of the 3D printing-related literature that is indexed, highlighting its significance in the field. The technology was patented in 1989. It is a type of 3D printing that falls under the category of extrusion-based printing [4, 15, 16]. It is a widely preferred 3D printing technology in biomaterial research owing to its affordability, compact size, and ability to create intricate structures without organic solvents. Moreover, FDM printing is a cost-effective method for manufacturing complex 3D objects in short printing times. This is because the process allows for the efficient use of materials, reducing waste and minimizing the need for added finishing processes. These advantages make FDM an attractive choice for researchers and developers working with biomaterials, as it offers a versatile and accessible method for creating complex structures with precision and control. It is a technique for constructing three-dimensional objects by layering thermoplastic material onto a surface using a printhead that regulates temperature. The process involves a heated printhead melting a thermoplastic filament, enabling the exact deposition of thin layers of partially melted polymers, such as polycaprolactone, to build the desired structure. In Fused Deposition Modelling (FDM), solid materials like filament or powder are heated close to their melting point and extruded through a heated nozzle in semi-molten form, following CAD patterns to create layers. After each layer is finished, the platform moves down by a set distance to begin printing the next layer [4, 7, 13, 17].

Hot-melt extrusion (HME) is commonly employed to produce FDM printer filaments. The matrix-forming polymer, drugs, and any extra excipients are combined in this procedure to generate a formulation that is well mixed, heated, and extruded through a die to produce filaments of the required size. Combining HME with FDM technologies increases the range of pharmaceutical-grade polymers suitable for 3D printing, offering new avenues for developing innovative drug delivery systems [18].

A typical commercially available FDM machine has a maximum operating temperature of approximately 300 °C. This temperature limitation makes materials possessing extremely sharp melting points unsuitable for this machine, and specifically, types of materials capable of melting at relatively moderate temperatures are considered appropriate. Among the most often used thermoplastic polymers in FDM are acrylonitrile butadiene styrene (ABS) and PLA. Other alternative polymers that can be used include polypropylene (PP), polymethyl methacrylate (PMMA), polyethylene (PE), polycarbonate (PC), and polyamide or nylon (PA). These materials have varying properties and are suitable for different applications. However, in recent years, there has been a focus on developing new FDM filaments with superior thermophysical properties to enhance the mechanical characteristics of printed components. This has been achieved by incorporating specialized engineering plastics, like polyetheretherketone (PEEK), polyetherimide (PEI), polyaryletherketone (PAEK), and polyetherketoneketone (PEKK), which have outstanding thermophysical characteristics and elevated melting points. This advancement is achieved by overcoming the limitations of the process and increasing the temperatures to 400–500 °C. It has opened new possibilities for using FDM technology in high-performance applications [19, 20].

11.4.3 Selective Laser Sintering

Powder bed fusion (PBF) techniques represent a widespread group of additive manufacturing (AM) methods for creating intricate 3D components with excellent mechanical characteristics. This approach involves depositing powder particles layer by layer and using a laser beam to selectively melt the powder within specific regions of each layer based on a digitally defined blueprint. SLS is also known as a side arm of PBF technology. This method employs a laser beam to melt and solidify a photopolymerizable polymer blend that includes therapeutic medication [21, 22].

It is a cutting-edge additive manufacturing technology that has revolutionized the production of complex and customized parts. This innovative process uses a high-powered laser to selectively fuse powdered materials, such as polymers, metals, or ceramics, stacking them sequentially in layers to generate three-dimensional structures with exceptional precision and detail. The technology stands out for its many benefits compared to other manufacturing methods. These advantages include diverse raw material options, the ability to produce overly complex components, rapid manufacturing capabilities, cost-effective production, and efficient resource use. As a result, SLS has appeared as a leading-edge technique in 3D printing, offering unparalleled versatility and efficiency in creating intricate and customized objects [23].

The technology was first patented in 1986 by researchers Carl Deckard and Joe Beaman at the University of Texas at Austin. This groundbreaking patent laid the foundation for developing and commercializing SLS technology. Following the first patent, the subsequent advancements were made by DTM, a company later acquired by 3D Systems in 2001 [24]. The commercially available materials for SLS 3D printing encompass a wide range, including metals and polymers like polystyrene, PC, PAEK, and PA. These materials are well-suited for SLS 3D printing applications [25].

11.4.4 Inkjet Bioprinting

Inkjet bioprinting is a variant of 3D printing technology that has gained popularity in biomedical device fabrication. It allows for the precise deposition of layered assembly of cells and biomaterials, creating complex three-dimensional structures. This technique uses heat-driven or sound-driven impulses to expel tiny "bioink" drops onto a substrate, enabling the fabrication of highly detailed structures with high-resolving cells and growth factors within the scaffolds during fabrication, promoting the formation of biomimetic tissue. Many uses of this technology, such as tissue engineering, regenerative medicine, drug research, and disease modelling, have proved considerable promise. In today's rapidly advancing field of bioprinting, inkjet bioprinting stands out as a promising technology due to its ability to precisely control the deposition of bioink, its versatility in printing a wide range of biomaterials, and its potential for high cell viability after printing. Inkjet bioprinting is a variant of 3D printing technology that allows for the precise deposition of cells and biomaterials layer-by-layer, creating complex three-dimensional structures. Inkjet bioprinting is revolutionizing the field of tissue engineering by allowing for the precise fabrication of scaffolds that mimic the native tissue microenvironment.

Inkjet bioprinting is a promising technique in the field of biomedical device fabrication, offering contactless printing for both hard and soft tissue regeneration. It is one of the four main categories of 3D bioprinting techniques, along with extrusion-based, laser-assisted, and stereolithography bioprinting [9]. It is also known for precisely depositing ink without contacting the printing surface. This non-contact approach enables meticulous command over the fabricated dose by regulating the volume of ink expelled from the printer nozzle. Moreover, inkjet printing (IJP) provides exceptional versatility for substrates, allowing formulated inks to be applied to diverse materials such as films, microneedles (MNs), stents used for clinical purposes, and medical devices. This highlights the adaptability of this method in creating distinctive pharmaceutical forms and drug delivery when it comes to medical devices.

Inkjet printing appeared in the late 1990s and has since gained widespread popularity owing to its capacity for generating high-resolution outputs. This is primarily attributed to the extremely small droplet volumes, ranging from pico-litres to even femtoliters, and the non-contact, "on-demand" delivery method employed in this technique. It offers many benefits, such as efficient high-throughput capacity, exceptional resolution, cost-effectiveness, reproducibility, and maintaining a high cell viability rate exceeding 90%. Additionally, this method enables the precise deposition of various cells or proteins onto specific spatial locations using multiple print heads. This capability eases the creation of intricate multicellular structures, making inkjet bioprinting a valuable tool in tissue engineering and regenerative medicine [15, 26].

11.4.5 Laser-Assisted Bioprinting

Laser-assisted bioprinting has its origins in laser direct-write technology. It represents an adaptation of the laser-induced forward transfer method initially introduced for the precise transfer of biological components like peptides, DNA, and cells. Figure 11.1 represents components of a Laser-Assisted Bioprinter.

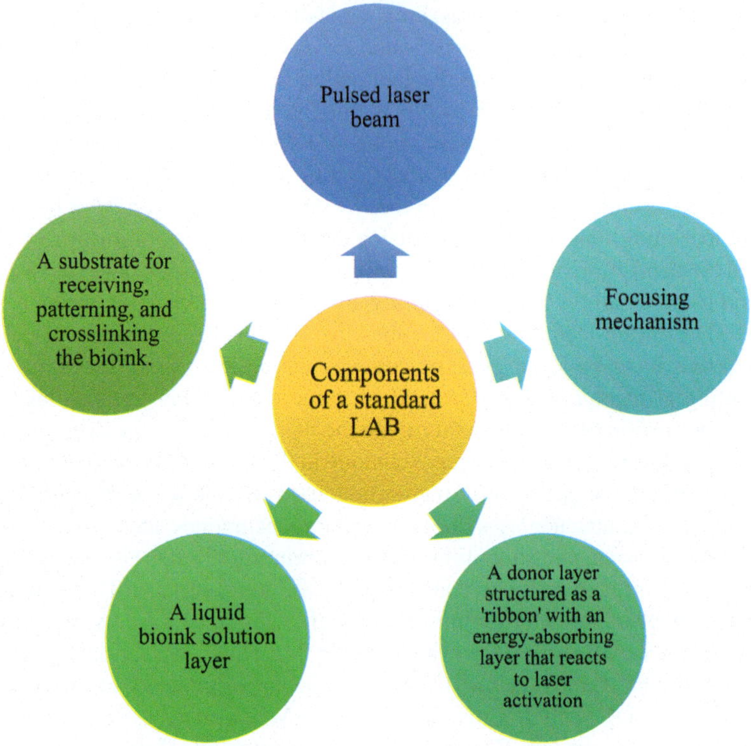

Fig. 11.1 Components of a laser-assisted bioprinter. (Created by using Biorender.com)

Laser-assisted bioprinting is a promising method for 3D printing cell-infused scaffolds for human tissues. This process avoids physical contact between the bio-ink and the dispenser, averting cell stress and elevating cell resilience (over 95%). Additionally, laser-assisted bioprinting is compatible with a wide range of bioink types and viscosities, from 1 to 300 mPa/s. These advantages make laser-assisted bioprinting well-suited for fabricating 3D tissue constructs with high cell survival rates [15, 27].

Laser-assisted bioprinting is a promising non-contact printing technique that does not require nozzles. It employs either a low-power ultraviolet pulsed laser, called Matrix-Assisted Pulsed Laser Evaporation Direct Writing, or a high-intensity laser source, called Laser-Induced Forward Transfer (LIFT) (MAPLE-DW) [6].

Laser-assisted bioprinting offers several benefits, such as the ability to command the placement of cells and biomaterials meticulously. However, the impact of laser exposure on the viability and function of printed cells is not yet fully understood. More research is needed to thoroughly investigate the biological effects of the laser on different cell types and to optimize printing parameters to minimize any potential damage. Another factor limiting the wider adoption of laser-assisted bioprinting is the prohibitive cost of the specialized equipment required. The laser printing system and the necessary control and monitoring components are significant investments. The complexity of programming and operating these systems also requires specialized ability, adding to the barriers for many researchers and applications [7].

Laser-assisted bioprinting represents a category of non-contact, nozzle-free printing processes that enable the precise deposition of biomaterials onto a substrate; which the most widely used method is laser-induced forward transfer, which is a non-contact printing technique that uses a laser pulse to deposit biomaterials onto a substrate. This method is similar to inkjet bioprinting in terms of droplet release, but it differs in the way the droplets are formed and propelled. In LIFT, a laser pulse is directed at the top layer of the donor slide, which is coated with the bioink. The laser pulse creates a bubble that expands and propels the bioink from the bottom layer of the donor slide onto the collection plate or substrate below. This process occurs without any direct contact between the nozzle and the bioink, eliminating the possibility of clogging and minimizing shear stress on the cells. One of the key advantages of LIFT is its ability to handle highly viscous bioinks, which can be challenging for other bioprinting techniques like extrusion and inkjet.

LIFT has also been shown to maintain high cell viability, often exceeding 95%, which is higher than extrusion and inkjet bioprinting methods. The non-contact nature of LIFT and its ability to handle viscous bioinks while maintaining high cell viability make it a promising technique for various applications in tissue engineering and regenerative medicine. The fundamental concept of LAB technology involves the creation of a three-layer structure utilizing a pulse laser irradiation, which is then focused on the absorption layer, causing a localized vaporization and expansion of the bioink layer at the irradiation point. This process generates a bioink jet deposited onto a receiving substrate. In the case of LIFT (laser-induced forward transfer), it also employs laser irradiation to transfer bioink. Still, it does so in continuous filaments, distinguishing it from LAB technology [9, 10, 13, 28–30].

Bioink plays a vital role in the success of laser-assisted bioprinting. It typically consists of cells enclosed in a biomaterial or a blend of different biomaterials in a hydrogel form. There are two main types of bioinks in current bioprinting practices: scaffold-based, which includes cells and scaffold-like hydrogels, microcarriers, or decellularized matrix, and scaffold-free, where cell aggregates are printed without additional biomaterials. Stem cells are widely used in bioprinting due to their versatility. These cells, known for their pluripotency, ability for cell renewal, and rapid proliferation, serve as an abundant and diverse cell source for 3D bioprinting applications [31]. Materials used in the additive manufacturing are listed in Table 11.1.

11.5 Additive Manufacturing *In Situ* Tissue Engineering

At a national science foundation (NSF) workshop in Granlibakken, California, the term "Tissue Engineering" first appeared. Skalak later published the meeting reports in 1988. Langer and Vacanti later reinforced this notion in 1993 [43]. It is split into three primary groups that concentrate on (a) implantation of isolated cells, (b) introducing growth factors to the cells to promote cell proliferation, and (c) introducing

Table 11.1 Materials used in additive manufacturing

Technique	Material class	Material example	Material used in	References
SLA	Photopolymer	Soybean oil epoxidized acrylate, PEGDA and Bisphenol A ethoxylate dimethacrylate (BPADMA), Methacrylated PCL, Aliphatic Poly(carbonate)Urethane, Bisphenoldiglycidylether and Graphene	Foldable scaffolds capable of supporting cardiomyogenic differentiation, Fabrication of trachea-mimetic tissue scaffolds.	[32]
SLA	Polymer	PEGDA, N-vinyl-pyrrolidone, PEG 300, Propylene glycol (PG).	Fabrication of immediate or Controlled release dosage forms, Osmotic or Floating Tablets. Even Gummies or Chocolate formulations designed explicitly for pediatrics use.	[33]
	Photo initiator	TPO, 2- hydroxy-2-methylpropiophenone (HMPP), 2-hydroxy-4-(2-hydroxy ethoxy)-2-methyl propiophenone (HHEMPP), riboflavin (RF), and triethanolamine, acetophenone (ACP) And camphorquinone (CQ)		
SLA	Natural bioinks, Synthetic bioinks	Gelatin methacryloyl (GelMA), allylated gelatin (GelAGE), hyaluronic acid methacrylate (HAMA), Methacryloyl-modified silk (Sil-MA), PEGDA, Poly (vinyl alcohol) methacrylate (PVA-MA), Poly(glycerol-co-sebacate) (PGS)-methacrylate (PGSM)	Cardiac tissue engineering, Fabrication of vascular and Retinal Implants, as well as Skin, Cartilage, And Neural regeneration	[14]
SLA	Ceramics	Hydroxyapatite, Zirconia	Bone tissue engineering, Dental implants	[34]
SLA	Photopolymer Resin	**Polyethylene glycol diacrylate (PEGDA), α, w-polytetrahydrofuranetherdiacrylate,**	**Tablets, Microneedles, Bio-Medical Scaffolds, Dental Prosthetics, and Medical Devices.**	[2]
FDM	Polymer, bio ceramic material	**PLA, Graphene nanoplatelets (GNPs)**	**Scaffold manufacturing technology**	**[4]**
FDM	Polymer	PLA	**Manufacturing medical devices and scaffolds**	**[4]**

FDM	Polymer	Alginate with hydroxyapatite as an additive, PLA, multiwalled carbon nanotubes	Glucose-responsive insulin, transdermal drug delivery, pain management (Microneedle patch, biodegradable polymeric microneedle arrays, microheater-integrated microneedle patch)	[4]
FDM	Polymer	Thermoplastic polyurethane (TPU), ABS, PLA, Nylon, High-impact polystyrene (HIPS), PVA, PETG, PC	Biodegradable implants	[35]
SLS	Ceramics	Hydroxyapatite, Zirconia.	Bone tissue engineering, dental implants.	[24, 34]
SLS	Composites	Graphene, Carbon fiber	Tissue engineering, implants, Drug delivery	[34]
SLS	Thermoplastics	Nylon	Tissue engineering (Trachea, stem cell model) Drug modelling for cancer therapy	[22, 23]
SLS	Synthetic biodegradable polymer material Polymer *semi-crystalline*	PCL	Tissue engineering applications	[11, 36]
SLS	Polymer	PEEK	Tissues and scaffolds	[11, 36]
SLS	Polymer	Polyetherimide (PEI)	Customized implants, surgical tools, prosthetics, etc	[11]
SLS	Powder mixture	Theophylline, Eudragit®	Controlled release	[37]
SLS	Polymer	PA, PC, PEI, etc.	Customized implants, surgical tools, prosthetics, etc.	[11]

(continued)

Table 11.1 (continued)

Technique	Material class	Material example	Material used in	References
IB	Edible Paper Gelatin Cellulose Polymer	Rice paper, Icing sheets, and wafer edible sheets. Gelatin HPC and HPMC Polyglycolic Acid (PGA), Polydimethylsiloxane	Oro dispersible films, Tissue engineering, Drug delivery systems, including oral, buccal, transdermal, and ocular routes. Tissue regeneration. Creating microdevices and drug-loaded microstructures	[38]
IB	Polymer	TPU and PLA	Construct structures, Capsules, and Scaffolds for bone growth	[39]
IB	Biomaterial	Collagen, Alginate, Chitosan	Tissue engineering, drug delivery, wound healing	[34]
IB	Ceramics Polymer	Hydroxyapatite, Alpha-TCP, beta-TCP, PVA, PEG, PEG hydrogel	Printing of biomolecules such as protein and nucleic acid	[40]
IB	Conductive inks Conducting polymers	Silver nanoparticles, gold nanoparticles. Polymer blend of poly(3,4-ethylenedioxythiophene) (PEDOT) and polystyrene sulfonate (PSS) - (PEDOT: PSS), Polyaniline (PANI), PP.	Printable devices for neurotechnology	[41]
LAB	Biomaterials	Collagen	Bone tissue engineering, Implantation of local bone defect, Repair and treatment of bone injury, Subchondral bone defect repair	[6]
LAB	Biomaterials Naturally, derived polymers	Stem cells, collagen, gelatin, decellularized ECM, fibrin, and alginate	For cell growth and tissue regeneration.	[31]
LAB	Biomaterials	Adipose-derived stem cells, bone-marrow-derived Stem Cells	Tissue regeneration, preparation of hydrogel scaffolds	[42]

cells on or within various scaffolds that have the goal of imitating the extracellular matrix. The latter method is the most widely used tissue engineering technique, which involves seeding living cells on extracellular scaffolds that are either synthetic or natural to create an implantable substrate [44]. This scaffold can serve as a functional 3D structural support system that fosters the formation of a new layer of cells (ECM) similar to the injured original tissue [45].

To guarantee cell survival, migration, proliferation, and differentiation, the environment needed to create a functioning tissue must resemble the organism's natural chemical signals and biophysical cues [46]. Due to limited anatomical site access and morbidity, TE faces numerous limitations even with a suitable scaffold, such as the inability to get a donor tissue biopsy [47].

A wide range of biomaterials, including artificial polymers featuring improved arbitrary, physio-chemical, and fabrication characteristics, and organic polymers that convey biological signals necessary for inducing the right kind of biological feedback, have been explored for use in tissue engineering scaffolds [48]. Scaffold micro- and nano-structural characteristics, in addition to materials science, influence cell binding, propagation, development, and rearrangement [49–53]. To maximize the final scaffolding outputs concerning the dimensions and forms of the exterior structure, along with interior design, many processing methods and strategies have been presented based on the tissue engineering technique and supporting material. Among other things, phase separation techniques, fiber bonding, particulate leaching, freeze drying, gas foaming, melt molding, fiber bonding, and additive manufacturing (am) techniques, also referred to as solid free-form creation techniques, all work in tandem with casting with solvent [48].

So basically, in tissue engineering, additive manufacturing refers to the operation of layer-by-layer combining components to produce products from the data of a three-dimensional model [54]. This method has been widely used to fabricate tissue engineering scaffolds using a variety of methods, such as stereolithography (SLA) and fused deposition modelling (FDM) [55]. The most prevalent based on the principles of fabrication, the four categories of additive manufacturing (am) methods employed in the creation of tissue engineering structures are as follows: i) stereolithography (SLA); ii) selective laser sintering (SLS); iii) three-dimensional printing (3DP); and iv) fused deposition modelling (FDM) [56].

Various scaffolds, such as monolithic, micro-porous, nanoparticle, fibrous, hydrogel, and 3D-printed scaffolds, are utilized for in situ tissue regeneration. Scaffolds can be made from biomaterials such as composites, metals, polymers, and ceramics. To promote regeneration, synthetic, natural, or a combination of both materials must react to biological cues and interact with endogenous cells and the immune system. These biophysical and biochemical characteristics of these responsive biomaterials allow them to interact with the body and change local tissue micro-environments by regulating the immune system and the rate and extent of endogenous cell healing (from scarring to complete regeneration). Via intracellular and intercellular signalling, the biophysical properties of biomaterials, such as stiffness, structure, topography, and degradation, can change the local tissue microenvironments. These alterations to the tissue microenvironment include adjusting the temperature or pH and regulating the number of cells, ions, enzymes, and radical species prevalent. The stiffness of the matrix determines the adherence, dissemination, and destiny of stem cells [57, 58].

Breakdown has an impact on scaffolds' in vivo biophysical and biochemical changes. The scaffold's structural integrity must offer a three-dimensional framework for infiltrating cells and tissues after implantation. Scaffolds should then be dismantled and fully resorbed to enable load transfer to develop new tissue and functional recovery. A delicate balance must be forged between the scaffold's mechanical integrity and degradation to prevent scarring and foster in situ tissue regeneration. Scaffold degradation kinetics depend on several factors, including structure (pore size and distribution, chemical functionalization, and biological milieu). Biomaterials should not degrade too quickly because it takes time for freshly created tissue to reorganize and endure in vivo stresses. Slow-degrading scaffolds, on the other hand, promote fibrosis and hinder regeneration while offering longer-lasting structural support. Therefore, to achieve both structural and functional regeneration, it is crucial to coordinate the scaffold's resorption pace with the new tissue creation rate. For instance, implanting a non-resorbable scaffold, such as hydroxyapatite (hap), for bone regeneration is not beneficial since hap is inert and does not break down quickly, restricting the amount of new tissue creation that may replace the scaffold. However, the resorption rate of this scaffold formulation can be increased by adding β-tricalcium phosphate (β-TCP) to HAp scaffolds [59].

Orthopaedics faces significant challenges in treating postoperative femoral nonunion with substantial segment mental bone defects [60]. The methods used now to treat such bone defects include applying distraction osteogenesis techniques and implanting autologous, allogeneic, or synthetic grafts. Nevertheless, all the strategies above have drawbacks and restrictions [61]. Lately, additive manufacturing (am) has become a potent method for precisely and freely constructing orthopaedic implants with intricate internal and external macro- and micro-architectures [62].

Additive manufacturing also plays a crucial role in gene therapy; techniques for additive manufacturing, such as 3D bioprinting, allow for the by-layer cell depositing, tissues, and related molecules, providing safe, biocompatible, efficient, and inert ways for the creation of biological structures and biomimetic scaffolds. Using successive layers, 3D printing technology advances the real-time creation of innovative drug distribution techniques and dosage formulations while remaining adaptable, stable, diversified, and rapid. This technique aids in enhancing traditional dosage forms, directing the medication to the active site, and managing the pharmacokinetic and dynamic parameters. Supporting the creation of patient-compliant procedures and individualized, controlled-release therapeutic systems further opens up new possibilities in the field of medicine [63–67] nucleic acids, growth factors, stem cells, and other compounds can be produced on scaffolds using 3D printing technology, providing a platform for the desired (localized, safe, and sustained) release of these molecules in the body. Moreover, improved or altered gene expression by 3D scaffolds opens new possibilities for skin rejuvenation, wound recovery, and tissue engineering. Specifically, biodegradable polymers have found extensive application as scaffolding components due to their limited use time. It is possible to

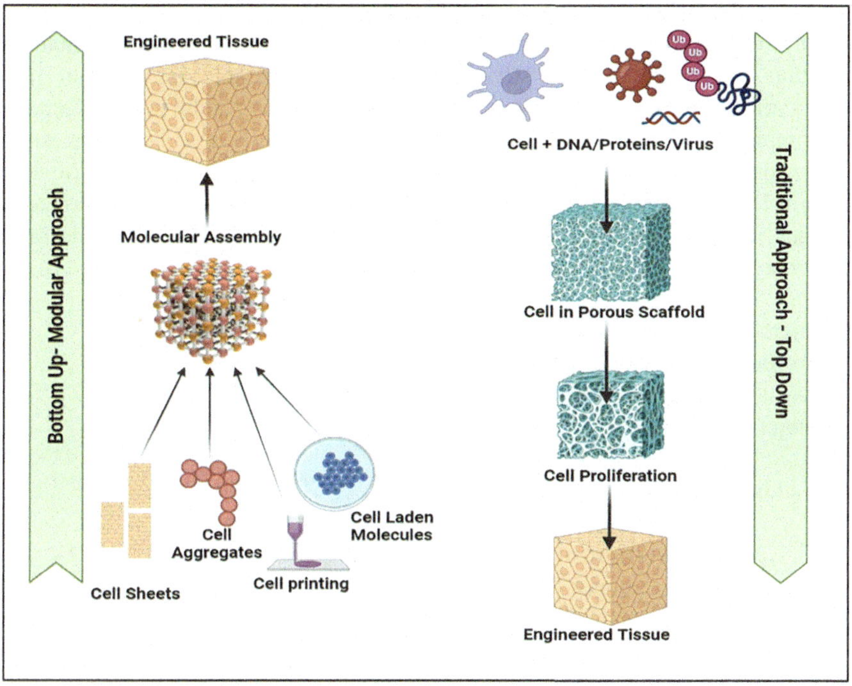

Fig. 11.2 Top-down and bottom-up approaches for additive manufacturing. (Created by using Biorender.com)

precisely tailor the deterioration process' timescale to a specific application [68, 69]. Figure 11.2 shows top-down and bottom-up approaches for additive manufacturing.

11.5.1 Cartilage Tissue

In today's world, osteochondral (OC) tissue damage is essential. It could be brought on by acute trauma, mechanical joint instability, or age-related degeneration. Owing to articular cartilage's (ac) restricted intrinsic regeneration ability; defects can worsen over time when its critical equilibrium is upset. In the end, this deterioration will lead to the loss of the synovial joint's articular surface, which often causes constant discomfort and a restriction in joint movement symptoms linked to the clinical manifestation of osteoarthritis [68, 70]. The majority of cartilage rejuvenation and repair methods have been demonstrated to be unsuccessful due to the cartilage's circumferential architecture and its intimate relationship to the subchondral bone layer underneath. Since cartilage lacks blood vessels, the synovial fluid supplies it with nutrition. At 65–80% of its wet weight, it is primarily an acellular tissue composed of 15–30% collagen, 5–10% proteoglycans, and water [71]. Collagens provide elasticity, and proteoglycans retain water molecules in the matrix.

11.5.1.1 Additive Manufacturing Techniques

By layering together materials, techniques for additive manufacturing facilitate the creation of three-dimensional structures. According to a software program, rapid prototyping (RP) or solid free-form fabrication (SFF) is the additive manufacturing of complex-formed scaffolds. In contrast with typical production techniques [56, 72], it provides an elevated level of autonomy on the construct's outward shape (macro-level) and interior design (micro-level: pore size, porosity, and surface-to-volume ratio), and it ensures that these characteristics will always be the same. Every layer depicts the model's cross-sectional shape at some point.

- Photo-polymerization method
- Powder bed fusion processes
- Extrusion-based process
- Inkjet printing.

One additional considerable point is,

11.5.1.2 Combined Additive Manufacturing Techniques

Hybrid scaffolding mimics the multi-composite nature of actual cartilage by combining cell-seeded, matrix-like hydrogels with a polymer for structural support. With this method, scaffold properties may be adjusted, cell distribution can be optimized, and signalling molecules and cells can even be distributed in specific directions. For the goal of cartilage [73, 74] and osteochondral [75], tissue engineering, hydrogel bioprinting and 3DF of supporting polymers have been developed together. The three main types of structures are fibrous scaffolds, hydrogels, and polymeric films [76, 77].

11.5.1.3 Hydrogel System

Natural resource-based hydrogel scaffolds have become increasingly popular among them. Because they are similar to native extracellular matrix (ECM) in structure and biology, allowing for easier cell transplantation, proliferation, and differentiation. In addition to their intriguing qualities, which include excellent biocompatibility biodegradation, high water content (usually 70–99%), variable elasticity and rigidity, etc., among these, hydrogel scaffolds derived from natural resources have gained popularity because they are similar to native extracellular matrix (ECM) in structure and biology, allowing for easier transplanting, growth, and division of cells [77, 78]. After physical or chemical cross-linking, their three-dimensional (3D) network architectures comprise synthetic polymers and natural macromolecules [79]. Recently, natural healing flexible frameworks with large distortion abilities have been developed using several cross-linking methods. These materials' wide range of uses in biomedical prostheses, electronics, and coatings has garnered much interest [80]. Materials with self-healing capabilities can fix faults and regain their mechanical and morphological characteristics. This property not only keeps a system going longer but also improves mechanical stability and keeps these materials from breaking down abruptly or irreversibly in delicate applications [81].

Depending on the damage's depth, there are three main types of articular osteochondral deficiencies: partial thickness chondral, complete thickness chondral, and osteochondral. In the initial partial thickness chondral defect, the calcified cartilage is undamaged, and only the articular hyaline cartilage zone is injured. When the whole architecture of the osteochondral tissue is destroyed, and the subchondral bone is entirely apparent, a full-thickness chondral defect can be classified as an osteochondral defect [82]. This defect will result in damage to the calcified cartilage layer.

Depending on the extent of restoration, three distinct groups can be used to categorize clinical therapy techniques for chondral or osteochondral lesions: reparative, regenerative/restorative, and palliative. Palliative care aims to relieve a patient's symptoms, such as arthritis and typically does not result in the replacement of damaged tissue. Reparative therapy aims to restore the damaged area by using biological materials or boosting the subchondral bone's blood flow to draw fresh tissues. Nevertheless, rejuvenation is the best technique for repairing damaged osteochondral tissue [83]. In regenerative treatment, autologous chondrocyte implantation (ACI) is a recognized clinical strategy for regenerative healthcare. That has developed over numerous generations for cartilage defects [84] during arthroscopy, from the cartilage biopsy specimens, the healthy homologous chondrocytes of the damaged knee are isolated. Following their culture in a controlled environment, the chondrocytes are injected into the lesion and covered by a periosteal covering. Ten to twenty years following ACI surgery, the long-term results have demonstrated excellent outcomes: the production of hyaline-like cartilage that is stable both structurally and operationally, and it blends nicely with the surrounding cartilage and subchondral bones [9, 85]. The autologous chondrocytes for the next-generation ACI are grown in vitro in a porcine type i/iii collagen matrix before inserting the cell-seeded matrix into the lesion. Furthermore, the process referred to as matrix-assisted autologous chondrocyte transplantation (MACI) minimizes surgical exposure along with operating time. Significantly improves chondrocyte proliferation, phenotype, and even distribution [86, 87]. However, achieving regeneration efficacy for osteochondral defects remains challenging if subchondral bone loss is present [88]. Essentially, two types of biomaterials are employed in cartilage tissue engineering: natural and synthetic. Natural materials having superior biological characteristics, minimal inflammatory reaction, biodegrading, and antigenic properties include collagen, a protein-based substance [89–91]; however, it is expensive, thrombogenic, weak mechanically, and difficult to modify [92, 93]. Then, gelatin is an inexpensive protein-based material with low immunogenicity, degrades quickly, and is compatible [94] but occasionally loses stability at high temperatures [95]. Next, there is silk fibroin, a protein-based material with excellent mechanical qualities, low immunogenicity [96], low thrombogenicity [97], but sluggish gelation [98] and sourcing issue. Then, polysaccharide-based glycosaminoglycans (gags), hyaluronic acid (ha), and chondroitin sulphate (cs) simulate the makeup of cartilage extracellular matrix (ECM), are biodegradable [99], reabsorbable [100], and bind signalling factors [101]. However, they degrade rapidly in vivo and need crosslinking to be stable [102, 103]. Alginate, on the other hand, is based on polysaccharides and has poor mechanical and cell adhesion qualities but is non-toxic and non-inflammatory and gels quickly [104, 105]. Then, polysaccharide-based chitosan has poor mechanical qualities [106] but is antibacterial, inexpensive, biocompatible, and has a tunable biodegradability [107]. Subsequently, decellularized hydrogels, which are protein-based and capable of delivering biochemical signals, are needed for complex processes and may be immunogenic [108].

Poly (ethylene glycol) (peg) is a synthetic substance with biodegradable and not immunogenic [109], has tunable characteristics, and has no adhesion support [110, 111]. Glutamic acid polymers are biodegradable through hydrolysis, thermoplasticity, mechanical qualities that can be adjusted, porosity, and localized inflammation [112–114]. However, weakly crosslinked gels and their hydrolysis products might cause irritation and rapid breakdown [114, 115]. Although the byproducts of hydrolysis might induce irritation, (PGA) poly (lactic acid) and copolymers are biodegradable by hydrolysis, have mechanical solid properties [116], are soluble in organic solvents [117], and strengthen hydrogels [118, 119].

11.5.2 Bone Tissue

Bone tissue is well known for its ability to mend itself. Thus, goes through controlled resorption, maturation, differentiation, and dynamic remodelling. Osteoclasts and osteoblasts, which preserve a healthy bone, dynamically participate in all these processes [120]. Patients' well-being is significantly compromised by bone loss. Consequently, the expenses and treatments associated with bone-related illnesses are rising. In addition, musculoskeletal because of an older worldwide population and increased life expectancies, diseases including osteoporosis, rheumatic disorders like osteoarthritis, congenital deformities, backache, scoliosis, cracks, infections, malignancies, and infections of the bones are fast growing [121]. Substantial bone abnormalities are not fully reparable by the body [122]. In such cases, outside intervention is required to return to standard functionality. Autografts, which use bone from the same patient, and allografts, which use bone from a deceased donor, are among the several therapy choices. However, novel biomaterials and scaffolding approaches are now being investigated since more efficient methods for healing massive bone lesions are required. Bone tissue engineering (BTE) is an essential field of study. Scaffolds are commonly utilized in tissue engineering (TE) procedures to offer structural reinforcement to the developing tissue. In addition to forming the extracellular matrix, scaffolds operate as transient templates for cell interaction [123]. Utilization of tailored to patients' analytical data to apply computational modelling, which, using additive manufacturing (am) techniques, subsequently achieved. Choosing the proper size, structural arrangement, and porosity to suit every distinct fault is made possible by computerized modelling.

11.5.3 Heart Tissue

Cardiac failure is a disorder that reflects a decrease in the heart's pumping efficiency. It can be brought on by some rudimentary illnesses, either with or without an occurrence of acute myocardial infarction, such as hypertensive heart disease, valvular heart disease, ischemic heart disease, and primary myocardial disease. Myocardial infarction is typically followed by myocyte displacement. Ventricular dilation and thinning of the heart wall are the outcomes of the collagen extracellular matrix's weakening. After a significant cell depletion following an infarction, the cardiac tissue loses a significant inherent regeneration capacity capable of replacing the missing cells, resulting in persistent weakening of the heart wall muscle [124].

The goals of pharmacological therapy are to guard against the overactivity of toxic humoral elements in cardiac failure and to reduce workload (with nitrates and diuretics). Interventional therapy is becoming increasingly common, especially for people who have limited activities and symptoms that are severe. It involves surgery or the insertion of pacing devices to correct electrical/mechanical asynchrony. Drugs and interventional therapy together are unable to stop the disease's progression to its final stage effectively [125–132]. Additive manufacturing in the production of polymeric frameworks is listed in Table 11.2 [133].

11.5.4 Liver Tissue

The liver is the most prominent internal organ, accounting for 2–5% of overall body weight. Its main metabolic centre is intricate, with numerous endocrine and exocrine capabilities. It also governs metabolic homeostasis [134, 135] of many hormones and nutrients in the bloodstream, the control of inflammation and tolerance, and mainly all plasma protein synthesis. In addition, the liver creates bile to aid digestion and serve as the primary method of excreting waste items from the liver. It also metabolizes medications, intrinsic hormones, xenobiotics, and waste elements. Even in severe damage, when fewer than one-third of the cells are still viable, the liver can regenerate itself, potentially restoring both mass and activity [136]. Partial or whole transplantation of the liver has become known as an achievable choice for tackling the physiological challenges that patients with a working liver face, yet it is ability to regenerate ability has been severely impaired in various pathological conditions [137]. Some practical constraints in liver transplantation include the low degree of transplantation and organ scarcity [138].

Due to lacking liver donations, tissue technology has become a viable substitute for existing liver-related treatments. Reconstructive medicine's main objective is to either help the sick liver heal itself or replace it with an ex-vivo bioengineered liver that can carry out regular hepatic activities [139]. therefore, to replicate the in-vivo microenvironment in an ex-vivo situation, it is imperative to develop suitable supporting biomaterials that mimic the liver extracellular matrix and exhibit certain hepatic features. Before a functional tissue is produced, the cells are cultivated in a bioreactor after being planted in the scaffold. Thus, the initial objective of this

Table 11.2 Additive manufacturing in the production of polymeric frameworks

Manufacturing technology	Essential qualities of the materials	Pore dimensions available (mm)	Porosity (%)	Porosity design
Melt molding Extrusion/particle leaching	Thermoplastic Thermoplastic	50–500 <100	<80 <84	Spherical spaces and bare interweaving Spherical spaces and bare interweaving
Solvent casting/ particle leaching Paraffin template Emulsion freeze drying Thermally induced phase separation/ freeze drying Electrospinning Membrane lamination	Soluble Soluble Soluble Soluble Soluble Soluble	30–300 100–700 <200 <200 20–100 30–300	20–50 >90% <97 <97 <95 <85	Variable pores Round pores Plenty of linked small pores Plenty of linked small pores Fibrous Variable pores
Chemical reactant gas foaming Physical induced gas foaming/ particle leaching	Amorphous Amorphous	100–700 Micropores <50	>90% <97	A large number of separate micropores Numerous interconnected macropores mixed with low-volume, non-interconnected micropores.
Rapid prototyping 3D printing Fused deposition modelling Direct rapid prototyping	Soluble	Macropores <400 45–150	<70	100% interconnected macrospores

discipline is to replicate liver constructions to repair damaged hepatic tissues in patients through transplantation [140, 141] different biomaterials are being employed to achieve this goal; among them, alginate scaffolds have endured the most popular because of their porosity, hydrophilic qualities, and biocompatibility. Alginate scaffolds have been shown to increase the lifespan of animal models with 70–80% partial hepatectomy when combined with hepatocytes or mesenchymal stem/stromal cells (MSC) [142, 143]because of its outstanding biodegradability, biocompatibility, and low cytotoxicity, chitosan is also utilized as a scaffold [144] collagen is a frequently employed molecule in this field because of its ubiquity, biocompatibility, and prevalence in the hepatic extracellular matrix. To cultivate a wide range of stem cells for numerous tissue engineering applications, 3D collagen scaffolds have been used. When cultivated as a monolayer on collagen gel, hepatocytes lost many of their liver-specific capabilities within a few days, but when cultured in-between two layers of gel, they preserve many of these functions for weeks [145, 146] furthermore, for a variety of pharmaceutical and medicinal applications, these tissue-engineered liver constructions serve as effective biological surrogates.

Thus, hepatic tissue bioengineering offers patients with incurable conditions some hope as well as new options and possibilities for treating various illnesses. The percentage of patients awaiting liver transplantation may be decreased with the use of extracorporeal liver support devices.

11.5.5 Skin

The skin, the most substantial part of the human body, accounts for more than 16% of our body mass. The epidermis, dermis, and an extra layer known as the hypodermis comprise the two main layers of skin. Every stratum possesses distinct configurations, amalgamations, and cooperative roles [147–149]. The epidermis comprises keratinocytes, melanocytes, Langerhans, Merkel, and inflammatory cells. The epidermis keeps the body hydrated, regulates temperature, and shields the body from harm [149]. Traditional skin constructs are primarily engineered using three cell types: dermal fibroblasts, melanocytes, and keratinocytes. Dermal fibroblasts are found in the dermis, while keratinocytes and melanocytes play significant roles in the epidermis. Globally, wound care and skin transplantation remain the two most pressing issues in healthcare. Statistics show that the global expenditure on skin repair exceeds 25 billion us dollars annually [150]. Human skin cannot fully regenerate when it is deeply charred, extensively injured in a big region, or afflicted with conditions such as diabetic foot ulcers and skin cancer. This prevents the regeneration of several skin appendages, such as sweat glands, sebaceous glands, and hair follicles [151, 152]. These days, a graft transplanting flawless skin in place of injured skin is a common procedure and the standard treatment for healing damaged skin. Synthetic tissues, autografts, allografts, and xenografts are examples of skin grafts [152, 153]. Slight infections can be treated with these techniques. About autografts despite their excellent compatibility, they have limited sources. Thus, immunological rejection is still a possibility [154, 155].

Animal skin tissues used for transplantation are called xenografts. Due to their longer healing durations, greater availability, and superior biocompatibility, they are preferred. However, they have disadvantages such as immunological rejection, increased infection risks, and particular tissue care and maintenance needs [152, 153]. Figure 11.3 represents additive manufacturing in tissue engineering showing scaffold and bioprinting.

11.6 Key Processing Parameters for 3D Bioprinting Using Hydrogel-Based Bioinks

11.6.1 Viscosity

The viscosity of hydrogel-based bioinks is a crucial factor affecting scaffolds' printability and structural integrity. Maintaining proper control over the viscosity of bioinks can help to ensure that cells are successfully extruded in their intended shape

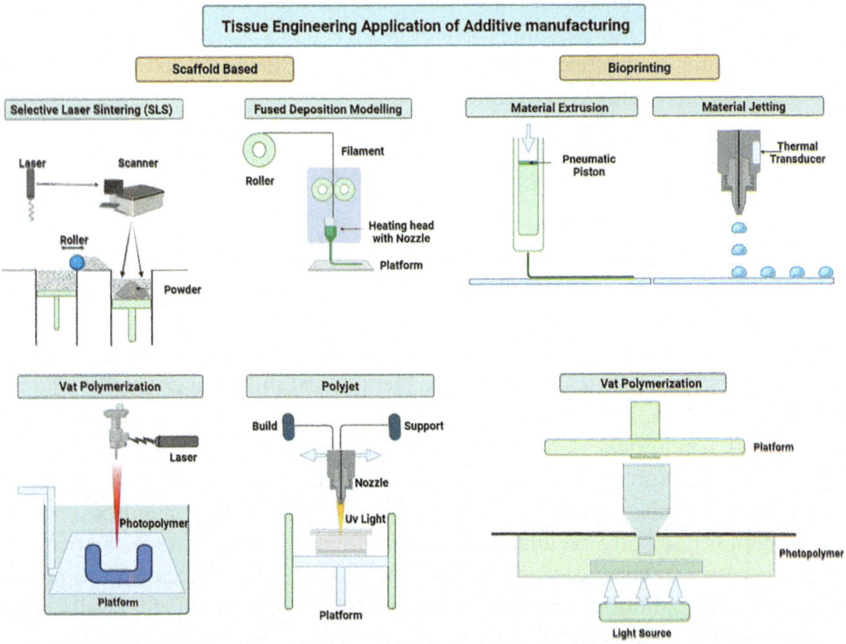

Fig. 11.3 Additive manufacturing in tissue engineering showing scaffold and bioprinting. (Created by using Biorender.com)

and with sufficient support. Viscosity is critically important for the stability and accuracy of scaffolds. High viscosity bioinks give better structural integrity, though they are apt to induce high shear stress that can damage cells during extrusion. However, low-viscosity bioinks are excellent for extrusion, although they often do not support maintaining the printed structure's structural stability, leading to buckling or collapse altogether [156]. Shear-thinning is a desired property for hydrogel-based bioinks. To account for the non-Newtonian nature of hydrogels, it allows bioinks' viscosity to decrease at high shear rates. Thus, the bioink can also be easily extruded from the nozzle during printing, and at the same time, increases in low shear rates support the scaffolds' structure. This behavior is significant for high-resolution bioprinting without hurting cells (Tian et al. 2021).

Maintaining a consistent viscosity for bioink across diverse batches is difficult due to variations in raw materials and environmental conditions. Viscosity is also significantly influenced by variables such as polymer concentration, weight of the molecules, temperature, and pre-cross-linking methods. For instance, hydrogels with lower acrylate content (0AC, 0.1AC, 0.2AC) exhibited higher viscosity and printability than those with higher acrylate content (0.3AC, 0.4AC), which had lower viscosity and appeared to have serious problems maintaining structural stability [157].

For extrusion-based bioprinting, the recommended viscosity typically falls within the range of 30–6000 mPa·s, depending on the application and cell. For example, the hydrogel bioinks' viscosity at various shear rates exhibited a shear-thinning behavior where their viscosities decreased with an increase in the rate of applied shear. The highest viscosities were detected in hydrogel compositions with higher acrylate content (0.5AC, 0.6AC) compared to other compositions [158]. Viscosity has further implications for the hydrogels' mechanical properties, like yield strength and yield strain at different temperatures.

The hydrogel of a 0AC composition had yield strength values of 2.78 kPa at room temperature and 2.03 kPa at 37 °C, with a yield strain of 17.71% at room temperature (R.T.) and 23.60% at 37 °C. In contrast, 0.4AC composition showed much higher yield strengths (50.22 kPa at R.T. and 72.77 kPa at 37 °C) and yield strain (44.32% at R.T. and 35.62% at 37 °C), pointing to the relationship between hydrogel composition, viscosity, and printability [157, 159].

11.6.2 Flow Behavior

The flow behavior of hydrogel-based bioinks is a critical factor in determining the ease of printing plus the accuracy of scaffolds. Suitable flow properties facilitate smooth extrusion through the printer nozzle, ensuring the maintenance of the intended shape and structure upon deposition. Consequently, these factors significantly impact the scaffolds' printability and biological functionality.

Hydrogel-based bioinks commonly display non-Newtonian, shear-thinning characteristics, characterized by a decline in viscosity with increasing shear rate. This attribute is essential in preventing nozzle blockages during printing and promoting consistent flow when subjected to shear stress. The bioink recovers its viscosity once the shear stress is relieved, aiding in upholding the integrity of the scaffold structure [157, 160]. Substances such as gelatin-methacryloyl and Gellan gum effectively leverage shear-thinning behavior for precise printing, ensuring printability and structural fidelity [157, 161].

The shear stress endured by cells during bioprinting significantly influences cell viability. Hydrogels with lower shear stress rates at moderate pressures are favored for maintaining printing accuracy while preserving cell viability. Studies on cell viability reveal that hydrogel bioinks extruded at pressures between 50 and 100 kPa exhibit higher cell viability than those extruded at elevated pressures, which can compromise cell health [161].

Flow curves, depicting the correlation between shear stress and shear rate, are instrumental in characterizing the flow behavior of bioinks. Desirable shear-thinning behavior is illustrated by a curve demonstrating a substantial viscosity drop as the shear rate increases. Furthermore, yield stress, representing the minimum stress required to initiate flow, is vital for upholding the structural integrity of the printed scaffold post-extrusion. Bioinks with yield stress exceeding 100 Pa are favored as they prevent the deformation of the printed layers under their weight [161, 162].

Additives like polyethylene oxide (PEO) and crosslinkers such as sodium persulfate (SPS) have a notable impact on the flow behavior and printability of bioinks. These enhancements enhance viscosity and printability by inhibiting droplet formation and ensuring a consistent jet formation during printing. For instance, incorporating 0.5% PEO into a gelatin-methacryloyl hydrogel elevates its viscosity by approximately 30%, enhancing its printability [161].

11.6.3 Flow Rate

The FR is a critical parameter that significantly impacts scaffolds' print fidelity, structural integrity, and cell viability. It is the volume or bulk of bio-ink traveling via a printer's nozzle per unit of time. It is affected by factors like nozzle diameter, applied pressure, bio-ink viscosity, and printing speed [157, 159, 161, 163, 164]. Lower flow rates result in finer filaments with improved resolution but necessitate higher pressure, which can impact cell viability. Conversely, higher flow rates increase the diameter of printed fibers, reducing resolution and potentially causing fiber coiling [157, 159].

Optimizing the FR is crucial for maintaining high print fidelity and balancing cell viability [157]. Achieving optimal print fidelity relies on calibrating the flow rate to ensure continuous filament formation and structural integrity. The printability range for different bio-inks, such as gelnor and silk fibroin, is contingent upon the applied pressure and flow rate [159]. Higher flow rates can result in increased shear stress, potentially harming the cells within the bioink. Ideal bio-inks display shear-thinning behavior, minimizing cell damage during extrusion [157]. The correlation between flow rate and cell viability is impacted by dispensing pressure and nozzle diameter, with optimal flow rates reducing shear stress and improving cell survival [164]. Cell viability can vary from 40% to 80%, potentially reaching up to 97% with optimized parameters [161].

The flow rate is also dependent on the type of bioprinter, and hydrogel utilized. Key aspects, like the diameter of the nozzle and applied pressure, have a critical role in influencing flow rates. In extrusion-based bioprinters, typical nozzle diameters range between 200 and 1000μm [157, 163]. The applied pressure must be optimized based on the viscosity of the bio-ink, with higher pressures necessary for bio-inks along with viscosities ranging from 30 to 6×10^7 mPa·s [157].

The characteristics of the rheological nature of the bio-ink, including viscosity and shear-thinning behavior, directly impact the flow rate. Bio-inks with higher viscosities require increased pressure for extrusion but offer superior structural support. Conversely, low-viscosity bio-inks are more accessible to extrude at lower pressures but may lack mechanical support [161]. Shear-thinning bio-inks, which exhibit a reduction in viscosity with higher shear rates, are favored for facilitating smoother extrusion at elevated flow rates without compromising cell viability [164].

Printability is significantly affected by the flow rate, influencing the bio-inks ability to create continuous, consistent filaments. Higher flow rates can enhance printing speed but may jeopardize resolution, whereas lower flow rates improve resolution at the expense of prolonging the printing process [159]. A steady flow rate is vital in fabricating scaffolds with the intended resolution, mechanical characteristics, and biological functionality [163].

11.6.4 Shear Stress

Shear stress refers to the force per unit area exerted by a fluid parallel to the surface it flows over, impacting the viability and functionality of cells within hydrogel-based bioinks. Maintaining optimal shear stress levels is crucial for cell viability, proliferation, and successful tissue formation [157].

The rheological characteristics of hydrogel-based bioinks significantly influence shear stress levels during bioprinting. Most bioinks exhibit non-Newtonian flow behavior, showing shear-thinning properties. This behavior, lowering viscosity under shear strain, aids in reducing shear stress during extrusion, preventing clogging, and enhancing scaffold fidelity. Bioinks like alginate and gelatin possess these attributes, making them suitable for bioprinting applications emphasizing cell viability [165]. Accurately measuring shear stress is vital for optimizing the bioprinting process. Low shear stress levels, achievable at moderate extrusion pressures, are ideal for preserving cell integrity. For optimal cell viability, shear stress levels typically should remain under 5 kPa, as excessive stress can significantly reduce viability rates [166].

Excessive shear stress during bioprinting can lead to cell deformation, rupture, apoptosis, or necrosis. Therefore, meticulous control of shear stress appears vital for sustaining the cell's high viability throughout and post-bioprinting. Notably, cell viability can plummet at elevated shear stress levels, with endothelial cells exhibiting a significant response at specific thresholds where the viability rate drops below 80% at shear stress levels exceeding 10 kPa [167].

The composition and concentration of hydrogels directly influences bioink viscosity and, consequently, the shear stress experienced during bioprinting. Adjusting polymer concentration and molecular weight and introducing shear-thinning agents can fine-tune viscosity to achieve desired shear stress levels. Moreover, different crosslinking methods, both physical and chemical, significantly impact bioink rheological properties and shear stress. Selecting a suitable crosslinking method is necessary for maintaining the needed balance between the elasticity and viscosity for effective bioprinting [168]. Balancing high-resolution printing with low-shear stress presents a notable challenge in 3D bioprinting. A precise adjustment of printing parameters and bioink properties is necessary to achieve this equilibrium. Proper management of shear stress is essential to prevent scaffold deformation and collapse, ensuring structural integrity. Techniques such as coaxial printing, the use of bioinks with optimized rheological properties, and employing computational fluid dynamics (CFD) simulations to predict SS (shear stress) can enhance bioprinting outcomes [169].

11.6.5 Viscoelasticity

Viscoelasticity is a fundamental characteristic of hydrogel-based bioinks used in 3D bioprinting, playing a crucial role in ensuring the printability, shape accuracy, and structural robustness of the scaffolds. This dual viscous and elastic nature is indispensable for facilitating proper material flow through the printing nozzle and preserving the deposited shape.

Assessment of the viscoelastic properties involves conducting oscillatory rheology tests such as amplitude and frequency sweeps. Amplitude sweeps demonstrate that bioinks demonstrate solid-like behavior, with the elastic modulus (G′) exceeding the viscous modulus (G″) up to a shear strain of approximately 10%. After this threshold, G″ exceeds G′, indicating a shift towards fluid-like behavior. For instance, the elastic modulus (G′) values for hydrogel bioinks are around 500 Pa without additives. Upon adding albumin, G′ values reduce to about 450 Pa with 2% albumin, 430 Pa with 4.5% albumin, and 410 Pa with 7% albumin. Frequency sweeps reveal that G′ remains higher than G″ across various frequencies (0.01–50 Hz), highlighting consistent gel-like behavior irrespective of the albumin concentration [170].

Furthermore, thixotropy and recovery are pivotal viscoelastic properties. Three-step thixotropy tests illustrate that hydrogel bioinks can recover their viscosity after high shear rates, which is crucial for upholding shape fidelity after printing. For example, bioinks exhibit initial viscosities of approximately 7800 Pa·s at a reduced shear rate (0.05 s^{-1}), with an eventual decrease to around 14 Pa·s under high shear (50 s^{-1}), recovering to about 550 Pa·s within 2 s of reducing the shear rate (Mainardi and Rezwan 2022). Viscoelastic behavior immensely impacts the fidelity and printability of shapes in bioinks. The shear-thinning property, characterized by decreasing viscosity with a rising shear rate, facilitates smoother extrusion and improved shape retention post-extrusion. An ideal bioink demonstrates a viscosity reduction from 1000 to 10 Pa·s as the shear rate escalates from 0.1 to 100 s^{-1} [171]. Furthermore, bioinks with higher storage modulus (G′) values are favored for their ability to uphold structure immediately after extrusion. A bioink possessing a G′ of 500 Pa and a yield stress of 50 Pa is anticipated to yield high-fidelity prints with minimal drooping or distortion [171].

Additives like polyethylene oxide (PEO) boost the viscosity of bioinks, ensuring consistent jet formation during processes like cell electro-writing. For instance, increased PEO concentrations (>3.4% v/v) notably elevate viscosity, preventing droplet formation and guaranteeing continuous fiber extrusion. This results in heightened electrical conductivity levels (0.9–1.3 mS/cm) compatible with the stable formation of jets under electric fields [159].

Furthermore, mechanical characteristics such as compressive modulus profoundly influence the performance of bioinks. Gelatin-alginate scaffolds with graphene oxide (GO) demonstrate compressive modulus values of 1.58 and 1.63 kPa for 1GO and 2GO scaffolds, respectively, compared to ~0.69 kPa for 0GO on day 1. Constructs with compressive moduli of 82.5 ± 18.9 MPa fall within the optimal range for bone tissue regeneration (10–1500 MPa), displaying their potential to address substantial bone defects (Pan et al. 2022). Optimal hydrogel formulations entail carefully balancing polymer concentration, crosslinking agents, and additives to achieve the desired viscoelastic properties. For example, combining 2% (w/v) alginate with 5% (w/v) gelatin can improve printability and structural integrity. Introducing nanoparticles or microgels can further enhance mechanical properties and regulate the release of bioactive compounds. For instance, adding 1% (w/v) nano-clay to a hydrogel formulation can significantly boost its mechanical strength and printing accuracy [157].

11.6.6 Nozzle Aperture Geometry, Size, Applied Pressure, and Geometry

The nozzle size aperture, geometry, and applied pressure are critical factors that significantly influence the scaffold's printing resolution, cell viability, and complete structural integrity. The diameter of the nozzle aperture is a fundamental factor that impacts the scaffold's resolution and the embedded cells' viability. Nozzle diameters ranging from 100 to 250μm are crucial in achieving finer details and higher-resolution prints. However, these smaller nozzle sizes can increase shear stress during extrusion, potentially harming cells. Research suggests that nozzle diameters between 150 and 250μm strike a balance between high resolution and cell viability, resulting in cell viability ranging from 70% to 95% [166, 172, 173]. However, larger nozzles exceeding 800μm decrease shear stress, promoting improved cell viability by reducing mechanical stress at the nozzle orifice. These larger nozzles compromise print resolution and may cause the formation of larger droplets and satellite drops. Nozzles of medium size, ranging from 250 to 800μm, offer a middle ground by providing adequate resolution while preserving a relatively high level of cell viability [174, 175]. The geometry of the nozzle, whether conical or cylindrical, significantly influences the shear stress experienced by both the bioink and the cells. Conical nozzles offer an advantage by concentrating the shear stress at the exit point, thereby reducing overall exposure time and potentially better preserving cell viability than cylindrical nozzles. Conical nozzles with diameters of 260μm or larger can sustain cell viability between 70% and 95% [166]. However, despite generating lower peak shear stresses, cylindrical nozzles prolong the passage length, leading to extended exposure to shear stress that can be detrimental to cells [174, 176].

Another critical parameter during the extrusion process is the applied pressure, which directly impacts shear stress and cell viability. Higher pressures are required to extrude bioinks through smaller nozzles, especially those with high viscosity. However, this elevated pressure also escalates shear stress, increasing cell damage. Optimal pressure ranges from 10 to 50 kPa for medium viscosity bioinks, balancing effective extrusion and preserving cell viability [173]. Applied pressures for extrusion-based bioprinting typically range from 5 to 200 kPa, depending on the bioink viscosity and nozzle diameter. Maintaining pressures within 30–50 kPa generally achieves a trade-off between flow rate and cell viability, where higher pressures enhance flow but also elevate shear stress, potentially decreasing cell viability to as low as 50–60% under extreme conditions [157, 176].

Optimizing the aperture size, geometry, and pressure applied to the nozzle is crucial for achieving high-quality 3D bio-printed structures. Using smaller nozzles can enhance resolution but requires careful management to prevent excessive cell damage from heightened shear stress. Conical nozzles are recommended for minimizing overall shear stress exposure, thereby safeguarding cell viability. It is essential to precisely regulate the applied pressure to ensure seamless extrusion without inflicting significant cell harm.

11.6.7 Volumetric Flow Rate (Q)

The Q serves as a critical processing parameter that notably affects the accuracy and quality of printed structures. It refers to the amount of bio-ink extruded per time unit, impacting extrusion speed, printed strand diameter, and overall print resolution. Precisely optimizing the volumetric flow rate is vital for achieving accurate and high-quality scaffolds [163].

The volumetric flow rate experiences a significant increase with applied pressure. For instance, elevating the pressure from 70 to 130 kPa can lead to a 15-fold rise in flow rate. This behavior is linked to the shear-thinning properties of hydrogels, where viscosity undergoes a decline as the shear rate increases, following the Power Law model:

$$\eta = Kn_{\gamma}^{\cdot} - 1$$

Where η is the viscosity, K is the consistency index, $\gamma^{\cdot}$ is the shear rate, and n is the power-law index. Constants such as $K = 222$ Pa · s^n and $n = 0.23$ can be used to facilitate the prediction of flow rates under varying pressures [163]. The eequation within the volumetric flow rate Q and the applied pressure ΔPcan be expressed as:

$$Q = \frac{\pi(3+n)}{(2K)^{1/n} L^{1/n}} \Delta P^{1/n} R^{(3+n)/n}$$

Where L is the nozzle length, and R is the nozzle radius. By way of an example, at an applied pressure of 100 kPa, the extrusion speed may be recorded at 2.6 mm/s, demonstrating the direct link between applied pressure and extrusion speed [163].

A very prominent role is played by temperature in managing the volumetric flow rate of hydrogel-based bioinks. For k-carrageenan gel systems, the optimal extrusion temperature hovers around 36 °C. This specific temperature facilitates a volumetric flow rate that yields filaments closely mirroring the nozzle diameter, a critical aspect for achieving top-notch prints. Filaments extruded at temperatures lower or higher than this threshold exhibit irregularities or structural weaknesses, underscoring the necessity for meticulous thermal regulation [177]. Precise management of the volumetric flow rate is imperative to guarantee that the printed structure aligns with the envisaged design, especially crucial for applications demanding high accuracy like tissue scaffolds. The rheological attributes of bio-inks, such as viscosity and shear stress, immediately affect the volumetric flow rate. The advantageous property of shear-thinning behavior, where the viscosity decreases under shear stress, facilitates the smooth flow of the bio-ink through the nozzle while upholding structural integrity during deposition [157]. Mathematical modeling stands as a cornerstone for optimizing the volumetric flow rate. The filament diameter (D) can be foreseen based on the correlation between the printing speed (v) and the volumetric flow rate:

$$D\alpha\sqrt{\frac{Q}{v}}$$

This correlation underscores the significance of harmonizing flow rate and printing speed to achieve the desired filament diameter without compromising structural integrity [157]. Optimizing the volumetric flow rate necessitates striking a balance among multiple factors encompassing nozzle diameter, bio-ink viscosity, and printing speed. This equilibrium is paramount in crafting bio printed structures with specified mechanical properties and cell viability while sidestepping challenges like clogging or inconsistent filament diameter. Theoretical models and empirical validations furnish a robust framework for predicting extrusion speeds and ensuring uniformity across diverse pressure settings, ultimately bolstering scaffold reliability [163].

11.7 Potential Biopolymers for Additive Manufacturing Scaffold Applications

11.7.1 Natural Polymers

11.7.1.1 Alginate

Alginate, a natural polymer obtained via brown seaweed, has attracted considerable attention for additive manufacturing in tissue engineering because of its distinctive characteristics. It is an anionic polysaccharide consisting of β-(1→4)-linked D-mannuronic acid (M block) and α-L-guluronic acid (G block) units arranged in GG, MG, and GM blocks [178, 179]. Alginate is valued for its biodegradability, non-toxicity, biocompatibility, ease of gelation and lack of antigenicity [178–180]. These attributes position alginate as a material of choice for fabricating biocompatible scaffolds in tissue engineering.

Alginate finds extensive usage in bioink formulations for 3D bioprinting, especially in applications like cartilage repair, wound healing, drug delivery and bone regeneration [178, 180, 181]. Its capacity to replicate the extracellular matrix (ECM) augments its applicability across diverse biomedical settings. By creating an ECM-mimetic environment conducive to cell integration and survival, alginate enhances the printability and mechanical strength of the fabricated structures [178, 179]. This feature is vital for preserving the capability and functionality of cells within tissue-engineered scaffolds. The properties of alginate-based bioinks can be tailored by adjusting factors such as the G/M ratio, molecular weight, solid content, and cell density to fulfill specific biomedical requisites [178, 182]. This capacity for customization enables the production of targeted scaffolds tailored for diverse tissue engineering applications.

Alginate undergoes ionic crosslinking with divalent cations like calcium ions (Ca^{2+}) to produce hydrogels. This crosslinking mechanism imparts structural integrity and mechanical rigidity, making it suitable for various bioprinting techniques [182]. Employing calcium ions (Ca^{2+}) for crosslinking enhances the architectural accuracy of printed structures, ensuring prolonged stability and mechanical durability [179] (Hurtado et al. 2022). The preservation of structural integrity is crucial for the practical application of alginate-based scaffolds in tissue engineering. Optimal mechanical properties and printability of alginate-based structures in crosslinking processes are commonly achieved by employing varying concentrations of calcium chloride ($CaCl_2$), typically around 100 mM [178, 182]. For instance, in CNF/alginate bioprinting, 100 mM $CaCl_2$ was used to fabricate cartilage tissue engineering scaffolds. Furthermore, adipose tissue engineering and cartilage lesion treatment have utilized hyaluronic acid CNF/alginate bioinks crosslinked with 100 mM $CaCl_2$ or 0.001% H_2O_2 [179].

Utilizing human chondrocyte-laden bioink formulated with CNF and alginate, crosslinked with 100 mM $CaCl_2$, has created printed grids, solid discs, and patient-specific auricular constructs. Such formulations highlight the versatility and effectiveness of alginate in generating intricate, biocompatible structures. The commercial CELLINK® bioink, a composite of CNF-alginate, has undergone in vitro and in vivo evaluations, demonstrating promising outcomes across various tissue engineering applications [179].

Alginate gel formation is primarily attributed to GG blocks interacting through ionic cross-linking with divalent cations such as Ca^{2+}, Zn^{2+}, and Mg^{2+} [178]. To address challenges like low mechanical strength, alginate is frequently blended with other substances like poly (ethylene oxide) (PEO) or utilized in composite structures with synthetic polymers [183]. These blends enhance the mechanical characteristics and overall functionality of alginate-based scaffolds.

Regarding mechanical traits, alginate-based scaffolds display favorable properties effective in bone tissue engineering when coupled with materials like poly (lactic-co-glycolic acid) and MgO [183]. However, utilizing alginate as a binder in hydroxyapatite (HAP) scaffolds have exhibited compatibility with cell growth and regeneration, highlighting the viability for bone tissue applications [181]. These scaffolds possess porosity and mechanical attributes that endorse their utility in biomedical scenarios.

Recent progressions in bio-fabrication methodologies, encompassing microfluidics, electrospinning, and 3D bioprinting, have broadened the utility of alginate in regenerative medicine. Alginate is a prevalent option for constructing cell-laden microgels utilizing microfluidic systems, offering precise regulation over the microgel size and shape, crucial for mimicking the natural cellular environment. Moreover, alginate/PEO blends are employed to manufacture cell-laden electro spun fibers containing encapsulated osteoblasts, displaying substantial cell viability and osteogenic differentiation potential [180].

11.7.1.2 Collagen

Collagen, a natural polymer, is highly valued in additive manufacturing for the engineering of tissue structure due to its unparalleled properties. It is a crucial biopolymer obtained from animal sources such as porcine or bovine tissues and is meticulously purified to ensure safety and remove immunogenic components [181, 183, 184]. Collagen fibrils typically have a diameter of around 500 nm and a length of approximately 1 cm, with its fundamental structural unit being a 300 nm protein consisting of three braided α-subunits [184].

Collagen stands out for its excellent biocompatibility, making it an ideal scaffold material for various applications in tissue engineering. It promotes proliferation, differentiation, and cell attachment, which is crucial for successful tissue regeneration [181, 183, 185]. Its bioactivity enhances cellular interactions and tissue integration, offering structural support and biochemical signals to cells that create effective tissue scaffolds [184, 195]. Moreover, collagen is biodegradable, adjusting its degradation rate to match tissue regeneration rates, ensuring the scaffold supports tissue growth without premature degradation [181, 184].

The porosity of collagen-based scaffolds is a considerable factor in facilitating cell infiltration and tissue growth. Optimal pore sizes typically fall within 200–800μm [181, 183, 185]. These scaffolds are engineered to mimic the extracellular matrix, creating a favorable environment for tissue growth [185]. The properties of the mechanical nature of the collagen scaffolds, including Young's modulus and tensile strength, can be modified using various crosslinking techniques. For example, glutaraldehyde crosslinking can increase tensile strength, while enzymatic crosslinking improves stability [184]. Notably, combining collagen with hydroxyapatite (HA) amplifies mechanical strength, making it suitable for applications in bone tissue engineering [181].

Various methods for fabricating collagen-based scaffolds include freeze-drying, electrospinning, and solvent casting. Each technique offers unique advantages in controlling scaffold architecture, porosity, and mechanical properties [184, 185, 196]. For instance, electrospinning can produce fibrous scaffolds that fabricate the architecture of the natural extracellular matrix, while freeze-drying can generate highly porous structures that benefit cell infiltration [184]. Moreover, surface modification techniques such as chemical grafting or coating with bioactive molecules can further enhance the functionality of collagen scaffolds. These alterations improve cell adhesion, proliferation, and differentiation, facilitating better integration with host tissues [182, 196]. Incorporating bioactive molecules like growth factors can significantly boost the scaffold's capacity to support tissue regeneration [184]. Despite its numerous benefits, collagen poses challenges such as potential immunogenicity and variability in mechanical properties. Addressing these issues requires meticulous material selection, processing, and cross-linking techniques to ensure consistent and reliable performance of the scaffold [182, 184] (Arifin et al. 2019). For instance, the immunogenicity of collagen can be mitigated through rigorous purification processes and the utilization of recombinant technologies for human collagen production [182].

The possibilities of collagen usage in tissue engineering are extensive, spanning from skin regeneration and bone repair to complex tissue reconstructions. Its adaptability and interplay with other biomaterials make it a suitable candidate for composite scaffolds, elevating their mechanical and biological attributes [181, 183]. Contemporary progress in bio-fabrication techniques and the emergence of innovative crosslinking methods have broadened the utility of collagen in regenerative medicine, underscoring its pivotal role in the advancement of functional tissue constructs [180, 184].

11.7.1.3 Chitosan

Chitosan is sourced from chitin in the exoskeletons of crustaceans like crabs and shrimp. It is a valuable biopolymer in additive manufacturing (AM) scaffold applications for regenerative medicine and tissue engineering. Known for its exceptional biocompatibility, antimicrobial properties and biodegradability, chitosan is highly suitable for various biomedical applications.

Chitosan demonstrates adequate strength for soft tissue engineering but can be further reinforced by blending with other materials like HA or polycaprolactone (PCL). For example, chitosan/HA composites find utility in bone tissue engineering due to their ability to replicate the natural bone matrix and enhance osteoconductive, creating a conducive environment for bone regeneration. Chitosan-based scaffolds support the adhesion and proliferation of chondrocytes, thereby assisting in cartilage regeneration.

The versatility of chitosan extends to its application in hydrogels and wound dressings. Chitosan hydrogels are efficacious in wound dressing applications as they sustain a moist environment, facilitate drug delivery, and foster cell proliferation and migration. Moreover, these hydrogels can create antibacterial wound dressings that shield against infections and expedite healing [186]. Recent progress in additive manufacturing has facilitated the creation of intricate chitosan-based scaffolds characterized by precise geometries and controlled porosity. Hybrid scaffolds comprising hydroxyapatite, chitosan, and silica developed through sol-gel techniques in combination with 3D plotting methods have exhibited improved mechanical attributes and bioactivity, rendering them suitable for bone tissue engineering purposes. Investigations on double-layer wound dressings composed of chitosan, polyvinyl alcohol, and copper have revealed notable antimicrobial properties and enhanced wound healing results both *in vivo* and *in vitro* [187].

Utilizing chitosan scaffolds with hydroxyapatite can yield compressive strengths reaching up to 110 MPa with a porosity of 34%, a feature well-suited for bone tissue engineering endeavors. The porosity levels within chitosan-based scaffolds may vary from 40% to 85%, facilitating cell infiltration and tissue growth. The degradation rate of chitosan-based scaffolds can be customized to promote the formation of new tissues, thereby preventing premature breakdown or excessively prolonged presence. Chitosan wound dressings have exhibited a capacity to reduce bacterial load by as much as 90%, underscoring their effectiveness in infection prevention.

Chitosan demonstrates compatibility with a range of additive manufacturing techniques, including solvent casting, particulate leaching, and electrospinning, enabling the fabrication of scaffolds with exact internal composition and porosity. Photopolymerization 3D printing has exhibited potential in producing high-resolution chitosan scaffolds characterized by finely controlled dimensions, enhancing mechanical properties and functionality. Innovations such as chitosan/polycaprolactone scaffolds intended for lung tissue engineering applications exhibit promise for deployment in studies related to COVID-19.

Chitosan possesses unique chemical and structural characteristics, such as a significant nitrogen content, hydrophilicity, crystallinity, and polyelectrolyte properties, which underlie its diverse applications. Alterations like cross-linking, graft copolymerization, carboxymethylation, etherification, esterification, and phosphorylation enhance its solubility, mechanical strength, and biological functions, broadening its use in biomedical sectors. Cross-linked chitosan hydrogels can absorb and hold generous amounts of water, facilitating swelling, while combining synthetic polymers through grafting can enhance solubility and antimicrobial qualities.

11.7.2 Synthetic Polymer

11.7.2.1 Polycaprolactone

Polycaprolactone (PCL) is a versatile synthetic polymer extensively utilized in additive manufacturing for scaffold applications in tissue engineering due to its biocompatibility. Despite its hydrophobic nature, which can restrict cell interactions, PCL's exceptional processability via electrospinning and 3D printing facilitates the fabrication of minute scaffold structures.

As a semicrystalline polymer with slow degradation kinetics, PCL typically requires up to 2 years for complete degradation, a feature advantageous for applications that necessitate long-term stability [188]. The polymer degrades via hydrolysis and has a melting point of approximately 60 °C [185]. The mechanical properties of PCL could be customized to meet various tissue requirements. PCL scaffolds exhibit compressive potencies ranging from 2.74 to 55.95 MPa and yield strengths from 1.17 to 5.03 MPa, rendering them suitable for bone tissue engineering applications where mechanical compatibility with bone is critical. PCL's Young's modulus also varies from 343.9 to 363.4 MPa [182].

Manufacturing techniques like fused deposition modeling (FDM) have confirmed the non-toxic nature of PCL scaffolds, further confirming their biocompatibility for tissue engineering purposes. PCL bone scaffolds developed through FDM have been granted FDA approval, highlighting their appropriateness for clinical utilization [181]. PCL's inherent hydrophobicity and modest water absorption capacity limit its cellular interactions, but these attributes can be advantageous in scenarios necessitating gradual degradation. Recent progress includes the innovation of functional polyesters incorporating aliphatic phenyl magnesium chloride (PhMgCl) into PCL, enhancing hydrophilicity to promote improved cell adhesion, proliferation, and differentiation [213]. Notably, PCL/HA composites bolster mechanical properties and biocompatibility by harnessing the osteoconductive potential of HA [213]. PCL also holds promise in skin and vascular tissue engineering applications, such as the fabrication of vascular grafts and skin tissue scaffolds (Shiva et al. 2023).

Various fabrication techniques optimize PCL's utility in scaffold applications. Electrospinning enables the creation of nanofiber scaffolds with notable mechanical stability and porosity, crucial for tissue regeneration. Electro spun PCL/HA-based scaffolds exhibit augmented osteogenic characteristics [181]. Approaches like FDM, selective laser sintering (SLS), and stereolithography (SLA) yield PCL-based scaffolds, tunable porosity, and tailored mechanical attributes. Integrating electrospinning and 3D printing methods, such as melt electrospinning writing, yields scaffolds featuring hierarchical structures and improved qualities for diverse tissue engineering applications [185].

The amalgamation of PCL with other materials notably enhances its characteristics. For instance, adding HA enhances osteoconductive and osteoinductive properties, while tricalcium phosphate (TCP) improves mechanical and biological attributes. Gelatin (Gel) boosts hydrophilicity, enhancing cell adhesion and proliferation (Shiva et al. 2023). Recent research has highlighted the efficacy of PCL gel-based scaffolds created through electrohydrodynamic printing (EHDP) in promoting neural tissue regeneration. Moreover, PCL/20 wt.% HA-based scaffolds have enhanced mechanical characteristics and biocompatibility, rendering them appropriate for bone tissue engineering endeavors [213].

11.7.2.2 Poly (Vinyl Alcohol)

Poly (vinyl alcohol) (PVA) is a versatile synthetic polymer extensively used in additive manufacturing for scaffold purposes in the engineering of tissues because of its helpful biocompatibility, biodegradability, and non-toxic attributes. Its hydrophilic nature enhances its applicability in biomedical settings like wound dressing and tissue engineering, which are crucial for maintaining a moist environment and offering mechanical support.

Regarding mechanical characteristics, PVA demonstrates significant properties suitable for scaffold applications. The tensile strength of PVA typically falls between 2.32 and 2.52 MPa, coupled with Young's modulus of approximately 97.8 MPa when blended with hydroxyethyl cellulose. The elongation at break for hydroxyethyl cellulose-PVA nanofibers is about 12.43%, indicating flexibility and durability in biomedical applications. Moreover, the common fiber diameter of PVA nanofibers ranges from 100 to 373 nm, depending on the specific blend and fabrication method, enabling precise control over scaffold architecture [189].

PVA's notable swelling capacity is highlighted by a swelling of around 108 ± 6.45% for sodium alginate-PVA nanofibers. This property ensures that PVA-based scaffolds can absorb significant amounts of fluid, which is beneficial for wound healing. The hydrophilicity of PVA is evident from its water contact angle of approximately 31.9°, aiding in enhanced cell attachment and proliferation. Additionally, PVA nanofibers combined with silver nanoparticles demonstrate antibacterial efficacy, showing inhibition zones of 1.1 ± 0.05 cm against E. coli and 1.3 ± 0.08 cm against *B. circulans*, indicating their potential to prevent infections in wound dressings (Alven and Aderibigbe 2021). Regarding biodegradability, PVA undergoes environmental degradation through photodegradation, biodegradation, and chemical degradation, rendering it a sustainable choice for medical purposes. Its hydrogels, three-dimensional networks of cross-linked hydrophilic macromolecules, exhibit an optimal water content (up to 90%), elasticity, and softness, making them suitable for applications requiring biocompatibility and low toxicity. PVA is extensively utilized in tissue engineering as it can imitate the extracellular matrix, supporting cell growth and proliferation. PVA-based composites such as PVA/alginate and PVA/chondroitin sulfate are employed in bone tissue engineering, while PVA/gelatin and PVA/poly L-lysine composites enhance wound dressing capabilities. These composites demonstrate notable mechanical and physical properties, rendering them effective for various biomedical applications. Recent developments have demonstrated that PVA composites can substantially boost their performance. For example, PVA/chitosan/polydopamine-functionalized graphene oxide composites are efficient at adsorbing ion elements, whereas PVA/alginate composites are utilized as slow-release fertilizers. These innovations underscore PVA's adaptability and potential for diverse applications in tissue engineering and beyond [190].

PVA's limitations, such as insufficient elasticity and rigidity, are commonly tackled by blending it with other polymers to improve its mechanical features, facets and stability. Despite these challenges, PVA remains a potent material for additive manufacturing scaffolds in tissue engineering because of its biocompatibility, ease of fabrication, and ability to promote wound healing. Introducing other biopolymers and bioactive agents into PVA nanofibers further boosts their mechanical properties and biological performance, hailing them extremely useful for various biomedical applications [195].

When employed as a thermoplastic binder within composite filaments, PVA retains the structural integrity of printed objects until their complete removal during the sintering process. This function has a pivotal contribution in the fabrication of intricate scaffold structures suitable for colonization with mesenchymal stem cells in applications of bone tissue engineering [191]. Furthermore, the contribution of PVA to improving the printability and antimicrobial characteristics of chitosan composites displays its efficacy in developing functional and efficient biomedical devices [192].

11.7.2.3 Polyurethane

Polyurethane (PU) is a highly versatile synthetic polymer widely used in additive manufacturing for scaffold applications in tissue engineering owing to its advantageous mechanical properties, biocompatibility, and biodegradability. Its tensile strength is 31–45 MPa, rendering it durable enough to cater to various biomedical applications, including load-bearing functions like bone tissue engineering (Shiva et al. 2023). Furthermore, PU's elastic modulus, varying from 1.14 to 1.55 GPa, offers the necessary balance of flexibility and rigidity to meet diverse tissue engineering needs (Shiva et al. 2023). With an approximate density of 0.905 g/cm^3 and an elongation at a break of around 11%, PU highlights flexibility and resilience under stress conditions (Shiva et al. 2023). Additionally, possessing a glass transition temperature (Tg) of approximately −25 °C, PU retains its properties under different environmental conditions. At the same time, its melting point of 160–163 °C plays a crucial role in processing and fabrication (Shiva et al. 2023).

When reprocessed, thermosetting PU exhibits minimal stiffness loss in the second cycle. Still, experiences diminished performance in the following cycles, with tensile strength decreasing from 15 MPa to below 12 MPa post the third cycle [193]. This mechanical degradation bears significance concerning the extended usage and recyclability of PU in biomedical fields. PU scaffolds boast extremely permeable structures with interconnected pore networks critical for cell proliferation and nutrient transportation. The macropore sizes can vary from 400 to 450μm, while the micropore sizes range from 5 to 50μm, depending on the specific application and fabrication technique [194]. PU's exceptional biocompatibility ensures it does not elicit adverse immune reactions upon implantation, promoting tissue regeneration while averting inflammation or rejection. Moreover, its biodegradability enables gradual scaffold degradation, facilitating seamless integration with host tissue as natural tissue gradually replaces it over time [195]. Conventional PU synthesis involving isocyanate monomers and organic solvents poses toxicity and environmental hazards. Nonetheless, alternatives like waterborne or water-based PU obviate toxic reagents, rendering the process more environmentally sustainable and safer for biomedical applications [193]. Various fabrication techniques heighten the efficacy of PU in scaffold applications. Solvent casting and particulate leaching yield extremely permeable scaffolds with interconnected pore structures crucial for cell viability and nutrient exchange. Combining solvent casting with centrifugation bolsters pore uniformity and scaffold durability. Furthermore, photo-cross-linkable PU enriched with additives like hyaluronic acid enhances PU scaffolds' mechanical attributes and biological functionality, making them ideal for cartilage tissue engineering [185].

PU's versatility in supporting the regeneration of hard and soft tissues renders it indispensable in tissue engineering. PU scaffolds are particularly favored in bone and cartilage tissue engineering for their adaptable mechanical properties and biocompatibility. Customizable to mimic bone rigidity or cartilage flexibility, these scaffolds give an abundant milieu for cell proliferation and differentiation. PU scaffold porosity and mechanical strength tunability allow tailored customization to meet diverse tissue requirements, a crucial advantage in crafting functional tissue-engineered constructs [195]. PU's versatility is evident in its application for developing hybrid constructs, such as a polyurethane-gelatin construct designed for producing implantable bioartificial livers. This use case underscores PU's pivotal role in sophisticated tissue engineering applications. Furthermore, PU has been employed in rapid prototyping to fashion vascular systems, displaying its capacity to fabricate elaborate and functionally crucial structures within tissue engineering. Integrating inorganic nanoparticles and other fillers during the processing phase substantially boosts PU's mechanical properties, enhancing compressive strength and modulus [194].

11.7.2.4 Polyether Ether Ketone

Polyether ether ketone (PEEK) is renowned for its exceptional mechanical properties, biocompatibility, and chemical resistance, making it an optimal choice for AM applications, especially in tissue engineering and biomedical fields.

PEEK demonstrates mechanical characteristics comparable to human bone, essential for its suitability in load-bearing implants. With a tensile strength ranging from 90 to 100 MPa and an elastic modulus of 3.5 to 4.0 GPa, PEEK helps minimize stress shielding post-implantation, thereby reducing the risk of osteopenia, and ensuring its viability for long-term use in biomedical settings [196]. Notably, PEEK's chemical robustness and biocompatibility are commendable, as it remains non-biodegradable, radiolucent, and resistant to heat and various chemicals, ensuring implant longevity and stability even in challenging physiological environments. Within additive manufacturing, PEEK finds extensive utility in FDM and SLS techniques. These approaches enable the precise and tailored fabrication of medical implants, such as bone replacements and dental prostheses. In bone reconstruction contexts, PEEK has demonstrated efficacy in statically and dynamically loaded implants like mandibles and cranial implants. The capacity for crafting patient-specific implants with bespoke shapes is particularly advantageous in craniofacial surgeries, offering personalized solutions that enhance patient outcomes [205]. Moreover, PEEK's application in dental prostheses capitalizes on its biocompatibility and mechanical properties, rendering it ideal for enduring dental usage.

Despite its benefits, PEEK's bioinert nature poses a challenge to osseointegration, prompting various modification techniques to enhance its bioactivity. The incorporation of HA into PEEK composites significantly boosts their biological characteristics. These PEEK/HA composites strike a balance between mechanical strength and bioactivity crucial for bone regeneration. Moreover, surface alterations such as sulfonation and the application of bioactive coatings like titanium oxide (TiO_2) or polydopamine (PDA) have proven effective in improving osseointegration and cellular responses [197].

For instance, a composite containing 70% PEEK, and 30% calcium hydroxyapatite (cHAp) exhibits a Young's modulus of 5.695 GPa and a shear modulus of 2.1454 GPa. These figures indicate that the composite not only upholds the strength and rigidity needed for load-bearing applications but also closely mimics the mechanical properties of natural bone compared to pure PEEK [198]. PEEK's thermal properties contribute to its suitability for additive manufacturing. With a glass transition temperature (Tg) of around 143 °C and a melting temperature (Tm) of approximately 343 °C, PEEK can maintain its properties under the elevated temperatures used in additive manufacturing processes [199]. The requirement for high printing head temperatures (up to 520 °C) and bed temperatures (up to 160 °C) ensures that the structural integrity of printed components remains intact under physiological conditions. PEEK's mechanical strength and capacity to fabricate patient-specific implants align with anatomical contours without compromising physiological functions in craniomaxillofacial bone reconstruction. Incorporating tantalum (Ta) nanoparticles into PEEK significantly enhances its osseointegration abilities. Specifically, composites containing 3% and 5% Ta content exhibited improved cell adhesion, collagen secretion, and the expression of genes related to osteogenesis. Likewise, the enhancement of carbon-fiber-reinforced PEEK (CFR-PEEK) with graphene improves its microstructural features and the average rate of mineral deposition, surpassing conventional CFR-PEEK implants [200].

11.7.2.5 Acrylonitrile Butadiene Styrene

Acrylonitrile butadiene styrene (ABS), a commonly used thermoplastic polymer in additive manufacturing, especially in FDM, is recognized for its advantageous mechanical properties, processing convenience, and cost efficiency. ABS plays a pivotal role in fabricating scaffolds for tissue engineering applications.

ABS demonstrates mechanical solid characteristics suitable for diverse uses. It boasts a tensile strength ranging from 22 to 50.5 MPa and an elastic modulus between 1.3 and 3.6 GPa, striking a balance between flexibility and rigidity crucial for load-bearing functions in tissue engineering projects [201, 202]. The material's impact resistance, vital for scaffold durability under mechanical strain, is approximately 15–20 kJ/m^2 [203]. Moreover, ABS elongates at a break of around 8%, highlighting its deformability before fracture, which is advantageous for components exposed to dynamic or cyclic loads [181].

With a glass transition temperature (Tg) of about 105 °C, ABS preserves its structural integrity at moderate temperatures encountered in tissue engineering processes [204]. Its melting temperature falls within the range of 220–260 °C, enhancing processing ease in FDM and fused filament fabrication (FFF) operations [205]. ABS's thermal conductivity of approximately 0.18 W/m·K aids in maintaining dimensional stability throughout the printing procedure [201]. ABS is not widely utilized in direct biological applications due to its lower native biocompatibility than biopolymers such as PCL or PLA. Nonetheless, the biocompatibility of ABS can be significantly enhanced through surface modifications. For example, photo grafting with polyethylene glycol (PEG) can boost hydrophilicity and reduce protein adhesion, rendering ABS more suitable for biomedical applications [181]. Furthermore, techniques like chemical dissolution using acetone can be applied to fabricate watertight microfluidic devices, augmenting the usability of ABS in biomedical fields [204].

ABS fabricates scaffolds that promote tissue growth; however, it typically necessitates surface modification or coating with biocompatible materials to meet medical standards (Shiva et al. 2023). Strengthening ABS with additives like carbon fibers can substantially enhance its properties. A composite containing 10 wt% carbon fiber can achieve a flexural modulus of up to 8 GPa and a flexural strength of 125 MPa, making the composite notably stiffer and more robust than pure ABS [202]. Additionally, graphene oxide/ABS composites exhibit potential for biomedical applications owing to improved mechanical and electrical properties, rendering them suitable for advanced applications [201].

Despite its adaptability, ABS encounters various challenges in biomedical applications. Its intrinsic biocompatibility is inferior to other biopolymers, necessitating surface modifications to meet requirements for direct biomedical applications [181].

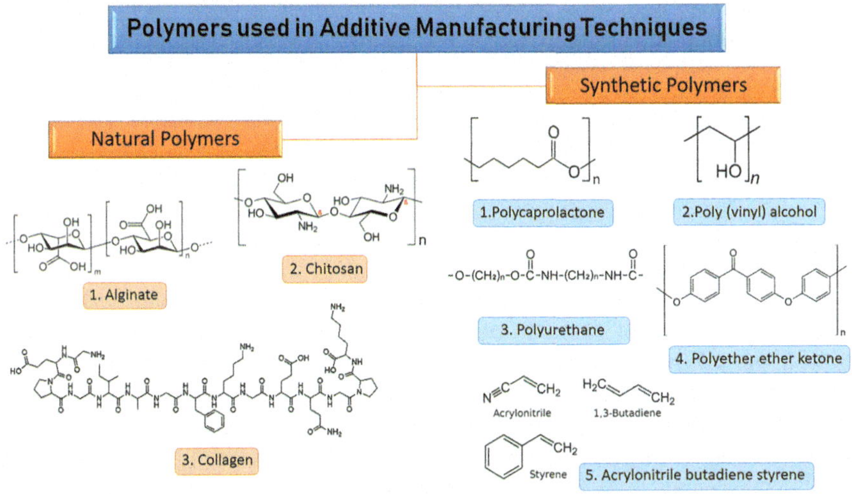

Fig. 11.4 Polymers used in different additive manufacturing techniques. (Created by using Biorender.com)

Additionally, ABS may experience warping during cooling due to its high shrinkage rate, necessitating controlled environmental conditions during the printing process [201]. ABS also emits harmful fumes; therefore, adequate ventilation is required during printing [206]. Polymers used in different additive manufacturing techniques (Fig. 11.4).

11.8 Current Challenges and Future Direction

Challenges faced in additive manufacturing usually occur due to limitations in creating intricate structures, slower speed compared to traditional methods, higher costs, and concerns about the quality and sustainability of additive manufacturing in the current scenario.

- (i) Speed: 3D printing is comparatively slower than traditional methods for mass production of materials. Automation can help initially, but traditional methods are still faster for large-scale production.
- (ii) Cost-effectiveness: 3D printing is found to be cost-effective only for small productions due to fewer machines and less waste, while for extensive production, traditional methods are found to be faster and more energy-efficient
- (iii) Quality: Structural flaws and uneven appearances can occur due to layering and require expensive high-resolution laser-based techniques to improve their quality
- (iv) Sustainability: The initial advantages of 3D printing based on sustainability are neutralized by their high energy consumption and requirement of material

refinement. Traditional methods are more environmentally friendly for mass production until green technology becomes more widely used [207].

11.8.1 Challenges with the Use of Biomaterials in Additive Manufacturing

(a) Tissue vascularization

Biomaterials used to replace large organs and tissues of larger dimensions have faced the challenge of vascularization. Without the appropriate amount of blood flow, tissue survival is threatened. Different methods, like creating tiny patterns in materials have been used to incorporate biomaterials to advance vascularization. Studies are continuously conducted to build 3d functional blood vessels, but it has not succeeded. Different methods, like creating tiny patterns in materials have been used to incorporate biomaterials to advance vascularization. This might be the reason tissue engineering is currently limited to only Cartilage, Skin and Ligaments. Studies are also being conducted on lab-grown blood vessels before they are inserted into the body and using unique proteins for their growth [208].

(b) Design of material

Researchers have been trying to build artificial materials that can mimic natural tissues in the body, which have the property to adapt and interact with other body tissues. Lab tests are being conducted to check the efficacy of biomaterials to replace these tissues and how well they interact with natural tissues. These tests will help us create tissues with suitable properties that can mimic the body properly without reacting to the body's immune system. Finding materials that are inert to the immune system is a challenge that must be overcome for proper utilization of biomaterials.

(c) Interface between tissues

Musculoskeletal function can be improved by enhancing the connection between several types of tissues. To enhance these connections, researchers have created scaffolds that encourage the growth of tissue as well as its integration. Using this data, methods to repair soft tissues are being developed by different strategies. Understanding the interface between tissues is currently a challenge, overcoming which can lead to better repair of tissues in the body.

d. Structural optimization of Components

Several factors like vibration and material fatigue must be considered for optimum biomaterial performance. Multiple materials and their fusions must be incorporated to achieve functional fusion of structures.

e. Real-time monitoring

For effective control of the process, real-time monitoring is required. The establishment of standards for 3d printing can ensure accurate prediction models as well as foreseeable outcomes.

(f) Complex scaffolds

Scaffolds needed for use require a variety of properties, like those found in natural tissues, which can pose a challenge to its use [209].

Apart from these, tissue engineering mainly faces two challenges: The first is the development of new bioinks and the second is Regulatory Hurdles. Ideally, research is ongoing for a universal bioink blending material that can support tissues and growth factors for angiogenesis. Additive manufacturing methods like 3d printing can assist in creating complex tissues. The challenge of vascularization, as mentioned earlier, can be overcome by incorporating some sacrificial biomaterials to create blood vessels. Regulatory clearance by the FDA is a requirement before these advancements can be used in patients [210].

11.8.2 Additive Manufacturing Challenges for Peripheral Nerve Conduits

(a) Developing novel biomaterial inks / bio-inks and crosslinking strategies

To improve their efficiency and performance, novel methods and crosslinking strategies are important for additive manufacturing. Traditionally, biopolymers like gelatin, polyethylene glycol and polyurethane have been used in varying combinations. These materials face limitations concerning their stability, brittleness, change in colour and potential toxicity at high concentrations and molecular weights. To overcome these limitations, further studies are required of natural and synthetic biopolymers and their crosslinking mechanisms. The data obtained from these studies will help identify suitable bio-inks for use and help us understand the necessary environmental conditions for regenerating nerve cells.

(b) Optimization challenges

3D scanning and defect-free nerve cell construction require further optimization to be of use to reconstruct nerves. Printing speed and accuracy can be optimized by collaboration of clinicians, engineers, and computer experts. Achieving appropriate resolution in 3D imaging techniques is a challenge. Challenges are also faced during implanting nerve cells into native cells and ensuring sterility.

(c) Multifunctional NGCs

Long-distance nerve injuries have posed a challenge for the generally used nerve conduits due to limited efficacy and complications like infections and stiffness, which lead to implant removal. Multifunctional nerve guidance conduits offer a solution for these issues. They can encapsulate live cells like stem and schwann cells and the drug and nerve factors within additively manufactured NGCs. This can lead to faster nerve growth and recovery without complications. This approach enhances both sensory and motor neuron recovery and minimizes the need for further interventions.

(d) Scaling up of 3D nerve constructs

It involves two main strategies: Vascularization of thicker constructs or in-situ bioprinting. In-situ bioprinting using a robotic arm is a direct approach to printing at the defective site based on scanned data [211].

11.9 Future Directions and Concluding Remark

Additive manufacturing has significantly impacted tissue engineering by helping create 3D tissue analogs, various cell types necessary for regenerative medicine and screening of drugs. Specifically, suppose we focus on nerve tissue engineering. In that case, we can observe that additive manufacturing has been used to build nerve-equivalent connections that mimic native tissue structures. Specific advantages of 3D printed/ bio-printed nerve connections include patient specificity, design flexibility, cost efficiency, and rapid manufacturing with high precision and reproducibility [211].

In recent years, the research on additive manufacturing has increased with a specific focus on tissue engineering for creating scaffolds and organs. Combining both these approaches, i.e., biomaterial scaffolds and tissues, can enhance the current healthcare systems. Studies in material-tissue interface and anatomical influences will shape the future generations of biomaterials used in additive manufacturing. Engineering surfaces to enhance physiological interactions with the help of additive manufacturing will play a crucial role. Additive manufacturing is expected to delve into many more areas like nanoparticles, hydrogels, biological systems, and cells. It is expected to evolve into hybrids and biomimetic structures that can incorporate various experimental materials and nano-patterns in different meta-biomaterial categories [209].

3D printing and nanotechnology have brought significant innovations into bone tissue engineering by creating large structures and customization options for these structures to adapt to the body quickly. It still faces the challenge of being expensive, having slower speeds compared to traditional methods, and limitations of materials to use. Bio-printing faces challenges while printing small vessels, lack of suitable bio-inks, and unstable cellular behaviour. These issues can be resolved with the help of extrusion and droplet-based bio-printing while also incorporating smart materials. Despite these challenges, additive manufacturing biomaterials have made considerable progress over the past years [212].

Although 3D printing with biopolymers is advancing, some challenges remain to overcome. Modern technology can mix materials for detailed scaffolds, but some issues still exist with intricate structures. We must understand how quickly these materials break down and their strength to create better materials. Collaboration might be the key to improving scaffolds. Material shortages slow progress in medicine. Innovative materials show promise but need more testing. Mimicking tissue movement is tricky. There's potential in printing things that change over time, but it is early days. We need better inks and an understanding of how things react. While promising, 3D printing with biopolymers needs more work to shine in medicine [213].

Advances in orthopaedic tissue engineering rely on improving manufacturing efficiencies to understand the relationships between product design and performance. Integrating artificial intelligence into material selection could facilitate the selection of ideal materials through a specialized database, with antimicrobial or anticancer properties added to orthopaedic properties to combat issues such as infection and allergies also require efficiency. Due to the harsh environment of the human body, it is crucial to ensure that bone grafts degrade rapidly. Factors such as microstructure can influence corrosion behaviour, and a thorough analysis of corrosion products is essential for long-term storage. Again, biological compatibility, texture, and growth factors are considered to contribute to and promote bone regeneration to improve the quality of the composites, thus improving the scaffold performance [200].

Additive manufacturing (AM) technology is essential in biomedical applications, especially in producing biomaterials for various needs such as bone repair, tissue regeneration, implants, prosthetics, and pharmaceuticals. However, the text also emphasizes that not all AM processes are produced or used equally in living organisms. Each method has its capabilities, limitations, and associated features such as print speed, resolution, material compatibility, and backend processing requirements, and the success of an AM application depends on the optimization equipment it has the power of this theory. Microarchitecture development is vital in biomaterial development, especially in making porous or mesh structures necessary for biomedical applications. These structures should facilitate the delivery of nutrients and oxygen to cells, where open pores together prevail.

Further research on cellular responses to immune rejection, reduction, and durability, are needed for the creation of better models of biomaterials. Future directions could include incorporating bioactive agents to stimulate desirable or inhibit adverse reactions [214].

This chapter was able to compile information on several topics like stereolithography, fused deposition modelling, selective laser sintering and so on. We can conclude by summarizing the critical topics covered: (1) Stereolithography: Stereolithography (SLA) utilizes UV/visible lasers to polymerize photopolymer resin layer-by-layer, producing high-precision 3D objects suitable for biomedical applications with smooth surfaces and high cell viability. (2) Fused deposition modelling (FDM): FDM is a widely used 3D printing technology in biomaterial research, utilizing filament extrusion to create complex structures with various materials, offering low cost and versatility. (3) Selective laser sintering: Selective laser sintering (SLS) is a competitive additive manufacturing method for producing biocompatible prostheses for bone replacement, including skull and hip joint implants. (4) Inkjet bio-printing: Inkjet bioprinting enables the creation of small batches of customized medical or other devices, beneficial for material studies, using a commercial 2D ink development system. (5) Additive manufacturing in (a) Cartilage tissue: Cartilage tissue engineering aims to replicate native cartilage's load-bearing function, structural performance, and resistance to stress, strain, and inflammation, addressing a significant medical need due to the prevalence of cartilage injuries and the growing demand for knee arthroplasty. (b) Bone tissue engineering: Bone tissue engineering aims to replicate the structure of bone tissue using bioactive biomaterials, such as metallic, ceramic, polymer, composite, or hybrid scaffolds, to promote bone regeneration and repair, with emerging strategies including in vitro pre-vascularization techniques. (c) Heart tissue/Cardiac tissue: Cardiovascular tissue engineering utilizes biomaterials such as metallic alloys, polymers, and biological materials for constructs like heart valves, pacemakers, and angioplasty devices, each with its associated drawbacks. (d) Liver/hepatic tissue: Liver ECM-based electro spun scaffolds enhance hepatocyte viability and functionality compared to conventional scaffolds by providing a native microenvironment rich in ECM components like collagen, laminin, and fibronectin. (e) Skin: Skin tissue engineering aims to create biomaterials supporting skin regeneration, facing challenges in mimicking native skin complexity, especially in establishing functional vascular networks and integrating synthetic materials at the dermis layer. Key processing parameters for 3D bioprinting using hydrogel- based bioinks are viscosity, flow behaviour, flow rate, shear stress, viscoelasticity, nozzle, volumetric flow rate. Potential biopolymers for additive manufacturing scaffold applications can be for Alginate: Alginate, derived from brown algae, forms hydrogels with divalent cations like Ca^{2+}, widely used in 3D bioprinting, with ongoing efforts to enhance cell adhesion and stability through chemical modifications and blending with bioactive materials and are applicable for chitosan, collagen etc.

Acknowledgments The corresponding author thanks the management of Bharati Vidyapeeth (Deemed to be University), Poona College of Pharmacy, Pune, for providing the necessary library facilities for conducting literature and writing.

Conflicts of Interest The authors report no conflicts of interest. The authors are responsible for the content and writing of the paper.

References

1. Shinde S, Mane R, Vardikar A, Dhumal A, Rajput A (2023) 4D printing: from emergence to innovation over 3D printing. Eur Polymer J 197:112356
2. Vyas, J., Raytthatha, N., Singh, S., & Prajapati, B. Next-generation computational automation-based additive manufacturing of pharmaceuticals: An approach to fabricate precise medicine. In Handbook of 3D printing in pharmaceutics (pp. 153-175). CRC Press
3. Shah SR, Modi CD, Singh S, Mori DD, Soniwala MM, Prajapati BG (2024) Recent advances in additive manufacturing of polycaprolactone-based scaffolds for tissue engineering applications: a comprehensive review. Regen Eng Transl Med 11:112–131
4. Singh, A., Mandal, P., Maheshwari, S., Akhtar, J., Prajapati, B. G., & Singh, S. Processing of sustainable biomaterials by additive manufacturing methods. In Advances in sustainable biomaterials (pp. 141-162). CRC Press
5. Vyas J, Raytthatha N, Singh S, Prajapati BG, Mohite P, Munde S (2024) Sustainable sources of raw materials as substituting biomaterials for additive manufacturing of dental implants: a review. Periodontal Implant Res 8(1):3
6. Xu H, Zhang Y, Zhang Y, Zhao Z, Xue T, Wang J, Li M, Zhao S, Zhang H, Ding Y (2024) 3D bioprinting advanced biomaterials for craniofacial and dental tissue engineering–a review. Mater Des 241:112886
7. Zhang J, Wehrle E, Rubert M, Müller R (2021) 3D bioprinting of human tissues: biofabrication, bioinks, and bioreactors. Int J Mol Sci 22(8):3971
8. Kantaros A (2022) 3D printing in regenerative medicine: technologies and resources utilized. Int J Mol Sci 23(23):14621
9. Veeravalli RS, Vejandla B, Savani S, Nelluri A, Peddi NC (2023) Three-dimensional bioprinting in medicine: a comprehensive overview of current progress and challenges faced. Cureus 15(7):e41624
10. Gu Z, Fu J, Lin H, He Y (2020) Development of 3D bioprinting: from printing methods to biomedical applications. Asian J Pharm Sci 15(5):529–557
11. Li Y, Zhang X, Zhang X, Zhang Y, Hou D (2023) Recent progress of the vat photopolymerization technique in tissue engineering: a brief review of mechanisms, methods, materials, and applications. Polymers 15(19):3940
12. Zhan S, Guo AX, Cao SC, Liu N (2022) 3D printing soft matters and applications: a review. Int J Mol Sci 23(7):3790
13. Zhang S, Chen X, Shan M, Hao Z, Zhang X, Meng L, Zhai Z, Zhang L, Liu X, Wang X (2023) Convergence of 3D bioprinting and nanotechnology in tissue engineering scaffolds. Biomimetics 8(1):94
14. Li W, Wang M, Ma H, Chapa-Villarreal FA, Lobo AO, Zhang YS (2023) Stereolithography apparatus and digital light processing-based 3D bioprinting for tissue fabrication. Iscience 26(2)
15. Zhang Y, Kumar P, Lv S, Xiong D, Zhao H, Cai Z, Zhao X (2021) Recent advances in 3D bioprinting of vascularized tissues. Mater Des 199:109398
16. Vanaei S, Parizi M, Salemizadehparizi F, Vanaei H (2021) An overview on materials and techniques in 3D bioprinting toward biomedical application. Engi Regen 2:1–18
17. Winarso R, Anggoro P, Ismail R, Jamari J, Bayuseno A (2022) Application of fused deposition modeling (FDM) on bone scaffold manufacturing process: a review. Heliyon 8(11):e11701
18. Shojaie F, Ferrero C, Caraballo I (2023) Development of 3D-printed bicompartmental devices by dual-nozzle fused deposition modeling (FDM) for colon-specific drug delivery. Pharmaceutics 15(9):2362
19. Acierno D, Patti A (2023) Fused deposition modelling (FDM) of thermoplastic-based filaments: process and rheological properties—an overview. Materials 16(24):7664
20. Mocerino D, Ricciardi MR, Antonucci V, Papa I (2023) Fused deposition modelling of polymeric auxetic structures: a review. Polymers 15(4):1008

21. Lupone F, Padovano E, Casamento F, Badini C (2021) Process phenomena and material properties in selective laser sintering of polymers: a review. Materials 15(1):183
22. Pavan Kalyan B, Kumar L (2022) 3D printing: applications in tissue engineering, medical devices, and drug delivery. AAPS PharmSciTech 23(4):92
23. Balasankar A, Anbazhakan K, Arul V, Mutharaian VN, Sriram G, Aruchamy K, Oh TH, Ramasundaram S (2023) Recent advances in the production of pharmaceuticals using selective laser sintering. Biomimetics 8(4):330
24. Gopal PM, Kavimani V, Gupta K, Marinkovic D (2023) Laser-based manufacturing of ceramics: a review. Micromachines 14(8):1564
25. Pal AK, Mohanty AK, Misra M (2021) Additive manufacturing technology of polymeric materials for customized products: recent developments and future prospective. RSC Adv 11(58):36398–36438
26. Barui S (2021) 3D inkjet printing of biomaterials: principles and applications. Med Dev Sensors 4(1):e10143
27. Dou C, Perez V, Qu J, Tsin A, Xu B, Li J (2021) A state-of-the-art review of laser-assisted bioprinting and its future research trends. ChemBioEng Rev 8(5):517–534
28. Keriquel V, Oliveira H, Rémy M, Ziane S, Delmond S, Rousseau B, Rey S, Catros S, Amédée J, Guillemot F (2017) In situ printing of mesenchymal stromal cells, by laser-assisted bioprinting, for in vivo bone regeneration applications. Sci Rep 7(1):1778
29. Tharakan S, Khondkar S, Ilyas A (2021) Bioprinting of stem cells in multimaterial scaffolds and their applications in bone tissue engineering. Sensors 21(22):7477
30. Zhu J, Wang Y, Zhong L, Pan F, Wang J (2021) Advances in tissue engineering of vasculature through three-dimensional bioprinting. Dev Dyn 250(12):1717–1738
31. Ventura RD (2021) An overview of laser-assisted bioprinting (LAB) in tissue engineering applications. Med Lasers Eng Basic Res Clin Appl 10(2):76–81
32. Kalogeropoulou M, Díaz-Payno PJ, Mirzaali MJ, van Osch GJ, Fratila-Apachitei LE, Zadpoor AA (2024) 4D printed shape-shifting biomaterials for tissue engineering and regenerative medicine applications. Biofabrication 16(2):022002
33. Curti C, Kirby DJ, Russell CA (2024) Systematic screening of photopolymer resins for stereolithography (SLA) 3D printing of solid oral dosage forms: investigation of formulation factors on printability outcomes. Int J Pharm 653:123862
34. Mamo HB, Adamiak M, Kunwar A (2023) 3D printed biomedical devices and their applications: a review on state-of-the-art technologies, existing challenges, and future perspectives. J Mech Behav Biomed Mater 143:105930
35. Karimi A, Rahmatabadi D, Baghani M (2024) Various FDM mechanisms used in the fabrication of continuous-fiber reinforced composites: a review. Polymers 16(6):831
36. Shirazi SF, Gharehkhani S, Mehrali M, Yarmand H, Metselaar HS, Adib Kadri N, Osman NA (2015) A review on powder-based additive manufacturing for tissue engineering: selective laser sintering and inkjet 3D printing. Sci Technol Adv Mater 16(3):033502. https://doi.org/10.1088/1468-6996/16/3/033502
37. Trenfield SJ, Xu X, Goyanes A, Rowland M, Wilsdon D, Gaisford S, Basit AW (2023) Releasing fast and slow: non-destructive prediction of density and drug release from SLS 3D printed tablets using NIR spectroscopy. Int J Pharm X 5:100148. https://doi.org/10.1016/j.ijpx.2022.100148
38. Carou-Senra P, Rodríguez-Pombo L, Awad A, Basit AW, Alvarez-Lorenzo C, Goyanes A (2024) Inkjet printing of pharmaceuticals. Adv Mater 36(11):e2309164. https://doi.org/10.1002/adma.202309164
39. Bas J, Dutta T, Llamas Garro I, Velázquez-González JS, Dubey R, Mishra SK (2024) Embedded sensors with 3D printing technology: review. Sensors (Basel) 24(6). https://doi.org/10.3390/s24061955
40. Sachdev A, Acharya S, Gadodia T, Shukla S, Harshita J, Akre C, Khare M, Huse S (2022) A review on techniques and biomaterials used in 3D bioprinting. Cureus 14(8):e28463. https://doi.org/10.7759/cureus.28463

41. Matta R, Moreau D, O'Connor R (2024) Printable devices for neurotechnology. Front Neurosci 18:1332827. https://doi.org/10.3389/fnins.2024.1332827
42. Tharakan S, Khondkar S, Ilyas A (2021) Bioprinting of stem cells in multimaterial scaffolds and their applications in bone tissue engineering. Sensors (Basel) 21(22). https://doi.org/10.3390/s21227477
43. Vacanti JP, Langer R (1999) Tissue engineering: the design and fabrication of living replacement devices for surgical reconstruction and transplantation. Lancet 354(Suppl 1):Si32–Si34. https://doi.org/10.1016/s0140-6736(99)90247-7
44. Mi H-Y, Jing X, Salick M, Cordie T, Peng X-F, Turng L-S (2013) Morphology, mechanical properties, and mineralization of rigid thermoplastic polyurethane/hydroxyapatite scaffolds for bone tissue applications: effects of fabrication approaches and hydroxyapatite size. J Mater Sci 49. https://doi.org/10.1007/s10853-013-7931-3
45. O'Brien F (2011) Biomaterials & scaffolds for tissue engineering. Mater Today 14:88–95. https://doi.org/10.1016/S1369-7021(11)70058-X
46. Ko IK, Lee SJ, Atala A, Yoo JJ (2013) In situ tissue regeneration through host stem cell recruitment. Exp Mol Med 45(11):e57. https://doi.org/10.1038/emm.2013.118
47. Ju YM, Atala A, Yoo JJ, Lee SJ (2014) In situ regeneration of skeletal muscle tissue through host cell recruitment. Acta Biomater 10(10):4332–4339. https://doi.org/10.1016/j.actbio.2014.06.022
48. Puppi D, Federica C, Piras A, Chiellini E (2010) Polymeric materials for bone and cartilage repair. Prog Polym Sci 35:403–440. https://doi.org/10.1016/j.progpolymsci.2010.01.006
49. Karageorgiou V, Kaplan D (2005) Porosity of 3D biomaterial scaffolds and osteogenesis. Biomaterials 26(27):5474–5491
50. Leong K, Cheah C, Chua C (2003) Solid freeform fabrication of three-dimensional scaffolds for engineering replacement tissues and organs. Biomaterials 24(13):2363–2378
51. Mota C, Puppi D, Dinucci D, Errico C, Bártolo P, Chiellini F (2011) Dual-scale polymeric constructs as scaffolds for tissue engineering. Materials 4(3):527–542
52. Puppi D, Piras AM, Detta N, Dinucci D, Chiellini F (2010) Poly (lactic-co-glycolic acid) electrospun fibrous meshes for the controlled release of retinoic acid. Acta Biomater 6(4):1258–1268
53. Stevens MM, George JH (2005) Exploring and engineering the cell surface interface. Science 310(5751):1135–1138
54. Standard A (2012) Standard terminology for additive manufacturing technologies. ASTM International F2792-12a 46:10918–10928
55. Woodruff MA, Hutmacher DW (2010) The return of a forgotten polymer—polycaprolactone in the 21st century. Prog Polym Sci 35(10):1217–1256
56. Hutmacher DW, Sittinger M, Risbud MV (2004) Scaffold-based tissue engineering: rationale for computer-aided design and solid free-form fabrication systems. Trends Biotechnol 22(7):354–362
57. Engler AJ, Sen S, Sweeney HL, Discher DE (2006) Matrix elasticity directs stem cell lineage specification. Cell 126(4):677–689
58. Reilly GC, Engler AJ (2010) Intrinsic extracellular matrix properties regulate stem cell differentiation. J Biomech 43(1):55–62
59. Jensen SS, Bornstein MM, Dard M, Bosshardt DD, Buser D (2009) Comparative study of biphasic calcium phosphates with different HA/TCP ratios in mandibular bone defects. A long-term histomorphometric study in minipigs. J Biomed Mater Res Part B Appl Biomater 90(1):171–181
60. Zhang Q, Zhang W, Zhang Z, Zhang L, Chen H, Hao M, Deng J, Tang P (2017) Femoral nonunion with segmental bone defect treated by distraction osteogenesis with monolateral external fixation. J Orthop Surg Res 12:1–7
61. Li Y, Xu J, Mi J, He X, Pan Q, Zheng L, Zu H, Chen Z, Dai B, Li X (2021) Biodegradable magnesium combined with distraction osteogenesis synergistically stimulates bone tissue regeneration via CGRP-FAK-VEGF signaling axis. Biomaterials 275:120984

62. Madrid APM, Vrech SM, Sanchez MA, Rodriguez AP (2019) Advances in additive manufacturing for bone tissue engineering scaffolds. Mater Sci Eng C 100:631–644
63. Beck R, Chaves P, Goyanes A, Vukosavljevic B, Buanz A, Windbergs M, Basit A, Gaisford S (2017) 3D printed tablets loaded with polymeric nanocapsules: an innovative approach to produce customized drug delivery systems. Int J Pharm 528(1–2):268–279
64. Tagami T, Ando M, Nagata N, Goto E, Yoshimura N, Takeuchi T, Noda T, Ozeki T (2019) Fabrication of naftopidil-loaded tablets using a semisolid extrusion-type 3D printer and the characteristics of the printed hydrogel and resulting tablets. J Pharm Sci 108(2):907–913
65. Tagami T, Yoshimura N, Goto E, Noda T, Ozeki T (2019) Fabrication of muco-adhesive oral films by the 3D printing of hydroxypropyl methylcellulose-based catechin-loaded formulations. Biol Pharm Bull 42(11):1898–1905
66. Wu M, Zhang Y, Huang H, Li J, Liu H, Guo Z, Xue L, Liu S, Lei Y (2020) Assisted 3D printing of microneedle patches for minimally invasive glucose control in diabetes. Mater Sci Eng C 117:111299
67. Yi H-G, Choi Y-J, Kang KS, Hong JM, Pati RG, Park MN, Shim IK, Lee CM, Kim SC, Cho D-W (2016) A 3D-printed local drug delivery patch for pancreatic cancer growth suppression. J Control Release 238:231–241
68. Hofmann D, Entrialgo-Castaño M, Kratz K, Lendlein A (2009) Knowledge-based approach towards hydrolytic degradation of polymer-based biomaterials. Adv Mater 21(32–33):3237–3245
69. Ratner BD, Bryant SJ (2004) Biomaterials: where we have been and where we are going. Annu Rev Biomed Eng 6:41–75
70. Swieszkowski W, Tuan BHS, Kurzydlowski KJ, Hutmacher DW (2007) Repair and regeneration of osteochondral defects in the articular joints. Biomol Eng 24(5):489–495
71. Castro NJ, Hacking SA, Zhang LG (2012) Recent progress in interfacial tissue engineering approaches for osteochondral defects. Ann Biomed Eng 40:1628–1640
72. Melchels FP, Domingos MA, Klein TJ, Malda J, Bartolo PJ, Hutmacher DW (2012) Additive manufacturing of tissues and organs. Prog Polym Sci 37(8):1079–1104
73. Kundu J, Shim JH, Jang J, Kim SW, Cho DW (2015) An additive manufacturing-based PCL–alginate–chondrocyte bioprinted scaffold for cartilage tissue engineering. J Tissue Eng Regen Med 9(11):1286–1297
74. Schuurman W, Khristov V, Pot MW, van Weeren PR, Dhert WJ, Malda J (2011) Bioprinting of hybrid tissue constructs with tailorable mechanical properties. Biofabrication 3(2):021001
75. Shim J-H, Lee J-S, Kim JY, Cho D-W (2012) Bioprinting of a mechanically enhanced three-dimensional dual cell-laden construct for osteochondral tissue engineering using a multi-head tissue/organ building system. J Micromech Microeng 22(8):085014
76. Zhang X, Li Z, Yang P, Duan G, Liu X, Gu Z, Li Y (2021) Polyphenol scaffolds in tissue engineering. Mater Horizons 8(1):145–167
77. Zou Y, Zhang L, Yang L, Zhu F, Ding M, Lin F, Wang Z, Li Y (2018) "Click" chemistry in polymeric scaffolds: bioactive materials for tissue engineering. J Control Release 273:160–179
78. Kwon JS, Yoon SM, Kwon DY, Tai GZ, Jin LM, Song B, Lee B, Kim JH, Han DK, Min BH (2013) Injectable in situ-forming hydrogel for cartilage tissue engineering. J Mater Chem B 1(26):3314–3321
79. Wei W, Ma Y, Yao X, Zhou W, Wang X, Li C, Lin J, He Q, Leptihn S, Ouyang H (2021) Advanced hydrogels for the repair of cartilage defects and regeneration. Bioact Mater 6(4):998–1011
80. Taylor DL, In Het Panhuis M (2016) Self-healing hydrogels. Adv Mater 28(41):9060–9093
81. Li Q, Liu C, Wen J, Wu Y, Shan Y, Liao J (2017) The design, mechanism and biomedical application of self-healing hydrogels. Chin Chem Lett 28(9):1857–1874
82. Lopa S, Madry H (2014) Bioinspired scaffolds for osteochondral regeneration. Tissue Eng A 20(15–16):2052–2076

83. Kwon H, Brown WE, Lee CA, Wang D, Paschos N, Hu JC, Athanasiou KA (2019) Surgical and tissue engineering strategies for articular cartilage and meniscus repair. Nat Rev Rheumatol 15(9):550–570
84. Chesterman P, Smith AU (1968) Homotransplantation of articular cartilage and isolated chondrocytes: an experimental study in rabbits. J Bone Joint Surg Br 50(1):184–197
85. Peterson L, Vasiliadis HS, Brittberg M, Lindahl A (2010) Autologous chondrocyte implantation: a long-term follow-up. Am J Sports Med 38(6):1117–1124
86. Behrens P, Bitter T, Kurz B, Russlies M (2006) Matrix-associated autologous chondrocyte transplantation/implantation (MACT/MACI)—5-year follow-up. Knee 13(3):194–202
87. Kon E, Verdonk P, Condello V, Delcogliano M, Dhollander A, Filardo G, Pignotti E, Marcacci M (2009) Matrix-assisted autologous chondrocyte transplantation for the repair of cartilage defects of the knee: systematic clinical data review and study quality analysis. Am J Sports Med 37(1_suppl):156–166
88. Marcacci M, Filardo G, Kon E (2013) Treatment of cartilage lesions: what works and why? Injury 44:S11–S15
89. Couet F, Rajan N, Mantovani D (2007) Macromolecular biomaterials for scaffold-based vascular tissue engineering. Macromol Biosci 7(5):701–718
90. Marelli B, Achilli M, Alessandrino A, Freddi G, Tanzi MC, Farè S, Mantovani D (2012) Collagen-reinforced electrospun silk fibroin tubular construct as small calibre vascular graft. Macromol Biosci 12(11):1566–1574
91. Pankajakshan D, Agrawal DK (2010) Scaffolds in tissue engineering of blood vessels. Can J Physiol Pharmacol 88(9):855–873
92. Achilli M, Lagueux J, Mantovani D (2010) On the effects of UV-C and pH on the mechanical behavior, molecular conformation and cell viability of collagen-based scaffold for vascular tissue engineering. Macromol Biosci 10(3):307–316
93. Ferreira AM, Gentile P, Sartori S, Pagliano C, Cabrele C, Chiono V, Ciardelli G (2012) Biomimetic soluble collagen purified from bones. Biotechnol J 7(11):1386–1394
94. Gyles DA, Castro LD, Silva JOC Jr, Ribeiro-Costa RM (2017) A review of the designs and prominent biomedical advances of natural and synthetic hydrogel formulations. Eur Polym J 88:373–392
95. Einerson NJ, Stevens KR, Kao WJ (2003) Synthesis and physicochemical analysis of gelatin-based hydrogels for drug carrier matrices. Biomaterials 24(3):509–523
96. Kapoor S, Kundu SC (2016) Silk protein-based hydrogels: promising advanced materials for biomedical applications. Acta Biomater 31:17–32
97. Rodriguez MJ, Brown J, Giordano J, Lin SJ, Omenetto FG, Kaplan DL (2017) Silk based bioinks for soft tissue reconstruction using 3-dimensional (3D) printing with in vitro and in vivo assessments. Biomaterials 117:105–115
98. Kumar M, Gupta P, Bhattacharjee S, Nandi SK, Mandal BB (2018) Immunomodulatory injectable silk hydrogels maintaining functional islets and promoting anti-inflammatory M2 macrophage polarization. Biomaterials 187:1–17
99. Armiento A, Stoddart M, Alini M, Eglin D (2018) Biomaterials for articular cartilage tissue engineering: learning from biology. Acta Biomater 65:1–20
100. Lapcık L Jr, Lapcık L, De Smedt S, Demeester J, Chabrecek P (1998) Hyaluronan: preparation, structure, properties, and applications. Chem Rev 98(8):2663–2684
101. Jackson RL, Busch SJ, Cardin AD (1991) Glycosaminoglycans: molecular properties, protein interactions, and role in physiological processes. Physiol Rev 71(2):481–539
102. Baier Leach J, Bivens KA, Patrick CW Jr, Schmidt CE (2003) Photocrosslinked hyaluronic acid hydrogels: natural, biodegradable tissue engineering scaffolds. Biotechnol Bioeng 82(5):578–589
103. Scalzone A, Bonifacio M, Cometa S, Cucinotta F, De Giglio E, Ferreira A, Gentile P (2020) pH-triggered adhesiveness and cohesiveness of chondroitin sulfate-catechol biopolymer for biomedical applications. Front Bioeng Biotechnol 8:712
104. Augst AD, Kong HJ, Mooney DJ (2006) Alginate hydrogels as biomaterials. Macromol Biosci 6(8):623–633

105. Rastogi P, Kandasubramanian B (2019) Review of alginate-based hydrogel bioprinting for application in tissue engineering. Biofabrication 11(4):042001
106. Gentile P, Ghione C, Ferreira A, Crawford A, Hatton P (2017) Alginate-based hydrogels functionalised at the nanoscale using layer-by-layer assembly for potential cartilage repair. Biomater Sci 5(9):1922–1931
107. Peers S, Montembault A, Ladavière C (2020) Chitosan hydrogels for sustained drug delivery. J Control Release 326:150–163
108. Xie Y, Liao X, Zhang J, Yang F, Fan Z (2018) Novel chitosan hydrogels reinforced by silver nanoparticles with ultrahigh mechanical and high antibacterial properties for accelerating wound healing. Int J Biol Macromol 119:402–412
109. Benders KE, Van Weeren PR, Badylak SF, Saris DB, Dhert WJ, Malda J (2013) Extracellular matrix scaffolds for cartilage and bone regeneration. Trends Biotechnol 31(3):169–176
110. Lin C-C, Anseth KS (2009) PEG hydrogels for the controlled release of biomolecules in regenerative medicine. Pharm Res 26:631–643
111. Zhu J (2010) Bioactive modification of poly (ethylene glycol) hydrogels for tissue engineering. Biomaterials 31(17):4639–4656
112. Fan Z, Cheng P, Liu M, Li D, Liu G, Zhao Y, Ding Z, Chen F, Wang B, Tan X (2017) Poly (glutamic acid) hydrogels crosslinked via native chemical ligation. New J Chem 41(16):8656–8662
113. Higgins SP, Solan AK, Niklason LE (2003) Effects of polyglycolic acid on porcine smooth muscle cell growth and differentiation. J Biomed Mater Res A 67(1):295–302
114. Liu B, Huang W, Yang G, An Y, Yin Y, Wang N, Jiang B (2020) Preparation of gelatin/poly (γ-glutamic acid) hydrogels with stimulated response by hot-pressing preassembly and radiation crosslinking. Mater Sci Eng C 116:111259
115. Munim SA, Raza ZA (2019) Poly (lactic acid) based hydrogels: formation, characteristics and biomedical applications. J Porous Mater 26(3):881–901
116. Gupta B, Revagade N, Hilborn J (2007) Poly (lactic acid) fiber: An overview. Prog Polym Sci 32(4):455–482
117. Dhandayuthapani B, Yoshida Y, Maekawa T, Kumar S (2011) Polymeric scaffolds in tissue engineering application: a review. Int J Polym Sci 2011. https://doi.org/10.1155/2011/290602
118. Basu A, Kunduru KR, Doppalapudi S, Domb AJ, Khan W (2016) Poly (lactic acid) based hydrogels. Adv Drug Deliv Rev 107:192–205
119. Kosik-Kozioł A, Costantini M, Bolek T, Szöke K, Barbetta A, Brinchmann J, Święszkowski W (2017) PLA short sub-micron fiber reinforcement of 3D bioprinted alginate constructs for cartilage regeneration. Biofabrication 9(4):044105
120. Mouriño V, Boccaccini AR (2010) Bone tissue engineering therapeutics: controlled drug delivery in three-dimensional scaffolds. J R Soc Interface 7(43):209–227
121. Agarwal R, García AJ (2015) Biomaterial strategies for engineering implants for enhanced osseointegration and bone repair. Adv Drug Deliv Rev 94:53–62
122. Jones AC, Arns CH, Sheppard AP, Hutmacher DW, Milthorpe BK, Knackstedt MA (2007) Assessment of bone ingrowth into porous biomaterials using MICRO-CT. Biomaterials 28(15):2491–2504
123. Mano J, Silva G, Azevedo HS, Malafaya P, Sousa R, Silva SS, Boesel L, Oliveira JM, Santos T, Marques A (2007) Natural origin biodegradable systems in tissue engineering and regenerative medicine: present status and some moving trends. J R Soc Interface 4(17):999–1030
124. Baig MK, Mahon N, McKenna WJ, Caforio AL, Bonow RO, Francis GS, Gheorghiade M (1998) The pathophysiology of advanced heart failure. Am Heart J 135(6):S216–S230
125. Chen FY, Cohn LH (2002) The surgical treatment of heart failure. A new frontier: nontransplant surgical alternatives in heart failure. Cardiol Rev 10(6):326–333
126. Hawkins NM, Petrie MC, MacDonald MR, Hogg KJ, McMurray JJ (2006) Selecting patients for cardiac resynchronization therapy: electrical or mechanical dyssynchrony? Eur Heart J 27(11):1270–1281
127. Kohli SK, Elliott P (2005) Cardiac resynchronization therapy: the procedure and progress so far. Br J Hosp Med 66(8):469–473

128. Leor J, Amsalem Y, Cohen S (2005) Cells, scaffolds, and molecules for myocardial tissue engineering. Pharmacol Ther 105(2):151–163
129. Loebe M, Soltero E, Thohan V, Lafuente JA, Noon GP (2003) New surgical therapies for heart failure. Curr Opin Cardiol 18(3):194–198
130. Martinez C, Tzur A, Hrachian H, Zebede J, Lamas GA (2006) Pacemakers and defibrillators: recent and ongoing studies that impact the elderly. Am J Geriatr Cardiol 15(2):82–87
131. Packer M (2002) The impossible task of developing a new treatment for heart failure. J Card Fail 8(4):193–196
132. Young JB, Mills RM (2004) Clinical management of heart failure. Professional Communications
133. Zimmermann W-H, Melnychenko I, Eschenhagen T (2004) Engineered heart tissue for regeneration of diseased hearts. Biomaterials 25(9):1639–1647
134. Rishi G, Subramaniam VN (2017) The liver in regulation of iron homeostasis. Am J Physiol Gastrointest Liver Physiol 313(3):G157–G165
135. Van den Berghe G (1991) The role of the liver in metabolic homeostasis: implications for inborn errors of metabolism. J Inherit Metab Dis 14(4):407–420
136. Mann RE, Smart RG, Govoni R (2003) The epidemiology of alcoholic liver disease. Alcohol Res Health 27(3):209
137. Alqahtani SA (2012) Update in liver transplantation. Curr Opin Gastroenterol 28(3):230–238. https://doi.org/10.1097/MOG.0b013e3283527f16
138. Agarwal T, Subramanian B, Maiti TK (2019) Liver tissue engineering: challenges and opportunities. ACS Biomater Sci Eng 5(9):4167–4182
139. Mason C, Dunnill P (2008) A brief definition of regenerative medicine. Regen Med 3(1):1–5. https://doi.org/10.2217/17460751.3.1.1
140. Abu-Absi SF, Hansen LK, Hu W-S (2004) Three-dimensional co-culture of hepatocytes and stellate cells. Cytotechnology 45:125–140
141. Inamori M, Mizumoto H, Kajiwara T (2009) An approach for formation of vascularized liver tissue by endothelial cell–covered hepatocyte spheroid integration. Tissue Eng A 15(8):2029–2037
142. Lin J, Meng L, Yao Z, Chen S, Yang J, Tang Z, Lin N, Xu R (2015) Use an alginate scaffold-bone marrow stromal cell (BMSC) complex for the treatment of acute liver failure in rats. Int J Clin Exp Med 8(8):12593
143. Shteyer E, Ya'acov AB, Zolotaryova L, Sinai A, Lichtenstein Y, Pappo O, Kryukov O, Elkayam T, Cohen S, Ilan Y (2014) Reduced liver cell death using an alginate scaffold bandage: a novel approach for liver reconstruction after extended partial hepatectomy. Acta Biomater 10(7):3209–3216
144. Shang Y, Tamai M, Ishii R, Nagaoka N, Yoshida Y, Ogasawara M, Yang J, Tagawa Y-i (2014) Hybrid sponge comprised of galactosylated chitosan and hyaluronic acid mediates the co-culture of hepatocytes and endothelial cells. J Biosci Bioeng 117(1):99–106
145. Glowacki J, Mizuno S (2008) Collagen scaffolds for tissue engineering. Biopolymers 89(5):338–344
146. Tahmasbi Rad A, Ali N, Kotturi HSR, Yazdimamaghani M, Smay J, Vashaee D, Tayebi L (2014) Conducting scaffolds for liver tissue engineering. J Biomed Mater Res A 102(11):4169–4181
147. Gao C, Lu C, Jian Z, Zhang T, Chen Z, Zhu Q, Tai Z, Liu Y (2021) 3D bioprinting for fabricating artificial skin tissue. Colloids Surf B: Biointerfaces 208:112041
148. Lai-Cheong JE, McGrath JA (2021) Structure and function of skin, hair and nails. Medicine 49(6):337–342. https://doi.org/10.1016/j.mpmed.2021.03.001
149. Yan W-C, Davoodi P, Vijayavenkataraman S, Tian Y, Ng WC, Fuh JY, Robinson KS, Wang C-H (2018) 3D bioprinting of skin tissue: from pre-processing to final product evaluation. Adv Drug Deliv Rev 132:270–295
150. Manita PG, Garcia-Orue I, Santos-Vizcaino E, Hernandez RM, Igartua M (2021) 3D bioprinting of functional skin substitutes: from current achievements to future goals. Pharmaceuticals 14(4):362

151. Tavakoli S, Klar AS (2021) Bioengineered skin substitutes: advances and future trends. Appl Sci 11(4):1493
152. Trabosh VA, Daher A, Divito KA, Amin K, Simbulan-Rosenthal CM, Rosenthal DS (2009) UVB upregulates the bax promoter in immortalized human keratinocytes via ROS induction of Id3. Exp Dermatol 18(4):387–395
153. Lima-Junior EM, de Moraes Filho MO, Costa BA, Fechine FV, de Moraes MEA, Silva-Junior FR, Soares MFAN, Rocha MBS, Leontsinis CMP (2019) Innovative treatment using tilapia skin as a xenograft for partial thickness burns after a gunpowder explosion. J Surg Case Rep 2019(6):rjz181
154. Gupta S, Mohapatra DP, Chittoria RK, Subbarayan E, Reddy SK, Chavan V, Aggarwal A, Reddy LC (2019) Human skin allograft: is it a viable option in management of burn patients? J Cutan Aesthet Surg 12(2):132–135
155. Wang C, Zhang F, Lineaweaver WC (2020) Clinical applications of allograft skin in burn care. Ann Plast Surg 84(3S):S158–S160
156. Vijayavenkataraman S (2023) 3D bioprinting: challenges in commercialization and clinical translation. J 3D Print Med 7(2):3DP8
157. Fatimi A, Okoro OV, Podstawczyk D, Siminska-Stanny J, Shavandi A (2022) Natural hydrogel-based bio-inks for 3D bioprinting in tissue engineering: a review. Gels 8(3):179
158. Sadeghzade S, Cao J, Yang R, Li Y, Li Y, Zhang D, Liu J, Yu Z, Fang L, Yuan H (2024) Highly stretchable alginate/methylcellulose hydrogels for 3D bio-printing: photopolymerization approach enhancing structural integrity. Giant 18:100280
159. Castilho M, Levato R, Bernal PN, de Ruijter M, Sheng CY, van Duijn J, Piluso S, Ito K, Malda J (2021) Hydrogel-based bioinks for cell electrowriting of well-organized living structures with micrometer-scale resolution. Biomacromolecules 22(2):855–866
160. Townsend JM, Beck EC, Gehrke SH, Berkland CJ, Detamore MS (2019) Flow behavior prior to crosslinking: the need for precursor rheology for placement of hydrogels in medical applications and for 3D bioprinting. Prog Polym Sci 91:126–140
161. Deo KA, Singh KA, Peak CW, Alge DL, Gaharwar AK (2020) Bioprinting 101: design, fabrication, and evaluation of cell-laden 3D bioprinted scaffolds. Tissue Eng Part A 26(5–6):318–338. https://doi.org/10.1089/ten.TEA.2019.0298
162. Antezana PE, Municoy S, Álvarez-Echazú MI, Santo-Orihuela PL, Catalano PN, Al-Tel TH, Kadumudi FB, Dolatshahi-Pirouz A, Orive G, Desimone MF (2022) The 3D bioprinted scaffolds for wound healing. Pharmaceutics 14(2):464
163. Arjoca S, Bojin F, Neagu M, Păunescu A, Neagu A, Păunescu V (2024) Hydrogel extrusion speed measurements for the optimization of bioprinting parameters. Gels 10(2). https://doi.org/10.3390/gels10020103
164. Karvinen J, Kellomäki M (2023) Design aspects and characterization of hydrogel-based bioinks for extrusion-based bioprinting. Bioprinting 32:e00274
165. Das S, Basu B (2019) An overview of hydrogel-based bioinks for 3D bioprinting of soft tissues. J Indian Inst Sci 99:405–428
166. Dufaud M, Solé L, Maumus M, Simon M, Perrier-Groult E, Subra G, Jorgensen C, Noël D (2022) 3D bioprinting of articular cartilage: recent advances and perspectives. Bioprinting 28:e00253
167. Gómez-Blanco JC, Pagador JB, Galván-Chacón VP, Sánchez-Peralta LF, Matamoros M, Marcos A, Sánchez-Margallo FM (2023) Computational simulation-based comparative analysis of standard 3D printing and conical nozzles for pneumatic and piston-driven bioprinting. Int J Bioprint 9(4):730
168. Gu Y, Forget A, Shastri VP (2022) Biobridge: an outlook on translational bioinks for 3D bioprinting. Adv Sci 9(3):2103469
169. Lima TP, Canelas CAA, Concha VO, Costa FA, Passos MF (2022) 3D bioprinting technology and hydrogels used in the process. J Funct Biomater 13(4):214
170. Condi Mainardi J, Bonini Demarchi C, Mirdrikvand M, Karim MN, Dreher W, Rezwan K, Maas M (2022) 3D bioprinting of hydrogel/ceramic composites with hierarchical porosity. J Mater Sci 57(5):3662–3677

171. Schwab A, Levato R, D'Este M, Piluso S, Eglin D, Malda J (2020) Printability and shape fidelity of bioinks in 3D bioprinting. Chem Rev 120(19):11028–11055
172. Kačarević ŽP, Rider PM, Alkildani S, Retnasingh S, Smeets R, Jung O, Ivanišević Z, Barbeck M (2018) An introduction to 3D bioprinting: possibilities, challenges and future aspects. Materials (Basel) 11(11). https://doi.org/10.3390/ma11112199
173. Naghieh S, Chen X (2021) Printability–a key issue in extrusion-based bioprinting. J Pharm Anal 11(5):564–579
174. Adhikari J, Roy A, Das A, Ghosh M, Thomas S, Sinha A, Kim J, Saha P (2021) Effects of processing parameters of 3D bioprinting on the cellular activity of bioinks. Macromol Biosci 21(1):2000179
175. Umur E, Bayrak E, Arslan F, Bulut SB, Baysoy E, Kaleli-Can G, Ayan B (2023) Advances in three dimensional bioprinting for wound healing: a comprehensive review. Appl Sci 13(18):10269
176. Xu H-Q, Liu J-C, Zhang Z-Y, Xu C-X (2022) A review on cell damage, viability, and functionality during 3D bioprinting. Mil Med Res 9(1):70
177. Russo Spena S, Grizzuti N, Tammaro D (2022) Linking processing parameters and rheology to optimize additive manufacturing of k-carrageenan gel systems. Gels 8(8):493
178. Sharma R, Malviya R, Singh S, Prajapati B (2023) A critical review on classified excipient sodium-alginate-based hydrogels: modification, characterization, and application in soft tissue engineering. Gels 9(5):430
179. Liu J, Sun L, Xu W, Wang Q, Yu S, Sun J (2019) Current advances and future perspectives of 3D printing natural-derived biopolymers. Carbohydr Polym 207:297–316
180. Puertas-Bartolomé M, Mora-Boza A, García-Fernández L (2021) Emerging biofabrication techniques: a review on natural polymers for biomedical applications. Polymers 13(8):1209
181. Veeman D, Sai MS, Sureshkumar P, Jagadeesha T, Natrayan L, Ravichandran M, Mammo WD (2021) Additive manufacturing of biopolymers for tissue engineering and regenerative medicine: an overview, potential applications, advancements, and trends. Int J Polym Sci 2021(1):4907027
182. Ianchis R, Marin MM, Alexa RL, Gifu IC, Alexandrescu E, Pircalabioru GG, Vlasceanu GM, Teodorescu GM, Serafim A, Preda S (2023) Nanoclay-reinforced alginate/salecan composite inks for 3D printing applications. Int J Bioprint 10(1):0967
183. Lett JA, Sagadevan S, Fatimah I, Hoque ME, Lokanathan Y, Léonard E, Alshahateet SF, Schirhagl R, Oh WC (2021) Recent advances in natural polymer-based hydroxyapatite scaffolds: properties and applications. Eur Polym J 148:110360
184. Chan WW, Yeo DCL, Tan V, Singh S, Choudhury D, Naing MW (2020) Additive biomanufacturing with collagen inks. Bioengineering 7(3):66
185. Arifin N, Sudin I, Ngadiman NHA, Ishak MSA (2022) A comprehensive review of biopolymer fabrication in additive manufacturing processing for 3D-tissue-engineering scaffolds. Polymers 14(10):2119
186. Mohite P, Shah SR, Singh S, Rajput T, Munde S, Ade N et al (2023) Chitosan and chito-oligosaccharide: a versatile biopolymer with endless grafting possibilities for multifarious applications. Front Bioeng Biotechnol 11:1190879
187. Kołodziejska M, Jankowska K, Klak M, Wszoła M (2021) Chitosan as an underrated polymer in modern tissue engineering. Nano 11(11):3019
188. Arif ZU, Khalid MY, Noroozi R, Sadeghianmaryan A, Jalalvand M, Hossain M (2022) Recent advances in 3D-printed polylactide and polycaprolactone-based biomaterials for tissue engineering applications. Int J Biol Macromol 218:930–968
189. Singh S, Supaweera N, Nwabor OF, Chaichompoo W, Suksamrarn A, Chittasupho C, Chunglok W (2024) Poly (vinyl alcohol)-gelatin-sericin copolymerized film fortified with vesicle-entrapped demethoxycurcumin/bisdemethoxycurcumin for improved stability, antibacterial, anti-inflammatory, and skin tissue regeneration. Int J Biol Macromol 258:129071
190. Feldman D (2020) Poly (vinyl alcohol) recent contributions to engineering and medicine. J Compos Sci 4(4):175

191. Thurzo A, Gálfiová P, Nováková ZV, Polák Š, Varga I, Strunga M, Urban R, Surovková J, Leško Ľ, Hajdúchová Z, Feranc J, Janek M, Danišovič Ľ (2022) Fabrication and in vitro characterization of novel hydroxyapatite scaffolds 3D printed using polyvinyl alcohol as a thermoplastic binder. Int J Mol Sci 23(23). https://doi.org/10.3390/ijms232314870
192. Tylingo R, Kempa P, Banach-Kopeć A, Mania S (2022) A novel method of creating thermoplastic chitosan blends to produce cell scaffolds by FDM additive manufacturing. Carbohydr Polym 280:119028. https://doi.org/10.1016/j.carbpol.2021.119028
193. Sanchez-Rexach E, Johnston TG, Jehanno C, Sardon H, Nelson A (2020) Sustainable materials and chemical processes for additive manufacturing. Chem Mater 32(17):7105–7119
194. Ali F, Kalva SN, Koç M (2022) Additive manufacturing of polymer/Mg-based composites for porous tissue scaffolds. Polymers 14(24):5460
195. Reddy MSB, Ponnamma D, Choudhary R, Sadasivuni KK (2021) A comparative review of natural and synthetic biopolymer composite scaffolds. Polymers 13(7):1105
196. Ana I (2022) Functional biomaterials. Drug delivery and biomedical applications. Springer Singapore
197. Su Q, Qiao Y, Xiao Y, Yang S, Wu H, Li J, He X, Hu X, Yang H, Yong X (2023) Research progress of 3D printed poly (ether ether ketone) in the reconstruction of craniomaxillofacial bone defects. Front Bioeng Biotechnol 11
198. Oladapo B, Zahedi A, Ismail S, Fernando W, Ikumapayi O (2023) 3D-printed biomimetic bone implant polymeric composite scaffolds. Int J Adv Manuf Technol 126(9):4259–4267
199. Shiva S, Asuwin Prabhu RG, Bajaj G, John AE, Chandran S, Kumar VV, Ramakrishna S (2023) A review on the recent applications of synthetic biopolymers in 3D printing for biomedical applications. J Mater Sci Mater Med 34(12):62. https://doi.org/10.1007/s10856-023-06765-9
200. Wu Y, Lu Y, Zhao M, Bosiakov S, Li L (2022) A critical review of additive manufacturing techniques and associated biomaterials used in bone tissue engineering. Polymers 14(10):2117
201. Daminabo SC, Goel S, Grammatikos SA, Nezhad HY, Thakur VK (2020) Fused deposition modeling-based additive manufacturing (3D printing): techniques for polymer material systems. Mater Today Chem 16:100248
202. Tan LJ, Zhu W, Zhou K (2020) Recent progress on polymer materials for additive manufacturing. Adv Funct Mater 30(43):2003062
203. Monfared V, Bakhsheshi-Rad HR, Ramakrishna S, Razzaghi M, Berto F (2021) A brief review on additive manufacturing of polymeric composites and nanocomposites. Micromachines 12(6):704
204. González-Henríquez CM, Sarabia-Vallejos MA, Rodriguez-Hernandez J (2019) Polymers for additive manufacturing and 4D-printing: materials, methodologies, and biomedical applications. Prog Polym Sci 94:57–116
205. Pepelnjak T, Stojšić J, Sevšek L, Movrin D, Milutinović M (2023) Influence of process parameters on the characteristics of additively manufactured parts made from advanced biopolymers. Polymers 15(3):716
206. Saleh Alghamdi S, John S, Roy Choudhury N, Dutta NK (2021) Additive manufacturing of polymer materials: progress, promise and challenges. Polymers 13(5):753
207. Mahmood A, Akram T, Chen H, Chen S (2022) On the evolution of additive manufacturing (3D/4D printing) technologies: materials, applications, and challenges. Polymers 14(21):4698
208. Arjun B, Alekya B, Hari R, Vikas V, Mahadevan A, Pandya HJ (2022) Electromechanical characterization of human brain tissues: a potential biomarker for tumor delineation. IEEE Trans Biomed Eng 69(11):3484–3493
209. Arjunan A, Baroutaji A, Robinson J, Praveen AS, Pollard A, Wang C (2021) Future directions and requirements for tissue engineering biomaterials. https://doi.org/10.1016/B978-0-12-815732-9.00068-1
210. Jammalamadaka U, Tappa K (2018) Recent advances in biomaterials for 3D printing and tissue engineering. J Funct Biomater 9(1):22

211. Zennifer A, Thangadurai M, Sundaramurthi D, Sethuraman S (2023) Additive manufacturing of peripheral nerve conduits–fabrication methods, design considerations and clinical challenges. SLAS Technol 28(3):102–126
212. Simorgh S, Alasvand N, Khodadadi M, Ghobadi F, Kebria MM, Milan PB, Kargozar S, Baino F, Mobasheri A, Mozafari M (2022) Additive manufacturing of bioactive glass biomaterials. Methods 208:75–91
213. Arif ZU, Khalid MY, Noroozi R, Hossain M, Shi HH, Tariq A, Ramakrishna S, Umer R (2023) Additive manufacturing of sustainable biomaterials for biomedical applications. Asian J Pharm Sci 18:100812
214. Mirzaali MJ, Moosabeiki V, Rajaai SM, Zhou J, Zadpoor AA (2022) Additive manufacturing of biomaterials—design principles and their implementation. Materials 15(15):5457
215. Tian et. al. : Yan, W.-C., P. Davoodi, S. Vijayavenkataraman, Y. Tian, W. C. Ng, J. Y. Fuh, K. S. Robinson and C.-H. Wang (2018). 3D bioprinting of skin tissue: From pre-processing to final product evaluation. Advanced drug delivery reviews 132: 270–295.
216. Mainardi and Rezwan (2022): Condi Mainardi, J., C. Bonini Demarchi, M. Mirdrikvand, M. N. Karim, W. Dreher, K. Rezwan and M. Maas (2022). 3D bioprinting of hydrogel/ceramic composites with hierarchical porosity. Journal of Materials Science 57(5): 3662–3677.
217. Pan et al. (2024): Pan L, Hu Y, Zhang Z, Yuan Y, Zhong Q, Yang ST. Reduced graphene oxide promotes the biodegradation of sulfamethoxazole by white-rot fungus Phanerochaete chrysosporium under cadmium stress. Water Research. 2024 Jun 1;256:121558.
218. Hurtado et al. (2022): Hurtado, A., A. A. Aljabali, V. Mishra, M. M. Tambuwala and Á. Serrano-Aroca (2022). Alginate: Enhancement strategies for advanced applications. International Journal of Molecular Sciences 23(9): 4486.
219. Arifin et al. (2019): Mohammad F, Arfin T, Bwatanglang IB, Al-lohedan HA. Starch-based nanocomposites: types and industrial applications. Bio-based polymers and nanocomposites: preparation, processing, properties & performance. 2019:157–81.
220. Alven and Aderibigbe (2021): Alven, S. and B. A. Aderibigbe (2021). Fabrication of hybrid nanofibers from biopolymers and poly (Vinyl alcohol)/poly (ε-caprolactone) for wound dressing applications. Polymers 13(13): 2104.

Biomaterials-Based Additive Manufacturing of Veterinary Products

12

Prashant L. Pingale, Shounak Y. Kulkarni, Sakshi P. Wani, Rupali A. Patil, Anjali P. Pingale, and Sachin N. Kothawade

Abstract

Over the past few years, substantial progress has been made in the development of biomaterials for a range of uses, spanning from veterinary to human health. They find applications in the areas of wound management, tissue regeneration and therapeutic delivery systems. The amount of research and explorations on biomaterials have surged significantly. Biodegradable polymers are promising candidates which find extensive applications in the field of biomaterials. A scaffold is of pivotal role as it is important in the regeneration of the dermis layer of the skin. Elements like Fibrinogen, Silk Fibroin, and Hyaluronic acid are instrumental in the fabrication of Scaffolds. Polymers like Chitin, Cellulose, Polylactic acid, Poly(ε-caprolactone) and certain polymers extracted from bacteria hold significant potential as biomaterials. In addition to tissue engineering, they serve as drug carriers in pharmaceutical delivery systems. This chapter illustrates the applicative potential of such biomaterials in Veterinary medicine. Although, extensive research is imperative for the ultimate usage of such materials, they yet prove to be of paramount importance in Veterinary medicine.

P. L. Pingale (✉) · S. Y. Kulkarni · S. P. Wani
Department of Pharmaceutics, GES's Sir Dr. M. S. Gosavi College of Pharmaceutical Education and Research, Nashik, India

R. A. Patil
Department of Pharmacology, Mahatma Gandhi Vidyamandir Pharmacy College, Nashik, India

A. P. Pingale
Department of Pharmacognosy, Adivasi Seva Samiti's Institute of Industrial and Pharmaceutical Technology, Nashik, India

S. N. Kothawade
Department of Pharmaceutics, SCSSS's Sitabai Thite College of Pharmacy, Shirur, India

S. Singh et al. (eds.), *Biomaterial-based Additive Manufacturing in Tissue Engineering and Regeneration*, Biomaterials, Bioengineering and Sustainability 7,
https://doi.org/10.1007/978-3-031-96070-3_12

Keywords

Biomaterials · Veterinary Medicine · Polymers · Regeneration · Tissues · Systems

Abbreviations

BC	Bacterial Cellulose
ECM	Extracellular Matrix.
PCL	Polycaprolactone
PHA	Polyhydroxyalkanoates
PHB	Poly(3-hydroxybutyrate)
PLA	Poly lactic acid
NO	Nitric Oxide

12.1 Introduction

Biomaterials are synthetic or natural substances that can augment or replace body tissues, organs, or functions to improve an organism's quality of life. They play a crucial role in the biomedical industry, supporting diagnosis, prevention, and therapy across several applications. Their applications include orthopaedics, vascular treatment, trauma, cosmetic surgery, ophthalmology, and maxillofacial surgery. Integrating implants, gadgets, prostheses, and drug-delivery systems can benefit face surgery, cardiology, urology, and neurology. Biomaterial science combines physics, chemistry, medicine, and materials science to develop and translate biomaterials for therapeutic use continuously. Research has led to a diverse range of biomaterials with medicinal applications on the market. Biomaterials might be natural or manufactured. Materials such as polymers, composites, and ceramics can be metallic or non-metallic. Along with all these applications, there has been an increasing demand of biomaterials for Veterinary applications [1].

Biomaterial development requires expertise from a variety of scientific areas. It encompasses various subjects such as material science, engineering, medical science, biophysics, chemistry, and biology. Biomaterials are primarily developed through research in material science and engineering. Researchers have analysed and describe metals, alloys, polymers, ceramics, and composites to uncover features that meet biomaterial criteria. Medical science drives the creation of biomaterials for their intended purpose [2]. It provides information on device development requirements, including physical, mechanical, chemical, and biological features. Biophysics supports the creation of biomaterials by assessing their compliance and physical properties, including magnetic, electrical, phase, and thermal properties. Biomaterials' growth is intricately linked to chemistry, making elemental reactivity a valuable predictor of their behaviour in their environment [3].

Biomaterials translation from bench (laboratory) to bed (patient) is a laborious and time-consuming process. In order to meet all the requirements, set forth by regulatory bodies like the European Conformity (CE) agency or the US Food and Drug Administration (FDA), one must first formulate specific needs, then synthesize materials, design, manufacture, and conduct clinical testing. The first generation of biomaterials simply offered physical support with the barest minimum of compatibility because they were intended to be inert. The idea of biomaterials has changed dramatically in the modern era. Biomaterials are now understood to be active, not only inert, but compounds also that aid in the healing process of damaged tissue. They can be found in many different forms, like intricate composite hybrids or metals coated in various polymers. Typically, one material is used for one particular purpose [4].

These Biomaterials also have numerous applications in Veterinary Medicine. They are mostly employed in wound management and treatment. Natural polysaccharides, like chitin, chitosan, as well as cellulose, certain polyesters are the most appealing biodegradable polymers that have veterinary applications. On a commercial level they are offered in a range of molecular weights and shapes (powder, beads, gel, fibres, films, etc.) that enable specialized uses and the management of polymer degradation when needed (drug delivery systems). In order to restore tissue, biomaterial cannot only function as a static replacement; it must interact with the tissue. Furthermore, biomaterials that are directly employed in tissue replacement or repair applications (like artificial skin) need to be more than just biocompatible; they also need to cause the right kind of biological response. As a result, controlling how cells interact with biomaterials is a primary goal of biomaterials for tissue engineering applications. This is frequently achieved by incorporating cellular response manipulation mechanisms into the supporting biomaterial as an essential design element.

Since every tissue engineering application requires a different environment for cell-cell interactions, the field requires to have a variety of biomaterial solutions accessible. Such applications involve:

- The avoidance of cellular reactions (when the growth of tissues is not desirable);
- Providing supportive framework for new tissue development (where the communication between cells and access to nutrients to cells, and the medicinally active constituents need to be optimized);
- Directed tissue reaction, which enhances specific cellular response while inhibits others;
- Preventing antibodies from targeting the cells of homograft as well as xenograft used in organ replacement treatments and blocking cellular adhesion or activation.

Biodegradable polymers are helpful in tissue engineering applications where the objective is remodelling or repair of tissue, but not in situations which require a material stability over a sustained period. Table 12.1 depicts the chemical structure and uses of various polymers used in biomaterials. Additionally, biodegradable polymers need to exhibit [5]:

Table 12.1 Various polymers used in biomaterials with their chemical structure and uses

Name of the polymer	Chemical structure	Uses
Polylactic Acid (PLA)	Lactic acid monomers, which undergo repetition to form a long linear chain. Available in D and L form.	Scaffold fabrication, Drug delivery vehicle, implant preparation
Polymers derived from Bacteria	Natural polyesters containing 3-, 4-, 5-, and 6-hydroxyalkanoic acids have molecular weights ranging from thousands to millions of Daltons.	Wound dressing, tissue engineering, artificial organ construction, drug delivery, and bio-implants.
Poly(ε-caprolactone)	Developed through ring-opening polymerization of ε-caprolactone with different catalysts.	Enhance biodegradation, improve adhesion, and stress crack resistance, or control medication release from PCL microcapsules.
Cellulose	The linear polymeric chain consists of hundreds to thousands of β (1→4) connected D-glucose units. It also has six hydroxyl groups.	Regenerative medicine, wound treatment, tissue engineering, implants, and drug delivery systems.
Chitin	N-acetyl glucosamine units which are linked by β-(1, 4) links.	Nutraceuticals as a supplement to food, pharmaceuticals, and 3D scans for biomedical applications, wound healing
Chitosan	Chitin derivative that has undergone partial deacetylation via alkaline deacetylation	Wound Management, Tissue Engineering.

- Adequate commercial viability and manufacturing capability.
- The capacity to shape the polymer in the desired outcome.
- Mechanical characteristics which do not impede long-term function and sufficiently address short-term function.
- Drug delivery compatibility in operations requiring the release or adhesion of active chemicals; low-level toxicity of breakdown products.

To repair damaged tissue, aid in tissue regeneration, or recover the function of tissue, biomaterials have long been employed in medicine in the form of devices, systems for delivering drugs, or implants (stents, ligaments, prostheses, screws, vascular grafts, heart valves, rods, plates, etc.). The choice of biomaterials is crucial for both the biological response and the outcome of clinical procedures. A biomaterial needs to meet several requirements to be approved, including high corrosion resistance, no cytotoxic effects, good mechanical properties, additionally, it must have the property of biocompatibility, biostability (chemical, biological, and structural resistance), bio functionality, and adherence to pathobiology as well as functional tissue structure when used as implants.

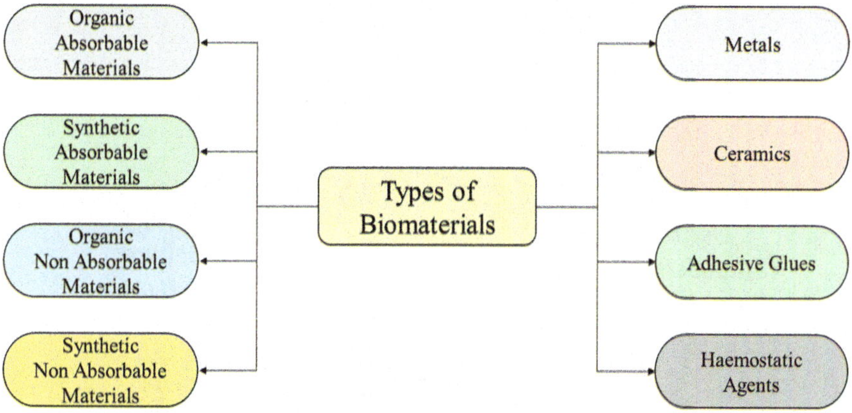

Fig. 12.1 Types of biomaterials. (Created by using Biorender.com)

12.2 Types of Biomaterials

Organic adsorbable materials include surgical catgut and collagen, whereas synthetic absorbable materials include polyglycolic acid and polyglactin 910. Silk, linen, and cotton are examples of organic non-absorbable materials, while synthetic non-absorbable materials include polyester as well as silicone rubber. Metals in this category include stainless steel 316L, alloys such as cobalt-chromium, and titanium. Alumina and zirconia are ceramic biomaterials. Autologous fibrin and cyanoacrylates constitute adhesive glues. Hemostatics agents include absorbable gelatin sponges [6] as presented in Fig. 12.1.

12.3 Biomaterials in Veterinary Medicine

In-vitro and *in-vivo* studies aim to create implantable biomaterials for wound dressing that integrate into the wound and promote dermal and epidermal healing. Diverse types of biomaterials employed in Veterinary Medicine are depicted in Fig. 12.2. These materials may contain growth factors and cytokines that promote healing. Three-dimensional scaffolds are required for dermis regeneration because they offer elasticity as well as rigidity to the epidermal graft while also nourishing the keratinocytes in the epidermis. In skin tissue engineering, a tissue scaffold acts as a template for cell reorganization as well as host infiltration of the graft. Over the last three decades, advancements in cell/molecular biology have resulted in successful skin tissue regeneration for the healing of wounds. The scaffold should support cell adhesion, infiltration, and proliferation for ECM formation. Additionally, it must be biodegradable, bioresorbable, and non-inflammatory [7]. Table 12.2 provides a comprehensive overview of several types of biomaterials, outlining their advantages, disadvantages, and applications [7].

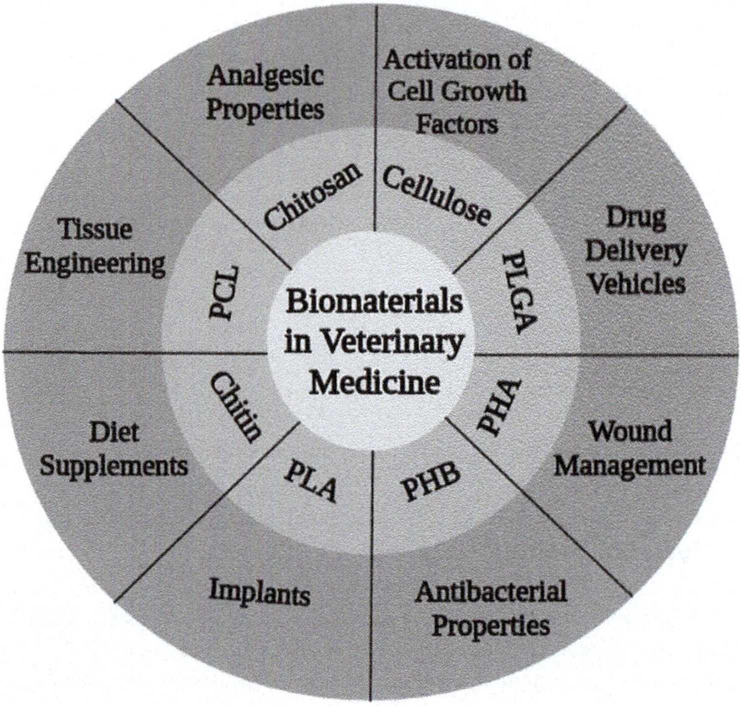

Fig. 12.2 Biomaterials in veterinary medicine. (Created by using Biorender.com)

Table 12.2 Examples of biomaterials with their pros, cons and uses

Type of biomaterial	Examples	Pros	Cons	Uses
Polymers	Silicone, polyurethane, polyesters, polyamides, polyacetals, polyolefins, rubber, polyether	Easy to Manufacture, Low density	Low Mechanical Strength, Oligomers may trigger tissue responses.	Cardiovasculaire and maxillofacial implants. Soft skeletal tissues, contact lenses, sutures, etc.
Metals	Stainless steel, cobalt-chromium alloys, titanium alloys, platinum, silver, gold	High conductivity and strength. Excellent wear resistance, and strong absorbency of shock.	Low biocompatibility, corrosive in the physiological environment, mechanical mismatch with surrounding soft tissues.	Orthopedic load-bearing and fixing devices, implants for teeth, and neuromuscular activation electrodes.

12.3.1 Elements That Can Be Used as Scaffolds

Fibrinogen and Fibrin Another family of naturally occurring compounds generated from tissue includes fibrinogen and fibrin glue, which are commonly used as biomaterials to create three-dimensional scaffold materials. Under the catalysis of calcium ions, the hemostatic coagulation processes require a mix of fibrinogen and thrombin. Its ability to be fabricated from autologous blood makes it appealing as a natural scaffold. Fibrin glue has been shown in studies using nude mice to be a useful biomaterial because the rate of breakdown and polymerization may be regulated. Fibrin glue can also be mixed with other polymers, such as polyurethane, to preserve cell viability, adhesion, and dispersion while expressing more type II collagen and aggrecan. Fibrin, the filament core protein, and sericin proteins, which resemble glue, make up silk fibroin. It is a naturally occurring biomaterial that is extracted from flies, spiders, scorpions, mites, and silkworm cocoons.

Silk Fibroin Using techniques like gas foaming or salt leaching, silk can be transformed into porous scaffolds. It is also frequently utilized as a suture material in surgical applications. This biomaterial has suitable mechanical properties and is highly biocompatible. Silk fibroin interacts with mammalian cells, making it an appealing option for tissue engineering purposes.

Hyaluronic Acid One of the main elements of the extracellular matrix and a very appealing natural biomaterial is hyaluronic acid, an acidic glycosaminoglycan. It exists in tissues as a gel-like material, but it can be effectively processed chemically to form fibers, membranes, or microspheres. Hyaff® is a modified form of hyaluronic acid that is sold commercially.

Gelatine Another biocompatible substrate that is generated from collagen is gelatine, which may be altered to become photopolymerizable gelatine by exposing it to visible light, which will cause the polymerization to occur and cause a crosslink. With enhanced proteoglycan spatial accumulation, gelatine/alginate gels can stimulate chondrocyte growth and the expression of type II collagen and hyaline matrix molecules. Hydrogels based on gelatin have good biocompatibility, are biodegradable, and allow for the introduction of ligands or functional groups.

Chitosan Chitosan is regarded as an excellent material for making scaffolds, either by itself or in conjunction with other polymers (such as hyaluronate). Preferred for its special qualities, it increases the activation of fibroblasts by amplifying growth factors including platelet-derived growth factor. It is mainly used in wound management [8].

12.3.2 Decellularization

As the biomaterials are instantly rejected on implantation, it is crucial to perform the process of 'Decellularization.' Decellularization aims to reduce the antigenicity of biological grafts by removing cellular and nuclear material while preserving the remaining ECM's composition, biological activity, and mechanical integrity. Decellularization can occur by physical, chemical, or enzymatic means, resulting in a substance primarily consisting of extracellular matrix (ECM) components. The acellular tissues maintained their mechanical qualities and helped the prosthesis undergo remodeling through neovascularization and recellularization by the host. Tissue engineering and regenerative medicine rely heavily on collagen-based biomaterials. Collagen is still a popular protein for biomaterials because of its great biocompatibility as well as low immunogenicity. It may be extracted from various tissues and mixed with other chemicals. It can be used in laboratories as a decellularized ECM for fundamental research or as a tissue substitution product in medical applications [7].

12.3.3 Collagen-Rich Extracellular Matrix as a Biomaterial

While there is a collagen-rich extracellular matrix (ECM) in all tissues and organs, only a small number of these may be extracted and used as a therapeutic scaffold. For tissue engineering, extracellular matrices have been generated from a range of tissues, including the skin [9], diaphragm [10, 11], pericardium, fish swim bladder [12, 13], tendons [14], and urinary bladder [15, 16]. Acellular dermal matrix was found to be an excellent means of directing the regeneration of normal skin morphology [17]. In rats, the revascularization capability of xenogeneic acellular dermal matrix (alloderm) [18].

Based on macroscopic and immunological observations the effectiveness of xenogeneic and allogeneic acellular dermal matrix in repairing the full thickness of skin lesions on specifically the dorsum of rats [19]. The mechanical and regenerative qualities of the cholecyst-derived matrix can be used in soft tissue regeneration [20]. The product that comes closest to the perfect wound dressing in humans is dermal replacements based on collagen accompanied by an epidermal autograft or, lately, bioengineered bilayer skin replacements [21, 22].

Research investigated the healing of stomach wounds in rabbits with absorbable sutures made from collagen found in fish intestines was investigated [23]. All test animals' stomach wounds healed by day 15 and showed no signs of irritation, allergies, toxic effects, leaking, or discontinuity of the suture line. Histologically, it was seen that the stomach mucosa had fully healed, and the suture material had dissolved by day 30. The healing tissue was not distinguished from typical healing tissue, and there was also epithelialization along with neovascularization over the suture site. Acellular swim bladders were studied for major skin wound clinical testing in animals following the evaluation of successful full healing in rabbits and rats [24, 25].

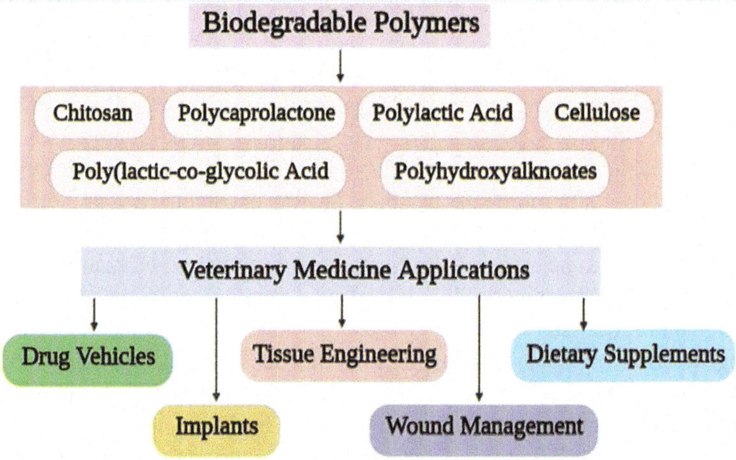

Fig. 12.3 Biodegradable Polymers have applications in veterinary medicine. (Created by using Biorender.com)

12.3.4 Polylactic Acid

Different biodegradable polymers are depicted in Fig. 12.3. Polylactic acid (PLA) consists of lactic acid monomers, which undergo repetition to form a long linear chain. They exist in two enantiomeric forms: the L and D isomers, with the L isomer being the more prevalent form that is created throughout the fermentation process [26, 27].

PLA possesses good mechanical resistance, biological safety, and processability profiles. As a result, it may be used in biomedical applications as an inexpensive substitute for biocompatible materials which are non-biodegradable. It finds applications in sutures that are biodegradable, orthopaedic devices, apparatus used for bone fixation purposes, surgical implantation substances, medication delivery frameworks, as well as scaffolds for applications in tissue engineering. However, the low hardness of PLA makes it unsuitable for various applications. Therefore, a lot of investigation is being undertaken to generate different co-polymers with superior efficacy to minimize this, PLA, hydroxyapatite as well as polycaprolactone are mixed [26, 28, 29].

12.3.4.1 Polylactic Acid for Veterinary Orthopaedic Surgeries

Conventional inserts made of metals are troublesome in veterinary applications and uses; they can lead to infections, resorption of the bone, and can also cause pain at the place where it is applied. Moreover, removing is risky and costly. As a result, new kinds of implants that are less expensive and safer are required, and PLA can be a potential element for these purposes. Furthermore, the advancement of 3D printing techniques makes it possible to create customized implants using biocompatible and biodegradable polymers that have anatomical structures similar to actual bones. This will lower the cost of orthopaedic procedures and allow implants to be customized for each patient.

In canine clinical trials, PLA implants have already been evaluated. They had high functional outcomes, were safe for veterinary patients, and were effective in treating tibial tuberosity progression. They displayed an acceptable rate of problems, comparable clinical recovery times to metallic implants, and quicker ossification, which allowed for more effective bone healing of osteotomy gaps. Furthermore, Bone fixation plates made from PLA or related materials of PLA, their screws, and nails have proven to be a superior substitute for their metallic counterparts, giving the fracture sufficient support while it heals and then gradually disintegrating when it is no longer required.

Regeneration of tissues in critical cases of bone abnormalities is an additional challenging issue in veterinary orthopaedics, and dependable osteoconductive materials are desperately needed for this use. It was discovered that PLA is appropriate for a method in 3D printing that allows the creation of scaffolds that might aid in the treatment of significant bone deformities, either alone or in combination with other biodegradable polymers. In Wistar rats with substantial bone defects, scaffolds consisting of polylactic acid, polycaprolactone, as well as hydroxyapatite supplied with stem cells from the bone marrow or unique bone cells were discovered to be effective in supporting bone repair. The usage of cell-loaded scaffolds exhibited identical consequences as the autograft category, and immunohistochemical, histological, histomorphometry, and biomechanical tests revealed that both scaffolds, as well as autograft groups, significantly enhanced bone healing when compared to the untreated category. PLA scaffolds can be loaded with osteoinductive drugs like purmorphamine or nandrolone to improve healing quality speed up bone regeneration and effectively cure keratoma-caused hoof wall defects in horses [5].

12.3.4.2 Polylactic Acid as Therapeutic Delivery Systems

Polymers like PLA appear like excellent carrier alternatives because of their prolonged biocompatibility and biodegradability. To address issues with drug absorption and associated side effects, animal medicine requires novel, lasting therapeutic compositions. Several clinical trials have previously been conducted to determine PLA's utility as a medicine delivery vehicle. Tylosin and florfenicol were administered intramuscularly to pigs to evaluate the medicinal effect of the microspheres made of PLA packed with lactones from Venenum Bufonis towards mycoplasma pneumoniae. The outcomes showed that the microspheres had a considerable medication slow-release impact, effectively lowering drug toxicity and lengthening its half-life.

Although metformin is useful in the treatment of diabetes, its application varies greatly in terms of gastrointestinal absorption as well as bioavailability, depending on the kind of the patient. Metformin-loaded PLA microparticles were evaluated in rabbits to see whether the pharmacokinetics of Metformin has been enhanced. Based on the outcomes, the generated microparticles exhibit the required physicochemical properties and exhibit sustained expulsion of metformin, indicating that they could be employed to treat diabetes in animals [30]. In guinea pigs who were induced with myocardial infarction, curcumin-nisin-based PLA nanoparticles have been effective for cardiovascular protection [31].

The manufacture of innovative biomaterials containing active substances for the treatment of both people and animals is made possible by the technologies of electrospinning and solution blow spinning. PLA/nano-hydroxyapatite/doxycycline fibers are involved in a number of these; they have shown effectiveness against common periodontal pathogens, such as *Aggregatibacter actinomycetem* compitans and *Porphyromonas gingivalis*, exhibiting stronger antimicrobial properties than antibiotics frequently used in similar cases and serving as a stand-in for non-surgical local therapy in periodontitis. Another example is progesterone-containing nanofibrous PLA mats, which offered an enticing option for the controlled transport of progesterone to control the animal oestrus cycle [32, 33].

Biocompatible PLA nanoparticles are potential adjuvants and carriers for the recently developed mucosal vaccination in human and veterinary medicine. Experiments on zebrafish have demonstrated their effectiveness and adaptability in delivering immune-promoting chemicals and antigens to cells those present antigens, indicating their considerable promise for the application of mucosal vaccines in aquaculture and human populations [5].

12.3.5 Polymers Derived from Bacteria

Biodegradable polymers that occur naturally include polyesters made from microorganisms. Polyhydroxyalkanoates (PHA) and polyhydroxybutyrate (PHB) are the two main types of polyesters in this category. They are produced by a variety of microorganisms as lipid inclusions, which they then store in their cells as a source of energy and carbon. Because different building monomers have different effects, PHAs have different chemical compositions and consequent properties. They are resistant to hydrolytic assault, insoluble in water, and biodegradable, however, the rate of degradation varies according to the microorganism doing the breaking down and the environment in which it occurs. In terms of production and recycling, bacterial polyesters are regarded as green plastics that have a beneficial social and environmental impact in contrast to traditional petrochemical-based plastic. These materials are distinguished by their nontoxicity, biocompatibility, biodegradability, tunable biodegradability, and variable thermal and mechanical attributes. But even with their exceptional qualities, they are still unable to completely replace conventional plastic due to their expensive production costs, which are five to ten times greater than those of polymers derived from petroleum.

12.3.5.1 Polyhydroxyalkanoates and Poly(3-Hydroxybutyrate) for Tissue Remodelling

Polyhydroxybutyrate-based composites, which have structural, and physicochemical properties have characteristics similar to cartilage or bone, can be an ideal material for the development of orthopaedic implants that are resorbable. PHB can help regenerate chondral, osteochondral, and osseous defects in articular cartilage and bones where their innate capacity to regrow is reduced. This is especially true when paired with 3-hydroxyvalerate to form P(HB-HV). At the beginning of the therapy, they provide good stability, and their biodegradability allows for the gradual removal of physical impedance and, in the end, implant resorption, which promotes the healing of defects.

Polyhydroxybutyrate/chitosan composites were one of the materials studied. Tested on an animal model of sheep with defective knee cartilage, implants constructed of this material were effective in speeding up the healing process, indicating that they may be used to repair traumatic chondral and osteochondral defects. Porous scaffolds containing mesenchymal stem cells have been successfully utilized to improve wound healing and cartilage-like tissue production in sheep with defective knee cartilage. Conversely, polyhydroxybutyrate/hydroxyapatite composites that were implanted in animals with bone defects demonstrated biocompatibility, biodegradability, osteoconductive properties, and successful integration into bone; yet, in certain instances, they elicited a persistent local inflammatory reaction [34–36].

12.3.5.2 Drug Carrier-Based on Polyhydroxyalkanoates and Poly(3-hydroxybutyrate)

Researchers are searching for new drug carriers among biodegradable polymers since veterinary medicine still struggles with safe, effective, and economical medication administration strategies and routes due to the differences among different animal species. Progesterone is a common steroid hormone utilized in the synchronization of cattle's oestrus. Recent research explores the use of ear implants containing progesterone encapsulated in biopolymer nanoparticles, focusing on the interaction between PHB-V nanoparticles and PCL for optimal drug distribution.

In vivo studies on cows using progesterone-loaded PHB/PCL implants demonstrated their effectiveness as drug carriers with adjustable release rates and biodegradation periods. Furthermore, it was demonstrated that P(HB-HV) and other PHB-based composite materials are excellent for drug delivery because surface erosion leads to deterioration. One of the most significant health problems affecting cattle is tuberculosis in milk; for this reason, the development of a vaccination that effectively prevents this illness is essential. This technology offers the potential to produce low-cost vaccinations to treat animal diseases like tuberculosis, as evidenced by the successful usage of PHB nanoparticles, which displayed antigens from mycobacterial species on the outside of the bio-beads.

Additionally, bacterial polyesters can be added to food. Effective disease prevention chemicals are needed in aquaculture to keep the aquatic animals that are farmed healthy and in good condition. Because PHB may be converted into short-chain fatty acids (SCFAs), which have been shown to have antibacterial properties, it is a perfect biocontrol agent. PHB, when used as a bioactive agent in bio flocs, significantly increased the presence of beneficial bacteria, such as Bacillus sp., beneficial for the health of farmed fish. An increase in the weight of the fish was seen with the incorporation of PHBs. When fed a diet enriched with PHB and butyrate, Pacific white shrimp demonstrated enhanced immunological function, growth, and digestion. PHB-fed Siberian sturgeons and Nile tilapias produced comparable results [5, 37–39].

12.3.6 Poly(ε-caprolactone)

Ring-opening polymerization of ε-caprolactone with different catalysts can yield the polymer. Because of its comparatively slow hydrolytic disintegration, this polymer is ideal for long-term (more than a year) medicinal applications. To improve adhesion, stress crack resistance, biodegradation, and drug release from PCL microcapsules, PCL is frequently utilized in conjunction with additional polymers, such as PLA and polylactic acid-co glycolic acid [40].

12.3.6.1 Poly(ε-caprolactone) Tissue Engineering Techniques

It can be used for scaffolds that are helpful in the regeneration of tissues, much like the polymers that were previously described. For this use, polycaprolactone tricalcium phosphate (PCL-TCP) is one of the promising composites. In a limb-sparing procedure, its implants were used and were effectively tested in canines suffering with distal radial osteosarcoma [41]. Polypropylene mesh modified with PCL nanofibers was studied in animal models as a framework for incisional hernia restoration or prophylaxis because of its structure, which is similar to that of natural extracellular matrix. A novel avenue meant for treating as well as preventing hernias has been opened up by promising results from experiments conducted on models of rabbits and minipigs [42]. These models showed that collagen maturation could be accelerated and that the scar left behind had increased elasticity and flexibility as compared to a normal mesh. Because of its biocompatibility and regenerative qualities, which promote excellent osteogenesis, in the rabbit model, a bilayer oxidized renewed cellulose/poly ε-caprolactone composite proved effective for dural closure after neurosurgery. Additionally, PCL has shown suitability for creating screws utilized in the surgical procedure of partial anterior cruciate ligament ruptures in animals during experimental procedures. The PCL-based scaffolds can have the right cells added to them to improve their capacity for regeneration. In tests conducted on rabbits, Poly(ε-caprolactone) scaffolds implanted with mesenchymal stem cells showed enhanced fibrocartilaginous regeneration of tissue, durability, and reduced cartilage degradation in comparison with PCL scaffolds without cells [43]. Furthermore, PCL-TCP scaffolds cured significant bone size deficiencies in a dog's jaw, suggesting the potential for bone regeneration [44].

12.3.6.2 Poly(ε-caprolactone) as a Therapeutic Delivery System

Due to its high porosity, biocompatibility, slow degradation, and lack of toxicity, poly(ε-caprolactone) is a material of choice for a variety of therapeutic delivery applications. This is especially true for difficult-to-deliver or site-specific drugs with high general toxicity or local action. For example, the toxicity, inadequate penetrability, and quick deterioration of free chloramphenicol can be avoided by using poly(ε-caprolactone)-Pluronic composite nanoparticles packed with novel chloramphenicol. This enables the therapy of MRSA-infected burn injuries, as demonstrated on a model of animals. The application of PCL nanoparticles reduced the toxicity and increased the bioavailability and therapeutic efficacy of chloramphenicol, demonstrating the significant promise of such a strategy for transporting strong but troublesome medicines in veterinary and human medicine [45]. An additional illustration would be a PLC-based intraocular implant with enhanced porosity that is packed with dexamethasone and dissolves quickly. Its capacity to administer drugs to treat retinal problems is demonstrated by its non-toxicity to retinal cells and its inability to change the structural and functional properties of the retina [46]. Moreover, poly(ε-caprolactone) demonstrated efficacy in creating intravaginal inserts that administered progesterone to cattle [47].

12.3.7 Cellulose

Cellulose is the major constituent of plants, some algae, and oomycetes, and it is present in a variety of bacterial secretions. While the walls of plants include a mixture of cellulose, hemicelluloses, as well as lignin, bacterial cellulose is relatively pure. There are now several BC-based biomedical products available, and bacterial cellulose (BC) is used expressly for medical applications. Due to its unique properties, *Gluconacetobacter xylinus* (*Acetobacter xylinum*) is the primary producer of bacterial cellulose. It is an ultrafine material that has excellent durability, adaptability, moldability, and water-holding capacity. It also has a high surface area, high polymerization, as well as elevated crystallinity degrees. Due to its great purity, this substance is ideal for a variety of uses in both human and veterinary medicine, such as wound care, implants, tissue engineering, regenerative medicine, and drug delivery [48–53]. It is also non-cytotoxic, non-genotoxic, and biocompatible. However, using a range of in-situ and ex-situ procedures, the cellulose from bacteria is additionally chemically and/or physically altered to enhance its porosity and the amount broken down as well as to promote cell interactions [51]. To alleviate its drawbacks, bacterial cellulose (BC) is frequently combined with other biomaterials of natural or artificial origin, such as chitosan, polylactic acid, graphene, or graphene oxide. This is because bacterial cellulose lacks the antioxidant as well as antibacterial properties needed for various biomedical applications, and natural cellulose hydrogels are challenging for handling, storing, and preserving [50].

12.3.7.1 Cellulose in Wound Management and Treatment

Pure crystalline cellulose in the form of a membrane has the potential to be used in veterinary medicine to cure lesions. It is simple to use, has been effectively used to treat skin wounds in Wistar rats, and has been shown to reduce discomfort and healing time while shielding the wound, preserving its humidity, and facilitating lesion visualization and management. Furthermore, a cellulose/graphene oxide nanocomposite that was cytocompatibility was evaluated on rats' dorsum wounds and shown an effective promotion of wound healing, enhancing neovascularization and re-epithelization as well as the accumulation of collagen as well as granulation tissue

[54]. Cellulose's limited ability to promote wound healing is due to its lack of antibacterial characteristics. As a result, it is frequently modified or mixed with different antimicrobial agents. Since bacterial cellulose is the purest form of cellulose and is therefore most frequently used in wound care, it undergoes the most changes.

A case in point is the thymol-enriched bacterial cellulose hydrogel, which demonstrated remarkable efficacy in the healing of third-degree burns in Wistar rats. In addition to shortening the duration required for burn site healing, it demonstrated remarkable antibacterial efficacy against microorganisms particular to burns [55]. Because of their possible antibacterial properties, silver nanoparticles are also frequently included in dressing products based on BC. They had strong antibacterial activity, were biocompatible, and efficiently sped up the healing process.

BC mixed with Cu/Na/Ca-modified montmorillonite creates a novel burn injury dressing material, promising for artificial skin substitute in burn wound therapy due to improved wound healing and antibacterial activity [56]. A unique modified Mohs paste was made by substituting CMC for zinc oxide starch powder, which reduces the amount of preparation work and metal-containing liquid waste produced during the scrubbing wound cleaning process. This paste is used to treat malignant wounds. It cured canine malignant lesions successfully, lessened bleeding and odour, and changed onto the wound surface to form a gel to facilitate excretion [57].

12.3.7.2 Cellulose in Tissue Engineering

Particularly well suited for this use is bacterial cellulose. Because of its shape and nanostructure resemblance to collagen, it is a desirable material for scaffolds made of soft and hard tissue that vary in length and have the potential to stimulate tissue regeneration. It can also be used alongside additional polymeric as well as non-polymeric materials to get antibacterial, adhesion of cells, and proliferation qualities [53, 58–60]. In mice and rabbits, pure BC scaffolds were effective in promoting bone repair [61].

The study demonstrated superior mechanical characteristics, biocompatibility, and osteoinductivity of scaffolds made of BC and gelatin, BC/gelatin/polylactic acid, or BC/gelatin/polylactic acid with hydroxyapatite coating [61]. This suggests an avenue for the research and development of artificial scaffolds for the regeneration of bones. A dog lateral femoral condyle injury model was used to study a carboxymethyl cellulose sheet loaded with calcium phosphate for bone regeneration [62].

Tissue-engineered blood arteries made of bacteria-cellulose offer a novel solution to the reconstructive issues that come with long-term vascular disorders. In multiple trials, small-diameter BC tubes were implanted into artificial carotid artery defects in rats or sheep, or they were utilized to substitute the carotid arteries in pigs and lambs. The investigations showed that the artificial vessels' suture retention strength and bursting strength were excellent. The study found no evidence of inflammatory capability in the migration of vascular smooth muscle cells into the synthetic matrix or the effective neoformation of a vascular wall network. The outcomes demonstrate that BC grafts are robust vascular conduits that support vascular wall neoformation and are viable building blocks for novel vascular surgery programs in the future [60, 63, 64].

To investigate the potential of potato starch/bacterial cellulose bilayer nanofibrous scaffolds as biomaterials for hollow organ regeneration, muscle cells were grown on them. To correct canine urethral abnormalities, they performed admirably. The composite scaffolds can be used to repair many hollow organ tissues, including the intestine, ureter, bladder, oesophagus, and vascular tissues, because of their adjustable pore size and porosity [65]. Furthermore, tested on rabbits, an engineered corneal stroma composed of polyvinyl alcohol hydrogel and bacterial cellulose had even superior qualities than a hydrogel formed of BC alone, indicating that corneal stroma can be replaced in both human and veterinary medicine. Rabbit corneal stroma replacement with BC scaffolds was successfully evaluated, and the scaffolds demonstrated good stability as well as biocompatibility [66].

12.3.8 Chitin

Long-chain polymers of N-acetylglucosamine, known as chitin, have been shown through experimentation to exist in both multicellular (such as sponges, corals, mollusks, worms, and arthropods) among unicellular (such as diatoms, protists, and fungus) organisms [67]. Chitin has been added to nutraceuticals as a supplement to food, pharmaceuticals, and 3D scans for biomedical applications [68, 69] in addition to technology uses [70–73]. Chitin is capable of being produced at elevated temperatures and has thermostability [74]. Chitin's high resistance to high chemical concentrations allows for the electrochemical deposition of metals like copper into it at room temperature [75]. Chitin is a linear polymer made up of (1,4)-N-acetyl-d-glucosamines joined by a glycosidic bond.

Tissue engineering has made extensive use of α-chitin [76, 77]. β-Chitin, on the other hand, is primarily present in squid pens. β-chitin has also been used in biological applications, such as improving the sense of saltiness [78], preservation agents for methylene blue [79], and wound healing [80, 81]. γ-Chitin was extracted from the moth's (*Orgyia dubia*) cocoon, and it was discovered that while its structure is comparable to that of chitin rather than chitin, its surface morphology is made up of microfibers, whereas α-chitin and β-chitin are made up of nanofibers.

12.3.8.1 Use of Chitin in Neurological and Musculoskeletal Disorders

Chitin was used as a scaffold for neural development in conjunction with carbon nanotubes [82]. Chitin, a physiologically absorbable tube or catheter, has been successfully used as a bridge for sural nerve transplants in a rat sciatic nerve insufficiency model [83, 84]. Chitin hydrogel effectively treats cartilage injury by preventing chondrocyte death and promoting chondrogenesis through macrophage immunomodulation [85]. Chitin nanofibers and platelet-rich plasma, including TEMPO-oxidized sacchachitin nanofibers and chitosan-activated platelet-rich plasma, have been found to enhance corneal injury by promoting cell growth and migration [86].

12.3.8.2 Use of Chitin in Cardiovascular and Haematological Disorders

Chitin, combined with pomegranate glucan and polyphenols, has been shown to restore endothelial dysfunction in apolipoprotein E defective animals by reducing inflammatory markers and increasing NO synthase. Additionally, extra bilirubin in the blood was effectively absorbed by chitin and graphene oxide mixed as aerogel beads [87].

In comparison to hemostatic products based on chitosan, chitin nanogel with rectorite nanocomposite halted bleeding in a rat tail vein in 121 s as well as increased haemostatic activity [88]. Additionally, chitin from the spider *Caribena versicolor* and the demosponge *Ianthella labyrinthus* worked well as 3D scaffolds for the cultivation of cardiomyocytes produced from induced pluripotent stem cells as well as commercial extracellular matrix [89, 90].

12.4 Conclusion

The extent of research and studies on Biomaterials has only grown in the last few years. They are specifically aimed in improvement in the quality of the treatment and thus the quality of life. Biomaterials have found widespread uses in the field of tissue engineering, wound management, and healing, as well as in drug delivery systems. Biomaterials can be metals, composites, polymers, or ceramics depending on their chemical makeup. Biodegradable polymers as well as metals remain the most widely employed biomaterials. Biodegradable polymers are widely used in veterinary treatment, benefiting both animals and the environment. It is essential for veterinary treatment that the biomaterials be both inexpensive and extremely effective. Polymers should be designed to achieve specific goals, such as promoting tissue growth, providing antimicrobial resistance, and addressing short-term functionality, minimizing toxicity, and delivering medications to particular locations after degradation. Additionally, it is paramount to select the type of material that will be used as a biomaterial. Choosing the right biomaterials is crucial for optimal biological response and clinical performance. Hence, Biomaterial development is currently on par with drug discovery research. Using them individually or in combination will benefit not only animals but also humans. As research progresses, we can expect to find a plethora of novel uses and new materials as biomaterials.

References

1. Biswal T, BadJena SK, Pradhan D (2020) Sustainable biomaterials and their applications: a short review. Mater Today Proc 30:274–282
2. Ratner BD, Hoffman AS, Schoen FJ, Lemons JE (2013) Biomaterials science: an evolving, multidisciplinary endeavor. Biomater Sci:xxv–xix
3. Mahyudin F, Hermawan H (2016) Biomaterials, and medical devices. Adv Struct Mater 58:207–234

4. Vallet-Regí M (2010) Evolution of bio ceramics within the field of biomaterials. C R Chim 13(1–2):174–185
5. Broda M, Yelle DJ, Serwańska-Leja K (2024) Biodegradable polymers in veterinary medicine—a review. Molecules 29(4):883
6. Sortino F, Lombardo C, Sciacca A (2008) Silk and polyglycolic acid in oral surgery: a comparative study. Oral Surg Oral Med Oral Pathol Oral Radiol Endod 105(3):e15–e18
7. Naveen K, Remya V, Kutty MV, Radhika S, Vineet K, Poonam S (2014) Biomaterials in veterinary world-with an emphasis on wound management. Int J Livestock Res 4(4):12–19
8. Mohite P, Shah SR, Singh S, Rajput T, Munde S, Ade N et al (2023) Chitosan and chito-oligosaccharide: a versatile biopolymer with endless grafting possibilities for multifarious applications. Front Bioeng Biotechnol 11:1190879
9. Hama R, Reinhardt JW, Ulziibayar A, Watanabe T, Kelly J, Shinoka T (2023) Recent tissue engineering approaches to mimicking the extracellular matrix structure for skin regeneration. Biomimetics 8(1):130
10. Perme H, Sharma AK, Kumar N, Singh H, Dewangan R, Maiti SK (2009) In-vitro biocompatibility evaluation of crosslinked cellular and acellular bovine pericardium. Trends Biomater Artif Organs 23(2):65–75
11. Kaarthick DT, Sharma AK, Kumar N, Gangwar AK, Mathew DD, Shrivastava S, Maiti SK, Kumar V, Singh R (2021) Evaluation of acellular pericardium and fibroblast seeded acellular pericardium for the repair of full thickness skin defects in rats. Trends Biomater Artif Organs 35(1):20–29
12. Trinh XT, Long NV, Van Anh LT, Nga PT, Giang NN, Chien PN, Nam SY, Heo CY (2022) A comprehensive review of natural compounds for wound healing: targeting bioactivity perspective. Int J Mol Sci 23(17):9573
13. Jones CL, Penney BT, Theodossiou SK (2023) Engineering Cell–ECM–Material Interactions for Musculoskeletal Regeneration. Bioengineering 10(4):453
14. Cartmell JS, Dunn MG (2000) Effect of chemical treatments on tendon cellularity and mechanical properties. J Biomed Mater Res 49(1):134–140
15. Chen F, Yoo JJ, Atala A (1999) Acellular collagen matrix as a possible "off the shelf" biomaterial for urethral repair. Urology 54(3):407–410
16. Dewangan R, Sharma AK, Kumar N, Maiti SK, Singh H, Kumar A, Shrivastava S, Singh R (2012) In vivo determination of biocompatibility of bladder acellular matrix in a rabbit model. Trends Biomater Artif Organs 26(1):43
17. Rennekampff HO, Kiessig V, Griffey S, Greenleaf G, Hansbrough JF (1997) Acellular human dermis promotes cultured keratinocyte engraftment. J Burn Care Rehabil 18(6):535–544
18. Erkin ÜR, Kerem M, Tug M, Orbay H, Sensöz Ö (2007) Prefabrication of a conjoint flap containing xenogenic tissues: a preliminary report on an experimental model. J Craniofac Surg 18(6):1451–1456
19. DeSagun EZ, Botts JL, Srivastava A, Hanumadass M, Walter RJ (2001) Long-term outcome of xenogenic dermal matrix implantation in immunocompetent rats. J Surg Res 96(1):96–106
20. Burugapalli K, Thapasimuttu A, Chan JC, Yao L, Brody S, Kelly JL, Pandit A (2007) Scaffold with a natural mesh-like architecture: isolation, structural, and in vitro characterization. Biomacromolecules 8(3):928–936
21. Shen JT, Falanga V (2003) Innovative therapies in wound healing. J Cutan Med Surg 7(3):217–224
22. Frame JD, Still J, Lakhel-LeCoadou A, Carstens MH, Lorenz C, Orlet H, Spence R, Berger AC, Dantzer E, Burd A (2004) Use of dermal regeneration template in contracture release procedures: a multicenter evaluation. Plast Reconstr Surg 113(5):1330–1338
23. Maiti SK, Hoque M, Kumar N, Kalicharan K, Singh GR (2001) Evaluation of a new absorbable suture fish-gut. Indian J Anim Sci 71(4)
24. Vellachi R, Kumar N, Sharma AK, Negi M, Maiti SK, Mathew DD, Shrivastava S, Kurade NP (2014) Acellular fish swin bladder biomaterial construct seeded with mesenchymal stem cells for full thickness skin wound healing in rats. Trends Biomater Artificial Organs 28(4):1–10

25. Ankomah AD, Boakye YD, Agana TA, Boamah VE, Ossei PP, Adu F, Agyare C (2022) Evaluation of dermal toxicity and wound healing activity of Cnestis ferruginea Vahl ex DC. Adv Pharmacol Pharm Sci 2022:5268613
26. Rathbone MJ, Bunt CR, Ogle CR, Burggraaf S, Macmillan KL, Pickering K (2002) Development of an injection molded poly (ϵ-caprolactone) intravaginal insert for the delivery of progesterone to cattle. J Control Release 85(1–3):61–71
27. de Albuquerque TL, Júnior JE, de Queiroz LP, Ricardo AD, Rocha MV (2021) Polylactic acid production from biotechnological routes: a review. Int J Biol Macromol 186:933–951
28. Sinha VR, Bansal K, Kaushik R, Kumria R, Trehan A (2004) Poly-ϵ-caprolactone microspheres and nanospheres: an overview. Int J Pharm 278(1):1–23
29. Chen Y, Geever LM, Killion JA, Lyons JG, Higginbotham CL, Devine DM (2016) Review of multifarious applications of poly (lactic acid). Polym-Plast Technol Eng 55(10):1057–1075
30. Chen X, Wang W, Wu Q, Shen X, Qiu D, Dong B, Liang Z, Fang B, Zeng Z, Chen J (2015) Preparation of polylactic acid microspheres containing lactones from Venenum Bufonis, its slow-release characteristics and therapeutic effects on mycoplasmal pneumonia of swine. Chin J Vet Sci 35(12):2014–2020
31. Bouriche S, Alonso-García A, Cárceles-Rodríguez CM, Rezgui F, Fernández-Varón E (2021) Potential of sustained release microparticles of metformin in veterinary medicine: an in vivo pharmacokinetic study of metformin microparticles as oral sustained release formulation in rabbits. BMC Vet Res 17(1):315
32. Nabofa WE, Alashe OO, Oyeyemi OT, Attah AF, Oyagbemi AA, Omobowale TO TO, Adedapo AA, Alada AR. Cardioprotective effects of curcumin-nisin based poly lactic acid nanoparticle on myocardial infarction in guinea pigs. Sci Rep 2018;8(1):16649
33. Oliveira JE, Medeiros ES, Cardozo L, Voll F, Madureira EH, Mattoso LH, Assis OB (2013) Development of poly (lactic acid) nanostructured membranes for the controlled delivery of progesterone to livestock animals. Mater Sci Eng C 33(2):844–849
34. Zinn M, Witholt B, Egli T (2001) Occurrence, synthesis, and medical application of bacterial polyhydroxyalkanoate. Adv Drug Deliv Rev 53(1):5–21
35. Raza ZA, Abid S, Banat IM (2018) Polyhydroxyalkanoates: Characteristics, production, recent developments, and applications. Int Biodeterior Biodegrad 126:45–56
36. Zhang J, Shishatskaya EI, Volova TG, da Silva LF, Chen GQ (2018) Polyhydroxyalkanoates (PHA) for therapeutic applications. Mater Sci Eng C 86:144–150
37. Krucińska I, Żywicka B, Komisarczyk A, Szymonowicz M, Kowalska S, Zaczyńska E, Struszczyk M, Czarny A, Jadczyk P, Umińska-Wasiluk B, Rybak Z (2017) Biological properties of low-toxicity PLGA and PLGA/PHB fibrous nanocomposite implants for osseous tissue regeneration. Part I: evaluation of potential biotoxicity. Molecules 22(12):2092
38. Rodríguez-Cendal AI, Gómez-Seoane I, de Toro-Santos FJ, Fuentes-Boquete IM, Señarís-Rodríguez J, Díaz-Prado SM (2023) Biomedical applications of the biopolymer poly (3-hydroxybutyrate-co-3-hydroxyvalerate) (PHBV): drug encapsulation and scaffold fabrication. Int J Mol Sci 24(14):11674
39. Leja K, Lewandowicz G (2010) Polymer biodegradation and biodegradable polymers-a review. Pol J Environ Stud 19(2):255–266
40. Shah SR, Modi CD, Singh S, Mori DD, Soniwala MM, Prajapati BG (2024) Recent advances in additive manufacturing of polycaprolactone-based scaffolds for tissue engineering applications: a comprehensive review. Regen Eng Transl Med 11:112–131
41. Conde G, De Carvalho JR, do Patrocínio Dias P, Moranza HG, Montanhim GL, de Oliveira Ribeiro J, Chinelatto MA, Moraes PC, Taboga SR, Bertolo PH, Funnicelli MI (2021) In vivo biocompatibility and biodegradability of poly (lactic acid)/poly (ε-caprolactone) blend compatibilized with poly (ε-caprolactone-b-tetrahydrofuran) in Wistar rats. Biomed Phys Eng Express 7(3):035005
42. Choi S, Oh YI, Park KH, Lee JS, Shim JH, Kang BJ (2019) New clinical application of three-dimensional-printed polycaprolactone/β-tricalcium phosphate scaffold as an alternative to allograft bone for limb-sparing surgery in a dog with distal radial osteosarcoma. J Vet Med Sci 81(3):434–439

43. Fedorová P, Srnec R, Pěnčík J, Dvořák M, Krbec M, Nečas A (2015) Intra-articular reinforcement of a partially torn anterior cruciate ligament (ACL) using newly developed UHMWPE biomaterial in combination with Hexalon ACL/PCL screws: ex-vivo mechanical testing of an animal knee model. Acta Chir Orthop Traumatol Cechoslov 82(3):222–228
44. Zhang ZZ, Wang SJ, Zhang JY, Jiang WB, Huang AB, Qi YS, Ding JX, Chen XS, Jiang D, Yu JK (2017) 3D-printed poly (ε-caprolactone) scaffold augmented with mesenchymal stem cells for total meniscal substitution: a 12-and 24-week animal study in a rabbit model. Am J Sports Med 45(7):1497–1511
45. Khojasteh A, Behnia H, Hosseini FS, Dehghan MM, Abbasnia P, Abbas FM (2013) The effect of PCL-TCP scaffold loaded with mesenchymal stem cells on vertical bone augmentation in dog mandible: a preliminary report. J Biomed Mater Res B Appl Biomater 101(5):848–854
46. Kalita S, Devi B, Kandimalla R, Sharma KK, Sharma A, Kalita K, Kataki AC, Kotoky J (2015) Chloramphenicol encapsulated in poly-ε-caprolactone–pluronic composite: nanoparticles for treatment of MRSA-infected burn wounds. Int J Nanomedicine 10:2971–2984
47. Boia R, Dias PA, Martins JM, Galindo-Romero C, Aires ID, Vidal-Sanz M, Agudo-Barriuso M, de Sousa HC, Ambrósio AF, Braga ME, Santiago AR (2019) Porous poly (ε-caprolactone) implants: A novel strategy for efficient intraocular drug delivery. J Control Release 316:331–348
48. Kendre PN, Lokwani D, Pote A, Singh S, Jayeoye TJ, Prajapati BG (2023) Potential technologies to develop cellulose beads and microspheres. In: Regenerated cellulose and composites: morphology-property relationship. Springer Nature Singapore, Singapore, pp 145–176
49. Singh S, Nwabor OF, Sukri DM, Wunnoo S, Dumjun K, Lethongkam S et al (2022) Poly (vinyl alcohol) copolymerized with xanthan gum/hypromellose/sodium carboxymethyl cellulose dermal dressings functionalized with biogenic nanostructured materials for antibacterial and wound healing application. Int J Biol Macromol 216:235–250
50. Dugan JM, Gough JE, Eichhorn SJ (2013) Bacterial cellulose scaffolds and cellulose nanowhiskers for tissue engineering. Nanomedicine 8(2):287–298
51. Silva IG, Pantoja BT, Almeida GH, Carreira AC, Miglino MA (2022) Bacterial cellulose and ECM hydrogels: an innovative approach for cardiovascular regenerative medicine. Int J Mol Sci 23(7):3955
52. Gorgieva S, Trček J (2019) Bacterial cellulose: Production, modification, and perspectives in biomedical applications. Nanomaterials 9(10):1352
53. Costa LM, de Olyveira GM, Basmaji P (2012) Bacterial cellulose towards functional medical materials. J Biomater Tissue Eng 2(3):185–196
54. Camargo MC, Nogueira RM, Sanches OD, Saab MD, Batista A, Vasconcelos D, Luvisotto LY, Lúcio MA (2014) Applicability of crystalline cellulose membrane in the treatment of skin wounds induced in Wistar rats. Acta Cir Bras 29:429–437
55. Soliman M, Sadek AA, Abdelhamid HN, Hussein K (2021) Graphene oxide-cellulose nanocomposite accelerates skin wound healing. Res Vet Sci 137:262–273
56. Song S, Liu Z, Abubaker MA, Ding L, Zhang J, Yang S, Fan Z (2021) Antibacterial polyvinyl alcohol/bacterial cellulose/nano-silver hydrogels that effectively promote wound healing. Mater Sci Eng C 126:112171
57. Sajjad W, Khan T, Ul-Islam M, Khan R, Hussain Z, Khalid A, Wahid F (2019) Development of modified montmorillonite-bacterial cellulose nanocomposites as a novel substitute for burn skin and tissue regeneration. Carbohydr Polym 206:548–556
58. Janmohammadi M, Nazemi Z, Salehi AO, Seyfoori A, John JV, Nourbakhsh MS, Akbari M (2023) Cellulose-based composite scaffolds for bone tissue engineering and localized drug delivery. Bioact Mater 20:137–163
59. Fukuyama Y, Maruo T, Nishiyama Y, Nemoto Y, Murayama K, Kayanuma H, Kawarai S (2021) Application of a novel carboxymethyl cellulose-based Mohs sol–gel on malignant wounds in three dogs. J Vet Med Sci 83(3):385–389
60. Bodea IM, Cătunescu GM, Stroe TF, Dîrlea SA, Beteg FI (2020) Applications of bacterial-synthesized cellulose in veterinary medicine–a review. Acta Vet Brno 88(4):451–471

61. Pang M, Huang Y, Meng F, Zhuang Y, Liu H, Du M, Ma Q, Wang Q, Chen Z, Chen L, Cai T (2020) Application of bacterial cellulose in skin and bone tissue engineering. Eur Polym J 122:109365
62. Huang Y, Wang J, Yang F, Shao Y, Zhang X, Dai K (2017) Modification, and evaluation of micro-nano structured porous bacterial cellulose scaffold for bone tissue engineering. Mater Sci Eng C 75:1034–1041
63. Sakamoto A, Qi P, Ohba S, Ohta S, Hara Y, Ogawa T, Tomokiyo M, Sasaki A, Takizawa H, Mochizuki M, Ito T (2019) Bone regeneration by calcium phosphate-loaded carboxymethyl cellulose nonwoven sheets in canine femoral condyle defects. J Biomed Mater Res B Appl Biomater 107(5):1516–1521
64. Wippermann J, Schumann D, Klemm D, Albes JM, Wittwer T, Strauch J, Wahlers T (2008) Preliminary results of small arterial substitutes performed with a new cylindrical biomaterial composed of bacterial cellulose. Thorac Cardiovasc Surg 56(S1):22
65. Chen PH, Liao HC, Hsu SH, Chen RS, Wu MC, Yang YF, Wu CC, Chen MH, Su WF (2015) A novel polyurethane/cellulose fibrous scaffold for cardiac tissue engineering. RSC Adv 5(9):6932–6939
66. Lv X, Yang J, Feng C, Li Z, Chen S, Xie M, Huang J, Li H, Wang H, Xu Y (2016) Bacterial cellulose-based biomimetic nanofibrous scaffold with muscle cells for hollow organ tissue engineering. ACS Biomater Sci Eng 2(1):19–29
67. Tsurkan MV, Voronkina A, Khrunyk Y, Wysokowski M, Petrenko I, Ehrlich H (2021) Progress in chitin analytics. Carbohydr Polym 252:117204
68. Wysokowski M, Machałowski T, Petrenko I, Schimpf C, Rafaja D, Galli R, Ziętek J, Pantović S, Voronkina A, Kovalchuk V, Ivanenko VN (2020) 3D chitin scaffolds of marine demosponge origin for biomimetic mollusk hemolymph-associated biomineralization ex-vivo. Mar Drugs 18(2):123
69. Mutsenko VV, Gryshkov O, Lauterboeck L, Rogulska O, Tarusin DN, Bazhenov VV, Schütz K, Brüggemeier S, Gossla E, Akkineni AR, Meißner H (2017) Novel chitin scaffolds derived from marine sponge Ianthella basta for tissue engineering approaches based on human mesenchymal stromal cells: Biocompatibility and cryopreservation. Int J Biol Macromol 104:1955–1965
70. Petrenko I, Khrunyk Y, Voronkina A, Kovalchuk V, Fursov A, Tsurkan D, Ivanenko VN (2020) Poriferan chitin: 3D scaffolds from nano-to macroscale. A review. Lett Appl NanoBioScience 9(2):1004–1014
71. Machałowski T, Wysokowski M, Petrenko I, Fursov A, Rahimi-Nasrabadi M, Moh'd MA, Meissner H, Joseph Y, Fazilov B, Ehrlich H, Jesionowski T (2020) Naturally pre-designed biomaterials: spider molting cuticle as a functional crude oil sorbent. J Environ Manag 261:110218
72. Machałowski T, Amemiya C, Jesionowski T (2020) Chitin of Araneae origin: structural features and biomimetic applications: a review. Appl Phys A 126:1–7
73. Khrunyk Y, Lach S, Petrenko I, Ehrlich H (2020) Progress in modern marine biomaterials research. Mar Drugs 18(12):589
74. Wysokowski M, Petrenko I, Stelling AL, Stawski D, Jesionowski T, Ehrlich H (2015) Poriferan chitin as a versatile template for extreme biomimetics. Polymers 7(2):235–265
75. Petrenko I, Bazhenov VV, Galli R, Wysokowski M, Fromont J, Schupp PJ, Stelling AL, Niederschlag E, Stöker H, Kutsova VZ, Jesionowski T (2017) Chitin of poriferan origin and the bioelectrometallurgy of copper/copper oxide. Int J Biol Macromol 104:1626–1632
76. Kumar PS, Srinivasan S, Lakshmanan VK, Tamura H, Nair SV, Jayakumar R (2011) Synthesis, characterization and cytocompatibility studies of α-chitin hydrogel/nano hydroxyapatite composite scaffolds. Int J Biol Macromol 49(1):20–31
77. Jayakumar R, Rani VD, Shalumon KT, Kumar PS, Nair SV, Furuike T, Tamura H (2009) Bioactive and osteoblast cell attachment studies of novel α-and β-chitin membranes for tissue-engineering applications. Int J Biol Macromol 45(3):260–264
78. Somsak P, Sriwattana S, Prinyawiwatkul W (2021) Ultrasonic-assisted chitin nanoparticle and its application as saltiness enhancer. Int J Food Sci Technol 56(2):608–617

79. Mulongo-Masamba R, El Hazzat M, El Hamidi A, Halim M, Arsalane S (2019) New functional β-chitin/calcium phosphate as promising support of copper nano catalyst for the reductive degradation of methylene blue. Int J Environ Sci Technol 16:8117–8128
80. Song S, Zhao Y, Yuan X, Zhang J (2019) β-Chitin nanofiber hydrogel as a scaffold to in situ fabricate monodispersed ultra-small silver nanoparticles. Colloids Surf A Physicochem Eng Asp 574:36–43
81. Jung HS, Kim MH, Shin JY, Park SR, Jung JY, Park WH (2018) Electrospinning, and wound healing activity of β-chitin extracted from cuttlefish bone. Carbohydr Polym 193:205–211
82. Singh N, Chen J, Koziol KK, Hallam KR, Janas D, Patil AJ, Strachan A, Hanley JG, Rahatekar SS (2016) Chitin, and carbon nanotube composites as biocompatible scaffolds for neuron growth. Nanoscale 8(15):8288–8299
83. Wang ZY, Wang JW, Qin LH, Zhang WG, Zhang PX, Jiang BG (2018) Chitin biological absorbable catheters bridging sural nerve grafts transplanted into sciatic nerve defects promote nerve regeneration. CNS Neurosci Ther 24(6):483–494
84. Wang Z, Fan J, Yang X, Zhang W, Zhang P, Jiang B (2018) The neural regeneration effect of chitin biological absorbable tubes bridging sciatic nerve defects with sural nerve grafts. Am J Transl Res 10(8):2362
85. Ji X, Lei Z, Yuan M, Zhu H, Yuan X, Liu W, Pu H, Jiang J, Zhang Y, Jiang X, Xiao J (2020) Cartilage repair mediated by thermosensitive photocrosslinkable TGFβ1-loaded GM-HPCH via immunomodulating macrophages, recruiting MSCs and promoting chondrogenesis. Theranostics 10(6):2872
86. Lin HL, Wu TH, Ho HO, Chao FC, Wu MH, Liu DZ, Chen LC, Sheu MT (2020) TEMPO-oxidized sacchachitin nanofibers (TOSCNFs) combined with platelet-rich plasma (PRP) for management of dry eye syndrome. Int J Nanomedicine 12:1721–1730
87. Song X, Huang X, Li Z, Li Z, Wu K, Jiao Y, Zhou C (2019) Construction of blood compatible chitin/graphene oxide composite aerogel beads for the adsorption of bilirubin. Carbohydr Polym 207:704–712
88. Zhang J, Xue S, Zhu X, Zhao Y, Chen Y, Tong J, Shi X, Du Y, Zhong Z, Ye Q (2019) Emerging chitin nanogels/rectorite nanocomposites for safe and effective hemorrhage control. J Mater Chem B 7(33):5096–5103
89. Schubert M, Binnewerg B, Voronkina A, Muzychka L, Wysokowski M, Petrenko I, Kovalchuk V, Tsurkan M, Martinovic R, Bechmann N, Ivanenko VN (2019) Naturally, prefabricated marine biomaterials: Isolation and applications of flat chitinous 3D scaffolds from Ianthella labyrinthus (Demospongiae: Verongiida). Int J Mol Sci 20(20):5105
90. Machałowski T, Wysokowski M, Żółtowska-Aksamitowska S, Bechmann N, Binnewerg B, Schubert M, Guan K, Bornstein SR, Czaczyk K, Pokrovsky O, Kraft M (2019) Spider Chitin. The biomimetic potential and applications of Caribena versicolor tubular chitin. Carbohydr Polym 226:115301

Hydrogel-Based Strategies: Tailoring Biomaterial Properties for Additive Manufacturing in Tissue Regeneration

13

Pijush Kumar Paul, Himanshu Paliwal, and Karuppasamy Prem Ananth

Abstract

Hydrogels are crucial materials in tissue engineering owing to their biocompatibility, tunable properties, and ability to mimic the extracellular matrix. Additive manufacturing, or 3D printing, revolutionizes tissue construct creation by precisely depositing biomaterials layer by layer. This requires adaptable hydrogels that can be fine-tuned for specific mechanical and biological needs. Tailoring the mechanical properties of hydrogels is crucial for their performance in tissue regeneration. By adjusting the cross-linking density and incorporating reinforcing agents, hydrogels can achieve the desired stiffness and elasticity to match different tissue types. Additionally, controlling the porosity and pore size distribution within hydrogels is essential for nutrient diffusion and cell proliferation, which are critical for tissue growth and repair. Various additive manufacturing techniques, including extrusion-based printing, inkjet bioprinting, and stereolithography, are employed to fabricate hydrogel-based constructs. Each method has its advantages and limitations, influencing the choice of hydrogel

P. K. Paul (✉)
Department of Nutrition & Health Sciences, University of Nebraska–Lincoln, Lincoln, NE, USA

Department of Pharmacy, Gono Bishwabidyalay (University), Dhaka, Bangladesh
e-mail: ppaul4@unl.edu

H. Paliwal
Marwadi University Research Center, Department of Pharmaceutics, Faculty of Pharmacy, Marwadi University, Rajkot, Gujarat, India

K. P. Ananth
Mechanical and Aerospace Department, New Mexico State University (NMSU), Las Cruces, New Mexico, USA

S. Singh et al. (eds.), *Biomaterial-based Additive Manufacturing in Tissue Engineering and Regeneration*, Biomaterials, Bioengineering and Sustainability 7,
https://doi.org/10.1007/978-3-031-96070-3_13

formulation and printing parameters. Hydrogel-based strategies for additive manufacturing hold immense potential in tissue regeneration. By precisely tailoring the properties of hydrogels, researchers can create customized, functional tissue constructs that address specific clinical needs. This synergy between advanced biomaterials and innovative manufacturing techniques heralds a new era in regenerative medicine, offering promising solutions for repairing and replacing damaged tissues.

Keywords

Hydrogel · Biomaterials · Additive manufacturing · Tissue regeneration · Three-dimensional printing · Extracellular matrix

13.1 Introduction to Hydrogel-Based Strategies

Hydrogel-based strategies for additive manufacturing (AM) include exploration of hydrogels as bioinks to fabricate 3D structures with superior biological compatibility and performance [1]. The hydrogels comprising of hydrophilic and cross-linked polymers render a wide variety of physico-biological characteristics appropriate for utilization in tissue engineering [2]. The inclusion of numerous organic and inorganic materials can be used in enhancement of the physical, mechanical, and biological features of hydrogels, conceiving them optimal for printing structures of living tissue or organs [3]. Moreover, progressions in functional hydrogels, like pH-responsive hydrogels, drug-delivery hydrogels, etc. have extended their biomedical usefulness, allowing accurate reproducibility of high-resolution structures into poly(dimethyl siloxane) (PDMS) elastomer for the application in tissue engineering [4]. Additionally, the incorporation of silica nanoparticles to microbial polysaccharide-based hydrogels has revealed potential outcomes in drafting printed hydrogel nanocomposites with exceptional biocompatibility, particularly appropriate for utility in regenerative medicine [5].

The advantages associated with hydrogels in AM, especially in biomedical applications, are quite significant. They hold the capacity to absorb more water, along with biocompatibility, and flexible mechanical properties, which confers them suitable for drug delivery and tissue engineering [1, 6]. AM techniques, like 3D printing, allow the production of complex structures with better resolution and modifiability, evading shortcomings of conventional polymerization techniques [3]. Hydrogels utilized as bioinks in bioprinting enable the manufacturing of biomimetic 3D assemblies that precisely be like natural tissues, enabling the investigation of active pharmaceutical ingredients (API) in more practical tissue models [7]. Moreover, advanced approaches, such as the use of hydrogels as transitional molds in double molding methods lessen the incorporation of toxic materials, subsequently increasing the utility of hydrogels in tissue engineering and biological applications [2].

The approach of using hydrogel for AM employs several methods such as combination of sacrificial templates and thermoresponsive hydrogels to fabricate engineered vasculature [8]. Direct ink writing (DIW) allows the requisite creation of efficient conductive hydrogels like poly(3,4-ethylenedioxythiophene): polystyrene sulfonate for applications in artificial skin and implantable bioelectronics [9]. Additionally, a double molding method concerning hydrogels enables the reproducibility of high-resolution stereolithographic prints into PDMS elastomer, which facilitates prompt design recapitulations and parallelized sample creation in tissue engineering [10]. A novel AM approach infusing 3D printed hydrogels with metallic precursors allowed the production of metallic micromaterials with microscale resolution and high-performance properties [3]. Such a variety of techniques display the flexibility and prospects of hydrogel-based methods in AM for several applications.

13.2 Fundamentals of Additive Manufacturing

AM is also referred to as 3D printing, is a pioneering manufacturing method that produces fragments by incorporating materials layer by layer, reforming conventional manufacturing approaches [11, 12]. The conventional manufacturing methods are mostly slow, expensive, and may pose harm to the environment, which makes the AM a preferable selection for fabricating composites with high precision and effectiveness [13]. It enables the fabrication of modified products, therefore AM can find its demand for innovative applications in several industries, like medicine, aerospace, automobile, and construction [11, 14]. AM has enables overcoming the limitations of traditional manufacturing processes by allowing the fabrication of complex individual parts instantly from 3D computer-aided design models, which leads to improved production and exceptional mechanical features in comparison with conventional methods [15]. Additionally, the AM with polymers and polymer composites is evolving at a fast pace, bringing forth new possibilities through the printing of architecture cellular materials and auxetic metamaterials.

AM has revolutionized the production of biomedical components, offering benefits such as patient-specific prosthetics and improved tissue engineering [16]. The technology's design principles and implementation for the fabrication of customized biomaterials and devices have been explored, with a focus on porous lattice structures and the use of metals, polymers, and ceramics [17]. Polymers, in particular, have become a key focus in additive manufacturing, with their synthetic versatility and adaptability making them widely employed in the field [18]. The technology's potential in the medical field is further emphasized by its ability to incorporate living cells, a process known as bioprinting, which has applications in tissue engineering and regenerative medicine [19].

Various techniques such as laser-assisted printing, extrusion printing, and inkjet printing have been developed to achieve optimal printability, cytocompatibility, and mechanical properties for tissue engineering applications [20]. Photopolymerization-based 3D bioprinting, specifically photocuring, has emerged as a promising method due to its simplicity, non-invasiveness, and spatial controllability, making it ideal for engineering biomimetic tissues [20, 21]. Additionally, the incorporation of micrometric magnesium particles into polylactic acid (PLA) matrices through fused deposition modeling (FDM) has shown enhanced mechanical, physical, chemical, and biological characteristics, making PLA-Mg composites promising for hard tissue engineering applications [22].

Recent advancements in tissue engineering and regenerative medicine have been driven by AM to create 3D-printed scaffolds. These scaffolds are often made from biopolymers, which offer biocompatibility, bioactivity, and biodegradability [23]. In bone tissue engineering, AM has enabled the fabrication of site-specific implants with tailored porosity and controlled chemistry, using natural biopolymers such as chitosan, alginate, collagen, and gelatin [24]. The incorporation of nanomaterials into biopolymeric matrices, followed by 3D printing, has further enhanced the properties and functionalities of these materials [25]. In tooth tissue engineering, the focus is on designing biomaterials that replicate the local extracellular environment and enable cell-scaffold interactions, often using a combination of natural and synthetic polymers [26].

13.3 Hydrogel Materials for Tissue Regeneration

Hydrogels have played a crucial role in tissue engineering due to their tunable properties and high similarity to human tissues [27]. Over time, researchers have focused on enhancing hydrogels to address challenges such as poor biocompatibility and non-degradability, leading to the development of tough hydrogels like gelatin-based tough hydrogels (GBTH) that mimic natural tissue components and promote tissue regeneration effectively [28]. Additionally, silk fibroin (SF) hydrogels have been extensively studied for their exceptional biological properties and regenerative effects in various tissues, making them a versatile material in tissue engineering applications [29]. Furthermore, recent advancements have seen the development of composite scaffolds combining GelMA hydrogels and electro spun polycaprolactone (PCL) with oxygen-releasing capabilities, enhancing cell survival and proliferation in bone-tissue engineering under hypoxic conditions [30].

Hydrogels offer several advantages over other biomaterials, making them a promising choice for various applications. Polysaccharide-based hydrogels stand out due to their biocompatibility, biodegradability, low cost, good mechanical properties, and thermal stability [31]. Organo gels, on the other hand, exhibit unique features like multi-stimuli responses, broad substance affinity, and electronic conductivity, enhancing their versatility and applicability [32]. Gelatin-based hydrogels derived from collagen present innovative characteristics, easy cross-linking, and biodegradability, making them sustainable and suitable for diverse uses, including biomedical applications and environmental remediation [33]. Hydrogels also play a crucial role in enhancing the bioavailability and solubility of hydrophobic drugs like curcumin, aiding in drug delivery systems for anti-cancer and wound healing applications [34]. Additionally, incorporating nanomaterials into hydrogels can address limitations such as burst release and poor bio adhesion, offering multifunctional properties for various biomedical applications like drug delivery, cancer treatment, and tissue regeneration [35].

Various types of hydrogels are utilized in tissue engineering applications, including multicomponent hydrogels that mimic the complexity of the extracellular matrix [27], human-derived hydrogels with unique physiochemical and biological properties supporting adipocyte and bone differentiation [36], silk fibroin hydrogels known for their mechanical strength and regenerative effects in cartilage, bone, skin, cornea, teeth, and eardrum [29], and functional hydrogels that provide a conducive microenvironment for cell proliferation, shape adaptation, intelligent drug delivery, and self-healing capabilities in repairing bone, cartilage, skin, muscle, and nerve tissues [37]. Additionally, hydrogels are increasingly studied for oral tissue engineering, with natural and synthetic hydrogels being explored for endodontic, periodontal, and oral soft tissue restorations, highlighting their potential in promoting tissue growth and combating bacterial infections [38]. Table 13.1 comprises of different types of hydrogels used in tissue engineering.

The synthesis and fabrication of hydrogels have seen remarkable advancements, including the development of quantum dots-hydrogel composites for bioimaging and biosensing [49], the creation of photodegradable double-network hydrogels with excellent mechanical properties and on-demand degradation capabilities for applications like skin-worn sensors [50], the incorporation of two-dimensional MXenes into hydrogels to enhance stability and conductivity for flexible strain sensors and healthcare monitoring [51], the utilization of hydrogels and nanocomposite hydrogels as smart wound dressing materials with self-healing properties and antimicrobial effects for wound healing applications [52], and the innovative combination of oxidized alginate and hyaluronan-based hydrogel systems using cost-effective domestic resources for tissue engineering purposes [53]. A range of hydrogels have been developed for tissue regeneration, each with unique properties. Tomić et al. synthesized superporous hydrogels with improved properties using 2-hydroxyethyl methacrylate and gelatin [54]. A self-healing hydrogel was fabricated by Tian et al. with on-demand antimicrobial activity and sustained biomolecule release, suitable for infected skin regeneration [55]. Perez-Puyana et al. evaluated type I collagen as a raw material for hydrogel fabrication, identifying optimal processing conditions [56]. Li et al. developed in situ forming biodegradable electroactive hydrogels with controlled properties, including conductivity and drug release, making them suitable for various tissue engineering applications [57].

Hydrogels are promising materials for tissue regeneration due to their unique properties such as:

(a) *Biocompatibility:* The ability of hydrogel's to be compatible with living tissues without causing any toxic or immunogenic response is quite significant. It allows the hydrogel to be used safely in the body, supporting cell growth and tissue regeneration without adverse effects [58].

Table 13.1 Summarizing the types of hydrogels employed in tissue engineering, along with their sources, properties, and typical application

Type of hydrogel	Source	Properties	Applications	References
Collagen	Natural (animal tissue)	Biocompatible, biodegradable, promotes cell adhesion and proliferation	Skin regeneration, cartilage repair, wound healing	[39]
Gelatine	Natural (denatured collagen)	Biocompatible, biodegradable, thermally reversible, promotes cell adhesion	Drug delivery, wound healing, tissue scaffolds	[40]
Alginate	Natural (seaweed)	Biocompatible, mild gelation conditions, adjustable mechanical properties	Cartilage repair, cell encapsulation, wound dressing	[4]
Hyaluronic acid	Natural (animal tissue, bacteria)	Biocompatible, biodegradable, promotes cell migration and proliferation	Cartilage repair, dermal fillers, wound healing	[41]
Chitosan	Natural (crustacean shells)	Biocompatible, biodegradable, antimicrobial, promotes cell adhesion	Bone regeneration, wound healing, drug delivery	[42]
Polyethylene Glycol	Synthetic	Biocompatible, non-biodegradable, tunable mechanical properties	Drug delivery, tissue scaffolds, cell encapsulation	[43]
Polyvinyl Alcohol	Synthetic	Biocompatible, non-biodegradable, high water content, good mechanical strength	Cartilage repair, wound dressings, contact lenses	[44]
Polyacrylamide	Synthetic	Biocompatible, non-biodegradable, high water content, tunable mechanical properties	Cell culture substrates, tissue scaffolds	[45]
Poly(N-isopropylacrylamide)	Synthetic	Thermoresponsive, biocompatible, non-biodegradable, reversible phase transition	Drug delivery, cell sheets, tissue engineering	[46]
Composite Hydrogels	Combination of natural and synthetic materials	Combined properties of individual components, tailored mechanical and biological properties	Various tissue engineering applications	[47]
Smart Hydrogels	Synthetic or natural	Stimuli-responsive (e.g., pH, temperature, light), biocompatible, tunable properties	Drug delivery, responsive tissue scaffolds	[48]

(b) *Biodegradability:* The biodegradable property is vital for temporary scaffolding applications where the hydrogel needs to provide initial support for tissue formation and then degrade, leaving behind only the newly formed tissue [59].
(c) *Mechanical Properties:* The mechanical properties include the strength, elasticity, and stiffness of the hydrogel, which determine its ability to withstand physical stresses. These properties must match those of the target tissue to provide appropriate mechanical support, ensuring the proper function and integration of the regenerated tissue [60].
(d) *Swelling Behavior:* The hydrogel's ability to absorb water and swell to a larger volume while retaining its structural integrity is critical for maintaining a hydrated environment, which is beneficial for cell viability, nutrient transport, and overall tissue regeneration [61].
(e) *Permeability and Diffusion:* Adequate permeability is essential for maintaining cell health and promoting efficient tissue regeneration, as it ensures that cells within the hydrogel scaffold can exchange necessary substances with their environment [62].
(f) *Structural and Functional Tunability:* Structural and functional tunability is the capacity to modify the hydrogel's properties to fit specific needs through changes in composition, crosslinking density, and other parameters. This adaptability allows the design of hydrogels tailored to various tissue engineering applications, enabling precise control over their mechanical and biological [63].

Novel approaches include the development of biofunctionalized hybrid hydrogels using biocompatible processes [64], the incorporation of extracellular matrix adhesion motifs to enhance biocompatibility [65], and the manipulation of adhesion through elastic heterogeneity for on-demand regulation [66]. Furthermore, the design of hydrogels with tunable mechanical properties, such as the addition of silk fibroin methacryloyl to promote cell adhesion, highlights advancements in 3D bioprinting for customized structures [67]. Additionally, the direct functionalization of PEG microparticles with cell-adhesion promoters like poly-l-lysine and GRGDS peptides demonstrates a simplified and effective method to enhance cell adhesion capabilities [68].

The usefulness of stimuli-responsive hydrogels in tissue regeneration have been explored due to their ability to mimic natural tissues' characteristics and respond dynamically to various stimuli, offering precise control over therapeutic agent delivery [69, 70]. These hydrogels, with their self-healing properties, can repair and restore function post-injury, making them attractive for applications in tissue engineering, drug delivery, and gene regulation [71]. In the realm of bone and cartilage regeneration, stimuli-responsive hydrogels have shown significant potential by responding to internal or external cues, aiding in controlled drug delivery and tissue repair [37]. Furthermore, hydrogels' physicochemical properties similar to the extracellular matrix of natural tissues provide an ideal microenvironment for cell growth and differentiation, making them valuable in the repair and regeneration of various tissue defects [61].

Designing hydrogels tailored for specific tissue types is a crucial aspect of tissue engineering, aiming to mimic the native extracellular matrix (ECM) complexity for effective tissue regeneration. Multicomponent hydrogels have emerged as promising biomaterials due to their tunable properties that can mimic the ECM's composition and structure [27]. Architected hydrogels, with native-like dynamic 3D environments, integrate with 3D cell culture and microfluidic technologies to support tissue formation and maturation [72]. Conductive hydrogels offer unique advantages in tissue repair by promoting cell activities and functions through electrical stimulation [73]. Decellularized tissue hydrogels play a critical role in generating organoids with biomolecular signatures for physiomimetic potential in tissue engineering [74]. The versatility and resemblance to the natural ECM make hydrogels excellent tools for tissue engineering, with advanced designs and processing methods enhancing cell-biomaterial interactions [75].

13.4 Tailoring Hydrogel Properties for Additive Manufacturing

Hydrogels are a type of material that has fascinated significant attention in additive manufacturing (AM), principally for biomedical uses like tissue engineering, drug delivery, and regenerative medicine. Customizing hydrogels for AM requires adjusting their composition and processing parameters to attain explicit mechanical properties, biocompatibility, printability, and functionality [76]. Since its initiation, the 3D printing industry has made extraordinary developments in sympathetic and optimizing the interaction between a component's structure, properties, and applications. Research has been paying attention to improving the accuracy, pace, and material capabilities of 3D printers. Innovations include multi-material printing and superior post-processing techniques to get better the strength and finish of printed parts. Hydrogels are a predominantly gifted class of materials for 3D printing. These are networks of polymer chains that are hydrophilic, often proficient in holding huge amounts of water while maintaining their structure [77]. Hydrogels, with their distinctive compatibility with living systems, are supremely appropriate for abundant life science applications. Obtainable in natural, synthetic, and hybrid forms, and in physically and chemically crosslinked varieties, hydrogels display adaptability and flexibility. Their high-water content and customizable mechanical properties make them well-suited for 3D printing, enabling precise and functional fabrication of biomaterials [78].

Hydrogels can be categorized based on their origin, structure, and responsiveness to ecological stimuli, allowing for tailored applications in medical, environmental, and industrial sectors. Their properties, such as high-water content, biocompatibility, and tunable mechanical strength, make them particularly appropriate for uses in drug delivery systems, tissue engineering, and wound care. The method of crosslinking, whether physical or chemical, further differentiates them. Additionally, their properties, configuration, and ionic charge play crucial roles in their classification, as depicted in Fig. 13.1 [79].

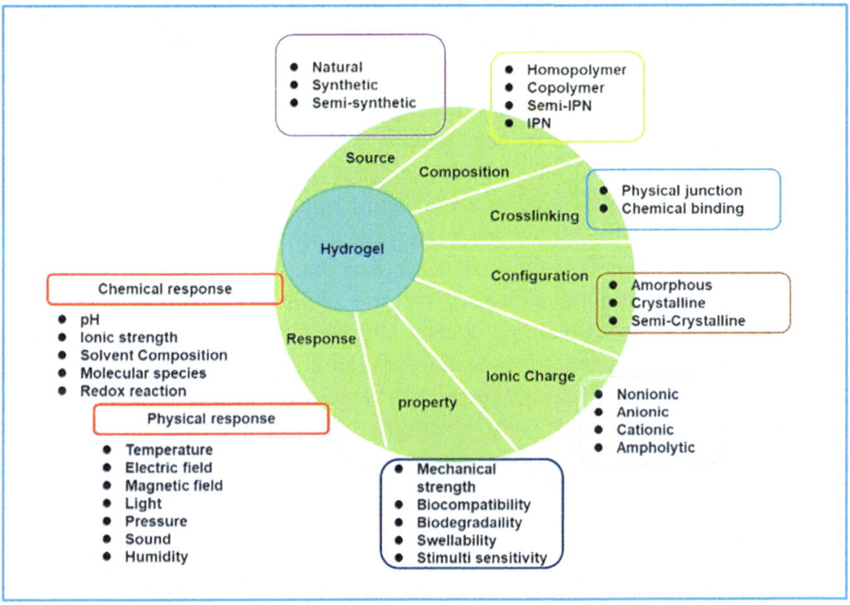

Fig. 13.1 Categories of hydrogels [79]. (Re-used under Creative Commons Attribution (CC BY 4.0))

Natural polymers, together with alginate, gelatin, hyaluronic acid, and chitosan, are extremely biocompatible, which makes them perfect for biomedical uses. Though, they frequently require modification to augment their mechanical strength. In contrast, synthetic polymers such as poly (ethylene glycol) (PEG), poly (vinyl alcohol) (PVA), and poly(acrylamide) can be engineered to assemble precise mechanical and degradation requirements. By combining natural and synthetic polymers, hybrid hydrogels are created, leveraging the advantages of both to get better their overall performance and flexibility in different applications [80].

Physical crosslinking, during ionic, hydrogen, and hydrophobic exchanges, is reversible and tender for cell encapsulation. Chemical crosslinking, via covalent bonds like photopolymerization and enzymatic reactions, yields constant structures proper for miscellaneous applications [81]. Incorporating nanoparticles, nanofibers, and microspheres into hydrogel materials improves the mechanical properties and adds functionalities like conductivity or antimicrobial property. Moreover, bioactive molecules such as growth factors, peptides, or drugs can be integrated for exact biological responses, contribution flexible applications in areas like tissue engineering and drug delivery [82]. By considering these factors, researchers and engineers can optimize hydrogel formulations for specific additive manufacturing techniques and applications, ensuring high-performance final printed structures.

13.4.1 Modulating Mechanical Properties of Hydrogels

Although the toughness of hydrogels is normally synchronized by shifting the cross-linking density and structure, or the polymer concentration, we have recognized a narrative approach to manage toughness by adjusting the side-chain length. By enhancement the side-chain length, we can attain an optimal balance between flexibility and strength, as long as a pioneering approach for designing advanced hydrogel materials. This method paves the way for creating hydrogels with customized mechanical properties for various applications [83]. Effect of Mechanical Properties and various applications are shown in Table 13.2.

The mechanical strength of hydrogels is vital for pharmaceutical and biomedical uses. Evaluating this strength is essential for their effectiveness in various applications, such as tendon and ligament repair, cartilage replacement, tissue engineering, wound dressing, and drug delivery. Hydrogels need to retain their structural integrity while releasing therapeutic agents over a specific duration [91]. Table 13.3. Shows a comprehensive overview of the mechanical properties of various hydrogels.

Table 13.2 Effect of Mechanical Properties and various application of hydrogels

Strategy	Effect on mechanical properties	Typical applications	References
Polymer concentration	Higher concentration increases stiffness and strength; lower concentration increases softness and flexibility	Tissue engineering, wound healing	[84]
Cross-linking density	Higher density increases rigidity and strength; lower density increases flexibility and reversibility	Drug delivery, tissue scaffolds, contact lenses	[85]
Type of polymer	Synthetic polymers offer tunability and predictability; natural polymers provide biocompatibility	Biomedical implants, drug delivery, tissue engineering	[86]
Composite hydrogels	Addition of nanoparticles or nanofibers enhances stiffness, strength, and elasticity	Bone regeneration, dental materials, soft robotics	[82]
Environmental conditions	pH, temperature, and ionic strength changes can modulate swelling and mechanical properties	Drug delivery, responsive coatings, biosensors	[87]
Dual-network hydrogels	Combining rigid and flexible networks increases toughness and elasticity	Cartilage replacement, load-bearing applications	[88]
Post-processing techniques	Freeze-thaw cycles, dehydration/ rehydration adjust strength and flexibility	Cryogels, tissue engineering scaffolds	[89]
Mechanical reinforcement	Applying stress during formation aligns polymer chains, increasing strength	Wearable devices, artificial muscles, flexible electronics	[90]

Table 13.3 Comprehensive overview of the mechanical properties of various hydrogels

Hydrogel	Young's modulus (kPa)	Tensile strength (MPa)	Elongation at break (%)	References
Polyacrylamide (PAAm)	10–1000	Up to 0.5	50–200	[92]
Polyethylene glycol (PEG)	10–100	0.1–0.3	10–50	[93]
Polyvinyl alcohol (PVA)	1–1000	0.01–1	50–200	[94]
Alginate	20–1000	0.05–0.3	10–40	[95]
Chitosan	100–500	0.01–0.2	10–30	[96]
Agarose	500–1000	0.02–0.1	5–20	[97]
Hyaluronic acid (HA)	0.1–10	0.01–0.1	50–150	[98]
Collagen	0.1–10	0.01–0.1	20–100	[99]
Gelatin	10–100	0.02–0.1	10–60	[100]
Poly(N-isopropylacrylamide) (PNIPAM)	100–1000	0.1–0.3	30–100	[101]
Poly (2-hydroxyethyl methacrylate) (PHEMA)	10–500	0.1–0.5	20–80	[102]
Dextran	10–100	0.01–0.1	10–50	[103]

13.4.2 Controlling Porosity and Pore Size Distribution

Material properties vary widely, especially pore sizes, which are crucial in cell culture and tissue engineering for oxygen transport, toxin removal, and cellular growth [104]. Understanding and controlling this pore size is crucial for tailoring the hydrogel's permeability, swelling behavior, and other key properties. Researchers often employ various techniques to manipulate and characterize these void spaces to optimize hydrogel performance for specific applications [105]. Table 13.4. Shows comprehensive overview of Porosity control methods and pore size distribution.

13.4.3 Strategies for Enhancing Bio-Functionality and Bioactivity

Bioactive agents, with nanoparticles, cells, and molecules, notably enhance the efficiency of hydrogels in wound healing by accelerating tissue regeneration and minimizing swelling. These agents can be accurately surrounded within hydrogels to deliver targeted therapeutic effects, facilitating a controlled liberate of bioactive substances at the wound site and improving the therapeutic process while reducing infection risk [118]. Embedding growth factors like VEGF, FGF, and BMPs in hydrogels improve cellular proliferation and differentiation. Functionalizing hydrogels with bioactive peptides such as RGD sequences promotes cell sticking together and signaling. Moreover, loading hydrogels with antibiotics, anti-inflammatory drugs, or other therapeutics enables localized and controlled release, improving their medical and therapeutic applications [119]. Natural polymers such as collagen, gelatin, and hyaluronic acid considerably develop biocompatibility and reproduce the ECM, facilitating cellular adhesion and proliferation. Collagen confers structural integrity, gelatin promotes cellular proliferation, and hyaluronic acid contributes to tissue hydration and regeneration [120].

Table 13.4 Comprehensive overview of Porosity control methods and pore size distribution

Hydrogel	Porosity control methods	Porosity percentage range	Pore size distribution control methods	Pore size range	Recent references
Polyacrylamide (PAAm)	Freeze-drying, sol-gel	50–95%	Templating, cross-linking	10 nm–100 μm	[106]
Polyethylene glycol (PEG)	Gas foaming, freeze-thaw cycling	40–80%	Surfactant addition, phase separation	1 nm–1 mm	[107]
Polyvinyl alcohol (PVA)	Freeze-thaw cycling, solvent casting	50–90%	Cross-linking density adjustment, templating	0.1 μm–5 mm	[108]
Alginate	Ionic cross-linking	50–80%	Emulsion templating, chemical cross-linking	1 nm–100 μm	[109]
Chitosan	Freeze-drying, ionic cross-linking	60–95%	Solvent casting, microfluidic templating	10 nm–50 μm	[110]
Agarose	Gelation	70–95%	Gelation temperature adjustment, mechanical compression	50 nm–500 μm	[111]
Hyaluronic acid (HA)	Cross-linking, sonication	50–90%	Enzymatic degradation, chemical modification	10 nm–100 μm	[112]
Collagen	Self-assembly, cross-linking	60–95%	Fibrillogenic, templating	50 nm–50 μm	[113]
Gelatin	Gelation, cross-linking	50–90%	Enzymatic degradation, physical cross-linking	50 nm–200 μm	[114]
Poly(N-isopropylacrylamide) (PNIPAM)	Freeze-drying, gas foaming	60–90%	Cross-linking density adjustment, solvent casting	10 nm–100 μm	[115]
Poly (2-hydroxyethyl methacrylate) (PHEMA)	Cryogelation, photopolymerization	60–95%	Salt leaching, electrospinning	100 nm–100 μm	[116]
Dextran	Cross-linking, sol-gel	50–85%	Templating, dialysis	10 nm–100 μm	[117]

Hydrophilic polymers, characterized by polar functional groups like hydroxyl (-OH), carboxyl (-COOH), and amino (-NH_2), are soluble in water or swell in its presence. Poly (vinyl alcohol) (PVA) is a notable hydrophilic polymer, ahead noteworthy consideration for its possible as a helpful biological drug-delivery matrix [121]. Hydrogel embedding of stem cells, backed by scientific research, holds assure for tissue repair. Co-culturing diverse cell types within these hydrogels enhances tissue model reliability. These advance mimics involved cellular interactions, advancing tissue engineering's therapeutic potential. The controlled microenvironment supports stem cell proliferation and differentiation, promotion effective tissue regeneration [122].

13.5 Additive Manufacturing Techniques for Hydrogel-Based Tissue Constructs

Recently, hydrogel materials utilized as bioink have demonstrated significant potential in the biomedical field due to their similarity to the extracellular matrix and their biocompatibility. These hydrogels can be leveraged in regenerative medicine, wound healing, and drug delivery. In tissue engineering, various hydrogels are employed to fabricate scaffolds through diverse additive manufacturing methodologies [123]. AM techniques for hydrogel-based tissue constructs are at the forefront of scientific research in tissue engineering and regenerative medicine. These techniques offer precise control over the architecture, composition, and mechanical properties of hydrogel constructs, which are crucial for mimicking the natural ECM and supporting cell growth and differentiation. Below are the main AM techniques utilized in scientific research for hydrogel-based tissue constructs [124].

13.5.1 Extrusion-Based Printing Methods

Extrusion-based printing techniques (EBB) for hydrogels have demonstrated significant potential in tissue engineering, allowing for the creation of intricate, biomimetic structures that promote cell growth and differentiation. Furthermore, improvements in the rheological properties of hydrogels used in these printing methods have enhanced print accuracy and structural stability, enabling the production of more precise and functional biomedical constructs [125]. Figure 13.2 shows the schematic illustration of extrusion-based printing techniques for hydrogels. Among rapid prototyping techniques, EBB is notable for its unique advantages. It enables precise layer-by-layer fabrication of 3D tissue constructs from bio-inks, supports higher cell seeding densities, faster printing speeds, and reduces cell damage compared to other methods. EBB can print continuous cylindrical filaments from various bio-inks, accommodating high cell density aggregates with diverse viscosities. Table 13.5 shows the recent studies of natural and synthetic hydrogels.

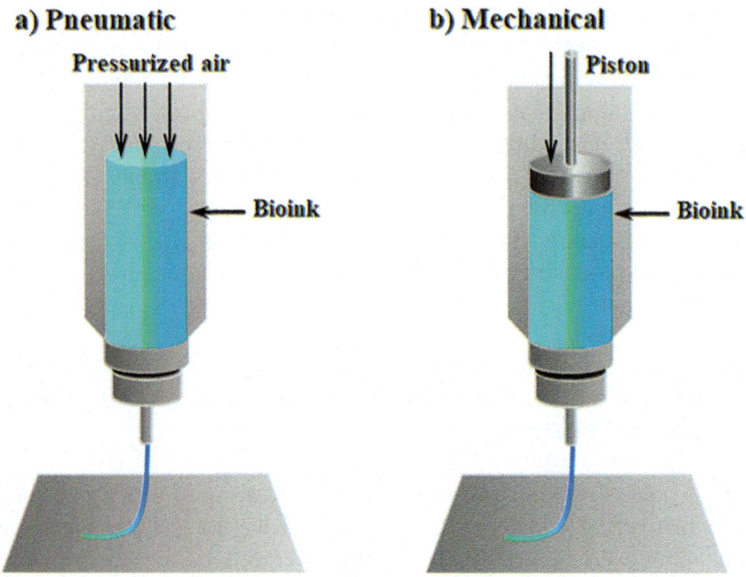

Fig. 13.2 Figure shows the schematic illustration of Extrusion-based printing techniques for hydrogels [126]. (**a**) Pneumatic. (**b**) Mechanical. (Re-used under Creative Commons license (CC BY 4.0))

Table 13.5 The recent studies on natural and synthetic hydrogels

Hydrogel materials	Applications	Description	References
Alginate	Tissue engineering, drug delivery	Derivative from seaweed, alginate hydrogels are biocompatible and gelate in the existence of calcium ions, making them valuable for tissue engineering scaffolds and drug delivery systems.	[127]
Collagen	Bioprinting, wound healing	As a major component of the extracellular matrix, collagen hydrogels mimic native tissue environments and endorse cell adhesion and proliferation, building them valuable in bioprinting and wound healing applications.	[128]
Gelatin	Tissue engineering, drug delivery	Gelatin hydrogels offer tunable mechanical properties and can be customized with bioactive molecules, making them versatile biomaterials for tissue engineering scaffolds and drug delivery vehicles.	[129]
Hyaluronic acid (HA)	Tissue engineering, ophthalmology	HA hydrogels exhibit high water content and biocompatibility, making them suitable for tissue engineering scaffolds, ophthalmic applications, and supplementation in osteoarthritis treatment.	[98]

(continued)

Table 13.5 (continued)

Hydrogel materials	Applications	Description	References
Chitosan	Wound healing, drug delivery	Chitosan hydrogels possess antibacterial properties and can be tailored for controlled drug release, making them effective wound dressings and drug delivery systems for various biomedical applications.	[130]
Polyethylene glycol (PEG)	Drug delivery, tissue engineering	PEG hydrogels offer excellent biocompatibility and tunable properties, making them suitable for drug delivery carriers, tissue engineering scaffolds, and regenerative medicine applications.	[131]
Poly(N-isopropylacrylamide) (PNIPAAm)	Drug delivery, cell culture	PNIPAAm hydrogels exhibit a LCST behavior, which allows for temperature-responsive drug delivery systems and cell culture substrates with reversible swelling properties.	[132]
Poly (acrylic acid) (PAA)	Drug delivery, wound dressing	pH-responsive PAA hydrogels swell or shrink in response to changes in pH, enabling controlled drug release and adaptable wound dressing materials for enhanced wound healing.	[133]
Poly (vinyl alcohol) (PVA)	Tissue engineering, drug delivery	PVA hydrogels offer good mechanical properties and biocompatibility, making them suitable for tissue engineering scaffolds, wound dressings, and drug delivery vehicles.	[134]
Polyacrylamide (PAAm)	Electrophoresis, cell culture	PAAm hydrogels are widely used in electrophoresis and cell culture applications due to their biocompatibility, high water content, and tunable mechanical properties.	[135]

13.5.2 Inkjet-Based Bioprinting Approaches

Figure 13.3 shows the schematic illustration. Inkjet-based bioprinting. Inkjet-based bioprinting is a superior method that utilizes inkjet printing technology to detail three-dimensional biological structures. This method works by precisely placing bioinks-suspensions that contain living cells and biocompatible materials. It has become a vital tool in tissue engineering and regenerative medicine, prominent for its high resolution, cost-effectiveness, and versatility [136].

The inkjet technique has the ability of generating droplets in the picoliter volume range, firing thousands of times in a few seconds, and printing in a noncontact manner. Since its appearance, inkjet technology has been extensively utilized in publishing engineering for the printing of text and images. As the expertise has urbanized, its applications have been stretched out from two-dimensional (2D) to

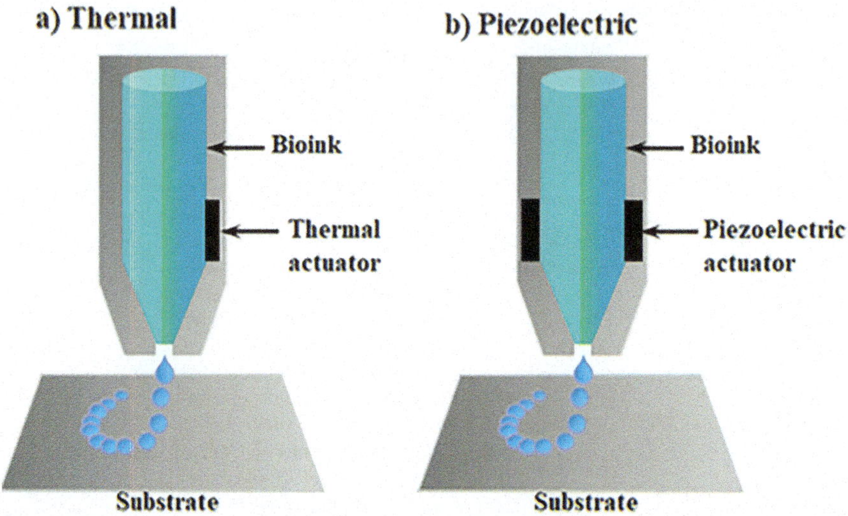

Fig. 13.3 Schematic illustration Inkjet-based bioprinting [126] (**a**) Thermal. (**b**) Piezoelectric. (Re-used under Creative Commons license (CC BY 4.0))

three-dimensional (3D) printing and have even been engaged in the manufacture of electronic device apparatus within scientific research [137]. The common methods can be further categorized into thermal and piezoelectric actuator methods, depending on the droplet actuation mechanism [138]. In thermal technology, ink droplets are fashioned by heating the ink, causing a magnified fizz to push the ink through a narrow nozzle and onto the substrates. This technology is affordable and widely utilized [139, 140]. An alternative method entails the use of thermal actuators, wherein localized heating of the bioink prompts quick development, facilitating droplet ejection. This advance necessitates accurate temperature organize to moderate impending harm to the biological material, as elucidated by scientific research [141] (Table 13.6).

13.5.3 Stereolithography and Digital Light Processing

Stereolithography (SLA) and digital light processing (DLP) are two types of additive manufacturing technologies that exploit photopolymerization to produce solid structures from liquid photopolymer resins (Fig. 13.4). SLA employs a focused ultraviolet laser to selectively cure the resin layer by layer, whereas DLP uses a digital micromirror device to project patterned light, curing whole layers concurrently. Both techniques are prominent for their high resolution and capability to produce complex and multifaceted geometries with excellent accuracy (Table 13.7) [153].

Bioinks for SLA/DLP bioprinting must cure quickly under light, limiting options to photocurable biomaterials [155]. Opacity affects resolution, and viscosity impacts

Table 13.6 Comprehensive overview of Inkjet-based bioprinting approaches

Hydrogel material	Type	Application in bioprinting	Thermal bioprinting	Piezoelectric bioprinting	References
Collagen	Natural	Tissue engineering, scaffolds	Limited due to temperature sensitivity	Suitable for bioinks	[70]
Alginate	Natural	Cell encapsulation, scaffolds	Generally, not used	Widely used, gelation with Ca ions	[142]
Chitosan	Natural	Wound healing, tissue regeneration	Limited due to thermal sensitivity	Suitable, good viscosity and gelation properties	[143]
Gelatin	Natural	Mimics extracellular matrix	Used in blends	Compatible, viscoelastic properties	[144]
Hyaluronic acid	Natural	Cartilage and soft tissue engineering	Sensitive to high temperatures	Suitable, supports cell encapsulation	[145]
Fibrin	Natural	Wound healing, vascular tissue engineering	Sensitive to high temperatures	Suitable, excellent cell adhesion	[146]
Polyethylene glycol (PEG)	Synthetic	Tunable hydrogels, tissue scaffolds	Used with temperature control	Suitable, controllable viscosity	[147]
Pluronic F127	Synthetic	Thermo-responsive, sacrificial templates	Widely used, temperature-sensitive	Limited use, primarily thermal	[148]
Polyvinyl alcohol (PVA)	Synthetic	Mechanical properties, biocompatibility	Can be used with temperature management	Suitable, viscoelastic properties	[149]
Polyacrylamide (PAAm)	Synthetic	High-strength hydrogels	Limited due to thermal instability	Suitable, structural integrity	[150]
Poloxamer	Synthetic	Drug delivery, tissue engineering	Thermo-responsive	Limited use, primarily thermal	[151]
Poly(N-isopropylacrylamide) (PNIPAAm)	Synthetic	Drug delivery, tissue scaffolds	Thermo-responsive	Limited use, primarily thermal	[152]

fabrication time. Natural and synthetic bioinks advantages and disadvantages are shown in Table 13.8.

SLA/DLP bioprinting offers fast fabrication, precise scaffold control, and high resolutions in biomedicine. By integrating patient-specific data and stem cell tech, it can create personalized tissue models for disease study and therapy development.

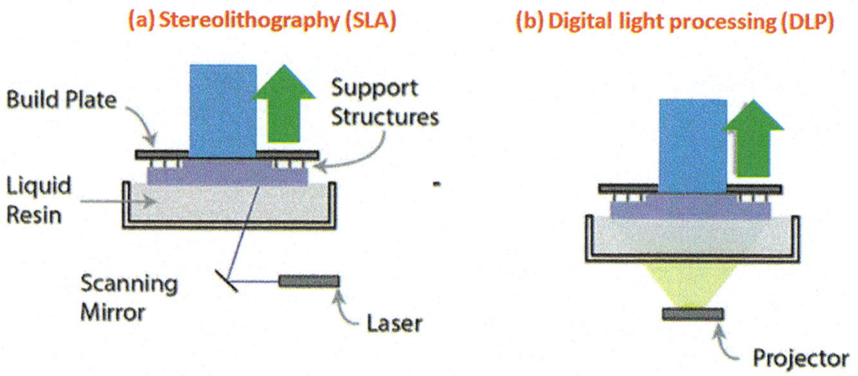

Fig. 13.4 Schematic illustration of Stereolithography (SLA) (**a**), Digital light processing (DLP) (**b**). (Reproduced with permission from [154])

Table 13.7 Outlines the key differences and similarities between SLA and DLP 3D printing technologies

Feature	Stereolithography (SLA)	Digital light processing (DLP)
Process	Uses a laser to cure liquid resin layer by layer	Uses a digital light projector to cure layers all at once
Components	Laser, resin tank, build platform, gantry system	Projector, resin tank, build platform, DMD
Speed	Slower due to point-by-point curing	Faster as entire layers are cured simultaneously
Resolution and detail	High precision and fine details	High resolution and fine details
Surface finish	Smooth surface finish	Smooth surface finish
Initial cost	Generally, more affordable initially	Generally higher due to projector technology
Ongoing costs	Potentially higher due to resin and maintenance	Lower in the long run due to faster print times
Material choices	Limited compared to FDM	Similar limitations in material choice
Build volume	Varies, often larger than DLP	Often limited build volume
Applications	Dental, jewelry, prototyping, complex geometries	Same as SLA, plus faster production needs
Post-processing	Requires post-curing and resin handling	Similar post-curing and resin handling

13.5.4 Emerging Additive Manufacturing Technologies for Hydrogels

Over the past two decades, considerable evolution has been made in rising strong and tough hydrogels. Shaping these hydrogels into precise structures for sensible applications remains demanding but vital. 3D printing, an emerging additive manufacturing technique, shows huge potential for engineering these tough hydrogels for dissimilar uses [166]. Emerging trends in hydrogel additive manufacturing contain the manufacture of smart hydrogels that respond to stimuli such as temperature, pH, and light, the combination with other materials like nanomaterials and bioactive

Table 13.8 Natural and synthetic bioinks advantages and disadvantages

Type of hydrogel	Source	Advantages	Disadvantages	References
GelMA	Gelatin derived from collagen	Biocompatible, tunable mechanical properties, cell adhesion	Limited control over degradation rate, potential immunogenicity	[156]
Alginate	Algae	Rapid gelation, biocompatibility, versatility	Limited cell adhesion, poor mechanical properties	[157]
Hyaluronic acid (HA)	Naturally occurring in the body	Excellent biocompatibility, promotes cell proliferation and migration	Limited mechanical strength, rapid degradation	[158]
PEG	Synthetic polymer	Highly tunable properties, biocompatible, minimal immunogenicity	Lack of inherent bioactivity, requires modification for cell adhesion	[159]
PEGDA	Synthetic polymer	High printing resolution, versatile, customizable properties	Limited biological cues, requires modification for cell attachment	[160]
PVA	Synthetic polymer	Biocompatible, easily modifiable, low cost	Limited cell adhesion, poor mechanical strength	[161]
Collagen	Animal connective tissue	Biocompatible, promotes cell adhesion, mimics native ECM	Limited mechanical strength, batch-to-batch variability	[162]
Fibrin	Blood protein	Mimics natural blood clotting process, promotes cell migration	Rapid degradation, potential immunogenicity	[163]
Chitosan	Shell of crustaceans	Biodegradable, antibacterial, promotes wound healing	Limited printability, poor mechanical properties	[164]
HAMA	Naturally occurring in the body	Biocompatible, promotes cell adhesion and proliferation	Limited mechanical strength, rapid degradation	[165]
PEGDMA	Synthetic polymer	High printability, versatile, tunable mechanical properties	Lack of inherent bioactivity, potential cytotoxicity	[159]

agents, and advancements in multi-material printing for complex, functional devices. Moreover, there is a rising importance on attractive mechanical properties and biocompatibility of hydrogels to widen their use in biomedical engineering. Efforts are also in progress to get better the accuracy and resolution of hydrogel 3D printing techniques for more involved and accurate structures [167]. Below table provides a complete summary of each additive manufacturing technology for hydrogels, along with their general applications and the types of hydrogels typically used [166] (Table 13.9).

Table 13.9 Additive manufacturing technology for hydrogels, along with their general applications

Technology	Description	Applications	Common hydrogels
3D bioprinting	Layer-by-layer deposition of bioinks (hydrogels mixed with living cells).	Tissue scaffolds, organ-on-a-chip, implants, and prosthetics	Alginate, Gelatin, collagen, hyaluronic acid
Inkjet bioprinting	Uses droplets of hydrogel-based bioinks via thermal or piezoelectric mechanisms.	High-resolution tissue constructs, drug testing models	Alginate, GelMA, fibrin
Extrusion bioprinting	Uses pneumatic or mechanical systems to extrude hydrogel strands.	Versatile tissue scaffolds, cell culture systems	Alginate, GelMA, Pluronic F127
Laser-assisted bioprinting	Uses laser-induced forward transfer to deposit bioinks with high precision.	Cell placement with minimal shear stress, complex tissue constructs	Alginate, GelMA, PEGDA
Stereolithography (SLA)	Photopolymerization of liquid hydrogel precursors using a laser or digital light projector.	Microfluidic devices, dental aligners, anatomical models	PEGDA, GelMA, hyaluronic acid
Digital light processing (DLP)	Uses a digital micromirror device to project entire layer patterns at once.	Faster printing of detailed structures, custom dental and medical devices	PEGDA, GelMA
Two-photon polymerization (2PP)	High-resolution SLA using femtosecond lasers for intricate microstructures.	Microdevices, high-precision scaffolds	PEGDA, hyaluronic acid
Direct ink writing (DIW)	Extrusion of hydrogel inks through a nozzle to form structures.	Flexible electronics, soft robotics, tissue engineering scaffolds	Alginate, Pluronic F127, GelMA
Fused deposition Modeling (FDM)	Adaptations for hydrogels using specialized nozzles and cooling systems.	Drug delivery systems, biocompatible supports, wearable hydrogel devices	Pluronic F127, PVA, GelMA
Electrospinning	Creates fine fibers from hydrogel solutions using a high-voltage electric field.	Wound dressings, filtration systems, tissue engineering scaffolds	PVA, gelatin, collagen
Hybrid printing techniques	Combines multiple AM technologies for complex structures.	Composite tissue constructs, multifunctional diagnostic and therapeutic devices	Alginate, PEGDA, GelMA

13.6 Applications of Hydrogel-Based Strategies in Tissue Regeneration

Constructing bioactive scaffolds for tissue regeneration offers promising treatment for osteochondral and cartilage defects. Polysaccharide hydrogels, mimicking the extracellular matrix, are highly biocompatible materials for this purpose. Combining polysaccharide hydrogels with 3D printing can precisely distribute cell-loaded biomaterials, creating complex 3D tissues with optimal structure and mechanical properties for osteochondral and cartilage repair [168]. Yan et al. used alginate to create injectable, biodegradable composite hydrogels that release calcium ions and are cross-linked with hydroxyapatite and gelatin microspheres, boosting osteoblast activity effectively [169]. Polysaccharides used in 3D printing mimic natural cartilage ECM but require appropriate cell types to effectively regenerate functional cartilage, promoting collagen and glycosaminoglycan production in the graft.

3D printing technology can create scaffolds with gradient porosity, using hydrogels or polymer scaffolds for the cartilage layer and biodegradable thermoplastics for the bone layer, enhancing subchondral bone tissue recovery. 3D-printed scaffolds with vertical porosity and pore diameter gradients, where 90–120 μm promotes cartilage formation in mesenchymal stem cells (MSCs) and around 300 μm enhances osteogenesis [170]. Recruiting endogenous MSCs, Kartogenin (KGN) in the upper layer promotes cartilage differentiation, while pro-osteogenesis aptamers in the lower layer promote osteogenic differentiation, significantly accelerating osteochondral defect repair in a rat model [171]. Bone morphogenetic protein-2 (BMP-2), an FDA-approved growth factor, is used in bone tissue engineering for sustained release and enhanced osteogenesis, as demonstrated in hydrogel-based 3D constructs by Poldervaart et al. [172]. Balakrishnan et al. developed a rapid-gelling injectable hydrogel using self-crosslinking of oxidized alginate and gelatin, enhanced chondrocyte activity and migration. They also innovated double cross-linked alginate microgels with oxidized and methacrylic acid alginate for 3D structures and long-term storage, maintaining stem cell viability like freshly treated cells, ideal for bioprinting [173, 174].

A double-layer silk fibroin scaffold was developed for bone and cartilage repair, featuring a biomimetic cartilage layer and a porous BMP-2-loaded layer, sealed with transforming growth factor-β3 (TGF-β3) loaded methacrylate silk fibroin sealant (Sil-MA) hydrogel. The BMP-2 porous network supports nutrient transport and stimulates BMSC osteogenic differentiation, while TGF-β3/Sil-MA bridges the scaffold to natural cartilage, promoting new cartilage growth. This composite material shows promise for osteochondral repair using endogenous cells [175].

In cartilage engineering, 3D printed PCL stents can be improved by adding polydopamine (PDA)and Poly (lactic-co-glycolic acid) (PLGA) nanoparticles for better cell affinity. Hybrid bioinks with alginate and high-density collagen hydrogel enhance mechanical stability and support cell growth, effectively promoting cartilage differentiation and tissue repair [176]. Researchers developed advanced bioink mixtures, such as alginate with gellan, alginate with nano-fibrillated cellulose, and PEG/alginate composites, which demonstrated superior performance and stability for 3D printed cartilage compared to natural cartilage [177, 178]. Hyaluronic acid (HA) enhances chondrocyte marker expression in 3D scaffolds; while combining chondroitin sulfate (CS) and HA improves glycosaminoglycan (GAG) accumulation and cell proliferation in cartilage matrices within 3D fibrin-alginate hydrogels [179, 180].

In addition, 3D scaffolds can release exosomes for joint tissue repair, enabling targeted drug delivery and using exosome-laden materials as cell-free therapeutic products for cartilage repair [181, 182]. By integrating growth factors and creating scaffolds with precise porosity and mechanical properties, 3D printing technology allows for the fabrication of complex structures that can support the regeneration of both cartilage and bone tissues. For instance, scaffolds with smaller pore sizes promote cartilage formation, while larger pores facilitate bone growth. This customization capability makes 3D-printed hydrogels a promising approach for personalized, effective treatments in regenerative medicine [183].

In wound healing, blood vessel regeneration is crucial for nutrient and oxygen transport needed for skin repair. Fibrin promotes endothelial cell adhesion and vascularization at injury sites. Chen et al. established an injectable hydrogel using SH-PEG and AgNO3 with the angiogenic drug deferoxamine. This hydrogel stimulated angiogenesis, resulting in a more extensive vascular network and longer blood vessels [184]. Similarly, Liu et al. developed an injectable, self-healing hydrogel combining chitosan (CS) with silver (Ag^+) and copper ions (Cu^{2+}) for effective healing of infectious and diabetic wounds. The CS-Ag-Cu hydrogel gradually delivered Ag^+ for antibacterial action and Cu^{2+} for angiogenesis, showing good adhesion, water absorption, antibacterial properties, and biocompatibility for skin regeneration [185].

Johnson and colleagues developed a 3D printing method using silicone to create complex nerve guidance channels. By collecting data on missing nerve pieces and printing a functionalized model, they successfully regenerated nerve injuries in rats over 10–12 weeks. This method, however, needs biodegradable alternatives due to the use of non-biocompatible silicone [186]. Hu and co-workers created a 3D-printed bio-conduit for peripheral nerve regeneration using cryoGelMA gel with adipose-derived stem cells (Fig. 13.5). These biodegradable conduits, with customizable structures, eliminate the need for removal surgery and advanced in vitro models are essential for studying various neurological phenomena [187].

13.7 Challenges and Future Perspectives

Despite advances in bioengineering hydrogel production, key parameters like degradation rates, reaction time, surface hybridization, microstructures, inflammation, and immunological response still require careful scientific research to advance cytocompatibility for tissue engineering. Future improvements in hydrogel function and structure will enhance our understanding of cell-material interactions, enabling modular tissue generation. These hydrogels also promise safer, more effective treatments for various diseases. Developing bioinks for 3D bioprinting that meet all clinical requirements remains a significant challenge due to the need for properties like in vivo stability, structural integrity, tissue-specific degradation, cell growth support, non-toxicity, and vascularization promotion, despite the potential impact on cell viability during printing.

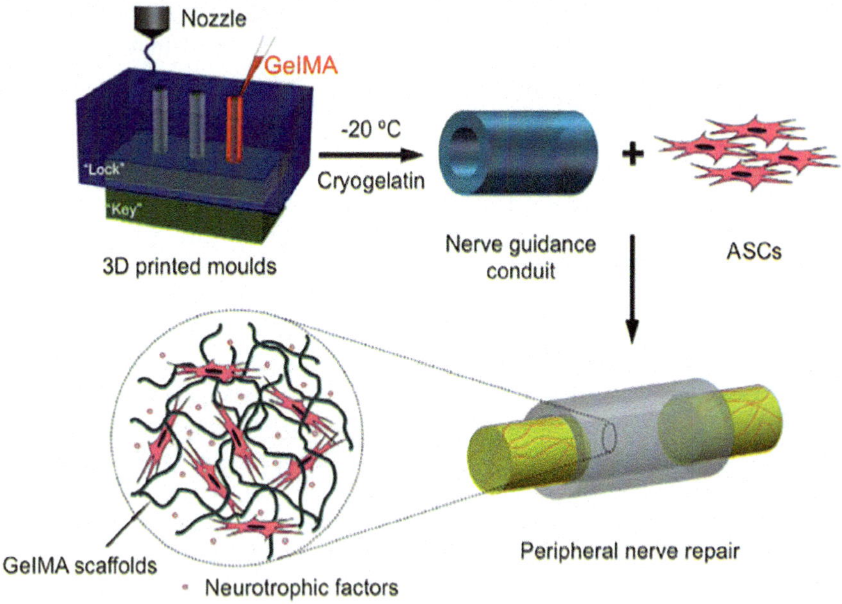

Fig. 13.5 Schematic illustration of the 3D-engineered bio-conduit for peripheral nerve regeneration [188]. (Re-used under Creative Commons Attribution (CC BY 4.0))

In situ bioprinting, the direct printing of living tissue onto surgical defects, remains a challenge due to the need to maintain a sterile environment, integrate printers into surgical procedures, and address ethical considerations around patient information disclosure. Regulatory compliance and cost-effectiveness are additional barriers that must be overcome for widespread clinical adoption. 3D bioprinting technologies have revolutionized tissue engineering by enabling precise control over scaffold characteristics and integrating biomaterials with imaging and computing. Challenges remain in developing suitable biological inks for tissue and organ printing. Recent advancements allow for the rapid creation of bioactive 3D structures using cells in a single step, offering promising applications in cartilage regenerative medicine through interdisciplinary collaboration.

13.8 Conclusion

Tissue engineering aims to restore tissue function using scaffold-based strategies involving cells, biomaterials, and growth factors. However, replicating the complex structures of native tissues, such as gradient pore sizes and spatial heterostructures, remains a significant challenge in achieving accurate tissue regeneration. Hydrogels are versatile for drug delivery and tissue engineering, providing spaces for cells and nutrient distribution. Bioprinting advances allow precise fabrication of hydrogel scaffolds with interconnected pores, mimicking native tissue structures for effective tissue regeneration. These bio-inks must balance biological properties, structural integrity, mechanical support, rheological behaviors, and degradation times to meet native tissue requirements.

Bioprinting material selection focuses on biocompatibility and printing properties like viscosity and stability. Analyzing ECM proteins in acellular scaffolds guides bio-ink development. Advanced chemical synthesis offers new avenues for functional materials that respond to external stimuli, restoring shape, size, and function. Recent advancements have enabled modifications in the physicochemical, rheological, mechanical, and biological properties of hydrogels through various techniques. Despite progress, challenges persist in cultivating diverse cell types, selecting suitable materials and cells for bio-inks, and ensuring stable constructs with appropriate properties for clinical use. Issues include the presence of animal-derived components in cell culture media, potential immunogenicity of biomaterials, and difficulties in managing swelling behaviors during storage and use. Additionally, achieving optimal degradation rates matching tissue regeneration remains crucial. Future advancements should focus on environmental responsiveness, self-healing properties, and multifunctionality of biomedical hydrogel-based bio-inks to enhance their clinical applicability in tissue engineering. Combining various natural or synthetic hydrogels and employing 3D printing techniques can significantly enhance the properties of hydrogel-based scaffolds. To advance these technologies, closer collaboration among scientists, engineers, and clinicians is essential. This collaboration should focus on assessing feasibility, practicality, and multifunctionality of hydrogel-based bio-inks, addressing unmet requirements, and prioritizing their design from laboratory development to clinical application.

References

1. Li J, Wu C, Chu PK, Gelinsky M (2020) 3D printing of hydrogels: rational design strategies and emerging biomedical applications. Mater Sci Eng R Rep 140:100543. https://doi.org/10.1016/j.mser.2020.100543
2. Singh S, Syukri DM, Ushir YV, Mishra A, Ontong JC, Nwabor OF et al (2025) Post-operative wound healing efficacy of Eucalyptus camaldulensis phenolic-rich extracts incorporated hydrogel with enhanced antioxidant, antibacterial, and anti-inflammatory activities. J Polym Environ 33:814–839
3. Mohite P, Asane G, Rebello N, Munde S, Ade N, Boban T et al (2024) Polymeric hydrogel sponges for wound healing applications: a comprehensive review. Regen Eng Transl Med 10:416–437
4. Sharma R, Malviya R, Singh S, Prajapati B (2023) A critical review on classified excipient sodium-alginate-based hydrogels: modification, characterization, and application in soft tissue engineering. Gels 9(5):430
5. Mohite P, Rahayu P, Munde S, Ade N, Chidrawar VR, Singh S et al (2023) Chitosan-based hydrogel in the management of dermal infections: a review. Gels 9(7):594
6. Vaupel S, Mau R, Kara S, Seitz H, Kragl U, Meyer J (2023) 3D printed and stimulus responsive drug delivery systems based on synthetic polyelectrolyte hydrogels manufactured via digital light processing. J Mater Chem B 11(28):6547–6559. https://doi.org/10.1039/D3TB00285C

7. Fatimi A, Okoro OV, Podstawczyk D, Siminska-Stanny J, Shavandi A (2022) Natural hydrogel-based bio-inks for 3D bioprinting in tissue engineering: a review. Gels 8(3):179. https://doi.org/10.3390/gels8030179
8. Yang X, Li S, Sun X, Ren Y, Qiang L, Liu Y, Wang J, Dai K (2023) Swelling compensation of engineered vasculature fabricated by additive manufacturing and sacrifice-based technique using thermoresponsive hydrogel. Int J Bioprint 9(5):749. https://doi.org/10.18063/ijb.749
9. Bao X, Meng J, Tan Z, Zhang C, Li L, Liu T (2024) Direct-ink-write 3D printing of highly-stretchable polyaniline gel with hierarchical conducting network for customized wearable strain sensors. Chem Eng J 491:151918. https://doi.org/10.1016/j.cej.2024.151918
10. Liu J, Garcia J, Leahy LM, Song R, Mullarkey D, Fei B, Dervan A, Shvets IV, Stamenov P, Wang W, O'Brien FJ, Coleman JN, Nicolosi V (2023) 3D printing of multifunctional conductive polymer composite hydrogels. Adv Funct Mater 33(37):2214196. https://doi.org/10.1002/adfm.202214196
11. Vyas J, Raytthatha N, Singh S, Prajapati B (2025) Next-generation computational automation-based additive manufacturing of pharmaceuticals: an approach to fabricate precise medicine. In: Handbook of 3D printing in pharmaceutics. CRC Press, Boca Raton, pp 153–175
12. Shah SR, Modi CD, Singh S, Mori DD, Soniwala MM, Prajapati BG (2024) Recent advances in additive manufacturing of polycaprolactone-based scaffolds for tissue engineering applications: a comprehensive review. Regen Eng Transl Med 11:112–131
13. Singh, A., Mandal, P., Maheshwari, S., Akhtar, J., Prajapati, B. G., & Singh, S. (2025) Processing of sustainable biomaterials by additive manufacturing methods. In Advances in sustainable biomaterials (pp. 141-162). CRC Press, Boca Raton
14. Vyas J, Raytthatha N, Singh S, Prajapati BG, Mohite P, Munde S (2024) Sustainable sources of raw materials as substituting biomaterials for additive manufacturing of dental implants: a review. Periodontal Implant Res 8(1):3
15. Vyas J, Singh S, Shah I, Prajapati BG (2023) Potential applications and additive manufacturing technology-based considerations of mesoporous silica: a review. AAPS PharmSciTech 25(1):6
16. Zaharin HA, Abdul Rani AM, Ginta TL, Azam FI (2018) Additive manufacturing technology for biomedical components: a review. IOP Conf Ser Mater Sci Eng 328(1):012003. https://doi.org/10.1088/1757-899X/328/1/012003
17. Mirzaali MJ, Moosabeiki V, Rajaai SM, Zhou J, Zadpoor AA (2022) Additive manufacturing of biomaterials—design principles and their implementation. Materials 15(15):5457. https://doi.org/10.3390/ma15155457
18. González-Henríquez CM, Sarabia-Vallejos MA, Rodriguez-Hernandez J (2019) Polymers for additive manufacturing and 4D-printing: materials, methodologies, and biomedical applications. Prog Polym Sci 94:57–116. https://doi.org/10.1016/j.progpolymsci.2019.03.001
19. Dhavalikar P, Lan Z, Kar R, Salhadar K, Gaharwar AK, Cosgriff-Hernandez E (2020) 1.4.8—Biomedical applications of additive manufacturing. In: Wagner WR, Sakiyama-Elbert SE, Zhang G, Yaszemski MJ (eds) Biomaterials science, 4th edn. Academic, pp 623–639. https://doi.org/10.1016/B978-0-12-816137-1.00040-4
20. Lim J, Bupphathong S, Huang W, Lin C-H (2023) Three-dimensional bioprinting of biocompatible photosensitive polymers for tissue engineering application. Tissue Eng Part B Rev 29(6):710–722. https://doi.org/10.1089/ten.teb.2023.0072
21. Martelli A, Bellucci D, Cannillo V (2023) Additive manufacturing of polymer/bioactive glass scaffolds for regenerative medicine: a review. Polymers 15(11):2473. https://doi.org/10.3390/polym15112473
22. Dadhich P, Kumar P, Roy A, Bitar KN (2023) Advances in 3D printing technology for tissue engineering. In: Chakravorty N, Shukla PC (eds) Regenerative medicine: emerging techniques to translation approaches. Springer Nature Singapore, Singapore, pp 181–206. https://doi.org/10.1007/978-981-19-6008-6_9
23. Veeman D, Sai MS, Sureshkumar P, Jagadeesha T, Natrayan L, Ravichandran M, Mammo WD (2021) Additive manufacturing of biopolymers for tissue engineering and regenerative

medicine: an overview, potential applications, advancements, and trends. Int J Polym Sci 2021:4907027. https://doi.org/10.1155/2021/4907027
24. Bose S, Koski C, Vu AA (2020) Additive manufacturing of natural biopolymers and composites for bone tissue engineering. Mater Horizons 7(8):2011–2027. https://doi.org/10.1039/D0MH00277A
25. Hassan M, Dave K, Chandrawati R, Dehghani F, Gomes VG (2019) 3D printing of biopolymer nanocomposites for tissue engineering: nanomaterials, processing and structure-function relation. Eur Polym J 121:109340. https://doi.org/10.1016/j.eurpolymj.2019.109340
26. Sugiaman VK, Jeffrey, Naliani S, Pranata N, Djuanda R, Saputri RI (2023) Polymeric scaffolds used in dental pulp regeneration by tissue engineering approach. Polymers 15(5):1082. https://doi.org/10.3390/polym15051082
27. Singh S, Syukri DM, Ushir YV, Mishra A, Ontong JC, Nwabor OF et al (2024) Post-operative wound healing efficacy of Eucalyptus camaldulensis phenolic-rich extracts incorporated hydrogel with enhanced antioxidant, antibacterial, and anti-inflammatory activities. J Polym Environ 1-26
28. Yuan X, Zhu Z, Xia P, Wang Z, Zhao X, Jiang X, Wang T, Gao Q, Xu J, Shan D, Guo B, Yao Q, He Y (2023) Tough gelatin hydrogel for Tissue Engineering. Adv Sci 10(24):2301665. https://doi.org/10.1002/advs.202301665
29. Lyu Y, Liu Y, He H, Wang H (2023) Application of silk-fibroin-based hydrogels in tissue engineering. 9(5):431. https://doi.org/10.3390/gels9050431
30. Augustine R, Nikolopoulos VK, Camci-Unal G (2023) Hydrogel-impregnated self-oxygenating electrospun scaffolds for bone tissue engineering. 10(7):854. https://doi.org/10.3390/bioengineering10070854
31. Berradi A, Aziz F, Achaby ME, Ouazzani N, Mandi L (2023) A comprehensive review of polysaccharide-based hydrogels as promising biomaterials. 15(13):2908. https://doi.org/10.3390/polym15132908
32. Kuzina MA, Kartsev DD, Stratonovich AV, Levkin PA (2023) Organogels versus hydrogels: advantages, challenges, and applications. 33(27):2301421. https://doi.org/10.1002/adfm.202301421
33. Andreazza R, Morales A, Pieniz S, Labidi J (2023) Gelatin-based hydrogels: potential biomaterials for remediation 15(4):1026. https://doi.org/10.3390/polym15041026
34. Madamsetty VS, Vazifehdoost M, Alhashemi SH, Davoudi H, Zarrabi A, Dehshahri A, Fekri HS, Mohammadinejad R, Thakur VK (2023) Next-generation hydrogels as biomaterials for biomedical applications: exploring the role of curcumin. ACS Omega 8(10):8960–8976. https://doi.org/10.1021/acsomega.2c07062
35. Huang S, Hong X, Zhao M, Liu N, Liu H, Zhao J, Shao L, Xue W, Zhang H, Zhu P, Guo R (2022) Nanocomposite hydrogels for biomedical applications. 7(3):e10315. https://doi.org/10.1002/btm2.10315
36. Belgodere JA, Lassiter HR, Robinson JT, Hamel KM, Rogers EL, Mohiuddin OA, Zhang L, Wu X, Gimble JM, Frazier TP, Monroe WT, Sanchez CG (2023) Biomechanical and biological characterization of XGel, a human-derived hydrogel for stem cell expansion and tissue engineering. 7(8):2200332. https://doi.org/10.1002/adbi.202200332
37. Li X, Xu M, Geng Z, Liu Y (2023) Functional hydrogels for the repair and regeneration of tissue defects. 11. https://doi.org/10.3389/fbioe.2023.1190171
38. Chen A, Deng S, Lai J, Li J, Chen W, Varma SN, Zhang J, Lei C, Liu C, Huang L (2023) Hydrogels for oral tissue engineering: challenges and opportunities. 28(9):3946. https://doi.org/10.3390/molecules28093946
39. Li Y, Liu Y, Li R, Bai H, Zhu Z, Zhu L, Zhu C, Che Z, Liu H, Wang J, Huang L (2021) Collagen-based biomaterials for bone tissue engineering. Mater Des 210:110049. https://doi.org/10.1016/j.matdes.2021.110049
40. Zheng M, Wang X, Yue O, Hou M, Zhang H, Beyer S, Blocki AM, Wang Q, Gong G, Liu X, Guo J (2021) Skin-inspired gelatin-based flexible bio-electronic hydrogel for wound healing promotion and motion sensing. Biomaterials 276:121026. https://doi.org/10.1016/j.biomaterials.2021.121026

41. Zhang K, Shi Z, Zhou J, Xing Q, Ma S, Li Q, Zhang Y, Yao M, Wang X, Li Q, Li J, Guan F (2018) Potential application of an injectable hydrogel scaffold loaded with mesenchymal stem cells for treating traumatic brain injury. J Mater Chem B 6(19):2982–2992. https://doi.org/10.1039/C7TB03213G
42. Mohite P, Shah SR, Singh S, Rajput T, Munde S, Ade N et al (2023) Chitosan and chito-oligosaccharide: a versatile biopolymer with endless grafting possibilities for multifarious applications. Front Bioeng Biotechnol 11:1190879
43. Kong X-b, Tang Q-y, Chen X-y, Tu Y, Sun S-z, Sun Z-l (2017) Polyethylene glycol as a promising synthetic material for repair of spinal cord injury. 12(6):1003–1008. https://doi.org/10.4103/1673-5374.208597
44. Singh S, Nwabor OF, Sukri DM, Wunnoo S, Dumjun K, Lethongkam S et al (2022) Poly (vinyl alcohol) copolymerized with xanthan gum/hypromellose/sodium carboxymethyl cellulose dermal dressings functionalized with biogenic nanostructured materials for antibacterial and wound healing application. Int J Biol Macromol 216:235–250
45. Geng Y, Liu T, Zhao M, Wei H, Yao X, Zhang Y (2024) Silk fibroin/polyacrylamide-based tough 3D printing scaffold with strain sensing ability and chondrogenic activity. Compos Part B 271:111173. https://doi.org/10.1016/j.compositesb.2023.111173
46. Xu X, Liu Y, Fu W, Yao M, Ding Z, Xuan J, Li D, Wang S, Xia Y, Cao M (2020) Poly(N-isopropylacrylamide)-based thermoresponsive composite hydrogels for biomedical applications. 12(3):580. https://doi.org/10.3390/polym12030580
47. Sheffield C, Meyers K, Johnson E, Rajachar RM (2018) Application of composite hydrogels to control physical properties in tissue engineering and regenerative medicine. 4(2):51. https://doi.org/10.3390/gels4020051
48. El-Husseiny HM, Mady EA, Hamabe L, Abugomaa A, Shimada K, Yoshida T, Tanaka T, Yokoi A, Elbadawy M, Tanaka R (2022) Smart/stimuli-responsive hydrogels: cutting-edge platforms for tissue engineering and other biomedical applications. Materials Today Bio 13:100186. https://doi.org/10.1016/j.mtbio.2021.100186
49. Ganguly S, Margel S (2023) Fluorescent quantum dots-based hydrogels: synthesis, fabrication and multimodal biosensing. Talanta Open 8:100243. https://doi.org/10.1016/j.talo.2023.100243
50. Fonseca RG, Kuster A, Fernandes PP, Tavakoli M, Pereira P, Fernandes JR, De Bon F, Serra AC, Fonseca AC, Coelho JFJ (2023) Facile synthesis of highly stretchable, tough, and photodegradable hydrogels. 12(22):2300918. https://doi.org/10.1002/adhm.202300918
51. Manzoor A, Jamil F, Salman AW, Khalid FA, Shoukat US, Iqbal MA (2023) Fabrication and structural design of MXene-based hydrogels. In: Rizwan K, Khan A, Ahmed Asiri AM (eds) Handbook of functionalized nanostructured MXenes: synthetic strategies and applications from energy to environment sustainability. Springer Nature Singapore, Singapore, pp 61–81. https://doi.org/10.1007/978-981-99-2038-9_5
52. Kumari S, Rathore R, Jaiswal R, Pandey S, Verma N, Sharma B, Sharma C, Aeshala LM, Singh S (2023) 1–Fundamental of hydrogels and nanocomposite hydrogels: synthesis, physiochemical characterization, and biomedical applications. In: Kumar A, Thakur VK (eds) Functional nanocomposite hydrogels. Elsevier, pp 1–24. https://doi.org/10.1016/B978-0-323-99638-9.00001-0
53. Sahoo DR, Biswal T (2021) Alginate and its application to tissue engineering. SN Appl Sci 3(1):30. https://doi.org/10.1007/s42452-020-04096-w
54. Tomić SL, Babić MM, Vuković JS, Perišić MD, Filipović VV, Davidović SZ, Filipović JM (2016) 2-hydroxyethyl metahcrylate/gelatin based superporous hydrogels for tissue regeneration. AIP Conf Proc 1736(1). https://doi.org/10.1063/1.4949668
55. Tian R, Qiu X, Yuan P, Lei K, Wang L, Bai Y, Liu S, Chen X (2018) Fabrication of self-healing hydrogels with on-demand antimicrobial activity and sustained biomolecule release for infected skin regeneration. ACS Appl Mater Interfaces 10(20):17018–17027. https://doi.org/10.1021/acsami.8b01740

56. Perez-Puyana V, Jiménez-Rosado M, Romero A, Guerrero A (2020) Fabrication and characterization of hydrogels based on gelatinised collagen with potential application in tissue engineering. 12(5):1146. https://doi.org/10.3390/polym12051146
57. Li L, Ge J, Guo B, Ma PX (2014) In situ forming biodegradable electroactive hydrogels. Polym Chem 5(8):2880–2890. https://doi.org/10.1039/C3PY01634J
58. Chyzy A, Plonska-Brzezinska ME (2020) Hydrogel properties and their impact on regenerative medicine and tissue engineering. 25(24):5795. https://doi.org/10.3390/molecules25245795
59. Guarino V, Raucci MG, Ronca A, Cirillo V, Ambrosio L (2014) 5—Multifunctional scaffolds for bone regeneration. In: Mallick K (ed) Bone substitute biomaterials. Woodhead Publishing, pp 95–117. https://doi.org/10.1533/9780857099037.2.95
60. Sarrigiannidis SO, Rey JM, Dobre O, González-García C, Dalby MJ, Salmeron-Sanchez M (2021) A tough act to follow: collagen hydrogel modifications to improve mechanical and growth factor loading capabilities. Materials Today Bio 10:100098. https://doi.org/10.1016/j.mtbio.2021.100098
61. Mantha S, Pillai S, Khayambashi P, Upadhyay A, Zhang Y, Tao O, Pham HM, Tran SD (2019) Smart hydrogels in tissue engineering and regenerative medicine. 12(20):3323. https://doi.org/10.3390/ma12203323
62. Sivaraj D, Chen K, Chattopadhyay A, Henn D, Wu W, Noishiki C, Magbual NJ, Mittal S, Mermin-Bunnell AM, Bonham CA, Trotsyuk AA, Barrera JA, Padmanabhan J, Januszyk M, Gurtner GC (2021) Hydrogel scaffolds to deliver cell therapies for wound healing. 9. https://doi.org/10.3389/fbioe.2021.660145
63. Xu F, Dawson C, Lamb M, Mueller E, Stefanek E, Akbari M, Hoare T (2022) Hydrogels for tissue engineering: addressing key design needs toward clinical translation. 10. https://doi.org/10.3389/fbioe.2022.849831
64. Mansur AAP, Rodrigues MA, Capanema NSV, Carvalho SM, Gomes DA, Mansur HS (2023) Functionalized bioadhesion-enhanced carboxymethyl cellulose/polyvinyl alcohol hybrid hydrogels for chronic wound dressing applications. RSC Adv 13(19):13156–13168. https://doi.org/10.1039/D3RA01519J
65. Morwood AJ, El-Karim IA, Clarke SA, Lundy FT (2023) The role of extracellular matrix (ECM) adhesion motifs in functionalised hydrogels. 28(12):4616. https://doi.org/10.3390/molecules28124616
66. Zhu H, Yin T, Hu X, Liu B, Yu H, Chen Z, Qu S, Yang W (2023) Mechanics of regulatable hydrogel adhesion with elastic heterogeneity. J Mech Phys Solids 175:105304. https://doi.org/10.1016/j.jmps.2023.105304
67. Li S, Huang C, Liu H, Han X, Wang Z, Huang J, Yan Y, Wang Z (2022) A silk fibroin methacryloyl-modified hydrogel promoting cell adhesion for customized 3D cell-laden structures. ACS Appl Polym Mater 4(10):7014–7024. https://doi.org/10.1021/acsapm.2c00952
68. Jang W, Kim DY, Mun SJ, Choi JH, Roh YH, Bong KW (2022) Direct functionalization of cell-adhesion promoters to hydrogel microparticles synthesized by stop-flow lithography. 60(11):1767–1777. https://doi.org/10.1002/pol.20210934
69. El-Husseiny HM, Mady EA, El-Dakroury WA, Doghish AS, Tanaka R (2023) Stimuli-responsive hydrogels: smart state of-the-art platforms for cardiac tissue engineering. 11. https://doi.org/10.3389/fbioe.2023.1174075
70. Amirthalingam S, Rajendran AK, Moon YG, Hwang NS (2023) Stimuli-responsive dynamic hydrogels: design, properties and tissue engineering applications. Mater Horizons 10(9):3325–3350. https://doi.org/10.1039/D3MH00399J
71. Ni X, Xing X, Deng Y, Li Z (2023) Applications of stimuli-responsive hydrogels in bone and cartilage regeneration. 15(3):982. https://doi.org/10.3390/pharmaceutics15030982
72. Maji S, Lee H (2022) Engineering hydrogels for the development of three-dimensional in vitro models. 23(5):2662. https://doi.org/10.3390/ijms23052662
73. Liang Y, Qiao L, Qiao B, Guo B (2023) Conductive hydrogels for tissue repair. Chem Sci 14(12):3091–3116. https://doi.org/10.1039/D3SC00145H

74. Moura BS, Monteiro MV, Ferreira LP, Lavrador P, Gaspar VM, Mano JF (2022) Advancing tissue decellularized hydrogels for engineering human organoids. 32(29):2202825. https://doi.org/10.1002/adfm.202202825
75. Vieira S, Silva-Correia J, Reis RL, Oliveira JM (2022) Engineering hydrogels for modulation of material-cell interactions. 22(10):2200091. https://doi.org/10.1002/mabi.202200091
76. Aghamirsalim M, Mobaraki M, Soltani M, Kiani Shahvandi M, Jabbarvand M, Afzali E, Raahemifar K (2022) 3D printed hydrogels for ocular wound healing. Biomedicines 10(7). https://doi.org/10.3390/biomedicines10071562
77. Pasini C, Pandini S, Ramorino G, Sartore L (2024) Tailoring the properties of composite scaffolds with a 3D-printed lattice core and a bioactive hydrogel shell for tissue engineering. J Mech Behav Biomed Mater 150:106305. https://doi.org/10.1016/j.jmbbm.2023.106305
78. Agrawal A, Hussain CM (2023) 3D-printed hydrogel for diverse applications: a review. Gels 9(12). https://doi.org/10.3390/gels9120960
79. Ho T-C, Chang C-C, Chan H-P, Chung T-W, Shu C-W, Chuang K-P, Duh T-H, Yang M-H, Tyan Y-C (2022) Hydrogels: properties and applications in biomedicine. Molecules 27(9). https://doi.org/10.3390/molecules27092902
80. Satchanska G, Davidova S, Petrov PD (2024) Natural and synthetic polymers for biomedical and environmental applications. Polymers 16(8). https://doi.org/10.3390/polym16081159
81. Salma-Ancane K, Sceglovs A, Tracuma E, Wychowaniec JK, Aunina K, Ramata-Stunda A, Nikolajeva V, Loca D (2022) Effect of crosslinking strategy on the biological, antibacterial and physicochemical performance of hyaluronic acid and ε-polylysine based hydrogels. Int J Biol Macromol 208:995–1008. https://doi.org/10.1016/j.ijbiomac.2022.03.207
82. Alshangiti DM, El-damhougy TK, Zaher A, Madani M, Mohamady Ghobashy M (2023) Revolutionizing biomedicine: advancements, applications, and prospects of nanocomposite macromolecular carbohydrate-based hydrogel biomaterials: a review. RSC Adv 13(50):35251–35291. https://doi.org/10.1039/D3RA07391B
83. Annabi N, Tamayol A, Uquillas JA, Akbari M, Bertassoni LE, Cha C, Camci-Unal G, Dokmeci MR, Peppas NA, Khademhosseini A (2014) 25th anniversary article: rational design and applications of hydrogels in regenerative medicine. Adv Mater 26(1):85–123. https://doi.org/10.1002/adma.201303233
84. Wang L, Yang Y, Chen Y, Majidi C, Iida F, Askounis E, Pei Q (2018) Controllable and reversible tuning of material rigidity for robot applications. Mater Today 21(5):563–576. https://doi.org/10.1016/j.mattod.2017.10.010
85. Kureha T, Takahashi K, Kino M, Kida H, Hirayama T (2023) Controlling the mechanical properties of hydrogels via modulating the side-chain length. Soft Matter 19(16):2878–2882. https://doi.org/10.1039/D3SM00134B
86. Prete S, Dattilo M, Patitucci F, Pezzi G, Parisi OI, Puoci F (2023) Natural and synthetic polymeric biomaterials for application in wound management. J Funct Biomater 14(9). https://doi.org/10.3390/jfb14090455
87. Drozdov AD, Christiansen J (2021) The effects of pH and ionic strength on the volume phase transition temperature of thermo-responsive anionic copolymer gels. Polymer 221:123637. https://doi.org/10.1016/j.polymer.2021.123637
88. Panyukov S (2020) Theory of flexible polymer networks: elasticity and heterogeneities. Polymers 12(4). https://doi.org/10.3390/polym12040767
89. Yıldız M, Soğancı AS (2012) Effect of freezing and thawing on strength and permeability of lime-stabilized clays. Sci Iranica 19(4):1013–1017. https://doi.org/10.1016/j.scient.2012.06.003
90. Thang NH, Chien TB, Cuong DX (2023) Polymer-based hydrogels applied in drug delivery: an overview. Gels 9(7). https://doi.org/10.3390/gels9070523
91. Bashir S, Teo YY, Ramesh S, Ramesh K, Mushtaq MW (2018) Rheological behavior of biodegradable N-succinyl chitosan-g-poly (acrylic acid) hydrogels and their applications as drug carrier and in vitro theophylline release. Int J Biol Macromol 117:454–466. https://doi.org/10.1016/j.ijbiomac.2018.05.182

92. Miao C, Li P, Yu J, Xu X, Zhang F, Tong G (2023) Dual network hydrogel with high mechanical properties, electrical conductivity, water retention and frost resistance, suitable for wearable strain sensors. Gels 9(3). https://doi.org/10.3390/gels9030224
93. Ying H, Shen C, Pan R, Li X, Chen Y (2023) Strategy insight: mechanical properties of biomaterials' influence on hydrogel-mesenchymal stromal cell combination for osteoarthritis therapy. Front Pharmacol 14. https://doi.org/10.3389/fphar.2023.1152612
94. Chen Y, Li J, Lu J, Ding M, Chen Y (2022) Synthesis and properties of poly(vinyl alcohol) hydrogels with high strength and toughness. Polym Test 108:107516. https://doi.org/10.1016/j.polymertesting.2022.107516
95. Patel, R. P., Patel, G. K., Patel, N., Singh, S., & Chittasupho, C. (2023). Alginate nanoparticles: a potential drug carrier in tuberculosis treatment. Tubercular drug Delivery systems: advances in treatment of Infectious diseases, pp 207–234
96. Begum RF, Singh S, Prajapati B, Sumithra M, Patel RJ (2024) Advanced targeted drug delivery of bioactive agents fortified with graft chitosan in management of cancer: a review. Curr Med Chem
97. Apte G, Lindenbauer A, Schemberg J, Rothe H, Nguyen T-H (2021) Controlling surface-induced platelet activation by agarose and Gelatin-based hydrogel films. ACS Omega 6(16):10963–10974. https://doi.org/10.1021/acsomega.1c00764
98. Fakhari A, Berkland C (2013) Applications and emerging trends of hyaluronic acid in tissue engineering, as a dermal filler and in osteoarthritis treatment. Acta Biomater 9(7):7081–7092. https://doi.org/10.1016/j.actbio.2013.03.005
99. Gachon E, Mesquida P (2022) Mechanical properties of collagen fibrils determined by buckling analysis. Acta Biomater 149:60–68. https://doi.org/10.1016/j.actbio.2022.06.044
100. Singh S, Supaweera N, Nwabor OF, Chaichompoo W, Suksamrarn A, Chittasupho C, Chunglok W (2024) Poly (vinyl alcohol)-gelatin-sericin copolymerized film fortified with vesicle-entrapped demethoxycurcumin/bisdemethoxycurcumin for improved stability, antibacterial, anti-inflammatory, and skin tissue regeneration. Int J Biol Macromol 258:129071
101. Ansari MJ, Rajendran RR, Mohanto S, Agarwal U, Panda K, Dhotre K, Manne R, Deepak A, Zafar A, Yasir M, Pramanik S (2022) Poly(N-isopropylacrylamide)-based hydrogels for biomedical applications: a review of the state-of-the-art. Gels 8(7). https://doi.org/10.3390/gels8070454
102. Moura D, Pereira AT, Ferreira HP, Barrias CC, Magalhães FD, Bergmeister H, Gonçalves IC (2023) Poly(2-hydroxyethyl methacrylate) hydrogels containing graphene-based materials for blood-contacting applications: from soft inert to strong degradable material. Acta Biomater 164:253–268. https://doi.org/10.1016/j.actbio.2023.04.031
103. Jung SA, Malyaran H, Demco DE, Manukanc A, Häser LS, Kučikas V, van Zandvoort M, Neuss S, Pich A (2023) Fibrin–dextran hydrogels with tunable porosity and mechanical properties. Biomacromolecules 24(9):3972–3984. https://doi.org/10.1021/acs.biomac.3c00269
104. Jayawardena I, Turunen P, Garms BC, Rowan A, Corrie S, Grøndahl L (2023) Evaluation of techniques used for visualisation of hydrogel morphology and determination of pore size distributions. Mater Adv 4(2):669–682. https://doi.org/10.1039/D2MA00932C
105. Aston R, Sewell K, Klein T, Lawrie G, Grøndahl L (2016) Evaluation of the impact of freezing preparation techniques on the characterisation of alginate hydrogels by cryo-SEM. Eur Polym J 82:1–15. https://doi.org/10.1016/j.eurpolymj.2016.06.025
106. Inphonlek S, Jarukumjorn K, Chumsamrong P, Ruksakulpiwat C, Ruksakulpiwat Y (2023) Preparation of crosslinked poly(acrylic acid-co-acrylamide)-grafted deproteinized natural rubber/silica composites as coating materials for controlled release of fertilizer. Polymers 15(7). https://doi.org/10.3390/polym15071770
107. Chiu Y-C, Larson JC, Isom A, Brey EM (2009) Generation of porous poly(ethylene glycol) hydrogels by salt leaching. Tissue Eng Part C Methods 16(5):905–912. https://doi.org/10.1089/ten.tec.2009.0646
108. Xie L, Jiang M, Dong X, Bai X, Tong J, Zhou J (2012) Controlled mechanical and swelling properties of poly(vinyl alcohol)/sodium alginate blend hydrogels prepared by freeze–thaw

followed by Ca2+ crosslinking. J Appl Polym Sci 124(1):823–831. https://doi.org/10.1002/app.35083
109. Zhang Y, Yang Y, Zhao X, Gao J (2023) Investigation on ionical cross-linking of alginate by monovalent cations to fabrication alginate gel for biomedical application. React Funct Polym 183:105484. https://doi.org/10.1016/j.reactfunctpolym.2022.105484
110. Patiño-Ruiz DA, De Ávila G, Alarcón-Suesca C, González-Delgado ÁD, Herrera A (2020) Ionic cross-linking fabrication of chitosan-based beads modified with FeO and TiO2 nanoparticles: adsorption mechanism toward naphthalene removal in seawater from Cartagena Bay Area. ACS Omega 5(41):26463–26475. https://doi.org/10.1021/acsomega.0c02984
111. Trudicova M, Smilek J, Kalina M, Smilkova M, Adamkova K, Hrubanova K, Krzyzanek V, Sedlacek P (2020) Multiscale experimental evaluation of agarose-based semi-interpenetrating polymer network hydrogels as materials with Tunable rheological and transport performance. Polymers 12(11). https://doi.org/10.3390/polym12112561
112. Kim M-H, Park J-H, Nguyen D-T, Kim S, Jeong DI, Cho H-J, Kim D-D (2021) Hyaluronidase inhibitor-incorporated cross-linked hyaluronic acid hydrogels for subcutaneous injection. Pharmaceutics 13(2). https://doi.org/10.3390/pharmaceutics13020170
113. Revell CK, Herrera JA, Lawless C, Lu Y, Kadler KE, Chang J, Jensen OE (2023) Modeling collagen fibril self-assembly from extracellular medium in embryonic tendon. Biophys J 122(16):3219–3237. https://doi.org/10.1016/j.bpj.2023.07.001
114. Campiglio CE, Contessi Negrini N, Farè S, Draghi L (2019) Cross-linking strategies for electrospun Gelatin scaffolds. Materials 12(15). https://doi.org/10.3390/ma12152476
115. Serizawa T, Wakita K, Akashi M (2002) Rapid deswelling of porous poly(N-isopropylacrylamide) hydrogels prepared by incorporation of silica particles. Macromolecules 35(1):10–12. https://doi.org/10.1021/ma011362+
116. Přádný M, Martinová L, Michálek J, Fenclová Tá, Krumbholcová E (2007) Electrospinning of the hydrophilic poly (2-hydroxyethyl methacrylate) and its copolymers with 2-ethoxyethyl methacrylate. 5(3):779-792. https://doi.org/10.2478/s11532-007-0021-0
117. Chen F-M, Zhao Y-M, Sun H-H, Jin T, Wang Q-T, Zhou W, Wu Z-F, Jin Y (2007) Novel glycidyl methacrylated dextran (Dex-GMA)/gelatin hydrogel scaffolds containing microspheres loaded with bone morphogenetic proteins: formulation and characteristics. J Control Release 118(1):65–77. https://doi.org/10.1016/j.jconrel.2006.11.016
118. Norahan MH, Pedroza-González SC, Sánchez-Salazar MG, Álvarez MM, Trujillo de Santiago G (2023) Structural and biological engineering of 3D hydrogels for wound healing. Bioact Mater 24:197–235. https://doi.org/10.1016/j.bioactmat.2022.11.019
119. Pishavar E, Luo H, Naserifar M, Hashemi M, Toosi S, Atala A, Ramakrishna S, Behravan J (2021) Advanced hydrogels as exosome delivery Systems for Osteogenic Differentiation of MSCs: application in bone regeneration. Int J Mol Sci 22(12). https://doi.org/10.3390/ijms22126203
120. Ye B, Wu B, Su Y, Sun T, Guo X (2022) Recent advances in the application of natural and synthetic polymer-based scaffolds in musculoskeletal regeneration. Polymers 14(21). https://doi.org/10.3390/polym14214566
121. Bustamante-Torres M, Romero-Fierro D, Arcentales-Vera B, Palomino K, Magaña H, Bucio E (2021) Hydrogels classification according to the physical or chemical interactions and as stimuli-sensitive materials. Gels 7(4). https://doi.org/10.3390/gels7040182
122. Tsou Y-H, Khoneisser J, Huang P-C, Xu X (2016) Hydrogel as a bioactive material to regulate stem cell fate. Bioact Mater 1(1):39–55. https://doi.org/10.1016/j.bioactmat.2016.05.001
123. Taneja H, Salodkar SM, Singh Parmar A, Chaudhary S (2022) Hydrogel based 3D printing: Bio ink for tissue engineering. J Mol Liq 367:120390. https://doi.org/10.1016/j.molliq.2022.120390
124. Omidian H, Chowdhury SD, Wilson RL (2024) Advancements and challenges in hydrogel engineering for regenerative medicine. Gels 10(4). https://doi.org/10.3390/gels10040238
125. Tejada Jacob G, Passamai VE, Katz S, Castro GR, Alvarez V (2022) Hydrogels for extrusion-based bioprinting: general considerations. Bioprinting 27:e00212. https://doi.org/10.1016/j.bprint.2022.e00212

126. Prem Ananth K, Jayram ND (2024) A comprehensive review of 3D printing techniques for biomaterial-based scaffold fabrication in bone tissue engineering. Ann 3D Print Med 13:100141. https://doi.org/10.1016/j.stlm.2023.100141
127. Farshidfar N, Iravani S, Varma RS (2023) Alginate-based biomaterials in tissue engineering and regenerative medicine. Mar Drugs 21(3). https://doi.org/10.3390/md21030189
128. Zhang Y, Wang Y, Li Y, Yang Y, Jin M, Lin X, Zhuang Z, Guo K, Zhang T, Tan W (2023) Application of collagen-based hydrogel in skin wound healing. Gels 9(3). https://doi.org/10.3390/gels9030185
129. Lukin I, Erezuma I, Maeso L, Zarate J, Desimone MF, Al-Tel TH, Dolatshahi-Pirouz A, Orive G (2022) Progress in gelatin as biomaterial for tissue engineering. Pharmaceutics 14(6). https://doi.org/10.3390/pharmaceutics14061177
130. Liu H, Wang C, Li C, Qin Y, Wang Z, Yang F, Li Z, Wang J (2018) A functional chitosan-based hydrogel as a wound dressing and drug delivery system in the treatment of wound healing. RSC Adv 8(14):7533–7549. https://doi.org/10.1039/C7RA13510F
131. Wang Z, Ye Q, Yu S, Akhavan B (2023) Poly ethylene glycol (PEG)-based hydrogels for drug delivery in cancer therapy: a comprehensive review. Adv Healthc Mater 12(18):2300105. https://doi.org/10.1002/adhm.202300105
132. Shaibie NA, Ramli NA, Mohammad Faizal NDF, Srichana T, Mohd Amin MCI (2023) Poly(N-isopropylacrylamide)-based polymers: recent overview for the development of temperature-responsive drug delivery and biomedical applications. Macromol Chem Phys 224(20):2300157. https://doi.org/10.1002/macp.202300157
133. Tan S, Han J, Yuan X, Song Z, Gao L, Gao J, Wang L (2023) Epigallocatechin gallate-loaded pH-responsive dressing with effective antioxidant, antibacterial and anti-biofilm properties for wound healing. Mater Des 227:111701. https://doi.org/10.1016/j.matdes.2023.111701
134. Teixeira MA, Amorim MTP, Felgueiras HP (2020) Poly(vinyl alcohol)-based nanofibrous electrospun scaffolds for tissue engineering applications. Polymers 12(1). https://doi.org/10.3390/polym12010007
135. Norris SCP, Soto J, Kasko AM, Li S (2021) Photodegradable polyacrylamide gels for dynamic control of cell functions. ACS Appl Mater Interfaces 13(5):5929–5944. https://doi.org/10.1021/acsami.0c19627
136. Vanaei S, Parizi MS, Vanaei S, Salemizadehparizi F, Vanaei HR (2021) An overview on materials and techniques in 3D bioprinting toward biomedical application. Eng Regen 2:1–18. https://doi.org/10.1016/j.engreg.2020.12.001
137. Li X, Liu B, Pei B, Chen J, Zhou D, Peng J, Zhang X, Jia W, Xu T (2020) Inkjet bioprinting of biomaterials. Chem Rev 120(19):10793–10833. https://doi.org/10.1021/acs.chemrev.0c00008
138. Boland T, Xu T, Damon B, Cui X (2006) Application of inkjet printing to tissue engineering. Biotechnol J 1(9):910–917. https://doi.org/10.1002/biot.200600081
139. Murphy SV, Atala A (2014) 3D bioprinting of tissues and organs. Nat Biotechnol 32(8):773–785. https://doi.org/10.1038/nbt.2958
140. Cui X, Boland T, D'Lima DD, Lotz MK (2012) Thermal inkjet printing in tissue engineering and regenerative medicine. Recent Pat Drug Deliv Formul 6(2):149–155. https://doi.org/10.2174/187221112800672949
141. Saunders RE, Gough JE, Derby B (2008) Delivery of human fibroblast cells by piezoelectric drop-on-demand inkjet printing. Biomaterials 29(2):193–203. https://doi.org/10.1016/j.biomaterials.2007.09.032
142. Gonzalez-Fernandez T, Tenorio AJ, Campbell KT, Silva EA, Leach JK (2020) Alginate-based bioinks for 3D bioprinting and fabrication of anatomically accurate bone grafts. Tissue Eng A 27(17–18):1168–1181. https://doi.org/10.1089/ten.tea.2020.0305
143. Madni A, Kousar R, Naeem N, Wahid F (2021) Recent advancements in applications of chitosan-based biomaterials for skin tissue engineering. J Bioresour Bioprod 6(1):11–25. https://doi.org/10.1016/j.jobab.2021.01.002
144. Jaipan P, Nguyen A, Narayan RJ (2017) Gelatin-based hydrogels for biomedical applications. MRS Commun 7(3):416–426. https://doi.org/10.1557/mrc.2017.92

145. Ding Y-W, Zhang X-W, Mi C-H, Qi X-Y, Zhou J, Wei D-X (2023) Recent advances in hyaluronic acid-based hydrogels for 3D bioprinting in tissue engineering applications. Smart Mater Med 4:59–68. https://doi.org/10.1016/j.smaim.2022.07.003
146. Wein S, Schemmer C, Al Enezy-Ulbrich MA, Jung SA, Rütten S, Kühnel M, Jonigk D, Jahnen-Dechent W, Pich A, Neuss S (2024) Fibrin-based hydrogels with reactive amphiphilic copolymers for mechanical adjustments allow for capillary formation in 2D and 3D environments. Gels 10:3. https://doi.org/10.3390/gels10030182
147. Lin C-C, Anseth KS (2009) PEG hydrogels for the controlled release of biomolecules in regenerative medicine. Pharm Res 26(3):631–643. https://doi.org/10.1007/s11095-008-9801-2
148. Chatterjee S, Hui PC, Kan C-w (2018) Thermoresponsive hydrogels and their biomedical applications: special insight into their applications in textile based transdermal therapy. Polymers 10(5). https://doi.org/10.3390/polym10050480
149. Jiang S, Liu S, Feng W (2011) PVA hydrogel properties for biomedical application. J Mech Behav Biomed Mater 4(7):1228–1233. https://doi.org/10.1016/j.jmbbm.2011.04.005
150. Buyanov AL, Gofman IV, Saprykina NN (2019) High-strength cellulose–polyacrylamide hydrogels: mechanical behavior and structure depending on the type of cellulose. J Mech Behav Biomed Mater 100:103385. https://doi.org/10.1016/j.jmbbm.2019.103385
151. Brambilla E, Locarno S, Gallo S, Orsini F, Pini C, Farronato M, Thomaz DV, Lenardi C, Piazzoni M, Tartaglia G (2022) Poloxamer-based hydrogel as drug delivery system: how polymeric excipients influence the chemical-physical properties. Polymers 14(17). https://doi.org/10.3390/polym14173624
152. Santhamoorthy M, Vy Phan TT, Ramkumar V, Raorane CJ, Thirupathi K, Kim S-C (2022) Thermo-sensitive poly (N-isopropylacrylamide-co-polyacrylamide) hydrogel for pH-responsive therapeutic delivery. Polymers 14(19). https://doi.org/10.3390/polym14194128
153. Lakkala P, Munnangi SR, Bandari S, Repka M (2023) Additive manufacturing technologies with emphasis on stereolithography 3D printing in pharmaceutical and medical applications: a review. Int J Pharm X 5:100159. https://doi.org/10.1016/j.ijpx.2023.100159
154. Yilmaz B, Al Rashid A, Mou YA, Evis Z, Koç M (2021) Bioprinting: a review of processes, materials and applications. Bioprinting 23:e00148. https://doi.org/10.1016/j.bprint.2021.e00148
155. Dionysios EM (2018) Advanced technologies in manufacturing 3d-layered structures for defense and aerospace. In: Charles AO (ed) Lamination. IntechOpen, Rijeka, p Ch. 5. https://doi.org/10.5772/intechopen.74331
156. Jakus AE, Rutz AL, Shah RN (2016) Advancing the field of 3D biomaterial printing. Biomed Mater 11(1):014102. https://doi.org/10.1088/1748-6041/11/1/014102
157. Chua CK, Yeong WY (2014) Bioprinting, vol 1. World Scientific Series in 3D Printing, World Scientific. https://doi.org/10.1142/9193
158. Lin L, Luo Y, Chen Q, Lai Q, Zheng Q (2020) Redox-modulated colorimetric detection of ascorbic acid and alkaline phosphatase activity with gold nanoparticles. Luminescence 35(4):542–549. https://doi.org/10.1002/bio.3749
159. Gopinathan J, Noh I (2018) Recent trends in bioinks for 3D printing. Biomater Res 22(1):11. https://doi.org/10.1186/s40824-018-0122-1
160. Rutz AL, Hyland KE, Jakus AE, Burghardt WR, Shah RN (2015) A multimaterial bioink method for 3D printing tunable, cell-compatible hydrogels. Adv Mater 27(9):1607–1614. https://doi.org/10.1002/adma.201405076
161. Pati F, Cho D-W (2017) Bioprinting of 3D tissue models using decellularized extracellular matrix bioink. In: Koledova Z (ed) 3D cell culture: methods and protocols. Springer New York, New York, pp 381–390. https://doi.org/10.1007/978-1-4939-7021-6_27
162. Lima TPL, Canelas CAA, Concha VOC, Costa FAM, Passos MF (2022) 3D bioprinting technology and hydrogels used in the process. Journal of functional. Biomaterials 13(4). https://doi.org/10.3390/jfb13040214
163. Ong CS, Yesantharao P, Huang CY, Mattson G, Boktor J, Fukunishi T, Zhang H, Hibino N (2018) 3D bioprinting using stem cells. Pediatr Res 83(1):223–231. https://doi.org/10.1038/pr.2017.252

164. Jiang T, Carbone EJ, Lo KWH, Laurencin CT (2015) Electrospinning of polymer nanofibers for tissue regeneration. Prog Polym Sci 46:1–24. https://doi.org/10.1016/j.progpolymsci.2014.12.001
165. Burdick JA, Prestwich GD (2011) Hyaluronic acid hydrogels for biomedical applications. Adv Mater 23(12):H41–H56. https://doi.org/10.1002/adma.201003963
166. Zhang XN, Zheng Q, Wu ZL (2022) Recent advances in 3D printing of tough hydrogels: a review. Compos Part B 238:109895. https://doi.org/10.1016/j.compositesb.2022.109895
167. Arif ZU, Khalid MY, Tariq A, Hossain M, Umer R (2024) 3D printing of stimuli-responsive hydrogel materials: literature review and emerging applications. Giant 17:100209. https://doi.org/10.1016/j.giant.2023.100209
168. Huang J, Huang Z, Xiong J, Xia J, Wang Y, Yang L, Liang Y (2023) 3D bioprinted scaffolds of polysaccharide hydrogels in osteochondral and cartilage tissue engineering. Des Monom Polym 26(1):258–272. https://doi.org/10.1080/15685551.2023.2284482
169. Yan J, Miao Y, Tan H, Zhou T, Ling Z, Chen Y, Xing X, Hu X (2016) Injectable alginate/hydroxyapatite gel scaffold combined with gelatin microspheres for drug delivery and bone tissue engineering. Mater Sci Eng C Mater Biol Appl 63:274–284. https://doi.org/10.1016/j.msec.2016.02.071
170. Karageorgiou V, Kaplan D (2005) Porosity of 3D biomaterial scaffolds and osteogenesis. Biomaterials 26(27):5474–5491. https://doi.org/10.1016/j.biomaterials.2005.02.002
171. Hu X, Wang Y, Tan Y, Wang J, Liu H, Wang Y, Yang S, Shi M, Zhao S, Zhang Y, Yuan Q (2017) A difunctional regeneration scaffold for knee repair based on aptamer-directed cell recruitment. Adv Mater 29(15). https://doi.org/10.1002/adma.201605235
172. Poldervaart MT, Wang H, van der Stok J, Weinans H, Leeuwenburgh SC, Öner FC, Dhert WJ, Alblas J (2013) Sustained release of BMP-2 in bioprinted alginate for osteogenicity in mice and rats. PLoS One 8(8):e72610. https://doi.org/10.1371/journal.pone.0072610
173. Distler T, Solisito AA, Schneidereit D, Friedrich O, Detsch R, Boccaccini AR (2020) 3D printed oxidized alginate-gelatin bioink provides guidance for C2C12 muscle precursor cell orientation and differentiation via shear stress during bioprinting. Biofabrication 12(4):045005. https://doi.org/10.1088/1758-5090/ab98e4
174. Balakrishnan B, Jayakrishnan A (2005) Self-cross-linking biopolymers as injectable in situ forming biodegradable scaffolds. Biomaterials 26(18):3941–3951. https://doi.org/10.1016/j.biomaterials.2004.10.005
175. Wu X, Zhou M, Jiang F, Yin S, Lin S, Yang G, Lu Y, Zhang W, Jiang X (2021) Marginal sealing around integral bilayer scaffolds for repairing osteochondral defects based on photocurable silk hydrogels. Bioact Mater 6(11):3976–3986. https://doi.org/10.1016/j.bioactmat.2021.04.005
176. Wei P, Xu Y, Gu Y, Yao Q, Li J, Wang L (2020) IGF-1-releasing PLGA nanoparticles modified 3D printed PCL scaffolds for cartilage tissue engineering. Drug Deliv 27(1):1106–1114. https://doi.org/10.1080/10717544.2020.1797239
177. Hong S, Sycks D, Chan HF, Lin S, Lopez GP, Guilak F, Leong KW, Zhao X (2015) 3D printing: 3D printing of highly stretchable and tough hydrogels into complex, cellularized structures. Adv Mater 27(27):4034. https://doi.org/10.1002/adma.201570182
178. Daly AC, Critchley SE, Rencsok EM, Kelly DJ (2016) A comparison of different bioinks for 3D bioprinting of fibrocartilage and hyaline cartilage. Biofabrication 8(4):045002. https://doi.org/10.1088/1758-5090/8/4/045002
179. Park JY, Choi JC, Shim JH, Lee JS, Park H, Kim SW, Doh J, Cho DW (2014) A comparative study on collagen type I and hyaluronic acid dependent cell behavior for osteochondral tissue bioprinting. Biofabrication 6(3):035004. https://doi.org/10.1088/1758-5082/6/3/035004
180. Little CJ, Kulyk WM, Chen X (2014) The effect of chondroitin sulphate and hyaluronic acid on chondrocytes cultured within a fibrin-alginate hydrogel. J Funct Biomater 5(3):197–210. https://doi.org/10.3390/jfb5030197
181. Liang Y, Xu X, Li X, Xiong J, Li B, Duan L, Wang D, Xia J (2020) Chondrocyte-targeted MicroRNA delivery by engineered exosomes toward a cell-free osteoarthritis therapy. ACS Appl Mater Interfaces 12(33):36938–36947. https://doi.org/10.1021/acsami.0c10458

182. Liang Y, Duan L, Lu J, Xia J (2021) Engineering exosomes for targeted drug delivery. Theranostics 11(7):3183–3195. https://doi.org/10.7150/thno.52570
183. Dell AC, Wagner G, Own J, Geibel JP (2022) 3D bioprinting using hydrogels: cell inks and tissue engineering applications. Pharmaceutics 14(12). https://doi.org/10.3390/pharmaceutics14122596
184. Chen H, Cheng R, Zhao X, Zhang Y, Tam A, Yan Y, Shen H, Zhang YS, Qi J, Feng Y, Liu L, Pan G, Cui W, Deng L (2019) An injectable self-healing coordinative hydrogel with antibacterial and angiogenic properties for diabetic skin wound repair. NPG Asia Mater 11(1):3. https://doi.org/10.1038/s41427-018-0103-9
185. Liu X, Zhou S, Cai B, Wang Y, Deng D, Wang X (2022) An injectable and self-healing hydrogel with antibacterial and angiogenic properties for diabetic wound healing. Biomater Sci 10(13):3480–3492. https://doi.org/10.1039/D2BM00224H
186. Johnson BN, Lancaster KZ, Zhen G, He J, Gupta MK, Kong YL, Engel EA, Krick KD, Ju A, Meng F, Enquist LW, Jia X, McAlpine MC (2015) 3D printed anatomical nerve regeneration pathways. Adv Funct Mater 25(39):6205–6217. https://doi.org/10.1002/adfm.201501760
187. Tao J, Hu Y, Wang S, Zhang J, Liu X, Gou Z, Cheng H, Liu Q, Zhang Q, You S, Gou M (2017) A 3D-engineered porous conduit for peripheral nerve repair. Sci Rep 7(1):46038. https://doi.org/10.1038/srep46038
188. Maiti B, Díaz Díaz D (2018) 3D printed polymeric hydrogels for nerve regeneration. Polymers (Basel) 10(9). https://doi.org/10.3390/polym10091041

Innovative Bioengineering Solutions for Otolaryngology Management: Customized Additive Manufacturing with Biomaterials

14

Rajdip Goswami, Subhajit Das, Sourav Ghosh, Biswajit Basu, Swarupananda Mukherjee, and Sudarshan Singh

Abstract

A novel approach to managing otolaryngology involves using additive manufacturing techniques with biomaterials to produce customized bioengineered items. The medical specialty of otolaryngology, which treats conditions affecting the ear, nose, and throat, frequently needs to provide patients with exact and customized therapy. The specific anatomical variances and needs of each patient may not be adequately addressed by traditional approaches. This novel technology provides new levels of customization and precision by utilizing modern biomaterials and additive manufacturing, also referred to as 3D printing. A patient's anatomy and pathology can be considered when creating custom implants, prosthetics, and other devices thanks to the integration of patient-specific data, such as medical imaging. Additionally, the danger of rejection or problems is

R. Goswami · S. Ghosh · B. Basu (✉)
Department of Pharmaceutical Technology, School of Health and Medical Sciences, Adamas University, Barasat, Kolkata, WB, India

S. Das
Department of Allied Health, School of Health and Medical Sciences, Adamas University, Barasat, Kolkata, WB, India

S. Mukherjee
Department of Pharmaceutical Technology, NSHM Knowledge Campus, Kolkata, West Bengal, India
e-mail: swarupananda.mukherjee@nshm.com

S. Singh
Faculty of Pharmacy, Chiang Mai University, Chiang Mai, Thailand

Office of Research Administration, Chiang Mai University, Mueang Chiang Mai, Thailand

S. Singh et al. (eds.), *Biomaterial-based Additive Manufacturing in Tissue Engineering and Regeneration*, Biomaterials, Bioengineering and Sustainability 7,
https://doi.org/10.1007/978-3-031-96070-3_14

reduced since biomaterials chosen for their biocompatibility and biomechanical qualities guarantee appropriate integration andj functionality within the patient's body. This new era in otolaryngology, where personalized solutions can result in better patient outcomes, shorter surgical times, and higher quality of life, is heralded by the synergistic combination of biology and technology. In the realm of otolaryngology, the chapter lays the groundwork for future research and development of biomaterials-based additive manufacturing, offering promising developments in patient care and treatment methods.

Keywords

Otolaryngology · Additive manufacturing · Biomaterials · Customization · Patient-specific data

Abbreviations

ENT	Ear, Nose, and Throat
AM	Additive Manufacturing
CT	Computed Tomography
MRI	Magnetic Resonance Imaging
CAD	Computer-Aided Design
CAM	Computer-Aided Manufacturing
FDA	Food and Drug Administration
PLA	Polylactic Acid
PCL	Polycaprolactone
PMMA	Polymethyl Methacrylate
PET	Positron Emission Tomography
BiPAP	Bilevel Positive Airway Pressure
CPAP	Continuous Positive Airway Pressure
QC	Quality Control
BAM	Bio-Additive Manufacturing
ECM	Extracellular Matrix
BPO	Bio-Printed Organs
DNP	Digital Near-infrared Photopolymerization
3D	Three-Dimensional
4D	Four-Dimensional

14.1 Introduction

Additive manufacturing process (3D and 4D) mark on a journey into the forefront of modern innovation with bio-additive manufacturing, where the realms of biology and technology intertwine to redefine the future [1]. Explore the captivating world of 3D and 4D printing, where each layer unveils the potential to create living, dynamic structures. Enter the dynamic world of bio-additive manufacturing, merging biology and technology on the frontier of innovation. Explore a future were life and synthetic blend seamlessly, unlocking endless potential with each layer. Bio-additive manufacturing, known as 3D and 4D printing, revolutionizes object creation. Unlike traditional methods, it builds layer by layer, incorporating biological materials for precise customization and pioneering bio-mimicry [2]. Bio-additive manufacturing thrives on biomaterials—derived from life or mimicking it. From proteins to living cells, they are sculpted with precision using 3D printing, crafting intricate and functional objects from intricate digital blueprints [3, 4]. The evolution extends beyond 3D printing: welcome to 4D printing, where time becomes a factor. Objects adapt to stimuli, offering self-repair and adaptive biomedical solutions, unlocking a realm of endless possibilities. Imagine a 4D-printed stent in a blood vessel, dynamically adapting to flow changes for peak efficiency. Envision scaffolds, seeded with cells, growing tissue on demand, heralding a transformative era for organ transplants and tissue repair [5]. Beyond healthcare, bio-additive manufacturing sparks innovation. Architects craft sustainable, self-healing biopolymer structures. Aerospace pioneers create lightweight, biomimetic materials, reshaping aircraft construction and fuel efficiency [6]. As we delve deeper into bio-additive manufacturing, ethical dilemmas emerge: the impact on life forms, human-machine boundaries, and identity. Yet, responsible innovation promises a future of sustainability, personalized medicine, and technological progress. Bio-additive manufacturing propels us into a boundless future, blurring biology and technology. With each printed layer, we approach a world where natural and artificial merge, unlocking limitless innovation [7, 8].

The history of additive manufacturing began in the 1980s, pioneering 3D printing for prototyping and rapid manufacturing [9]. Key patents in the 1990s laid the foundation for diverse techniques. The 2010s saw the emergence of 4D printing, adding time as a dimension for dynamic object transformation. Today, additive manufacturing spans industries, from automotive to healthcare, driven by material science and software advancements. With its promise of innovation and creativity, additive manufacturing continues to revolutionize design and production processes, shaping a future where customization and sustainability thrive [10]. Table 14.1 illustrates summarizes the history of additive manufacturing [11].

The tissue engineering is a cutting-edge technology, where the potential for ground-breaking medical breakthroughs is there with scaffolds and bio-printed

Table 14.1 History of additive manufacturing

Decade	Milestones
1980s	Pioneering of 3D printing for prototyping and rapid manufacturing.
1990s	Key patents laying the foundation for diverse additive manufacturing techniques.
2000s	Expansion of additive manufacturing applications across various industries.
2010s	Emergence of 4D printing, introducing time as a dimension for dynamic object transformation.
Present	Additive manufacturing spans industries, driven by material science and software advancements.
Future	Promise of continued innovation and creativity, revolutionizing design, and production processes.

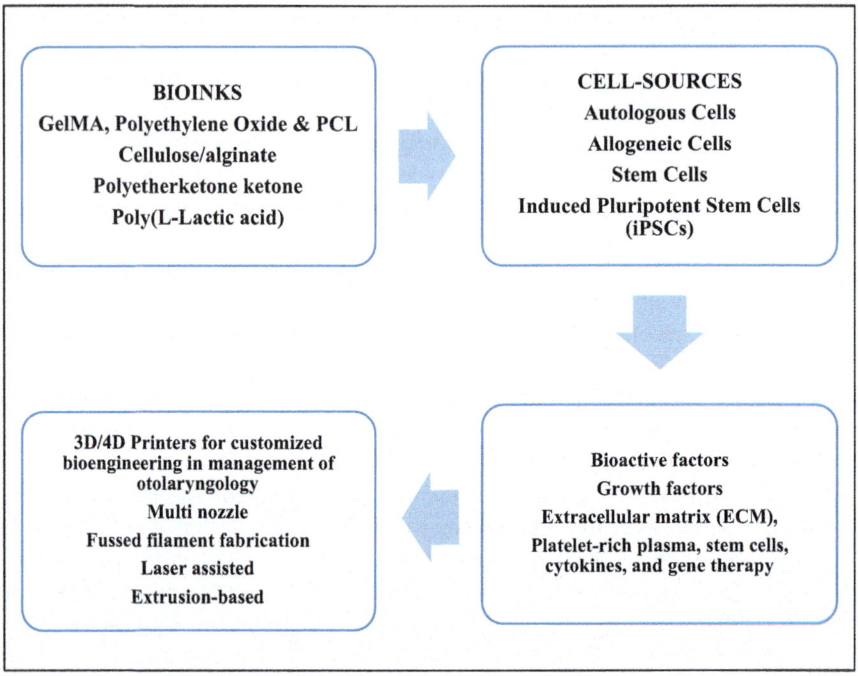

Fig. 14.1 Overview on additive manufacturing process (3D and 4D). (Created by using Biorender.com)

organs. Tight structures called scaffolds, which mimic the extracellular matrix found naturally in the body, offer a supportive environment for tissue regeneration and cell proliferation. Biological materials are precisely layered to build customized scaffolds that are suited to tissues or organs by utilizing cutting-edge 3D printing technology [11]. These scaffolds provide the building blocks for the creation of bio-printed organs, which imitate the structure and capabilities of real tissues. Bio-printed organs, which overcome the obstacles of donor scarcity and rejection, provide hope for patients in need of transplants, from skin grafts to heart valves. Step into the future of medicine, where customized regenerative medicines and improved quality of life are made possible by scaffold-based tissue engineering [12]. Figure 14.1 throws light on additive manufacturing process (3D and 4D)-bioprinting.

14.2 Bio-inks

Embark on a journey with bio-ink and biological printing, revolutionizing medicine. Mimicking native tissues, bio-ink combines cells and polymers for 3D printing living tissues and organs. This breakthrough holds promise for personalized solutions in regenerative medicine, shaping the future of healthcare and biotechnology [13].

14.3 Advantage of Bio-printed Product over Conventional Additive Manufacturing Product

Comparing bio-printed items to traditional additive manufacturing, there are clear benefits. They can produce more lifelike and useful tissue and organ structures by closely resembling the intricate microenvironments of natural tissues [14]. Furthermore, bio-printed goods can be customized to fit each patient's anatomical and physiological traits, improving compatibility, and lowering the possibility of rejection. Additionally, living cells and growth hormones can be incorporated via bio-printing, which facilitates tissue integration and regeneration after implantation. Because of these benefits, bio-printed goods are extremely significant in the field of regenerative medicine [15]. They provide specialized, more accurate, and effective solutions for tissue healing, organ replacement, and drug testing.

Bio-printed products represent a significant advancement in the realm of tissue healing, organ replacement, and drug testing. These products offer a range of advantages over traditional additive manufacturing methods:

A significant benefit lies in their capacity to emulate the intricate structure and operational characteristics of native tissues and organs. This biological mimicry allows bio-printed products to closely resemble their biological counterparts, providing a more effective solution for medical applications [16]. Biocompatibility is another crucial aspect of bio-printed products. They are often crafted from materials that are compatible with living organisms, reducing the risk of rejection or adverse reactions when implanted [17]. Bio-printed products also excel in tissue integration, facilitating better healing processes and lowering the chances of complications. This integration with surrounding tissues enhances the overall effectiveness of these products in medical treatments [18]. The ability to personalize bio-printed products based on individual patient needs is a significant benefit. This customization allows for tailored medical treatments, leading to improved outcomes for patients [19]. Moreover, bio-printed products hold regenerative potential by stimulating tissue repair and regeneration. This capability opens new avenues for treating injuries and degenerative diseases effectively [20]. In the realm of drug testing, bio-printed tissues offer a more accurate platform for studying how drugs interact with human biology. This advancement provides valuable insights into drug efficacy and safety, benefiting the field of pharmaceutical development [21]. Advantages of bio printed products over conventional additive manufacturing product has been summarized in Table 14.2 Overall, bio-printed products promise to advance regenerative medicine, personalized healthcare, and drug discovery. Their superior performance and functionality compared to traditional additive manufacturing products make them a compelling option for various medical applications [28, 29].

Table 14.2 Advantage of bio-printed product over conventional additive manufacturing product

Advantage	Bio-printed products	Traditional additive manufacturing products	References
Mimicry of natural tissues and organs	Closely resemble the complexity and functionality of natural tissues and organs.	Limited ability to replicate the intricate microenvironments of biological structures.	[22]
Biocompatibility	Crafted from materials compatible with living organisms, reducing the risk of rejection or adverse reactions.	May use materials that are not inherently biocompatible, increasing the risk of rejection.	[23]
Tissue integration	Facilitate better healing processes and lower the chances of complications due to improved integration with surrounding tissues.	Integration with surrounding tissues may be limited, potentially leading to complications.	[24]
Customization	Can be personalized based on individual patient needs, leading to tailored medical treatments and improved outcomes.	Limited ability to customize products to fit specific patient anatomical and physiological traits.	[25]
Regenerative potential	Stimulate tissue repair and regeneration, offering promising solutions for treating injuries and degenerative diseases.	May lack the ability to promote tissue regeneration, limiting their effectiveness in regenerative medicine.	[26]
Drug testing	Provide a more accurate platform for studying drug interactions with human biology, enhancing drug efficacy and safety assessments.	Less effective in mimicking human biology for drug testing purposes, potentially leading to inaccurate results.	[27]

14.4 Advancements and Challenges in Bio-printed Additive Manufacturing for Tissue Engineering

In the realm of sophisticated bio-printed additive manufacturing, bio-ink assumes the role of the future tissue regeneration architect. This review aims to highlight the significant advancements, The practical uses, and captivating hurdles of this groundbreaking method in tissue engineering and regenerative medicine, with an emphasis on the state-of-the-art in nasal, tracheal, and craniofacial restorations [30]. Our initial task is to delve deeply into the state-of-the-art additive manufacturing bioprinting industry. Come explore the most recent advancements in this fascinating topic and the complex mechanisms that provide the groundwork for the miracles of tissue engineering to come [31]. We next focused on using bio-ink in the field of craniofacial, tracheal, and nasal bone restorations [32]. See how bio-ink can revolutionize reconstructive surgery by providing customized solutions for these critical anatomical areas. Not only that, but we are still on the path, carefully evaluating the many advantages and constraints of bio-printed constructions in tissue engineering. We analyse the benefits and the challenges of better biocompatibility and regulatory obstacles.

Step into the future of innovation where bio-printed additive manufacturing is the way to go, providing a novel approach to the creation of complex tissue structures. Imagine these biomaterials are meticulously applied layer by layer to generate three-dimensional formations that accurately mirror both the structure and operational capabilities of human tissues. With its accuracy and potential, this bio-printed additive manufacturing technique is redefining the area of tissue engineering [33–35].

14.5 Applications in Nasal, Tracheal, and Craniofacial Bone Reconstructions

In the arena of nasal, tracheal, and craniofacial bone reconstructions, the use of bio-printed additive manufacturing with bio-ink opens doors to exciting possibilities. Consider this: in nasal reconstructions, personalized nasal scaffolds can now be crafted to precisely fit each patient's anatomy. This not only enhances the aesthetic results but also improves functionality significantly. The accurate creation of bio-printed tracheal stents or grafts aids in tracheal restorations by offering support and encouraging tissue regeneration. Furthermore, bio-printed structures provide a customized method for mending abnormalities originating from trauma, congenital deformities, or surgical treatments in craniofacial bone reconstructions [36].

14.6 Techniques Involved with Bio-printed Additive Manufacturing in Management of Otolaryngology

14.6.1 Bioinks

Bioinks are an exciting innovation in the field of tissue engineering and regenerative medicine. These are biocompatible materials, often in the form of hydrogels or polymers, designed to act as supportive scaffolds for the growth and differentiation of cells into functional tissues. The primary purpose of bio-inks is to create an environment that closely mimics the natural tissue structure, providing the necessary cues and support for cells to organize and develop into tissues such as cartilage, bone, or skin [37].

This is achieved by incorporating cells and growth factors into the bioink and customizing it to the specific needs of the tissue being regenerated. One of the remarkable features of bioinks is their ability to respond to external stimuli, leading to what is known as "4D printing" in the field of tissue engineering. This concept expands upon 3D printing by incorporating the element of time, where the printed structure can change its shape, properties, or function over time in response to stimuli [38]. The current uses of bioinks in the various medical industries are highlighted in Fig. 14.2.

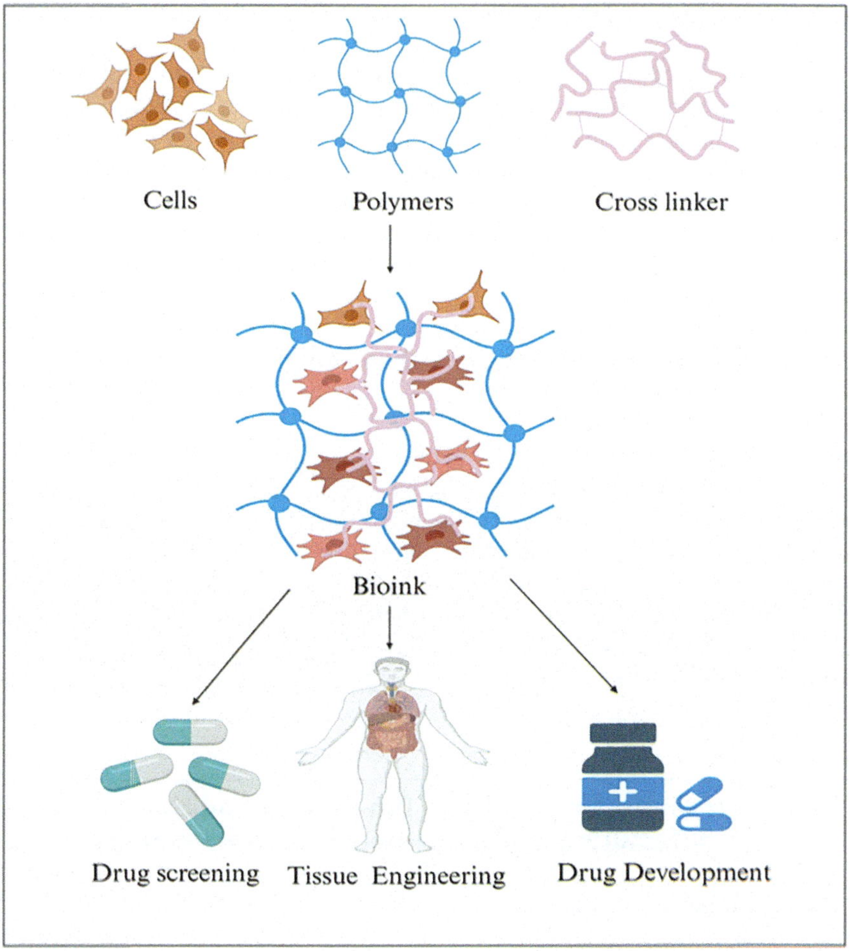

Fig. 14.2 Current applications of bioinks across various medical industries. (Created by using Biorender.com)

The stimuli's can be classified into three main categories as highlighted in Fig. 14.3.

A. *Chemical stimuli-responsive materials*
B. *Biological stimuli-responsive materials*
C. *Physical stimuli-responsive materials*

Chemical Stimuli-Responsive Materials These bio-inks are designed to change in response to specific chemical signals. For example, they might respond to changes in pH, the presence of certain ions, or the introduction of biochemical molecules. This can trigger processes such as cell differentiation or the release of growth

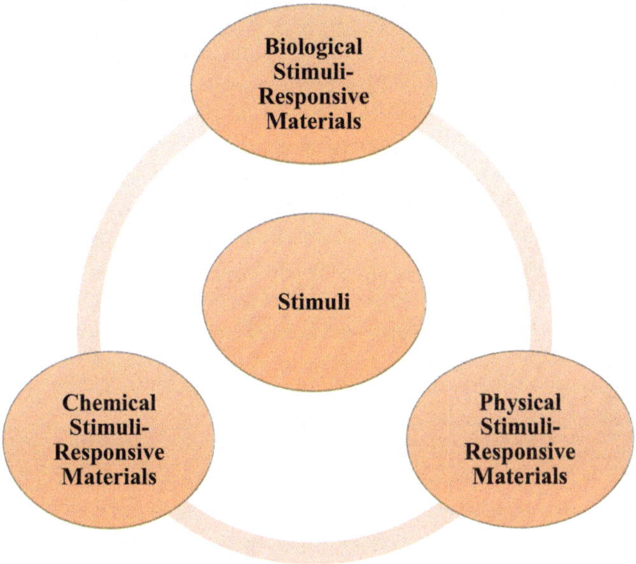

Fig. 14.3 Stimuli involved with bio-printed additive manufacturing. (Created by using Biorender.com)

factors at specific times and locations [39]. Responsive materials refer to substances, Instances include hydrogels responsive to ions and polymers altering their structure, characteristics, or behaviour upon exposure to chemical stimuli, such as fluctuations in pH [40]. Figure 14.4 highlights Chemical as well as mechanical stimulation can promote tissue development.

Ion-Sensitive Hydrogels Various research studies have demonstrated the feasibility of creating scaffolds capable of supporting structures filled with cells on a large clinical scale. In order to produce scaffolds with adequate strength and adaptable mechanical properties, bioprinting allows for the use of crosslinking agents like Zn^{2+} and Ca^{2+}. The manipulation of poly (acrylonitrile) chains through bidirectional dipole-dipole interactions with Zn^{2+} led to the development of A newly developed shape memory hydrogel, characterized by its remarkable strength and suitability for cellular applications, was created through the synthesis of imidazole-zinc ion bonding. By employing chelating agents to remove the zinc ions, it is possible to restore the enduring forms of these constructed configurations demonstrate their ability to retain memory bidirectionally. To maintain the integrity of the hydrogel layer containing cells, a temporary cylindrical form can be applied, allowing for cell culture alongside zinc ions. These hydrogels, which react to ion-sensitive crosslinking, offer novel avenues for producing shape-shifting constructs integrated with cells, particularly within the fields of tissue engineering and 4D bioprinting advancements [11, 40–43].

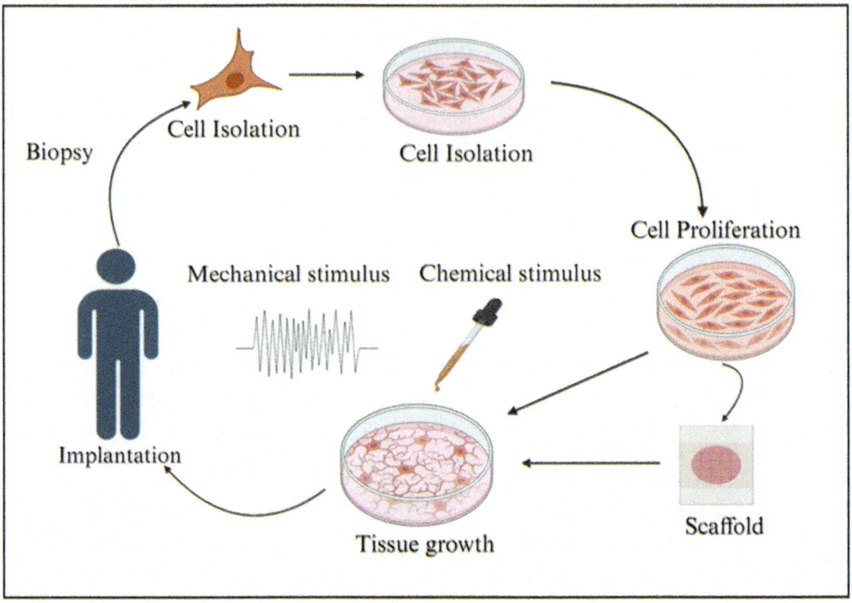

Fig. 14.4 Tissue growth is stimulated by both mechanical and chemical stimuli. (Created by using Biorender.com)

pH-Responsive Materials Investigations by researchers focus on chemical groups like carboxyl, pyridine, sulfone, and phosphate, which exhibit responsiveness to pH changes. These groups are harnessed in the creation of self-assembling structures, including pH-responsive materials capable of transitioning from spherical to spiral shapes within specific pH ranges. For instance, the polymer poly (N-isopropyl acrylamide) undergoes a shift from an electrostatic-repellent phase to forming globules upon neutralization of its functional group. Moreover, chitosan/TPP scaffolds with varying levels of cross-linked primary amine are developed for bone regenerative therapies, where their adsorption and release behaviour are pH-sensitive and contingent upon the degree of amine cross-linking. These newly engineered scaffolds, crafted from pH-responsive materials, demonstrate malleable shapes and mechanical properties, presenting promising alternatives to 4D bioprinting in biological applications [44].

Physical Stimuli-Responsive Materials Bio-inks in this category react to physical changes in their environment. This could include variations in temperature, light exposure (photo responsive), or mechanical forces. For instance, a bio-ink might change its stiffness or porosity When illuminated with light of a specific wavelength, allowing for precise control over tissue formation [45].

Temperature-Responsive Materials Commonly employed materials for inducing shape changes in bio-printed objects are activated by exposure to light of specific wavelengths. Examples of such materials encompass poly (N-isopropyl acrylamide), polyethylene glycol, chitosan, gelatine, and poly(caprolactone) (PCL) [46]. When subjected to appropriate conditions, these materials exhibit shape alterations, which are particularly advantageous for fabricating self-adjusting implants for minor bone defects. For instance, a tracheal stent could safely expand post-implantation, reducing potential procedural risks. Researchers like Apsite et al. have devised multi-layer scaffolds utilizing temperature-sensitive polymers such as poly (N-isopropyl acrylamide) and PCL. These scaffolds naturally assume tubular structures with PCL serving as the inner layer in aqueous environments at 37 °C, facilitating cell encapsulation. Coating these polymers with collagen enhances cell adhesion and viability, expanding the scaffold's applicability across diverse tissue engineering contexts [47].

Magnetic Responsive Materials Wei and colleagues devised a method involving poly (lactic acid) polymer solution infused with magnetic iron oxide nanoparticles to fabricate tubular structures with magnetically modifiable and recoverable attributes using additive manufacturing (AM) technology. The application of a resistive magnetic field heated the iron oxide, exerting sufficient force to revert the temporary shape to its original form [48]. By adjusting the strength and alignment of the magnetic field during tissue engineering, it became feasible to manipulate the scaffold's structure and shape. This technology holds promise for various applications including aligning structural components, administering mechanical stimulation, and guiding stem cell differentiation [1]. Utilizing hydrogels manipulated through magnetic fields presents encouraging prospects for biomaterial printing due to their exceptional rheological characteristics. In an independent investigation, Zhu and collaborators pioneered the development of 4D-printed bioproducts responsive to magnetic fields by integrating Fe-based nanoparticles into poly (dimethyl siloxane). During the bioprinting process, which yields scaffolds with anisotropic features, the magnetized bio-prints have the potential to influence the orientation of the nanoparticles [49].

Electro-responsive materials Electro-responsive materials, primarily composed of polyelectrolytic polymers, demonstrate the ability to expand, compress, or bend under the influence of an electric field, with their properties being affected by both the intensity and direction of the field. Certain hydrogels, enriched with electrically conductive polymers such as poly(thiophene), poly(aniline), and poly(pyrrole), exhibit favourable biocompatibility and suitability for printing, opening avenues for potential applications in 4D printing. For example, the integration of 3D printing techniques with polypyrrole interfacial polymerization has led to the development of a conductive electroactive hydrogel. These printed structures are being employed in the creation of innovative neuro-prosthetic devices and bioelectronic designs, owing to their distinctive mechanical properties, advantageous electrical conductivity, and adaptable surface chemistry [50, 51].

Carbon-based nano-biomaterials like carbon nanotubes and graphene have attracted considerable attention in recent times for their role in exploring and controlling the behaviour and destiny of stem cells. Harnessing electrically conductive and biocompatible properties of carbon-based nano-biomaterials to fabricate stimuli-responsive 4D structures offers promising prospects for improving the regeneration of bone and brain tissues [52, 53].

Photo-Responsive Materials Experience mechanical alterations upon exposure to direct optical pulses, with light in near-infrared, infrared, and ultraviolet spectra triggering these biomaterials, frequently employed in areas such as tissue engineering and regulated drug delivery. The primary pathways for the response of photosensitive materials, which are extensively used to create active, shape-shifting 4D structures, involve photo-isomerization and photo-degradation of polymer chains [54–56].

In a particular investigation, Wei and associates employed a UV cross-linking agent to produce a tubular poly (lactic acid) structure with shape memory capabilities. Simultaneously, Kuang et al. explored a UV light-assisted bioink comprising a linear semi-crystalline polymer and urethane diacrylate, demonstrating remarkable shape memory and self-healing attributes essential for the advancement of 4D bioprinting methodologies [57].

Arakawa and colleagues conducted a separate study displaying the tailored photo-degradation of biomaterials for generating multicellular 3D endothelial vascular networks within hydrogels filled with cells. Employing a multiphoton lithography technique with adjustable 4D manipulation, they swiftly fabricated microchannel networks resembling the dimensions of actual human vasculature. These photo-responsive 4D bio-inks can authentically mimic the dynamic degradation characteristics of the extracellular matrix in practical settings [58].

Water-Responsive Materials Experience alterations in response to humidity levels, triggered by the presence of moisture. One approach to 4D printing of humidity-responsive scaffolds involves utilizing a composite material comprising a hydrogel matrix embedded with particles or fibres sensitive to humidity, enabling the scaffold to demonstrate controlled dimensional changes when exposed to varying humidity conditions. Among the commonly utilized materials for 4D printing, poly (ethylene glycol) (PEG) and hydrogels are favoured choices due to their responsiveness to moisture. An exemplary instance involves a nanocomposite that is both humidity-sensitive and thermo-responsive, crafted from a blend of PCL, PEG, and cellulose nanocrystals (CNCs). When submerged in water at 37 °C, this nanocomposite strip returns to its original shape by absorbing water and swelling, demonstrating its sensitivity to changes in environmental humidity [59]. Figure 14.5 highlights the various modes of stimuli-responsive materials, illustrating the different ways these

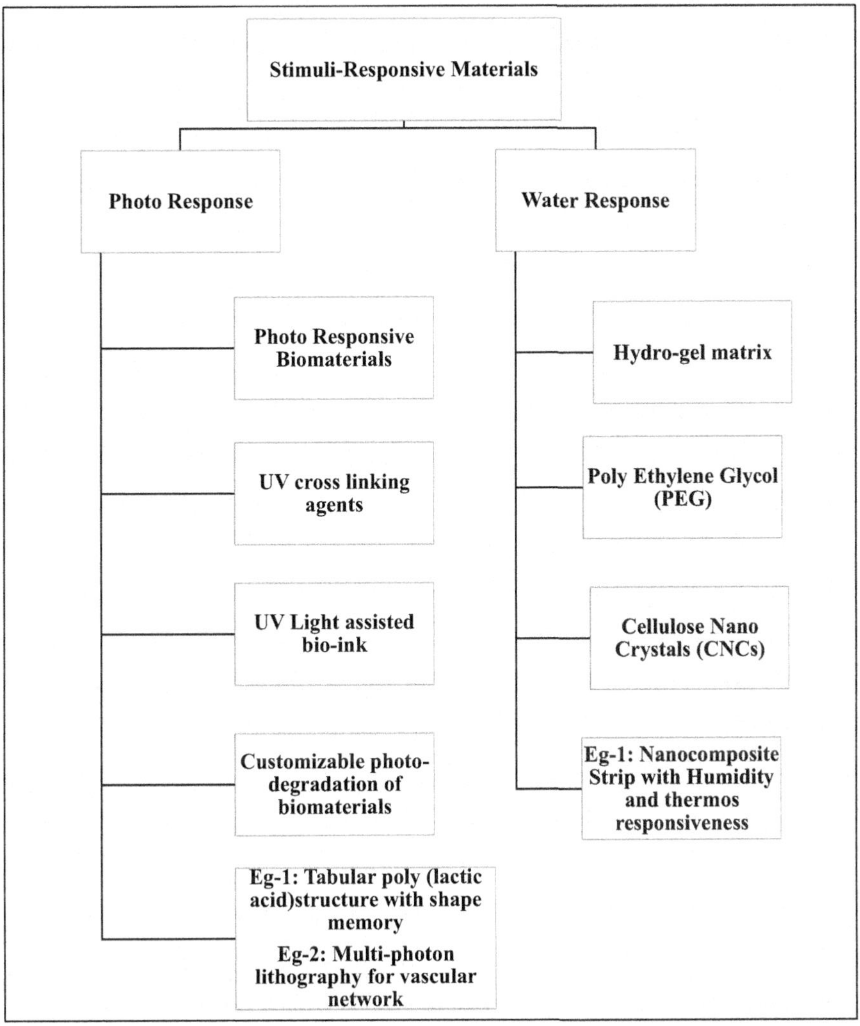

Fig. 14.5 Different mode of stimuli-responsive material. (Created by using Biorender.com)

materials can react to external stimuli and Different types of stimuli-responsive materials are showcased in Fig. 14.6.

Biological Stimuli-Responsive Material These bio-inks respond to biological cues present in the surrounding tissue or are introduced deliberately. This could involve interactions with proteins, enzymes, or other molecules naturally found in the body [60]. Figure 14.6 illustrates the different modes of stimuli-responsive material. By incorporating these biological triggers, bio-inks can create a microenvironment that supports cell adhesion, migration, and tissue development. In the exploration

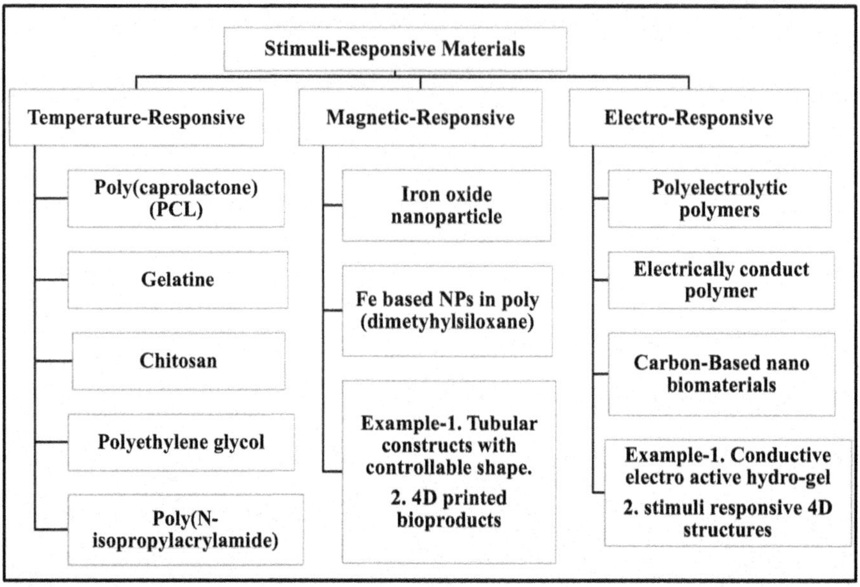

Fig. 14.6 Different stimuli-responsive material. (Created by using Biorender.com)

discussed by El-Husseiny and collaborators (2022), biologically responsive materials are developed utilizing bio-enzymes, glucose, and other biological triggers. Specific bio-enzymes have the capability to degrade biomolecules and peptide chains, triggering changes within polymers and resulting in increased swelling. Ionic interactions with substances such as polypeptides and polynucleotides, along with reversible hydrogen bonds, contribute to the gelation of shape memory hydrogels through ionic crosslinking [61]. A novel concept in 4D structure involves the incorporation of biologically active molecules like enzymes and antibodies within three-dimensional objects. Enzymes play a crucial role in various biological reactions, with those naturally occurring in the human body utilized to activate the shape memory properties of hydrogels. For optimal functionality of hydrogels, enzyme substrates are necessary to serve as crosslinkers or functional groups, as elucidated by Wang (2018). An important enzyme involved in breaking down extracellular matrix components is matrix metalloproteinase. Studies on hyaluronic acid-based hydrogels sensitive to matrix metalloproteinase have demonstrated notable properties, including significant cell attachment, as well as adjustable swelling and degradation characteristics [62].

14.6.2 Cell Sources

Here are several essential methods for bio-printing otolaryngological tissues for regeneration, with an emphasis on cell0 sources.

Autologous Cells These are the patient's own cells that were taken out. In bioprinting, autologous cells are desired to reduce the possibility of immunological rejection. Cells from a variety of sources, such as cartilage, bone marrow, adipose tissue, and the mucosal linings of the nose, throat, and ears, can be used in otolaryngology [62].

Allogeneic Cells These are not the patient's cells; they came from another donor. Obtaining allogenic cells has benefits in terms of availability and scalability. To avoid rejection, immunological compatibility concerns must be properly addressed [63].

Stem Cells These cells were taken from a donor who was not the patient. In terms of availability and scalability, allogenic cell sources can be beneficial. Nonetheless, to avoid rejection, immunological compatibility concerns must be properly handled [64].

Induced Pluripotent Stem Cells These are pluripotent somatic cells—like embryonic stem cells—those have undergone reprogramming. The benefit of patient-specificity that comes with embryonic stem cells is absent withiPSCs. They can develop into diverse cell types related to otolaryngology, like epithelial cells for mucosal repair or chondrocytes for cartilage regeneration [65].

Extracellular Matrix (ECM) Bioinks Synthetic or natural biomaterials that replicate the natural milieu of tissues make up ECM bioinks. They offer biochemical cues and structural support that are necessary for cell adhesion, proliferation, and differentiation. Otolaryngology bioprinting can benefit from the use of extracellular matrix (ECM) bioinks produced from decellularized tissues, such as decellularized cartilage or extracellular matrix from the ear, nose, or throat [66].

14.6.3 Bioactive Factors

Bioactive substances like growth factors and cytokines are crucial for controlling cell behavior and tissue development. In otolaryngology, these substances are vital for enhancing tissue regeneration using bio-printed structures. They help cells differentiate properly, promote blood vessel growth, and encourage nerve development in ear, nose, and throat tissues. By combining advanced cell sourcing techniques, including stem cells and extracellular matrix bioinks, with bioactive factors, bioprinting offers promising ways to repair and reconstruct damaged tissues in otolaryngology, leading to more personalized treatments and better outcomes [67]. In otolaryngology, bioactive substances are essential for bio-printed tissue regeneration.

Growth Factors These proteins control the growth and migration of cells. Growth factors such as TGF-β, FGF, and EGF are added to bio-inks in otolaryngology to aid in the regeneration of injured tissues, such as the vocal cords and inner ear [68].

Components of the Extracellular Matrix Collagen and hyaluronic acid provide signals and support for cell division and proliferation. By mimicking the natural surroundings of the target tissues, these components are included into bio-inks to produce printed structures that more effectively facilitate tissue regeneration [64].

Chemicals Called Chemokines and Cytokines It regulate immune responses and tissue repair. Their incorporation into bio-inks promotes tissue regeneration, reduces inflammation, and alters the immune environment [69].

Antibacterial Agents To stop infections, bio-printed structures can be infused with antibiotics or antimicrobial peptides. This holds significance for otolaryngology procedures where infections pose a risk [69].

Angiogenic Factors These factors, which include VEGF, encourage the growth of blood vessels, which gives the tissues undergoing regeneration enough oxygen and nourishment. Tissue integration and survival are enhanced when they are included into bio-inks. Bio-printed constructs, which offer promising options for tissue repair and reconstruction in otolaryngology, provide an environment that encourages cell proliferation, differentiation, and tissue regeneration by adding these bioactive elements into bio-ink formulations [70]. These Several critical techniques for bioprinting otolaryngological tissues for regeneration—with a focus on cell sources—are presented in Fig. 14.7.

14.7 Current Methods of 3D Bioprinting

3D bio-printing has emerged as a groundbreaking technology in tissue engineering and regenerative medicine, offering promising avenues for creating complex, functional tissues, and organs. Several methods are utilized in 3D bio-printing, each with its unique approach and advantages. Here, we will briefly describe three current methods: micro extrusion, droplet 3D bio-printing and light-based 3D bio-printing, including laser-assisted and stereo-lithography bio-printing [71]. Several crucial techniques for bio-printing otolaryngological tissues for regeneration are shown in Fig. 14.7, with a focus on cell sources.

Micro Extrusion Bio-printing Micro extrusion bio-printing operates on the principle of depositing bio-inks through a small nozzle onto a substrate, layer by layer,

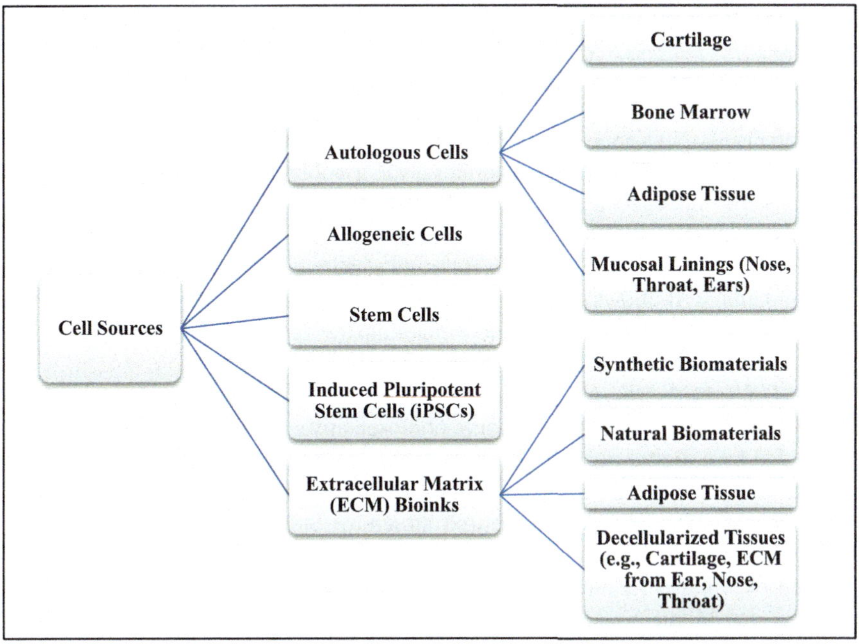

Fig. 14.7 Several essential methods for bio-printing otolaryngological tissues for regeneration, with an emphasis on cell sources. (Created by using Biorender.com)

to build three-dimensional structures. This method typically involves pneumatic or mechanical forces to push the bio-ink through the nozzle, allowing precise control over the deposition process. One of the main advantages of micro extrusion bio-printing is its compatibility with a wide range of biomaterials, including hydrogels, cell-laden inks, and bioactive factors. Additionally, it offers high printing speed and scalability, making it suitable for fabricating large-scale tissue constructs [72].

Droplet 3D Bio-printing Droplet 3D bio-printing, also known as inkjet bioprinting, utilizes a technology like traditional inkjet printing but with bio-inks containing living cells and biomaterials. In this method, tiny droplets of bio-ink are ejected from a print head onto a substrate, forming patterns and layers according to a digital design. Droplet bioprinting offers high resolution and precise control over cell deposition, allowing the creation of intricate tissue architectures. Moreover, it enables the co-printing of multiple cell types and bioactive agents within the same construct, facilitating the development of complex, heterogeneous tissues [73].

Light-Based 3D Bio-printing Light-based 3D bio-printing encompasses various techniques that utilize light to solidify or crosslink photosensitive biomaterials, forming desired structures layer by layer. Two prominent methods within this category are laser-assisted bio-printing and stereo-lithography bio-printing [74].

Laser-Assisted Bio-printing: Laser-assisted bio-printing involves the use of a laser to precisely focus energy onto a donor substrate coated with a bio-ink layer. This localized energy absorption generates a vapor bubble, which propels small droplets of bio-ink onto a receiving substrate, forming the desired pattern. Laser-assisted bio-printing offers high resolution and the ability to print a wide range of biomaterials, including living cells and sensitive bioactive factors, with minimal damage. It allows precise control over cell positioning and spatial distribution within the printed construct, enabling the fabrication of complex tissue architectures with cellular precision.

Stereo-Lithography Bio-printing: Stereo-lithography bio-printing utilizes a photopolymerization process to solidify photosensitive biomaterials layer by layer, guided by a digital light projection system. In this method, a UV light source is directed onto a photosensitive resin, triggering a chemical reaction that solidifies the material and forms the desired structure. Stereo-lithography bio-printing offers high resolution and the capability to fabricate intricate tissue geometries with excellent structural integrity. It allows the incorporation of multiple cell types and biomaterials within the same construct, enabling the development of heterogeneous tissues with precise control over composition and spatial arrangement [75].

In summary, micro extrusion, droplet 3D bio-printing, and light-based 3D bio-printing, including laser-assisted and stereo-lithography bio-printing, represent three prominent methods in current 3D bio-printing technologies. Each method offers distinct advantages and capabilities for fabricating complex, functional tissues, and organs for various biomedical applications, including tissue engineering, regenerative medicine, and drug screening. Continued advancements in these technologies hold the potential to revolutionize healthcare by enabling the production of patient-specific tissues and organs for transplantation and personalized medicine [76].

14.8 Challenges after Post Printing

Challenges in Customized Bioengineering in Otolaryngology Management via Post-Printing Biomaterials-Based Additive Manufactured Products because biomaterials-based additive manufacturing allows for the fabrication of customized bioengineered products that are suited to the specific demands of each patient, it has completely changed the management of otolaryngology [77]. Nevertheless, several issues that surface post-printing affect how well these items integrate and perform. The major post-printing challenges—biocompatibility, sterilization, accuracy, and integration with existing tissues—are covered in this work. When it comes to additively created items based on biomaterials, biocompatibility is still a major challenge after printing [78]. Although biocompatibility is a need for 3D printing materials, biocompatibility can also be influenced by surface roughness, porosity, and degradation kinetics. To improve biocompatibility, surface modification methods like chemical coating or plasma treatment are used. Furthermore, biocompatibility may be harmed by the release of harmful compounds during printing or post-processing procedures, requiring extensive testing and assessment [79]. Making sure 3D-printed biomaterials are sterile is essential to avoiding patient responses and infections. Sterilization procedures are complicated by the intricate geometries and porous materials of additively built goods. Intricate structures may be too small for traditional sterilization techniques like autoclaving or gamma irradiation to penetrate, leaving gaps in the sterilization process. The suitability of other methods, such as vapor sterilization using hydrogen peroxide or ethylene oxide, for use with 3D-printed materials is investigated. To expedite sterilization processes, new developments in in-situ sterilization techniques are being developed and incorporated into the printing process [80, 81]. For additive manufacturing to produce biomaterials-based goods with complex geometries and patient-specific designs, high precision and accuracy are necessary. However, issues including inconsistent material quality, nozzle clogging, and layer thickness can all compromise printing accuracy. To improve accuracy, techniques like slicing software optimization, printer performance monitoring, and parameter calibration are used. Moreover, the precision of additive manufacturing processes is enhanced by developments in printing technologies, such as high-resolution printing methods and multi-material printing [82]. The functional efficacy and long-term stability of 3D-printed biomaterials depend on their successful integration with native tissues. The mechanical qualities, surface features, and tissue reactions of printed implants and native tissues differ, posing challenges. Tissue integration is encouraged, and host responses are reduced by bio-mimetic design, surface changes, and bioactive substances. Furthermore, methods like tissue engineering and regenerative medicine aid in the creation of biomaterials that actively promote tissue remodeling and regeneration, improving integration with neighboring tissues. Otolaryngology poses problems in the selection of appropriate biomaterials for additive manufacturing because of the various functional requirements and intricate anatomical features of the targeted tissues. Important factors to take into account while choosing a material include biocompatibility, mechanical attributes, degradation dynamics, and manufacturability [83]. Furthermore, in order to avoid material mismatches and encourage smooth integration, compatibility between the various materials used in multi-material printing processes must be guaranteed [84]. These issues are addressed by developments in material science, such as the creation of innovative biomaterials and composite materials, which provide specialized solutions for particular otolaryngology applications. Significant problems exist in navigating regulatory procedures and securing authorization for devices based on 3D-printed biomaterials in otolaryngology management. Strict rules are imposed by regulatory organizations like the FDA on the development, production, and clinical application of medical devices, including items made by additive manufacturing. Satisfaction with ISO 13485 and ASTM F2503 standards is necessary to guarantee the effectiveness, safety, and quality of the product [85–87]. Moreover, regulatory clearance for 3D-printed goods requires proof that they are equivalent to their traditionally made equivalents through extensive testing and validation studies. Overcoming obstacles regarding cost-effectiveness, scalability, and clinical acceptability is necessary to move 3D-printed biomaterials-based solutions from research to clinical practice. In otolaryngology, the limited reimbursement regulations and high initial investment costs impede the widespread adoption of additive manufacturing technologies. Furthermore, to promote acceptance and confidence in these cutting-edge solutions, patients and healthcare professionals must be informed about the advantages and constraints of 3D-printed goods. In order to effectively solve these issues and enable the smooth integration of 3D-printed biomaterials into otolaryngology practice, cooperation between academic institutions, industry, and healthcare organizations is required. The creation and application of biomaterials-based additive manufactured products for tailored bioengineering in otolaryngology management faces many obstacles due to post-printing procedures [85–91]. To guarantee the safety, effectiveness, and clinical translation of 3D-printed items, addressing these issues calls for interdisciplinary cooperation, technological developments, and regulatory compliance. If these obstacles are overcome, additive manufacturing has enormous potential to transform the practice of otolaryngology by offering unique and customized patient care solutions [92].

14.9 Application of Additive Manufacturing in Management of Otolaryngology

Additive manufacturing, also known as 3D printing, has revolutionized various industries, including healthcare. In otolaryngology (ear, nose, and throat), additive manufacturing finds several applications beyond the commonly cited examples of customized implants, surgical planning, training, patient education, medical devices, research, and temporal bone models.

Auricular cartilage scaffolds have been fabricated using laser sintered 3D printing technology, as shown in Fig. 14.8.

Customized Implants Otolaryngology is being revolutionized by additive manufacturing, which creates customized implants based on each patient's individual anatomy. 3D-printed implants guarantee a flawless fit, improving surgical outcomes for anything from trauma to tumor removal. With better functionality and aesthetics promised for patients with congenital malformations or post-operative needs, these customized solutions herald a new era of precision in reconstructive treatments [93, 94].

Surgical Planning and Simulation Surgeons use 3D printing to create realistic models of their patients' throats, noses, and ears using medical images. Surgical teams can plan and practice complex procedures with these detailed models, honing their accuracy and efficiency. This technique promises to revolutionize

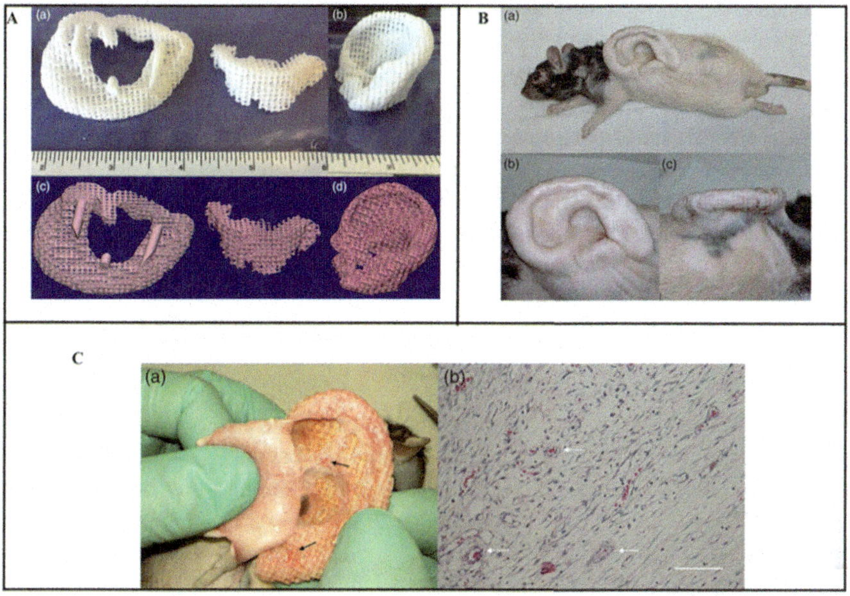

Fig. 14.8 Auricular cartilage scaffolds have been made via 3D printing with laser sintered technology. The methods consist of a single-stage PCL design (**A**b) and a two-stage lock-and-key design with PCL (**A**a). Furthermore, STL pictures depict the single-stage design (**A**d) and the two-stage lock-and-key design (**A**c). Side images (**B**c) complement lateral views (**B**a, **B**b) in showing the anatomical integrity of the ear scaffold. When the scaffold was harvested after 8 weeks, angiogenesis was evident (**C**a, **C**b). (This material is reprinted with permission under CCBY from [1])

otolaryngology's pursuit of excellence by streamlining operations, reducing risks, and optimizing results through pre-simulation of surgeries [95, 96].

Training and Education Vibrant anatomical models are moulded by innovative 3D printing for otolaryngology trainees. Immersion learning opportunities are provided by these replicas, which cover a wide range of diseases and subtle anatomic details. Aspiring ENT specialists accelerate their path to expertise by honing their abilities in a secure, controlled environment by imitating real-world conditions. Additive manufacturing, with its minutely detailed components, ushers in a new era of hands-on education in the field of ear, nose, and throat medicine [76, 83].

Patient Education 3D printed models provide patients with a clear understanding of their internal mechanisms. By demystifying medical language, these tactile reproductions encourage active participation and informed decision-making. Through gaining a visual understanding of their disease and suggested therapies, patients feel more empowered and take charge of their healthcare experience.

Understanding grows in this tactile domain, fostering patient-provider partnerships and opening doors to informed decision-making and ideal results [1, 63, 68].

Medical Devices and Instruments Using additive manufacturing to create custom instruments for ENT procedures opens new possibilities. Every gadget, from customized prosthesis to finely honed scalpels, is painstakingly made to meet the specific needs of each patient. With its revolutionary technologies, surgery becomes more precise and improves patient outcomes. Otolaryngologists confidently traverse difficult situations with personalized solutions at their disposal, ushering in a new era of individualized care and revolutionary outcomes [1, 63, 68].

Research and Development 3D printing spurs creativity and expedites otolaryngology research and development. This technology encourages quick prototyping and iterative improvement for everything from revolutionary implants to cutting-edge medicine delivery. With seamless collaboration, researchers incorporate patient and clinician feedback into each iteration. The future of ENT care is evolving with each print, driven by innovation and teamwork. Within this world of boundless potential, additive printing surfaces as the driving force behind revolutionary developments in otornolaryngoscopic and throat medicine [72].

Temporal Bone Models *3*D printing is used in otology to create complex temporal bone models. With the help of these lifelike replicas, ENT surgeons may now perform delicate surgeries like mastoidectomy and cochlear implantation with unprecedented precision. As each model painstakingly replicates the intricacies of the temporal bone, physicians engage in intense training and precise surgical planning. Expertise thrives in this tactile environment, leading to improved results and a new benchmark for care in ear surgery [59]. Pioneering otolaryngology, additive manufacturing unveils a realm of possibilities. From personalized treatments to surgical mastery, its precision reshapes care paradigms. As technology marches forward, so does our capacity for innovation, propelling patient care to unprecedented heights. With each advancement, additive manufacturing cements its place as a cornerstone of ENT excellence, promising a future where every patient receives tailored care and optimal outcomes [89]. The utilization of 3D and 4D bioprinting technologies is elucidated in Table 14.3.

14.9.1 Hearing Aids and Prosthetics

Additive manufacturing allows for the creation of custom-fit hearing aids and ear prosthetics. Traditional manufacturing methods often result in products that may not fit perfectly, leading to discomfort or reduced functionality. With 3D printing, audiologists can scan a patient's ear and produce a hearing aid that fits precisely,

Table 14.3 3D/4D bioprinting for otologic application using various polymers

Anatomic Region	3D/4D Bioprinting Techniques	Polymeric Materials Used	Cell Source	Animal Model Used in Assessment	Implant Position in Human	References
Otologic	Extrusion-based bioprinting, Inkjet-based bio-printing	Polycaprolactone (PCL), Polyethylene Glycol (PEG), Polylactic acid	Autologous chondrocytes, Mesenchymal stem cells (MSCs), Fibroblasts	Rodents (mice, rats), Canine models	Middle ear, Cochlear implants	[1, 92]
Nasal	Inkjet-based bioprinting, Laser-assisted bioprinting	Polycaprolactone (PCL), Polyethylene Glycol (PEG), Polyvinyl Alcohol (PVA)	Autologous chondrocytes, Nasal epithelial cells, Mesenchymal stem cells (MSCs)	Rodents (mice, rats), Porcine models	Nasal cavity, Nasal septum, Nasal cartilage	[1]
Craniofacial Bones	Extrusion-based bioprinting, Inkjet-based bioprinting	Polycaprolactone (PCL), Hydroxyapatite (HA), Poly (lactic-co-glycolic acid) (PLGA)	Autologous bone marrow-derived mesenchymal stem cells (BM-MSCs), Osteoblasts, Chondrocytes	Rodents (mice, rats), Canine models, non-human primates	Cranial bones, Facial bones	[1, 75]
Tracheal	Extrusion-based bioprinting, Laser-assisted bioprinting	Polyethylene Glycol (PEG), Polycaprolactone (PCL), Collagen	Autologous chondrocytes, Fibroblasts, Mesenchymal stem cells (MSCs)	Rodents (mice, rats), Canine models, Porcine models	Tracheal cartilage, Tracheal epithelial cells	[1, 2, 97]

improving comfort and performance. Additionally, this technology enables the production of aesthetic ear prosthetics that match the patient's skin tone and shape, enhancing both appearance and confidence [72].

14.9.2 Bio-fabrication of Tissue Scaffold

3D bioprinting can be used to create tissue scaffolds that support the growth of new tissue. This is particularly useful in reconstructive surgery where damaged or diseased tissues need to be replaced. In otolaryngology, bio fabrication techniques can produce scaffolds that mimic the extracellular matrix of ear, nose, and throat tissues, promoting cellular growth and tissue regeneration. These scaffolds can be impregnated with growth factors and other bioactive molecules to enhance healing and integration with existing tissues [76, 83].

14.9.3 Development of Drug Delivery Systems

Additive manufacturing is being used to develop advanced drug delivery systems specifically designed for ENT applications. For example, 3D-printed nasal sprays and inhalers can be customized to deliver medication more effectively to the sinuses or respiratory tract. These systems can be designed to release drugs at a controlled rate, targeting specific areas to maximize therapeutic effects while minimizing side effects [76, 83].

14.9.4 Regenerative Medicine and Stem Cell Applications

In regenerative medicine, 3D printing is utilized to create scaffolds that support the growth and differentiation of stem cells into functional tissues. This is particularly promising for repairing damage caused by diseases or injuries in the ENT region. Researchers are exploring the use of stem cells combined with 3D-printed scaffolds to regenerate complex structures such as the trachea or vocal cords [79].

14.9.5 Innovative Surgical Tools and Devices

Beyond custom instruments, additive manufacturing allows for the rapid prototyping and production of innovative surgical tools and devices tailored to specific procedures in otolaryngology. Surgeons can collaborate with engineers to design tools that enhance precision and reduce invasiveness, potentially leading to faster recovery times and better outcomes [71].

14.9.6 Patient-Specific Anatomical Models for Complex Cases

For complex surgical cases that require extensive planning, 3D printing can produce detailed, patient-specific anatomical models. These models help surgeons understand unique anatomical variations and plan interventions with greater accuracy. This is particularly useful for surgeries involving tumours, congenital abnormalities, or intricate structures like the inner ear [71].

14.9.7 Cochlear Implants and Auditory Brainstem Implants

Additive manufacturing facilitates the development and customization of cochlear implants and auditory brainstem implants. By precisely tailoring these devices to the patient's anatomy, the functionality and integration of the implants can be significantly improved, leading to better auditory outcomes for patients with severe hearing loss [75].

14.9.8 Innovation in Biocompatible Materials

Advances in additive manufacturing have led to the development of new biocompatible materials that can be used in ENT applications. These materials are designed to be more compatible with human tissues, reducing the risk of rejection and improving the integration of implants and prosthetics [79].

14.9.9 Customized Facial Prosthetics

3D printing technology enables the creation of highly customized facial prosthetics for patients who have undergone surgery for head and neck cancers or trauma. These prosthetics can be designed to match the patient's facial features precisely, improving both functionality and aesthetic appearance [75].

Additive manufacturing continues to transform otolaryngology by providing highly customized, patient-specific solutions that enhance surgical outcomes, improve patient comfort, and drive innovation in treatment methodologies. As technology advances, its integration into ENT practice will expand, further improving the standard of care and patient quality of life [79] (Fig. 14.9).

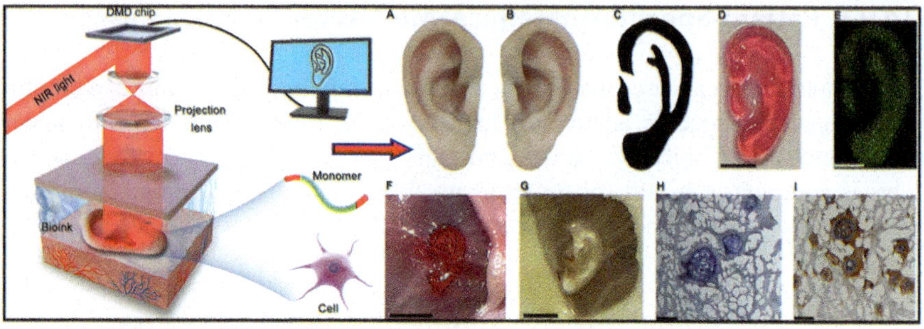

Fig. 14.9 A non-invasive 3D bioprinting method using digital near-infrared photopolymerization (DNP) is shown in Fig. 14.9. Tissue resembling ears can be bio printed using this technique. An optimized ear-outline image (**c**), a mirrored version of a representative image of a natural ear (**a**), and a printed ear-like construct made of bioink and coated in skin using the DNP method (**d**) are all included in the sequence. Live/dead staining was applied to the skin-covered ear constructions that were encapsulated with chondrocytes and bio printed with bioink after a 7-day cell culture (**e**). The DNP technique is used to create an ear-shaped construct that is bio printed in vivo noninvasively (**f**). The construct is printed subcutaneously on BALB/c nude mice (**g**, **h**). An image of the recovered ear-shaped construct at 1 month is displayed using collagen type II immunostaining (I). (The CC BY-NC license has been used to reprint this content with permission from [1])

14.10 Future Prospects and Challenges Associated with Bio-printing for Tissue Regeneration in Management of Otolaryngology

Advancing bio-printed additive manufacturing holds promise in enhancing printing speed and resolution, optimizing bio ink formulations for improved cell viability, and integrating state-of-the-art imaging technologies for personalized designs [78, 98]. Challenges persist in establishing standardized methods, ensuring regulatory compliance, and scaling up production for clinical applications. However, the potential benefits of personalized treatment plans, reduced surgical complications, and enhanced patient outcomes underscore the importance of continued research and development in this field [1, 79, 99]. Anticipated breakthroughs in repairing nasal, tracheal, and craniofacial bones through bio-printed additive manufacturing are significant. Future studies will focus on developing bio-ink compositions for increased strength and flexibility, creating hybrid materials for enhanced adaptability, and refining printing techniques for unparalleled precision [1, 100, 101]. With tissue engineering and regenerative medicine poised for expansion, there are ample opportunities to drive innovation and address current challenges. Bio-printed additive manufacturing offers exciting prospects for improving patient outcomes and advancing regenerative medicine [75, 84].

14.11 Conclusion

In conclusion, the application of biomaterials-based additive manufacturing offers significant promise in transforming the area of otolaryngology. By utilizing the capabilities of 3D printing technology and the plasticity of biomaterials, bespoke solutions can be tailored to match the anatomical and functional needs of patients. From tailored implants for reconstructive surgery to finely engineered scaffolds for tissue regeneration, the potential applications are vast. Furthermore, the combination of modern imaging techniques such as CT and MRI with additive manufacturing enables precise modeling and manufacture of patient-specific devices. This not only boosts surgical outcomes but also saves operative time and reduces the risk of complications. Furthermore, some biomaterials' biocompatibility and bioresorbable qualities have the benefit of encouraging tissue integration and regeneration, which eventually improves patient recovery and long-term results. However, despite the enormous improvements made in this industry, there are still hurdles to solve, including regulatory considerations, material optimization, and cost-effectiveness. Collaborative efforts among researchers, clinicians, and business partners are required to overcome these challenges and fully realize the potential of biomaterials-based additive manufacturing in otolaryngology. In conclusion, there is a great chance to revolutionize the treatment of otolaryngological illnesses through the convergence of biomaterials and additive manufacturing technologies. This will enable the development of targeted, individualized interventions that enhance patient care and quality of life.

References

1. Vyas J, Shah I, Singh S, Prajapati BG (2023) Biomaterials-based additive manufacturing for customized bioengineering in management of otolaryngology: a comprehensive review. Front Bioeng Biotechnol 11(September):1–26. https://doi.org/10.3389/fbioe.2023.1234340
2. Vyas J, Raytthatha N, Singh S, Prajapati B (2025) Next-generation computational automation-based additive manufacturing of pharmaceuticals: an approach to fabricate precise medicine. In: Handbook of 3D printing in pharmaceutics. CRC Press, Boca Raton, pp 153–175
3. Shah SR, Modi CD, Singh S, Mori DD, Soniwala MM, Prajapati BG (2024) Recent advances in additive manufacturing of polycaprolactone-based scaffolds for tissue engineering applications: a comprehensive review. Regen Eng Transl Med 11:112–131
4. Singh A, Mandal P, Maheshwari S, Akhtar J, Prajapati BG, Singh S (2025) Processing of sustainable biomaterials by additive manufacturing methods. In: Advances in sustainable biomaterials. CRC Press, Boca Raton, pp 141–162
5. Vyas J, Raytthatha N, Singh S, Prajapati BG, Mohite P, Munde S (2024) Sustainable sources of raw materials as substituting biomaterials for additive manufacturing of dental implants: a review. Periodontal Implant Res 8(1):3
6. Vyas J, Singh S, Shah I, Prajapati BG (2023) Potential applications and additive manufacturing technology-based considerations of mesoporous silica: a review. AAPS PharmSciTech 25(1):6
7. Jacofsky D, Jacofsky EM, Jacofsky M (2020) Understanding antibody testing for COVID-19. J Arthroplast 35(7):S74–S81. https://doi.org/10.1016/j.arth.2020.04.055

8. Schwarzkopf R, Maher NA, Slover JD, Strauss EJ, Bosco JA, Zuckerman JD (2020) The response of an orthopedic department and specialty hospital at the Epicenter of a pandemic: the NYU Langone Health experience. J Arthroplast 35(7):S3–S5. https://doi.org/10.1016/j.arth.2020.04.041
9. Pelt CE et al (2020) The rapid response to the COVID-19 pandemic by the arthroplasty divisions at two academic referral centers. J Arthroplast 35(7):S10–S14. https://doi.org/10.1016/j.arth.2020.04.030
10. Stambough JB et al (2020) The past, present, and future of orthopedic education: lessons learned from the COVID-19 pandemic. J Arthroplast 35(7):S60–S64. https://doi.org/10.1016/j.arth.2020.04.032
11. Ibrahim AMS, Jose RR, Rabie AN, Gerstle TL, Lee BT, Lin SJ (2015) Three-dimensional printing in developing countries. Plast Reconstr Surg Glob Open 3(7):1–8. https://doi.org/10.1097/GOX.0000000000000298
12. Wellman S et al (2020) Contents S1 July 2020 reports from the frontlines S3 the rapid response to the COVID-19 pandemic by the arthroplasty divisions at two academic S10 S19 S32 S37 S42 S60 S65 S68 Orthopedic Community Response American Association of Hip and Knee Surgeons Advo 35(7). https://doi.org/10.1016/S0883-5403(20)30527-1
13. Athey AG et al (2020) Survey of AAHKS international members on the impact of COVID-19 on hip and knee arthroplasty practices. J Arthroplast 35(7):S89–S94. https://doi.org/10.1016/j.arth.2020.04.053
14. Subburaj K, Nair C, Rajesh S, Meshram SM, Ravi B (2007) Rapid development of auricular prosthesis using CAD and rapid prototyping technologies. Int J Oral Maxillofac Surg 36(10):938–943. https://doi.org/10.1016/j.ijom.2007.07.013
15. Staiger MP, Pietak AM, Huadmai J, Dias G (2006) Magnesium and its alloys as orthopedic biomaterials: a review. Biomaterials 27(9):1728–1734. https://doi.org/10.1016/j.biomaterials.2005.10.003
16. Sharafeldin M et al (2021) Detecting cancer metastasis and accompanying protein biomarkers at single cell levels using a 3D-printed microfluidic immunoarray. Biosens Bioelectron 171(September 2020):112681. https://doi.org/10.1016/j.bios.2020.112681
17. Sames WJ, List FA, Pannala S, Dehoff RR, Babu SS (2016) The metallurgy and processing science of metal additive manufacturing. Int Mater Rev 61(5):315–360. https://doi.org/10.1080/09506608.2015.1116649
18. Yang J et al (2022) An integrative review on the applications of 3D printing in the field of in vitro diagnostics. Chinese Chem Lett 33(5):2231–2242. https://doi.org/10.1016/j.cclet.2021.08.105
19. Roskies M et al (2016) Improving PEEK bioactivity for craniofacial reconstruction using a 3D printed scaffold embedded with mesenchymal stem cells. J Biomater Appl 31(1):132–139. https://doi.org/10.1177/0885328216638636
20. Meneghini RM (2020) Resource reallocation during the COVID-19 pandemic in a suburban hospital system: implications for outpatient hip and knee arthroplasty. J Arthroplast 35(7):S15–S18. https://doi.org/10.1016/j.arth.2020.04.051
21. Rankin TM, Giovinco NA, Cucher DJ, Watts G, Hurwitz B, Armstrong DG (2014) Three-dimensional printing surgical instruments: are we there yet? J Surg Res 189(2):193–197. https://doi.org/10.1016/j.jss.2014.02.020
22. Ponader S et al (2010) In vivo performance of selective electron beam-melted Ti-6Al-4V structures. J Biomed Mater Res Part A 92(1):56–62. https://doi.org/10.1002/jbm.a.32337
23. Peter SJ, Kim P, Yasko AW, Yaszemski MJ, and Mikos AG (1998) Crosslinking characteristics of an injectable poly(propylene fumara.pdf
24. Fillingham YA, Grosso MJ, Yates AJ, Austin MS (2020) Personal protective equipment: current best practices for orthopedic teams. J Arthroplast 35(7):S19–S22. https://doi.org/10.1016/j.arth.2020.04.046
25. S. A. J. L. No- and W. Bengal (2023) Tax invoice grand total, no. 90, p. 66904500
26. Noble JW, Moore CA, Liu N (2012) The value of patient-matched instrumentation in total knee arthroplasty. J Arthroplast 27(1):153–155. https://doi.org/10.1016/j.arth.2011.07.006

27. Mishra S (2016) Application of 3D printing in medicine. Indian Heart J 68(1):108–109. https://doi.org/10.1016/j.ihj.2016.01.009
28. Leoncini E et al (2014) Adult height and head and neck cancer: a pooled analysis within the INHANCE Consortium. Head Neck 36(10):1391. https://doi.org/10.1002/HED
29. Chisari E, Krueger CA, Barnes CL, Van Onsem S, Walter WL, Parvizi J (2020) Prevention of infection and disruption of the pathogen transfer chain in elective surgery. J Arthroplast 35(7):S28–S31. https://doi.org/10.1016/j.arth.2020.04.049
30. Mertz L (2013) Dream it, design it, print it in 3-D: what can 3-D printing do for you? IEEE Pulse 4(6):15–21. https://doi.org/10.1109/MPUL.2013.2279616
31. Mazzoni S, Bianchi A, Schiariti G, Badiali G, Marchetti C (2015) Computer-aided design and computer-aided manufacturing cutting guides and customized titanium plates are useful in upper maxilla waferless repositioning. J Oral Maxillofac Surg 73(4):701–707. https://doi.org/10.1016/j.joms.2014.10.028
32. Mazzoli A, Germani M, Raffaeli R (2009) Direct fabrication through electron beam melting technology of custom cranial implants designed in a PHANToM-based haptic environment. Mater Des 30(8):3186–3192. https://doi.org/10.1016/j.matdes.2008.11.013
33. Liu X et al (2012) Inkjet printing assisted synthesis of multicomponent mesoporous metal oxides for ultrafast catalyst exploration. Nano Lett 12(11):5733–5739. https://doi.org/10.1021/nl302992q
34. Linhart W et al (2001) Biologically and chemically optimized composites of carbonated apatite and polyglycolide as bone substitution materials. J Biomed Mater Res 54(2):162–171. https://doi.org/10.1002/1097-4636(200102)54:2<162::AID-JBM2>3.0.CO;2-3
35. Li Y, Chen H, Liu D, Wang W, Liu Y, Zhou S (2015) PH-responsive shape memory poly(ethylene glycol)-poly(ε-caprolactone)-based polyurethane/cellulose nanocrystals nanocomposite. ACS Appl Mater Interfaces 7(23):12988–12999. https://doi.org/10.1021/acsami.5b02940
36. Kaye R, Goldstein T, Zeltsman D, Grande DA, Smith LP (2016) Three dimensional printing: a review on the utility within medicine and otolaryngology. Int J Pediatr Otorhinolaryngol 89:145–148. https://doi.org/10.1016/j.ijporl.2016.08.007
37. Kai D et al (2016) Biocompatible electrically conductive nanofibers from inorganic-organic shape memory polymers. Colloids Surf B Biointerfaces 148:557–565. https://doi.org/10.1016/j.colsurfb.2016.09.035
38. Jovic TH, Stewart K, Kon M, Whitaker IS (2020) Auricular reconstruction: a sociocultural, surgical and scientific perspective. J Plast Reconstr Aesthetic Surg 73(8):1424–1433. https://doi.org/10.1016/j.bjps.2020.03.025
39. Spirig JM, Golshani S, Farshad-Amacker NA, Farshad M (2021) Patient-specific template-guided versus standard freehand lumbar pedicle screw implantation: a randomized controlled trial. J Neurosurg Spine 35(2):147–153. https://doi.org/10.3171/2020.10.SPINE201383
40. Lee HY, Lee JW (2023) Current status and future outlook of additive manufacturing technologies for the reconstruction of the trachea. J Funct Biomater 14(4). https://doi.org/10.3390/jfb14040196
41. Kaboodkhani R, Mehrabani D, Karimi-Busheri F (2021) Achievements and challenges in transplantation of mesenchymal stem cells in otorhinolaryngology. J Clin Med 10(13). https://doi.org/10.3390/jcm10132940
42. Javaid M, Haleem A (2018) Additive manufacturing applications in medical cases: a literature based review. Alexandria J Med 54(4):411–422. https://doi.org/10.1016/j.ajme.2017.09.003
43. Khanuja HS, Chaudhry YP, Sheth NP, Oni JK, Parsley BS, Morrison JC (2020) Humanitarian needs: the arthroplasty Community and the COVID-19 pandemic. J Arthroplast 35(7):S85–S88. https://doi.org/10.1016/j.arth.2020.04.054
44. Warth LC, Noiseux NO, Duncan ST, Daines SB, Mahoney CR (2020) How has COVID-19 affected our orthopedic implant industry partners? Implications for the surgeon-industry relationship in 2020 and beyond. J Arthroplast 35(7):S56–S59.e10. https://doi.org/10.1016/j.arth.2020.04.063

45. Erickson MM, Richardson ES, Hernandez NM, Bobbert DW, Gall K, Fearis P (2020) Helmet modification to PPE with 3D printing during the COVID-19 pandemic at Duke University Medical Center: a novel technique. J Arthroplast 35(7):S23–S27. https://doi.org/10.1016/j.arth.2020.04.035
46. Gavenis K, Schneider U, Groll J, Schmidt-Rohlfing B (2010) BMP-7-loaded PGLA microspheres as a new delivery system for the cultivation of human chondrocytes in a collagen type I gel: the common nude mouse model. Int J Artif Organs 33(1):45–53. https://doi.org/10.1177/039139881003300107
47. Gao B et al (2019) Long-segmental tracheal reconstruction in rabbits with pedicled tissue-engineered trachea based on a 3D-printed scaffold. Acta Biomater 97:177–186. https://doi.org/10.1016/j.actbio.2019.07.043
48. Fedorovich NE et al (2012) Biofabrication of osteochondral tissue equivalents by printing topologically defined, cell-laden hydrogel scaffolds. Tissue Eng Part C Methods 18(1):33–44. https://doi.org/10.1089/ten.tec.2011.0060
49. Fantino E et al (2018) 3D printing/interfacial polymerization coupling for the fabrication of conductive hydrogel. Macromol Mater Eng 303(4):1–8. https://doi.org/10.1002/mame.201700356
50. Bedard NA, Elkins JM, Brown TS (2020) Effect of COVID-19 on hip and knee arthroplasty surgical volume in the United States. J Arthroplast 35(7):S45–S48. https://doi.org/10.1016/j.arth.2020.04.060
51. Saxena A et al (2020) Educating surgeons to educate patients about the COVID-19 pandemic. J Arthroplast 35(7):S65–S67. https://doi.org/10.1016/j.arth.2020.04.037
52. Browne JA et al (2020) Ed board. J Arthroplast 35(7):A1. https://doi.org/10.1016/s0883-5403(20)30525-8
53. O'Connor CM, Anoushiravani AA, DiCaprio MR, Healy WL, Iorio R (2020) Economic recovery after the COVID-19 pandemic: resuming elective orthopedic surgery and Total joint arthroplasty. J Arthroplast 35(7):S32–S36. https://doi.org/10.1016/j.arth.2020.04.038
54. D'Apolito R, Faraldi M, Ottaiano I, Zagra L (2020) Disruption of arthroplasty practice in an orthopedic center in Northern Italy during the coronavirus disease 2019 pandemic. J Arthroplast 35(7):S6–S9. https://doi.org/10.1016/j.arth.2020.04.057
55. Bini SA et al (2020) Digital orthopaedics: a glimpse into the future in the midst of a pandemic. J Arthroplast 35(7):S68–S73. https://doi.org/10.1016/j.arth.2020.04.048
56. Cutright DE, Bhaskar SN, Brady JM, Getter L, Posey WR (1972) Reaction of bone to tricalcium phosphate ceramic pellets. Oral Surg Oral Med Oral Pathol 33(5):850–856. https://doi.org/10.1016/0030-4220(72)90457-4
57. Cui X, Boland T, D'Lima DD, Lotz MK (2012) Thermal inkjet printing in tissue engineering and regenerative medicine. Recent Pat Drug Deliv Formul 6(2):149–155. https://doi.org/10.2174/187221112800672949
58. Cuellar JS, Smit G, Zadpoor AA, Breedveld P (2018) Ten guidelines for the design of non-assembly mechanisms: the case of 3D-printed prosthetic hands. Proc Inst Mech Eng Part H J Eng Med 232(9):962–971. https://doi.org/10.1177/0954411918794734
59. Culp BM, Frisch NB (2020) COVID-19 impact on young arthroplasty surgeons. J Arthroplast 35(7):S42–S44. https://doi.org/10.1016/j.arth.2020.04.058
60. Correction to Lancet Oncol 2019; 20: e262–73 (The Lancet Oncology (2019) 20(5) (e262–e273), (S1470204519301494), (https://doi.org/10.1016/S1470-2045(19)30149-4)). Lancet Oncol 20(6):293, 2019. https://doi.org/10.1016/S1470-2045(19)30294-3
61. Correction to Lancet Oncol 2019; 20: 849–61 (The Lancet Oncology (2019) 20(6) (849–861), (S1470204519300270), (https://doi.org/10.1016/S1470-2045(19)30027-0)). Lancet Oncol 20(6):e293, 2019. https://doi.org/10.1016/S1470-2045(19)30283-9
62. Correction to Lancet Oncol 2019; 20: 827–36 (The Lancet Oncology (2019) 20(6) (827–836), (S1470204519300889), (https://doi.org/10.1016/S1470-2045(19)30088-9)). Lancet Oncol 20(6):e293, 2019, doi: https://doi.org/10.1016/S1470-2045(19)30293-1

63. Correction to Lancet Oncol 2019; 20: 297–310 (The Lancet Oncology (2019) 20(2) (297–310), (S1470204518307782), (https://doi.org/10.1016/S1470-2045(18)30778-2)). Lancet Oncol 20(6):e293, 2019. https://doi.org/10.1016/S1470-2045(19)30348-1
64. Correction to Lancet Oncol 2018; 19: 1530–42 (The Lancet Oncology (2018) 19(11) (1530–1542), (S1470204518306181), (https://doi.org/10.1016/S1470-2045(18)30618-1))', Lancet Oncol 20(6):e293, 2019. https://doi.org/10.1016/S1470-2045(19)30280-3
65. Correction to Lancet Oncol 2017; 18: 1061–75 (The Lancet Oncology (2017) 18(8) (1061–1075), (S1470204517304163), (https://doi.org/10.1016/S1470-2045(17)30416-3)). Lancet Oncol 20(6):e293, 2019. https://doi.org/10.1016/S1470-2045(19)30297-9
66. Correction to Lancet Oncol 2016; 17: 747–56 (The Lancet Oncology (2016) 17(6) (747–756), (S147020451600111X), (https://doi.org/10.1016/S1470-2045(16)00111-X)). Lancet Oncol 20(6):e293, 2019. https://doi.org/10.1016/S1470-2045(19)30347-X
67. Cohen A, Laviv A, Berman P, Nashef R, Abu-Tair J (2009) Mandibular reconstruction using stereolithographic 3-dimensional printing modeling technology. Oral Surg Oral Med Oral Pathol Oral Radiol Endod 108(5):661–666. https://doi.org/10.1016/j.tripleo.2009.05.023
68. Correction to Lancet Oncol 2019; 20: 816–26 (The Lancet Oncology (2019) 20(6) (816–826), (S147020451930097X), (https://doi.org/10.1016/S1470-2045(19)30097-X)). Lancet Oncol 20(6):e293, 2019. https://doi.org/10.1016/S1470-2045(19)30292-X
69. The Lancet Oncology (2019) Clinical trial registry reporting: a transparent solution needed. Lancet Oncol 20(6):741. https://doi.org/10.1016/S1470-2045(19)30350-X
70. Canzi P et al (2018) New frontiers and emerging applications of 3D printing in ENT surgery: a systematic review of the literature. Acta Otorhinolaryngol Ital 38(4):286–303. https://doi.org/10.14639/0392-100X-1984
71. Wellman S, Moskal J, Barnes CL, Parvizi J (2020) Business unusual: COVID-19 ramifications for arthroplasty. J Arthroplast 35(7):S1–S2. https://doi.org/10.1016/j.arth.2020.04.062
72. Bandyopadhyay A, Espana F, Balla VK, Bose S, Ohgami Y, Davies NM (2010) Influence of porosity on mechanical properties and in vivo response of Ti6Al4V implants. Acta Biomater 6(4):1640–1648. https://doi.org/10.1016/j.actbio.2009.11.011
73. Bakarich SE, Gorkin R, In Het Panhuis M, Spinks GM (2015) 4D printing with mechanically robust, thermally actuating hydrogels. Macromol Rapid Commun 36(12):1211–1217. https://doi.org/10.1002/marc.201500079
74. Ayoub AF et al (2014) A novel approach for planning orthognathic surgery: the integration of dental casts into three-dimensional printed mandibular models. Int J Oral Maxillofac Surg 43(4):454–459. https://doi.org/10.1016/j.ijom.2013.08.016
75. Attar H, Bönisch M, Calin M, Zhang LC, Scudino S, Eckert J (2014) Selective laser melting of in situ titanium-titanium boride composites: processing, microstructure and mechanical properties. Acta Mater 76:13–22. https://doi.org/10.1016/j.actamat.2014.05.022
76. Apsite I, Stoychev G, Zhang W, Jehnichen D, Xie J, Ionov L (2017) Porous stimuli-responsive self-folding electrospun mats for 4D biofabrication. Biomacromolecules 18(10):3178–3184. https://doi.org/10.1021/acs.biomac.7b00829
77. Huddleston JI, Iorio R, Bosco JA, Kerr JM, Bolognesi MP, Barnes CL (2020) American Association of Hip and Knee Surgeons Advocacy Efforts in response to the SARS-CoV-2 pandemic. J Arthroplast 35(7):S82–S84. https://doi.org/10.1016/j.arth.2020.04.050
78. Ahmed A, Arya S, Gupta V, Furukawa H, Khosla A (2021) 4D printing: fundamentals, materials, applications and challenges. Polymer (Guildf) 228(May):123926. https://doi.org/10.1016/j.polymer.2021.123926
79. Ahmed S, VanKoevering KK, Kline S, Green GE, Arts HA (2017) Middle cranial fossa approach to repair tegmen defects assisted by three-dimensionally printed temporal bone models. Laryngoscope 127(10):2347–2351. https://doi.org/10.1002/lary.26438
80. Ahadian S et al (2016) Carbon nanotubes and graphene-based nanomaterials for stem cell differentiation and tissue regeneration. J Nanosci Nanotechnol 16(9):8862–8880. https://doi.org/10.1166/jnn.2016.12729

81. Zeegen EN, Yates AJ, Jevsevar DS (2020) After the COVID-19 pandemic: returning to normalcy or returning to a new normal? J Arthroplast 35(7):S37–S41. https://doi.org/10.1016/j.arth.2020.04.040
82. Adel-Khattab D et al (2018) Development of a synthetic tissue engineered three-dimensional printed bioceramic-based bone graft with homogenously distributed osteoblasts and mineralizing bone matrix in vitro. J Tissue Eng Regen Med 12(1):44–58. https://doi.org/10.1002/term.2362
83. Adamzyk C et al (2016) Bone tissue engineering using polyetherketoneketone scaffolds combined with autologous mesenchymal stem cells in a sheep calvarial defect model. J Cranio-Maxillofacial Surg 44(8):985–994. https://doi.org/10.1016/j.jcms.2016.04.012
84. Piplani P, Danta CC (2015) University Institute of Pharmaceutical Sciences. Bioorg Chem Vd:87–93 [Online]. Available: https://doi.org/10.1016/j.bioorg.2015.04.004
85. Fernández D et al (2022) Economics of dry needling and botulinum toxin type A for treatment of post-stroke spasticity: a review. Explor Neuroprotective Ther:131–140. https://doi.org/10.37349/ent.2022.00024
86. Choi Y-J, Park JH, Jang J, Cho D-W (2018) 3D bioprinting technologies and bioinks for therapeutic and tissue engineering applications. J 3D Print Med 2(4):187–203. https://doi.org/10.2217/3dp-2018-0014
87. Banerjee SS, Burbine S, Shivaprakash NK, Mead J (2019) 3D-printable PP/SEBS thermoplastic elastomeric blends: preparation and properties. Polymers (Basel) 11(2). https://doi.org/10.3390/polym11020347
88. Di Gesù R, Acharya AP, Jacobs I, Gottardi R (2020) 3D printing for tissue engineering in otolaryngology. Connect Tissue Res 61(2):117–136. https://doi.org/10.1080/03008207.2019.1663837
89. Lin Y et al (2019) Combination of polyetherketoneketone scaffold and human mesenchymal stem cells from temporomandibular joint synovial fluid enhances bone regeneration. Sci Rep 9(1):1–13. https://doi.org/10.1038/s41598-018-36778-2
90. Vanaei S, Parizi MS, Vanaei S, Salemizadehparizi F, Vanaei HR (2021) An overview on materials and techniques in 3D bioprinting toward biomedical application. Eng Regen 2(November 2020):1–18. https://doi.org/10.1016/j.engreg.2020.12.001
91. Balçik C et al (2007) Early weight bearing of porous HA/TCP (60/40) ceramics in vivo: a longitudinal study in a segmental bone defect model of rabbit. Acta Biomater 3(6):985–996. https://doi.org/10.1016/j.actbio.2007.04.004
92. Niermeyer WL, Rodman C, Li MM, Chiang T (2020) Tissue engineering applications in otolaryngology—the state of translation. Laryngoscope Investig Otolaryngol 5(4):630–648. https://doi.org/10.1002/lio2.416
93. Brennan JR et al (2021) Preclinical assessment of clinically streamlined, 3D-printed, biocompatible single- and two-stage tissue scaffolds for ear reconstruction. J Biomed Mater Res Part B Appl Biomater 109(3):394–400. https://doi.org/10.1002/jbm.b.34707
94. Tan D, Nokhodchi A, Maniruzzaman M (2019) 3D and 4D printing technologies: innovative process engineering and smart additive manufacturing. 3D 4D Print Biomed Appl:25–52. https://doi.org/10.1002/9783527813704.ch2
95. Xu Y, Han J, Chai Y, Yuan S, Lin H, Zhang X (2018) Development of porous chitosan/tripolyphosphate scaffolds with tunable uncross-linking primary amine content for bone tissue engineering. Mater Sci Eng C 85(June 2017):182–190. https://doi.org/10.1016/j.msec.2017.12.032
96. Pettersson ABV, Salmi M, Vallittu P, Serlo W, Tuomi J, Mäkitie AA (2020) Main clinical use of additive manufacturing (three-dimensional printing) in Finland restricted to the head and neck area in 2016–2017. Scand J Surg 109(2):166–173. https://doi.org/10.1177/1457496919840958
97. Albini A, Bruno A, Gallo C, Pajardi G, Noonan DM, Dallaglio K (2015) Cancer stem cells and the tumor microenvironment: interplay in tumor heterogeneity. Connect Tissue Res 56(5):414–425. https://doi.org/10.3109/03008207.2015.1066780

98. Schonberg DL, Lubelski D, Miller TE, Rich JN (2013) Molecular aspects of medicine brain tumor stem cells: molecular characteristics and their impact on therapy. J Mol Asp Med:1–20. https://doi.org/10.1016/j.mam.2013.06.004
99. Ellis AJ, Hendrick VM, Williams R, Komm BS (2015) Selective estrogen receptor modulators in clinical practice: a safety overview. Expert Opin Drug Saf 14(6):921–934. https://doi.org/10.1517/14740338.2015.1014799
100. Elmahallawy EK, Mohamed Y, Abdo W, Yanai T Melatonin and mesenchymal stem cells as a key for functional integrity for liver cancer treatment, pp 1–19
101. Wahl GM, Spike BT (2017) Cell state plasticity, stem cells, EMT, and the generation of intra-tumoral heterogeneity. NPJ Breast Cancer 3(1):1–12. https://doi.org/10.1038/s41523-017-0012-z

Biomaterials-Based Additive Manufactured Constructs for Drug Testing

15

Renjith P. Nair, Helen Jalaja Shibu, Hega Vincent, Anugya Bhatt, and Naresh Kasoju

Abstract

The constantly evolving domain of pharmaceutical research necessitates the discovery, development, and screening of novel drugs. Most of the drugs that successfully pass preclinical drug screening encounter failure during clinical trials, primarily attributed to the inadequacies of prevailing drug screening systems. The conventional two-dimensional (2D) drug screening models and animal studies are inadequate mainly due to their inability to mimic the *in vivo* architecture of human organs. Three-dimensional (3D) constructs with human cells exhibit notable advantages over traditional 2D screening systems, making them a more promising model for drug screening. The development of more sophisticated and innovative additive manufacturing technology led to the 3D printing of live tissue constructs, known as 3D bioprinting where the tissue organisation of the human body can be replicated. Various 3D bio-printed organ models, including liver, heart, lungs, kidney, skin, intestine, and bone, are currently being developed using additive manufacturing technology and evaluated for testing the efficacy of drugs for various applications. Disease models are also being constructed for the preclinical examination of drugs or for the personalised therapies. In this chapter, an overview of the utilisation of additively manufactured products will be

R. P. Nair (✉) · H. J. Shibu · H. Vincent · A. Bhatt
Division of Thrombosis Research, Department of Applied Biology, Biomedical Technology Wing, Sree Chitra Tirunal Institute for Medical Sciences and Technology, Thiruvananthapuram, Kerala, India
e-mail: renjith@sctimst.ac.in; anugyabhatt@sctimst.ac.in

N. Kasoju
Division of Tissue Culture, Department of Applied Biology, Biomedical Technology Wing, Sree Chitra Tirunal Institute for Medical Sciences and Technology, Thiruvananthapuram, Kerala, India
e-mail: naresh.kasoju@sctimst.ac.in

S. Singh et al. (eds.), *Biomaterial-based Additive Manufacturing in Tissue Engineering and Regeneration*, Biomaterials, Bioengineering and Sustainability 7,
https://doi.org/10.1007/978-3-031-96070-3_15

discussed with particular emphasis on the bioinks and techniques employed for the development of different 3D bio-printed tissue constructs for drug testing. The challenges faced by the technique, as well as the critical outlooks and future advancements, will be discussed.

Keywords
Additive manufacturing · 3D bioprinting · Drug testing · Bioink · Disease modelling

15.1 Introduction

With the global prevalence of numerous illnesses and the alarming increase in their occurrence rates, there is an urgent need to develop innovative technologies and novel drugs capable of combating these diseases [1]. Successful development of a new drug is a long-term process that takes about 12–15 years, where the drug is scrutinized not only to determine its safety and efficacy but also to check whether it is in accordance with the existing regulatory standards [2]. Before a drug is marketed, it must undergo an extensive screening process comprising two major phases, namely, preclinical trials and clinical trials. Preclinical trials involve *in vitro, in vivo,* and *in silico* studies on two-dimensional models such as monolayer cultures and animal models. It evaluates the toxicity, pharmacokinetics and pharmacodynamics of the drug before it is sent for clinical trials on human subjects, which further assesses its safety, effectiveness, dosage, and potential side effects [3]. However, two-dimensional models are unable to completely mimic *in vivo* conditions like complex microenvironments, cellular anchorage, and heterogenicity [4]. Animal models are also not greatly reliable due to the differences in physiology and microenvironment of test animals from that of humans, together with the arising ethical concerns. To address this, three-dimensional constructs closely resembling the *in vivo* conditions are being developed to serve as an effective alternative, thereby providing a precise platform for drug screening [5].

Additive manufacturing technology is an emerging technique that collaborates computers, materials, and other machinery for the rapid prototyping of equipment and three-dimensional substitutes. It involves precise printing of biomolecules sequentially to form a three-dimensional matrix resembling *in vivo* architecture in terms of structure and function [6]. Cells and tissues are printed in such a way that they can be held together in a specific pattern, forming the respective tissue construct, thus mimicking their natural architecture and complexity. Layer-by-layer bioprinting of biological materials creates interconnected pores ideal for the exchange of nutrients and gases in addition to providing excellent inter and intracellular communication [7, 8]. 3D bioprinting is an advanced method based on additive manufacturing technology, gaining importance and driving major innovations in all fields of medicine, particularly tissue engineering and regenerative medicine. Cells encapsulated in ink are used for bioprinting, which enables spatial arrangement, scaffold adjustment, and other specific formulations to create an ideal multiscale tissue architecture [9].

Recently, pharmaceutical industries have been testing their drugs on 3D bio-printed substitutes to identify potential drugs that reduce the risk of financial losses due to drug failure in the clinical phase of testing [4]. Several constructs have been developed for drug testing using 3D bioprinting, including liver organoids, heart tissues, kidney organoids, skin, vascular grafts, and lungs. 3D bio-printed substitutes also allow patient-specific formulations through precision medicine, which involves the fabrication of individualized constructs representing the specific illness of the patient by utilising their own cells to incorporate individual mutations and genetic variations corresponding to the patient onto the model [6, 10]. This is greatly advantageous as genetic information is crucial for determining pharmacological action, progression of disease, and drug response. Customised constructs through additive manufacturing, thus enable personalized treatments by exploiting patient-specific genetic, proteomic, and phenotypic characteristics and provide better therapies in addition to financial benefits [11].

This chapter highlights the advancement of 3D bioprinting based on additive manufacturing technology as a breakthrough in the realm of drug testing. Emphasis is given to the commonly developed three-dimensional tissue constructs through layer-by-layer deposition of biomaterials and the techniques employed for their fabrication. Also, insights into the bioinks utilised for the purpose, and potential future perspectives of the technology are detailed.

15.2 Additive Manufacturing Techniques for the Development of 3D Bio-printed Constructs

Layer-by-layer deposition of biomaterials, growth factors, and cells in 3D bioprinting is achieved through various techniques which differ based on resolution, choice of material, and time taken for manufacture, and enables the precise production of organs and organoids with potential applications in regenerative medicine, drug testing, and disease modelling. These techniques, along with their principles, are briefly described as follows.

15.2.1 Extrusion-Based Method

The extrusion-based method deposits bioink in the form of multiple layers using a nozzle, enabling versatility, rapid pattering, and differential density deposition of biomaterials [12]. Extrusion and printing are controlled by a fluid-dispensing system connected to an automated robotic system, which in turn regulates the flow of bioink through a deposition system attached to a computer-aided design (CAD) [13]. The technique utilises both the pneumatic method where the bioink is pushed by compressed air, and the mechanical method, which exploits mechanical force produced by a screw or piston. A tubular-shaped fibre construct using simultaneous yet separate ejection of bioink and cross-linking agent through a coaxial nozzle is one of the recent advancements of this method [11]. Extrusion-based bioprinting is versatile, enabling multifarious bioinks such as cell aggregates, hydrogels infused with cells, decellularized matrix, and micro-carriers for the construction of a wide variety of organs, organoids and tissue/organ-on-a-chip constructs [14], with an added advantage of printing anatomically accurate porous models at high speed and least cell-damage during the process [15]. However, poor resolution, formulation of different bioinks for each tissue type, and difficulty in cell delivery into a biomimetic microenvironment are major drawbacks concerning cell patterning and organisation. Additionally, when a large number of cells are passed through the nozzle, cell viability is compromised [16].

15.2.2 Inkjet-Based Method

The inkjet-based method involves non-contact bioprinting where the bioink droplets are prompted by piezoelectric input, thermal input or electrostatic actuator and sprayed onto a patterned substrate through a nozzle [17]. First used by Kleeb in 1988 to develop 2D and 3D synthetic tissues using collagen and fibronectin, the technique is now reaching out extensively to various dimensions such as toxicology, tissue regeneration, and drug testing because of its ability to produce droplets in the picolitre range. Ink formation on an inkjet printer can be either in a continuous manner or on demand [18]. The picolitre droplets increase the resolution and enable easy dispensing of cells and cell aggregates [19]. It dispenses ink precisely with predetermined cell numbers, making it suitable for the development of drug testing models. However, dispensing may cause a loss of cell integrity, which was solved by the development of a stress-free inkjet bioprinter that dispenses the cells directly onto the media [20]. There are numerous commercially available inkjet printers that have been transformed into customised 3D bioprinters for printing tissues of varying sizes for drug testing and development [21].

15.2.3 Laser-Assisted Method

The laser-assisted method is a scaffold-free technique that holds promise in the bioengineering of intricate tissue and organ scaffolds for drug testing due to its precision and accuracy [22]. Three systems collectively constitute this technology, namely, a pulsated laser source, a ribbon for coating the liquid samples, which are further deposited onto a metallic film, and a substrate containing biopolymer or culture media to sustain cell adhesion and growth [23]. It works on the principle of matrix-assisted laser-induced vaporisation of donor slide coated with metals such as gold or titanium, which push the material onto the receiver slide in the form of a high-pressure liquid bubble, thus enabling bioprinting at higher resolution [24]. Usually, the lasers will be in the wavelengths of 193 nm, 248 nm, or near UV wavelength, such as 1064 nm, to obtain a pulse energy of 1–20 μJ per pulse [25]. The method promotes cell-scaffold interactions, tissue and organ regeneration, and construction of artificial cell niches, all of which could not be achieved by traditional fabrication methods without deteriorating cell and material integrity [22].

15.2.4 Light-Assisted Method

Light-assisted methods are much more advanced than the above-mentioned methods. Two main types of light-assisted methods in use are digital light processing (DLP) and two-photon polymerisation (TPP). DLP uses light emitted from a digitalised micromirror device (DMD chip) to produce a solidified monomer, whereas TPP is a direct writing method based on laser rasterization, which generates photopolymerisable and photosensitive monomers [26]. Crosslinking of the biomaterials is made possible through the free-radical polymerisation of the bioink [27]. The method enables superior printing quality and resolution in addition to high cell viability rates after bioprinting. It also comes in multiple variations which allows the fabrication of constructs by adjusting mechanical and chemical parameters along with cell and biomaterial distribution. However, a major challenge arises when the cells are repeatedly exposed to UV radiations of very high intensities [28].

15.3 Bioinks for 3D Bioprinting of Tissue Constructs

Appropriate selection of bioink is a prerequisite for the accurate representation of pathological and physiological conditions of tissues to facilitate effective drug screening or testing. Bioink refers to the cell-laden printable fluid biomaterials used for the fabrication of tissue or organ constructs with specific biochemical or physiological properties [29]. The biomaterials used in bioink should possess tailored properties, including biocompatibility, mechanical strength, and degradation kinetics [29]. Commonly used cell sources and biomaterials will be discussed in detail, giving importance to their applications in the landscape of drug screening.

15.3.1 Cell Source

Choice of cell source demands contemplation of the type, viability, and compatibility of the cells, along with the technique employed for bioprinting and target organ [30–32]. Human-derived cells such as primary cells, cell lines, and stem cells are preferred over animal cells for drug testing and disease modelling platforms to minimise the errors caused by the differences in their genetic constitution and cellular arrangements [33] (Table 15.1). Primary cells, isolated directly from the tissues, could mimic tissue characteristics at specific stages but face challenges due to individual variability and limited lifespan. Cell lines are immortalised cells having homogenous genotypic features and unlimited proliferation due to genetic modifications. However, they lack inherent characteristics of the tissue of interest and, thus, are inefficient for precision engineering. Stem cells, being self-renewable and differentiable, are beneficial for *in vitro* disease modelling and tissue regeneration as they could portray individual phenotypic variability but face difficulty in the regulation of differentiation pathway and require extensive cell characterisation [34, 35]. Among the different classes of stem cells, mesenchymal stem cells (MSCs) are

Table 15.1 Some of the commonly used cells for the development of 3D bio-printed tissue constructs

Cell type	Biomaterial	Printing method	Tissue model constructed	Reference
Adipose mesenchymal stem cells	Fibrinogen and sodium alginate	Extrusion-based method	Neural tissue model	[37]
Primary human breast fibroblasts	Decellularized extracellular matrix	Extrusion-based method	Breast cancer model	[39]
NIH-3T3 fibroblasts	GelMA and PEDGA	Digital light processing stereolithography	Intestinal tissue model	[40]
ReNcell VM neural progenitor cells	Sodium alginate	Extrusion-based method	Neural tissue model	[41]
Bone marrow mesenchymal stem cells	Acrylated peptides and PEG	Ink jet-based-method	Peptide-conjugated PEG scaffold	[36]
Primary human foreskin fibroblasts	Collagen	Extrusion-based method	Fibrosis model	[42]
iPSCs derived cardiomyocyte	GelMA	Extrusion-based method	Cardiac model	[43]
Hepatocyte-like cells	Alginate hydrogel matrix	Microvalve bioprinter	Mini liver model	[44]
Human dermal fibroblasts	Gelatin, alginate and fibrinogen	Scaffold-free technique	Skin tissue model	[45]
Ovine endothelial cells	Acrylated polyethylene glycol	Laser assisted method	Multicellular construct	[46]
HUVECs, HHSCs, and HLCs	Decellularized extracellular matrix	Extrusion-based method	Liver lobule model	[47]
Keratinocyte, melanocyte, fibroblast	Collagen	Inkjet based method	Skin tissue model	[48]
Fibroblasts	GelMA, PEDGA, LAP and tartazine	Digital light processing stereolithography	Intestinal tissue model	[49]
Endothelium, cardiomyocytes	Alginate and GelMA	Extrusion and photocuring method	Cardiac model	[50]
Human mesenchymal stem cells	PLGA and hydroxyapatite	Extrusion-based method	Bone model	[51]

most widely investigated due to its ease of isolation and multilineage differentiation potential [36, 37]. Induced pluripotent stem cells (iPSCs) are also considered a promising stem cell source due to their unlimited proliferation potential and ability to differentiate into all three germ layers, thereby facilitating the development of any organ and mimicking its structural and functional properties [38].

15.3.2 Biomaterial

Unlike the traditional printing components, bioink is equipped with live cells embedded in a hydrogel biomaterial for cell-cell adhesion, proliferation, and differentiation [52]. Conventional biomaterials used in bioink for drug testing models can be classified as natural, synthetic, and composite materials. Natural biomaterials include collagen, gelatin, alginate, fibrin, chitosan, silk, and hyaluronic acid, which closely resemble the *in vivo* composition of native tissues, thereby exhibiting the desirable features including exceptional cell affinity and biocompatibility [53–55]. Decellularised extracellular matrix (dECM) is gaining recognition nowadays due to its exceptional biocompatibility and synergistic effect on the encapsulated cells, thereby providing niche microenvironment and tissue-specific characteristics by maintaining *in vivo* ECM components [56]. However, natural biomaterials are limited by weak mechanical properties and degradation profiles [57].

Synthetic biomaterials are characterised by tunable degradation and strong mechanical strength. Some of the synthetic biomaterials used for bioprinting and drug testing include polyethylene glycol (PEG), polyvinylpyrrolidone (PVP), poly(l-lactic) acid (PLA), gelatin methacrylate (GelMA), polylactic-co-glycolic acid (PLGA), polycaprolactone (PCL), acrylonitrile butadiene styrene (ABS) and polyether ether ketone (PEEK) [28, 57, 58]. These biomaterials lack the signals required for cell proliferation and differentiation in addition to the absence of cellular recognition sites [59]. To overcome the above-mentioned limitations, natural and synthetic polymers were combined to develop composite biomaterials with enhanced mechanical properties and degradation. For instance, a composite bioink developed using a combination of GelMA, AlgMA (methacrylated sodium alginate), and fibrin was found to possess the synergistic characteristics of all the components, including high stability, printability, and cell viability as well as resistance to degradation [60]. Similarly, a composite bioink developed by copolymerisation of polyethylene glycol with methacrylated gelatin demonstrated enhanced printability, biocompatibility, and stiffness [61, 62].

15.4 3D Bio-printed Tissue Constructs for Drug Testing and Disease Modelling

3D bio-printed tissue constructs have emerged as a potent tool for evaluating the therapeutic potential and safety of drug candidates as well as for disease modelling. These precisely developed tissue constructs closely mimic the microenvironment and functions of living tissues, offering valuable insights into drug effects and mechanisms. Compared to traditional 2D cell cultures, they provide a more realistic testing platform, enhancing predictions of drug behaviour within the body. This section briefly elaborates on the successfully developed 3D bio-printed tissue constructs and their application in drug testing.

15.4.1 Cardiac Tissue Constructs

The hierarchical architecture of the heart presents a challenge for the replicative construction of myocardial models, which is necessary for the development of novel drugs to treat various cardiovascular diseases [50, 63]. 3D bioprinting is a promising alternative for replicating complex cardiac tissue architecture by utilising cells such as primary myocardial cells, neonatal human cardiac progenitor cells, and stem cell-derived cardiomyocytes (CMs), which can proliferate into cardiac tissues [64]. Human iPSC-derived cardiac micro-tissue was constructed by encapsulating human iPSC-CMs and human cardiac fibroblasts in GelMA followed by printing using a micro continuous optical printing system to obtain similar sarcomere arrangement and mRNA expression markers of *in vivo* cardiac tissues. Validation using two benchmark drugs, isoproterenol, and verapamil, at varying concentrations through observation of the changes in beating frequency and displacement after drug administration proved the reliability of the model [65]. Another approach involved the bioprinting of endothelial cells onto micro-fibrous hydrogel scaffolds followed by the seeding of cardiomyocytes and embedding it onto a perfusion bioreactor to form a contractible and synchronous endothelialized-myocardium-on-a-chip. It was evaluated for cardiovascular toxicity using doxorubicin at two varying concentrations demonstrating the models' potential as a promising innovation for drug screening, disease modelling and regenerative medicine [50].

Researchers also developed cardiac tissue spheroids using scaffold-independent bioprinting, where cardiomyocytes, fibroblasts, and endothelial cells were cocultured to form spheroids and arranged on a needle array using a 3D bioprinter. Spontaneously beating spheroid cardiac patches could be obtained after 3 days, exhibiting mechanical assimilation similar to *in vivo* cardiac tissues. This model can also serve as a transformative technology to screen the efficacy of a drug before clinical trials [66]. In another study, some researchers employed a novel strategy wherein the cardiac tissue developed using human iPSC-CMs, endothelial cells, and fibroblasts encapsulated in a bioink exhibited conduction velocities and calcium transient in addition to the physiological responses, which was used to evaluate the efficacy of the drugs isoproterenol and verapamil [67]. Also, a cardiac model using a combination of methacrylated gelatin and collagen onto which human iPSC-derived CMs were encapsulated was proved to possess tensile strength, gap junctions, calcium utilisation, and patient-specific properties. The same model, when fabricated using human iPSC-CM derived from a patient possessing a mutated gene for polymorphic ventricular tachycardia, followed by adrenalin treatment, showed a similar response to the disease-specific original phenotype, which portrays the model's application in disease modelling and personalised medicine [43].

15.4.2 Liver Constructs

The critical role of the liver in drug metabolism necessitates an understanding of drug-induced liver damage, also known as hepatotoxicity, and drives continuous research into developing robust and reliable *in vitro* liver models [68]. 3D models created with additive manufacturing techniques and perfusion culture are now demonstrating excellent capability in accurately replicating the intricate microenvironment of the liver and actual metabolic conditions experienced by drugs, enabling the study of drug response *in vitro* [69]. Researchers developed a novel bio-printed human liver tissue construct comprising patient-derived hepatocytes and non-parenchymal cells with distinct intercellular hepatocyte junctions, CD31+ endothelial networks, and essential hepatic functions, such as ATP production, albumin production, and cytochrome P450 enzyme activity. It was employed to study drug-induced liver injury (DILI), where significant toxicity of trovafloxacin, a known hepatotoxic agent, was revealed at clinically relevant doses (<4 uM) compared to structurally related non-toxic drug Levofloxacin [70].

Extrusion bioprinting was employed to develop various 3D liver tissue constructs for hepatotoxicity assessment. One model containing primary human hepatocytes and liver stellate cells used thiolated hyaluronic acid and methacrylated collagen bioink to evaluate the toxicity of common hepatotoxicant, acetaminophen (APAP) (100 uM). It revealed the model's ability to sustain hepatocyte viability and respond to hepatotoxicity through decreased LDH (liver damage indicator) and α-GST (detoxification protein) activities and decreased urea and albumin levels [71]. Another model used Hepa RG cells in alginate-gelatin-matrigel hydrogel to print liver construct, through which hepatotoxicity of the drug, aflatoxin B1 (10 or 20 μM) was evaluated, where increased resistance to the drug and dose-dependent effect was observed on cell viability, compared to 2D cultures [72]. Further, a novel hepatocyte construct was fabricated with HUVECs, HHSCs, and HLCs in liver ECM-based bioink, mimicking the native hepatocyte environment with increased albumin production, urea synthesis, and cytochrome P450 activity, that served as an efficient platform to evaluate the hepatotoxicity of drugs like aspirin, dexamethasone, trovafloxacin mesylate, acetaminophen, and troglitazone [47].

Fatal conditions like liver cancer or hepatocellular carcinoma (HCC) drive the urgent development and screening of novel anti-cancer drugs. Researchers are investigating 3D hepatocyte culture through bioprinting to create sophisticated liver cancer models accurately mimicking the tumor microenvironment to evaluate *in vitro* drug responses. For instance, an endothelialized liver lobule-like construct was developed by printing hepatocyte-laden GelMA hydrogel microbeads and HUVEC-laden gelatin microbeads through dot extrusion printing (DEP) technology, which displayed high expression of hepatic markers and metabolic product secretion and served as a platform to screen sorafenib (25 uM) [73]. Researchers also successfully created a reliable patient-derived three-dimensional bio-printed HCC (3DP-HCC) model, using primary HCC cells isolated from patients combined with gelatin and sodium alginate to form bioink for printing. This model accurately predicted patient-specific drug responses and expanded the scope of drug screening for personalized therapeutic interventions and pharmaceutical innovations in HCC treatment [74]. Extrusion-based bioprinting was also employed to construct a 3D liver cancer model using HepG2 cells suspended in gelatin/alginate through which anti-tumor response to several drugs like cisplatin, sorafenib, and regorafenib was assessed. The model showcased increased expression of tumor-related genes and higher half-maximal inhibitory concentration (IC50) values for tested drugs compared to 2D models, suggesting the capability of the 3D scaffold to accurately replicate the tumor environment and enable more efficient drug screening [75].

Another breakthrough in 3D bioprinting technology was the fabrication of a viable 3D bio-printed liver system through NovoGen™ technology, which is a technology patented by Organovo, Inc., USA, that works on the principle of extrusion printing. The model was developed by incorporating primary hepatocytes, hepatic stellate cells, and liver endothelial cells with enhanced complexity and critical liver functions to mimic *in vivo* physiological features for drug screening and liver biology investigations [76]. This advancement paved the way for the development of the commercially available 3D printed exVive3D™ Human Liver Tissue by Organovo. The model was printed with a hexagonal configuration using primary hepatocytes, stellate cells, and endothelial cells to form microcapillaries and was exclusively used for the screening and toxicity testing of newly developed hepatotoxic and non-hepatotoxic drugs, as well as for cancer disease modelling [77].

15.4.3 Kidney Constructs

Several drugs used for treatment have inexorable nephrotoxic effects, which are discovered during clinical trials, leading to economic loss and necessitating kidney models for nephrotoxicity testing. Despite the multi-faceted framework of nephrons, researchers have been able to develop authentic models for drug testing using 3D bioprinting technology [78]. An *in vitro* cellular model of the human proximal convoluted tubule interstitial interface was developed using renal fibroblasts, epithelial cells from primary human renal proximal tubule and endothelial cells mixed with thermosensitive NovoGel on polyester membranes. Cisplatin treatment induced dose-dependent tissue viability loss, and the added TGFβ induced fibrotic response. Exact reproduction of tubule architecture was achieved as observed by histological characterization, making the construct ideal for nephrotoxicity studies [79]. A similar experiment with cisplatin was performed using bio-printed renal epithelial cell spheroids through piezo-actuated drop on demand, where the bioink positioned on a Matrigel substrate was encapsulated using a second Matrigel layer to form spheroids. Heightened sensitivity was observed when the cell death rate was assessed through a dose-dependent lactate dehydrogenase assay [80].

3D bioprinting was further employed to develop kidney organoids from induced pluripotent stem cells derived nephron progenitor cells on polycarbonate membranes for nephrotoxicity screening. With merely 8000 cells for its fabrication, the model exhibited expression markers for podocytes, proximal and distal tubules, and endothelial cells [81]. Also, a vascularised 3D model of proximal convoluted tubule attached with a perfusion system was constructed using renal proximal tubular and glomerular microvascular endothelial cells in Pluronic F127-polyethylene oxide bioink on gelatin-fibrinogen hydrogel. Albumin uptake, glucose reabsorption, and epithelium-endothelium crosstalk could be studied using this construct after inducing hyperglycemic conditions, making the model ideal for disease modelling and pharmacology [82].

Pathological conditions like clear cell renal cell carcinoma and papillary renal cell carcinoma demand the exact replication of the tumour microenvironment in drug testing [83]. Three-dimensional spheroids of patient derived papillary renal carcinoma cells surrounded by a human fibroblast layer were successfully developed by incorporating magnetic nanoparticles, which induce cell aggregation into spheroids. When treated with capmatinib, a MET oncogene inhibitor, and compared to the conventional monolayers, the model was found to exhibit increased biological dose-dependent responses [84]. Also, due to high rates of mortality due to metastasis, researchers have fabricated a basic metastasis model having a central nephroblastoma encircled by kidney fibroblasts through extrusion method for comparative analysis of drug sensitivity using panobinostat which targets cancerous cells and cytotoxicity using blasticidin which acts on all cells. It was found that the fibroblasts of the model exhibited less sensitivity to the drugs when compared to monolayers, which implies the impact of tumour microenvironment on drug testing [85].

15.4.4 Lung Constructs

Lungs are complex organs with a sophisticated hierarchical organization that traditional methods become inefficient in accurately replicating its microenvironment, making 3D bioprinting crucial for creating precise *in vitro* models for drug development and cytotoxicity studies [86]. A reproducible and automated air-blood tissue barrier analogue was constructed where endothelial and epithelial cells, along with basement membranes, were used to produce thin and uniform cell layers. The constructed model satisfied the properties of the air-blood tissue barrier *in vivo* and thus could be used for drug testing [87]. A 3D bio-printed lung tissue model for drug testing was developed by extrusion printing using a multicomponent bioink consisting of A549 cells, alginate, gelatin, diethylaminoethyl cellulose, and collagen peptide. They compared the efficacy of this 3D model with a monolayer of A549 cells and observed dose-dependent drug response with different IC50 values [88]. Using the drop-on-demand method, a 3D bio-printed triple-layer alveolar lung model was fabricated using polyvinylpyrrolidone-collagen-based hydrogel and human lung epithelial cells, human endothelial cells, and human lung fibroblasts [89].

Advancements in 3D bioprinting enabled the development of a reliable alveolar barrier model for testing drug efficacy against pulmonary fibrosis, a common lung disorder. The model exhibited fibrosis markers upon treatment with pro-fibrotic cytokines and was used to screen two approved anti-fibrotic drugs and to study the changes in mRNA expression markers [90]. In another experiment, A549 cell line was encapsulated in a gelatin-albumin bioink and was bio-printed onto matrigel that enabled proper spatial arrangement and sustainability of the construct. When tested with the influenza A virus strain, the model was found to possess an *in vivo* virus infection pattern, which enables the proper environment for drug testing [91]. Also, an array of lung cancer organoids was developed through droplet-based method using a bioink consisting of sodium alginate, hyaluronic acid, and arginine-glycine-aspartic acid peptide, which exhibited increased anti-cancer drug resistance and reduced IC50 values in addition to increased cellular activity and presence of expression markers including P-CK, ProSP-C, MUC1, and Caveolin-1 [92].

Patient-derived lung cancer organoids and lung fibroblasts, along with perfusable vessels, were printed with porcine lung-derived decellularized extracellular matrix hydrogel and developed a fibrotic niche using idiopathic pulmonary fibrosis-derived lung fibroblasts to mimic *in vivo* conditions. The success of this model proved that it could be utilised for targeted therapies and the identification of fibrosis biomarkers [93]. Further, by the incorporation of perfusion systems combined with digital light processing, an easily reproducible and small non-cell lung cancer model using GelMa-H358 bioink incorporating methacrylated gelatin-crosslinked vascular structures was fabricated. Comparative analysis of 2D monolayers and 3D bio-printed model using the anti-cancer drug gemcitabine demonstrated a threefold increase in IC50 values in the 3D bio-printed model [94].

15.4.5 Bone Constructs

Development of *in vitro* bone models for discovering therapeutic drugs for various metabolic bone disorders, as well as their screening and toxicity evaluation, is essential. Several studies on 3D bioprinting technology have effectively produced defect-matched or custom-designed 3D-printed bone constructions that mimic the complex architecture of native bone tissue by utilizing bioinks containing osteoblasts, growth factors, and biomaterials [95]. 3D bio-printed bone framework with differentiated bone marrow-derived mesenchymal stem cells (BM-MSC) was fabricated to evaluate the efficiency of two drugs, namely, Icariin and Purmorphamine, both of which promote osteogenesis and bone repair [96]. This construct, when treated with Purmorphamine (2 μM) and Icariin (1 μM), showed increased ALP activity, mineralization, and expression of bone formation markers (*BGLAP*, *SSP1*, and *COL1A1*), signalling molecules (*MAPK1*, *WNT1*, and *AKT1*), and transcription factors (*RUNX2* and *GLI1*), which confirms its ability to mimic the niche of native bone, and to serve as a relevant drug screening platform [97–99].

Researchers have also been integrating ceramic materials like hydroxyapatite and beta-tricalcium phosphate into 3D-printed scaffolds to enhance osteogenesis and to develop vascularized tissue constructs that better mimic physiological conditions for effective drug screening [100]. Human mesenchymal stem cells (MSCs) were seeded on the combined formulation of PLGA and hydroxyapatite to develop a hyperelastic bone (HB) ink, which was used to print 3D bone constructs and observed enhanced upregulation of pro-osteogenic genes such as collagen type 1, osteopontin, and osteocalcin [51]. However, ceramic materials could not mimic the complex extracellular matrix (ECM) essential for osteogenesis, and this issue was addressed by printing PCL/PLGA/TCP scaffolds coated with bone-decellularized ECM, which increased calcium deposition and gene expression [95]. Further, 3D bioprinting technology was successfully employed to address the problem of bone metastasis in advanced breast cancer and detect the progression of disease by fabricating a heterogeneous tumor-vessel-bone construct using gelatin-methacrylate

(GelMA), polyethylene glycol diacrylate (PEGDA) and hydroxyapatite nanoparticles (HAP), to which breast cancer cells (MDA-MB-231), osteoblasts (h-OB) and human umbilical vein endothelial cells (HUVECs) were incorporated to mimic the natural environment of bone metastasis of breast cancer cells [101]. The 3D co-culture model will help in evaluating the fundamental processes of breast cancer cell migration and offers an effective platform for personalized drug testing.

15.4.6 Skin Constructs

Dermatological disorders are rapidly increasing due to environmental and lifestyle changes, necessitating the development of novel therapeutic drugs and a deeper understanding of fundamental skin biology [102]. Researchers have been trying to create intricate skin substitutes and complex skin tissue models for treating skin injuries and to examine the efficacy of drugs [103, 104]. Recently, 3D bioprinting has emerged as an advanced technique for developing skin tissue substitutes to evaluate and screen various cosmetic products, skin irritants, and therapeutic drugs and has become an alternative to animal models. Several test guidelines (TG) have been provided by the Organization for Economic Cooperation and Development (OECD) for conducting toxicology tests on skin substitutes. The major guidelines include OECD TG 428, OECD TG 431, OECD TG 439, and OECD TG 442D for skin absorption, skin corrosion, skin irritation, and skin sensitization tests, respectively [105]. According to a new study, bioprinted reconstructed human epidermis (B-RHE) and manually reconstructed human epidermis (M-RHE) models were developed to serve as artificial skin for testing cosmetics and drugs, as per OECD 439 guidelines [106]. Researchers found that both models showed well-stratified, multi-layered epidermis and equivalent barrier functions, mimicking natural skin's structure and effectively responding to irritants and non-irritants. This development indicates the potential of 3D bioprinting to reconstruct epidermal models and offers a reliable alternative to the Draize test, which is a commonly employed toxicity test that directly applies the chemical to be tested on rabbit skin [106].

A 3D bio-printed skin model resembling the dermis layer was created by printing a bioink consisting of a blend of human dermal fibroblasts, gelatin, alginate, and fibrinogen. This model was then utilized for drug screening purposes due to its ability to mimic natural human dermis [45]. In another study, a full-thickness skin construct was developed using a bioink formulation consisting of alginate, gelatin, diethylaminoethyl cellulose, primary human fibroblasts, and keratinocytes. The construct expressed skin-specific markers and distinct epidermal-dermal features [107]. The utilization of 3D bioprinting technology has also revolutionized the development of diseased skin models like wounded skin models, diabetic skin models, skin cancer models etc., which enables the exploration of unknown pathological mechanisms and the development of efficient drug screening platforms. A 3D-bioprinted diabetic skin construct was developed where normal keratinocytes were co-cultured with a diabetic dermal compartment, and a vascularized diabetic hypodermis was added to enhance its resemblance to natural diabetic skin. The construct possessed typical hallmarks of type 2 diabetes, such as adipocyte hypertrophy, insulin resistance, and inflammatory reactions. It was used to test metformin and eicosapentaenoic acid (EPA), which are the first-line basic agents for treating diabetes [108].

Researchers also fabricated a pigmented skin model that represented the physiological features of vitiligo, a disorder caused by reduced melanin production by epidermal melanocytes. Skin pigmentation was induced by printing melanocytes and keratinocytes onto the dermal layer, which contains multiple layers of fibroblast-containing hydrogel. The model closely resembled the natural skin tissue due to its well-formed, layered epidermis and provides a platform for testing potential treatment options for individuals with vitiligo [48, 109]. Similarly, 3D bioprinting technology has been applied to create models for psoriasis, an inflammatory immune-related condition characterized by hardened, inflamed skin patches. This technique allows precise deposition of immune cells, growth factors, psoriatic keratinocytes, and fibroblasts to develop collagen-based skin substitutes that mimic the pro-inflammatory conditions of psoriasis-like increased cell proliferation and expression of pro-inflammatory cytokines, thus enabling effective drug screening for the disease [110, 111].

15.4.7 Intestinal Tissue Constructs

Successful drug development is often restricted by bioavailability and toxicity issues, which are linked to the intestinal mucosa, the major site of absorption, distribution, metabolism, and excretion. The development of intestinal mucosa models is inevitable for the pre-clinical screening of drugs, and the complex phenotypic organisation of intestinal villi can be reproduced only by 3D models [112]. A 3D intestinal tissue was fabricated by additive manufacturing technique using human primary intestinal epithelial cells and myofibroblasts, which was found to express functional and inducible CYP450 enzymes, tight junctions, and polarised epithelial cell types in addition to responding to inflammation and toxicity. Testing using a gastrointestinal toxicant NSAID indomethacin proved the model's reliability by demonstrating concentration dependent barrier function reduction, decreased TEER, and increased LDH release [113]. Another study employed the use of a digital light processing stereolithography system with a bioink consisting of a cross-linked mixture of GelMA, PEDGA, lithium phenyl-2,4,6-trimethylbenzoyl phosphinate, tartrazine, and fibroblasts for developing an intestinal construct with and without microstructures. Testing for differential marker expression and quantification, along with permeability assays using the drugs colchicine, atenolol, and metoprolol, revealed barrier function maintenance and other characteristics of the intestinal construct [49]. A colitis-like 3D model was developed through the inkjet method using Caco-2 and HT-29 cell lines for determining the barrier crossing efficiency of albumin nano-encapsulated anti-inflammatory drugs, roxadustat and caffeic acid phenethyl ester. Characterisation and comparison with monolayers emphasised the model's compatibility in preclinical assays, along with its relevance in drug testing [114].

Mortality due to colorectal cancers is increasing worldwide, calling for novel medications and suitable 3D models for their *in vitro* testing. An affordable and flexible colorectal cancer model was developed by bioprinting Caco-2 cells with alginate containing covalently bound arginine-glycine-aspartic acid and nanofibrillar cellulose. The construct showed granular architecture and upregulated genes for hypoxia, EGFR/KRAS signalling and cell adhesion. The efficacy of the construct was tested with three known chemotherapeutics, 5-fluoruracil, oxaliplatin, and irinotecan [115]. Also, a multicellular concentric 3D model using SW480 cells, tumor-associated macrophages, and endothelial cells was constructed through layer-by-layer extrusion, which also exhibited elevated expression levels of tumor-related genes. The model, when tested for antitumor drug sensitivity against fluorouracil, oxaliplatin, and irinotecan, showed increased chemotherapy resistance [116].

15.4.8 Other 3D Bio-printed Constructs for Disease Modelling

Drug resistance is a serious challenge faced during many cancer treatments, wherein cancer cells become resistant to chemotherapy, anti-tumour drugs, and radiation, leading to the progression of the disease and failure of treatment. In order to study this issue, drug-resistant breast cancer spheroids were created in a gelatin-alginate hydrogel using 3D bioprinting. These spheroids maintained drug-resistant phenotypes and exhibited higher levels of resistance markers, also with high EC50 values for the anti-tumour agents, camptothecin, and paclitaxel, indicating high drug resistance [117]. A similar approach was employed to fabricate multicellular tumour spheroids with controllable size for more reliable drug testing. A blood vessel layer containing fibroblasts and endothelial cells in gelatine, alginate, and fibrinogen was printed onto which glioblastoma cells (U87 MG) were seeded to mimic tumour microenvironment (TME) and proved to be an efficient platform for screening the combined effect of temozolomide (anti-cancer drug) and sunitinib (angiogenic inhibitor) [118].

A 3D bio-printed human biliary tract cancer model was developed through the sacrificial extrusion technique, where tubular structures were created inside GelMA gel, into which cholangiocarcinoma cells were cultured. The model closely mimicked *in vivo* conditions, replicating bile duct stenosis and demonstrating higher sensitivity to anti-tumor drugs than conventional 2D cell models, offering a novel and effective platform for drug screening in bile duct cancer treatment [119]. Researchers also developed a grid-shaped 3D bio-printed cervical cancer model by combining gelatin/alginate/fibrinogen hydrogel with Hela cells. The model facilitated molecular diffusion and exhibited higher synthesis of matrix metalloproteinases and lower sensitivity against paclitaxel, as compared to traditional models. It will also serve as an efficient platform to investigate cancer initiation, progression, and screening of anticancer agents [120].

Thus, drug testing using 3D bioengineered tissue constructs developed using additive manufacturing provides new insights for pharmaceutical innovation, offering a transformative platform for accelerated drug development and highlighting their potential to reduce reliance on animal testing that aligns with ethical considerations and regulatory demands. By closely replicating the intricate microenvironment of human tissues, these constructs provide more accurate and reliable results compared to traditional methods, thereby reducing the likelihood of adverse drug reactions and speeding up the discovery of new treatments. A summary of the above-mentioned 3D bio-printed tissue constructs developed for drug testing is provided in Table 15.2.

Table 15.2 Organs in human body and some of their 3D bio-printed constructs developed for drug testing

Organs	3D Bioprinted tissue constructs	Reference
Heart	Human iPSC-derived cardiac micro-tissue	[65]
	Endothelialized-myocardium-on-a-chip	[50]
	Cardiac tissue spheroids on a needle array	[66]
	Cardiac tissue construct using human iPSC-CMs, endothelial cells and fibroblasts	[67]
	Human iPSC derived cardiomyocyte model for polymorphic ventricular tachycardia	[43]
Liver	Construct with patient-derived hepatocytes, non-parenchymal cells, hepatocyte junctions	[70]
	Construct using primary human hepatocytes and liver stellate cells	[71]
	HepaRG cultures in alginate-gelatin-Matrigel hydrogel	[72]
	Construct with HUVECs, HHSCs, and HLCs in ECM-based bioink	[47]
	Endothelialized liver lobule-like HCC construct	[73]
	Patient-derived HCC (3DP-HCC) model	[74]
	Liver cancer model using HepG2 cells in gelatin/alginate	[75]
	Viable 3D bioprinted liver system through NovoGen™ technology	[76]
	exVive3D™ Human Liver Tissue by Organovo	[77]
Lung	Alveolar barrier model using epithelial and endothelial cells	[87]
	Lung tissue model using multicomponent bioink	[88]
	Triple-layer alveolar lung model	[89]
	Alveolar barrier model for pulmonary fibrosis	[90]
	Construct for testing of Influenza A virus strain	[91]
	Cancer organoids in sodium alginate, hyaluronic acid, and Arg-Gly-Asp acid peptide	[92]
	Patient-derived lung cancer organoids with perfusable vessels	[93]
	Non-cell lung cancer model using GelMa-H358 bioink	[94]
Kidney	Human PCT interstitial interface model	[79]
	3D Renal epithelial cell spheroids on Matrigel	[80]
	PCT model in Pluronic F127 PEO bioink on gelatin-fibrinogen hydrogel	[82]
	Patient-derived papillary renal carcinoma spheroids surrounded by human fibroblast layer	[84]
	3D Metastasis model with neuroblastoma encircled by kidney fibroblasts	[85]
	Kidney organoids from iPSC derived nephron progenitor cells	[81]
Skin	Human epidermis models	[45, 107]
	Diabetic skin construct with diabetic dermal compartment and vascularized diabetic hypodermis	[108]
	Pigmented skin construct in fibroblast-containing hydrogel– vitiligo disease model	[109]
	Collagen-based construct with pro-inflammatory conditions—psoriasis model	[110, 111]

(continued)

Table 15.2 (continued)

Organs	3D Bioprinted tissue constructs	Reference
Bone	Bone framework with differentiated bone marrow-derived mesenchymal stem cells	[96]
	Tumor-vessel-bone construct using GelMA, PEGDA and HAP nanoparticles	[101]
	Bone construct using hyperelastic bone (HB) bioink	[51]
	PCL/PLGA/TCP scaffolds, coated with bone decellularized ECM	[95]
Intestine	Construct with human primary intestinal epithelial cells and myofibroblasts	[113]
	Construct using crosslinked mixture of fibroblasts, GelMA, PEDGA, LAP and tartrazine	[49]
	Colitis-like model using Caco-2 and HT-29 cell lines	[114]
	Colorectal cancer model with Caco-2 cells in alginate	[115]
	Multicellular concentric 3D model using SW480 cells, tumor-associated macrophages, and endothelial cells	[116]

15.5 Challenges and Future Prospectives

Additive manufacturing technology holds significant promise for the fabrication of 3D tissue constructs that closely resemble the *in vivo* conditions of human organs. Advancements in control systems and formulation of new bioink combinations have enabled the development of more realistic tissue and organ substitutes with great potential to serve as drug screening and disease modelling platforms. Despite these contributions, there are numerous challenges that need to be addressed to ensure the successful evolution of this technology in the near future.

The primary challenge of 3D bioprinting is its inability to print tissue constructs at a macro scale, resulting in substitutes that are often much smaller in size than the actual organs in the human body [121–123]. In order to develop large-sized constructs, challenges such as time consumption, loss of cell viability and stability of the construct, and molecular transportation between cells in the construct should be solved [124]. The damage to cells by dehydration during long printing times can be mitigated by incorporating a humid chamber system with low printing temperature, which can preserve the viability of cells [125, 126]. Molecular diffusion is the next challenge that has to be solved for the advancement of 3D bioprinting. Even 3D-printed constructs with a porous polymeric framework are unable to mimic the native structure of the capillary network, which facilitates the diffusion of oxygen, nutrients, and waste products between cells [127–129]. Another limitation that needs to be addressed is the development of tissues and organs with multiple cell types. The currently existing biomaterials, bioinks, and culture mediums are insufficient to induce differentiation and optimise co-culture conditions to maintain microenvironments for various cell types [130, 131]. To an extent, this issue can be solved by the utilisation of tissue-specific bioink and decellularized extracellular matrix (dECM), which is capable of promoting cell maturation and differentiation [132]. Also, the technological difficulties in 3D bioprinting include the need for higher resolution, faster printing speeds, improved biocompatibility, and large-scale production of tissue constructs for clinical and commercial applications [28, 122].

Looking to the future, 3D bioprinting aims to revolutionize tissue engineering and regenerative medicine by evolving and expanding its capabilities. The development of a multi-organ system is crucial to resemble the actual interconnected network of organs in the human body and to study their realistic functionality [133]. Though conventional machining processes and MEMs-based multi-organ-on-a-chip systems have demonstrated improved functionality for pharmacology investigations, there's a lack of research on 3D cell printing-based multi-organ-on-a-chip systems. Researchers are trying to fabricate a vascular endothelial network, focusing on methods such as sacrificial bioink for hollow structure formation and coaxial nozzle printing to construct intricate pre-vascularized muscles containing both endothelial and muscle components [27]. Personalized medicine through 3D bioprinting is another emerging area of research, which aims to recognize unique genetic and environmental factors among patients and provide them with individualized treatment options [134]. However, achieving the complete functional maturity of cells still remains a major hurdle, especially due to the limited availability of primary cells from patients, highlighting the need for further research to improve iPSC differentiation and maturation. This may pave the way for personalized disease modelling using primary patient cells in co-culture or tri-culture platforms, offering promising avenues for future applications [28, 135].

15.6 Conclusion

Additive manufacturing technique has revolutionized the field of drug testing by developing highly precise and biologically relevant three-dimensional constructs that closely resemble the complex architecture of human tissues and organs. In recent years, the emergence of new bioink combinations, along with innovative printing technologies, has transformed additive manufacturing into a novel engineering tool capable of fabricating three-dimensional tissue constructs through layer-by-layer deposition, resulting in more reliable and accurate models for evaluating drug efficacy. Although 2D-based cell cultures and animal models were used traditionally to screen drugs and discover the mechanisms underlying various disease states, they were inefficient in accurately resembling the *in vivo* conditions of the human body due to differing physiology, microenvironments, and genetic discrepancies. However, the ability of 3D construct to mimic living tissues, made it possible to overcome the limitations of existing models and improve disease modelling as well as drug response assessment, which can reduce financial losses from drug failures in clinical trials.

Pharmaceutical companies are increasingly using 3D bio-printed constructs for drug testing and disease modelling, by employing various techniques like extrusion-based, inkjet-based, laser-assisted, and light-assisted methods. The commonly developed 3D bio-printed constructs for drug testing and disease modelling include liver, kidney, lung, heart, skin, intestinal tissues, and cancer spheroids. Furthermore, the potential of 3D bioprinting to recognize unique genetic and environmental factors among patients enables the creation of customized models for each patient and aids the development of highly personalized prevention strategies and treatment options, paving the way for personalized medicine. In conclusion, 3D bioprinting represents a significant breakthrough in drug testing and holds great promise for revolutionizing drug development and personalized medicine, leading to safer and more effective therapies.

References

1. Lage OM, Ramos MC, Calisto R, Almeida E, Vasconcelos V, Vicente F (2018) Current screening methodologies in drug discovery for selected human diseases. Mar Drugs 16:279. https://doi.org/10.3390/md16080279
2. Frankowski J, Kurzątkowska M, Sobczak M, Piotrowska U (2023) Utilization of 3D bioprinting technology in creating human tissue and organoid models for preclinical drug research—state-of-the-art. Int J Pharm 644:123313. https://doi.org/10.1016/j.ijpharm.2023.123313
3. Huang W, Percie Du Sert N, Vollert J, Rice ASC (2019) General principles of preclinical study design. In: Bespalov A, Michel MC, Steckler T (eds) Good research practice in non-clinical pharmacology and biomedicine. Springer International Publishing, Cham, pp 55–69
4. Pagnotta G, Kalia S, Di Lisa L, Cicero AFG, Borghi C, Focarete ML (2022) Progress towards 3D bioprinting of tissue models for advanced drug screening: in vitro evaluation of drug toxicity and drug metabolism. Bioprinting 27:e00218. https://doi.org/10.1016/j.bprint.2022.e00218
5. Edmondson R, Broglie JJ, Adcock AF, Yang L (2014) Three-dimensional cell culture systems and their applications in drug discovery and cell-based biosensors. ASSAY Drug Dev Technol 12:207–218. https://doi.org/10.1089/adt.2014.573
6. Murphy SV, Atala A (2014) 3D bioprinting of tissues and organs. Nat Biotechnol 32:773–785. https://doi.org/10.1038/nbt.2958
7. Agarwal S, Saha S, Balla VK, Pal A, Barui A, Bodhak S (2020) Current developments in 3D bioprinting for tissue and organ regeneration–a review. Front Mech Eng 6:589171. https://doi.org/10.3389/fmech.2020.589171
8. Knowlton S, Anand S, Shah T, Tasoglu S (2018) Bioprinting for neural tissue engineering. Trends Neurosci 41:31–46. https://doi.org/10.1016/j.tins.2017.11.001
9. Mirzaali MJ, Moosabeiki V, Rajaai SM, Zhou J, Zadpoor AA (2022) Additive manufacturing of biomaterials—design principles and their implementation. Materials 15:5457. https://doi.org/10.3390/ma15155457
10. Devarasetty M, Mazzocchi AR, Skardal A (2018) Applications of bioengineered 3D tissue and tumor organoids in drug development and precision medicine: current and future. BioDrugs 32:53–68. https://doi.org/10.1007/s40259-017-0258-x
11. Gao G, Ahn M, Cho W-W, Kim B-S, Cho D-W (2021) 3D printing of pharmaceutical application: drug screening and drug delivery. Pharmaceutics 13:1373. https://doi.org/10.3390/pharmaceutics13091373
12. Malekpour A, Chen X (2022) Printability and cell viability in extrusion-based bioprinting from experimental, computational, and machine learning views. J Funct Biomater 13:40. https://doi.org/10.3390/jfb13020040
13. Khoda AKMB, Ozbolat IT, Koc B (2011) Engineered tissue scaffolds with variational porous architecture. J Biomech Eng 133:011001. https://doi.org/10.1115/1.4002933
14. Ozbolat IT, Hospodiuk M (2016) Current advances and future perspectives in extrusion-based bioprinting. Biomaterials 76:321–343. https://doi.org/10.1016/j.biomaterials.2015.10.076
15. Rider P, Kačarević ŽP, Alkildani S, Retnasingh S, Barbeck M (2018) Bioprinting of tissue engineering scaffolds. J Tissue Eng 9:204173141880209. https://doi.org/10.1177/2041731418802090

16. Gou M, Qu X, Zhu W, Xiang M, Yang J, Zhang K, Wei Y, Chen S (2014) Bio-inspired detoxification using 3D-printed hydrogel nanocomposites. Nat Commun 5:3774. https://doi.org/10.1038/ncomms4774
17. Li X, Liu B, Pei B, Chen J, Zhou D, Peng J, Zhang X, Jia W, Xu T (2020) Inkjet bioprinting of biomaterials. Chem Rev 120:10793–10833. https://doi.org/10.1021/acs.chemrev.0c00008
18. Azizi Machekposhti S, Mohaved S, Narayan RJ (2019) Inkjet dispensing technologies: recent advances for novel drug discovery. Expert Opin Drug Discov 14:101–113. https://doi.org/10.1080/17460441.2019.1567489
19. Hagenbuchner J, Nothdurfter D, Ausserlechner MJ (2021) 3D bioprinting: novel approaches for engineering complex human tissue equivalents and drug testing. Essays Biochem 65:417–427. https://doi.org/10.1042/EBC20200153
20. Yumoto M, Hemmi N, Sato N, Kawashima Y, Arikawa K, Ide K, Hosokawa M, Seo M, Takeyama H (2020) Evaluation of the effects of cell-dispensing using an inkjet-based bioprinter on cell integrity by RNA-seq analysis. Sci Rep 10:7158. https://doi.org/10.1038/s41598-020-64193-z
21. Angelopoulos I, Allenby MC, Lim M, Zamorano M (2020) Engineering inkjet bioprinting processes toward translational therapies. Biotechnol Bioeng 117:272–284. https://doi.org/10.1002/bit.27176
22. Ventura RD (2021) An overview of laser-assisted bioprinting (LAB) in tissue engineering applications. Med Lasers 10:76–81. https://doi.org/10.25289/ML.2021.10.2.76
23. Papaioannou TG, Manolesou D, Dimakakos E, Tsoucalas G, Vavuranakis M, Tousoulis D (2019) 3D bioprinting methods and techniques: applications on artificial blood vessel fabrication. Acta Cardiol Sin 35. https://doi.org/10.6515/ACS.201905_35(3).20181115A
24. Chang J, Sun X (2023) Laser-induced forward transfer based laser bioprinting in biomedical applications. Front Bioeng Biotechnol 11:1255782. https://doi.org/10.3389/fbioe.2023.1255782
25. Li J, Chen M, Fan X, Zhou H (2016) Recent advances in bioprinting techniques: approaches, applications and future prospects. J Transl Med 14:271. https://doi.org/10.1186/s12967-016-1028-0
26. Li H, Dai J, Wang Z, Zheng H, Li W, Wang M, Cheng F (2023) Digital light processing (DLP)-based (bio)printing strategies for tissue modeling and regeneration. Aggregate 4:e270. https://doi.org/10.1002/agt2.270
27. Kolesky DB, Truby RL, Gladman AS, Busbee TA, Homan KA, Lewis JA (2014) 3D bioprinting of vascularized, heterogeneous cell-laden tissue constructs. Adv Mater 26:3124–3130. https://doi.org/10.1002/adma.201305506
28. Ma X, Liu J, Zhu W, Tang M, Lawrence N, Yu C, Gou M, Chen S (2018) 3D bioprinting of functional tissue models for personalized drug screening and in vitro disease modeling. Adv Drug Deliv Rev 132:235–251. https://doi.org/10.1016/j.addr.2018.06.011
29. Whitford WG, Hoying JB (2016) A bioink by any other name: terms, concepts and constructions related to 3D bioprinting. Future Sci OA 2:FSO133. https://doi.org/10.4155/fsoa-2016-0044
30. Dufour A, Gallostra XB, O'Keeffe C, Eichholz K, Von Euw S, Garcia O, Kelly DJ (2022) Integrating melt electrowriting and inkjet bioprinting for engineering structurally organized articular cartilage. Biomaterials 283:121405. https://doi.org/10.1016/j.biomaterials.2022.121405
31. Mahdavi SS, Abdekhodaie MJ, Kumar H, Mashayekhan S, Baradaran-Rafii A, Kim K (2020) Stereolithography 3D bioprinting method for fabrication of human corneal stroma equivalent. Ann Biomed Eng 48:1955–1970. https://doi.org/10.1007/s10439-020-02537-6
32. Moncal KK, Ozbolat V, Datta P, Heo DN, Ozbolat IT (2019) Thermally-controlled extrusion-based bioprinting of collagen. J Mater Sci Mater Med 30:55. https://doi.org/10.1007/s10856-019-6258-2
33. Maltman DJ, Przyborski SA (2010) Developments in three-dimensional cell culture technology aimed at improving the accuracy of *in vitro* analyses. Biochem Soc Trans 38:1072–1075. https://doi.org/10.1042/BST0381072

34. Ong CS, Yesantharao P, Huang CY, Mattson G, Boktor J, Fukunishi T, Zhang H, Hibino N (2018) 3D bioprinting using stem cells. Pediatr Res 83:223–231. https://doi.org/10.1038/pr.2017.252
35. Ma Y, Deng B, He R, Huang P (2024) Advancements of 3D bioprinting in regenerative medicine: exploring cell sources for organ fabrication. Heliyon 10:e24593. https://doi.org/10.1016/j.heliyon.2024.e24593
36. Gao G, Yonezawa T, Hubbell K, Dai G, Cui X (2015) Inkjet-bioprinted acrylated peptides and PEG hydrogel with human mesenchymal stem cells promote robust bone and cartilage formation with minimal printhead clogging. Biotechnol J 10:1568–1577. https://doi.org/10.1002/biot.201400635
37. Restan Perez M, Sharma R, Masri NZ, Willerth SM (2021) 3D bioprinting mesenchymal stem cell-derived neural tissues using a fibrin-based bioink. Biomolecules 11:1250. https://doi.org/10.3390/biom11081250
38. Chang E-A, Jin S-W, Nam M-H, Kim S-D (2019) Human induced pluripotent stem cells : clinical significance and applications in neurologic diseases. J Korean Neurosurg Soc 62:493–501. https://doi.org/10.3340/jkns.2018.0222
39. González-Callejo P, Vázquez-Aristizabal P, García-Astrain C, Jimenez De Aberasturi D, Henriksen-Lacey M, Izeta A, Liz-Marzán LM (2023) 3D bioprinted breast tumor-stroma models for pre-clinical drug testing. Mater Today Bio 23:100826. https://doi.org/10.1016/j.mtbio.2023.100826
40. Torras N, Zabalo J, Abril E, Carré A, García-Díaz M, Martínez E (2023) A bioprinted 3D gut model with crypt-villus structures to mimic the intestinal epithelial-stromal microenvironment. Biomater Adv 153:213534. https://doi.org/10.1016/j.bioadv.2023.213534
41. Zhang Y, Chen H, Long X, Xu T (2022) Three-dimensional-engineered bioprinted in vitro human neural stem cell self-assembling culture model constructs of Alzheimer's disease. Bioact Mater 11:192–205. https://doi.org/10.1016/j.bioactmat.2021.09.023
42. Petrachi T, Portone A, Arnaud GF, Ganzerli F, Bergamini V, Resca E, Accorsi L, Ferrari A, Delnevo A, Rovati L, Marra C, Chiavelli C, Dominici M, Veronesi E (2023) Novel bioprinted 3D model to human fibrosis investigation. Biomed Pharmacother 165:115146. https://doi.org/10.1016/j.biopha.2023.115146
43. Lappi H, Kauppila M, Aalto-Setälä K, Mörö A (2023) The 3D bioprinted human induced pluripotent stem cell-derived cardiac model: toward functional and patient-derived in vitro models for disease modeling and drug screening. Bioprinting 36:e00313. https://doi.org/10.1016/j.bprint.2023.e00313
44. Faulkner-Jones A, Fyfe C, Cornelissen D-J, Gardner J, King J, Courtney A, Shu W (2015) Bioprinting of human pluripotent stem cells and their directed differentiation into hepatocyte-like cells for the generation of mini-livers in 3D. Biofabrication 7:044102. https://doi.org/10.1088/1758-5090/7/4/044102
45. Pourchet LJ, Thepot A, Albouy M, Courtial EJ, Boher A, Blum LJ, Marquette CA (2017) Human skin 3D bioprinting using scaffold-free approach. Adv Healthc Mater 6:1601101. https://doi.org/10.1002/adhm.201601101
46. Ovsianikov A, Gruene M, Pflaum M, Koch L, Maiorana F, Wilhelmi M, Haverich A, Chichkov B (2010) Laser printing of cells into 3D scaffolds. Biofabrication 2:014104. https://doi.org/10.1088/1758-5082/2/1/014104
47. Janani G, Priya S, Dey S, Mandal BB (2022) Mimicking native liver lobule microarchitecture in vitro with parenchymal and non-parenchymal cells using 3D bioprinting for drug toxicity and drug screening applications. ACS Appl Mater Interfaces 14:10167–10186. https://doi.org/10.1021/acsami.2c00312
48. Ng WL, Qi JTZ, Yeong WY, Naing MW (2018) Proof-of-concept: 3D bioprinting of pigmented human skin constructs. Biofabrication 10:025005. https://doi.org/10.1088/1758-5090/aa9e1e
49. Macedo MH, Torras N, García-Díaz M, Barrias C, Sarmento B, Martínez E (2023) The shape of our gut: dissecting its impact on drug absorption in a 3D bioprinted intestinal model. Biomater Adv 153:213564. https://doi.org/10.1016/j.bioadv.2023.213564

50. Zhang YS, Arneri A, Bersini S, Shin S-R, Zhu K, Goli-Malekabadi Z, Aleman J, Colosi C, Busignani F, Dell'Erba V, Bishop C, Shupe T, Demarchi D, Moretti M, Rasponi M, Dokmeci MR, Atala A, Khademhosseini A (2016) Bioprinting 3D microfibrous scaffolds for engineering endothelialized myocardium and heart-on-a-chip. Biomaterials 110:45–59. https://doi.org/10.1016/j.biomaterials.2016.09.003
51. Jakus AE, Rutz AL, Jordan SW, Kannan A, Mitchell SM, Yun C, Koube KD, Yoo SC, Whiteley HE, Richter C-P, Galiano RD, Hsu WK, Stock SR, Hsu EL, Shah RN (2016) Hyperelastic "bone": a highly versatile, growth factor–free, osteoregenerative, scalable, and surgically friendly biomaterial. Sci Transl Med 8. https://doi.org/10.1126/scitranslmed.aaf7704
52. Ashammakhi N, Hasan A, Kaarela O, Byambaa B, Sheikhi A, Gaharwar AK, Khademhosseini A (2019) Advancing frontiers in bone bioprinting. Adv Healthc Mater 8:1801048. https://doi.org/10.1002/adhm.201801048
53. Axpe E, Oyen M (2016) Applications of alginate-based bioinks in 3D bioprinting. Int J Mol Sci 17:1976. https://doi.org/10.3390/ijms17121976
54. Gomes S, Leonor IB, Mano JF, Reis RL, Kaplan DL (2012) Natural and genetically engineered proteins for tissue engineering. Prog Polym Sci 37:1–17. https://doi.org/10.1016/j.progpolymsci.2011.07.003
55. Paxton N, Smolan W, Böck T, Melchels F, Groll J, Jungst T (2017) Proposal to assess printability of bioinks for extrusion-based bioprinting and evaluation of rheological properties governing bioprintability. Biofabrication 9:044107. https://doi.org/10.1088/1758-5090/aa8dd8
56. Yeleswarapu S, Dash A, Chameettachal S, Pati F (2023) 3D bioprinting of tissue constructs employing dual crosslinking of decellularized extracellular matrix hydrogel. Biomater Adv 152:213494. https://doi.org/10.1016/j.bioadv.2023.213494
57. Peng W, Datta P, Ayan B, Ozbolat V, Sosnoski D, Ozbolat IT (2017) 3D bioprinting for drug discovery and development in pharmaceutics. Acta Biomater 57:26–46. https://doi.org/10.1016/j.actbio.2017.05.025
58. Vanaei S, Parizi MS, Vanaei S, Salemizadehparizi F, Vanaei HR (2021) An overview on materials and techniques in 3D bioprinting toward biomedical application. Eng Regen 2:1–18. https://doi.org/10.1016/j.engreg.2020.12.001
59. Khoeini R, Nosrati H, Akbarzadeh A, Eftekhari A, Kavetskyy T, Khalilov R, Ahmadian E, Nasibova A, Datta P, Roshangar L, Deluca DC, Davaran S, Cucchiarini M, Ozbolat IT (2021) Natural and synthetic bioinks for 3D bioprinting. Adv NanoBiomed Res 1:2000097. https://doi.org/10.1002/anbr.202000097
60. García-Lizarribar A, Villasante A, Lopez-Martin JA, Flandez M, Soler-Vázquez MC, Serra D, Herrero L, Sagrera A, Efeyan A, Samitier J (2023) 3D bioprinted functional skeletal muscle models have potential applications for studies of muscle wasting in cancer cachexia. Biomater Adv 150:213426. https://doi.org/10.1016/j.bioadv.2023.213426
61. Ashammakhi N, Ahadian S, Xu C, Montazerian H, Ko H, Nasiri R, Barros N, Khademhosseini A (2019) Bioinks and bioprinting technologies to make heterogeneous and biomimetic tissue constructs. Mater Today Bio 1:100008. https://doi.org/10.1016/j.mtbio.2019.100008
62. Hutson CB, Nichol JW, Aubin H, Bae H, Yamanlar S, Al-Haque S, Koshy ST, Khademhosseini A (2011) Synthesis and characterization of tunable poly(ethylene glycol): gelatin methacrylate composite hydrogels. Tissue Eng Part A 17:1713–1723. https://doi.org/10.1089/ten.tea.2010.0666
63. Polonchuk L, Gentile C (2021) Current state and future of 3D bioprinted models for cardiovascular research and drug development. ADMET DMPK. https://doi.org/10.5599/admet.951
64. Wang Z, Wang L, Li T, Liu S, Guo B, Huang W, Wu Y (2021) 3D bioprinting in cardiac tissue engineering. Theranostics 11:7948–7969. https://doi.org/10.7150/thno.61621
65. Miller KL, Xiang Y, Yu C, Pustelnik J, Wu J, Ma X, Matsui T, Imahashi K, Chen S (2021) Rapid 3D BioPrinting of a human iPSC-derived cardiac micro-tissue for high-throughput drug testing. Organs---Chip 3:100007. https://doi.org/10.1016/j.ooc.2021.100007
66. Ong CS, Fukunishi T, Nashed A, Blazeski A, Zhang H, Hardy S, DiSilvestre D, Vricella L, Conte J, Tung L, Tomaselli G, Hibino N (2017) Creation of cardiac tissue exhibiting

mechanical integration of spheroids using 3D bioprinting. J Vis Exp 55438. https://doi.org/10.3791/55438
67. Finkel S, Sweet S, Locke T, Smith S, Wang Z, Sandini C, Imredy J, He Y, Durante M, Lagrutta A, Feinberg A, Lee A (2023) FRESH™ 3D bioprinted cardiac tissue, a bioengineered platform for *in vitro* pharmacology. APL Bioeng 7:046113. https://doi.org/10.1063/5.0163363
68. Soldatow VY, LeCluyse EL, Griffith LG, Rusyn I (2013) In vitro models for liver toxicity testing. Toxicol Res 2:23–39. https://doi.org/10.1039/C2TX20051A
69. Lee J-H, Ho K-L, Fan S-K (2019) Liver microsystems in vitro for drug response. J Biomed Sci 26:88. https://doi.org/10.1186/s12929-019-0575-0
70. Nguyen DG, Funk J, Robbins JB, Crogan-Grundy C, Presnell SC, Singer T, Roth AB (2016) Bioprinted 3D primary liver tissues allow assessment of organ-level response to clinical drug induced toxicity in vitro. PLoS One 11:e0158674. https://doi.org/10.1371/journal.pone.0158674
71. Mazzocchi A, Devarasetty M, Huntwork R, Soker S, Skardal A (2018) Optimization of collagen type I-hyaluronan hybrid bioink for 3D bioprinted liver microenvironments. Biofabrication 11:015003. https://doi.org/10.1088/1758-5090/aae543
72. Schmidt K, Berg J, Roehrs V, Kurreck J, Al-Zeer MA (2020) 3D-bioprinted HepaRG cultures as a model for testing long term aflatoxin B1 toxicity in vitro. Toxicol Rep 7:1578–1587. https://doi.org/10.1016/j.toxrep.2020.11.003
73. Fan Z, Wei X, Chen K, Wang L, Xu M (2023) 3D bioprinting of an endothelialized liver lobule-like construct as a tumor-scale drug screening platform. Micromachines 14:878. https://doi.org/10.3390/mi14040878
74. Xie F, Sun L, Pang Y, Xu G, Jin B, Xu H, Lu X, Xu Y, Du S, Wang Y, Feng S, Sang X, Zhong S, Wang X, Sun W, Zhao H, Zhang H, Yang H, Huang P, Mao Y (2021) Three-dimensional bioprinting of primary human hepatocellular carcinoma for personalized medicine. Biomaterials 265:120416. https://doi.org/10.1016/j.biomaterials.2020.120416
75. Sun L, Yang H, Wang Y, Zhang X, Jin B, Xie F, Jin Y, Pang Y, Zhao H, Lu X, Sang X, Zhang H, Lin F, Sun W, Huang P, Mao Y (2020) Application of a 3D bioprinted hepatocellular carcinoma cell model in antitumor drug research. Front Oncol 10:878. https://doi.org/10.3389/fonc.2020.00878
76. Robbins JB, Gorgen V, Min P, Shepherd BR, Presnell SC (2013) A novel in vitro three-dimensional bioprinted liver tissue system for drug development. FASEB J 27. https://doi.org/10.1096/fasebj.27.1_supplement.872.12
77. Nelson B (2015) 3-dimensional bioprinting makes its mark: new tissue and organ printing methods are yielding critical new tools for the laboratory and clinic. Cancer Cytopathol 123:203–204. https://doi.org/10.1002/cncy.21543
78. Yu P, Zhu H, Bosholm CC, Beiner D, Duan Z, Shetty AK, Mou SS, Kramer PA, Barroso LF, Liu H, Cheng K, Ihnat M, Gorris MA, Aloi JA, Woldemichael JA, Bleyer A, Zhang Y (2023) Precision nephrotoxicity testing using 3D in vitro models. Cell Biosci 13:231. https://doi.org/10.1186/s13578-023-01187-0
79. King SM, Higgins JW, Nino CR, Smith TR, Paffenroth EH, Fairbairn CE, Docuyanan A, Shah VD, Chen AE, Presnell SC, Nguyen DG (2017) 3D proximal tubule tissues recapitulate key aspects of renal physiology to enable nephrotoxicity testing. Front Physiol 8. https://doi.org/10.3389/fphys.2017.00123
80. Tröndle K, Miotto G, Rizzo L, Pichler R, Koch F, Koltay P, Zengerle R, Lienkamp SS, Kartmann S, Zimmermann S (2022) Deep learning-assisted nephrotoxicity testing with bioprinted renal spheroids. Int J Bioprinting 8:528. https://doi.org/10.18063/ijb.v8i2.528
81. Shin J, Chung H, Kumar H, Meadows K, Park S, Chun J, Kim K (2024) 3D bioprinting of human iPSC-derived kidney organoids using a low-cost, high-throughput customizable 3D bioprinting system. Bioprinting 38:e00337. https://doi.org/10.1016/j.bprint.2024.e00337
82. Lin NYC, Homan KA, Robinson SS, Kolesky DB, Duarte N, Moisan A, Lewis JA (2019) Renal reabsorption in 3D vascularized proximal tubule models. Proc Natl Acad Sci 116:5399–5404. https://doi.org/10.1073/pnas.1815208116

83. Xu K, Han Y, Huang Y, Wei P, Yin J, Jiang J (2022) The application of 3D bioprinting in urological diseases. Mater Today Bio 16:100388. https://doi.org/10.1016/j.mtbio.2022.100388
84. Rosette KA, Lander SM, VanOpstall C, Looyenga BD (2021) Three-dimensional coculture provides an improved in vitro model for papillary renal cell carcinoma. Am J Physiol-Ren Physiol 321:F33–F46. https://doi.org/10.1152/ajprenal.00141.2021
85. Wu D, Berg J, Arlt B, Röhrs V, Al-Zeer MA, Deubzer HE, Kurreck J (2021) Bioprinted cancer model of neuroblastoma in a renal microenvironment as an efficiently applicable drug testing platform. Int J Mol Sci 23:122. https://doi.org/10.3390/ijms23010122
86. Da Rosa NN, Appel JM, Irioda AC, Mogharbel BF, De Oliveira NB, Perussolo MC, Stricker PEF, Rosa-Fernandes L, Marinho CRF, De Carvalho KAT (2023) Three-dimensional bioprinting of an in vitro lung model. Int J Mol Sci 24:5852. https://doi.org/10.3390/ijms24065852
87. Horváth L, Umehara Y, Jud C, Blank F, Petri-Fink A, Rothen-Rutishauser B (2015) Engineering an in vitro air-blood barrier by 3D bioprinting. Sci Rep 5:7974. https://doi.org/10.1038/srep07974
88. Geevarghese R, Somasekharan LT, Bhatt A, Kasoju N, Nair RP (2022) Development and evaluation of a multicomponent bioink consisting of alginate, gelatin, diethylaminoethyl cellulose and collagen peptide for 3D bioprinting of tissue construct for drug screening application. Int J Biol Macromol 207:278–288. https://doi.org/10.1016/j.ijbiomac.2022.02.191
89. Ng WL, Ayi TC, Liu Y-C, Sing SL, Yeong WY, Tan B-H (2024) Fabrication and characterization of 3D bioprinted triple-layered human alveolar lung models. Int J Bioprinting 7:332. https://doi.org/10.18063/ijb.v7i2.332
90. Kang D, Lee Y, Kim W, Lee H-R, Jung S (2023) 3D pulmonary fibrosis model for anti-fibrotic drug discovery by inkjet-bioprinting. Biomed Mater 18:015024. https://doi.org/10.1088/1748-605X/aca8e3
91. Berg J, Hiller T, Kissner MS, Qazi TH, Duda GN, Hocke AC, Hippenstiel S, Elomaa L, Weinhart M, Fahrenson C, Kurreck J (2018) Optimization of cell-laden bioinks for 3D bioprinting and efficient infection with influenza A virus. Sci Rep 8:13877. https://doi.org/10.1038/s41598-018-31880-x
92. Dong Q, Su X, Li X, Zhou H, Jian H, Bai S, Yin J, You Q (2023) In vitro construction of lung cancer organoids by 3D bioprinting for drug evaluation. Colloids Surf Physicochem Eng Asp 666:131288. https://doi.org/10.1016/j.colsurfa.2023.131288
93. Choi Y, Lee H, Ann M, Song M, Rheey J, Jang J (2023) 3D bioprinted vascularized lung cancer organoid models with underlying disease capable of more precise drug evaluation. Biofabrication 15:034104. https://doi.org/10.1088/1758-5090/acd95f
94. Mei Y, Wu D, Berg J, Tolksdorf B, Roehrs V, Kurreck A, Hiller T, Kurreck J (2023) Generation of a perfusable 3D lung cancer model by digital light processing. Int J Mol Sci 24:6071. https://doi.org/10.3390/ijms24076071
95. La W-G, Jang J, Kim BS, Lee MS, Cho D-W, Yang HS (2016) Systemically replicated organic and inorganic bony microenvironment for new bone formation generated by a 3D printing technology. RSC Adv 6:11546–11553. https://doi.org/10.1039/C5RA20218C
96. Breathwaite E, Weaver J, Odanga J, Dela Pena-Ponce M, Lee JB (2020) 3D bioprinted osteogenic tissue models for in vitro drug screening. Molecules 25:3442. https://doi.org/10.3390/molecules25153442
97. Lou D, Ye J, Yang L, Wu Z, Zheng W, Zhang H (2019) Icariin stimulates differentiation of bone marrow-derived mesenchymal stem cells (BM-MSCs) through activation of cAMP/PKA/CREB. Braz J Pharm Sci 55:e18300. https://doi.org/10.1590/s2175-97902019000218300
98. Wöltje M, Böbel M, Heiland M, Beck-Broichsitter B, Al-Dam A, Gröbe A, Friedrich RE, Hanken H, Smeets R (2015) Purmorphamine and oxysterols accelerate and promote osteogenic differentiation of mesenchymal stem cells in vitro. Vivo Athens Greece 29:247–254
99. Yao X, Jing X, Guo J, Sun K, Deng Y, Zhang Y, Guo F, Ye Y (2019) Icariin protects bone marrow mesenchymal stem cells against iron overload induced dysfunction through mitochondrial fusion and fission, PI3K/AKT/mTOR and MAPK pathways. Front Pharmacol 10:163. https://doi.org/10.3389/fphar.2019.00163

100. Mondrinos M, Dembzynski R, Lu L, Byrapogu V, Wootton D, Lelkes P, Zhou J (2006) Porogen-based solid freeform fabrication of polycaprolactone–calcium phosphate scaffolds for tissue engineering. Biomaterials 27:4399–4408. https://doi.org/10.1016/j.biomaterials.2006.03.049
101. Cheng S, Li Y, Yu C, Deng Z, Huang J, Zhang Z (2023) 3D bioprinted tumor-vessel-bone co-culture scaffold for breast cancer bone metastasis modeling and drug testing. Chem Eng J 476:146685. https://doi.org/10.1016/j.cej.2023.146685
102. Jee MH, Mraz V, Geisler C, Bonefeld CM (2020) γδ T cells and inflammatory skin diseases. Immunol Rev 298:61–73. https://doi.org/10.1111/imr.12913
103. Ng WL, Yeong WY, Naing MW (2014) Potential of bioprinted films for skin tissue engineering. In: Proceedings of the 1st international conference on progress in additive manufacturing. Research Publishing Services, pp 441–446
104. Vijayavenkataraman S, Lu WF, Fuh JYH (2016) 3D bioprinting of skin: a state-of-the-art review on modelling, materials, and processes. Biofabrication 8:032001. https://doi.org/10.1088/1758-5090/8/3/032001
105. Ng WL, Yeong WY (2019) The future of skin toxicology testing—three-dimensional bioprinting meets microfluidics. Int J Bioprinting 5:237. https://doi.org/10.18063/ijb.v5i2.1.237
106. Bagatin JDT, Camarena DEM, Osaki LH, Freitas VM, Da Silva RO, Nold JCL, Maria-Engler SS (2023) Bioprinted and manual human epidermis production: a compared performance for skin irritation tests. Bioprinting 29:e00251. https://doi.org/10.1016/j.bprint.2022.e00251
107. Somasekharan LT, Raju R, Kumar S, Geevarghese R, Nair RP, Kasoju N, Bhatt A (2021) Biofabrication of skin tissue constructs using alginate, gelatin and diethylaminoethyl cellulose bioink. Int J Biol Macromol 189:398–409. https://doi.org/10.1016/j.ijbiomac.2021.08.114
108. Kim BS, Ahn M, Cho W-W, Gao G, Jang J, Cho D-W (2021) Engineering of diseased human skin equivalent using 3D cell printing for representing pathophysiological hallmarks of type 2 diabetes in vitro. Biomaterials 272:120776. https://doi.org/10.1016/j.biomaterials.2021.120776
109. Min D, Lee W, Bae I, Lee TR, Croce P, Yoo S (2018) Bioprinting of biomimetic skin containing melanocytes. Exp Dermatol 27:453–459. https://doi.org/10.1111/exd.13376
110. Ahn M, Cho W-W, Park W, Lee J-S, Choi M-J, Gao Q, Gao G, Cho D-W, Kim BS (2023) 3D biofabrication of diseased human skin models in vitro. Biomater Res 27:80. https://doi.org/10.1186/s40824-023-00415-5
111. Chandra A, Ray A, Senapati S, Chatterjee R (2015) Genetic and epigenetic basis of psoriasis pathogenesis. Mol Immunol 64:313–323. https://doi.org/10.1016/j.molimm.2014.12.014
112. Creff J, Malaquin L, Besson A (2021) In vitro models of intestinal epithelium: toward bioengineered systems. J Tissue Eng 12:204173142098520. https://doi.org/10.1177/2041731420985202
113. Madden LR, Nguyen TV, Garcia-Mojica S, Shah V, Le AV, Peier A, Visconti R, Parker EM, Presnell SC, Nguyen DG, Retting KN (2018) Bioprinted 3D primary human intestinal tissues model aspects of native physiology and ADME/tox functions. iScience 2:156–167. https://doi.org/10.1016/j.isci.2018.03.015
114. Almutary AG, Alnuqaydan AM, Almatroodi SA, Bakshi HA, Chellappan DK, Tambuwala MM (2023) Development of 3D-bioprinted colitis-mimicking model to assess epithelial barrier function using albumin nano-encapsulated anti-inflammatory drugs. Biomimetics 8:41. https://doi.org/10.3390/biomimetics8010041
115. Sbirkov Y, Molander D, Milet C, Bodurov I, Atanasov B, Penkov R, Belev N, Forraz N, McGuckin C, Sarafian V (2021) A colorectal cancer 3D bioprinting workflow as a platform for disease Modeling and chemotherapeutic screening. Front Bioeng Biotechnol 9:755563. https://doi.org/10.3389/fbioe.2021.755563
116. Wang P, Sun L, Li C, Jin B, Yang H, Wu B, Mao Y (2023) Study on drug screening multicellular model for colorectal cancer constructed by three-dimensional bioprinting technology. Int J Bioprinting 9:694. https://doi.org/10.18063/ijb.694
117. Hong S, Song JM (2022) 3D bioprinted drug-resistant breast cancer spheroids for quantitative in situ evaluation of drug resistance. Acta Biomater 138:228–239. https://doi.org/10.1016/j.actbio.2021.10.031

118. Han S, Kim S, Chen Z, Shin HK, Lee S-Y, Moon HE, Paek SH, Park S (2020) 3D bioprinted vascularized tumour for drug testing. Int J Mol Sci 21:2993. https://doi.org/10.3390/ijms21082993
119. Liu Q, Mille LS, Villalobos C, Anaya I, Vostatek M, Yi S, Li W, Liao J, Wu H, Song Y, Xiong L, Zhang YS (2023) 3D-bioprinted cholangiocarcinoma-on-a-chip model for evaluating drug responses. Bio-Des Manuf 6:373–389. https://doi.org/10.1007/s42242-022-00229-9
120. Lan PX, Lee JW, Seol Y-J, Cho D-W (2009) Development of 3D PPF/DEF scaffolds using micro-stereolithography and surface modification. J Mater Sci Mater Med 20:271–279. https://doi.org/10.1007/s10856-008-3567-2
121. Choi Y-J, Yi H-G, Kim S-W, Cho D-W (2017) 3D cell printed tissue analogues: a new platform for theranostics. Theranostics 7:3118–3137. https://doi.org/10.7150/thno.19396
122. Levato R, Jungst T, Scheuring RG, Blunk T, Groll J, Malda J (2020) From shape to function: the next step in bioprinting. Adv Mater 32:1906423. https://doi.org/10.1002/adma.201906423
123. Wu CA, Zhu Y, Woo YJ (2023) Advances in 3D bioprinting: techniques, applications, and future directions for cardiac tissue engineering. Bioengineering 10:842. https://doi.org/10.3390/bioengineering10070842
124. Xu H-Q, Liu J-C, Zhang Z-Y, Xu C-X (2022) A review on cell damage, viability, and functionality during 3D bioprinting. Mil Med Res 9:70. https://doi.org/10.1186/s40779-022-00429-5
125. Lee J-S, Kim BS, Seo D, Park JH, Cho D-W (2017) Three-dimensional cell printing of large-volume tissues: application to ear regeneration. Tissue Eng Part C Methods 23:136–145. https://doi.org/10.1089/ten.tec.2016.0362
126. Ricci G, Gibelli F, Sirignano A (2023) Three-dimensional bioprinting of human organs and tissues: bioethical and medico-legal implications examined through a scoping review. Bioengineering 10:1052. https://doi.org/10.3390/bioengineering10091052
127. Cai B, Kilian D, Ramos Mejia D, Rios RJ, Ali A, Heilshorn SC (2024) Diffusion-based 3D bioprinting strategies. Adv Sci 11:2306470. https://doi.org/10.1002/advs.202306470
128. Kang H-W, Lee SJ, Ko IK, Kengla C, Yoo JJ, Atala A (2016) A 3D bioprinting system to produce human-scale tissue constructs with structural integrity. Nat Biotechnol 34:312–319. https://doi.org/10.1038/nbt.3413
129. Kang T-Y, Hong JM, Jung JW, Kang H-W, Cho D-W (2016) Construction of large-volume tissue mimics with 3D functional vascular networks. PLoS One 11:e0156529. https://doi.org/10.1371/journal.pone.0156529
130. Baldwin J, Antille M, Bonda U, De-Juan-Pardo EM, Khosrotehrani K, Ivanovski S, Petcu E, Hutmacher D (2014) In vitro pre-vascularisation of tissue-engineered constructs A co-culture perspective. Vasc Cell 6:13. https://doi.org/10.1186/2045-824X-6-13
131. Belk L, Tellisi N, Macdonald H, Erdem A, Ashammakhi N, Pountos I (2020) Safety considerations in 3D bioprinting using mesenchymal stromal cells. Front Bioeng Biotechnol 8:924. https://doi.org/10.3389/fbioe.2020.00924
132. Kim BS, Das S, Jang J, Cho D-W (2020) Decellularized extracellular matrix-based bioinks for engineering tissue- and organ-specific microenvironments. Chem Rev 120:10608–10661. https://doi.org/10.1021/acs.chemrev.9b00808
133. Zhao Y, Kankala RK, Wang S-B, Chen A-Z (2019) Multi-organs-on-chips: towards long-term biomedical investigations. Molecules 24:675. https://doi.org/10.3390/molecules24040675
134. Shopova D, Yaneva A, Bakova D, Mihaylova A, Kasnakova P, Hristozova M, Sbirkov Y, Sarafian V, Semerdzhieva M (2023) (bio)printing in personalized medicine—opportunities and potential benefits. Bioengineering 10:287. https://doi.org/10.3390/bioengineering10030287
135. Scognamiglio C, Soloperto A, Ruocco G, Cidonio G (2020) Bioprinting stem cells: building physiological tissues one cell at a time. Am J Physiol-Cell Physiol 319:C465–C480. https://doi.org/10.1152/ajpcell.00124.2020

Challenges and Future Directions Biomaterial-Based Additive Manufacturing in Tissue Engineering and Regeneration

16

Amol D. Gholap, Pankaj R. Khuspe, Dipali V. Mane, Vishal Galave, Audumbar Mali, Popat Mohite, Abhijeet V. Puri, and Md. Jasim Uddin

Abstract

In the field of tissue engineering and regeneration, biomaterial-based additive manufacturing (AM) has shown great promise in providing accurate scaffolds with specific features. The difficulties and potential applications of biomaterial-based AM in tissue engineering and regeneration are covered in this chapter. The difficulties of recreating the natural tissue microenvironment, the lack of

A. D. Gholap (✉)
Department of Pharmaceutics, AETs St. John Institute of Pharmacy and Research, Palghar, Maharashtra, India

P. R. Khuspe · D. V. Mane
Department of Pharmaceutics, Shriram Shikshan Sanstha's College of Pharmacy, Paniv, Solapur, Maharashtra, India

V. Galave
Department of Pharmaceutics, Galave Research Consultancy, Sangola, Solapur, Maharashtra, India

A. Mali
Department of Pharmaceutics, Sahyadri College of Pharmacy, Methwade, Solapur, Maharashtra, India

P. Mohite
Department of Pharmaceutical Chemistry, AETs St. John Institute of Pharmacy and Research, Palghar, Maharashtra, India

A. V. Puri
Department of Pharmacognosy, AETs St. John Institute of Pharmacy and Research, Palghar, Maharashtra, India

M. J. Uddin
Faculty of Pharmacy, Department of Pharmaceutical Technology, Universiti Malaya, Kuala Lumpur, Malaysia

S. Singh et al. (eds.), *Biomaterial-based Additive Manufacturing in Tissue Engineering and Regeneration*, Biomaterials, Bioengineering and Sustainability 7,
https://doi.org/10.1007/978-3-031-96070-3_16

biocompatible materials appropriate for printing, and the requirement for faster and more precise printing techniques are some of the difficulties associated with biomaterial-based additive manufacturing. Innovations in material science, bio-ink creation, and printing technologies are necessary to overcome these obstacles. The creation of novel biomaterials with improved biocompatibility and bioactivity, the integration of additive manufacturing with other bio fabrication methods (such as bioprinting and electrospinning), and the application of innovative imaging methods for patient-specific scaffold design are some future directions for biomaterial-based AM in tissue engineering and regeneration. Furthermore, the fabrication of functional tissue constructs for drug screening and transplantation offers considerable potential when biomaterial-based AM is combined with stem cell technologies and bioreactor systems. Translating these advances from the lab to clinical practice will need cooperation between investigators, medical professionals, and regulatory agencies. In conclusion, by overcoming present obstacles and pursuing novel avenues, biomaterial-based AM can transform tissue engineering and regeneration. Biomaterial-based AM can lead to better patient outcomes and individualized regenerative medicine by tackling these issues and looking toward the future.

Keywords

Biomaterial-based additive manufacturing · Tissue engineering · Regeneration · Challenges · Future directions

Abbreviations

2D	Two-dimensional
2PP	Two-photon polymerization
3D	Three-dimensional
AFM	Atomic force microscopy
AM	Additive manufacturing
CT	Computed tomography
DLP	Digital light processing
ECM	Extracellular matrix
FDA	Food and Drug Administration
HA	Hydroxyapatite
micro-CT	Micro-computed tomography (micro-CT)
MRI	Magnetic resonance imaging
PCL	Polycaprolactone
PEG	Polyethylene glycol
PLGA	Poly(lactic-co-glycolic acid)
SEM	Scanning electron microscopy
TCP	Tricalcium phosphate
VEGF	Vascular endothelial growth factor

16.1 Introduction

In the realm of tissue engineering and regeneration, biomaterial-based additive manufacturing has become a notable change, providing hitherto unseen possibilities to tackle pressing issues. This chapter offers a thorough summary of the state of biomaterial-based additive manufacturing today, emphasizing the role it plays in overcoming present obstacles and examining potential future paths [143]. Researchers and clinicians can create advanced tissue constructs with biomimetic properties, promising improved therapeutic outcomes for patients, by utilizing the special qualities of additive manufacturing, such as precision, customization, and scalability. The complexity of simulating the native tissue microenvironment presents another challenge [145]. The extracellular matrix, many cell types, and signaling molecules that make up bodily tissues interact in a highly coordinated way to preserve tissue homeostasis. It is an extremely difficult endeavor to replicate this intricacy in vitro; it calls for a thorough grasp of tissue biology as well as the development of advanced bio fabrication techniques. Additionally, there are difficulties in combining additive manufacturing with other bio fabrication methods, such electrospinning and bioprinting, to produce multi-scale, multi-material structures with hierarchical architectures [123].

Notwithstanding these obstacles, the science of biomaterials has made great strides that have resulted in the creation of innovative biomaterials with improved biocompatibility and bioactivity [87, 117, 124]. These substances, which include scaffold materials and bioinks, are made to offer the biochemical cues and mechanical support required for tissue regeneration and cell proliferation. In order to create hybrid constructs with improved structural and functional properties, researchers are also investigating the integration of additive manufacturing with other bio fabrication techniques, such as electrospinning and bioprinting [130]. Collaboration among researchers, clinicians, and regulatory bodies is crucial for the advancement of biomaterial-based additive manufacturing in tissue engineering and regeneration. In order to guarantee that the created tissue constructs fulfill clinical specifications and are secure for transplantation, researchers must collaborate closely with physicians [146]. Furthermore, cooperation with regulatory agencies is required to guarantee that the manufactured structures adhere to the applicable laws and specifications [48, 50, 52]. To help patients in need of tissue regeneration therapies, researchers, doctors, and regulatory agencies can collaborate to expedite the translation of advances in biomaterial-based additive manufacturing to clinical practice. For tissue engineering and regeneration, biomaterial-based additive manufacturing has enormous potential and presents hitherto unheard-of chances to tackle critical issues in the area [144]. Researchers, physicians, and regulatory agencies may collaborate to transform tissue engineering and regeneration and, in the process, enhance patient outcomes and quality of life by tackling these obstacles and investigating novel avenues. The future of regenerative medicine, in which intricate tissues and organs can be precisely and effectively built, is glimpsed through the combination of biomaterial-based additive manufacturing with improved imaging tools, stem cell technologies, and bioreactor systems [4].

16.1.1 Biomaterial-Based Additive Manufacturing in Tissue Engineering and Regeneration

The three-dimensional (3D) bioprinting, also known as biomaterial-based additive manufacturing, is a innovative technique that combines materials science, engineering, and regenerative medicine. This novel method allows for the layer-by-layer deposition of biomaterials, cells, and bioactive compounds to precisely fabricate 3D structures [99]. In contrast to conventional manufacturing techniques like molding and casting, additive manufacturing provides unmatched control over the material's spatial distribution, enabling the construction of intricate objects with exact design and functionality. Biomaterial-based additive manufacturing has enormous potential to meet the increasing need for functioning tissues and organs for drug screening, transplantation, and disease modeling in the context of tissue engineering and regeneration [61].

Fundamentally, biomaterial-based additive manufacturing depends on the use of biomaterials that can enable cell adhesion, proliferation, and differentiation while remaining in harmony with living systems [134]. The term "biomaterials" refers to a broad class of materials that include ceramics (like hydroxyapatite, tricalcium phosphate), synthetic polymers (like polycaprolactone, polyethylene glycol), natural polymers (like collagen, gelatin, and alginate), and composite materials (such polymer-ceramic hybrids) [47, 50, 102]. The target tissue or organ's unique requirements, along with other parameters including the desired mechanical qualities, degradation kinetics, and bioactivity, all influence the choice of biomaterials [25].

Biomaterials are combined with additive manufacturing to create printable inks or scaffolds that may be layered to create intricate three-dimensional structures. For biomaterial-based additive manufacturing, several printing techniques have been developed, including stereolithography, extrusion, inkjet, and laser printing. Regarding resolution, speed, scalability, and suitability for various biomaterials, each of these modalities has specific benefits and drawbacks [107]. For instance, printing very fluid biomaterials like hydrogels is a good fit for extrusion-based printing, which also produces structures with excellent structural stability. Conversely, inkjet printing provides fine control over the size and location of droplets, making it possible to deposit cells and bioactive compounds with a high degree of spatial resolution [139].

Cell sources are essential to biomaterial-based additive manufacturing for tissue engineering and regeneration, in addition to biomaterials and printing techniques [44, 49, 51]. To produce cell-filled structures with tissue-specific functionality, the printing method can be combined with primary cells, stem cells, and cell lines obtained from diverse tissue sources [17]. Because they may differentiate into several cell types, stem cells provide specific features that make them suitable for creating complex tissues and organs. Moreover, the incorporation of bioactive compounds into the printed structures, like growth factors, cytokines, and extracellular matrix proteins, can improve tissue regeneration and cell behavior [118].

The creation of functional tissues and organs that closely resemble their natural counterparts in terms of shape, content, and function is the goal of biomaterial-based additive manufacturing in tissue engineering and regeneration. Numerous uses for these created tissues exist, such as organ transplantation, tissue replacement and repair, drug screening, and disease modeling [69]. To encourage regeneration and repair, for instance, 3D-printed scaffolds infused with patient-derived cells can be inserted into diseased or damaged tissues. In a similar vein, 3D-printed microstructures integrated into organ-on-a-chip platforms can replicate the physiological intricacy of human organs, facilitating in vitro research on medication responses and disease causes [90].

To sum up, biomaterial-based additive manufacturing is a revolutionary method to tissue engineering and regeneration, providing unmatched precision and efficacy in the fabrication of complex tissues and organs [39–41]. Researchers and clinicians can advance the field of regenerative medicine and ultimately improve the quality of life for patients suffering from tissue damage, organ failure, and degenerative diseases by utilizing the synergistic interplay between biomaterials, printing modalities, cells, and bioactive molecules [154]. Nonetheless, there are still a few issues that need to be resolved, such as creating biocompatible and useful biomaterials, streamlining printing procedures, and integrating laboratory-based research into clinical settings. Biomaterial-based additive manufacturing holds the promise of transforming healthcare and launching a new chapter in the history of regenerative medicine if it can overcome these obstacles and continue to push artistic boundaries [101].

16.1.2 Importance of Addressing Challenges and Exploring Future Directions

In the field of tissue engineering and regeneration, biomaterial-based additive manufacturing has gained traction as a potential technology that can produce intricate, patient-specific tissue constructions [43, 45, 100]. The traditional tissue engineering method is presented in Fig. 16.1. Even though this field has seen tremendous advancements, several issues still need to be resolved before this technology can reach its full potential. To further the field and reach its maximum therapeutic potential, it is imperative to investigate future directions. The scarcity of biocompatible materials that are suitable for printing is one of the main obstacles facing biomaterial-based additive manufacturing [114]. While numerous biomaterials, such as hydrogels, polymers, and ceramics, have been employed in additive manufacturing, many of these materials lack the mechanical qualities and biocompatibility required to promote tissue regeneration and cell proliferation. In order to overcome this obstacle, new biomaterials with improved biocompatibility and bioactivity must be created, and current materials must be optimized to make them more suitable for additive manufacturing [12].

The requirement for faster and higher resolution printing processes presents another difficulty. Creating intricate, functioning tissue constructions requires great precision and speed, which is sometimes beyond the capabilities of current additive printing processes. This restriction makes it more difficult to create tissues that

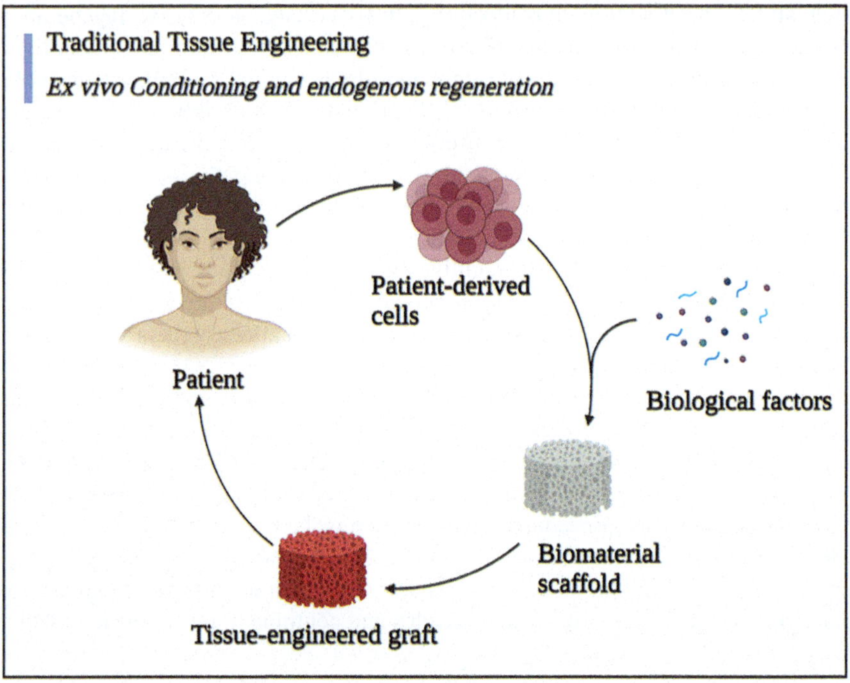

Fig. 16.1 The traditional tissue engineering method contains *ex vivo* conditioning and endogenous regeneration. (Created by using Biorender.com)

closely resemble natural tissues in terms of microarchitecture and cellular organization. The creation of novel printing methods and technologies that can increase speed and resolution without sacrificing the viability and functionality of printed tissues is necessary to overcome this obstacle [151].

The difficulty of simulating the natural tissue microenvironment is another concern. The extracellular matrix, many cell types, and signaling molecules that make up bodily tissues interact in a highly coordinated way to preserve tissue homeostasis [9]. It is extremely difficult to replicate this in vitro, necessitating a thorough grasp of tissue biology and the creation of advanced bio fabrication methods [42, 46, 110]. To overcome this difficulty, it is necessary to incorporate various biomaterials, cells, and bioactive chemicals into the printing process in order to produce structures that closely resemble the natural tissue microenvironment [82].

Biomaterial-based additive manufacturing has enormous potential for tissue engineering and regeneration, despite these obstacles. To fulfill this potential and take the profession to new heights, it is imperative that future directions in this area be investigated [22]. The incorporation of innovative imaging methods for patient-specific scaffold construction is one potential path forward. Researchers can produce extremely accurate 3D models of a patient's anatomy by combining additive manufacturing with methods like MRI and CT scans. This enables the creation of scaffolds that exactly match the patient's tissue structure. Better patient results could result from this tailored strategy, which has the potential to greatly increase the efficacy of tissue engineering and regeneration therapies [88].

The combination of biomaterial-based additive manufacturing with bioreactor systems and stem cell technology is another potential avenue for the future [15, 64, 65, 129]. Because stem cells may develop into many types of cells, they are perfect for tissue regeneration. Researchers may generate complex tissue constructs that closely resemble the original tissue milieu by merging stem cells with biomaterial-based scaffolds made using additive manufacturing. Additionally, the physiological cues required for cell growth and tissue maturation such as mechanical stresses and biochemical signals can be supplied via bioreactor systems. This coordinated strategy may quicken the creation of functional tissue constructs for use in medication screening and transplantation [152].

To advance the field and reach the full clinical potential of biomaterial-based additive manufacturing for tissue engineering and regeneration, it is imperative to tackle obstacles and investigate new avenues for future research [30, 34, 38, 142]. Through the resolution and speed of printing processes, the difficulty of simulating the native tissue microenvironment, the scarcity of biocompatible materials, and other issues, researchers can get past the obstacles that stand in the way of developing novel and creative approaches to tissue engineering and regeneration [74]. Furthermore, researchers can progress the discipline and create new methods for producing intricate, functional tissue structures by investigating future avenues like integration with stem cell technologies, advanced imaging tools, and bioreactor systems. Biomaterial-based additive manufacturing holds the potential to transform tissue engineering and regeneration, as well as enhance the quality of life for numerous patients experiencing tissue damage and organ failure, by tackling obstacles and investigating novel avenues [132].

16.1.3 Challenges in Biomaterial-Based Additive Manufacturing

3D bioprinting, or biomaterial-based additive manufacturing, has become a meaningful change in tissue engineering and regenerative medicine. To fully reap the benefits of this state-of-the-art technology, several obstacles must be overcome, despite its enormous potential. These issues cover a wide range of topics related to additive manufacturing, such as the difficulty of simulating the natural tissue microenvironment and material selection as well as printing resolution [105]. To progress the field and get over the present obstacles in the creation of functioning tissue constructions, several issues must be resolved. We will delve into these issues in-depth in this chapter and talk about solutions that will help biomaterial-based additive manufacturing become widely used in tissue engineering and regeneration [12].

16.1.4 Limited Availability of Biocompatible Materials Suitable for Printing

The scarcity of biocompatible materials that are suitable for printing is one of the main obstacles facing biomaterial-based additive manufacturing. Many natural and synthetic polymers, ceramics, and composite materials are among the many biomaterials that have been investigated for use in additive manufacturing; however, many of these materials are deficient in the mechanical, biocompatible, and bioactive qualities needed for successful tissue engineering applications. This restriction makes it more difficult to create viable tissue constructs with the appropriate levels of cell viability, structural integrity, and tissue regeneration [109].

The quality and functionality of the created tissue constructs are directly impacted by the choice of appropriate biomaterials for additive manufacturing [27, 28, 35, 137]. To enable cell adhesion, proliferation, and differentiation, biomaterials used in additive manufacturing should ideally have a combination of qualities, such as biocompatibility, mechanical strength, degradation kinetics, and bioactivity [26, 31, 32, 113]. Finding biomaterials that satisfy each of these requirements is still quite difficult, though [8]. Natural polymers that are similar to the extracellular matrix (ECM) of biological tissues and biocompatible, such collagen, gelatin, and alginate, are frequently utilized in additive manufacturing. For instance, the primary structural protein in the ECM, collagen, has been applied to tissue engineering for the regeneration of skin, bone, and cartilage. However, natural polymers frequently have weak mechanical qualities and quickly deteriorate, which makes them unsuitable for long-term tissue regeneration and load-bearing applications [133].

In comparison to natural polymers, synthetic polymers like polycaprolactone (PCL), poly (lactic-co-glycolic acid) (PLGA), and polyethylene glycol (PEG) have superior mechanical properties and degradation kinetics. One biodegradable polymer that works well in additive manufacturing is PCL, which possesses good mechanical strength and flexibility. Nevertheless, it is possible that synthetic polymers do not have the biological cues that encourage tissue regeneration and cell attachment, thus adding bioactive chemicals to the printing process will be important [83]. Because of their superior osteoconductive qualities and biocompatibility, ceramics like hydroxyapatite (HA) and tricalcium phosphate (TCP) are frequently utilized in additive manufacturing. Because of their ability to encourage bone regeneration and integration with surrounding tissues, these materials are frequently used in bone tissue engineering applications. Ceramics, on the other hand, are brittle materials that can be difficult to work with using additive manufacturing techniques, necessitating the development of specific printing procedures and tools [115].

Composite materials present a viable way to get beyond the drawbacks of individual materials by combining two or more types of biomaterials [33, 80, 112]. For instance, scaffolds with improved mechanical and biological functioning can be created by combining the flexibility of polymers with the mechanical strength of ceramics in polymer-ceramic composites. To guarantee uniformity and compatibility, however, the development of composite materials necessitates meticulous component selection and processing parameter optimization. Significant advancements in the creation of biocompatible materials for additive manufacturing have been accomplished in spite of these obstacles [67]. Researchers have investigated several techniques, including surface modification, crosslinking, and the addition of bioactive compounds, to improve the characteristics of biomaterials. For example, to build cell-laden constructions with enhanced biological functionality, researchers have developed bioinks, which are printed materials containing cells and bioactive chemicals. Bioinks provide a flexible foundation for tissue engineering applications since they may be adapted to certain tissue types and printing methods [91].

The variety of materials available for additive manufacturing has also increased as a result of developments in material science, including the creation of novel polymers, ceramics, and composite materials [29, 36, 37, 78, 79]. For application in tissue engineering, for instance, scientists have created biodegradable polymers with adjustable mechanical characteristics and degradation kinetics. Comparably, the potential of innovative ceramic materials like bioactive glasses and glass ceramics for tissue repair and bone regeneration has been investigated. Extracellular Matrix-Based Strategy for Intervertebral Disc Regeneration is presented in Fig. 16.2. One major obstacle to biomaterial-based additive manufacturing is the scarcity of biocompatible materials that can be printed. However, novel, and creative biomaterials for tissue engineering and regeneration have been made possible by recent developments in material science and bio fabrication methods. Researchers can get over the obstacles in the way of fully realizing the potential of biomaterial-based

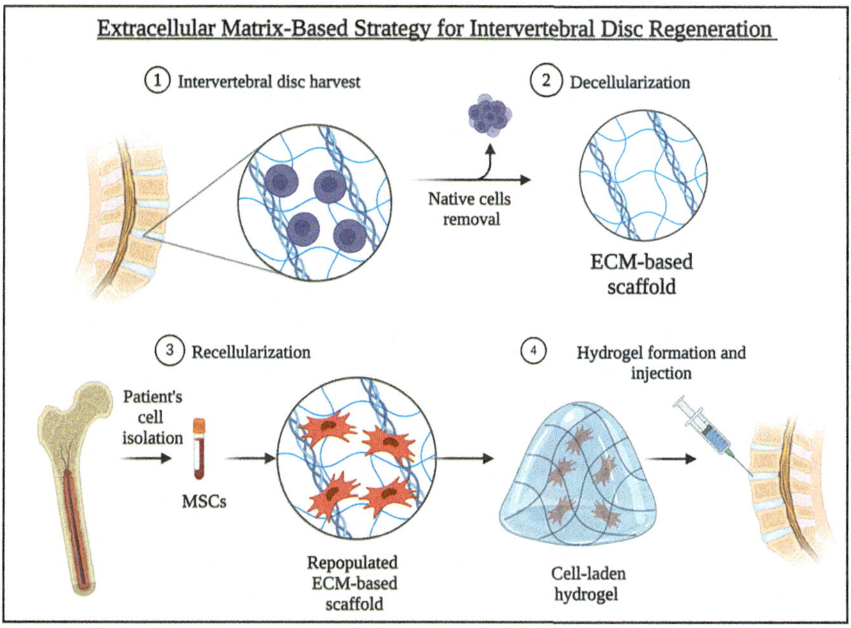

Fig. 16.2 Extracellular matrix-based strategy for intervertebral disc regeneration. (Created by using Biorender.com)

additive manufacturing in tissue engineering and regenerative medicine by tackling the difficulties related to material development and selection [95].

16.1.5 Need for Improved Resolution and Speed in Printing Processes

The resolution and speed of printing techniques are critical factors in defining the quality and usefulness of the generated tissue structures in biomaterial-based additive manufacturing, or 3D bioprinting. Even though biocompatible materials and printing methods have advanced significantly, there is still a rising need to increase printing processes' speed and resolution in order to satisfy the needs of intricate tissue engineering applications. This chapter examines the benefits of increased resolution and speed in additive manufacturing and the obstacles and possibilities that must be overcome to reach this goal. The quality and performance of printed tissue constructs are directly impacted by two crucial factors: resolution and speed. Speed is the rate at which the printing process can be finished, whereas resolution is the capacity to precisely reproduce minute details and complex structures in the printed forms. To create tissues with complex geometries, hierarchical structures, and high cell density which is essential for simulating the natural tissue milieu and stimulating tissue regeneration additive manufacturing resolution and speed must be improved [136].

The printing technology's limitations are a major obstacle to attaining high resolution in additive manufacturing. Most additive manufacturing methods, such inkjet and extrusion-based printing, depend on the layer-by-layer deposition of material, which can cause "layering," a staircase effect that lowers resolution and surface roughness. This restriction is especially problematic when printing delicate details that need high resolution to function, such blood arteries and nerve fibers [20]. The scarcity of printing materials with high resolution is another difficulty. Although many different biomaterials have been employed in additive manufacturing, some materials' viscosity, mechanical characteristics, and suitability for printing techniques prevent them from being used to achieve high resolution. For additive manufacturing to advance, novel biomaterials with the qualities needed to achieve high resolution must be created [96].

The intricacy of the printing process, the requirement for exact control over material deposition, and the limits of the printing apparatus all make it difficult to achieve fast speeds in additive manufacturing [103]. For instance, the continuous extrusion of material through a nozzle is the basis of extrusion-based printing, which is frequently employed in tissue engineering. This can impede printing speed because it necessitates maintaining a constant flow rate and deposition pattern. An additional difficulty is the requirement for post-processing procedures like cell culture and crosslinking, which can slow down manufacturing and increase the total printing time. Accelerating additive manufacturing requires boosting the effectiveness of post-processing stages and creating novel printing techniques that can shorten the time needed for these procedures [5].

The resolution and speed of additive manufacturing have improved greatly despite these obstacles. In comparison to conventional printing methods, researchers have created novel printing processes like digital light processing (DLP) and two-photon polymerization (2PP), which offer improved resolution and speed. For instance, DLP enables the quick construction of high-resolution structures by selectively curing a photosensitive resin using a digital light projector [3].

Furthermore, novel biomaterials with improved qualities for additive manufacturing have been developed because of advances in material science. To fabricate tissues with intricate geometries and hierarchical architectures, for instance, scientists have created bioinks with great resolution and mechanical stability. Furthermore, improvements in post-processing methods like the use of microfluidic devices for cell culture have allowed researchers to shorten the time needed for post-processing procedures, which has accelerated additive manufacturing as a whole [111].

Tissue engineering and regenerative medicine are fields that depend on increased resolution and speed in additive manufacturing to progress. Researchers can create novel and creative methods for creating tissues with intricate geometries and hierarchical structures by tackling the difficulties in reaching high resolution and speed. This will benefit patients in need of tissue regeneration therapies. The achievement of high-resolution, high-speed additive manufacturing in tissue engineering will depend on ongoing developments in material science, printing technology, and post-processing methods [59].

16.1.6 Complexity of Mimicking the Native Tissue Microenvironment

The intricate and ever-changing local tissue microenvironment is a vital component in controlling tissue function and cellular behavior. Replicating the complex interactions between cells, ECM, and signaling molecules present in real tissues in vitro presents a substantial difficulty in biomaterial-based additive manufacturing. This chapter examines the intricate characteristics of the natural tissue microenvironment, the difficulties in reproducing its intricacy in additive manufacturing, and the approaches being developed to address these difficulties [76].

The intricate web of cells, ECM proteins, growth factors, and signaling molecules that makes up the native tissue microenvironment interacts in a highly coordinated way to preserve tissue homeostasis and control cellular activity. In addition to giving cells structural support, the ECM stores growth factors and signaling chemicals. ECM elements and signaling molecules are secreted by cells in the tissue microenvironment, such as immune cells, stromal cells, and parenchymal cells. These chemicals aid in the regulation of cell adhesion, migration, proliferation, and differentiation [93].

It is difficult to replicate in vitro the intricacy of the natural tissue milieu for several reasons. Replicating the temporal and spatial arrangement of cells and extracellular matrix elements present in natural tissues is one of the primary challenges. The ECM is an intricate web of proteins and polysaccharides that gives cells structural support and controls their motility. Precise control over the deposition of cells and ECM components is necessary to replicate this intricate architecture in vitro, and this is difficult to accomplish with conventional additive manufacturing methods [120]. Replicating the dynamic character of the tissue microenvironment presents another difficulty. Since cells in the native tissue microenvironment are always interacting with the extracellular matrix and reacting to environmental changes, it is a dynamic system. It will take the creation of biomaterials and culture systems that can imitate the natural tissue microenvironment to replicate this dynamic behavior in vitro. Furthermore, there are many different cell types in the native tissue microenvironment, each with special characteristics and roles. It is difficult to replicate this cellular diversity in vitro and necessitates the creation of culture techniques that can sustain the growth and differentiation of several cell types at once [92].

Despite these obstacles, several techniques have been developed to replicate in biomaterial-based additive manufacturing the intricacy of the native tissue microenvironment. Using bioinks, printable materials with cells and ECM components that can replicate the mechanical characteristics and biochemical makeup of the natural ECM, is one method. Bioinks provide a flexible foundation for tissue engineering applications since they may be adapted to certain tissue types and printing methods. Utilizing sophisticated bio fabrication methods, such as electrospinning and bioprinting, which enable the exact deposition of cells and ECM components to produce intricate tissue constructions, is an additional strategy. For instance, multicellular spheroids that replicate the cellular relationships and architecture of natural tissues can be produced via bioprinting processes. Similarly, nanofibrous scaffolds with a structure similar to the original ECM can be made via electrospinning procedures [84].

A difficult but crucial objective for tissue engineering and regenerative medicine is to mimic the complexity of the natural tissue microenvironment in biomaterial-based additive manufacturing. Researchers can create novel and creative approaches to creating tissues with improved functionality and regenerative potential by tackling the issues related to signaling control, cell architecture, and tissue composition. To imitate the native tissue milieu in additive manufacturing and enhance the results of tissue regeneration therapies, further developments in biomaterials, bio fabrication methods, and tissue engineering strategies will be essential [89].

16.2 Advancements in Material Science

Material science has played a vital role in the evolution of tissue engineering and regenerative medicine, particularly in the context of biomaterial-based additive manufacturing. This chapter analyzes the recent breakthroughs in material science that are moving the industry forward, concentrating on their implications for overcoming hurdles and setting future directions in biomaterial-based additive manufacturing for tissue engineering and regeneration. The discovery of innovative biomaterials sits at the heart of these breakthroughs, with researchers continuously aiming to better biocompatibility, bioactivity, and mechanical qualities [63]. Novel biomaterials offer personalized solutions for tissue engineering applications, delivering scaffolds that closely replicate the original ECM and enable cell adhesion, proliferation, and differentiation. Additionally, breakthroughs in material science have led to the creation of bioactive materials that may actively interact with cells and biological systems, further improving their regenerative potential [135]. Furthermore, the integration of additive manufacturing with material science has opened new possibilities to produce complex tissue constructions. Additive manufacturing techniques, such as 3D bioprinting and electrospinning, allow for the exact deposition of biomaterials, cells, and growth factors to construct things with specific topologies and features. These techniques enable the fabrication of tissue constructs with hierarchical architecture and regulated mechanical properties, closely approximating real tissues. In addition to biomaterial creation, developments in material characterization techniques have also been crucial in developing the area. Techniques such as atomic force microscopy (AFM), scanning electron microscopy (SEM), and confocal microscopy allow researchers to observe and investigate the structure and properties of biomaterials at the nanoscale [104]. This level of detail is critical for understanding how biomaterials interact with cells and tissues, driving the creation of optimal scaffolds for tissue engineering applications. Looking ahead, the future of biomaterial-based additive manufacturing in tissue engineering and regeneration is bright, with continued developments in material science driving innovation. By tackling present problems and exploring new pathways for study, researchers can unlock the full potential of biomaterials for tissue engineering applications, opening the way for the development of next-generation therapeutics for tissue repair and regeneration [122].

16.2.1 Development of Novel Biomaterials with Enhanced Biocompatibility and Bioactivity

In the field of biomaterial-based additive manufacturing, material science is essential, especially for tissue engineering and regenerative medicine. To create functional tissue constructs that can replicate the natural tissue microenvironment and encourage tissue regeneration, new biomaterials with improved biocompatibility and bioactivity must be developed. The creation of innovative biomaterials for additive manufacturing and their possible uses in tissue engineering and regenerative medicine are the main topics of this chapter's exploration of current developments in material science [140]. Two important characteristics that establish a biomaterial's appropriateness for tissue engineering applications are biocompatibility and bioactivity. The term "biocompatibility" describes a material's capacity to interact with living tissues without producing unfavorable effects like toxicity or inflammation. Contrarily, bioactivity describes a substance's capacity to trigger biological reactions in live tissues, such as encouraging cell adhesion, proliferation, and differentiation. In order to create functional tissue constructs that can completely integrate with the host tissue and assist tissue regeneration, biomaterials with enhanced biocompatibility and bioactivity are crucial [66].

It is difficult to create new biomaterials with improved biocompatibility and bioactivity for many reasons. The requirement to replicate the intricate mechanical and biological characteristics of the native ECM is one of the primary obstacles. The ECM is an intricate web of proteins and polysaccharides that gives cells structural support and controls their motility. It takes a profound grasp of tissue biology and the creation of novel biomaterials to replicate this intricacy in synthetic biomaterials. The requirement to create biomaterials that can be processed by additive manufacturing methods presents another difficulty [149]. Materials that can be accurately deposited layer by layer are needed for additive manufacturing to generate complicated tissue constructions. To enable accurate deposition and maintain the structural integrity of the printed structures, biomaterials used in additive manufacturing need to have the appropriate rheological characteristics, such as viscosity and shear-thinning behavior [1].

The creation of innovative biomaterials with improved biocompatibility and bioactivity for additive manufacturing has advanced significantly despite these obstacles. Using natural polymers that closely mimic the makeup of the native extracellular matrix, such as collagen, gelatin, and hyaluronic acid, is one strategy. For tissue engineering applications, these natural polymers can be mixed or altered to improve their mechanical characteristics and bioactivity. The creation of artificial polymers with specific qualities for tissue engineering is an additional strategy [14]. For instance, scientists have created artificial polymers with adjustable mechanical characteristics and kinetics of breakdown, which enables the creation of tissue constructions with certain attributes for various uses. Furthermore, the production of nanomaterials such as nanoparticles and nanofibers that can increase the bioactivity of biomaterials by encouraging cell adhesion, proliferation, and differentiation has been made possible by advances in nanotechnology [108].

Enhancing the bioactivity and functionality of biomaterials for tissue engineering applications has been the focus of recent developments in biomaterial development. To promote tissue integration and cell adhesion, for instance, scientists have created biomimetic scaffolds that closely resemble the structure and makeup of the natural extracellular matrix. Moreover, it has been demonstrated that adding bioactive molecules to biomaterials, such as cytokines and growth hormones, increases their bioactivity and encourages tissue regeneration [77]. For the area of biomaterial-based additive manufacturing to advance, new biomaterials with improved biocompatibility and bioactivity must be developed. It is possible for researchers to produce functioning tissue constructs that can replicate the natural tissue milieu and encourage tissue regeneration by tackling the difficulties involved in material synthesis and processing. Novel biomaterials for tissue engineering and regenerative medicine applications will require ongoing developments in material science, nanotechnology, and bio fabrication methods [126].

16.2.2 Integration of Additive Manufacturing with Other Bio-fabrication Techniques

One intriguing way to get beyond the drawbacks of individual approaches and improve tissue engineering's potential is to combine additive manufacturing with other bio fabrication technologies like electrospinning and bioprinting. This chapter examines the prospects and challenges of merging additive manufacturing with other bio fabrication techniques to advance tissue engineering and regenerative medicine. It also looks at the synergistic potential of this integration [131]. Using precise deposition of cells, biomaterials, and bioactive substances, bioprinting is a specialized form of additive manufacturing that produces three-dimensional tissue constructions. Researchers can precisely regulate the spatial distribution of cells and biomaterials to generate complex tissue structures through the combination of additive manufacturing and bioprinting. The creation of tissues with diverse cell populations, vascular networks, and biochemical cue gradients is all necessary to replicate the natural tissue microenvironment which is made possible by this integration [18].

Keeping the printing materials and biological components compatible is one of the biggest obstacles to combining bioprinting and additive manufacturing. Cells must continue to be viable and functioning during the printing process, and biomaterials utilized in additive manufacturing must be both biocompatible and compatible with the printing process. To obtain the required tissue architecture and functionality, bioprinting processes must also be adjusted in terms of resolution and speed [141]. A flexible method for creating nanofibrous scaffolds that closely mimic the composition of the natural extracellular matrix is electrospinning. Researchers can produce hybrid scaffolds with improved mechanical characteristics and hierarchical structures by fusing additive manufacturing and electrospinning. By supporting cells physically and encouraging adhesion, proliferation, and differentiation, these hybrid scaffolds can improve the structure and functionality of tissues [94].

Achieving a seamless connection between the electro spun scaffolds and printed constructs is a major problem in the merging of additive manufacturing with electrospinning. To guarantee that the electro spun fibers are deposited and distributed properly, printed constructions must be carefully monitored during the deposition process. In order to prevent negative impacts on cell behavior and tissue regeneration, it is also important to carefully assess the compatibility between the printing materials and the electro spun fibers [21]. Tissue engineering and regenerative medicine seem to benefit greatly from the combination of additive manufacturing and other bio fabrication methods. Researchers can produce tissue constructions with improved structural complexity, biological functioning, and regenerative potential by combining the advantages of several approaches. By making it possible to create tissues with characteristics that closely resemble genuine tissues, these hybrid creations have the potential to completely transform the field. The integration of additive manufacturing with other bio fabrication techniques will be fueled by future developments in material science, bio fabrication methods, and tissue engineering approaches. Researchers can create functional tissue constructs for a variety of purposes, including as drug screening, organ replacement, and tissue regeneration, by tackling the integration's obstacles, which include material compatibility, resolution, and speed [153].

16.2.3 Future Directions in Imaging Techniques

Tissue engineering has seen a revolution thanks to imaging tools that make it possible to see organs and tissues at the cellular and molecular levels. The use of innovative imaging methods in the creation of patient-specific scaffolds for tissue engineering applications has gained popularity in recent years. This chapter highlights the opportunities and problems in this quickly developing discipline as it examines the future trends in imaging techniques with an emphasis on their application for patient-specific scaffold construction [16]. A key component of tissue engineering is patient-specific scaffold design, which enables the creation of scaffolds that closely resemble the anatomical and physiological traits of individual patients. By increasing the scaffold's integration with the host tissue, encouraging tissue regeneration, and lowering the possibility of implant rejection, this customized strategy may enhance the results of tissue engineering treatments. The utilization of advanced imaging techniques is crucial in the creation of scaffolds that are specifically tailored to the needs of individual patients, as they offer comprehensive anatomical information [19].

Although patient-specific scaffold design can benefit from the application of imaging techniques, there are several issues that need to be resolved. One of the primary obstacles is the existing imaging methods' poor resolution, which can make it challenging to precisely record the intricate anatomical details of tissues and organs. The time and expense involved in gathering and analyzing imaging data presents another difficulty, especially for extensive research or medical applications. Researchers are looking into a range of cutting-edge imaging methods that provide better resolution, speed, and affordability in an effort to address these issues [150]. Using high-resolution imaging modalities, such as micro-computed tomography (micro-CT), computed tomography (CT), and magnetic resonance imaging (MRI), which can produce detailed 3D images of tissues and organs with submicron resolution, is one strategy. These methods enable more precise and customized scaffold construction by enabling researchers to see the inside structure of tissues and organs with never-before-seen detail [55].

Combining imaging data with computer modeling methods is another new trend in imaging for scaffold design. Through the integration of mathematical models and imaging data, scientists may replicate the behavior of various tissues and organs, facilitating the refinement of scaffold design factors like porosity, stiffness, and degradation rate. With its thorough understanding of how scaffolds interact with host tissue and how to tune them to promote tissue regeneration, this integrated technique, called image-based modeling, has the potential to completely transform the field of tissue engineering. By enabling the creation of scaffolds that are specifically customized to each patient's needs, the application of sophisticated imaging techniques for patient-specific scaffold design holds the potential to revolutionize the field of tissue engineering. Through the enhancement of imaging technologies' resolution, speed, and cost-effectiveness, scientists can open up new avenues for personalized medicine and regenerative therapies. Future developments in imaging technology, in conjunction with creative computational modeling techniques, will propel the creation of scaffolds that are tailored to each patient's needs, ultimately improving the prognosis of patients undergoing tissue engineering treatments [2].

16.2.4 Integration with Stem Cell Technology and Bioreactor Systems

The integration of additive manufacturing with stem cell technology and bioreactor systems offers a substantial development in tissue engineering and regenerative medicine. This chapter addresses the current state of integration, obstacles, and future perspectives in utilizing these technologies for biomaterial-based additive manufacturing in tissue engineering and regeneration [81]. Stem cell science has enormous promise for tissue engineering, having the possibility to restore damaged tissues and organs. By integrating stem cells with additive manufacturing techniques, researchers may produce complex tissue constructs with the ability to develop into specific cell types, closely imitating the underlying tissue microenvironment. This integration permits the manufacture of tissues with greater regenerating potential, making them ideal for transplantation and other regenerative therapies [97].

Bioreactor systems serve a significant role in tissue engineering by providing a regulated environment for the maturation and development of tissue constructions. By merging additive manufacturing with bioreactor systems, researchers may build dynamic culture settings that mirror the physiological parameters encountered in vivo. This integration enables the production of tissues with better structural and functional features, making them more suited for transplantation and therapeutic applications. Despite the potential benefits of merging additive manufacturing with stem cell technologies and bioreactor systems, various hurdles must be solved. One of the primary issues is ensuring the viability and functionality of stem cells during the production process. Stem cells are sensitive to environmental cues, such as mechanical stresses, oxygen levels, and nutrition availability, which can impact their behavior and differentiation potential. Therefore, it is vital to optimize the printing parameters and culture conditions to retain the viability and functionality of stem cells in the printed constructs [121].

Another hurdle is the incorporation of stem cells into the tissue constructions in a regulated and reproducible manner. The spatial distribution of stem cells within the construct might dramatically affect its regeneration ability, making it necessary to develop ways for precise cell insertion and patterning. Additionally, the interaction between stem cells and the biomaterials utilized in additive manufacturing must be carefully managed to enable the optimal biological response and tissue regeneration [155]. Despite these challenges, the integration of additive manufacturing with stem cell technologies and bioreactor systems presents significant potential for developing tissue engineering and regenerative medicine. By integrating these technologies, researchers can generate tissue constructs with greater regeneration potential and functionality, making them more suited for therapeutic applications. Continued research and development in this area will be vital for unlocking the full potential of these technologies and improving the field of tissue engineering and regenerative medicine [98].

16.3 Potential Applications in Functional Tissue Constructs for Transplantation and Drug Screening

Utilizing stem cell technologies in conjunction with bioreactor systems is a viable method for generating functional tissue constructs that may find use in drug screening and transplantation. Because of their capacity for self-renewal and differentiation, stem cells can be made to differentiate into cell types in a bioreactor setting. The construction of intricate the 3D structures that closely resemble native tissues is the result of this process, which imitates the natural evolution of tissues. The growth and maturation of these tissue constructions are greatly aided by bioreactors. They offer a regulated setting with all the necessary elements for cell growth, including food delivery, oxygenation, pH control, and mechanical stimulation. Bioreactors facilitate the creation of functional tissues with improved cellular organization and ECM synthesis by simulating the physiological conditions seen in the human body [75].

Transplantation is one of the main uses for these modified tissues. Functional tissue constructs present a promising treatment option for a variety of illnesses and injuries since they can be utilized to replace or repair damaged tissues. For instance, cartilage structures have shown promise in the healing of injured joints, and engineered skin grafts have been developed for the treatment of burn injuries. The development and screening of drugs is a significant additional application. When it comes to drug testing, engineered tissues can be more predictive than conventional two-dimensional (2D) cell cultures [106]. These tissues' 3D structure and cellular makeup more closely resemble the in vivo environment, which improves the accuracy of determining the toxicity and efficacy of drugs. This can enhance the effectiveness of the medication development process and lessen the need for animal testing. There is much potential for tissue engineering to advance with the integration of stem cell technologies and bioreactor systems. This method has the potential to revolutionize drug screening and transplantation by producing functioning tissue constructs that closely resemble native tissues. This could lead to novel approaches for treating diseases and enhancing human health [58].

16.3.1 Enhanced Scaffold Design

Scaffold design is a critical aspect of tissue engineering, influencing the success of tissue regeneration and transplantation. The emergence of biomaterial-based additive manufacturing has revolutionized scaffold fabrication, offering unprecedented control over scaffold architecture and properties. This chapter explores the current challenges and future directions in scaffold design using biomaterial-based additive manufacturing techniques, highlighting innovations that enhance scaffold functionality and regenerative potential [125]. The traditional approach to scaffold design involves the use of natural or synthetic materials to create porous structures that support cell attachment, proliferation, and differentiation. While these scaffolds have shown promise in promoting tissue regeneration, they often lack the complexity and functionality of native tissues. Biomaterial-based additive manufacturing has overcome these limitations by enabling the fabrication of scaffolds with precise control over pore size, shape, and interconnectivity [147].

One of the key challenges in scaffold design is achieving a balance between mechanical properties and biocompatibility. Scaffolds must be strong enough to withstand mechanical forces in vivo while also being biocompatible to support cell growth and tissue regeneration. Biomaterial-based additive manufacturing allows for the creation of scaffolds with tunable mechanical properties, enabling researchers to tailor scaffolds to match the mechanical properties of native tissues [57]. Another challenge is the incorporation of bioactive molecules into scaffolds to enhance their regenerative potential. Biomaterial-based additive manufacturing offers a solution to this challenge by allowing for the precise deposition of bioactive molecules, such as growth factors and cytokines, within the scaffold structure. This enables the creation of scaffolds that can actively promote cell adhesion, proliferation, and differentiation, leading to enhanced tissue regeneration [10]. Furthermore, the integration of advanced imaging techniques with biomaterial-based additive manufacturing has opened new possibilities for scaffold design. By using imaging data to create patient-specific scaffolds, researchers can tailor scaffolds to match the anatomical and physiological characteristics of individual patients, improving the success rate of tissue regeneration and transplantation. Looking ahead, the future of scaffold design lies in the development of multifunctional scaffolds that can mimic the complexity of native tissues. These scaffolds may incorporate multiple types of biomaterials, bioactive molecules, and cells to create tissues with enhanced functionality and regenerative potential [154]. Additionally, advancements in scaffold fabrication techniques, such as 4D printing and bio fabrication, will further enhance the capabilities of biomaterial-based additive manufacturing in tissue engineering and regeneration. Biomaterial-based additive manufacturing offers exciting opportunities for enhancing scaffold design in tissue engineering and regeneration. By addressing current challenges and exploring new avenues for innovation, researchers can develop scaffolds with enhanced functionality and regenerative potential, paving the way for the development of next-generation tissue engineering therapies [128].

16.3.2 Design Optimization for Improved Tissue Regeneration

Tissue engineering relies heavily on improved scaffold design since it offers a framework that promotes cell adhesion, proliferation, and differentiation. By optimizing the mechanical properties, surface features, and bioactivity of the scaffold, design optimization strategies seek to improve tissue regeneration outcomes. Using innovative fabrication methods, such additive printing, which enables exact control over scaffold architecture at the micro- and nanoscale, is one way to improve scaffold design [62]. Through the optimization of the scaffold's pore size, interconnectivity, and surface roughness, scientists can enhance tissue integration, vascularization, and cell infiltration. Furthermore, adding bioactive substances to the scaffold, such peptides and growth factors, can improve its capacity for regeneration. These chemicals have the ability to trigger biological reactions, including cell migration, proliferation, and differentiation, which can enhance the results of tissue regeneration [116]. Furthermore, improved cell-scaffold interactions and tissue regeneration can be facilitated by the adoption of biomimetic scaffold designs, which are modeled after the composition and structure of native extracellular matrix. These scaffolds can give cells cues to organize and differentiate, resulting in the development of functional tissues, by imitating the natural extracellular matrix [54]. A multidisciplinary strategy combining materials science, engineering, and biology is needed to optimize scaffold design for better tissue regeneration. Through the utilization of sophisticated manufacturing methods, bioactive compounds, and biomimetic designs, scientists can create scaffolds that more closely resemble the natural tissue environment, resulting in improved outcomes for tissue regeneration [148].

16.3.3 Collaboration Between Researchers, Clinicians, and Regulatory Bodies

For advances in biomedical research to be transferred from the laboratory to clinical practice, cooperation between investigators, practitioners, and regulatory agencies is essential. Through this partnership, patients will always have access to safe, efficient, and innovative technologies, medicines, and treatments. Through fundamental and applied research, scientists are instrumental in the creation of novel technology and therapeutic approaches. They seek to comprehend the fundamental causes of illnesses, pinpoint possible intervention targets, and create novel treatments and technological advancements. On the other hand, clinicians are at the frontline of patient care and have direct knowledge of the difficulties and complications involved in providing care [11]. Their feedback is crucial to ensuring that novel technology and therapeutic approaches are realistic and practicable in a clinical context, as well as meeting patients' needs. The Food and Drug Administration (FDA) in the United States (US) is one of the regulatory authorities that is vital to guaranteeing the efficacy and safety of novel technology and medical interventions. They assess information from preclinical and clinical research to ascertain whether a novel technology or therapy satisfies the prerequisite regulatory requirements for authorization [13].

For several reasons, these three groups must work together. First, to help researchers concentrate on areas with the highest effect potential, clinicians offer insightful information about the clinical significance of modern technologies and treatments. Second, to make decisions and guarantee the safety and efficacy of modern technologies and therapies, regulatory agencies consult with researchers and clinicians. Lastly, cooperation between these organizations can aid in recognizing and resolving roadblocks to the adoption of novel technologies and therapies, such as difficulties with reimbursement or regulatory obstacles. Translating advances in biomedical research from the laboratory to clinical practice, in general, requires cooperation between investigators, practitioners, and regulatory agencies [60]. Together, these organizations can guarantee that novel medical procedures and technology are risk-free, efficacious, and easily available to patients, eventually enhancing patient outcomes and propelling the medical domain forward [71].

16.3.4 Regulatory Aspects and Standards

When developing and utilizing biomaterial-based AM for tissue engineering, regulatory issues are crucial. With the ability to create patient-specific constructions with intricate geometries, customized mechanical qualities, and bioactive cues to encourage tissue regeneration, this technique has enormous promise. However, regulatory frameworks need to be carefully traversed and followed in order to guarantee the quality, safety, and efficacy of these constructions [85]. Essential components of regulatory compliance are biomaterial selection and characterization. Biomaterials utilized in AM must be biocompatible, which means the body should not react negatively to them or become poisonous. Several in vitro and in vivo assays that measure how the biomaterial interacts with cells or tissues can be used to determine biocompatibility. In addition, biomaterials need to have the right mechanical characteristics to sustain physiological loads and give the regenerated tissue structural support. Therefore, one of the most important steps in the regulatory process to make sure the biomaterial satisfies the requirements is mechanical testing [53].

To guarantee their quality and consistency, biomaterials utilized in AM need to be thoroughly described in addition to having mechanical qualities and being biocompatible. This entails characterizing the surface characteristics, molecular weight, chemical content, and rate of deterioration of the material. These factors need to be regulated in order to guarantee the final product's repeatability and dependability since they have the potential to affect the material's performance in vivo [127]. Validation of the AM process is also necessary to make sure that the constructed objects fulfill predefined requirements. This covers the verification of the tools, supplies, and software employed throughout the procedure. Equipment needs to be maintained and calibrated in accordance with manufacturer guidelines, and any departures from SOPs need to be supported by documentation. The materials used in the AM process must be of the highest caliber and adhere to strict purity and uniformity regulations. To guarantee that it generates accurate and trustworthy results, software used to develop and manage the AM process must also be tested [138].

Another crucial component of regulatory compliance for biomaterial-based AM in tissue engineering is biological evaluation. To evaluate the biological safety of the produced constructs, biocompatibility testing including cytotoxicity, genotoxicity, and immunogenicity assays is crucial. These tests aid in confirming that the constructions are appropriate for their intended purpose and do not hurt the body. Apart from biocompatibility testing, in vivo investigations could be carried out to assess the efficacy of the creations in animal models. These investigations can yield important insights into the constructs' capacity to support tissue regeneration, biocompatibility, and biodegradability [23]. Regulatory approval procedures differ by nation or area, but they usually require submitting a regulatory dossier with extensive information on the product's quality, safety, and effectiveness. Preclinical and clinical study data, as well as details on the manufacturing procedure, quality assurance protocols, and labeling, must all be included in this dossier. These data are examined by regulatory bodies to ascertain if the product satisfies the requirements for approval [6].

Another crucial component of regulatory compliance for biomaterial-based AM in tissue engineering is post-market surveillance. Manufacturers must keep an eye on the product's performance and safety once on the market to identify and resolve potential problems. This entails gathering and examining information on unfavorable incidents, defective products, and other safety-related issues [70]. Additionally, manufacturers are required to notify regulatory bodies of any modifications made to the product or manufacturing process that would compromise its efficacy or safety. Regulatory factors play a crucial role in the creation and implementation of biomaterial-based AM in tissue engineering. Respecting legal requirements guarantees the safety, efficacy, and excellence of the created constructs, which eventually improves patient outcomes and promotes tissue engineering. Working together, researchers, producers, authorities, and medical experts can help you manage these obstacles and get safe, effective items into the market [86].

16.4 Clinical Trials and Patents

The clinical trials and patents for biomaterials involved in tissue engineering are tabulated in Tables 16.1 and 16.2.

16.5 Current Status and Future Prospects

Material, method, and application developments in biomaterial-based AM for tissue engineering are noteworthy at this point. Advances in the development of biocompatible and bioactive materials that replicate native tissue properties have made it possible to create sophisticated, patient-specific structures. Improved cell

Table 16.1 The clinical trial updates for use of biomaterials into tissue engineering

NCT number	Study status	Conditions	Interventions	Phases	Start date	Completion date
NCT01742260	Unknown	Surgically created resection cavity	Repair of cranial defects by tissue engineering	PHASE1	2013-07	2017-12
NCT03766217	Completed	Cleft lip and palate	Mesenchymal stem cells associated with biomaterials for iliac crest autogenous bone graft	PHASE3	05-2019	15-12-2019
NCT03102879	Completed	Periapical periodontitis	Regenerative endodontic for conventional root canal treatment	NA	09-2016	30-09-2018
NCT01932164	Completed	Cleft lip and palate	Maxillary alveolar graft by tissue engineering through bone tissue engineering using mesenchymal stem cells	NA	2013-05	16-12-2015
NCT05599087	Recruiting	Periapical periodontitis	Acellular system	PHASE1	10-2022	30-12-2023
NCT01813188	Completed	Pseudoarthrosis	ABM seeded onto a porous TCP and autologous bone graft	PHASE2	2011-04	2014-06

Data obtained from https://clinicaltrials.gov/search?intr=biomaterial%20tissue%20engineering

survivability, tissue integration, and functionality have resulted from the exact control over the architecture and composition of these constructs made possible by additive manufacturing methods like 3D bioprinting and electrospinning [7]. A major obstacle in biomaterial-based additive manufacturing is obtaining the mechanical characteristics needed for applications requiring load bearing. To increase the mechanical robustness of printed materials, researchers are looking into innovative techniques like using gradient structures or adding reinforcing strands. Improvements in cell adhesion, proliferation, and differentiation on printed scaffolds are also being researched by surface modification techniques including plasma treatment or coating with bioactive compounds [68].

Table 16.2 The patents updates for use of biomaterials into tissue engineering

Patent ID	Title	Inventor/author	Publication date	Grant date
US-9631171-B2	Microstructure biomaterials and fabrication methods therefor	Pranav Soman, Shaochen Chen, David Fozdar	25-04-2017	25-04-2017
US-11767505-B2	Biomaterial substrates, cell culture systems comprising the same and uses thereof in cell screening applications	Patrick van Rijn, Martin Conrad Harmsen, Qihui Zhou, Philipp till Kuhn, Gabriel Liguori	26-09-2023	26-09-2023
US-10828396-B2	Degradable magnesium-containing calcium phosphate-calcium sulfate porous composite biological scaffold	Yaping Li, Sang Li, Haijiao Mao, Tingting Tang, Jun Fu, Hanfeng Xiong, Lei Nie, Jianmin Zheng, Guofeng Zhang	10-11-2020	10-11-2020
CA-2645764-C	Oxidized microbial cellulose and use thereof	Jeremy J. Harris, Gonzalo Serafica, Christopher J. Damien, Heather R. Nonnenmann	03-11-2015	03-11-2015
US-11179243-B2	Implantable devices	Ryan K. Roeder, Gabriel L. Converse, Stephen M. Smith	23-11-2021	23-11-2021
US-9155818-B2	Cartilaginous and osteochondral substitute comprising multilayer structure and use thereof	Anna Tampieri, Daniele Pressato, Claudio De Luca, Sergio Di Fede, Elena Landi	13-10-2015	13-10-2015
EP-2358407-B1	Bone implant matrix and method of preparing the same	Gianni Pertici	09-12-2015	09-12-2015
EP-3338818-B1	Bone repair reusable composite material based on acellular biological tissue matrix material and preparation method therefor	Hua Liao, Chunhua Yu, Tao Jiang, Xiaohua Yu	28-07-2021	28-07-2021
EP-1973554-B1	Cartilage repair methods	Mitchell S. Seyedin, Matthew Matava	14-01-2015	14-01-2015
US-8992967-B2	Poly (diol-co-citrate) hydroxyapatite composite for tissue engineering and orthopaedic fixation devices	Guillermo Ameer, Hongjin Qiu, Jian Yang	31-03-2015	31-03-2015

(continued)

Table 16.2 (continued)

Patent ID	Title	Inventor/author	Publication date	Grant date
EP-3678714-B1	Tissue-engineered medical device	Simon P. Hoerstrup, Maximilian Y. Emmert, Frank Baaijens, Anita Driessen-Mol	22-02-2023	22-02-2023
EP-3338819-B1	Injectable composite material for bone repair, and preparation method thereof	Chunhua Yu, Hua Liao, Xiaohua Yu, Tao Jiang	28-07-2021	28-07-2021
US-10098681-B2	Segmented delivery system	Guobao Wei, Todd M. Boyce, Robert Cohen, Cristy Richards, Lawrence A. Shimp	16-10-2018	16-10-2018
EP-2182054-B1	Process for identification and isolation of mesenchymal cells with marked bone regenerative ability	Antonio Crovace, Edda Giuseppina Francioso, Giulio Alessandri	13-07-2016	13-07-2016
DK-2841114-T3	Self-harding bioactive cement compositions with partial deacetylated chitin as bone transplant replacements	Johannes Gislason, Jon M Einarsson, Chuen How Ng	24-09-2018	24-09-2018
US-10420857-B2	High density fibrous polymers suitable for implant	Timothy A Ringeisen, William Christopher Wattengel	24-09-2019	24-09-2019
US-9902932-B2	Methods, apparatus, and systems for fabrication of polymeric nano- and micro-fibers in aligned configurations	Amrinder Singh Nain	27-02-2018	27-02-2018
ES-2714782-T3	Procedure to create a custom implant that is activated by genes to regenerate bone tissue	Roman Vadimovich Deev, Artur Aleksandrovich Isaev, Ilya Yadigerovich Bozo, Vladimir Sergeevich Komlev, Alexey Yurevich Drobyshev	30-05-2019	30-05-2019
US-9492279-B2	Bioabsorbable plug implants and method for bone tissue regeneration	Swee Hin Teoh, Kim Cheng Tan, Dietmar Hutmacher, Thiam Chye Lim, Jan-Thorsten Schantz, Ning Chou	15-11-2016	15-11-2016

(continued)

Table 16.2 (continued)

Patent ID	Title	Inventor/author	Publication date	Grant date
EP-1896056-B2	Reinforced collagen scaffold	Chunlin Yang, Joseph Hammer, Ziwei Wang	30-05-2018	30-05-2018
EP-2243749-B1	Resorbable and biocompatible fibre glass compositions and their uses	Timo Lehtonen, Jukka Tuominen, Fredrik Ollila	08-04-2015	08-04-2015
EP-2967880-B1	Conduits for enhancing tissue regeneration	Thomas D. McGee	06-11-2019	06-11-2019
US-10271938-B2	Multi-phasic solid implants for tissue repair	Nir Altschuler	30-04-2019	30-04-2019
EP-3419678-B1	Trizonal membranes for periosteum regeneration	Ennio Tasciotti, Francesca Taraballi, Silvia Minardi	29-01-2020	29-01-2020
EP-3423120-B1	Porous polymer scaffold and preparation method thereof	Sarindr Bhumiratana, Eric Jeffries, Melanie Reschke	18-11-2020	18-11-2020
AU-2011341420-B2	Tissue engineered blood vessels	Iksoo Chun, Kevin Cooper, Carrie H. Fang	12-05-2016	12-05-2016
US-8936638-B2	Coral bone graft substitute	Ohad Schwartz, Itzhak Binderman	20-01-2015	20-01-2015
US-9168327-B2	Bone/collagen composites and uses thereof	Sherry Hsiu-Ying Wang, Michael Jaffe, Lawrence A. Shimp, Nels J. Lauritzen, Charles M. Schiff	27-10-2015	27-10-2015
US-10729813-B2	Extracellular matrix-derived gels and related methods	Stephen F. Badylak, Donald Freytes	04-08-2020	04-08-2020
US-9364584-B2	Demineralized cancellous bone scaffolds	Eric A. Nauman, Darryl A. Dickerson	14-06-2016	14-06-2016
US-10046084-B2	Solid forms for tissue repair	Nir Altschuler, Razi Vago	14-08-2018	14-08-2018
US-11642835-B2	Additive manufacturing methods utilizing a robotic arm	David J. Hoelzle, Desmond M. D'Souza, Andrej Simeunovic, Ali Asghari Adib	09-05-2023	09-05-2023
US-9066996-B1	Method to control the biodegradation rates of polymer structures	Chaoyin Zhou, Alan J. Jacobsen	30-06-2015	30-06-2015
EP-3180043-B1	A method for making a porous scaffold suitable for use in repair of osseous, chondral, or osteochondral defects in a mammal	Daniel John Kelly, Grainne Cunniffe, Henrique Almeida, Rajalakshmanan Eswaramoorthy, Conor Buckley, Pedro Diaz Peyno, David Browe	26-12-2018	26-12-2018

(continued)

Table 16.2 (continued)

Patent ID	Title	Inventor/author	Publication date	Grant date
US-10729816-B2	Customized load-bearing and bioactive functionally graded implant for treatment of osteonecrosis	Yunzhi Yang, Stuart B. Goodman, Yaser Shanjani	04-08-2020	04-08-2020
CA-2661548-C	Macroporous and highly resorbable apatitic calcium-phosphate cement	Ibrahim Khairoun, Pierre Weiss, Jean-Michel Bouler	01-12-2015	01-12-2015
US-9771410-B2	Foam-formed collagen strand	Nels Lauritzen, Lawrence A. Shimp, Brent S. Mitchell	26-09-2017	26-09-2017
US-11903612-B2	Bioprinter and methods of using same	Ibrahim T. Ozbolat, Howard Chen, Yin Yu, Yahui Zhang	20-02-2024	20-02-2024
US-9095391-B2	Osseointegration and biointegration coatings for bone screw implants	Laurence M. McKinley	04-08-2015	04-08-2015
EP-3717027-B1	Composition of a calcium phosphate and an additive compound containing a phosphate or phosphonate group	Philip Procter, Michael Pujari-Palmer, Gerard Insley, Hã¥kan Engqvist	07-06-2023	07-06-2023
EP-2608778-B1	Compositions and methods for treating bone defects	Michael J. Voor, Jr. Robert L. Burden	01-11-2017	01-11-2017
US-5916265-A	Method of producing a biological extracellular matrix for use as a cell seeding scaffold and implant	Jie Hu	29-06-1999	29-06-1999
AU-2018271341-B2	Anti-thrombogenic grafts	Sashka Dimitrievska, Laura Niklason	03-09-2020	03-09-2020
CA-2770490-C	Compositions and methods for implantation of processed adipose tissue and processed adipose tissue products	Zayna Nahas, Jennifer H. Elisseeff, Iwen Wu	16-04-2024	16-04-2024
US-10857264-B2	Multi-dimensional biomaterial and method for producing the same	Denis Dufrane, Christian Delloye	08-12-2020	08-12-2020

(continued)

Table 16.2 (continued)

Patent ID	Title	Inventor/author	Publication date	Grant date
CA-2962223-C	Articular cartilage repair	Farshid Guilak, Bradley T. Estes, Franklin Thomas Moutos	25-08-2020	25-08-2020
US-9814803-B2	Allografts combined with tissue derived stem cells for bone healing	Yaling Shi	14-11-2017	14-11-2017
US-9849211-B2	Dimensionally stable molded bone replacement element with residual hydraulic activity	Berthold Nies	26-12-2017	26-12-2017
EP-2129406-B1	Collagen products and methods for producing collagen products	Nels J. Lauritzen, Lawrence A. Shimp, Brent Mitchell	12-08-2015	12-08-2015
US-11013858-B2	Nonjacketing side-entry connectors and prosthetic disorder response systems	David S. Goldsmith	25-05-2021	25-05-2021
US-10300167-B2	Functionalized calcium phosphate artificial bone and joint compositions and methods of use and manufacture	Kenneth A. McGowan, Ellen S. Gawalt, Rachelle Palchesko	28-05-2019	28-05-2019
US-11389171-B2	Integrated system for the infixion and retrieval of implants	David S. Goldsmith	19-07-2022	19-07-2022
US-9480773-B2	Mineralization and biological modification of biomaterial surfaces	William L. Murphy, Martin C. Peters, David J. Mooney	01-11-2016	01-11-2016
US-10028837-B2	Delivery system attachment	Guobao Wei, Keyvan Behnam, Nanette Forsyth, John Winterbottom, James Beisser, Todd Boyce, Mohamed Attawia, Cristy J. Richards, Lawrence A. Shimp	24-07-2018	24-07-2018
US-11911535-B2	Implantable medical device with thermoplastic composite body and method for forming thermoplastic composite body	Douglas Snell, Robert Ball, Ryan K. Roeder	27-02-2024	27-02-2024

(continued)

Table 16.2 (continued)

Patent ID	Title	Inventor/author	Publication date	Grant date
US-9289534-B2	Biocompatible composite and its use	Timo Lehtonen, Jukka Tuominen	22-03-2016	22-03-2016
US-10524916-B2	Resorbable macroporous bioactive glass scaffold and method of manufacture	Qiang Jie, Jiang Chang, Weiming J. Gu, Jipin Zhong, Gregory J. Pomrink	07-01-2020	07-01-2020
US-10004587-B2	Macroporous 3-D scaffolds for tissue engineering	Fan Yang, Li-Hsin Han, Xinming Tong	26-06-2018	26-06-2018
US-9414939-B2	Delivery systems containing bioactive materials	Daniel A. Shimko, Kerem N. Kalpakci	16-08-2016	16-08-2016
CA-2682453-C	Method for obtaining three-dimensional structures for tissue engineering	Alvaro Meana Infiesta, Eva Garcia Perez, Veronica Garcia Diaz, Jose Luis Jorcano Noval, Marcela Del Rio Nechaevsky, Fernando Larcher Laguzzi, Blanca Duarte Gonzalez, Almudena Holguin Fernandez	12-07-2016	12-07-2016
ES-2959245-T3	Preparation and applications of bioinks for 3D bioprinting for the repair of bone defects, based on cellulose nanofibril hydrogels with natural or synthetic calcium phosphate particles	Paul Gatenholm, Hector Martinez, Michela Schettino, Erik Gatenholm	22-02-2024	22-02-2024
US-11357889-B2	Native soft tissue matrix for therapeutic applications	Hani Awad, Bradley T. Estes, Farshid Guilak	14-06-2022	14-06-2022
US-10463770-B2	Bone tissue engineering by ex vivo stem cells ongrowth into three-dimensional trabecular metal	Zou Xuenong, Haisheng Li, Cody Bunger	05-11-2019	05-11-2019

(continued)

Table 16.2 (continued)

Patent ID	Title	Inventor/author	Publication date	Grant date
US-10894895-B2	Two-component bioink, 3D biomaterial comprising the same and method for preparing the same	Kyung Sook Kim, Da Yeon Kim, Seung Hun Park, Ja Yong JANG, Moon Suk Kim	19-01-2021	19-01-2021
DK-2766061-T3	Published prosthesis	Laura Niklason, Angela Huang, Shannon Dahl, Liping Zhao	29-04-2019	29-04-2019
US-9603968-B2	Methods of making high-strength NDGA polymerized collagen fibers and related collagen-prep methods, medical devices, and constructs	Thomas J. Koob, Douglas Pringle, Daniel Hernandez	28-03-2017	28-03-2017

Data obtained from https://patents.google.com/?q=(biomaterial+tissue+engineering)&oq=biomaterial+tissue+engineering

Researchers are concentrating on enhancing the repeatability and scalability of biomaterial-based AM in addition to advances in materials and processes. To guarantee the consistent generation of high-quality constructions, efforts are being made to establish standardized procedures and quality control mechanisms. Biomaterial-based AM in tissue engineering has a bright future ahead of it. One of the emerging trends is the creation of patient-specific models for more precise and customized tissue constructs with the use of modern imaging techniques like CT and MRI. It is anticipated that the incorporation of bioactive compounds, such cytokines or growth factors, into printed scaffolds may improve the results of tissue regeneration [72]. Moreover, the creation of complex, heterogeneous tissues with a variety of functions will be made possible by the advancement of multi-material and multi-modal printing techniques. Regenerative medicine is projected to undergo a revolution because of these developments, which will propel biomaterial-based AM from the lab into clinical practice. For biomaterial-based AM in tissue engineering to reach its full potential and overcome the remaining obstacles, cooperation between researchers, doctors, and regulatory agencies will be crucial [119].

The promise of biomaterial-based AM in tissue engineering has been shown by recent studies and research. A study by Wang et al., for instance, that was published in Advanced Materials demonstrated a novel way to improve the mechanical properties of 3D-printed constructions by adding cellulose nanocrystals to the polymer matrix. The resultant scaffolds demonstrated the potential of nanocomposite materials in tissue engineering applications by demonstrating greater mechanical strength and encouraging stem cell proliferation and differentiation [73]. Using bioactive compounds in 3D-printed scaffolds to improve tissue regeneration is another fascinating breakthrough. For example, a work by Smith et al. published in Biomaterials (2020) showed how vascular endothelial growth factor (VEGF) may be released from 3D-printed scaffolds in a regulated manner to enhance vascularization and encourage angiogenesis in engineered tissues. This method may improve the survivability and performance of tissue constructs, especially in situations where significant vascularization is necessary, such bone or cardiac tissue engineering [143].

The development of very precise and customized tissue constructs has been made possible by improvements in imaging techniques. In Nature Biotechnology, for instance, Chen et al.'s study showed how to combine high-resolution MRI and CT images to make patient-specific models for 3D printing of intricate bone structures. With this method, implants could be precisely customized, which resulted in a better fit and functionality than with conventional off-the-shelf implants [56]. These illustrations show how biomaterial-based AM has advanced recently and how it has the potential to completely transform tissue engineering. The development of next-generation tissue constructs with improved functioning and therapeutic utility is being facilitated by researchers pushing the envelope in regenerative medicine through the use of innovative materials, bioactive chemicals, and better imaging tools [24].

16.6 Conclusion

Advancements in materials, techniques, and applications have created tremendous possibilities for tissue engineering and regeneration through biomaterial-based AM. With an emphasis on mechanical characteristics, quality control procedures, and biocompatibility, the selection and characterization of biomaterials are crucial. The ultimate products' uniformity and reproducibility are guaranteed by quality assurance and process validation. In vivo research and biocompatibility testing are two crucial components of biological evaluation that determine the safety and effectiveness of produced constructions. Ensuring the quality and safety of products requires strict respect for regulations, which are of utmost importance. The potential of AM based on biomaterials in tissue engineering has been shown by recent investigations. Personalized tissue constructs made with innovative imaging techniques, the controlled release of bioactive compounds to encourage tissue regeneration, and the use of nanocomposite materials to improve mechanical qualities are a few examples. These developments highlight how biomaterial-based AM has the potential to transform tissue engineering by making it possible to create intricate, functioning tissues that closely resemble natural tissues. In general, the translation of biomaterial-based AM from the laboratory to clinical practice depends heavily on cooperation between researchers, physicians, regulatory agencies, and industrial stakeholders. Biomaterial-based AM can transform tissue engineering and regeneration by using these developments and encouraging teamwork, providing fresh approaches to illness treatment and better patient outcomes.

References

1. Aamodt JM, Grainger DW (2016) Extracellular matrix-based biomaterial scaffolds and the host response. Biomaterials 86:68–82. https://doi.org/10.1016/j.biomaterials.2016.02.003
2. Adel IM, ElMeligy MF, Elkasabgy NA (2022) Conventional and recent trends of scaffolds fabrication: a superior mode for tissue engineering. Pharmaceutics 14:306. https://doi.org/10.3390/pharmaceutics14020306
3. Agrawaal H, Thompson JE (2021) Additive manufacturing (3D printing) for analytical chemistry. Talanta Open 3:100036. https://doi.org/10.1016/j.talo.2021.100036
4. Al-Shalawi FD, Mohamed Ariff AH, Jung D-W et al (2023) Biomaterials as implants in the orthopedic field for regenerative medicine: metal versus synthetic polymers. Polymers (Basel) 15:2601. https://doi.org/10.3390/polym15122601
5. Altıparmak SC, Yardley VA, Shi Z, Lin J (2022) Extrusion-based additive manufacturing technologies: state of the art and future perspectives. J Manuf Process 83:607–636. https://doi.org/10.1016/j.jmapro.2022.09.032
6. Anderson JM (2016) Future challenges in the in vitro and in vivo evaluation of biomaterial biocompatibility. Regen Biomater 3:73–77. https://doi.org/10.1093/rb/rbw001
7. Arif ZU, Khalid MY, Noroozi R et al (2023) Additive manufacturing of sustainable biomaterials for biomedical applications. Asian J Pharm Sci 18:100812. https://doi.org/10.1016/j.ajps.2023.100812
8. Arifin N, Sudin I, Ngadiman NHA, Ishak MSA (2022) A comprehensive review of biopolymer fabrication in additive manufacturing processing for 3D-tissue-engineering scaffolds. Polymers (Basel) 14:2119. https://doi.org/10.3390/polym14102119
9. Bahraminasab M (2020) Challenges on optimization of 3D-printed bone scaffolds. Biomed Eng Online 19:69. https://doi.org/10.1186/s12938-020-00810-2
10. Bertsch C, Maréchal H, Gribova V et al (2023) Biomimetic bilayered scaffolds for tissue engineering: from current design strategies to medical applications. Adv Healthc Mater 12:e2203115. https://doi.org/10.1002/adhm.202203115
11. Bhatia D, Paul S, Acharjee T, Ramachairy SS (2024) Biosensors and their widespread impact on human health. Sens Int 5:100257. https://doi.org/10.1016/j.sintl.2023.100257
12. Bishop ES, Mostafa S, Pakvasa M et al (2017) 3-D bioprinting technologies in tissue engineering and regenerative medicine: current and future trends. Genes Dis 4:185–195. https://doi.org/10.1016/j.gendis.2017.10.002
13. Bohr A, Memarzadeh K (2020) The rise of artificial intelligence in healthcare applications. In: Artificial intelligence in healthcare. Elsevier, London, pp 25–60
14. Bose S, Koski C, Vu AA (2020) Additive manufacturing of natural biopolymers and composites for bone tissue engineering. Mater Horiz 7:2011–2027. https://doi.org/10.1039/D0MH00277A
15. Bothiraja C, Gholap AD, Shaikh KS, Pawar AP (2014) Investigation of ethyl cellulose microsponge gel for topical delivery of eberconazole nitrate for fungal therapy. Ther Deliv 5:781–794. https://doi.org/10.4155/tde.14.43
16. Chandra PK, Soker S, Atala A (2020) Tissue engineering: current status and future perspectives. In: Principles of tissue engineering. Elsevier, London, pp 1–35
17. Cidonio G, Glinka M, Dawson JI, Oreffo ROC (2019) The cell in the ink: improving biofabrication by printing stem cells for skeletal regenerative medicine. Biomaterials 209:10–24. https://doi.org/10.1016/j.biomaterials.2019.04.009
18. Dalton PD, Vaquette C, Farrugia BL et al (2013) Electrospinning and additive manufacturing: converging technologies. Biomater Sci 1:171–185. https://doi.org/10.1039/C2BM00039C
19. De Chiara F, Ferret-Miñana A, Fernández-Costa JM, Ramón-Azcón J (2024) The tissue engineering revolution: from bench research to clinical reality. Biomedicines 12:453. https://doi.org/10.3390/biomedicines12020453

20. Diener S, Zocca A, Günster J (2021) Literature review: methods for achieving high powder bed densities in ceramic powder bed based additive manufacturing. Open Ceram 8:100191. https://doi.org/10.1016/j.oceram.2021.100191
21. Ejiohuo O (2023) A perspective on the synergistic use of 3D printing and electrospinning to improve nanomaterials for biomedical applications. Nano Trends 4:100025. https://doi.org/10.1016/j.nwnano.2023.100025
22. Flores-Rojas GG, Gómez-Lazaro B, López-Saucedo F et al (2023) Electrospun scaffolds for tissue engineering: a review. Macromolecules 3:524–553. https://doi.org/10.3390/macromol3030031
23. Frisch E, Clavier L, Belhamdi A et al (2023) Preclinical in vitro evaluation of implantable materials: conventional approaches, new models and future directions. Front Bioeng Biotechnol 11. https://doi.org/10.3389/fbioe.2023.1193204
24. Gao W, Wang C, Li Q et al (2022) Application of medical imaging methods and artificial intelligence in tissue engineering and organ-on-a-chip. Front Bioeng Biotechnol 10. https://doi.org/10.3389/fbioe.2022.985692
25. Geevarghese R, Sajjadi SS, Hudecki A et al (2022) Biodegradable and non-biodegradable biomaterials and their effect on cell differentiation. Int J Mol Sci 23:16185. https://doi.org/10.3390/ijms232416185
26. Gholap AD (2012) Translational nanomedicine: status assessment and opportunities. Int J Res Pharm Biomed Sci 4:1454–1463
27. Gholap AD, Dighe SB (2012) Role of sebum in human health & acne. Inven Rapid Cosmeceuticals 1:1–3. [ISSN 0976-3864]
28. Gholap AD, Dighe SB (2012) Dandruff: pathogenesis and treatment. Inven Rapid Cosmeceuticals 1:1–4. [ISSN 0976-3864]
29. Gholap AD, Musmade DS (2011) Advances in the field of nanooncology: an overview. Inven Impact 4:1–3. NDDS [ISSN 0976-3791]
30. Gholap AD, Rajurkar VG (2012) Potential applications of superparmagnetic nanoparticles technology. Inven Rapid 1:1–3. NDDS [ISSN 0976-3791]
31. Gholap AD, Rajurkar VG (2012) New threat of clinical toxicities evoked from nanocarrier systems. Inven Rapid 1:1–4. NDDS [ISSN 0976-3791]
32. Gholap AD, Rajurkar VG (2012) Lipid nanocapsules: a novel promising tool as drug delivery system. Inven Rapid 1:1–4. NDDS [ISSN 0976-3791]
33. Gholap AD, Rajurkar VG (2012) Potential applications of capillary electromigration methods for the improvement of separation efficiency. Inven Rapid Pharm Anal Qual Assur 1:1–3. [ISSN 0976-3813]
34. Gholap AD, Tupe D (2012) Artificial viral envelopes: emerging vehicle for gene delivery. PharmaTutor-ART-1439 1:1–3
35. Gholap AD, Tupe D (2012) Potential applications of botanical extract. Inven Rapid Ethnopharmacol. 1:1–4. [ISSN 0976-3805]
36. Gholap AD, Borude SS, Mahajan AM (2011) Smart crystals technology: a review. Pharmacologyonline 1:238–243
37. Gholap AD, Borude SS, Mahajan AM, Gholap MAD (2011) Aquasomes: a potential drug delivery carrier. Pharmacologyonline 1:230–237
38. Gholap AD, Rajurkar VG, Dighe SB (2012) Phonophoresis: ultrasound enhanced transdermal drug delivery. Inven Rapid 1:1–3. NDDS [ISSN 0976-3791]
39. Gholap A, Pardeshi S, Giram P (2023) Preparation, antibacterial and antiviral activity measurements and detection methods. In: Antibacterial and antiviral functional materials. ACS Publications, Wasington, DC, pp 33–64
40. Gholap AD, Bhowmik DD, Deshmukh AY, Hatvate NT (2023) Quintessential impact of dendrimer bioconjugates in targeted drug delivery. In: Bioconjugated materials part 2 applications in drug delivery, vaccine formulations and important conjugates for cancer therapy. Elsevier, Amsterdam, pp 257–302
41. Gholap AD, Gupta JS, Kamandar PA et al (2023) Antibody-drug conjugates for cancer therapy: an up-to-date review on the chemistry and pharmacology. In: Bioconjugated materials

part 2 applications in drug delivery, vaccine formulations and important conjugates for cancer therapy. Elsevier, Amsterdam, pp 105–190
42. Gholap AD, Hatvate NT, Khuspe PR et al (2023) Proteomics in oncology: retrospect and prospects. In: Novel technologies in biosystems, biomedical & drug delivery. Springer Nature, Singapore, pp 243–269
43. Gholap AD, Hatvate NT, Sawant SD et al (2023) Unlocking the potential of nanobiohybrids to combat environmental pollution. In: Nanobiohybrids for advanced wastewater treatment and energy recovery. IWA Publishing, London, pp 169–190
44. Gholap AD, Said RP, Pawar RD et al (2023) Importance of carbohydrate-drug conjugates in vaccine development: a detailed review. In: Bioconjugated materials part 2 applications in drug delivery, vaccine formulations and important conjugates for cancer therapy. Elsevier, Amsterdam, pp 191–256
45. Gholap AD, Sawant SD, Sayyad SF et al (2023) Food biopackaging for human benefits: status and perspectives. In: Biobased packaging materials. Springer Nature, Singapore, pp 1–35
46. Gholap AD, Sayyad SF, Hatvate NT et al (2023) Drug delivery strategies for avobenzone: a case study of photostabilization. Pharmaceutics 15:1008. https://doi.org/10.3390/pharmaceutics15031008
47. Gholap AD, Gupta J, Kamandar P et al (2024) Harnessing nanovaccines for effective immunization—a special concern on COVID-19: facts, fidelity, and future prospective. ACS Biomater Sci Eng 10:271–297. https://doi.org/10.1021/acsbiomaterials.3c01247
48. Gholap AD, Hatvate NT, Khuspe PR et al (2024) Decoding of nanoneutraceuticals and their delivery. In: Sustainable green nanotechnology. CRC Press, New York, pp 284–309
49. Gholap AD, Kapare HS, Pagar S et al (2024) Exploring modified chitosan-based gene delivery technologies for therapeutic advancements. Int J Biol Macromol 260:129581. https://doi.org/10.1016/j.ijbiomac.2024.129581
50. Gholap AD, Pardeshi SR, Hatvate NT et al (2024) Environmental implications and nanotechnological advances in octocrylene-enriched sunscreen formulations: a comprehensive review. Chemosphere 358:142235. https://doi.org/10.1016/j.chemosphere.2024.142235
51. Gholap AD, Rojekar S, Kapare HS et al (2024) Chitosan scaffolds: expanding horizons in biomedical applications. Carbohydr Polym 323:121394. https://doi.org/10.1016/j.carbpol.2023.121394
52. Gholap AD, Uddin MJ, Faiyazuddin M et al (2024) Advances in artificial intelligence for drug delivery and development: a comprehensive review. Comput Biol Med 178:108702. https://doi.org/10.1016/j.compbiomed.2024.108702
53. Giannitelli SM, Mozetic P, Trombetta M, Rainer A (2015) Combined additive manufacturing approaches in tissue engineering. Acta Biomater 24:1–11. https://doi.org/10.1016/j.actbio.2015.06.032
54. Golebiowska AA, Intravaia JT, Sathe VM et al (2024) Decellularized extracellular matrix biomaterials for regenerative therapies: advances, challenges and clinical prospects. Bioact Mater 32:98–123. https://doi.org/10.1016/j.bioactmat.2023.09.017
55. Grottkau BE, Hui Z, Yao Y, Pang Y (2020) Rapid fabrication of anatomically-shaped bone scaffolds using indirect 3D printing and perfusion techniques. Int J Mol Sci 21:315. https://doi.org/10.3390/ijms21010315
56. Hussain S, Mubeen I, Ullah N et al (2022) Modern diagnostic imaging technique applications and risk factors in the medical field: a review. Biomed Res Int 2022:1–19. https://doi.org/10.1155/2022/5164970
57. Hutmacher DW (2001) Scaffold design and fabrication technologies for engineering tissues—state of the art and future perspectives. J Biomater Sci Polym Ed 12:107–124. https://doi.org/10.1163/156856201744489
58. Jain A, Bansal R (2015) Applications of regenerative medicine in organ transplantation. J Pharm Bioallied Sci 7:188. https://doi.org/10.4103/0975-7406.160013
59. Jammalamadaka U, Tappa K (2018) Recent advances in biomaterials for 3D printing and tissue engineering. J Funct Biomater 9:22. https://doi.org/10.3390/jfb9010022

60. Javaid M, Haleem A, Singh RP (2023) ChatGPT for healthcare services: an emerging stage for an innovative perspective. BenchCouncil Trans Benchmarks Stand Eval 3:100105. https://doi.org/10.1016/j.tbench.2023.100105
61. Jayanth N, Venkata Roshan M, Sakthi Balaji S et al (2023) Additive manufacturing of biomaterials: a review. Mater Today Proc 113:287–298. https://doi.org/10.1016/j.matpr.2023.09.190
62. Johnston A, Callanan A (2023) Recent methods for modifying mechanical properties of tissue-engineered scaffolds for clinical applications. Biomimetics 8:205. https://doi.org/10.3390/biomimetics8020205
63. Kantaros A, Ganetsos T (2023) From static to dynamic: smart materials pioneering additive manufacturing in regenerative medicine. Int J Mol Sci 24:15748. https://doi.org/10.3390/ijms242115748
64. Khedkar A, Tupe D, Gholap A (2012) Pharmacosomes: an emerging vesicular system. Inven Rapid 1:1–3. NDDS [ISSN 0976-3791]
65. Khedkar A, Tupe D, Gholap A (2012) Cubosomes: emerging drug delivery system. Inven Rapid 1:1–3. NDDS [ISSN 0976-3791]
66. Koo S, Santoni SM, Gao BZ et al (2017) Laser-assisted biofabrication in tissue engineering and regenerative medicine. J Mater Res 32:128–142. https://doi.org/10.1557/jmr.2016.452
67. Krishani M, Shin WY, Suhaimi H, Sambudi NS (2023) Development of scaffolds from bio-based natural materials for tissue regeneration applications: a review. Gels 9:100. https://doi.org/10.3390/gels9020100
68. Kumar M, Sharma V (2021) Additive manufacturing techniques for the fabrication of tissue engineering scaffolds: a review. Rapid Prototyp J 27:1230–1272. https://doi.org/10.1108/RPJ-01-2021-0011
69. Leberfinger AN, Dinda S, Wu Y et al (2019) Bioprinting functional tissues. Acta Biomater 95:32–49. https://doi.org/10.1016/j.actbio.2019.01.009
70. Lee EJ, Kasper FK, Mikos AG (2014) Biomaterials for tissue engineering. Ann Biomed Eng 42:323–337. https://doi.org/10.1007/s10439-013-0859-6
71. Légaré F, Adekpedjou R, Stacey D et al (2018) Interventions for increasing the use of shared decision making by healthcare professionals. Cochrane Database Syst Rev 2018:CD006732. https://doi.org/10.1002/14651858.CD006732.pub4
72. Li W, Mille LS, Robledo JA et al (2020) Recent advances in formulating and processing biomaterial inks for vat polymerization-based 3D printing. Adv Healthc Mater 9:e2000156. https://doi.org/10.1002/adhm.202000156
73. Li N, Cui J, Chi M et al (2023) Building a better bone: the synergy of 2D nanomaterials and 3D printing for bone tissue engineering. Mater Des 234:112362. https://doi.org/10.1016/j.matdes.2023.112362
74. Liu T-M (2022) Additive manufacturing and removal of biomaterials. In: Ultrafast imaging and spectroscopy for biomedicine. AIP Publishing, Melville, pp 1–24
75. Liu G, David BT, Trawczynski M, Fessler RG (2020) Advances in pluripotent stem cells: history, mechanisms, technologies, and applications. Stem Cell Rev Rep 16:3–32. https://doi.org/10.1007/s12015-019-09935-x
76. Liu C, Pei M, Li Q, Zhang Y (2022) Decellularized extracellular matrix mediates tissue construction and regeneration. Front Med 16:56–82. https://doi.org/10.1007/s11684-021-0900-3
77. Liu S, Yu J-M, Gan Y-C et al (2023) Biomimetic natural biomaterials for tissue engineering and regenerative medicine: new biosynthesis methods, recent advances, and emerging applications. Mil Med Res 10:16. https://doi.org/10.1186/s40779-023-00448-w
78. Magar TV, Gholap AD, Musmade DS (2011) Elastic vesicles as drugs carriers through the skin. Inven Rapid 1:1–2. NDDS [ISSN 0976-3791]
79. Mahajan AM, Gholap AD, Borude SS (2011) Recent updates on Q fever: diagnosis and treatment. Pharmacologyonline 1:254–264
80. Mahajan AM, Gholap AD, Borude SS (2011) POEMS syndrome: a review. Pharmacologyonline 1:265–274
81. Mahla RS (2016) Stem cells applications in regenerative medicine and disease therapeutics. Int J Cell Biol 2016:1–24. https://doi.org/10.1155/2016/6940283

82. Mahmood A, Akram T, Chen H, Chen S (2022) On the evolution of additive manufacturing (3D/4D printing) technologies: materials, applications, and challenges. Polymers (Basel) 14:4698. https://doi.org/10.3390/polym14214698
83. Makadia HK, Siegel SJ (2011) Poly lactic-co-glycolic acid (PLGA) as biodegradable controlled drug delivery carrier. Polymers (Basel) 3:1377–1397. https://doi.org/10.3390/polym3031377
84. Mao H, Yang L, Zhu H et al (2020) Recent advances and challenges in materials for 3D bioprinting. Prog Nat Sci Mater Int 30:618–634. https://doi.org/10.1016/j.pnsc.2020.09.015
85. Martelli A, Bellucci D, Cannillo V (2023) Additive manufacturing of polymer/bioactive glass scaffolds for regenerative medicine: a review. Polymers (Basel) 15:2473. https://doi.org/10.3390/polym15112473
86. Mateu-Sanz M, Fuenteslópez CV, Uribe-Gomez J et al (2024) Redefining biomaterial biocompatibility: challenges for artificial intelligence and text mining. Trends Biotechnol 42:402–417. https://doi.org/10.1016/j.tibtech.2023.09.015
87. Mesut B, Al-Mohaya M, Gholap AD et al (2024) Demystifying the potential of lipid-based nanocarriers in targeting brain malignancies. Naunyn Schmiedeberg's Arch Pharmacol 397:9243–9279. https://doi.org/10.1007/s00210-024-03212-6
88. Miar S, Shafiee A, Guda T, Narayan R (2018) Additive manufacturing for tissue engineering. In: 3D printing and biofabrication. Springer, Cham, pp 1–52
89. Miklosic G, Ferguson SJ, D'Este M (2024) Engineering complex tissue-like microenvironments with biomaterials and biofabrication. Trends Biotechnol 42:1241–1257. https://doi.org/10.1016/j.tibtech.2024.03.008
90. Mirshafiei M, Rashedi H, Yazdian F et al (2024) Advancements in tissue and organ 3D bioprinting: current techniques, applications, and future perspectives. Mater Des 240:112853. https://doi.org/10.1016/j.matdes.2024.112853
91. Monia T, Ridha BC (2024) Polymer-ceramic composites for bone challenging applications: materials and manufacturing processes. J Thermoplast Compos Mater 37:1540–1557. https://doi.org/10.1177/08927057231190066
92. Morwood AJ, El-Karim IA, Clarke SA, Lundy FT (2023) The role of extracellular matrix (ECM) adhesion motifs in functionalised hydrogels. Molecules 28:4616. https://doi.org/10.3390/molecules28124616
93. Muncie JM, Weaver VM (2018) The physical and biochemical properties of the extracellular matrix regulate cell fate. Curr Top Dev Biol 130:1–37
94. Murphy SV, Atala A (2014) 3D bioprinting of tissues and organs. Nat Biotechnol 32:773–785. https://doi.org/10.1038/nbt.2958
95. Nath SD, Nilufar S (2020) An overview of additive manufacturing of polymers and associated composites. Polymers (Basel) 12:2719. https://doi.org/10.3390/polym12112719
96. Ngo TD, Kashani A, Imbalzano G et al (2018) Additive manufacturing (3D printing): a review of materials, methods, applications and challenges. Compos Part B Eng 143:172–196. https://doi.org/10.1016/j.compositesb.2018.02.012
97. Odeleye AOO, Baudequin T, Chui C-Y et al (2020) An additive manufacturing approach to bioreactor design for mesenchymal stem cell culture. Biochem Eng J 156:107515. https://doi.org/10.1016/j.bej.2020.107515
98. Pacelli S, Basu S, Whitlow J et al (2017) Strategies to develop endogenous stem cell-recruiting bioactive materials for tissue repair and regeneration. Adv Drug Deliv Rev 120:50–70. https://doi.org/10.1016/j.addr.2017.07.011
99. Panja N, Maji S, Choudhuri S et al (2022) 3D bioprinting of human hollow organs. AAPS PharmSciTech 23:139. https://doi.org/10.1208/s12249-022-02279-9
100. Pardeshi S, Gholap A, More M et al (2023) Dendrimers based antibacterial and antiviral materials. In: Wasington DC (ed) Antibacterial and Antiviral Functional Materials. ACS Publications, pp 139–169
101. Parihar A, Pandita V, Kumar A et al (2022) 3D printing: advancement in biogenerative engineering to combat shortage of organs and bioapplicable materials. Regen Eng Transl Med 8:173–199. https://doi.org/10.1007/s40883-021-00219-w

102. Parit S, Manchare A, Gholap AD et al (2024) Antibody-drug conjugates: a promising breakthrough in cancer therapy. Int J Pharm 659:124211. https://doi.org/10.1016/j.ijpharm.2024.124211
103. Patel A, Taufik M (2024) Extrusion-based technology in additive manufacturing: a comprehensive review. Arab J Sci Eng 49:1309–1342. https://doi.org/10.1007/s13369-022-07539-1
104. Paul AA, Aladese AD, Marks RS (2024) Additive manufacturing applications in biosensors technologies. Biosensors 14:60. https://doi.org/10.3390/bios14020060
105. Pavan Kalyan B, Kumar L (2022) 3D printing: applications in tissue engineering, medical devices, and drug delivery. AAPS PharmSciTech 23:92. https://doi.org/10.1208/s12249-022-02242-8
106. Peng W, Datta P, Ayan B et al (2017) 3D bioprinting for drug discovery and development in pharmaceutics. Acta Biomater 57:26–46. https://doi.org/10.1016/j.actbio.2017.05.025
107. Prem Ananth K, Jayram ND (2024) A comprehensive review of 3D printing techniques for biomaterial-based scaffold fabrication in bone tissue engineering. Ann 3D Print Med 13:100141. https://doi.org/10.1016/j.stlm.2023.100141
108. Puertas-Bartolomé M, Mora-Boza A, García-Fernández L (2021) Emerging biofabrication techniques: a review on natural polymers for biomedical applications. Polymers (Basel) 13:1209. https://doi.org/10.3390/polym13081209
109. Pugliese R, Beltrami B, Regondi S, Lunetta C (2021) Polymeric biomaterials for 3D printing in medicine: an overview. Ann 3D Print Med 2:100011. https://doi.org/10.1016/j.stlm.2021.100011
110. Puri AV, Gholap AD (2018) Comparative evaluation of papaya and banana starch as disintegrant in diclofenac sodium tablet formulation. Int J Pharm Biol Sci 8:512–522
111. Rahman M, Islam KS, Dip TM et al (2023) A review on nanomaterial-based additive manufacturing: dynamics in properties, prospects, and challenges. Prog Addit Manuf 9:1197–1224. https://doi.org/10.1007/s40964-023-00514-8
112. Rajurkar VG, Gholap AD (2012) Recent technologies in nanomedicine for treatment of cancer. Inven Rapid Pharm Tech 4:1–4
113. Rajurkar VG, Gholap AD (2012) Recent technologies in nanomedicine for treatment of cancer. Inven Rapid Pharm Tech 4:1–3
114. Ramezani M, Mohd Ripin Z (2023) 4D printing in biomedical engineering: advancements, challenges, and future directions. J Funct Biomater 14:347. https://doi.org/10.3390/jfb14070347
115. Reddy MSB, Ponnamma D, Choudhary R, Sadasivuni KK (2021) A comparative review of natural and synthetic biopolymer composite scaffolds. Polymers (Basel) 13:1105. https://doi.org/10.3390/polym13071105
116. Rodríguez-Vázquez M, Vega-Ruiz B, Ramos-Zúñiga R et al (2015) Chitosan and its potential use as a scaffold for tissue engineering in regenerative medicine. Biomed Res Int 2015:1–15. https://doi.org/10.1155/2015/821279
117. Rojekar S, Gholap AD, Togre N et al (2024) Current status of mannose receptor-targeted drug delivery for improved anti-HIV therapy. J Control Release 372:494–521. https://doi.org/10.1016/j.jconrel.2024.06.002
118. Saini G, Segaran N, Mayer J et al (2021) Applications of 3D bioprinting in tissue engineering and regenerative medicine. J Clin Med 10:4966. https://doi.org/10.3390/jcm10214966
119. Schot M, Araújo-Gomes N, van Loo B et al (2023) Scalable fabrication, compartmentalization and applications of living microtissues. Bioact Mater 19:392–405. https://doi.org/10.1016/j.bioactmat.2022.04.005
120. Scodellaro C, Pina RR, Ferreira FC et al (2024) Unlocking the potential of stem cell microenvironments in vitro. Bioengineering 11:289. https://doi.org/10.3390/bioengineering11030289
121. Selden C, Fuller B (2018) Role of bioreactor technology in tissue engineering for clinical use and therapeutic target design. Bioengineering 5:32. https://doi.org/10.3390/bioengineering5020032
122. Seo C, Zhang C, Liu H (2022) A review on the 3D printing of composite scaffolds for bone tissue engineering. In: Musculoskeletal tissue engineering. Elsevier, Cambridge, pp 201–241

123. Shah SR, Modi CD, Singh S, Mori DD, Soniwala MM, Prajapati BG (2024) Recent advances in additive manufacturing of polycaprolactone-based scaffolds for tissue engineering applications: a comprehensive review. Regen Eng Transl Med 11:1–20
124. Shaikh MS, Faiyazuddin M, Khan MS et al (2024) Chikungunya virus vaccine: a decade of progress solving epidemiological dilemma, emerging concepts, and immunological interventions. Front Microbiol 15. https://doi.org/10.3389/fmicb.2024.1413250
125. Shimojo AAM, Rodrigues ICP, Perez AGM et al (2020) Scaffolds for tissue engineering: a state-of-the-art review concerning types, properties, materials, processing, and characterization. In: Racing for the surface. Springer, Cham, pp 647–676
126. Shin H, Jo S, Mikos AG (2003) Biomimetic materials for tissue engineering. Biomaterials 24:4353–4364. https://doi.org/10.1016/S0142-9612(03)00339-9
127. Shiva S, Asuwin Prabu RG, Bajaj G et al (2023) A review on the recent applications of synthetic biopolymers in 3D printing for biomedical applications. J Mater Sci Mater Med 34:62. https://doi.org/10.1007/s10856-023-06765-9
128. Singh G, Pandey PM (2020) Role of imaging data in additive manufacturing for biomedical applications. In: 3D printing in biomedical engineering. Materials horizons: from nature to nanomaterials. Springer, Singapore, pp 69–94
129. Singh T, Ravi Kumar V, Gholap AD, Sanap G et al (2016) In vitro antipyrial activity of psidium guajava leaf extract. Eur J Pharm Med Res 3:312–315
130. Singh A, Mandal P, Maheshwari S, Akhtar J, Prajapati BG, Singh S (2024) Processing of sustainable biomaterials by additive manufacturing methods. In: Advances in sustainable biomaterials. CRC Press, Boca Raton, pp 141–162
131. Smith JA, Mele E (2021) Electrospinning and additive manufacturing: adding three-dimensionality to electrospun scaffolds for tissue engineering. Front Bioeng Biotechnol 9. https://doi.org/10.3389/fbioe.2021.674738
132. Socci MC, Rodríguez G, Oliva E et al (2023) Polymeric materials, advances and applications in tissue engineering: a review. Bioengineering 10:218. https://doi.org/10.3390/bioengineering10020218
133. Sta. Agueda JRH, Chen Q, Maalihan RD et al (2021) 3D printing of biomedically relevant polymer materials and biocompatibility. MRS Commun 11:197–212. https://doi.org/10.1557/s43579-021-00038-8
134. Szwed-Georgiou A, Płociński P, Kupikowska-Stobba B et al (2023) Bioactive materials for bone regeneration: biomolecules and delivery systems. ACS Biomater Sci Eng 9:5222–5254. https://doi.org/10.1021/acsbiomaterials.3c00609
135. Tabata Y (2009) Biomaterial technology for tissue engineering applications. J R Soc Interface 6:S311–S324. https://doi.org/10.1098/rsif.2008.0448.focus
136. Tripathi S, Mandal SS, Bauri S, Maiti P (2023) 3D bioprinting and its innovative approach for biomedical applications. MedComm 4:e194. https://doi.org/10.1002/mco2.194
137. Tupe D, Gholap AD (2012) The threat of photoaging: clinical conditions, evaluation and treatment. Inven Rapid Cosmeceuticals 1:1–3. [ISSN 0976-3864]
138. Vach Agocsova S, Culenova M, Birova I et al (2023) Resorbable biomaterials used for 3D scaffolds in tissue engineering: a review. Materials (Basel) 16:4267. https://doi.org/10.3390/ma16124267
139. Vanaei S, Parizi MS, Vanaei S et al (2021) An overview on materials and techniques in 3D bioprinting toward biomedical application. Eng Regen 2:1–18. https://doi.org/10.1016/j.engreg.2020.12.001
140. Veeman D, Sai MS, Sureshkumar P et al (2021) Additive manufacturing of biopolymers for tissue engineering and regenerative medicine: an overview, potential applications, advancements, and trends. Int J Polym Sci 2021:1–20. https://doi.org/10.1155/2021/4907027
141. Vijayavenkataraman S (2023) 3D bioprinting: challenges in commercialization and clinical translation. J 3D Print Med 7. https://doi.org/10.2217/3dp-2022-0026
142. Vilas Bhagat AG (2012) Gold nanorods: a promising drug delivery vehicle. Inven Rapid 1:1–4. NDDS [ISSN 0976-3864]

143. Vyas J, Shah I, Singh S, Prajapati BG (2023) Biomaterials-based additive manufacturing for customized bioengineering in management of otolaryngology: a comprehensive review. Front Bioeng Biotechnol 11. https://doi.org/10.3389/fbioe.2023.1234340
144. Vyas J, Singh S, Shah I, Prajapati BG (2023) Potential applications and additive manufacturing technology-based considerations of mesoporous silica: a review. AAPS PharmSciTech 25(1):6
145. Vyas J, Raytthatha N, Singh S, Prajapati B (2024) Next-generation computational automation-based additive manufacturing of pharmaceuticals: an approach to fabricate precise medicine. In: Handbook of 3D printing in pharmaceutics. CRC Press, Boca Raton, pp 153–175
146. Vyas J, Raytthatha N, Singh S, Prajapati BG, Mohite P, Munde S (2024) Sustainable sources of raw materials as substituting biomaterials for additive manufacturing of dental implants: a review. Periodontal Implant Res 8(1):3
147. Wang P, Sun Y, Shi X et al (2021) 3D printing of tissue engineering scaffolds: a focus on vascular regeneration. Bio-Design Manuf 4:344–378. https://doi.org/10.1007/s42242-020-00109-0
148. Wang F, Cai X, Shen Y, Meng L (2023) Cell–scaffold interactions in tissue engineering for oral and craniofacial reconstruction. Bioact Mater 23:16–44. https://doi.org/10.1016/j.bioactmat.2022.10.029
149. Williams DF (2022) Biocompatibility pathways and mechanisms for bioactive materials: the bioactivity zone. Bioact Mater 10:306–322. https://doi.org/10.1016/j.bioactmat.2021.08.014
150. Woo HN, Cho YJ, Tarafder S, Lee CH (2021) The recent advances in scaffolds for integrated periodontal regeneration. Bioact Mater 6:3328–3342. https://doi.org/10.1016/j.bioactmat.2021.03.012
151. Xie Z, Gao M, Lobo AO, Webster TJ (2020) 3D bioprinting in tissue engineering for medical applications: the classic and the hybrid. Polymers (Basel) 12:1717. https://doi.org/10.3390/polym12081717
152. Xu Y, Chen C, Hellwarth PB, Bao X (2019) Biomaterials for stem cell engineering and biomanufacturing. Bioact Mater 4:366–379. https://doi.org/10.1016/j.bioactmat.2019.11.002
153. Xue J, Wu T, Dai Y, Xia Y (2019) Electrospinning and electrospun nanofibers: methods, materials, and applications. Chem Rev 119:5298–5415. https://doi.org/10.1021/acs.chemrev.8b00593
154. Zadpoor AA, Malda J (2017) Additive manufacturing of biomaterials, tissues, and organs. Ann Biomed Eng 45:1–11. https://doi.org/10.1007/s10439-016-1719-y
155. Zakrzewski W, Dobrzyński M, Szymonowicz M, Rybak Z (2019) Stem cells: past, present, and future. Stem Cell Res Ther 10:68. https://doi.org/10.1186/s13287-019-1165-5

Regulatory Aspects, Future Prospects, and Challenges Associated with Biomaterials-Based Additive Manufacturing for Tissue Engineering

17

Popat Mohite, Abhijeet Puri, Amol D. Gholap, Yogesh Chaudhari, Pooja V. Nagime, and Sudarshan Singh

Abstract

Tissue engineering has been transformed by biomaterials-based additive manufacturing, which makes it possible to precisely fabricate intricate structures with customized characteristics. This chapter examines the opportunities, difficulties, and regulatory environment around the use of additive manufacturing in tissue engineering applications. The development and commercialization of additive manufacturing technologies based on biomaterials are significantly influenced by regulatory factors. The European Union's Medical Device Regulation (MDR) and the FDA's regulations for 3D-printed medical devices are only two examples of the regulatory processes and standards covered in this chapter that govern these technologies. It is important to comprehend and adhere to these rules to guarantee the safety and effectiveness of tissue-engineered products. Tissue engineering has great promise for the use of biomaterials-based additive manufacturing in the future. Innovation in this field is being driven by developments in bio-ink development, printing technologies, and materials science. The chapter covers new developments in the field, including stem cell applications in additive manufacturing, bio-fabrication of vascularized tissues, and multi-material

P. Mohite (✉) · A. Puri · A. D. Gholap
AETs St. John Institute of Pharmacy and Research, Palghar, Maharashtra, India

Y. Chaudhari
Dr. L. H. Hiranandani College of Pharmacy, Ulhasnagar, Maharashtra, India

P. V. Nagime
Faculty of Agro-Industry, Centre of Excellence in Innovative Biotechnology for Sustainable Utilization of Bioresources, Prince of Songkhla University, HatYai, Thailand

S. Singh
Office of Research Administration, Chiang Mai University, Mueang Chiang Mai, Thailand

Faculty of Pharmacy, Chiang Mai University, Chiang Mai, Thailand

S. Singh et al. (eds.), *Biomaterial-based Additive Manufacturing in Tissue Engineering and Regeneration*, Biomaterials, Bioengineering and Sustainability 7,
https://doi.org/10.1007/978-3-031-96070-3_17

printing. These advancements have enormous potential for producing organoids and functional tissue substitutes for use in drug testing and regenerative medicine. The book chapter provides an overview of regulatory considerations that must be carefully considered, technological developments must continue, and present obstacles must be overcome. Also, insights into the regulatory environment, potential opportunities, and difficulties related to biomaterials-based additive manufacturing for tissue engineering are provided in this chapter, which will be helpful to researchers and industry participants in this quickly developing subject.

Keywords

Additive manufacturing · Bioinks · Tissue engineering · Regulatory landscape · Regulations · Regulatory pathways

17.1 Introduction

Additive manufacturing techniques offer significant advantages over traditional methods in tissue engineering. These techniques, such as photopolymerization-based 3D bioprinting and Light-Emitting-Diode-based 3D printing, provide high-resolution structures at a lower cost, with faster fabrication times and smaller equipment sizes compared to established methods like Multi-Photon Polymerization and Selective Laser Melting [1, 2]. Additionally, additive manufacturing allows for the creation of personalized and intricate three-dimensional neural constructs with native-like features, enhancing the potential for successful nerve regeneration [3]. The versatility of additive manufacturing platforms, including laser-assisted printing and extrusion printing, enables precise construction of complex tissue engineering products with tailored printability, cytocompatibility, and mechanical properties, addressing the limitations of traditional fabrication methods.

Additive manufacturing facilitates the creation of intricate tissue structures with precise property control through various printing technologies and material advancements [4, 5]. Technologies like laser-assisted printing, extrusion printing, and inkjet printing offer tailored printability, cytocompatibility, and mechanical characteristics for diverse tissue engineering applications [6]. Temperature monitoring and controlled printing environments are crucial to ensure the quality of scaffolds, with innovations like compact heating jackets and semi-closed printing setups enhancing uniformity and efficiency. The evolution towards composite materials in additive manufacturing has further improved mechanical features and expanded application possibilities for tissue engineering products. Additionally, additive manufacturing methods, such as stereolithography, have enabled rapid geometric modifications in microscale tissues, enhancing design flexibility and replication fidelity for tissue engineering models (Fig. 17.1).

Regulatory requirements significantly influence the design and development process of tissue-engineered products (TEPs). These requirements, such as those set by

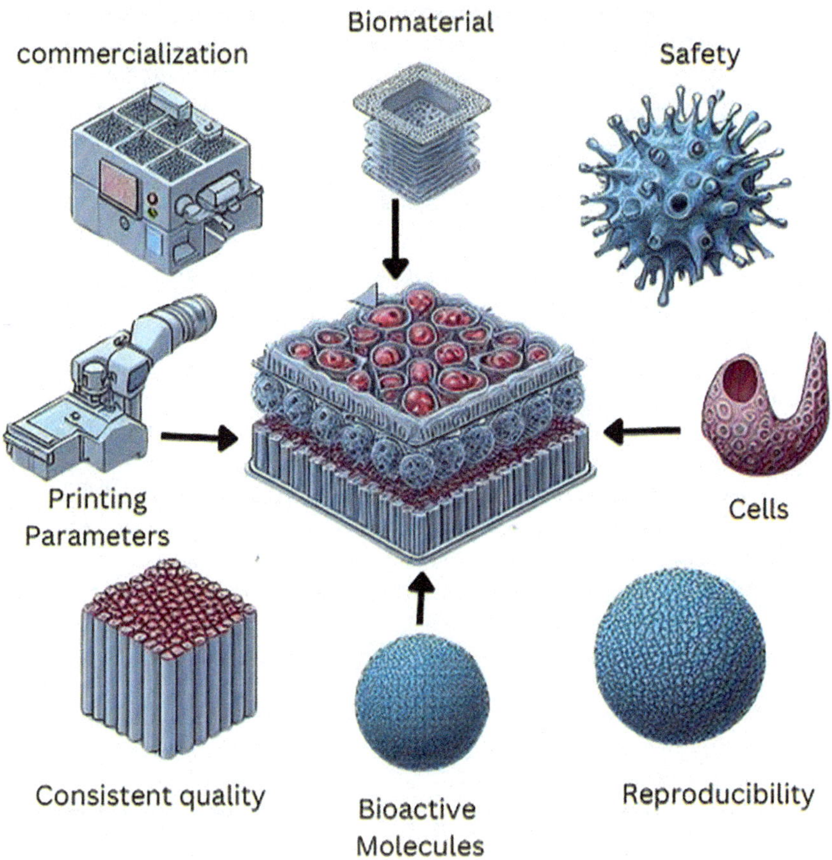

Fig. 17.1 Tissue engineering with additive manufacturing. (Created by using Biorender.com)

the US FDA and other regulatory bodies, dictate the quality, safety, and efficacy standards that TEP manufacturing processes must meet [7, 8]. Compliance with these regulations necessitates thorough process development to ensure quality, scalability, and cost-effectiveness. The regulatory landscape for TEPs is evolving, with different pathways for approval like Biologics License Applications (BLA) and Premarket Approvals, each with specific data requirements [9]. Failure to meet regulatory standards can pose challenges and increase costs, emphasizing the need for robust, reproducible, and compliant manufacturing processes. Overall, regulatory requirements play a crucial role in shaping the design and development of TEPs, guiding manufacturers towards producing safe and effective products for patients.

Non-compliance with regulatory standards in tissue engineering can have significant consequences. Failure to adhere to good laboratory practices (GLP) in non-clinical safety studies may impede the efficient clinical translation of innovative therapies [10]. The lack of European Union regulation for autologous tissue-engineering products poses a risk to the future of this technology in Europe. New regulatory regimes for advanced treatments like cell therapies and gene therapies, if not carefully implemented, could stifle innovation in life extension technologies [11]. The evolving nature of biotechnology blurs traditional boundaries, necessitating a flexible approach to regulatory governance and technical standardization processes to accommodate these changes. Exciting developments in tissue engineering present a complex challenge for regulations and standards, highlighting the importance of compliance to ensure safety and efficacy in this field.

Standardization plays a crucial role in enhancing the reliability and reproducibility of tissue-engineered products manufactured through additive manufacturing (AM) processes. The development of industry-accepted standards and guidelines is essential to ensure consistency in AM parts across various sectors [12]. Specifically, in the medical field, where AM is utilized for producing medical devices and implants, the need for specific standards is highlighted to address issues like anisotropy and process parameters, ensuring the quality of the final products [13]. Moreover, the control of printing environment temperature and the development of innovative cooling mechanisms significantly impact the quality and efficiency of tissue scaffolds, highlighting the importance of standardized processes in AM for tissue engineering applications [5]. Overall, standardization efforts in AM contribute to the quality, safety, and reproducibility of tissue-engineered products, fostering advancements in the field [14].

Implementing standardized protocols across various additive manufacturing platforms and materials presents challenges due to the complexity of combining models at multiple scales and multi-physics [15]. The differences between additive manufacturing and conventional processes pose extraordinary challenges for industrial implementation [16]. Lack of specific testing and inspection procedures for additive manufacturing inhibits the development of industry-accepted standards. Furthermore, the absence of comprehensive standards hinders the broad adoption of additive manufacturing, despite its potential advantages like weight reduction and design freedom [17]. Addressing these challenges requires harmonizing terminology, integrating semantic descriptions, and developing dedicated domain ontologies to ensure consistent use of terms and definitions across technologies and industries.

17.2 Regulatory Landscape

Regulatory agencies play a crucial role in overseeing tissue engineering products by establishing comprehensive regulatory frameworks [7, 18]. These agencies are responsible for evaluating the safety, efficacy, and quality of tissue-based products (TBPs) through various regulatory pathways such as Biologics License Applications (BLA), Premarket Approvals, and human cells, tissues, and cellular and tissue-based products designations. They require sponsors to provide detailed data, including clinical trials, to support the approval or clearance of these innovative products. Additionally, regulatory bodies like the US Food and Drug Administration [19] oversee the development of cell-based products, ensuring compliance with regulations related to chemistry, manufacturing, controls, pharmacology/toxicology, and clinical testing. By setting stringent standards and guidelines, regulatory agencies ensure the safety and efficacy of tissue engineering products for patient use.

Regulatory agencies ensure the safety and efficacy of tissue-engineered products through stringent frameworks and processes. These agencies, such as the U.S. Food and Drug Administration [19] and regulations in Japan, oversee the approval or clearance of these products [9, 20]. The regulatory pathways involve different classifications like Biologics License Application (BLA), 510(k), Premarket Approval, or human cells, tissues, and cellular and tissue-based products designation, each requiring specific data to support the filing [7]. Manufacturing processes are also closely monitored to meet technical demands, ensure quality, scalability, and cost-effectiveness [21]. Additionally, preclinical investigations, toxicity testing, immunogenicity assessments, and in vivo remodeling evaluations are conducted to determine the effectiveness of tissue-engineered products [22]. Regulatory agencies play a crucial role in setting standards, evaluating data, and ensuring the safety and efficacy of these innovative products before they are introduced to the market.

The regulations governing the use of additive manufacturing (AM) technologies in tissue engineering applications vary across different countries and regions. These regulations stipulate specific requirements for medical device manufacturers to bring their products to market. The regulatory frameworks address aspects such as CE marking, US FDA regulations, and guidelines from countries like Australia, Canada, Brazil, and Japan [23]. Challenges arise due to the disruptive nature of AM and 3D bioprinting technologies, which introduce complexities like incorporating living cells in the fabrication process. Regulators face difficulties in adapting traditional frameworks designed for mass-manufactured therapies to accommodate bespoke solutions in regenerative medicine [24, 25]. Additionally, the legal regime faces challenges in areas such as bioethics, safety, and compliance with regulations like the EU Advanced Therapy Medicinal Products Regulation and the Medical Device Regulation [26].

Regulatory frameworks for healthcare products vary across regions like the United States and the European Union. The FDA in the US and the EMA in the EU have distinct approaches [27, 28]. The FDA emphasizes expedited review programs, leading to shorter approval times compared to the EMA [29]. In terms of biosimilar medicines, the EU market has seen faster growth than the US due to patent challenges and exclusivity periods [30]. Additionally, the ICH public consultation process shows differences in the distribution speed and consultation periods among the FDA, EMA, and Japanese authority, with the EMA being quicker and allowing longer consultation times [31]. These variations highlight the diverse regulatory landscapes and priorities in different regions, impacting the approval processes and market access for healthcare products.

Adherence to regulatory standards plays a vital role in ensuring the safety and efficacy of tissue-engineered products manufactured using additive techniques. Regulatory frameworks and robust standards are essential to validate the quality of parts produced through additive manufacturing (AM) [32]. Temperature control during the printing process is crucial for maintaining the properties of biomaterial inks and enhancing scaffold quality [5]. Additionally, the implementation of good laboratory practices (GLP) in non-clinical studies for cell-, gene-, and tissue-engineered therapies is imperative for meeting pharmaceutical quality management standards [10]. By following these regulatory guidelines, manufacturers can demonstrate the reliability and effectiveness of tissue-engineered products produced through additive manufacturing techniques, ensuring their safety and efficacy for use in various applications.

Non-compliance with regulatory requirements in the development and commercialization of additive-manufactured tissue products can have significant consequences. Failure to meet regulatory standards can lead to increased costs due to the need for robust and compliant processes [25]. In the context of 3D bioprinting, which involves living cells, challenges include classification, risk assessment, standardization, quality control, and technical issues related to materials and cells [33]. Regulatory frameworks and standards are crucial for ensuring the safety and efficacy of additively manufactured medical devices, especially in the healthcare industry, which relies on personalized and complex structures produced through additive manufacturing [32]. Non-compliance can hinder confidence in the use of such products, emphasizing the importance of adhering to regulatory requirements for successful development and commercialization.

Collaborative efforts to harmonize regulatory standards in additive manufacturing for tissue engineering have been a focus of organizations like the International Organization for Standardization (ISO) and the American Society for Testing and Materials (ASTM) [34]. These efforts have led to the establishment of joint ISO/ASTM standards to address the diverse needs of various industrial sectors requiring product certification. Additionally, ISO Technical Committee 261 (ISO/TC 261) plays a crucial role in standardizing additive manufacturing processes, including data, materials, processes, software, and quality parameters [35]. By defining test procedures, supply agreements, and ensuring environmental health and safety standards, these collaborative initiatives aim to facilitate the adoption and regulation of additive manufacturing technologies in tissue engineering applications.

Global harmonization in the development of tissue-engineered products benefits manufacturers and researchers by facilitating standardized procedures, enhancing comparability, and streamlining regulatory processes. By aligning definitions, regulations, and jurisdictional assignments internationally [36], manufacturers can operate more efficiently across borders, reducing redundancies and costs [37]. Standardization efforts also extend to critical quality assays, aiming to harmonize identity, purity, safety, and potency assessments for cell therapies [38]. Moreover, harmonized culture conditions for mesenchymal stromal cells (MSCs) have shown reproducible results across different laboratories, emphasizing the importance of standardized protocols in research [39, 40]. Overall, global harmonization fosters collaboration, accelerates innovation, and ensures the safety and efficacy of tissue-engineered products on a worldwide scale.

17.3 Regulatory Pathways

The key differences between regulatory pathways for tissue-engineered products lie in the requirements for approval or clearance. In the United States, the FDA regulates these products through pathways like Biologics License Application (BLA), 510(k), Premarket Approval, or human cells, tissues, and cellular and tissue-based products designation, each with varying data requirements [7]. Similarly, the European Union and the USA have established specific regulatory schemes for tissue-based products, while other jurisdictions face challenges in oversight [18]. In Latin American countries, regulatory policies for medical devices, including tissue-engineered products, are slowly evolving, with a top-down diffusion pattern influencing current regulations [41]. Japan has introduced a new conditional approval pathway for regenerative medicine products, emphasizing post-marketing studies to address limitations of existing systems [42]. Overall, navigating these regulatory frameworks is crucial for ensuring the safety and efficacy of tissue-engineered products [43].

17.3.1 Different Regulatory Pathways for Tissue-Engineered Products

The choice of regulatory pathway significantly influences the timeline and cost of bringing a product to market. Different regulatory authorities, such as the US FDA and EMA, have specific pathways for approval [43]. Expedited regulatory pathways, like priority review and provisional approval, aim to accelerate the review and approval process for novel medicines, impacting timelines and costs [44]. The complexity of regulatory requirements can lead to delays and increased expenses in product development, with costs ranging from 2.8 billion and the time to market up to 15 years [45]. Academic institutions face challenges in navigating these pathways due to a lack of understanding of regulatory requirements, necessitating tools like Regulatory Readiness (RRL) for better integration of regulatory science into product development processes.

17.3.2 Challenges and Considerations for Electing the Appropriate Regulatory Pathway

When selecting a regulatory pathway for additive manufacturing-based tissue engineering products, companies should consider various factors. These include the specific regulatory requirements set by different countries and regions [46], the type of product being developed (living/nonliving, genetically modified, etc.) [7], the complexity of the product (dermal, epidermal, composite, etc.) [24], and the incorporation of living cells in the fabrication process, which adds complexity [18]. Additionally, companies must take into account the different regulatory pathways available, such as Biologics License Application (BLA), 510(k), Premarket

Approval, or human cells, tissues, and cellular and tissue-based products designation [25]. Understanding the regulatory frameworks in the European Union and the United States, which have clear pathways for tissue-based products, is crucial. Overall, companies need to navigate these factors to ensure compliance with regulations and successful market approval for their innovative tissue engineering products.

The complexity and novelty of additive manufacturing (AM) technologies significantly impact the regulatory pathway selection process [25, 47]. AM's unique capabilities, such as mass personalization and intricate structures, pose challenges for traditional regulatory frameworks designed for mass production. The need to upgrade regulatory frameworks to accommodate high-innovation products and processes, like those in the pharmaceutical and healthcare industries, is emphasized. Engineers working in regulated industries, like the space sector, face a conflict between safety and innovation when adopting AM, highlighting the challenges of creativity within regulated environments. The regulatory framework for AM medical devices is crucial to ensure safety and efficacy, requiring robust standards and ongoing development efforts to address gaps and ensure compliance across different regions.

17.3.3 Considerations for Combination Products

Developing combination products using additive manufacturing techniques presents unique regulatory challenges due to the complexity of merging drugs and medical devices [25]. The rapid advancements in additive manufacturing technologies, such as 3D bioprinting, introduce complexities in regulatory frameworks designed for mass-produced therapies [25]. The lack of clear classification criteria for combination products with evolving modes of action further complicates regulatory processes [47]. Additionally, the intricate nature of innovative manufacturing processes like additive manufacturing requires regulatory frameworks to adapt to ensure quality standards and public health protection [48]. Addressing these challenges necessitates multidisciplinary collaborations within the health sector and regulatory bodies to navigate the intricate classification and approval processes for personalized 3D-printed combination products [49].

Regulatory strategies for additive manufacturing involving multiple components, such as custom-made 3D-printed scaffolds with incorporated medicinal substances, require adaptation due to the complexity of these products. The challenges include classification based on the principal mode of action (PMOA) [25], the inclusion of living cells in the fabrication process [50], and the need for reliable process monitoring and control for quality assurance [49]. Current regulations do not fully address these complexities, necessitating multidisciplinary collaborations within the health sector and with regulators to overcome classification challenges [51]. Recommendations include a risk-based approach in the legal regulation of bioprinting and the introduction of a decision-tree scheme to assist in classification.

Addressing gaps in regulation and adapting to innovative healthcare solutions are crucial to effectively regulate additive manufacturing involving multiple components.

17.4 Future Prospective

Recent developments in bio-ink formulation for additive manufacturing in tissue engineering include the utilization of decellularized extracellular matrices (dECM) [52, 53], hyaluronic acid (HA) [54], and composite biomaterial inks incorporating fish scale (FS) and alginate dialdehyde (ADA)-gelatin (GEL) [55]. dECM-based bioinks offer complex internal structures and bioactive factors for tissue regeneration [56], while HA bioinks provide stability and printability for tissue constructs. The incorporation of FS in ADA-GEL biomaterial inks enhances mechanical strength and osteogenic differentiation, making it a promising additive for bone tissue engineering applications. These advancements highlight the diverse approaches in bioink formulation, ranging from natural biomaterials like dECM to chemically modifiable compounds such as HA, aiming to improve the efficacy and clinical translation of bio printed tissue constructs.

Researchers are addressing the challenge of developing biocompatible materials suitable for additive manufacturing applications by focusing on the utilization of novel materials and composites tailored for specific biomedical purposes [57]. Efforts are being made to produce biodegradable polymers and bioplastics derived from renewable resources to replace conventional plastics, thus contributing to a sustainable and circular economy [58]. In healthcare, particularly in dentistry, additive manufacturing is being utilized to create patient-specific implants, dentures, and prosthetic aids, with a keen emphasis on optimizing surface quality, dimensional accuracy, and mechanical properties of the biocompatible materials produced [59, 60]. The design and preparation of biodegradable polymer composites for biomedical additive manufacturing applications are being explored as cost-effective and environmentally friendly alternatives to traditional materials like metals and ceramics (Fig. 17.2).

Multi-material printing in tissue engineering holds great promise for various applications. One significant application lies in creating complex tissue constructs with hierarchical architectures [61, 62]. This approach enables the fabrication of multilayered structures that can mimic the native tissue properties, promoting enhanced cell alignment and functionality [63]. Additionally, multi-material printing allows for the development of bioinks with high structural fidelity and bioactivity, essential for in situ tissue engineering [64]. Moreover, the use of light-responsive microgels as bio resins for volumetric bioprinting offers a novel way to spatially pattern multiple cell types, facilitating the creation of physiologically relevant models [65]. Overall, multi-material printing in tissue engineering shows promise in producing regenerative grafts with enhanced functionality, developing engineered living systems, and modeling metabolic diseases.

Researchers are addressing the technical challenges of vascularized tissue fabrication through additive manufacturing techniques by focusing on key aspects such

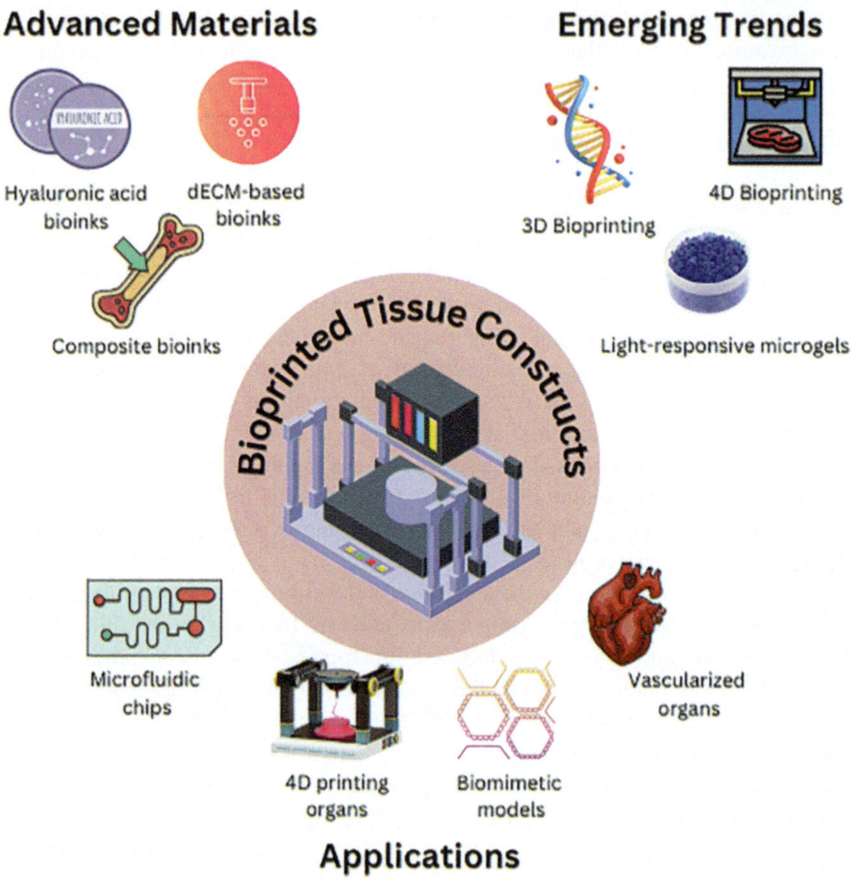

Fig. 17.2 Future prospects of additive manufacturing in tissue engineering. (Created by using Biorender.com)

as bioink selection, cell integration, and printing methods. By integrating stem and endothelial cells into prints, selecting bioinks based on their physical properties, and choosing appropriate printing methods, successful vascularization in 3D bioprinted tissues can be achieved [66, 67]. Techniques like 4D bioprinting are emerging as revolutionary methods for fabricating vascular constructs that can integrate with the host vasculature, offering new possibilities in vascular tissue engineering [63]. Additionally, significant progress has been made in 3D printing vascularized organs like the liver, heart, bone, and pancreas, indicating a promising future for bioartificial organ manufacturing using additive manufacturing technologies [68]. The advancements in bio-fabrication methods for blood vascular and vascularized tissues, including 3D bioprinting, assembled scaffolds, and electrospinning, further contribute to overcoming challenges in creating vascularized tissues [69].

Additive manufacturing, particularly 3D and 4D printing, plays a crucial role in advancing regenerative medicine applications like organoid production and personalized tissue constructs. These technologies enable the fabrication of patient-specific tissue scaffolds with high precision [70]. By incorporating smart materials that respond to stimuli, such as changing shape or becoming bioactive, 4D printing allows for the creation of dynamic structures, enhancing tissue regeneration [71]. Furthermore, 3D bioprinting facilitates the production of vascularized organs and tissues in vitro, in situ bioprinting directly onto defect sites, and 3D in vitro tissue models, offering personalized solutions for patients with chronic diseases and neurodegenerative disorders [72, 73]. These advancements have significantly impacted medical interventions, prosthetics, drug testing, and personalized treatments, displaying the potential of additive manufacturing in revolutionizing regenerative medicine.

Additive manufacturing platforms, particularly in combination with microfluidics, offer significant potential for enhancing drug testing processes [74, 75]. These platforms enable the creation of organ-on-chips and 3D bio-printed human organoids, which mimic physiological systems for more reliable drug screening [76, 77]. By utilizing 3D bioprinting, researchers can develop biomimetic models in organ-on-chips, improving the efficiency of drug investigation [78]. Additionally, the integration of 3D bio-printed hydrogel-based microfluidic chips within high-throughput screening platforms allows for continuous and seamless manufacturing, addressing challenges such as large cell volumes and high reagent consumption. These advanced techniques facilitate the modeling of in vitro tissue, providing a more holistic view on cell biology and increasing predictability in drug response.

17.5 Challenges

Materials intended for tissue engineering applications must meet specific criteria to be considered both biocompatible and bioresorbable. Biocompatibility is crucial for ensuring the material's ability to interact favorably with the host tissue, while bioresorb ability allows for gradual degradation without causing harm. These criteria are essential for successful tissue regeneration [79, 80]. Biocompatible materials should not elicit adverse immune responses and should support cell adhesion and proliferation [81]. On the other hand, bioresorbable materials should degrade gradually in a controlled manner, ensuring compatibility with the body's natural processes [82]. By meeting these criteria, materials can effectively integrate into the host tissue, promote regeneration, and eventually be replaced by the newly formed tissue, making them ideal for tissue engineering applications.

Researchers ensure the safety and effectiveness of biocompatible and bioresorbable materials used in additive manufacturing through rigorous biocompatibility testing, which includes assessing cytotoxic effects and acute systemic toxicity [59]. This testing involves various cell viability assays and cytotoxicity tests to evaluate the impact of these materials on living cells, guiding the optimization of cell culture conditions and drug candidates. Additionally, the properties of the materials, such as surface quality, dimensional accuracy, and mechanical properties, are thoroughly analyzed to meet the requirements of specific applications like tissue engineering [83–85]. By understanding the interactions between these materials and biological systems, researchers aim to minimize adverse reactions and improve patient outcomes, ensuring the successful integration of these materials into clinical practice.

Various strategies are being explored to replicate the mechanical properties of native tissues in additive manufacturing constructs. One approach involves designing scaffold geometries with specific pore areas and angles to tailor mechanical behavior [86]. Another strategy includes utilizing complex hierarchical scaffold designs inspired by nature to mimic the multiscale structure-function relationship found in natural tissues [87]. Additionally, the use of material jetting additive manufacturing allows for the fabrication of full-scale anatomic models with materials that resemble relevant tissues, enabling patient-specific surgical planning or training models. Furthermore, the development of material microstructures based on a micromechanical model can replicate the viscoelastic properties of soft tissues through additive manufacturing [88, 89]. These strategies collectively aim to enhance the biomimicry of mechanical properties in additive manufacturing constructs.

Researchers incorporate cell signaling cues into additive manufacturing scaffolds for tissue regeneration by utilizing various strategies. One approach involves the synthesis of dissolvable and conductive polyurethane copolymers processed into scaffolds with different patterns to regulate electrical signal transmission and cell behaviors concerning bone tissue [90]. Additionally, the incorporation of bioactive materials into 3D printed scaffolds aims to optimize bioprinting performance, including vascularization, antibacterial properties, immunomodulation, and collagen regulation [91]. Furthermore, nanomaterials like carbon-family nanomaterials have been integrated into polymeric hydrogels to enhance stability and shape fidelity of printed gels for tissue engineering applications [92]. Moreover, the development of 3D-printed scaffolds with specific topographical patterns, such as "lines and ridges," and the inclusion of graphene have shown promising results in promoting cell alignment, neurite orientation, and cell differentiation for various tissue types [93].

What are the primary scalability challenges faced when transitioning from laboratory-scale additive manufacturing to large-scale production of tissue-engineered constructs?

The primary scalability challenges encountered when moving from laboratory-scale additive manufacturing to large-scale production of tissue-engineered constructs include ensuring quality, meeting regulatory demands, process scalability, and managing manufacturing costs [3]. Additionally, challenges involve achieving successful electrochemical/mechanical cell coupling, efficient cardiomyocyte maturation, functional vascularization, immune response balance, and large-scale construct generation [94]. Addressing conflicting requirements like cell placement, structural integrity, and maintaining a supportive cell culture environment throughout the scaffold thickness is crucial for large, anatomically accurate scaffold design [95]. Furthermore, the choice of printable biomaterials, 3D microstructural design, conductivity, permeability, degradation, mechanical properties, and sterilization are essential parameters for successfully fabricating large-scale additive-manufactured constructs [96]. These challenges necessitate a combination of techniques and considerations to achieve optimal tissue-engineered constructs for clinical translation.

Advancements in process engineering and automation play a crucial role in addressing scalability concerns in additive manufacturing for tissue engineering applications [97, 98]. Automation solutions, such as the use of liquid handlers and robotics, enable the efficient fabrication of complex tissue structures, leading to higher throughput and reproducibility. By automating processes like tissue ring retrieval, quality control, and transfer, researchers can streamline production and reduce manual labor costs. Additionally, the standardization of characterization techniques and the development of new materials and techniques through process engineering contribute to the advancement of the field, enhancing translation potential and aiding in scalability. Overall, integrating automation into additive manufacturing processes can significantly improve the scalability and commercial viability of tissue engineering technologies.

The Effects of Automation on Additive Manufacturing for Tissue Engineering are enlisted in Table 17.1.

17.6 Addressing Manufacturing Scale-Up

Scaling up additive manufacturing processes from laboratory prototypes to commercially viable production faces several challenges. These include the need for cost-effective production, shorter lead times, improved aesthetics, precise geometrical accuracy, and enhanced mechanical properties [99]. Additionally, transitioning from R&D to commercial manufacturing involves overcoming cultural gaps, ensuring quality control, and implementing GMP and quality systems [100]. In the context of recycling waste plastics through additive manufacturing, barriers to implementation include the lack of flexibility in achieving circular economy goals, as well as other obstacles identified through techniques like the Delphi method and Interpretative Structural Modeling [101]. Overcoming these hurdles is crucial for realizing the full potential of additive manufacturing in achieving sustainable and efficient production processes.

Regulatory frameworks face challenges in adapting to the scale of additive manufacturing for tissue engineering applications due to the unique complexities involved. The rapid development of 3D bioprinting technologies introduces regulatory hurdles related to classification, risk assessment, standardization, and quality control, especially when incorporating living cells [25]. Legal aspects also play a crucial role, necessitating adjustments to accommodate the advancements in bioprinting, such as in transplant ology, implant manufacturing, and drug testing [102]. Moreover, ethical, legal, and social concerns surrounding bioprinting technology require a comprehensive policy approach, including addressing privacy issues related to digital watermarks in 3D prints [49, 71, 103]. To effectively regulate the scaling of additive manufacturing for tissue engineering, a multidisciplinary

Table 17.1 Effects of automation on additive manufacturing for tissue engineering

Aspect	Effects of automation	Examples	Advantages
Process efficiency	1. Improved process control and repeatability	Automated parameter adjustment	Consistent production quality
	2. Reduced human error	Closed-loop feedback systems	Faster production cycles
	3. Faster production cycles	Automated build platforms	Lower labor costs
Quality control	1. Enhanced consistency and precision in fabrication	In-line monitoring systems	Decreased defect rates
	2. Real-time monitoring and adjustment	Automated inspection tools	Improved product reliability
	3. Minimized defects	AI-based defect detection	Enhanced product safety
Scalability	1. Facilitated scale-up of production	Robotic handling and post-processing systems	Increased production capacity
	2. Streamlined workflow for mass production	Automated material replenishment	Efficient resource utilization
	3. Integration with robotic systems for continuous operation	Automated build farm	Seamless transition to large-scale manufacturing
Complexity of structures	1. Ability to produce intricate and complex designs	Multi-axis robotic arms	Customized and personalized products
	2. Multi-axis printing for varied geometries	Adaptive slicing algorithms	Advanced design capabilities
	3. Layer-by-layer precision for detailed constructs	AI-driven support structure generation	Enhanced design freedom
Material handling	1. Automated material loading and dispensing	Material handling robots	Reduced material waste
	2. Optimized material usage and waste reduction	Automated powder sieving	Improved material traceability
	3. Integration with material characterization and tracking	Barcode scanning systems	Enhanced inventory management
Data integration and analytics	1. Utilization of data-driven insights for optimization	IoT sensors for real-time data collection	Predictive maintenance
	2. Predictive maintenance and fault detection	Cloud-based analytics platforms	Continuous process improvement
	3. Integration with CAD/CAM software for design workflows	Digital twin simulations	Enhanced design collaboration

(continued)

Table 17.1 (continued)

Aspect	Effects of automation	Examples	Advantages
Regulatory compliance	1. Enhanced traceability and documentation	Blockchain technology	Automated compliance reporting
	2. Automated compliance with regulatory standards	Electronic batch records	Simplified audit processes
	3. Audit trails and reporting functionalities	Automated validation protocols	Improved regulatory oversight
Cost efficiency	1. Reduction in labor costs and manual interventions	Automated scheduling systems	Lower production costs
	2. Improved resource utilization and waste management	Energy-efficient production processes	Cost-effective manufacturing solutions
	3. Economies of scale with automated batch processing	Automated maintenance scheduling	Increased return on investment

international panel and national-level risk-benefit analysis panels are recommended to ensure proper oversight and alignment with evolving technological capabilities.

17.7 Conclusion

The future of biomaterials-based additive manufacturing for tissue engineering is highlighted in this book chapter as depending on three key factors: technological breakthroughs, regulatory compliance, and interdisciplinary collaborations. Strict legal frameworks guarantee the efficacy and safety of innovations, promoting confidence and quickening their clinical acceptance. Advancements in technology propel the creation of more accurate, effective, and adaptable treatments, expanding the realm of tissue engineering possibilities. In the meanwhile, multidisciplinary teams make use of a wide range of skills, combining knowledge from engineering, biology, materials science, and medicine to solve challenging problems and spur innovation. When combined, these factors will not only improve the field but also guarantee that newly developed technologies are successfully transferred from laboratory to clinical settings, ultimately leading to improved patient outcomes and a revolutionary shift in medical practices. Rigorous regulatory frameworks ensure that innovations are safe and effective, fostering trust and accelerating clinical adoption. Technological advancements drive the development of more precise, efficient, and customizable solutions, pushing the boundaries of what is possible in tissue engineering and achieving transformative impacts in regenerative medicine.

References

1. Vyas J, Shah I, Singh S, Prajapati BG (2023) Biomaterials-based additive manufacturing for customized bioengineering in management of otolaryngology: a comprehensive review. Front Bioeng Biotechnol 11:1234340
2. Vyas J, Raytthatha N, Singh S, Prajapati B (2024) Next-generation computational automation-based additive manufacturing of pharmaceuticals: an approach to fabricate precise medicine. In: Handbook of 3D printing in pharmaceutics. CRC Press, Boca Raton, pp 153–175
3. Shah SR, Modi CD, Singh S, Mori DD, Soniwala MM, Prajapati BG (2024) Recent advances in additive manufacturing of polycaprolactone-based scaffolds for tissue engineering applications: a comprehensive review. Regen Eng Transl Med 11:1–20
4. Singh A, Mandal P, Maheshwari S, Akhtar J, Prajapati BG, Singh S (2024) Processing of sustainable biomaterials by additive manufacturing methods. In: Advances in sustainable biomaterials. CRC Press, Boca Raton, pp 141–162
5. Vyas J, Raytthatha N, Singh S, Prajapati BG, Mohite P, Munde S (2024) Sustainable sources of raw materials as substituting biomaterials for additive manufacturing of dental implants: a review. Periodontal Implant Res 8(1):3
6. Vyas J, Singh S, Shah I, Prajapati BG (2023) Potential applications and additive manufacturing technology-based considerations of mesoporous silica: a review. AAPS PharmSciTech 25(1):6
7. Belsky K, Smiell J (2021) Navigating the regulatory pathways and requirements for tissue-engineered products in the treatment of burns in the United States. J Burn Care Res 42(4):774–784
8. Joyce K, Buljovcic Z, Rosic G, Kaszkin-Bettag M, Pandit A (2023) Issues with tissues: trends in tissue-engineered products in clinical trials in the European union. Tissue Eng Part B Rev 29(1):78–88
9. Claes E, Heck T, Sonnaert M, Donvil F, Schaschkow A, Desmet T, Schrooten J (2023) Product and process design: scalable and sustainable tissue-engineered product manufacturing. In: Tissue engineering. Elsevier, London, pp 689–716
10. Lopez-Navas L, Torrents S, Sánchez-Pernaute R, Vives J (2022) Compliance in non-clinical development of cell-, gene-, and tissue-based medicines: good practice for better therapies. Stem Cells Transl Med 11(8):805–813
11. Sethe SC (2010) The implications of "advanced therapies" regulation. Rejuvenation Res 13(2–3):327–328
12. Gunasegaram DR, Murphy A, Barnard A, DebRoy T, Matthews M, Ladani L, Gu D (2021) Towards developing multiscale-multiphysics models and their surrogates for digital twins of metal additive manufacturing. Addit Manuf 46:102089
13. Monzón MD, Paz R, Bordón P (2023) Present and future of standardization of additive manufacturing in the medical field. In: 3D printing in medicine. Elsevier, London, pp 361–378
14. García-Domínguez A, Claver J, Camacho AM, Sebastián MA (2020) Analysis of general and specific standardization developments in additive manufacturing from a materials and technological approach. IEEE Access 8:125056–125075
15. Gouttebroze S, Friis J, Hovig EW, Boivie K (2023) Toward semantic standard and process ontology for additive manufacturing. IOP Conf Ser Mater Sci Eng 1:012014
16. Rohde J, Jahnke U, Lindemann C, Kruse A, Koch R (2019) Standardised product development for technology integration of additive manufacturing. Virtual Phys Prototyp 14(2):141–147
17. Koch C (2017) Standardization in emerging technologies: The case of additive manufacturing. In: 2017 ITU kaleidoscope: challenges for a data-driven society (ITU K). Piscataway, IEEE, pp 1–8
18. Oberweis CV, Marchal JA, López-Ruiz E, Gálvez-Martín P (2020) A worldwide overview of regulatory frameworks for tissue-based products. Tissue Eng Part B Rev 26(2):181–196

19. Samadder A, Bhattacharjee B, Dey S, Chakrovorty A, Dey R, Sow P, Tarafdar D, Biswas M, Nandi S (2024) Enhanced drug carriage efficiency of curcumin-loaded PLGA nanoparticles in combating diabetic nephropathy via mitigation of renal apoptosis. J Pharm 27(1):1
20. Joly J, Marechal M, Van Assche D, Moos M Jr, Luyten FP (2023) Clinical translation. In: Tissue engineering. Elsevier, London, pp 717–746
21. Fujiwara Y, Maruyama Y, Honda F (2021) Balancing safety and efficacy with early availability in the regulation of regenerative medicine product. Clin Pharmacol Ther 109(5):1182–1185
22. Shankhwar N, Saha P, Zurek-Mortka M, Thomas S, Unal A (2022) Tissue engineered products—translational avenues. In: Tissue engineering. Elsevier, London, pp 575–625
23. Kara A, Distler T, Polley C, Schneidereit D, Seitz H, Friedrich O, Tihminlioglu F, Boccaccini AR (2022) 3D printed gelatin/decellularized bone composite scaffolds for bone tissue engineering: fabrication, characterization and cytocompatibility study. Mater Today Bio 15:100309
24. McDonald J (2019) Regulatory considerations for devices manufactured using additive manufacturing technologies. In: Polymer-based additive manufacturing: biomedical applications. Springer, Cham, pp 243–254
25. Mladenovska T, Choong PF, Wallace GG, O'Connell CD (2023) The regulatory challenge of 3D bioprinting. Regen Med 18(8):659–674
26. Phoebe L, Faulkner A (2017) 3D bioprinting regulations: a UK/EU perspective. Eur J Risk Regul 8(2):441–447
27. Fink M, Akra B (2023) Comparison of the international regulations for medical devices – USA versus Europe. Injury 54:110908
28. Bhambere DS, Kale AS, Jinturkar FK, Mali RN, Gawali DM, Ahirrao SP, Kshirsagar SJ (2024) Comparative analysis of market authorization approval processes for medical devices in USA and Brazil. JUSST Journal 26(5):300–308
29. Joppi R, Bertele V, Vannini T, Garattini S, Banzi R (2020) Food and Drug Administration vs European Medicines Agency: review times and clinical evidence on novel drugs at the time of approval. Br J Clin Pharmacol 86(1):170–174
30. Gherghescu I, Delgado-Charro MB (2020) The biosimilar landscape: an overview of regulatory approvals by the EMA and FDA. Pharmaceutics 13(1):48
31. Torqui A, Macau AM (2018) Regional differences during the ICH regulatory consultation process between the EU, US, and Japan. Ther Innov Regul Sci 52(4):438–441
32. Rafi K, Zhonghong Liu A, Di Prima M, Bates P, Seifi M (2022) Regulatory and standards development in medical additive manufacturing. MRS Bull 47(1):98–105
33. Parpworth N (2022) Consenting to prosecutions: the legal consequences of non-compliance with the requirement: R v Lalchan [2022] EWCA Crim 736. J Crim Law 86(5):368–371
34. Pei E, Seidel C (2023) Standardisation efforts of ISO/TC 261 "additive manufacturing" 21st plenary meeting of ISO/TC 261 "additive manufacturing". Prog Addit Manuf 8(6):1143–1144
35. Moroni G, Petrò S, Shao H (2020) On standardization efforts for additive manufacturing. In: Proceedings of 5th international conference on the Industry 4.0 model for advanced manufacturing: AMP 2020. Springer, Cham, pp 156–172
36. Calcat-i-Cervera S, Rendra E, Scaccia E, Amadeo F, Hanson V, Wilm B, Murray P, O'Brien T, Taylor A, Bieback K (2023) Harmonised culture procedures minimise but do not eliminate mesenchymal stromal cell donor and tissue variability in a decentralised multicentre manufacturing approach. Stem Cell Res Ther 14(1):120
37. Hernandez Pichardo A, Wilm B, Liptrott NJ, Murray P (2023) Intravenous administration of human umbilical cord mesenchymal stromal cells leads to an inflammatory response in the lung. Stem Cells Int 2023(1):7397819
38. Burlington DB (1997) FDA's approach to international harmonization of biomaterials and tissue-engineered products. Tissue Eng 3(1):79–83
39. Brown P, Dombal G, Dornbusch D, Eslami B, Lawton A, Stewart S, Vasconcelles M, Tambuyzer E (2001) Human cell-and tissue-based products: progress, promise and regulatory issues. J Commer Biotechnol 7(4). https://doi.org/10.5912/JCB395

40. Karanu F, Ott L, Webster DA, Stehno-Bittel L (2020) Improved harmonization of critical characterization assays across cell therapies. Regen Med 15(5):1661–1678
41. Molina CP (2018) Regulation of tissue engineered devices in some Latin American countries: development and external influences. Univ Odontol 37(79). https://doi.org/10.11144/Javeriana.uo37-79.rted
42. Jokura Y, Yano K, Yamato M (2018) Comparison of the new Japanese legislation for expedited approval of regenerative medicine products with the existing systems in the USA and European Union. J Tissue Eng Regen Med 12(2):e1056–e1062
43. Zhang N, Baume A, Payne R, Allickson J (2016) Regulatory aspects: regulation of cell-free biomaterial implants. In: In situ tissue regeneration. Elsevier, London, pp 383–403
44. Franco P, Jain R, Rosenkrands-Lange E, Hey C, Koban MU (2023) Regulatory pathways supporting expedited drug development and approval in ICH member countries. Ther Innov Regul Sci 57(3):484–514
45. McGowran E, Harris E (2020) Regulatory readiness level: a tool to enhance early regulatory adoption in academic discovery. Level 3 15(2):6
46. Adamo JE, Grayson WL, Hatcher H, Brown JS, Thomas A, Hollister S, Steele SJ (2018) Regulatory interfaces surrounding the growing field of additive manufacturing of medical devices and biologic products. J Clin Transl Sci 2(5):301–304
47. Minghetti P, Musazzi UM, Rocco P (2023) Regulatory aspects and barriers in using groundbreaking technologies. In: Nano-and microfabrication techniques in drug delivery: recent developments and future prospects. Springer, Cham, pp 467–491
48. Cannon N, Biegalski S, Erickson A (2022) Additive manufacturing: a challenge to nuclear nonproliferation. J Radioanal Nucl Chem 331(12):4997–5002
49. Romanovskaya OV, Romanovskiy GB (2023) Legal regulation of additive technologies in modern biomedicine. RUDN J Law 27(1):21–40
50. Reis ME, Bettencourt A, Ribeiro HM (2022) The regulatory challenges of innovative customized combination products. Front Med 9:821094
51. Sahin A, Rey P, Panoutsos G (2022) Self-tuning multi-model statistical process control for process monitoring in additive manufacturing. In: 2022 8th international conference on control, decision and information technologies (CoDIT). Piscataway, IEEE, pp 1049–1054
52. Liu H, Gong Y, Zhang K, Ke S, Wang Y, Wang J, Wang H (2023) Recent advances in decellularized matrix-derived materials for bioink and 3D bioprinting. Gels 9(3):195
53. Zhe M, Wu X, Yu P, Xu J, Liu M, Yang G, Xiang Z, Xing F, Ritz U (2023) Recent advances in decellularized extracellular matrix-based bioinks for 3D bioprinting in tissue engineering. Materials 16(8):3197
54. Rijal G (2023) Bioinks of natural biomaterials for printing tissues. Bioengineering 10(6):705
55. Özenler AK, Distler T, Tihminlioglu F, Boccaccini AR (2023) Fish scale containing alginate dialdehyde-gelatin bioink for bone tissue engineering. Biofabrication 15(2):025012
56. Sekar MP, Suresh S, Zennifer A, Sethuraman S, Sundaramurthi D (2023) Hyaluronic acid as bioink and hydrogel scaffolds for tissue engineering applications. ACS Biomater Sci Eng 9(6):3134–3159
57. Martínez G, Vázquez J, Begines B, Alcudia A (2023) Emerging strategies to improve the design and manufacturing of biocompatible therapeutic materials. Pharmaceutics 15:1938
58. Andanje MN, Mwangi JW, Mose BR, Carrara S (2023) Biocompatible and biodegradable 3D printing from bioplastics: a review. Polymers 15(10):2355
59. Krishna AV, Reddy VV, Dexter DW, Wälivaara D-Å, Abrahamsson P, Rosén BG, Anderud J (2023) Quality assurance of stereolithography based biocompatible materials for dental applications. Surf Topogr Metrol Prop 11(1):014008
60. Kumar S, Krishnan S, Prabakaran K (2023) Advance biodegradable polymer composite materials for biomedical additive manufacturing applications. In: Development, properties, and industrial applications of 3D printed polymer composites. Hershey, IGI Global, pp 107–129
61. Ribezzi D, Gueye M, Florczak S, Dusi F, de Vos D, Manente F, Hierholzer A, Fussenegger M, Caiazzo M, Blunk T (2023) Shaping synthetic multicellular and complex multimaterial tissues via embedded extrusion-volumetric printing of microgels. Adv Mater 35(36):2301673

62. Ribezzi D, Pinos R, Bonetti L, Cellani M, Barbaglio F, Scielzo C, Farè S (2023) Design of a novel bioink suitable for the 3D printing of lymphoid cells. Front Biomater Sci 2:1081065
63. Isik M, Karakaya E, Arslan TS, Atila D, Erdogan YK, Arslan YE, Eskizengin H, Eylem CC, Nemutlu E, Ercan B (2023) 3D printing of extracellular matrix-based multicomponent, all-natural, highly elastic, and functional materials toward vascular tissue engineering. Adv Healthc Mater 12(20):2203044
64. Vellayappan MV, Duarte F, Sollogoub C, Dirrenberger J, Guinault A, Frith JE, Parkington HC, Molotnikov A, Cameron NR (2023) Fabrication of architectured biomaterials by multilayer co-extrusion and additive manufacturing. Adv Funct Mater 33(31):2301547
65. Yeleswarapu S, Dash A, Chameettachal S, Pati F (2023) 3D bioprinting of tissue constructs employing dual crosslinking of decellularized extracellular matrix hydrogel. Biomater Adv 152:213494
66. Mir A, Lee E, Shih W, Koljaka S, Wang A, Jorgensen C, Hurr R, Dave A, Sudheendra K, Hibino N (2023) 3D bioprinting for vascularization. Bioengineering 10(5):606
67. Zeenat L, Zolfagharian A, Sriya Y, Sasikumar S, Bodaghi M, Pati F (2023) 4D printing for vascular tissue engineering: progress and challenges. Adv Mater Technol 8(23):2300200
68. Li S, Liu S, Wang X (2022) Advances of 3D printing in vascularized organ construction. Int J Bioprint 8(3):588
69. Zhang Z, Yang Q, Wang C (2022) Techniques for biofabrication of vascular and vascularized tissue. In: Biofabrication for orthopedics: methods, techniques and applications, vol 2. Wiley, Newark, pp 461–479
70. Pushparaj K, Balasubramanian B, Pappuswamy M, Anand Arumugam V, Durairaj K, Liu W-C, Meyyazhagan A, Park S (2023) Out of box thinking to tangible science: a benchmark history of 3D bio-printing in regenerative medicine and tissues engineering. Life 13(4):954
71. Gupta RK (2023) 3D printing: fundamentals to emerging applications. CRC Press, Boca Raton
72. Akilbekova D, Turlybekuly A (2023) Patient-specific 3D bioprinting for in situ tissue engineering and regenerative medicine. In: 3D printing in medicine. Elsevier, London, pp 149–178
73. Subramanian B, Das P, Biswas S, Roy A, Basak P (2023) Polymers for additive manufacturing and 4D-printing for tissue regenerative applications. In: Advances in biomedical polymers and composites. Elsevier, London, pp 159–182
74. Tabatabaei Rezaei N, Kumar H, Liu H, Lee SS, Park SS, Kim K (2023) Recent advances in organ-on-chips integrated with bioprinting technologies for drug screening. Adv Healthc Mater 12(20):2203172
75. Parihar A, Pandita V, Khan R (2022) 3D printed human organoids: high throughput system for drug screening and testing in current COVID-19 pandemic. Biotechnol Bioeng 119(10):2669–2688
76. Berry BJ, Luttrell SM, Moerk CT, Macadangdang J, Perez J, Gray K, Ghazizadeh H, Kharoufeh S, Nelsen B, Geisse NA (2023) Preclinical drug testing in scalable 3D engineered muscle tissues. J Vis Exp 194:e64399
77. Bhusal A, Dogan E, Nieto D, Mousavi Shaegh SA, Cecen B, Miri AK (2022) 3D bioprinted hydrogel microfluidic devices for parallel drug screening. ACS Appl Bio Mater 5(9):4480–4492
78. Engel M, Belfiore L, Aghaei B, Sutija M (2022) Enabling high throughput drug discovery in 3D cell cultures through a novel bioprinting workflow. SLAS Technol 27(1):32–38
79. Vach Agocsova S, Culenova M, Birova I, Omanikova L, Moncmanova B, Danisovic L, Ziaran S, Bakos D, Alexy P (2023) Resorbable biomaterials used for 3D scaffolds in tissue engineering: a review. Materials 16(12):4267
80. Wang X, Li S, Yu H, Lv J, Fan M, Wang X, Wang X, Liang Y, Mao L, Zhao Z (2022) The biocompatibility of multi-source stem cells and gelatin-carboxymethyl chitosan-sodium alginate hybrid biomaterials. Tissue Eng Regen Med 19(3):491–503
81. Saska S, Pilatti L, Blay A, Shibli JA (2021) Bioresorbable polymers: advanced materials and 4D printing for tissue engineering. Polymers 13(4):563

82. Raut HK, Das R, Liu Z, Liu X, Ramakrishna S (2020) Biocompatibility of biomaterials for tissue regeneration or replacement. Biotechnol J 15(12):2000160
83. Sukumaran A, Sweety VK, Vikas B, Joseph B (2023) Cytotoxicity and cell viability assessment of biomaterials. In: Cytotoxicity-understanding cellular damage and response. IntechOpen, London
84. Lores NJ, Hung X, Talou MH, Ballarre J, Abraham GA, Caracciolo PC (2022) Additive manufacturing of bioresorbable poly (ester-urethane)/glass-ceramic composite scaffolds. Polym Compos 43(8):5611–5622
85. Rekowska N, Huling J, Brietzke A, Arbeiter D, Eickner T, Konasch J, Riess A, Mau R, Seitz H, Grabow N (2022) Thermal, mechanical and biocompatibility analyses of photochemically polymerized PEGDA250 for photopolymerization-based manufacturing processes. Pharmaceutics 14(3):628
86. Pickering E, Paxton NC, Bo A, O'Connell B, King M, Woodruff MA (2022) 3D printed tubular scaffolds with massively tailorable mechanical behavior. Adv Eng Mater 24(11):2200479
87. Haag H, Dalton PD, Bloemen V (2022) The synergy of biomimetic design strategies for tissue constructs. Adv Funct Mater 32(32):2201414
88. Bezek LB, Chatham CA, Dillard DA, Williams CB (2022) Mechanical properties of tissue-mimicking composites formed by material jetting additive manufacturing. J Mech Behav Biomed Mater 125:104938
89. Estermann S-J, Pahr DH, Reisinger A (2022) Material design of soft biological tissue replicas using viscoelastic micromechanical modelling. J Mech Behav Biomed Mater 125:104875
90. Zhang Y, Tang J, Fang W, Zhao Q, Lei X, Zhang J, Chen J, Li Y, Zuo Y (2023) Synergetic effect of electrical and topographical cues in aniline trimer-based polyurethane fibrous scaffolds on tissue regeneration. J Funct Biomater 14(4):185
91. Fang W, Yang M, Liu M, Jin Y, Wang Y, Yang R, Wang Y, Zhang K, Fu Q (2023) Review on additives in hydrogels for 3D bioprinting of regenerative medicine: from mechanism to methodology. Pharmaceutics 15(6):1700
92. Raja IS, Kang MS, Hong SW, Bae H, Kim B, Hwang Y-S, Cha JM, Han D-W (2023) State-of-the-art techniques for promoting tissue regeneration: combination of three-dimensional bioprinting and carbon nanomaterials. Int J Bioprint 9(1):635
93. Gasparotto M, Bellet P, Scapin G, Busetto R, Rampazzo C, Vitiello L, Shah DI, Filippini F (2022) 3D printed graphene-PLA scaffolds promote cell alignment and differentiation. Int J Mol Sci 23(3):1736
94. Qiu T, Hanna E, Dabbous M, Borislav B, Toumi M (2020) Regenerative medicine regulatory policies: a systematic review and international comparison. Health Policy 124(7):701–713
95. Chang Y-C, Mirhaidari G, Kelly J, Breuer C (2021) Current challenges and solutions to tissue engineering of large-scale cardiac constructs. Curr Cardiol Rep 23:1–6
96. Forrestal DP, Klein TJ, Woodruff MA (2017) Challenges in engineering large customized bone constructs. Biotechnol Bioeng 114(6):1129–1139
97. Lindner N, Blaeser A (2022) Scalable biofabrication: a perspective on the current state and future potentials of process automation in 3D-bioprinting applications. Front Bioeng Biotechnol 10:855042
98. Manning KL, Feder J, Kanellias M, Murphy J III, Morgan JR (2020) Toward automated additive manufacturing of living bio-tubes using ring-shaped building units. SLAS Technol 25(6):608–620
99. Minary-Jolandan M (2022) Formidable challenges in additive manufacturing of solid oxide electrolyzers (SOECs) and solid oxide fuel cells (SOFCs) for electrolytic hydrogen economy toward global decarbonization. Ceramics 5(4):761–779
100. Kumar A, Kumar P, Mittal RK, Singh H (2023) Preprocessing and postprocessing in additive manufacturing. In: Advances in additive manufacturing. Elsevier, London, pp 141–165
101. Alfnes E, Romsdal A, Strandhagen JO, Von Cieminski G, Romero D (2023) Advances in production management systems. Production management systems for responsible manufacturing, service, and logistics futures: IFIP WG 5.7 international conference, APMS

2023, Trondheim, Norway, September 17–21, 2023, proceedings, part II, vol 690. Springer Nature, Cham

102. Trommelmans L, Selling J, Dierickx K (2009) Is tissue engineering a new paradigm in medicine? Consequences for the ethical evaluation of tissue engineering research. Med Health Care Philos 12:459–467
103. Dutta A, Roy T, Ray PG, Rajasekaran R, Banerjee M, Chattopadhyay S, Gupta S, Dhara S (2020) 3D printing: challenges and its prospect in futuristic tissue engineering applications. In: 3D printing in biomedical engineering. Springer, Singapore, pp 1–22

Index

S. Singh et al. (eds.), *Biomaterial-based Additive Manufacturing in Tissue Engineering and Regeneration*, Biomaterials, Bioengineering and Sustainability 7,
https://doi.org/10.1007/978-3-031-96070-3

H

I

L

M

N

O

P

R

S

T

The manufacturer's authorised representative in the EU is Springer Nature Customer Service Centre GmbH, Europaplatz 3, 69115 Heidelberg, Germany. If you have any concerns regarding our products, please contact ProductSafety@springernature.com

Printed and bound by CPI Group (UK) Ltd, Croydon, CR0 4YY
30/09/2026
02225454-0015